RANCHO SANTIAGO COLLEGE
OR MAIN   QH366.2 .H37
The origin and early evolution o
Hanson, Earl D.

3 3065 00041 8537

D1614139

QH   Hanson, Earl D.
366.2   The origin and
.H37   early evolution of
             animals
    35.00   221458

DATE DUE

Rancho Santiago College
Orange Campus Library

The Origin and Early Evolution of Animals

The Origin and Early Evolution of Animals

by Earl D. Hanson

Wesleyan University Press
Middletown, Connecticut

Pitman
London • Melbourne

Copyright © 1977 by Wesleyan University

All rights reserved

Library of Congress Cataloging in Publication Data

Hanson, Earl D
    The origin and early evolution of animals.

    Bibliography: p.
    Includes index.
        1. Evolution.   I. Title.
QH366.2.H37   591.3'8   76-41480
US ISBN 0 8195 5008 6
UK ISBN 0 273 01132 4

Published simultaneously by
Wesleyan University Press
Middletown, Connecticut 06457

Pitman Publishing Limited
39 Parker Street, London WC2B 5PB

Associated Companies: Pitman Publishing Co. SA [Pty] Ltd, Johannesburg • Pitman Publishing New Zealand Ltd, Wellington • Pitman Publishing Pty Ltd, Melbourne • Sir Isaac Pitman Ltd, Nairobi

Manufactured in the United States of America
First edition

*Dedicated to
the memory of
Erwin Goodenough
and to
Donald D. Jensen and Stanley R. Hanson
For their encouragement*

Table of Contents

Preface — ix

Part One
Concepts and Methodology

Chapter 1   The Evolutionary Process and Phylogeny   5
2   The Informational Content of Organisms   23
3   Phylogenetic Analysis   68

Part Two
Phylogeny of the Protozoa and Early Metazoa

4   The Protozoa   127
5   Rhizoflagellates: The Zooflagellates   155
6   Rhizoflagellates: The Ameboid Forms   202
7   Heliozoans   242
8   Radiolarians   265
9   Acantharians   290
10   Ciliates   326
11   Sponges   419
12   Cnidarians   454
13   Flatworms: Turbellarians   493

Part Three
Perspectives

14   Phylogenetic Methodology   555
15   Phylogenetic Conclusions   570

References   594
Author Index   638
Organism Index   645
Subject Index   656

Preface

"Men love to wonder and that is the seed of our science." — EMERSON

NOT only do we love to wonder but we are genuinely agitated when presented with unsolved problems. If such problems are seen with special clarity, the urge towards their solution becomes compulsive. Such is the case today with regard to phylogeny. It is one of contemporary biology's most pressing problems, for the reality of evolution is conceptually central in the life sciences and we stand informed in convincing detail on the nature of the process; but yet there is little effective agreement on the actual historical sequence of the major forms produced by evolution. This volume addresses the problem of phylogeny in three steps. First, there is developed in Part I a consistent set of phylogenetic principles deriving from both the classical awareness of homology as being a prerequisite for phylogenetic conclusions and our current knowledge of speciation through natural selection. Next, in Part II, these principles are applied to an area which has continually provoked some of our most heated and diverse opinions on the course of evolution, namely, the origin and early evolution of animals. Finally, in Part III, an overview is provided of the conclusions which emerge from the preceding two parts.

In that phylogeny is concerned with the full biological life of the organisms it studies, there must, of necessity, be attention paid to fine structure and to gross anatomy, to genetics of chromosomes and of populations, to functions of organs and groups of organisms, and to development, ecology, and paleontology as well. One writer cannot cover these various aspects with equal balance and care, and this book surely suffers in that regard.

The necessary breadth of coverage also results in another regrettable limitation. The material reviewed is felt to be quite complete up to a few years ago and then is less complete thereafter. The reason is that updating the whole volume takes at least a year (added to teaching, committees, etc.), and by then things are naturally one year behind their most current status. The added work of revising bibliography, completing illustrations (with permissions), and the work of publication, all adds further delays. Hence, one must finally call a halt to further revisions and get the book out. Nonetheless, having halted, it is still worth using this space to identify more recent work which readers could profitably consult. These include research reviews on

Tetrahymena (Elliott 1973, with an addendum in Corliss 1973), *Paramecium* (van Wagtendonk 1974), and *Blepharisma* (Giese 1973); a symposium volume on the Turbellaria (Riser and Morse 1974); the discovery of uniflagellate sperm in a species of *Nemertoderma* by Tyler and Rieger (1975) (see comments in Hanson 1976); and collected research reviews from the Fourth International Congress of Protozoology (Puytorac and Grain 1974). There are, of course, new treatments of phylogenetic procedures and two useful reviews are those of Estabrook (1972) and Cracroft (1974). Most important, however, are the recent papers by Corliss, which revise the systematics of ciliates and include certain phylogenetic proposals of Jankowski. Some of this is incorporated in the present text, but papers not discussed there include Corliss (1974b, 1974c, 1974d, 1975). Finally, one must inevitably acknowledge the continuing surge of new knowledge in cellular fine structure relating especially to membranes, microtubules, and microfilaments; unhappily much that was written for Chapter 2 bears revision in the light of new knowledge in these areas.

Despite these obvious limitations, the basic reasons for attempting a volume of this sort, stated above, are compelling, and the major goal of developing a consistent theory for phylogenetic work has been carried out and tested by its application to protozoans and certain metazoans. That effort has been evaluated in Chapters 14 and 15 and nothing more need be said here.

Lastly, there are those who deserve special thanks for their help in bringing this study to completion. Apart from the author, the two who spent most time on this effort are Mrs. Anne Wrubel and Ms. Jane T. Sibley. Mrs. Wrubel, with unfailing patience and diligent care, transformed an untidy, handwritten manuscript into legible typescript and Ms. Sibley completed the illustrations with thoughtful competence and cheerful cooperation. Next, because of the essential support they provided I must thank my own institution Wesleyan University, for its generous sabbatical program, and the Zoological Institute of Tübingen University, under the directorship of Professor Karl Grell, for its congenial hospitality during the year (1967–68) when the first draft was written. Not only are the Director and staff of that Institute to be deeply thanked, but also the staff of the Geological and Paleontological Institute of that University, with special acknowledgments due to Professors Otto Schindewolf and Adolf Seilacher, and Privatdozent Jost Wiedman, for their intellectual hospitality. Over the years there have been many colleagues and students whose questions, ideas, and comments have been essential in forcing me to rethink and rework the ideas developed in this work. These include Drs. Joseph Frankel, Tracy Sonneborn, Donald Jensen, and John Corliss. John Corliss also deserves special thanks for reading the entire manuscript in his typically careful and constructive manner. To all of these I owe and render deep thanks.

October 15, 1975
Middletown, Conn.

Earl D. Hanson

Part One
Concepts and Methodology

Concepts and Methodology

THE study of phylogeny has two goals: first, to define the course of evolution and, second, to understand the underlying innovations. To achieve the first goal we construct phylogenetic diagrams that indicate genetic relationships in the process of organismic diversification. Conclusions are usually most convincing at the level of the lower taxa and become progressively less convincing among the higher taxa. But it is among the higher taxa that the most fascinating problems are found. Broadly speaking, these problems are the origins of major evolutionary innovations.

The essential constraint on the phylogenetic inquirer is to envisage only those lines of evolutionary radiation or development that preserve evolutionarily fit systems. Static, diagrammatic phylogeny that disregards natural selection is today unacceptable; to derive a chordate from an annelid simply by turning the latter over so that the ventral nerve cord is now dorsal, is to totally ignore the functional integrity of an organism and, hence, to demand the impossible in terms of survival in nature. Evolutionary phylogeny demands that we understand the conservative character of the organism along with its innovations, and our problems then become those of identifying the old and new characters and of interpreting them in view of functionally adapted organisms. In the process, we will discover the pattern of genetic relations during evolutionary diversification. But also, and more exciting, we will be able to see where evolutionary innovations arise and understand these new potentials, not just as morphological or physiological facts, but as selectively advantageous endowments available to all descendants of that line. And, looking in the other direction, we will understand better the limitations and specializations of the ancestral line. Phylogeny, in this sense, is an analysis of the evolutionary potential of different organizational modes of living systems.

The elucidation of phylogenetic history must be preceded by agreement on a method for its analysis. Therefore, this book starts (Part I) with the development of an analytical rationale for phylogeny which is derived from current theories on species formation by natural selection. It proceeds (Part II) to an application of this methodology to the problems of the origin and early evolution of animals. This consists of proceeding from the first unicellular living systems deserving the appellation of animal to the emergence of the first multicellular animals. The volume ends (Part III) with ideas on the general validity of this approach, its applications, and possible conclusions.

Chapter 1

The Evolutionary Process and Phylogeny

"EVOLUTION is a change in the genetic composition of populations." (Dobzhansky 1951, p. 16). To expand this extraordinarily succinct summary of very complex natural phenomena so as to see its implications for phylogeny, this chapter will examine the functional basis of evolution and species formation. The point of view throughout is to see the organism as a product of the forces shaping the natural population of which it is a part.

The Functional Basis of Evolution

ORGANIZATIONAL UNITS

Organisms, however defined, must be alive and capable of evolution. And organic evolution, as we know it today, will occur when there is a finite environment; that is, the space available for supporting these living systems is limited, and, hence, only the more highly adapted or fit variants will persist. Furthermore, these survivors are most often seen to organize themselves as groups of individuals pooling their genetic resources through sexual reproduction to insure the survival of the group as a whole. Consequently, the two organizational units that emerge of paramount interest are the individual organism and the interbreeding population to which it belongs, the species. Both entities need to be carefully characterized for the purposes of this study.

ORGANISM

The difficulties in defining an organism are legion. Definitions based on structure bog down in the diversity of detail that lies between submicroscopic viruses and macroscopic whales and redwood trees. Definitions emphasizing function usually resolve themselves into chemical substances and their reactions, and these pay no special recognition to living systems, and therefore at this level the problem of distinguishing life from nonlife is very difficult (Pirie 1937; Kluyver and Van Niel 1956).

Organisms as evolvable systems. One answer to this problem is to recognize that the universal and unique property of living systems is their evolvability (Muller 1955). One then asks, what is the biological foundation of this property? The answer is that evolvable systems are ones that take matter and energy from their environment to maintain and reproduce themselves and at least certain of their variants, and they do so in a limited environment. Such systems have three functional aspects in their nature: (a) an ecological or exploitive function, which relates the organism to the resources in its environment; (b) a physiological or homeostatic function, which uses the energy and matter obtained from the organism's surroundings to maintain the system so that it can reproduce; and (c) the genetic or reproductive function, which assures more copies of the original system including some variants of it. By accumulation of variations that allow the system to exploit more efficiently its environment, maintain and reproduce itself, we get that sequence of changes that constitutes evolution. To the extent these functions and the changes in them depend on the genetic materials of the system, then evolution is, indeed, essentially a matter of temporal change in the genetic resources of the system.

However, focusing only on the genetic aspect of the problem excludes attention from another, equally fascinating aspect, which is that evolving systems develop into more and more complex systems while the world—or, better, universe—in which they exist is moving, thermodynamically speaking, in the other direction. This is not to argue that organisms are exceptions to the second law of thermodynamics, for they are not. It is to point to a unique character they possess as systems, which is to become less random, or more highly organized, in time (Blum 1951; Grobstein 1964). This, for our purposes, is another way of saying organisms evolve.

A more rigorous statement of these ideas is Morowitz's (1968) lucid development of the proposition that whenever energy flows from a source to a sink and through an intervening system, the latter will gain in order. And therefore, if that system can make more systems like itself and even a few that differ in terms of greater efficiency in utilizing the energy flow or can use a different energy flow, then they will persist and increase in their turn. Organisms are such systems and their ability to become progressively more highly organized is truly a distinctive feature when seen against the backdrop of the rest of nature which is—all of it—tending to greater disorder or entropy. Evolvability is the distinctive property of organisms.

Organismic function and natural selection. Natural selection, the main driving force of evolution, acts on functions. This is sometimes obscured by direction of our attention to changes in structure, such as the diversity of mammalian forelimbs or suture patterns in ammonoid shells, or even to changes in gene frequencies. The interconnections of genes, form, and function are basically clear. Any bodily function, be it exploitive, homeostatic, or reproductive, depends on some organized matrix for its expression, and this organization is ultimately genetically determined. And so, when we refer to natural selection as favoring a certain allele or alleles, we are, more accurately speaking, saying that a given function, arising from a certain material organization that is dependent on a specified gene or genes, is a genetically determined function that renders the organism possessing it better adapted or more fit than the organism that lacks it.

Most phylogenetic work starts from anatomical descriptions of given organisms or their parts. If we are concerned with the genetic basis of phylogeny, then the genetic determinants of the structures in question must be worked out. If one chooses to explore the biological basis of selective fitness, then one must understand the functional significance of the morphological features that are at issue. We are today better able to move from form to function than we are from form to genes. The interdependence of form and function, clearly enunciated by Cuvier in the 18th century in his "theory of correlations," has been actively explored by both physiologists and morphologists in the intervening decades and even stands today as a formal principle in biology (van Potter 1971). Its practical importance to phylogeny is the assurance it provides in precisely imputing function to form and thus allowing the investigator to transform the static data of anatomy into the dynamics of life.

But so much for the formal principle of the Cuvierian idea. As will be seen in Part II of this volume, there still remains a significant body of descriptive anatomy for which human knowledge has no functional correlates. In those cases, conjectures, labelled as such, will be used when trying to understand the selective pressures to which groups of organisms are responding as inferred from their anatomy. And when one tries to understand morphological change in terms of accumulated gene mutations, more conjecture may be necessary. There are still very few organisms whose genetic natures are so public as those of the T-even bacteriophages, *Escherichia coli*, *Neurospora*, *Zea*, *Paramecium*, and *Drosophila*. The most precise way around these difficulties comes from the amino acid sequence studies of proteins. Here direct reference to gene structure is possible, but there are two limitations to this approach. One is that few such studies have been done on the organisms of special concern to this study (Part II). The other is that the relation of proteins to higher levels of organization, such as specific cell organelles, cells, or tissues, is very complex and still incompletely known.

In brief, attempts to thoroughly understand the action of selection in terms of visible structure must be handled cautiously. The outlines of the process are clear. Natural selection winnows out more or less efficient ways of functioning, which in turn select the correlated structures, and this finally means certain genes are or are not favored. But the practical use of this conception, starting from specific anatomical features will, in many cases, generate problems of interpretation.

Varieties of forms and functions. It is commonplace to refer to the "endless variety of nature," "the multitude of living things," "the boundless diversity of life," and so on. But what are the quantitative parameters of organismic diversity? What is its organizational and functional basis? The point of these questions is to perceive the ways in which evolutionary history is conserved in organismic structure and function.

One measure of diversity is to count species. It has been estimated (Mayr 1969) that there are close to 1,071,000 described species in the major animal groups. The total number of species, described and undescribed, plants and animal, may be as high as 10 million (Handler 1970). Presumably this means there may be as many as 10^7 different ways to exploit this earth for the matter and energy needed to maintain and reproduce a species. (This assumes the so-called Gausean principle that no two species totally occupy the same niche.) The question as to why species diversity reaches this number rather than, say, 10^{10} or only 10^3, no one knows as yet (Hutchinson 1959).

Underlying species diversity is an extraordinary unity whose discovery is one of the great intellectual victories of biology in this century. First, there is the universality of the genetic code: all living things use the same pattern of specificity to transfer the linear information of the nucleic acids into that of the amino acid backbone of proteins. Second, there is the great uniformity of metabolic patterns: all living things make and use proteins, nucleic acids, lipids, and carbohydrates in much the same way and have similar catalytic and energetic requirements in doing so. Third, there is universal dependence on cells as the organizational unit of life: all living things are cellular or, as in the case of viruses, depend on them for life, and all cells consist of nuclear material lying in a cytoplasm that is bounded by at least a unit membrane. The seeming paradox of species diversity superimposed on genetic-metabolic-cellular unity must mean that the latter is a wonderfully flexible way to meet the varied exploitive, homeostatic, and reproductive needs of a living system.

The contrast between species diversity and the underlying unity of living systems can be placed in another informative context. For this analysis we are again indebted to Morowitz (1971). He has estimated that the energy levels per unit mass of living material is about 5 kcal/mole higher than that same material in a ground or nonliving state. Furthermore, this difference between living and nonliving organizations of matter is quite constant, regardless of whether we take a unit of mammalian tissue cells or the same unit of bacterial cells. The conclusion is that evolutionary complexity does not come from building more complex organisms in the energetic sense, but in taking units of a certain energetic level and exploiting all possible combinations of their parts that can survive under the conditions found on this earth.

It appears, then, that the diversity of terrestrial life lies compressed between two constraints: on one side a certain complexity is needed to reach that minimal organization which is the genetic-metabolic-cellular basis of life as we know it; on the other side is the fact that it takes work (an energy flow through the system) to maintain this system. Any system indulging in a profligate use of energy may be able to build a different, even a totally different, pattern of life but will expend much effort to maintain and reproduce itself. In a world where materials for survival are finite and energy flow is restricted basically to what arrives in the form of sunlight, the more efficient users of matter and energy are going to be favored. Profligate experimentation will not survive the rigors of natural selection.

This leads to two further points. We should expect to see different solutions to the problems of survival arising simply from a different juxtaposition of similar building blocks—bricks can build a doghouse, a factory or a mansion, as cells can build bacteria, protozoans, and metazoans. Species diversity is a reflection of the multiple ways in which a certain energetic level of molecular constituents can be used to sustain evolvable systems. The other point, one of great importance to phylogeny, is the fact that different organisms, quite independently of each other, will arrive at the same structural solution for a given functional need. This is what Pantin (1951) has referred to as limitations of "organic design". He emphasizes that for organisms, just as for engineers, there is a limited number of ways to build a structure to serve a given function. And, hence, similar structures in different organisms need not be a consequence of common evolutionary descent but of the special properties of matter and energy. Add to this the facts as accumulated and interpreted by D'Arcy Thompson (1942) "that throughout the whole range of organic morphology there are innumerable phenomena of form which are not peculiar to living things, but which are more or less simple manifestations of ordinary physical law" (p. 732), and we can see that phylogenetic interpretations of similarity between organisms are going to be more, much more, than just sorting out the successive steps of divergence starting from that common ancestral form that first achieved the beginnings of the modern genetic-metabolic-cellular mode of life.

POPULATIONS

The capacity to interbreed among themselves and the incapacity to breed with other such groups circumscribes those organisms which belong to a single species (Dobzhansky 1965, 1970; Mayr 1963, 1969). A species so defined is the fundamental unit of evolution because, though composed of individual organisms, it is the population as a whole which defines the niche in which the individual can survive rather than the reverse. Put another way, it turns out that forms with the greatest possibility of persisting through successive generations are the ones which share or pool their genetic resources by means of sexual processes. The term "gene pool" (Dobzhansky 1950) is the conceptual abstraction that goes to the heart of the problem.

Of the extensive literature on the nature and operation of gene pools (e.g., Dobzhansky 1965; 1970; Mayr 1963; Ehrlich and Holm 1963; Lewontin

1968, etc.), our interest at present is on the variability expressed by its members. For it is through an understanding of the origin, preservation, and further diversification of these variations that we come to understand the process of species formation or speciation. And speciation is the fundamental process of phyletic divergence.

Variability of the gene pool. The sources of genetic variability are well known. Changes in the base sequence of DNA, or point mutations, are the most fundamental source of new genetic material. Reshuffling of genes on a chromosome through crossing-over creates new alignments of genes, as does reshuffling of whole chromosomes during meiotic segregations. These are intraorganismic changes. Interorganismic changes, changes at the population level, come about by migration of individuals from one population to another. The effect of this gene flow is much like that of mutation, especially when the frequency of a single set of alleles is being studied (e.g., Li 1955; Crow and Kimura 1970). Natural selection and certain chance events, collectively called "genetic drift," are the determinants of the fate of the new genes and gene combinations, with selection playing much the most important role (Mayr 1963).

The retention of this variability in the natural population can take various patterns, depending on such factors as homogeneity or heterogeneity of the environment, size of the population, and breeding habits. For our present purposes it will be sufficient to review briefly various patterns of variability and to comment on the role of subpopulations, or demes. Finally, we must understand what is meant by the cohesiveness of a species.

It is a truism that organisms are perishable objects. They persist not because they possess the durability of inorganic crystalline solids but by a totally different strategy. They replace old organisms by new ones. It is therefore the *sine qua non* of biological persistance that an organism or group of organisms must persist long enough to form more of themselves (Wagner and Mitchell 1964). Fundamentally, selection favors those changes in exploitive and homeostatic functions which insure the survival of the individual long enough for it to engage successfully in reproduction (Hanson 1966). Add to this a second truism: the environment is variable. It is variable in the short run—there are day-night fluctuations of temperature and light, changes in rainfall, and seasonal changes—and also in the long run—continental uplift and sinking, movements of the axis of earthly rotation, which change the positions of polar regions, and long-range changes in mean temperature. These fluctuations are especially pronounced for terrestrial forms, but they also affect aquatic ones too. The problem, then, is for successive replacements or generations of organisms to inherit genetic endowments that adapt them to a probable range of environmental exigencies. Obviously, a population can store more such variability than any individual, and if the population is widely dispersed, natural calamities may affect only part of the population.

How does a population generate the variability for long and short term needs? How can it disperse itself and still remain adapted and able to tap the resources of the larger gene pool? It is obvious that species are not equally successful in the evolutionary sweepstakes. One model system that illustrates this quite effectively is the peak-valley model of population structure (Wright 1932; Dobzhansky 1965). What this model tells us is that small populations are lucky to survive in the long run because there is little probability that their gene pool will reach the optimally adaptive peak that helps insure survival. The problem is that a limited number of individuals provides only a very limited source of new mutations, and the tendency to inbreeding in small groups offsets the action of selection by the operation of genetic drift. Intermediate-sized populations do better, for they have more new mutations and inbreeding is less likely. Therefore, selection has a better chance of driving the population to its adaptive peak.

The most successful populations are thought to be large ones with local differentiations into subpopulations characterized by gene frequencies that differ from the population as a whole. The local populations or demes represent responses to local selection pressures. If gene flow between demes is low, it will be possible to develop significant genetic differences in different geographic localities of the whole area inhabited by the species in question. Gene flow insures influx of new genes as a supplement to mutation and insures, if local conditions change, that the resources of the larger population are available to achieve adaptation to the new conditions. Such large populations are well adapted to current local environments and are also buffered against future changes in the environment both through variability and dispersal.

Of course the nature of the habitat has much to do with how well-differentiated a deme can become. It is clear that relatively homogeneous environments, such as grasslands, harbor less variability in their endemic species than do discontinuous and heterogeneous habitats such as island archipelagoes (Mayr 1940, 1963). The situation for aquatic forms is not well studied.

Sharing in the gene pool has been thought to be the chief restraint on progressively greater and greater differentiation of demes (Mayr 1963, and others including Dobzhansky 1965, 1970, and Stebbins 1950, who laid the foundations of population genetics through integrated field and laboratory work). The restraint comes about in several ways. There is a continual influx, by migration, of genes from other demes that continually dilutes the force of local selection pressures. Mayr (1963, p. 521) says, "... a population cannot change drastically as long as it is exposed to the normalizing effects of gene flow." There is also a set of developmental and physiological restraints, referred to by Mayr (1963, p. 523) as "epigenetic systems" and "homeostatic devices." The former refers particularly to Waddington's (1960) concept of canalization in embryonic development, wherein there is a developmental norm for the species, a rather fixed and rigid pattern whereby adults are formed. Selection seems to favor stabilizing a successful epigenetic pathway—canalizing it—rather than experimenting with it. The result is a fixed pattern of development that can mask much genotypic variability, to be sure, but results in phenotypic stability. Similarly, homeostatic devices that perpetuate a stable internal milieu can be stabilized by selection for dominant genetic determinants that create a very uniform set of bodily functions throughout the species. Demic variation is whatever can squeeze by these restraints.

Another point of view has been espoused more recently by Ehrlich and his coworkers (Ehrlich and Raven 1969). They argue, in essence, that the cohesiveness of species is poorly demonstrated to be a consequence of gene flow and the sharing of common, rigid, epigenetic and homeostatic mechanisms. They argue that in demes of butterflies, for example, where gene flow is almost negligible, the uniformity of phenotype from deme to deme must be due to selection. This controversy cannot be resolved here. It does bring out the point that those few species, such as many free-living amebas, which have no known sexual processes and hence have not shared genetic resources, nonetheless appear on morphological grounds to be good species. Sonneborn (1957) has cited this as an important exception to the definition of species as a set of interbreeding individuals. Mayr (1963) has suggested that in such cases selection has driven certain groups of amebas to certain discrete adaptive peaks that give them the appearance of species cohesiveness. Ehrlich would presumably agree with this interpretation. It is certainly consistent with the energetic and organizational constraints discussed earlier. If there is an efficient way to serve the environmental needs, maintain, and reproduce an organism within the energetic and organizational limits enforced by natural selection, then it probably will be done, given the flexibility of the building materials. In the amebas, sexuality has probably been lost, as we shall see, so as to remove the variability coming from chromosomal rearrangements that would disturb the genotype of an otherwise well-adapted system. The asexual amebas provide the extreme example, but it may well be that in all the protozoa where reproduction and sexuality are separate processes, and where the latter is much rarer than the former, that in these forms as a whole, with the possible exception of the ciliates, selection is the predominant force in determining phenotypic uniformity.

Returning again to the question of variability, the point to be stressed is that regardless of its origin through mutation or recombination or gene flow, maintenance of variability in a population is due, in the main, to selection. When demes differ from the species norm, it is because of adaptive responses to local conditions. This type of variation within the species is the starting point for speciation. Henceforth our references to species, unless otherwise stated, will imply the larger, more genetically heterogeneous interbreeding populations, for these are the ones on which evolutionary development is dependent.

The Formation of Species and Higher Taxa

One species changes into other species in two ways. Either it transforms itself over time with a new set of gene frequencies replacing the ancestral one, or it diversifies in space and develops into more than one gene pool. It is the continuation of this process, with different rates of change in different species and with extinction in some and invasion of new habitats by others, that in the long run of history produces the diversity of life that we know today. How that change progresses is just what determines the ancestral traits that are preserved and those that are changed. This process must be understood explicitly if we are to interpret correctly the record of its past that is inherent in every organism.

SPECIATION

Darwin (1872) referred to species transformation as "descent with modifications" and diversification as the "origin of species." Romanes (1897) provided another pair of emphases: "transformation of species in time" and "multiplication of species in space," though recognizing the latter also took time. According to Mayr (1963, p. 429) "... it has become increasingly customary in recent times to restrict 'speciation' to the process that leads to multiplication of species through branching of phyletic lines." This is surely the neontologist speaking and reflecting his concern with that which he can best observe, i.e., the ongoing process of diversification arising from intraspecific variability. The paleontologist takes, perforce, a longer view and refers to "phyletic evolution" (Simpson 1953a) as the totality of change that occurs with or without species diversification. It seems best for our present purposes to refer to all changes of species into other species as speciation and refer to the two subprocesses as transformation and diversification. We start with the former and then turn to the latter, which is more complex.

Transformation. A species changes to another one as a result of changes in the environment, through the agency of changed selection pressures. If these changes are abrupt and drastic, i.e., demanding such extensive phenotypic changes for survival that the existing gene pool cannot effectively meet these demands, then extinction could well be the consequence. But lasting changes in selection pressure that can be met by the genetic resources of the population will eventually result in the predominance of new phenotypes and new gene frequencies.

Change will be conservative; the old characters will be more conspicuous than the innovative ones, and the change from one species to another will be gradual. The fossil record, when relatively complete, reveals a continuum, and only by using various somewhat arbitrary conventions can the paleontologist provide an operationally meaningful separation of a parent from a descendant species (Simpson 1953a).

If the change in the surroundings is very slow and the species well adapted to it, detectable change will be negligible, and the low evolutionary rate that may be documented in a fossil sequence can be termed bradytelic (Simpson 1953a) (Fig. 1–1). The brachiopod *Lingula* would be an example. On the other hand, significant change, encompassing several to many specific forms, would document other rates of change and probably would show some diversification of species, too. These faster or horotelic rates of evolution (Fig. 1–1) are exemplified by the evolution of land snails and mammalian carnivores.

Diversification. A necessary requirement for species diversification is geographic isolation. (Here we will omit any discussion of sympatric vs. allopatric speciation, isolation by food plant selection and so forth. The issue has been convincingly reviewed by Mayr (1963) and his conclusion that spatial isolation is the key to speciation is unavoidable.) Spatial isolation provides the essential opportunity for escaping the sexual tethers of the parent population. As has been made clear, an interbreeding population is a cohesive unit; it is an amalgam of mutually adapted genotypes. To escape the swamping effects of the larger gene pool, the local deme must irrevocably cut off genetic intercourse with it.

There are two ways in which geographic isolation can be initiated. Reduction in the numbers of a species may leave behind isolated pockets of locally adapted individuals, the remnants of previously interbreeding demes. Or, at the edge of a species, fortuitous circumstances could remove a few or even one member of the species to an environment that would support survival; and the chances of repeating the circumstances of removal being essentially zero, it means that there is effective isolation of the emigrants. If the single individual in this extreme case were a pregnant female capable of producing progeny of both sexes, then a new population could be inaugurated by what is now termed the "founder principle" (Mayr 1942, 1963).

It is important to realize that in both cases—constricted distribution and chance removal—the new individuals carry but a small sample of the genetic pool from which they came. This is both a liability and an advantage. The former is clear from what has been said earlier on the desirability of significant genetic variability in populations. The latter point needs some elaboration. The reduction in genetic resources is a kind of purge; the new population is starting from a relatively clean slate. At least it is free from the cohesive restraints of a fully evolved gene pool, and the way is effectively open for building its own constellation of genes in response to its own local environment. This is not to say the purge is a revolution in the sense of introducing vastly changed phenotypes within a generation or two of leaving the parent population. Drastic phenotypic change would be lethal. As with

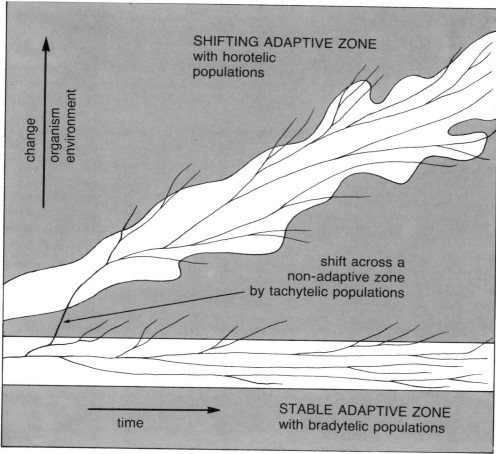

Fig. 1-1 Types of evolutionary change. Bradytelic populations show little change in time. Horotelic populations show steady change in time. Both of these are exploiting recognizable adaptive zones; the former is in a stable zone and the latter in a changing zone. Tachytelic populations show rapid change and are characteristic of forms moving from one adaptive zone to another. (From Hanson (1972) by permission, originally adapted from Simpson (1953a).)

species transformation, the new individuals are more conservative than innovative. The old developmental patterns must be retained to ensure the formation of functional new organisms, but within the developmental genotype or epigenotype (Waddington 1960) there will be stored new patterns of variability, and as time goes on there can be that subtle remodeling of the phenotype that visibly betokens a new biological species. The genetic revolution is, literally, by evolution.

One further point must be added about these newly isolated populations. It is obvious in the case of reduced distribution that the pocket of survivors is already significantly adapted to its local environment. A similar situation holds for the situation described by the founder principle, though it is not so obvious. The organisms arriving at a new locale cannot survive there unless they carry adaptions permitting them to exploit that area. These may be old adaptations used in the old way or old adaptations used in a new way. The latter are most important evolutionarily; they are the entering wedge into new niches and have been called preadaptations or prospective adaptions (Simpson 1953a). Preadaptations can be refined and extended by subsequent selection pressure, as can be seen in the classic transformation of the swim bladder of fishes to the lungs of terrestrial vertebrates. Their initial and fundamental importance, however, is that they make the transition to a new niche possible by endowing their possessors with that adaptive edge that can mean the difference between extinction and survival.

In isolation now, the new population can become so differentiated that if contact is reestablished with the parent population there will be no effective interbreeding. This can range from formation of hybrids with reduced fertility through hybrids with reduced viability to reduction in fertilization, reduction in mating, to, finally, no mating attempts whatsoever—to mention only a few of the many possible blocks to efficient sexual reproduction. In these instances where reproductive efforts are wasted, selection will favor those individuals who mate within their own populations, forming productive unions, and the process of isolation will be pushed rapidly to completion. Of course, if the spatially isolated populations reunite before isolating mechanisms (Dobzhansky 1950) are fully evolved, then differences will be minimized and the temporarily isolated deme will be incorporated again into the parent species. But where barriers to genetic intercourse are real, the new species will remain distinct from the old; and even if their range does again become sympatric, any tendency to competition will be removed by the evolution of characters emphasizing the distinctness of the two groups, thus removing them from a wasteful interspecies competition.

In such cases of speciation, evolutionary change can occur at great speed. Such tachytelic evolution may proceed twice or even ten times as fast as horotelic events (Simpson 1953a) (Fig. 1-1). This of course reduces the absolute numbers of the intermediate forms and, from the point of view of a fossil record, could be expected to result in gaps. Such apparently is one reason why certain paleontologists (e.g., Schindewolf 1950), geneticists (e.g., Goldschmidt 1952) and systematists (e.g., Petrunkevitch 1952) have favored a saltational or macromutational view of the emergence of new taxa, especially above the species level. All that seems left to the saltational view today is the genetic revolution described above, where new groups are founded with much reduced genetic resources. But this purge is distinctly not what has been proposed by those favoring macrogenetic changes. This brings us to a fuller and explicit account of the broader trends of evolutionary change, the formation of the higher taxonomic categories or transspecific evolution (Rensch 1954, 1960).

TRANSSPECIFIC EVOLUTION

The evidence is overwhelming today that the evolutionary development of diversity above the species level is by continuation of the processes that formed species initially, that is, by the further accumulation of differences between members of different gene pools (Simpson 1953a, Rensch 1960, Mayr 1963).

Evolutionary conservatism and continued adaptedness. The conservative forces of evolutionary change simply do not allow the establishment of major taxonomic differences between parent and progeny species. Where selection pressures differ between two recently isolated populations, there will be the first evolutionary divergence. The functions that allow an adaptive response to these different pressures, and the structures and genes determining these functions, will change first, and the rest, continuing to respond to the old pressures, will remain largely unchanged. But each population may transform and diverge again, and hence the differences will accumulate and become greater. Initially, therefore, the species that eventually gave rise to, say, a distinct phylum was not recognizable as a member of that final taxonomic group.

"The higher category is higher because it *became* distinctive, varied, or both to a higher degree and not directly because of characteristics it had when it was arising." (Simpson 1953a, p. 342). One need only consider the rare but decisive cases of intermediate forms such as *Archaeopteryx*, a feathered reptile, or the transition in the mammals of condylarths to artiodactyls, where the difference in the intermediates is essentially restricted to the tarsal bones (Simpson 1953a, Mayr 1963). In hindsight, *Archaeopteryx* will be called a bird because of its obvious importance as an example of the progenitor of a major order of vertebrates. But without the subsequent largely adaptive radiation of birds, *Archaeopteryx* would in all probability have to be considered simply an aberrant reptile.

Major evolutionary diversifications and trends must, therefore, always be seen in the context of continued niche exploitation. And in the perspective of geological time, it is clear that certain niches were the entering wedge on as yet empty or incompletely exploited adaptive zones (Simpson 1953a). Hence, the first amphibians were the forerunners in the zone subsequently exploited by vertebrate tetrapods as terrestrial predators. This permitted a vast array of specializations, including various herbivores and carnivores, some of which returned to the sea and some went on to become winged in several independent evolutionary experiments. In all changes, in these movements from one adaptive zone to another, in the further specialization of a zone, in its refinement into subzones, or even its narrowing into a single small zone—in all of these—the continuous, unremitting maintenance of adaptedness in both ancestral and descendant populations is an absolutely necessary requisite to survival. The transition from one zone to another can be materially aided by prospective adaptations, but that in nowise lessens the necessity for adaptedness. As we will see in Chapter 3, this unremitting demand for functional organisms puts very important restraints on phylogenetic speculation. It means explicitly that at no time are we permitted to envisage morphological transitions that would cripple the exploitive abilities of an organism nor disrupt its normal and ongoing homeostatic and reproductive functions. Natural selection does not accommodate arbitrary reductions in the fitness of members of a natural population. Conservatism in evolutionary change is necessary to preserve adaptedness.

Anagenesis and cladogenesis. At this point we insert two terms which will subsequently be important, i.e., anagenesis and cladogenesis. The terms as originally proposed by Rensch (1954, 1960) refer to the long-term results of species transformation and species diversification, respectively. Cladogenesis contained also an anagenetic dimension, for it included the appearance and rise of new structures in the process of diversification. Undoubtedly, in actual fact, the emergence of new parts does accompany the emergence of new taxa, so from a purely descriptive view the inclusion of diversification and trends together under cladogenesis is understandable. For analytical reasons, however, it is clearer to separate the two (Simpson 1953a), for above the species level the whole diversity of surviving lines of organic descent shows only two essential elements: separation and continuation. Huxley (1957) also has separated the two terms, whereby he refers to taxa that show significant anagenetic development and others that are predominantly cladogenetic, showing much proliferation of species. (He also recognizes stasigenetic, or evolutionarily quiescent, groups.) The emphasis we find useful in these terms is the following.

When a major taxonomic group, e.g., a phylum, shows within it the ability to evolve new modes of exploitation, homeostasis, or reproduction (the latter two will be especially important), then we can describe it as anagenetic. Or better, its anagenetic potential is demonstrably high. When, on the other hand, a major taxon shows a high degree of variation on a given adaptive theme, i.e., its variation is in the direction of diversifying a given exploitive mode with little or no change in homeostatic and reproductive mechanisms, then this is a cladogenetic group. Its cladogenetic potential is great. More simply put, cladogenetic groups show variations on a limited theme (typically exploitive) and anagenetic ones show new themes (importantly, reproductive and homeostatic ones).

We cannot, however, insist on a clean separation between these two sets of potentials. It is obvious that anagenetic groups must be able to exploit their environment to survive, and, hence, any new innovations that they produce, if successful, will demonstrate their success by at least a certain amount of cladogenesis. Conversely, extended cladogenetic development is never completely impoverished in terms of uncovering new potentials for functional and structural innovation. Hence, there will be some anagenetic development in these lines. What we come to is a judgment that will, on balance, argue that such-and-such a group is predominantly cladogenetic and showing little anagenesis in its evolutionary development, and another group shows very significant anagenesis with concomitant cladogenesis.

Simpson (1953a) recognizes four patterns of anagenetic change: (1) arrested, (2) trends, (3) casual or episodic, and (4) quantum changes. *Arrested change* will be Huxley's stasigenesis or any bradytelic groups. Rates of change are very slow; the overall pattern is often hard to discern, for evidently the group in question is well adapted to a very stable environment. This combined with the known mechanisms of genetic cohesion and canalization of development all cooperate to suppress phenotypic variability. *Trends* refer to changes in groups showing a constant, moderate rate of evolution. In them there is apparent change in response to a changed environment, or change as further and further exploitation of a broad adaptive zone, or both. These changes are relatively common in the fossil record, outstanding examples being the various trends found in horse evolution (broadly defined). Also relatively common is *casual or episodic development*. This differs from the preceding category in its variable rates of progress; that is, successive forms may evolve at different speeds. It is perhaps needless to emphasize that these rates are not inherent to the organisms but in all probability reflect the vagaries of environmental change and the chance exploitation of new niches and subsequent zones of adaptation. *Quantum change*—an undesirable term because of its connotations for saltational processes—refers to periods of very rapid, tachytelic, evolution that result in final forms significantly different from the parental form. This is not a common sort of process, both because fossil evidence for it would be expected to be, and is, rare, and because it represents rare evolutionary opportunities—the reaching out to what are very often major new adaptive zones. However, because they do so often represent highly significant innovations when they occur, these bursts of rapid evolution "have had an extremely important role in the whole of evolution" (Simpson 1953a, p. 386).

To summarize the foregoing and provide a transition to a phylogenetic view of these problems, let us briefly look at what stasigenetic and predominantly cladogenetic or predominantly anagenetic tendencies could be expected to produce. A stasigenetic group would be expected to be restricted in numbers because of its stable adaptation to a restricted adaptive zone. It may be highly specialized, which renders improbable its ability to exploit any new major adaptive zone. Hence, it is probably an evolutionary dead end. Probably no examples of stasigenetic taxa exist above the generic level. Thus examples of stasigenesis, such as the brachiopod *Lingula* or the mollusc *Ostrea* or the bony fish *Latimeria*, all are restricted groups within phyla that have shown considerable diversification and general evolutionary development.

Cladogenetic taxa show numerous variations on a major theme, as, for example the sponges with their ubiquitous flagellated chambers, the brachiopods with their dorsal-ventral valves and lophophorate mode of feeding, or the echinoderms with their pentaradiate body plan and water vascular system. These forms, as they exploit and refine the potentialities of a basic body plan, of necessity specialize its exploitive features. There is little within each group that is significantly anagenetic except very early. Indeed their origins may show tachytelic anagenetic development, or such may be inferred from the suddenness with which these groups appear in the fossil record with their basic form well evolved. These groups, with their emphasis on the exploitive potential of their epigenotype, are probably limited to no further evolution than the refinements that can be extracted from their basic organizational type.

Finally, groups which are notable for their anagenetic development, such as molluscs, arthropods and chordates, all show a basic body plan but, in addition, they also show a great variety of innovative developments which have also given rise to significant cladogenesis. Here again specific origins are unclear, and very high initial rates of evolution are probable. But between fossil evidence and the extensive data from contemporary forms, it is possible to make meaningful inferences about ancestral and a very significant succession of descendant forms. A comparison with predominantly cladogenetic phyla makes this point especially clear. In these latter phyla the presumed ancestral form is much more similar to subsequent forms than is the case for the tunicate larvae and mammals, for example, or for chitonlike forms and squids. The anagenetic phyla retain a plasticity in their organizational capabilities which result in significantly more truly innovative developments than is found in cladogenetic phyla. It clearly indicates that in them there has been achieved an epigenotype capable of new patterns of efficient development resulting in various new exploitive characters of formidable efficiency. It further suggests that if a given organizational mode or epigenotype—the terms are broadly interchangeable—is of selective advantage (and this can only be seen in the phenotypes it produces), that there is the possibility of far-reaching evolutionary innovations. Whereas, in the other cases, if an epigenotype's only advantage is in the exploitive characters it determines, the outcome is limited anagenesis until the exploitive character is fully developed and then cladogenesis predominates thereafter. This is our first view of the thesis that will be fully developed later, that the anagenetic phyla are the keys to broad-scale phylogenetic investigation.

Phylogenetic Implications

The organism as a member of a species is the product of two universal sets of constraints. One is that all organisms are evolvable systems; they are capable of exploiting their surroundings to maintain and make more of themselves and their variants. The second is that there are limited resources available for doing this. In that these evolvable or living systems are not equally proficient at perpetuating themselves under a given set of conditions, natural selection occurs which brings the potential for evolution to expression. The outcome of this is that organisms (1) depend on a cellular-genetic-metabolic organization for life (from subcellular viruses to supracellular multicellular systems), (2) form discrete gene pools through interbreeding, and (3) the members of different pools specialize in their utilization of the environment.

These facts are as important to the ambitions of phylogeny as firm foundations are to the aspirations of an architect or builder. They permit the conclusion that though there are many patterns today for exploiting the environment there is but one basic mode of functional organization for doing it; that is, there are many unique niches but only the cellular type of organism occupies them. This permits the student of phylogeny to consider seriously the possibility of fulfilling his or her most exciting goal: the unravelling of evolutionary history starting from the minimum of origins, which is one primordial type of cellular system (Pirie 1937; Hanson 1966).

This can be seen more clearly if we look for a moment at the contrasting possibility in its most extreme form, namely, a really different type of organism for each different niche. In this case phylogeny would be no more than niche enumeration. Or even if there were only 50, or even 5, modes of building organisms—each with its own type of molecular building blocks, its own independent source of energy, and its own three- or two- or one-dimensional pattern of organization—the problems of phylogeny would be different, perhaps demanding as many methodologies as there were types of organisms. But with the knowledge that though niches are unique but organismic organization is not unique, the phylogenist is allowed and encouraged to aim rationally at a unitary scheme of evolutionary descent.

However, there is one more requirement that has to be met to confirm the possibility of a unitary, rational approach to phylogeny. This requirement is that as organisms change in time their innovations not be so pervasive as to wipe out all traces of preexisting form and function. The extreme view would be that from one primordial cell all diversity arose by major extensive changes, leaving little or no trace of intermediates—the phylogenetic tree would be all leaves with no hints of twigs, branches, or trunks and their interconnections. That such is not the case is quickly inferred by anyone familiar with plant and animal classification. But that it contains disturbingly far more than a germ of truth is also common knowledge. For that reason we must know just how the process of diversification in time actually works.

The felicitous term "mosaic evolution" (Mayr 1963) goes to the heart of the matter for the phylogenist. Its meaning is immediately apparent by considering *Archaeopteryx* again. Feathers are the organizational innovation that resulted in a whole new array of exploitive possibilities. In regard to feathers, *Archaeopteryx* is a bird; in other regards, it appears to be a reptile. Here the response to selection pressures favoring some sort of thermoregulation also turned out to be a preadaptation for flight. The aerial niche was not completely filled by other forms, and birds as aerial vertebrates took on their distinctive form, moving significantly away from the forms of their reptilian ancestors. This was a mosaic process, some parts and their functions evolving much faster than others—what has been called in more technical terms "lineage allomorphosis" (Simpson 1953a)—with their evolutionary progress always determined by selection pressures. To have changed all reptilian characters at once to avian ones would necessitate a vastly improbable concatenation of genetic mutations and recombinations. To do it halfway or one quarter of the whole way would still be improbable statistically and almost certainly lethal developmentally. Epigenotypes, as highly integrated systems, cannot be juggled with impunity without a drastic, i.e., lethal, reduction in fitness of the organism as a whole. Hence, at any step in the evolutionary process—and the process is, emphatically, stepwise—evolution is far more conservative than innovative, as we have seen in reviewing the processes of speciation.

The central problem for the phylogenist is, therefore, to decipher the historical accumulation of information that lies within each organism. That this is theoretically a reasonable aspiration has been the aim of the preceding discussion. In what follows, we deal with the practical dimensions of its realization.

Chapter 2
The Informational Content of Organisms

MODERN evolutionary theory is to the phylogenist as the heliocentric theory of planetary motions was to Kepler, or as Kepler's laws of motion were to Newton, for foundational concepts clarify much, but, equally, they initiate new inquiries. Using an evolutionary theory which integrates genetics, population ecology, systematics, and paleontology, we can more clearly analyze and comprehend the similarities and differences in classical organ and organismic morphology, which largely comprise the older resource materials of phylogeny. But this theory tells us more: that what evolved ranges from nucleic acids to species, or even communities of species (Lewontin 1970). These are the new and definitive nadir and zenith of the phylogenist's purview. Our problem is to look at the array of functions and structures extending, in living systems, from the molecular to the populational level and see them as products of evolutionary forces, as an accumulation of historical information. The first step, which is the explicit aim of this chapter, is to identify that informational material. Its interpretation is relegated to the succeeding chapter.

Very simply put, the informational content of a system depends on the number of its subunits and how they interrelate. Hence, a recurrent question will be: What are the compositional units of a given structure or function and how are these units interrelated? And in the context of phylogeny a further question must be asked: How many different sets of possible interrelations actually occur, i.e., how diverse is a given structure or function? Answers to these two questions will be sought at the molecular, cellular, multicellular, and populational levels.

Molecular Information

The base line of our inquiry is the genetic information stored in the nucleic acids. Molecules smaller than the nucleic acids can provide information about the composition and diversity of living systems, but they do so as a consequence of acting with or being acted on by proteins which receive their specificity from a nucleic acid source. Size is not the criterion of the informational base line; the ultimate source of genetic specficity is that base line. "Molecular evolution stems from events that take place in DNA molecules" (Jukes 1966, p. 265).

NUCLEIC ACIDS

There are only two kinds of nucleic acids, deoxyribonucleic and ribonucleic acid, or DNA and RNA, respectively. Except for the RNA of certain viruses, DNA is accepted as being the final source of genetic specificity in an organism. Therefore, DNA is predominantly the molecular foundation for organismic specificity.

Deoxyribonucleic acid. It is now known that DNA resides in at least three places in the cell, i.e., in nuclei, mitochondria, and plastids. We shall be primarily concerned with nuclear DNA and will discuss it here. Mitochondrial and plastid DNA will be commented on later, in the context of examining those organelles.

The model of DNA as two complementary, helical, polydeoxynucleotide strands (Watson and Crick 1953a, 1953b) is now universally accepted and is thought to occur in all DNA-containing organisms except for certain bacterial viruses that carry only one polydeoxynucleotide strand. In bacteria and some viruses, judging from certain carefully studied forms, the DNA is a circular molecule (Cairns 1963a; DuPraw 1970). In the chromosomes of eukaryotic cells, though, it is linear and two ended, regularly possessing two or more double helices per chromosome, oftenly highly coiled in places, and intimately associated with RNA and proteins (DuPraw 1970).

The storage of genetic information in DNA, regardless of its configuration or associated molecules, is invariably dependent on the linear sequence of the purine and pyrimidine bases present in the successive nucleotides. Through genetic analysis we now know that the DNA, which extends throughout the length of a chromosome, can be subdivided into various units. Starting from triplet codons, these include the progressively larger units of cistrons or genes, operons (in prokaryotes), areas of crossover suppression (in eukaryotes), and arms of chromosomes (again in eukaryotes). These organizational units are better discussed as parts of chromosomes rather than in terms of DNA, for no chromosomal length of DNA has been so finely dissected as to demarcate all its genetic components in terms of its nucleotide components. Furthermore, the mode of comparing DNA molecules from different organisms, rather than relying on any of the foregoing genetic concepts, has made use of various physicochemical approaches. There are four of them.

The first is to identify the base sequence, base-by-base, for the whole length of a given DNA molecule. The second is to infer that sequence by indirect means, such as from known sequences in gene products, namely, from RNA or polypeptides. Third, there is the possibility of comparing two polynucleotide strands from separate DNA molecules by seeing how precisely they pair. And fourth, there are some general features of DNA molecules that permit certain comparative statements. These four approaches pose four different sets of technical problems and provide answers of differing degrees of usefulness from a phylogenetic viewpoint. We must discuss each point further, in some detail.

At present, base sequence determination by direct analysis is impractical. The essential problem is that DNA is an extremely long thread relative to its diameter. Normally the Watson-Crick double helix is about 20 Å thick in cross section, and a single DNA molecule from the bacterium *Escherichia coli* is about 700–900 μm long (Cairns 1963b). The length : width ratio in this case is 4×10^5. To isolate such a long, slender thread intact and then resolve it into its constituent nucleotides is technically a futile task today. This is not to say stepwise removal of purine and pyrimidine bases cannot be accomplished, for it can be. Holley and his coworkers (1965) and others have done it with transfer RNA molecules, but these molecules carry only about 77 nucleotides, whereas the bacterial chromosome carries about 2×10^7 nucleotides per DNA strand (Sager and Ryan 1961).

An indirect analysis of DNA base sequence is in some degree possible using known sequences in gene products to deduce the structure of the DNA parent molecule. Holley's work, just cited, illustrates the possibilities and limitations. In this case the complete length of a transfer, or tRNA, molecule was resolved into its nucleotide subunits. Sonneborn (1965a) promptly pointed out that this allowed the first unambiguous, end-to-end specification of the base sequence of a gene, i.e., the gene which specified the tRNA. This conclusion is based on the widely accepted assumption that RNA is formed or transcribed as a complementary copy of one of the two DNA polynucleotide strands. And knowing one of these strands from the RNA which it specifies, the other DNA strand can also be specified. The rules for this are straightforward. Each adenine, guanine, and cytosine in RNA designates a complementary thymine, cytosine, and guanine, respectively, in DNA; and for uracil in RNA, we read adenine into the DNA. And then, within DNA, adenine-thymine and guanine-cytosine pairing will account for the total structure of the double helix (Watson 1970).

The foregoing type of analysis depends, therefore, on complete end-to-end knowledge of nucleotide composition of given RNA molecules. Thus far, we know the base sequence of a mere handful of tRNA molecules. The larger messenger, or mRNA, molecules have not been deciphered. Even though these molecules have fewer nucleotides by many orders of magnitude than DNA—they possess on the average about 300 nucleotides—they still pose serious technical problems in manipulating a long delicate molecular thread so as to achieve its controlled degradation.

Another indirect method might appear to be the use of polypeptides whose amino acid sequence has been totally deciphered. These, too, are gene products, and progressing backwards from polypeptide, through mRNA to

DNA, it would seem possible to specify the base sequence of a complete cistron. This would be true except for the degeneracy of the code, wherein a given amino acid can be specified by more than one codon (Table 2-1) (Crick 1963; Watson 1970). This blocks attempts to deduce exactly the DNA base sequence responsible for a given polypeptide (Jukes 1965). As will be seen shortly, polypeptides, when studied in themselves, are capable of providing significant phylogenetic information without recourse to the nucleic acids.

The third and most successful approach to determining the informational content of DNA is to bypass entirely an explicit determining of base sequence and go directly to comparisons of similarity between DNA from different sources in terms of their ability to pair or hybridize with each other. This is, of course, the point of phylogenetic research—to specify the degree of similarity (or difference) between some aspect of two different organisms. Being able to make such a specification with regard to DNA would seem, then, to drive to the heart of the matter. The problem comes, however, in the pre-

Table 2-1 The genetic code. The first base in a given triplet of the code is at the left of the table, the second base at the top, and the third on the right. Hence, GCU specifies the amino acid alanine. Abbreviations of the amino acids: Ala—alanine; Arg—arginine; Asn—asparagine; Cys—cysteine; Gln—glutamine; Glu—glutamic acid; Gly—glycine; His—histidine; Ile—isoleucine; Leu—leucine; Met—methionine; Phe—phenylalanine; Pro—proline; Ser—serine; Thr—threonine; Trp—tryptophan; Tyr—tyrosine; Val—valine. Three triplets, Non-1, Non-2, and Non-3, specify no amino acid but are otherwise functional in reading the code.

	U	C	A	G	
U	Phe	Ser	Tyr	Cys	U
	Phe	Ser	Tyr	Cys	C
	Leu	Ser	Non2	Non3	A
	Leu	Ser	Non1	Trp	G
C	Leu	Pro	His	Arg	U
	Leu	Pro	His	Arg	C
	Leu	Pro	Gln	Arg	A
	Leu	Pro	Gln	Arg	G
A	Ile	Thr	Asn	Ser	U
	Ile	Thr	Asn	Ser	C
	Ile	Thr	Lys	Arg	A
	Met	Thr	Lys	Arg	G
G	Val	Ala	Asp	Gly	U
	Val	Ala	Asp	Gly	C
	Val	Ala	Glu	Gly	A
	Val	Ala	Glu	Gly	G

cision with which the degree of pairing can be specified. To make this clear, the technique for inducing and measuring pairing must be made explicit (Hoyer, McCarthy, and Bolton, 1964; and earlier papers).

The two strands of a DNA molecule can be separated by gentle heating and upon cooling they will rejoin, reconstituting the initial configuration. To make possible the pairing of DNA from different sources—heterologous reconstitutions—it is first necessary to prevent homologous unions. This is done as follows. The separated DNA strands from one donor source are embedded in agar (and this necessitates using DNA strands whose molecular weight is at least 10^7) to minimize subsequent leaching out. This DNA-agar preparation is then passed through a sieve and washed, and the particulate matter so produced is, in effect, a finely particulate gel with trapped DNA. To this is added specially prepared DNA from sources selected as desired—either homologous or heterologous. This added DNA has been radioactively labelled, then sheared, and, finally, the remaining fragments have had their double helices separated by heating to give single strands of DNA whose average molecular weight is about 5×10^5. These are small enough to diffuse through the agar gel, reach the entrapped DNA, and combine with it, if possible. Conditions allowing such combining can be relatively lax or stringent. Hoyer, McCarthy, and Bolton (1964) use routinely a "rather restrictive condition", which permits unions between homologous and heterologous materials, resulting in " ... properties expected of well-formed helical structures." The amount or extent of pairing is determined by first washing away any DNA not bound to trapped DNA and then freeing the hybridized material by elution, using gentle heating or change in ionic strength of the buffer solution or both. The fractions collected are then assayed for radioactivity. The value of recovered radioactivity in comparison to that initially added is a measure of the bound DNA, usually expressed as a percentile.

Combining nucleotide strands can be RNA and DNA as well as the DNA-DNA combinations just described. In the former case it is necessary that the RNA represent the total genome of its source. This is easily done in bacterial material, where the total genome is apparently wholly transcribed in each cell cycle, and so RNA from a random sample of log phase cells will provide RNA representatives of essentially the whole set of genetic information which is present. With this techinque, 90% of the binding sites have been found to be filled by complementary RNA. But where RNA does not adequately represent the genome, as in multicellular systems where differentiation results in transcription of only selected parts of the DNA, depending on the tissue and stage of development, then DNA-DNA bindings must be used. In these cases, for various reasons, the maximum pairing found, even between homologous DNA strands, never exceeds 30–40% of the added materials. This then is the base line against which heterologous pairings have to be measured. This has now been done using various vertebrate materials, and the results have encouraged *a priori* hopes that DNA hybridization work would show significant differences among the DNA studied, and that these differences could be interpreted in ways useful to phylogeny. The most critical way to make comparisons is to add unlabelled heterologous DNA along with homologous labelled DNA, the former being in concentrations up to a thousand times greater than the labelled material. The point of this is clear if one considers a few examples. If the unlabelled fragments are from a source very different from the labelled material, for example chicken and rhesus monkey DNA, it is clear one might expect little detectable competition between the chicken and monkey fragments for trapped monkey DNA. Indeed, the data show that in this case the measured homologous binding is reduced by only a few percentiles due to the presence of the heterologous materials. As the source of the heterologous material comes closer, in the taxonomic or systematic sense, to the monkey DNA, the amount of competition increases until, with unlabelled baboon and labelled homologous rhesus monkey DNA as an example, the inhibition of binding is very extensive. These same studies have been done with mouse DNA competing with salmon and bacterial (*Bacillus subtilis*) DNA for homologous mouse DNA (Fig. 2–1). The fish material reduces the homologous binding only slightly (a few percentiles) and the bacterial material failed altogether to inhibit the homologous reaction. We see then that the technique of DNA-DNA hybridization does not discriminate well between closely related nor very distantly related levels, designated systematically as below generic and above ordinal levels, respectively.

Furthermore, an additional complication has arisen in this type of analysis with discovery of extensive repetition or redundancy of genetic information in eukaryotic DNA (Britten and Kohne 1967). However, it seems possible to separate the repetitive DNA sequences from the unique ones on the basis of their rates of reassociation or reannealing following separation of polynucleotide strands by heating (Britten and Kohne 1967) and to concentrate on just the unique sequences. What these sequences represent genetically remains to be cleared up, and whether they insert a sampling bias into the DNA being examined is another unresolved question.

Before turning to RNA there is the final phylogenetic use of DNA analyses that deserves brief comment. Gross characterization of the DNA from a significant variety of organisms in terms of its guanine-cytosine (G-C) content (Sueoka 1965) raises phylogenetic problems of a very broad sort. For example, the range of the G-C content is 25–75% of the total DNA and is considerably more variable in lower forms like the bacteria than in higher plants and animals. Attempts to see if the observed values are indistinguishable from a random distribution leads to the conclusion that they are nonrandom and best explained in terms of natural selection acting on the proteins and thus determining the codons. Marmur, Falkow, and Mandel (1963) have used the

G-C content of bacterial DNA as an additional taxonomic trait in classifying this group of organisms. They find that it correlates well with other characters such as genetic behavior and DNA hybridization and to that extent is useful. These data on base ratios of different DNA molecules raise issues fundamental to the nature of the code and to gross distinctions between prokaryote and eukaryote cells (Antonov 1971) but they do not as yet answer questions regarding the finer details of evolutionary development. The same is true for certain other studies in DNA. For example, Hinegardner (1968) has shown that in certain teleost fishes the smaller DNA genomes correlate with the more advanced fishes and larger genomes with less advanced ones. Here phylogenetic determination of more or less advanced forms is putting genomic size in an interesting light. This DNA study, as Hinegardner fully understands, poses intriguing questions for the evolution of DNA rather than finding that DNA resolves phylogenetic problems.

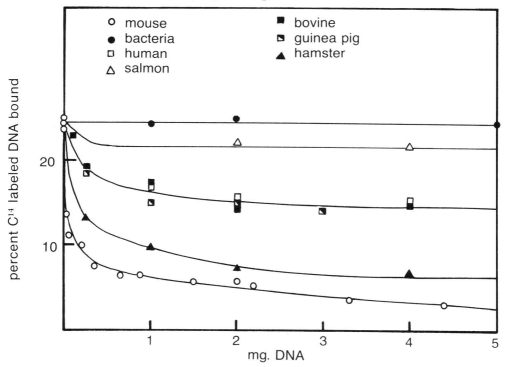

Fig. 2–1 Complementary relationships among single strands of DNA as seen in competition of homologous and nonhomologous DNA fragments for binding onto DNA immobilized in agar. Single-stranded, C^{14}-labeled DNA fragments from mouse cells were incubated with agar containing unlabeled, single-stranded DNA fragments from other species. The percent of the labeled fragments remaining bound to the immobilized DNA is plotted against amount of unlabeled DNA fragments. (Redrawn, by permission, from Hoyer, McCarthy, and Bolton (1964). Copyright 1964 by the American Association for the Advancement of Science.)

Ribonucleic acid. There are three kinds of RNA, each with its own physical and biological properties. At present, it appears that in terms of phylogenetic questions, these RNA molecules will tell us little that cannot also be learned from DNA or protein. The purpose in discussing them is to see them as adjuncts to studies on DNA and proteins.

Ribosomal RNA consists of several major components but none has been characterized in terms of its base sequence. From other properties, such as functioning rather nonspecifically as a part of the protein synthesizing apparatus, combining with protein into ribosomes, and existing as folded, single stranded polynucleotides with constant sedimentation coefficients (among their various physical properties), it is only the latter that provides some information on certain broad differences among ribosomes to indicate possible evolutionary variation. Loening (1968) reports that bacteria, actinomycetes, blue-green algae, and higher plant chloroplasts all have RNA molecules of 23S and 16S sizes; that is, whose molecular weights are 1.1×10^6 and 5.6×10^5, respectively. In higher plants, ferns, algae, and certain protozoa the molecular weights are 1.3×10^6 and 7.0×10^5. The latter component is common to all animals but the larger, the 28S, component increases in size, showing examples of 1.4×10^6 and 1.75×10^6. It is proposed by Loening that there are selective reasons, presently unknown, for these differences. (A somewhat similar story is made for whole ribosomes by Reisner, Rowe, and Macindoe (1968); see below.) Beyond these broad differences and general speculation, little can be said now about the composition and variability of rRNA.

Soluble or transfer RNA is more variable than rRNA; it is estimated that there are as many as 61 different molecules of tRNA in an organism (64 possible combinations of triplets, minus the three nonsense or punctuation triplets). If the genetic code is completely degenerate (Sonneborn 1965b) then all 64 available codons should be used either for punctuation or for specifying amino acids. The universality of the code puts very severe restrictions on the variability of the tRNA at the sites of the anticodon, which shows complementary pairing with the mRNA codon, and at the attachment to the appropriate amino acid (Hinegardner and Engelberg 1963). The latter seems most inflexible because of the limitation of twenty essential amino acids; degeneracy of the code allows the anticodon some variational latitude. Whether or not the rest of the molecule is free to vary depends, of course, on its function. If, as seems probable, it has a double helical configuration along parts of its length (Arnott 1971), then this may constrain random changes in those areas, the severity of constraints being determined by selection pressures to maintain a functional molecule.

The complete analysis of the sequence of nucleotide bases in the tRNA for alanine (Holley *et al.* 1965) and certain other amino acids (Madison, Everett, and Kung, 1966, and others) opens the possibility of comparisons between tRNA molecules regarding their structural variability (Dayhoff and

McLaughlin 1969). However, complete sequential analyses are still time-consuming, tedious operations, and it seems it will be some time before an extensive array of end-to-end analyses will be available for comparative study. When this happens, we will of course also have the same information on the genes that specified these molecules, which may or may not prove illuminating. However, it has been argued that tRNA is arguably a very conservative molecule (Hinegardner and Engelberg 1963; Dayhoff 1971) and this may mean we cannot ever expect much evolutionary information from it.

If pure samples of one kind of mRNA could be isolated, it too would provide information like that being obtained from base sequence studies on tRNA. At present such is not the case. There are as many mRNA molecules produced as there are cistrons transcribing them. Estimates range from 2500 for *E. coli* to as many as 10^6 for man (Monod 1971). They may differ little or significantly among themselves in their properties and under certain conditions are transient in their stability; under other situations they are stable and intimately bound to ribosomes. These circumstances combine to make it extremely difficult to obtain pure samples of the transcriptional product of a single gene. It has been done (Hall and Spiegelman 1961), but the amount obtained has apparently been insufficient for base sequence studies. It would seem simpler to go to the desired protein and analyze it directly.

In sum, studies on RNA will tell us little that is not also determinable from studying DNA or protein. Certain RNA molecules, the tRNA, do allow deducing the end-to-end sequence of bases in their respective DNA templates. This may be a way to study directly at least certain segments of the genetic material.

POLYPEPTIDES AND PROTEINS

Proteins are built from amino acid subunits into complex, three dimensional macromolecules. Between the amino acid level of organization and that of functional macromolecule are three and sometimes four levels of interaction among the constituent amino acid residues. The primary level is determined by a linear, unbranched sequence of covalently bonded (the peptide bond) amino acids. In some parts of this polypeptide chain, the residues can fold to form a characteristic helical structure, termed the α-helix, which is the secondary structure of the protein. This single-stranded coil is stabilized through weak intramolecular bonding—largely hydrogen bonds. Tertiary structure refers to the folding of the remaining parts of the amino acid chain onto itself or onto the secondary structure, or both. Weak bonds and occasional covalent bonds, notably disulfide linkages (-S-S-), provide the atomic interactions stabilizing this level of configuration. Up to this point, a single polypeptide chain is the organizational unit whose subunits are amino acids. Many proteins are made up of two or more polypeptide subunits or monomers: the interaction of these, primarily through many intermolecular weak bonds, defines the quaternary structure of proteins. The final result, at either the tertiary or quaternary level, is a three-dimensional molecular complex with specific properties, which are determined by the location and nature of open bonding sites, the placement of polar and nonpolar amino acid side chains, and the effective attachment, in certain proteins, to other atoms or molecules. For enzymes, the surface configuration must generate an active site for binding with substrates, and also there may be sites for combining with molecules that regulate enzymatic activities through allosteric properties (Monod, Changeux, and Jacob 1963). Enzymes may also combine with inorganic molecules such as the coenzymes CoA or certain flavins, these unions being necessary for proper functioning of the enzyme. Proteins may also conjugate with larger molecules to form the lipoproteins of membranes or mucoproteins of cell walls, or they may combine with prosthetic groups like heme and chlorophyll. And, finally, proteins will interact with themselves to build up membranes and fibers. (For more details on protein structure see references such as Lehninger (1970).)

It is obvious that the one-dimensional array of information specified by the nucleic acids is transformed through folding of the polypeptide chains and their combining with other chemicals, both like and unlike themselves, into more highly ordered complexes with new capacities for interaction. This suggests various parameters that can be measured to compare proteins from different sources. (a) In that amino acid sequence is a direct reflection of codon information, comparative sequence analyses speak directly to degrees of genetic difference between the polypeptides being compared. (b) Gross amino acid analyses permit less precise statements regarding the basic genetic information but can describe differences between proteins with some accuracy, the usefulness of which depends, naturally, on the questions being asked. (c) Enzymatic function can be described in terms of rates of reactions under specified conditions, which include temperature, pH, and specificity towards substrate. (d) Various physical properties can also be used to characterize proteins, such as stability under varying conditions, size, shape, and molecular weight. (e) Finally, immunological studies can be used to characterize the proteins by defining their similarities in terms of serological cross-reactions. Each of these parameters has yielded information on the evolution of proteins, but with varying degrees of relevance to phylogenetic questions; thus far, the analysis of amino acid sequences has been the most informative line of inquiry.

Certain globins and cytochromes c have had the most extensive analysis, and their stories illustrate, as well as any, the trends in such studies.

There are at least 43 hemoglobins and myoglobins, all from vertebrates, that have been analyzed completely in terms of their amino acid sequence (Dayhoff 1969). It is possible, moreover, to assign homologous positions to all the residues, thus permitting precise identification of differences—additions, deletions, substitutions—for the total primary structure of each molecule. Of the 141 amino-acid residues in the globin polypeptides, only a few are invariant. In their survey of this problem, Zuckerkandl and Pauling (1965) report eleven invariant residues, among which are those, or the one, necessary for binding the heme group. But even this is changeable, for in humans a hemoglobin designated Hemoglobin Gun Hill (HbGH) lacks the heme groups and also five amino acid residues located near it (Bradley, Wohl, and Rieder 1967). This reduces the number of invariant sites to six, or 4% of the total, and there seems reason to think even these can change. The question must be raised, however, whether HbGH would persist in the evolutionary sense, for it is one thing to find a newly arisen variant, as HbGH apparently is, and another to know whether or not it will survive in the face of natural selection. It seems that humans with HbGH are less fit than others and that this form of hemoglobin would be selected against. In which case, the number of residues that are invariant in the longer perspective of evolutionary time may well be the larger figure of eleven, or 8% of the total.

What are the patterns of substitution that are found to occur in the 92% of the residues that do show variability? The amount of substitution varies from one site to another, from one to a maximum of eight. For the whole myoglobin molecule, the average number of substitutions per site is three. Furthermore, it must be asked: At those sites where heavy substitution occurs, is there a regular pattern to the changes? For example, do polar amino acids tend to be substituted for polar amino acids more often than would be expected on the basis of random substitution? The problem is to see if conservation of protein structure or function can be inferred, from present knowledge of the properties of amino acids, to be a basis for directing single residue substitutions. The surprising answer is, No. There are apparently quite unexpected substitutions; it is unexpected that somewhat polar methionine would be a common replacement for nonpolar leucine (Zuckerkandl and Pauling 1965). Or to express it more vividly, "Apparently chemist and protein molecules do not share the same opinions regarding the definition of the most prominent properties of a residue." (Zuckerkandl and Pauling 1965, p. 129–130). And it is concluded that, for these globins, there is no, or almost no, single amino acid residue that is specifically needed for stabilizing a given conformation of the molecule. The stability of the whole molecule has to be explained, then, as a coordinated overlapping of the various properties of the many residues.

> "It appears that the basis for extensive changes in amino acid sequence without any radical change in tertiary structure and protein function is furnished by the fact that each amino acid residue has several important functional properties [—charge, hydrogen, bond forming capacity, polar-

ity, bulk and perhaps others (p. 134)], and that the set of amino acids that are coded for is chosen so that changes in one or more properties can occur while one or more other properties are maintained constant. By a propitious choice of the properties to be maintained constant, in the case of each particular residue, the amino acid sequence of the polypeptide chain can be transformed, and yet its basic pattern of intramolecular and intermolecular interactions remain the same." (Zuckerkandl and Pauling 1965, p. 127.)

A striking example of this is the fact that 107 of the 141 residues are different between whale myoglobin and human α-hemoglobin, but the two chains are very similar in their tertiary structure and their function (Zuckerkandl and Pauling 1965). It appears, therefore, that there is a kind of degeneracy in protein structure in that more than one primary structure can determine very nearly the same tertiary pattern and function. This point raises serious questions regarding the role of natural selection in determining the primary structure of proteins; we will return to this problem in the next chapter.

These results with the globins are in sharp contrast to what is found with cytochrome c, where about 30% (34/113) of the amino acid residues are invariant (Margoliash and Smith 1965; Dayhoff and Park 1969). Here, complete sequence analyses have been made on materials from 21 vertebrates, 4 insects, 3 different fungi, and wheat (Dayhoff 1969). (Cytochrome c is also found in bacteria but is thought to be of a different type than the so-called "mammalian type", which forms the basis of the present comments.) Regarding the variable portion of the molecule it appears that

". . . an exact study of quantitative aspects of cytochrome c evolution in relation to the evolution of species will be possible only if and when appreciably accurate knowledge will become available of the structural-functional role of each amino acid residue, alone and in conjunction with others." (Margoliash and Smith 1965. p. 235.)

This conclusion, very like that coming from the globin work, reflects the present inability to make detailed sense of the known patterns of residue substitution in these cytochromes. A beginning towards resolving this problem is indicated by Dayhoff and Eck (1969) where five groups of mutational substitutions in various proteins have been described.

It seems, then, that two choices are available for further work. One is to make a careful analysis aimed at elucidating the function of the whole structure of the proteins and the role played by each amino acid by itself and cooperatively. This promises to be a gargantuan task. The other approach is to short-cut the detailed analytical approach and derive empirically some measure of the variability between proteins that is of phylogenetic import. Let us try to summarize where we are with both approaches. First, the guidelines that have emerged from analyses of amino acid sequences.

Invariant sites are to be expected, but their extent is not known in detail for any molecule and, therefore, firm rules for their definition and delimitation are absent. Logically, invariance is related to conservation of function. (But as we will see below, this is not the only way to conserve function.) Some functions, such as binding to prosthetic groups like heme, may involve only one amino acid residue. Others, such as attachment to membranes, may involve extensive areas of the protein molecule. This has been suggested as the reason for the large number of invariant residues of cytochrome c, which implies the membranes themselves are largely invariant. This could be so, especially because of the very stable lipid component of unit membranes. Indeed, one would expect that the more invariant the material to which the protein binds, the more conservative are the protein binding sites associated with it. Other sites, which involve more than the one or two residues of prosthetic groups and fewer than the postulated multitudinous bonds to membranes, are active sites for substrate binding by enzymes. However, with dehydrogenase, where the residue sequence of the active site is known, only one residue appears invariant (Kaplan 1965). In general, however, in terms of interacting with other substances, invariant sites can apparently range in number from one to dozens.

Invariance can also be due to preservation of those residues that, though not directly concerned with active or binding sites, nonetheless, through their influence on tertiary structure, can affect the integrity of those sites. For although it is now argued (see citations above) that cooperative association of residues is most important for understanding protein structure and function, there is evidence that single substitutions can seriously change the property of these macromolecules; the difference between sickle cell hemoglobin and normal hemoglobin comes down to one amino acid residue (Ingram 1957). Critical sites that can affect the integrity of the molecule are the following (Zuckerkandl and Pauling 1965): 1. inside the molecule: increase in bulk or change in polarity; there is little or no room for the former and the latter could destabilize the molecule; 2. at the surface: changes in polarity can affect the solubility of the protein; 3. at bends: changes could occur that would destabilize such sites and tend to straighten out that region; and 4. in helices: changes could eliminate the helical pattern.

Finally, invariance could be the result of inability to reach a stable new configuration except by going through intermediates which are selectively disadvantageous and which may require many mutational steps to pass through them (Zuckerkandl and Pauling 1965). Invariance in this sense is a kind of adaptive peak.

Conversely, variant sites would be those not lying at the critical patterns just discussed or would represent changes that can be made, even at these critical sites, providing the substitution does not adversely affect the essential structure and function of the molecule. It is still one of the enigmas of protein structure that so much of the molecule can be changed or lost without affecting the detectable function of the molecule. In the adrenocorticotrophic hormone (ACTH), 16 or perhaps 24 of the 39 amino acid residues are necessary for hormonal activity. With the enzyme papaine, 80 of 180 amino acid residues of the enzyme can be removed without impairment of enzymatic function (examples cited in Anfinsen 1955). If, indeed, there are large areas of the molecule with no real function—silent areas in terms of detectable mutational effects—then random mutational events can occur here, and divergences in such areas between two or more species can be used for what has been termed a "molecular evolutionary clock" (Zuckerkandl and Pauling 1965) that measures the absolute rate of mutation of the cistrons coding for these neutral areas, and where degree of change would be closely correlated with phyletic distance.

This question of the selective advantage of mutational events at the macromolecular level will be examined more fully in the next chapter. Here we need only make the point that due to degeneracy of the code it will obviously be possible to change the third triplet in many codons and have no detectable change in amino acid content of a polypeptide. As King and Jukes (1969) put it, "Specifically, there are 61 amino acid-specifying codons. Since each of the three base pairs can mutate three ways, each codon can mutate nine ways by single substitution. Of the 549 possible single-base substitutions, 134 (one fourth) are to synonymous codons. These are heritable changes in the genetic material, and hence true mutations. As far as is known, synonymous mutations are truly neutral with respect to natural selection." (p. 789) The important question regarding the role of selection lies beyond these mutations and, as we shall see, opinion is divided as to the best answer.

Turning to the empirical approaches to phylogenetic relations as seen in proteins, it is convenient to start with the work on various dehydrogenases (Kaplan 1965).

The amino acid composition of triosephosphate dehydrogenase from eleven different sources has been compared. The most striking fact seems simply to be that there is a similar number of residues (1083–1117) in all of these proteins, arguing that the cistrons coding for these molecules have changed little in size. Looking at the catalytic characteristics of these enzymes largely in terms of the coenzymes used, it would appear that some relationships are determinable to the extent that the insect enzymes are more like crustacean ones than they are like those from arachnids. This suggestion says little by itself; it is one more item to be added to phylogenetic inquiries within the Arthropoda. Comparisons of physical properties—stability at various temperatures and electrophoretic mobility—were also made but encouraged no new conclusions; they were simply in general agreement with already established views.

Immunological studies on serum albumins (Hafleigh and Williams 1966) as well as on the dehydrogenases show rather sensitive abilities to discriminate between the proteins of various groups. However, since the full details of complex antigen-antibody reactions are not clear, it appears that the use of the detected array of cross-reactions must be evaluated by other than purely analytical means. That is, since we have no ideas such as the mutational changes in sequence analyses to guide our interpretations of serological differences, we are forced to compare the serological results with phylogenies based on classical morphological and paleontological data. When this is done (Leone 1964; Hafleigh and Williams 1966), it can be seen that, in general, broad serological differences correlate well with phyletic differences. But whether one can now extrapolate these serological results to areas where no other independent phylogenetic work is available and be confident of the conclusions indicated from the immunological data, is still an open question.

BIOSYNTHETIC PATHWAYS AND PRODUCTS

There remains now to be considered the evolutionary information contained in the ways organisms build up and break down their molecular constituents. Much of this information is not useful to phylogenetic problems, for it forms an ubiquitous set of biochemical processes unable to distinguish a virus from an elephant or a fungus from an octopus. These have been alluded to earlier (Chapter I) as the biochemical unity of living systems. Arising from that common substratum there are, however, a variety of metabolic capabilities that do mirror, in varying degree, organismic diversity. What are the organizational units of these special processes and what are their interrelations? What differences can we see in the processes as they exist in different forms? The discussion will first consider specialized metabolic pathways and special metabolic products, and then control mechanisms—genetic and metabolic.

Biochemical specializations. To see most clearly the nature of biochemical specializations, it is best to display them against the backdrop of general trends in comparative and evolutionary biochemistry. Generally it is agreed that the first living system was a chemoheterotroph, taking building blocks of relatively complex organic compounds—formed spontaneously without the intervention of living systems—and using energy released from chemical reactions (Horowitz 1945; Needham 1959; Oparin 1962) to make more of itself and its variants. Initial biochemical evolution is postulated to be the result of selective pressures leading organisms to exploit new resources when ones initially present were exhausted. This means not so much the turning to a new category of molecule as building substance, but, rather, gaining the ability to supply the depleted material from its naturally-occurring precursors, and when these precursors are depleted, developing the resources to build them anew from their precursors. The essential method of gaining new synthetic abilities is through mutations of genes forming enzymatic proteins. In this way, gene-controlled biosynthetic pathways emerge, and the direction of biochemical evolution is from relatively unsophisticated biochemical abilities to very complex ones in terms of using relatively simple sources for building blocks—CO_2, NH_3, CH_4, for example—and sunlight for energy. In brief, chemoheterotrophs evolved towards photoautotrophs (Horowitz 1945; Calvin 1961). There was a limit to this evolution, which relies solely on the inorganic environment as a source of materials. Eventually, life supported from this source would reach a limit for its natural resources are not infinite, and much material would be tied up in the products of organisms, living and dead—the accumulation of metabolic efforts of life up to that point. With the emergence of biochemical pathways capable of exploiting these products of metabolic activities and, especially, of cooperative utilization of resources in the sense of carbon and nitrogen cycles and of ecosystems in general, evolution, broadly speaking, and biochemical evolution in particular, stepped into patterns of adaptive cooperation that characterize ecological interactions to this day. Some organisms retained their photosynthetic activities and reliance on the simpler compounds of their environment, i.e., the producers; but others, living off of these photoautotrophs, were no longer under selective pressures to maintain a wide variety of biosynthetic capacities, since so much—amino acids, sugars, fats, various coenzymes (vitamins)—was already formed for them by the producers. With no selection to oppose the eroding effect of random mutation, various synthetic abilities were lost, with the consequence that chemoheterotrophs reappeared, no longer exploiting complex abiogenetic organic compounds, which were no longer abundant and never would be again, but depending exclusively on biogenetically formed organic materials. In simplified outline, early animal evolution is a story of the progressive loss of biosynthetic capabilities (Lwoff, A. 1943).

Looking further at animals, unicellular and multicellular, what are the sorts of metabolic specialization that can be distinguished above and beyond the processes that are their common heritage as chemoheterotrophs derived from photoautotrophs? They can be classified, roughly, as of two types: adaptations to special food sources and the building of special products. The former can be seen especially clearly in parasitic forms, which, again through unopposed mutation pressure, lose those characters not necessary to survival within their hosts. But also, responding to unique selection pressures arising from exploitation of very specialized habitats, they also show significant innovations. Under such conditions there arise the distinctive metabolic patterns of the blood-flagellates (M. Lwoff 1951) or of various insects (Dadd 1970). But a combination of special selection pressures is potentially a source of specialized adaptation, and biochemical characters, like other characters of living systems, will be responsive to these pressures. In addition to anabolic capabilities there are also catabolic ones. As an example, there exist various patterns of breakdown of nitrogenous materials largely in relation to water conservation, since such materials are largely eliminated by excretion (Schmidt-Nielsen 1964). In all of these situations, anabolic and catabolic, organisms are bringing the utilization of sometimes exotic chemicals into conjunction with basic metabolic pathways such as those of protein, nucleic acid, lipid, or carbohydrate syntheses or are developing pathways to eliminate efficiently catabolic waste products derived from these pathways.

The building of special products is seen in the formation of structural parts such as skeletal parts, e.g., the strontium sulfate spicules of radiolarians (Trégouboff 1953a) or the quinones of chitinous exoskeletons (Baldwin 1967). Or again, special compounds appear with regard to special functions, such as those controlled by molting hormones in insects (Sehnal 1971) or sexual hormones in vertebrates (Barrington 1968). And there is also a special list of those chemicals used as defensive agents in various insects (Pavan and Dazzini 1971).

In general, in the animals, biochemical innovations are almost certainly adaptations allowing for more effective exploitation of a niche. Not unexpectedly, then, we find these adaptations to have a spotty distribution. Hemoglobin appears in the flatworms, annelids, certain molluscs and insects, some echinoderms and in the vertebrates (Baldwin 1966). Presumably these represent groups where selection pressures have been met by appearance of this molecule; its origin is multiple and independent and for that reason is of less phylogenetic use than would have been the case if there had been one origin with subsequent elaboration of the function of the molecule. That elaboration, as far as it has been followed, has occurred within the molecule as it occurs among the vertebrates; the story is contained in the amino acid sequence rather than at the supramolecular level of the whole complex business of oxygen transportation. We conclude, therefore, that special metabolic products and pathways, though possessing subunit complexity in terms of their molecular structure or of their biosynthesis, nevertheless show little variation of these structural or functional patterns. It would appear that this sort of molecular information is either adaptive, and thereby gains a straightforward selective value, or it is not and is not preserved.

Control mechanisms. By genetic control is meant not only the genic loci concerned with elaborating the enzymes that determine a given sequence of biosynthetic steps, but also the pattern of chromosomal organization of the loci in question. The bacterial pattern shows the loci controlling a given series of biosynthetic reactions as all lying linked to each other (Demerec and Hartmann 1956), but this is not found in eukaryotic systems (Demerec 1965). In higher forms, even though the same synthetic steps are present, the loci can be scattered and lie in different chromosomes. There is evidence, however, that if two polypeptides interact to form a single functional unit, their respective cistrons may be converted to a single genetic unit by translocation (if they were on separate chromosomes) with subsequent elimination of code punctuation between them, so that they are transcribed, in tandem, forming a single mRNA, and this is translated into a single polypeptide (Bonner, DeMoss, and Mills 1965). This raises the possibility of genetic analyses to describe the pattern of distribution in a genome of the loci concerned with sequential biosynthetic reactions. In the case of eukaryotes, the present data are equivocal (Lewis and John 1970). In the most intensively studied forms—the fungi, *Neurospora*, *Ascobolus*, and *Saccharomyces*—there is both evidence of the scattering of metabolically-related loci among different chromosomes, even though their activity can be coordinated, and also evidence for some operon-like clusterings of genetic loci (Giles *et al.* 1967; Berlyn 1967; Berlyn, Ahmed, and Giles 1970). The functional analysis of the eukaryotic chromosome is just getting under way.

Lastly, there is the suggestion that metabolic control mechanisms in biosynthetic pathways may be useful in phylogenetic studies. This work comes from studying the formation of β-ketoadipyl-CoA in bacteria (Cánovas, Ornston, and Stanier 1967). Despite the fact that various bacteria have a metabolic pathway for β-ketoadipate, it is clear there are two different patterns for regulating it, which correlate clearly with the two different groups of bacteria in which they are found. These patterns are seen essentially in terms of the inducers of gene action, and from these differences it is inferred that despite apparent metabolic homology—the two systems produce the same end product—they do so as a result of convergent evolution, not descent from a common ancestor. Here, in this case, control mechanisms have been useful in broadly clarifying the evolutionary development of a given synthetic pathway.

Control mechanisms do not, at this stage of our knowledge, appear as a rich source of phylogenetic information. First, where suggestive data are found, they seem limited to the bacteria. Second, when eukaryotes are involved, the data are fragmentary and at best only suggestive of evolutionary changes.

Structure and Functional Organization of Cells

It has been conjectured (Setlow and Pollard 1962) that to build a typical bacterial cell, starting from atoms, there would have to be specified 10^{12} decisions of the "yes" or "no" sort. That is, the information content of a prokaryote cell is thought to be of the order if 10^{12} bits of information. To use this high degree of order phylogenetically we need, as stated earlier, to specify the subunits and their interrelations. Our inability to do so, to take a given part of the cell and accurately map its constituent atoms, renders much of the informational content of cell systems unavailable to phylogenetic inquiry. Where can we start?

For the present we have no choice but to proceed from cell organelles and their functions to the compounding and complexing of their forms and their functions into cells of various sorts. Though we know that organelles are composed of macromolecules, and these in turn are made up of smaller molecules, and finally there are atoms, nonetheless we are only at the beginning of understanding how organelles are assembled. The bacterial flagellum can come together *in vitro* from its constituent macromolecules (Abram and Koffler 1964) as can certain viruses (Oosawa *et al.* 1966; Casper and Klug 1962) or microtubules (Stephens 1968; Tilney 1968), but these isolated examples are insufficient for the comparative data needed to study the path of evolution. More importantly, they may never provide that information, for it seems increasingly certain that the major cell organelles are very conservative systems and may be largely invariant throughout all living systems. This is, of course, the unity of cellular structure discussed in the preceding chapter. If this is so, or even if the variation is limited to the neutral changes that can occur in proteins due to changes in degenerate codons, then we must look for a second base line from which to specify another organizational level, another set of information. It seems highly likely that the organizing of covalently bonded atoms into their various molecules and the organization of molecules into macromolecular aggregates is the chief organizational effort of living systems. It represents an as yet largely unexplored area for biology. Be that as it may, it is clear that provided with certain macromolecular aggregates, i.e., cell organelles, cellular systems show an astounding diversity in the way they intermingle these elements. Though we no longer take seriously the idea of a typical cell (Wilson 1928), we do agree that there are typical cell organelles; these are the subunits of cellular organization.

Because the essential problems of this study lie with animal cells, we will concentrate on them in the following discussion, referring to prokaryotes and plant cells only in selected instances. A general treatment of this topic, but stressing tissue cells, is the work by Threadgold (1967).

NUCLEAR STRUCTURES AND FUNCTIONS

Of the four major parts of the nucleus—nuclear membrane, nucleoplasm, nucleoli, and the chromosomes—the chromosomes are the most complex in terms of structure and functional behavior. For that reason they must be considered in some detail, which will be done after more limited comments on the other structures. Lastly, there will be considered the coordinated behavior of these nuclear components, as in mitosis and meiosis.

Nuclear membrane. Structurally, the nuclear membrane is a double unit membrane whose most striking further feature is the presence of annuli (Du Praw 1970). A typical annulus represents a localized fusion and thinning of the two unit membranes, surrounded by a raised ring of material, sometimes consisting of electron dense spheroids. On occasion the center of the annulus appears open. The size of annuli is somewhat variable: outside diameters range from 600–1450 Å and inside diameters are from 200–800 Å. Center-to-center spacings are from 900 to 1400 Å and the numbers of annuli per square micrometer vary from seven up to 145. There seem to be as yet no patterns to this variability, i.e., none is correlated with any particular kind of cell, tissue, or species. The variability more likely reflects function, for there is evidence of the movement of materials between nucleus and cytoplasm via annuli. Possibly more active cells possess more and larger annuli, but this remains to be established.

On the inside of the nuclear membrane there can be specialized structures such as the peculiar honeycomb structure seen in *Amoeba* (Pappas 1956) or the densely packed fibers of certain vertebrate cells (Fawcett 1966). The inside of the nuclear membrane can also serve as an attachment site for chromosomes. This may be related to processes of DNA replication (Becker and Hurwitz 1971) or of chromosomal movement, as seen in certain flagellates (Cleveland 1956b and earlier papers).

All eukaryotic cells possess a nuclear membrane (in contrast to prokaryotes which uniformly lack such a structure) and except for rare instances like the marine dinoflagellate *Noctiluca* (Afzelius 1963) it is always an annulated, double-layered structure.

Nucleoplasm. The nuclear sap or nucleoplasm is probably best defined negatively as being that material which is not nuclear membrane, chromosomal, or nucleolar material. In these terms it is usually seen as a relatively homogeneous, probably fluid area with occasional instances of inclusions. These may be metabolic products such as hemoglobin (Moses 1964) or bodies of as yet undefined chemical nature and biological function, variously termed karyosomes and endosomes (Wilson 1928; Grell 1968; Raikov 1969).

Nucleoli. These are RNA-rich bodies, usually 2–5 μm in diameter which are attached to specific chromosomal loci (Du Praw 1970). Not all cells possess nucleoli, but those that do often have one nucleolus per haploid set of chromosomes: thus diploids have two nucleoli, and triploids have three; however, nucleoli can fuse. Details of nucleolar structure, as seen in the electron microscope, have led to distinguishing three parts: an outer *pars granulosa*, an inner *pars fibrosa* within which lies the DNA thread of the *pars chromosoma* of nucleonema (Hay 1968). The outer, granular portion contains densely packed ribosome-like granules, rich in RNA, and often about 200 Å in diameter. The fibrous part is less dense and contains filaments about 100 Å in diameter. The nucleonema is continuous with the DNA of the chromosome to which the nucleolus is attached and is, in effect, the nucleolar gene or, in older terms, the nucleolus organizer. It is this segment of DNA that transcribes the ribosomal or rRNA (Brown and Gurdon 1964).

The function of nucleoli is the controlled formation and release of rRNA, possibly in the form of protein-rRNA complexes (pre-ribosomal particles), which eventually find their way to the cytoplasm, presumably through the annuli of the nuclear membrane. Whether the nucleoli are also sites for the synthesis of ribosomal protein remains to be determined. Also, in certain nucleoli there are multiple copies of the organizer DNA, and, therefore, a sort of special regulation of DNA synthesis occurs in the nucleus termed gene amplification, because of the multiple copying. Finally, it should be mentioned that the organizer region, at least in certain forms, consists of two regions, one for the 28S RNA and one for the 16S RNA, and these are repeated, in tandem fashion, along the length of the nucleolonema for as many as 450 times, as in the case of the toad oocyte (Brown and Dawid 1968).

Nucleoli are, therefore, quite uniform structures. Their variation lies largely in their occurrence, and this depends largely on the functional state of the cell. If the cell is actively synthesizing protein (a secretory cell) there is the need for many ribosomes and nucleoli are conspicuously present. Otherwise, they may even be absent. Hence different cells of the same organism may or may not have nucleoli.

Chromosomes. The eukaryotic chromosome is a long, continuous, two-ended, highly-coiled thread, whose backbone is a Watson-Crick helix of DNA. Attached to the DNA, in ways only poorly understood at present, are RNA and certain proteins, notably histones (DuPraw 1970) (Fig. 2-2). This structural unit has various distinguishing features: (a) size, (b) internal differentiations and organization, and patterns of (c) coiling and uncoiling, (d) replication and (e) transcription. Additionally, chromosomes exist as sets

Fig. 2-2 Idealized drawing of a eukaryotic chromosome. To the left is a randomly coiled chromosome composed of proteins, RNA, and DNA. The more detailed view on the right shows the coiling and supercoiling known to be present and suggests how the DNA double helix might be packed with the help of wedges of protein. (Redrawn, by permission, from Du Praw (1970) *DNA and Chromosomes.* Freeman & Co.)

distinguished by (a) number of individual chromosomes, (b) number of sets or ploidy, and (c) aggregate behavior at mitosis and meiosis. Clearly, chromosomes are highly complex organelles, especially in contrast to other parts of the nucleus.

Chromosomal size can refer to linear length at some time in the cell cycle or, thanks to recent techniques of measurement, can refer to the mass of individual chromosomes (Du Praw 1970). The former provides the more common data. The central point is that a given chromosome is a very constant unit of organization. It will, of course, duplicate and divide at precise times in the cell cycle and will engage in transcription at other times. But when a given chromosome, e.g., the Y chromosome in *Drosophila melanogaster* or a human chromosome, is taken at comparable times from different cells that contain it, and its length or mass is determined, this is found to be very stable. Highly coiled metaphase chromosomes can be as short as one or two micrometers (micra or microns) in length, and highly extended polytene chromosomes can be more than 100 micrometers long. The DNA core itself can be as much as one metre (10^6 μm) long in the chromosomes of *Lillium* (Taylor 1957). This poses, dramatically, the packing problem facing each chromosome in reducing such a linear dimension to the one observed at metaphase (ca. 5 μm).

Differentiations along the length of the chromosome include the two ends, some point in between where the centromere lies (which includes the kinetochore as attachment site to spindle fibers), possible patterns of more or less dense coiling of the DNA, and possibly the presence of a nucleolar organizer. Any one chromosome will show a fixed organizational pattern with regard to centromere, coiling patterns, etc. The classic example of coiling patterns are the bands of the giant polytene chromosomes of dipteran salivary gland cells. These have such constant detail as to permit correlations with genetic mapping. The positions of centromeres are often marked as localized constrictions. Similar constrictions can occur elsewhere on the same chromosome, having nothing to do with the centromere, and can define a terminal knob or satellite on a chromosome. Nucleolus organizers can also occur as constrictions.

In addition to these visible landmarks, genetic analyses reveal another set of subdivisions. The smallest such unit is a point mutation, which can be a change in a single purine or pyrimidine base. At the next higher level are the triplet codons—clusters of three bases, which specify amino acids or act as punctuation at the termini of genes. Genes, or more precisely cistrons, consist of discrete sequences of codons and determine polypeptides. Superimposed on these units, at least in prokaryotes, are operons—clusters of coordinately acting cistrons and various regulatory loci (reviewed in Beckwith and Zipser 1970 and Stent 1971). In eukaryotes, blocks of genes can be identified through the minimal occurrence of recombination within them. Presumably they represent an adaptive complex, and chromosomal inversions on homologous chromosomes are the device which preserves them (Wallace 1966; Dobzhansky 1970). Additionally, there are various rearrangements of eukaryotic chromosomes that become apparent during meiotic synapsis, which is a kind of hybridizing through intimate point-by-point pairing of homologous chromosomes. These rearrangements include, in addition to inversions, various deletions, duplications, and translocations (White 1954; Swanson 1957). In brief, it is possible to identify many individual chromosomes by direct microscopic observation of their morphology, or by means of breeding techniques, or by observing homologous pairings.

Patterns of uncoiling and coiling, of replication and transcription, all necessitate an awareness of chromosomal function that is closely coordinated with the functioning of the cell as a whole. It will be convenient to look at chromosomal activity in the context of the cell cycle—the progression of events from one cell division to the next. In terms of DNA synthesis, the cell cycle has been divided into two G periods (G_1 and G_2) with an S period in between (Howard and Pelc 1951). And between G_2 and G_1, some workers now recognize an M period when the cell is undergoing mitosis and dividing. The duration of these periods differs, it turns out, with the culture conditions being used and the species being examined. During G_1 the chromosomes are maximally extended (uncoiled). The evidence is that in this state transcription (formation of mRNA) occurs; all three types of RNA are synthesized, because the extended DNA thread is now open to read-out by RNA-polymerase. Before coiling, some important changes occur and DNA-polymerase acts to replicate the extended genetic thread and the cell switches from G_1 to the S period. Following completion of DNA synthesis, the chromosomes start to coil in anticipation of the onset of division. In this G_2 period the cell as a whole is preparing for division. The highly coiled chromosomes now give visual evidence of being doubled, since each consists of two chromatids. With the onset of division (M period) the metabolically almost inert chromosomes—tightly coiled packages of genetic information—are passively distributed to daughter cells. At the completion of division they uncoil and commence a new G_1.

The foregoing recapitulates an obviously complex series of events, and it is understandable that considerable variation exists from one cell system to another as to how the actual sequence of events runs its course. By and large, under given conditions or with a given type of cell, there is a distinctive pattern of $G_1 \rightarrow S \rightarrow G_2 \rightarrow M$; distinctive in terms of duration of each period and the behavior of each chromosome during these periods.

Mitosis and meiosis. This leads next to the distributional behavior of chromosomes during cell division.

First, the major features of the mitotic apparatus need to be described. In animal cells, this apparatus consists typically of a central spindle, at each of whose poles lies a centriole surrounded by an aster (Fig. 2-3). The spindles

Fig. 2-3 The mitotic apparatus. Chromosomes lie within the spindle fibers, which are all formed of microtubules, as are the aster fibers. Two centrioles lie at the center of each aster. (Redrawn, by permission, from Du Praw (1970) *DNA and Chromosomes.* Freeman & Co.)

composed of fibers as seen in light microscopy, have been resolved by electron microscopy into microtubules, and these into regularly organized rows of protein subunits. Initially, workers believed that this subunit was one type of protein with a molecular weight of 34,700, a diameter of 33 Å, to mention only certain of its characteristics (Sakai 1966; Kiefer *et al.* 1966). Studies on microtubules from brain tissue and sperm tails, to be discussed below, now point to a dimeric unit and the presence of two different proteins. Returning to the organelle level, it is seen that the mitotic microtubules are very uniform except for some variation in outside diameter (140–250 Å) (DuPraw 1970). Also, there are two kinds of tubules, those that extend from pole to pole of the spindle, and those that extend from pole to kinetochore of a given chromosome or chromatid. The fibers of the asters seem to be constructed on exactly the same plan as the spindle microtubules. They, however, are arranged differently in that they typically radiate in all directions outward from the vicinity of the polar centrioles.

The centrioles are highly ordered structures. Electron microscopy reveals that they are cylinders whose walls are formed of nine sets of triplet tubules (Fig. 2-4). This structure also occurs as the basal body of cilia and flagella, where it is designated a kinetosome. Centrioles vary somewhat in outside diameter and greatly in length (Du Praw 1970). Their constituent tubules also seem to be reducible to protein dimers or monomers (Ross 1968).

Of special interest regarding centrioles is the pattern of formation of new ones. These appear at right angles to one end of a preexisting centriole in the majority of cases (Dippell 1968). Under special conditions, as in the ameba to flagellate transformation in *Naegleria*, they appear to arise *de novo*, for none is seen in the ameba and they are present as kinetosomes in the flagellate (Schuster 1963). The pattern of appearance of new centrioles is varied in different cells and is, therefore, a source of some comparative information.

The classical steps in mitosis and meiosis—prophase, metaphase, anaphase, and telophase—are familiar to all biologists and will not be reviewed here. What needs to be done is to remind ourselves of the ways in which the constituent or unit events in these processes can change from one cell or organism to another. Starting with the chromosomes, the older cytological literature is replete with various patterns of coiling in the early stages of prophase (Wilson 1928). This includes the fate of heterochromatic regions, the different behavior of different chromosomes in the same cell, the course of synapsis, the appearance of chiasmata, and movement of chromosomes to the spindle, on the spindle, and to the poles. Add to these events those of spindle formation, which includes separation and migration of centrioles, the appearance of intranuclear or extranuclear spindle fibers and aster fibers. Further, there is the behavior of the nuclear membrane—if and when it breaks down and how it is reformed—and of the nucleolus.

Before going further, the obvious point emerging from this should be made explicit: The greater the number of subunits in a structure or function,

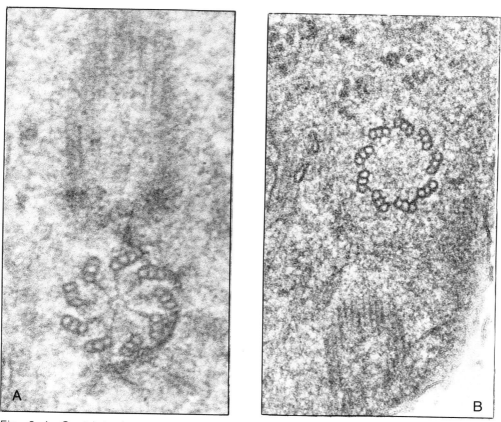

Fig. 2-4 Centriole from the sporozoan *Dermocystidium marinum*. A. A proximal cross section showing the presence of specialized structures within the circle of nine triplet microtubules. (57,000X) B. A distal cross section showing the relative absence of specialized structures within the triplets. (58,000X) (Originals courtesy of Dr. Frank O. Perkins, Virginia Institute of Marine Science.)

the greater its complexity and informational content; and here, in the eukaryote nucleus, we see that the coordinated activity of organelles such as unit membranes, microtubules and microtubular complexes (e.g., centrioles), and DNA strands with their associated macromolecules, all add up to wealth of organizational detail, both structurally and functionally. Indeed, it is the material that kept generations of cytologists busy for the past century in sorting out the constancy of chromosomal organization and distribution, which underlies a wealth of unique detail as one goes from species to species. Clearly, this area will be of extensive phylogenetic use.

Returning to organizational patterns in nuclear structure and function, there is a last major feature to be commented upon: the alternation of mitotic and meiotic events, and the relationships between haplo- and diplophases. This consideration is of crucial importance to the discussions which will follow in Part II of this volume, because animal forms in all probability arose from predominantly haploid algae and evolved into predominantly diploid metazoans. This transition will necessitate recurrent reference to where in a life cycle meiosis occurs and where mitosis perpetuates a haploid and where a diploid chromosomal set. Indeed, there will be the need to examine polyploid sets and patterns of chromosomal distribution that are even amitotic.

In that these matters deal with the manner in which an organism treats its genetic material—its most precious resource, in effect—it is most probable that any pattern of nuclear behavior is not a random concurrence of certain possible events, but a carefully chosen product of natural selection. A given pyramiding of such details, is, understandably, often unique and unlikely to have arisen independently more than once, and therefore similarities in these details between different organisms clearly suggest a common ancestry, just as differences point to divergence or even to no common ancestor for the trait in question. Here, then, is a complex set of information that is removed some distance from the genetic material. We cannot say what genes, much less what base sequences, are involved in specifying a given chromosomal cycle. Nevertheless, as we shall see, it will allow us to see the possibility of phylogenetic conclusions that can apply across broad taxonomic distances, i.e., at the level of classes or even phyla. It is, of course, just these kinds of data that have traditionally been of special use to phylogenists, though usually derived from multicellular structures such as organs and organ systems.

CYTOPLASMIC STRUCTURES AND FUNCTIONS

It will be useful to simplify the complexity of cytoplasmic organization by viewing it broadly as composed of membraneous, fibrillar (including tubular), or particulate elements lying in a fluid matrix. Membrane systems will be divided into two categories—those within the cytoplasm and those bounding the cytoplasm. The latter discussion will bring us to fibrous systems and their most prominent representatives, the peripherally located cilia and flagella, but will also include other organelles. Lastly there will be mentioned particulate elements and special organelles that either are of rare occurrence or simply do not fit neatly into the membrane-fiber-particle rubric.

Included in the cytoplasmic membrane systems are the endoplasmic reticulum, the Golgi apparatus, mitochondria, lysosomes, and, in plants, plastids. All of these entities are vastly more constant in their structure and function than they are variable. This is especially true of their function and of their structure as seen by the electron microscope, even though in the light microscope there is notable variability, as in the mitochondria (Wilson 1928; Lehninger 1965).

Endoplasmic reticulum. The endoplasmic reticulum varies in its elaboration within a given cell in terms of that cell's function, and so can vary as much from one cell to another in a single multicellular body as it will vary from one organism to another.

The endoplasmic reticulum is not conspicuous in rapidly dividing cells or cells with minimal protein production; embryonic blastomeres and protozoa generally show only a poorly developed reticulum. On the other hand, cells which are differentiated for secretion of proteinaceous products, like the acinar cells of the pancreas, show a very extensive development of the endoplasmic reticulum and associated ribosomes (Porter and Bonneville 1963). Though ribosomes are particulate structures, their intimate association with the endoplasmic reticulum must be recognized at this point. That association has often led to classifying the reticulum as rough or smooth depending on the presence of ribosomes. It is further to be noted that the attachment of ribosomes is asymmetrical; they attach to only one side of the endoplasmic reticulum. At one time the side of ribosome attachment was considered to be the "inside" of the cell in the special sense that it was continuous with annuli of the nuclear membrane and not continuous with the outside of the cell via pores in the cell or plasma membrane. This interpretation reflects the triphasic interpretation of cell membranes, namely that they are all continuous and arbitrarily divided into innermost nuclear membrane, outermost plasma membrane, and connected between by the tortuous maze of the endoplasmic reticulum. This view also incorporated the notion that all membranes were universally composed of the same unit structure (Robertson 1959), which was two outer, electron-dense protein layers with a lipid, less dense layer between. Though this pattern of organization is universal, it is now known that there are various differentiations in membranes. These are most conspicuous in the plasma membrane, and will be commented on shortly. Within the cell, however, the endoplasmic reticulum is a unit membrane, with or without ribosomes attached to one side; it is usually 75–100 Å across, and often lies folded or enclosing flattened vesicular spaces.

Golgi apparatus. The Golgi apparatus, which functions as a repository for materials (proteins) formed by ribosomal activity, is a system of flattened and stacked vesicles. Each vesicle is bounded by a unit membrane. There is evidence that these vesicles are formed from the endoplasmic reticulum and, after accumulating proteinaceous products, they move to the cell boundary, fuse with the cell membrane and discharge their contents externally (Porter 1964). Variations in the Golgi bodies are therefore a reflection of the activity of a given cell rather than of special genetic differences, though, of course, different activities can be genetically determined. The interactions of the various intracellular membrane systems is yet to be fully elucidated. One possible interpretation favoring significant interconversions of membrane systems is given in Threadgold's monograph (1967, see Fig. 9.1).

Lysosomes. These are usually seen as small (0.25–0.5 μm), rounded vesicles, bound by a single unit membrane. De Duve *et al.* (1955) were the first to identify them and found them rich in hydrolytic enzymes. Subsequently there has been some confusion between these and so-called endocytotic vesicles containing acid phosphate activity. This problem was clearly formulated by Bennett (1956) and has been resolved, at least terminologically (De Duve 1964), by referring to primary lysosomes (the ones originally identified) and secondary lysosomes, which are the endocytotic vesicles. In terms of function, the primary lysosomes are known to empty their contents into food vacuoles and aid in digestion. This is especially clear from the work of Hirsch (1962) on macrophages. Also, upon cell death, lysosomes are known to rupture, releasing their contents, which attack and destroy the cell and aid in its elimination from the host organism. Speculation ties lysosomal activity into a complex set of other events: they are perhaps pinched off from the Golgi apparatus as vesicles loaded with their potentially lethal load of enzymes, they fuse with pinocytotic vesicles or food vacuoles (perhaps to form endocytotic vesicles) and aid in digestion, and finally their vestige is eliminated at the cell surface as small vesicles containing undigestible debris (Novikoff, Essner, and Quintana 1964). This complex cycle is not well established for any cell and so permits of no comments regarding its variability. At present, the picture is that lysosomes are much more stable in their structure and function than they are variable (De Duve 1969).

Mitochondria. The stability of mitochondrial form is conspicuous in its fine structure. One exception is the folding of the inner unit membrane—the cristae—as tubules rather than the more common flattened vesicles. The protozoa typically have tubular cristae (Pitelka 1963). At the level of the light microscope the varieties of mitochondria or of the chondriome system (Wilson 1928) are numerous, the individual organelles ranging in size from spheres several micrometers in diameter to long threadlike structures, which may branch, ten or more micrometers in length. A given size and morphology is typical for a given cell. There is, however, no rational basis for classifying the various shapes and certainly no basis for evolutionary speculations based on the various forms. Margulis (1970) presents certain speculations regarding promitochondria and present day mitochondria, including some evidence of regression in blood flagellates, about which there will be more later; but this does not add up to anything that can be designated as rich in variability. In some cases the mitochondria can show specializations, such as the special aggregation of DNA-containing kinetoplasts of the blood flagellates (Steinert 1960), but these are very rare. In all forms where they are found, the mitochondria are the site of breakdown of acetyl CoA, derived from various sources, to water and CO_2 with concomitant generation of ATP. In cells where these processes go on anaerobically, such as certain parasitic flagellates like *Trichomonas vaginalis*, there are no typical mitochondria (Inoki, Nakanishi, and Nakabayashi 1959). There has always been extensive speculation as to how mitochondria arise (Wilson 1928; Lehninger 1965; Roodyn and Wilkie 1968). Present evidence clearly suggests that they are self-reproducing and insert new membraneous material into the old (Luck 1965), thus presumably growing in anticipation of subsequent division. Considerable work has been done on the genetics of mitochondria, starting with the discovery of mutants in yeast (Ephrussi 1949, 1951) and continuing on to discoveries of intracellular mitochondrial diversity in certain strains (Avers, Rancourt, and Lin 1965). Other mutants have been found in the other eukaryote cells. (Prokaryotes lack mitochondria.) These mutants can depend on changes in nuclear genes or changes in the hereditary material of the mitochondria themselves. In recent years intensive work has successfully demonstrated the ubiquitous occurrence of DNA in mitochondria. Interestingly, it is in all cases a naked Watson-Crick double helix and there is considerable evidence that at least in some forms it is circular (Wolstenholme and Dawid 1968; Borst and Kroon 1969). This is one of the main lines of evidence arguing for the idea (Brucke 1861; Mereschkowsky 1910; Wallin 1923; reviewed in Minchin 1915; Wilson 1928; Margulis 1970) that the eukaryotic cell evolved as a symbiotic association of prokaryotic cells. Be that as it may, within the eukaryotes we see the mitochondrion as a key organelle in meeting the energetic needs of the cell and possessing a distinctive fine structure, a somewhat variable gross structure, with a heredity system of its own in addition to dependence on nuclear genes.

Plastids. These are analogous to mitochondria in several respects. Their fine structure, as seen in the electron microscope, is rather uniform except that the innermost membranes—the lamellae—can show significant variability. In the algae there are lamellar differences that correlate well with major systematic groups (Granick 1959; Ueda 1961). In the green algae (Chlorophyta) and the higher plants (Bryophyta and Tracheophyta) there is a common pattern of organization in lamellar detail (Granick 1959). However, light microscopy reveals a great variety in size, number, and shape of plastids, but these characters are constant in a given species. In the algae, plastids are capable of division and, hence, are self-reproducing; but in the higher plants they arise from small, colorless, usually spherical vesicles, called proplastids (Gibor and Granick 1964; Gibor 1967). As with the mitochondria, there has been established the presence of DNA in plastids as a naked, double-stranded polynucleotide, which may be circular. And also, there is evidence of nuclear-dependent and autonomous inheritance of these organelles. It may be for plant phylogeny that plastids will offer some useful data in conjunction with other characters, but they are obviously of no use in animal phylogeny. Our only concern with them here will be to understand the consequences of their loss as phototrophs evolve into chemotrophs.

Plasma membrane. The membrane of the cell surface, the plasma membrane, must be discussed from two points of view: on the one hand, this unit membrane is the site of various specializations having to do with passage of materials into and out of the cell, or with points or areas of contact with and adhesion to other cells or substances, or with the placement of vibratile elements; and, on the other hand, it may remain relatively unspecialized in appearance yet carry extensive depositions of materials external to it, such as cell walls, shells and tests, or other extracellular substances, or aggregate there inorganic materials taken up from the environment. In that many of these activities and functions are a direct response to a given cell's external environment, they reflect directly on a cell's functioning in a given niche. Hence, their diversity is great.

The organizational basis of the plasma membrane is the unit membrane. A major variation in its structure is the fact that in prokaryotes the lipid component contains no sterols—highly unsaturated fatty acids—but these are always present in eukaryote membranes. It is speculated that this may be the basis of the special development of membranes in the eukaryotes (Block 1965). Certainly at the cell surface the plasma membrane evidences a remarkable variety of formations. Are these useful in phylogeny? Yes and no. No, if we look only at the shapes the membrane takes as it wraps around the somewhat arbitrary forms of pseudopodia or indents for pinocytotic vesicles or other ingestatory structures, or shows special thickenings or thinnings or foldings at special sites of contact with other objects. But the answer is Yes, if we examine these shapes in the context of the whole cell, where they indicate three-dimensional form that is a locomotory or ingestive or attachment unit contributing to the functional and organizational nature of the complete cells. Thus we see a larger pattern of organization whose increased detail is interpretable in a variety of ways, among which are evolutionary ones. More simply put, the uniform nature of the plasma membrane itself has little information even when seen as a variously formed plane surface—as a sheet draped over a statue has little significance when viewed simply as a variously deformed piece of cloth. But the membrane in conjunction with information about the cell as a living unit provides opportunities for meaningful inferences about the organizational and functional complexity of a cell and allows us rich details for comparison—just as awareness of the statue under the cloth immediately throws solid meaning into the conformations of the enfolding drapery. However, we are not much past the beginning of understanding the plasma membrane as a physiological component of the cell.

The extracellular secretions and accumulations of material present a different story, in its main emphasis, but first there are two minor emphases. One is that the extracellular products may be built of, or at least contain, some unique chemical products—recall the examples of strontium sulfate in radiolarian spicules—and these may have some use in defining biochemical diversity, as has already been discussed. Second, the form of the test or cell wall can delimit cellular form very precisely and is another way of seeing or defining cellular shape, but, as we have just said, this information is best used in connection with general information on the cell as a whole. These are the so-called minor emphases. Of special importance is the possibility of finding, particularly in tests, a story of evolutionary change as is done with fossil materials all the time, be they radiolarian, acantharian or foraminferan tests, valves of brachiopods or molluscs, exoskeletons of corals or arthropods, or vertebrate endoskeletons. Here the invaluable dimension of time, through geological stratigraphy, is added to our stock of information. Knowledge of the mechanisms of population transformations and speciation, plus careful discrimination as to the nature and degree of similarities and differences, all this on a time axis reveals very detailed and often convincing stories of the evolution of forms with preservable hard parts. The more detailed the patterns of preserved material—for example, a vertebrate skeleton, with its great detail of different parts and the precise articulation of these with each other, is more critically appraisable than a clam valve—the more convincing the evolutionary arguments. As we will see, this is possible among certain of the protozoa and lower metazoan groups.

Microfilaments. The fibrous organelles of the cell include a variety of structures extending from microfilaments and unit microtubules about 50 to 300 Å in diameter up through the complex structures of the cilia and their associated parts to the very complex fiber system of the mitotic apparatus, which we have already considered in some detail.

Microfilaments are 40–50 Å in diameter and of variable length. They have been most extensively studied as components of striated muscle. However, more recently they are being found in a wide variety of cells ranging from amebas to embryonic vertebrate tissue cells (Wohlman and Allen 1968; Wessells et al. 1971). The function of these filaments is not entirely clear. Present speculation favors the view that cell contractility and motility are dependent on microfilament interaction, perhaps as in models of muscle contraction. That they function also as a cytoskeleton is also possible. There seems to be little variation in the structure of filaments. In addition to the filamentous components of bacterial flagella (Abram and Koffler 1964) there is information on the molecular makeup of microfilaments from the squid axon (Huneeus and Davison 1970). They find a monomer whose molecular weight is around 80,000.

Microtubules. The microtubule is emerging as a common building unit of many fibrous elements, and also it is one of the few instances where organization in terms of component proteins is beginning to be understood. As seen in the sea urchin mitotic apparatus (Kiefer et al. 1966), the disrupted microtubule is a cylinder formed from the polymerization of a single unit, about 33 Å in diameter, and which may well represent a single protein monomer. In this material, these units form thirteen rows, parallel to the long axis of the microtubule. More recent work is tending to favor a somewhat modified view of microtubular substructure. Shelanski and Taylor (1968) report that a protein subunit of 60,000 molecular weight is found in sea urchin sperm tails and, more recently, material from chick-embryo brains suggests that two such units consisting of different monomers might combine to form a heterodimer as the basic subunit of tubule organization (Bryan and Wilson 1971).

In the cases studied thus far (Stephens 1968) it is clear that some preexisting tubular material is necessary as a "seed" or nucleation site for the further assembly of microtubules from their macromolecular components. Understanding of this process is under intensive investigation (Oosawa and Higashi 1967; Tilney 1968), and this may be the organelle whose mechanism of development from molecular subunits will be the first to be understood.

The pattern of microtubule arrangement in a cilium or flagellum of a eukaryote cell is largely invariant, though not so much so as formerly believed. The basic picture, as seen in cross sections, is the now familiar circle of nine doublets with attached arms plus a central doublet (Gibbons and Grimstone 1960; Pitelka 1963). Grain (1969) has presented an especially informative review of the kinetosome in ciliates and establishes, in the process, a standardized set of terms for describing its details and derivatives. There are some variations in the internal fine structure as one proceeds from the outer tip of these organelles down to the cell body, but these are relatively minor and occur with no presently identifiable patterns in their distribution from one organism to another. Externally the flagella, especially, show various elaborations (Pitelka 1963). These are usually delicate, hairlike protrusions of the outer flagellar membrane, and there may be a few or scores of them per flagellum. They are variously arranged on the flagellum but show a pattern that is fixed for whatever species is examined. They are most common in the unicellular algae.

Just below the plasma membrane the cilia and flagella terminate in a basal body. In the older literature this was termed a blepharoplast when found in conjunction with a flagellum and kinetosome when with a cilium. The present convention is to use kinetosome as the generic term (Pitelka 1963) and that convention will be followed here. The bounding membrane of the cilium and flagellum is continuous with the plasma membrane. The nine bundles of peripheral tubules continue through to the kinetosome but the central tubules do not. The latter end on or near a kind of plate at about the level of the cell surface. Kinetosomal structure is indistinguishable from that of a centriole. In certain groups, notably the flagellated and ciliated protozoa, there are quite complex structures associated with each kinetosome, extending in precise patterns into the cytoplasm or parallel to the cell surface (Pitelka 1963). These cases represent just that kind of information needed for phylogenetic studies, for they show complex structures varying in different directions from a common organizational theme and basically similar subunits.

In those cells which carry many cilia or flagella, there can be very diverse patterns in their distribution on the cell surface. This is especially true of free-swimming cells. Here these structures serve for locomotion and as aides to feeding, and they can be modified into a variety of structures. Some of the more obvious changes are compounding of a group of cilia into membranelles of various sorts, or using microtubules, possibly of flagellar origin, as the stiffening element in pseudopodia or using flagella or cilia as attachment devices. In that these modes of organization are closely associated with an organism's peculiar mode of life, they are informative of selection pressures, and their diversity, then, is a clue to organismic diversity itself. As is well known for the protozoa, the distribution of locomotory devices is a time-honored method of classifying these forms (Kudo 1954) and differences and similarities in the organization of cilia and flagella are useful indicators of phyletic relationships. In higher forms, bulk necessitates the development of more effective means of locomotion—based on muscular contractions as a source of mechanical force—and so external ciliation is lost except on small embryonic stages. Internal ciliation is usually of uniform occurrence as a heavy brush border on epithelia lining a tubule along which fluid substances must be moved. Thus the variety of external ciliation is useful only in a restricted group of organisms, i.e., the unicellular and very small multicellular ones.

In addition to patterns of organization of the surface fiber systems, there are also complex patterns in their development. These are most clearly seen at the level of the compounding of these organelles, for at the level of the single flagellum or cilium there seems to be a very uniform pattern, which is that kinetosomes form the new organelles. The exact course of events is not clear, and, hence, there remain questions as to the role of the basal body as a kind of seed. It could serve as a site of initiation of fiber growth, with polymerization of protein units occurring distally, or the new units could be inserted proximally and thus push the fibrous structures out. Work on the green alga *Chlamydomonas* clearly indicates that flagellar growth occurs at the distal tip (Rosenbaum and Child 1967). One further question is, where do new kinetosomes themselves come from? This is a problem that has had a long history in the protozoa, especially the ciliated protozoa (Lwoff 1950; Tartar 1961; Sonneborn 1963). There is no final answer as yet. In the ciliates, new kinetosomes, in most cases—and certainly it is true for the most intensively studied cases (*Paramecium:* Dippell 1968; *Tetrahymena:* R.D. Allen 1969)—always arise next to preexisting ones. But there is no basis as yet to argue for a template-like relationship existing between the old and new parts (Grimstone 1961; Sonneborn 1963; Fulton 1971).

The pattern of development of the total flagellation or ciliation of free-swimming cells has received extensive attention, again especially in the ciliated protozoa (Lwoff 1950; Tartar 1961, 1967; Sonneborn 1963; Hanson 1967a). In that this material will, of necessity, be discussed in considerable detail later, it suffices now simply to point out that development, as a process, runs on a time axis and is analyzable in detail by determining the precise sequence of events that achieve a given final form or function. In the development of complex ciliary patterns, for example, it is clear that detailed histories of the formation of two new ciliary patterns from one old one—as in fission—will provide documents rich in detail in proportion to the number and diversity of pattern of the various cilia. In that genetic information controls development and the form of an organism is responsive to selection pressure, it is clear that organizational patterns as reflected in ciliary developmental patterns can reveal a great deal about evolutionary history.

Obviously, then, at the level of the whole cell there is going to be available an extensive residue of evolutionary information, even though the base line of our information rests at the organelle rather than the molecular level.

Ribosomes. It is time to examine next the last large category of cytoplasmic inclusions or specializations, i.e., the particulate or granular elements. The most important example of this category are the ribosomes, which were necessarily alluded to in conjunction with the rough endoplasmic reticulum. These ubiquitous elements of all cellular organization are separable into two parts, and their size is most commonly given in terms of their sedimentation constant. There are some gross differences in these sizes, and they correlate broadly with phylogenetic position. Reisner, Rowe, and Macindoe (1968) point out that prokaryotes have 23 S and 16 S particles, algae and fungi 24 S and 16 S particles, certain protozoa 25 S and 18 S, and metazoa have 28 S and 18 S particles. In addition to these differences, there have also been reported differences in the amount of

RNA and protein present. These range from 40 to 60%. Work is still under way characterizing the nature of the RNA and protein which is present. Spirin and Gavrilova (1969) indicated that there are at least three different RNA molecules and 20 or more proteins, many of which are rich in basic residues and that picture is already being modified further. The function of ribosomes is the all-important one of translation of mRNA into polypeptide. That the same ribosomal preparation can translate mRNA from different sources (Watson 1970, Stent 1971) argues for a certain lack of specificity in the ribosomes. They are workbenches for the formation of specific polypeptides from specific messengers. It is not so surprising, then, that they show such little diversity throughout the diversity of living things.

Metabolic reserves. The other large group of granular inclusions in the cytoplasm are the metabolic products, such as starch or glycogen reserves, lipid droplets, and various crystals. These show no significant specificity relative to the cells which contain them. In fact, their presence seems variable depending on the physiological state of the cell under examination.

Special organelles. It is not fruitful to catalog here the various special organelles that can be found scattered amongst various cell types, but, rather, to mention selected examples to illustrate how such specializations might provide information on the evolutionary change. One example is the contractile vacuoles. These are special membrane bounded systems, perhaps better referred to as expulsive vesicles (Wigg, Bovee, and Jahn 1967), whose special function is control of the water content of cells. They are restricted to freshwater forms, and by their nature as expulsive devices they periodically fill and empty. Though variable in size, due to their function, the various sequences of forms that they manifest are fixed for a given species.

Another membrane-bounded vacuole is the food vacuole. It is of transitory existence, as are those around special metabolic products, and so they are somewhat difficult to use as stable characters, if that were to be attempted. Interest in these systems is probably best limited to an awareness that they reflect food-gathering adaptations of the cell.

Another example of cellular specialization is the occurrence of extrusible elements, which may appear as explosively extruded materials, such as trichocysts or nematocyts, or simply as secreted materials that are sometimes toxins. As toxins, or secretions, or chemically characterizable microprojectiles, these entities can be treated in the category of specialized metabolic products. The cellular organelles that produce them are often characterized by unique forms, but as they are somewhat uncommon and are not entities that are compounded or complexed into higher levels of organization, little can be done with them except to treat them as their products would be treated, that is, as unit bits of information to be added to a larger set of information.

CELLULAR DIVERSITY

By contrast with their organelles, cells are exuberantly diverse. We can deal with this seeming paradox in the following terms: it is their exploitive functions that render cells diverse; it is their homeostatic functions that determine organellar uniformity. Behind this perhaps too-neat generalization is the assumption that natural selection is at work, as it surely is. What is being expressed here can be restated as follows. First, as regards organelles, their special structures are commonly assumed to be correlated with the discrete functions they perform; the mitochondrion is "the powerhouse of the cell", chromosomes are linear units of genetic information, plastids are concerned with photosynthesis, ribosomes with protein synthesis, and so on. And so starting with the macromolecular building blocks of living systems, there are conceivably only a very limited number of optimal engineering solutions to given functional problems, especially when there are homeostatic restraints on the intracellular environment into which these solutions must fit. By contrast, cells fit into the highly variable niches that they must exploit as unicells, or they meet diverse functional roles as components of multicellular systems.

An alternative explanation for this cellular diversity superimposed on organellar uniformity is to postulate that the first eukaryote appeared with organelles similar to those present today, and because the system could not survive significant change in any component of its interacting parts, the primordial components have persisted largely unchanged. (These arguments parallel those relating to the universality of the genetic code: Is the present code the most efficient one in terms of natural selection? Or is it the first functional code to have evolved and all subsequent organisms are stuck with it, so to speak? (See Hinegardner and Engelberg 1963; Sonneborn 1965b; Woese 1967.)) The only way to resolve these contending views is to build alternative organelles and insert them in cells in place of the original ones and measure their performance. Unfortunately, we are of course nowhere near able to perform such experimental feats, except that mutant mitochondria have been transferred by microinjection (Beale, Knowles, and Tait 1972).

The reality remains that a given set of rather stereotyped organelles can be assembled into an enormously diverse array of cells: there are typical organelles, but, certain textbooks notwithstanding, there are no typical cells.

It is the reproductive function of cells that adds an often overlooked, or at least insufficiently utilized, further dimension of information. This refers to the cell and life cycles of cellular systems (Hanson 1963a, 1967a).

The cell cycle can be described using the G_1, S, G_2 and M periods, which helped characterize nuclear division. Here, of course, not only does one follow nuclear events, but also cytoplasmic events — the formation of enzymes and of metabolic products, of structural proteins and assembly of new organelles, the distribution of organelles in anticipation of cleavage, cleavage itself, and appearance of the two cell products where initially there was one. It is sufficient to recall the number of subunits involved in this process to realize, at least intuitively, that there will be variety in this process. Additionally, the fact that the process must succeed in a variety of environments says that the cell cycles we observe will be the surviving products of natural selection and therefore will show evolutionary change. Hence, cell cycles can be of great phylogenetic interest.

Superimposed on cell cycles will be cell life cycles. These are sequences of division or suppression of division with insertion of other processes such as spore or cyst formation, sexual events, passage from one environment (or host) to another (as in symbionts) with concomitant changes in form and function, or even simple cumulative changes associated with the normal growth curve. All of these add up to yet another pattern of processes which, again because they are liable to selection, can provide yet another set of relevant phylogenetic data.

COMPLEXITY AND DIVERSITY IN MULTICELLULAR SYSTEMS

From general knowledge and as an obvious inference from the foregoing sections, it is clear that multicellular plants and animals are complex and diverse; the problem now is to formulate that complexity and diversity in ways that reveal information pertinent to the study of evolutionary descent. It seems easiest to comment on anatomical and associated functional details first and then turn to developmental ones. Our treatment can be relatively brief because this level of organization is the most generally familiar one, and also because the multicellular organisms that we will examine later on are among the simpler ones and anatomical details are relatively uncomplicated.

Anatomy and Function. The constituent parts of many-celled organisms, when examined above the cellular level, are seen to be hierarchies of tissues, organs, and organ systems. Tissues are defined usually as groups of cells similar in structure and function. Various tissues are found in an organ such as the liver or brain. Organs, in turn, are anatomically delimitable groups whose cooperative action contributes a more or less specific function to the organism, predominantly in conjunction with one organ system. The brain, for example is made up of various types of nerve cells and contributes a coordinatory and information storage and retrieval service to the organism through the central nervous system. The liver is anatomically discrete, made up of various tissues, and associated intimately in its function with the digestive system, but is also closely coordinated with the circulatory system. Indeed, the brain too is dependent on an adequate blood supply for its proper functioning. Nonetheless, the general definition of an organ being largely related to one organ system does hold despite our awareness of cooperative interactions with other systems.

Organ systems are most simply defined as various organs that cooperate to perform a major bodily function, such as digestion or coordination, and so forth. Except for the case of widely separated parts in the endocrine system, organ systems show significant anatomical continuity from one organ to another of the same system. This is a result, in large part, of their mode of development.

Especially from an anatomical point of view, it can be shown that there is a very significant variation in the tissues of an organ, and the organs of a system, as one examines different species of organisms. Thus, it is easy to see that there is a great variety of subparts in organ systems and that they have various types of interrelationships. Critical scrutiny of these relationships can reveal changes that are quite analogous to those found in the macromolecules; that is, there have been deletions of parts in one system that still remain in other systems; there have been additions, also, and substitutions and innovations.

Furthermore, there can be found broad trends. One of these is progressive complexity, and this is often correlated with increase in size. There can be, also, decrease in complexity, which is not always associated with decrease in size, but more often simply accompanies changes in selection pressures. For example, sessile forms, evolved from free-swimming ones, will lose those specializations necessary for a free-swimming mode of life since these are now unnecessary for sedentary habits.

In the brief review of organ systems and their parts which follows, it is useful to keep in mind that organ systems are specialized ways of performing functions that are common to all organisms, even unicellular ones. Hence we will rarely find the evolutionary emergence of completely new functions, but we will commonly find new anatomical bases for old functions, and these will show adjustment to the demands made on the function often by increasing size and multicellular complexity of the total system they serve and less commonly by changes in the habits of the organism of which they are a part.

We will concern ourselves here only with animal systems. An *integumentary system* surrounds all metazoans. This ranges from epithelia one cell layer thick to elaborate skins replete with hard cuticles, scales, feathers, or hair. Associated with these are various glands and other secretory devices which augment the essentially protective nature of this outer covering. The integument is broken only by sites for extrusion of material, or by openings associated with the digestive, excretory, reproductive, respiratory, or sensory systems.

The *digestive system* has either one external opening attached to an internal cavity, or two external openings attached to the two ends of a tube which has various differentiations along its length. The single opening attached to a gastrovascular cavity is found only in the simpler metazoan phyla. The limitations of this system are lack of specialization and difficulty in supplying all parts of the body with the products of digestion. This latter function is solved by the appearance of a true vascular or circulatory system. The former difficulty is solved by "one-way traffic" (Buchsbaum 1948) in a tubular digestive system with two openings, thus allowing different parts of the gut to act differently as food passed by, instead of all parts doing all things when food lay simply in a common enclosure in the body. In fact, the two-opening digestive system can become highly differentiated along its length, with special organs being present that discharge their products into the gut to aid in digestion.

Musculature is necessary to move food substances along the digestive tube by peristalsis, but that muscle tissue is distinct from the musculature involved in contraction of the outer body wall or that which is attached to skeletal parts. One can see contractile material in its simplest form in the contractile activities of cilia and flagella, or in pseudopodial formation, all of which are concerned with locomotion of unicells. In higher forms, the muscular system acts by controlling contractions of the body wall or activating undulatory swimming motion or movement of appendages (Clark 1964). Also it plays an important role in opening and closing body apertures—oral and anal sphincters, gill openings, and siphons of various sorts. The tissue responsible for this is one of the most intensively studied of all, because of interest in molecular mechanisms for converting chemical into mechanical energy, and seems to be much the same in all animals. Its variety comes from the variety of uses that reversible contractility is put to in an organism.

A *coordinatory-sensory system* is essential for animals to allow effective exploitation of the predaceous mode of life, where there is a premium on sensing, capturing, and ingesting food. Such activities necessitate complex behavioral integrations on the part of the predator. There are important new ideas appearing on the nature of the primitive nervous system (Pantin 1956; Passano 1963; Bullock and Horridge 1965) and on the origins of animal behavior (Jensen 1959). The first nervous systems appear as nets of special cells in the lower metazoa and gradually, as sensory structures evolve and bodies become larger and more complex, recognizable central nervous systems are apparent, which take inputs from sensory systems and release outputs to effectors such as muscle cells.

The other aspect of coordination is *endocrine* or chemical coordination. The use of hormones is highly developed in higher forms such as the arthropods with their developmental hormones and the vertebrates with their varied hormonal activities. The occurrence of such substances has not been satisfactorily shown in the lower phyla, though there is evidence of the effectiveness of substances released into the medium, such as glutathione inducing development of sexual organs of *Hydra* (Lenhoff and Loomis 1961). It is highly probable that there are such chemicals in widespread use among the lower invertebrates. The fact they work interorganismically rather than intraorganismically, as in the larger higher forms, is an obvious functional difference.

Reproductive organs are either a testis, for forming sperm, or an ovary, for egg formation. Some forms (hermaphrodites) carry both organs. Where fertilization takes place externally to the parent organisms and gametes are shed into the surrounding environment, usually aqueous, there is usually little development of accessory sexual structures. However, when internal fertilization occurs with associated copulation and perhaps courtship behavior, which may be very elaborate, the accessory structures can become very complex. These include vesicles for storage of gametes and ducts for their transportation and a variety of copulatory devices. Where courtship is elaborate, there are secondary sexual characters which can serve as displays in one manner or another.

The foregoing systems are all represented by at least some specialized tissues in the simplest metazoans and, of course, are liable to evolutionary elaboration as complexity and size increase. The remaining systems arise as tissues and organs in forms above the simplest of the multicellular animals.

Circulatory systems are necessary as transportation devices, to take nutrients to cells lying beyond the ability of the digestive system to serve them adequately by cell-to-cell transport, and to remove from the cells various metabolic wastes, including CO_2. This system is dependent on vessels as conducting tubes and some sort of pumping system. Initially, peristaltic motion of the body was probably effective in moving materials along the primitive vessels, which may simply have been body cavities, but this crude force was replaced by the system's own motive force derived from the pumping action of a specialized organ, the heart. And crude vessels were replaced, with few exceptions, by elaborate systems of arteries, capillaries, and veins serving all parts of the body.

To pick up O_2 and release CO_2, there had to be, also, access to the external environment. This was supplied by bringing the vessels close to the body surface and allowing diffusion of the small molecules to solve the last step of the transportation problem. The site of this exchange with the environment could be special external appendages, including gills, or the gills could be located internally, as are the special *respiratory organs*, the lungs.

Excretory needs were initially taken care of by simple diffusion, but eventually special cells (flame cells) and tubules were evolved. As these simple protonephridia became more specialized, the inner ends of the tubules opened into body cavities (metanephridia) or became intimately associated with the circulatory system (kidneys) and removed from the blood the materials to be excreted. The activities of such *excretory systems* have come to be intimately related to the water conservation needs of the organism. For example, desert animals excrete almost no fluids but aquatic organisms have the opposite problem of pumping out excess fluids. The nature of nitrogenous wastes is, in turn, largely determined by whether there is much or little water available in which to dissolve these wastes, for some, such as ammonium ions, are highly toxic and must be held in dilute solution. Others, like urea, are essentially harmless even when highly concentrated (Schmidt-Nielsen 1964; Baldwin 1967).

Finally, there is the *skeletal system*. As an external structure, being both supportive and protective in function, it is variously formed. The corals secrete a rocky, calcium carbonate exoskeleton. Various worms have a highly impermeable, waxy exoskeleton, termed a cuticle. Molluscs and brachiopods are furnished with calcium carbonate shells or valves. The arthropods have a tough chitinous exoskeleton. In the starfish there are calcareous plates embedded in the integument, which act as protective plates, for the most part. As endoskeletons there are the hydraulic ones of the corals and certain worms (Clark 1964). Sponges show an endoskeleton variously formed of protein, calcium carbonate, or siliceous materials. And especially familiar are the cartilaginous or bony endoskeletons of the chordates. Some of these skeletal elements are very complex, as was mentioned during the earlier brief discussion of fossils, and for that reason are valuable sources of information for evolutionary studies.

Development. Developmental biology has, in the form of the biogenetic law—ontogeny recapitulates phylogeny (Haeckel 1874)—been overused as a solution to evolutionary problems but has been put into perspective again by various workers (Sewertzoff 1931; de Beer 1940; Hyman 1940; Gelei 1950). There remains in this section on multicellular forms to outline the sort of information provided by development. In that the events between gamete formation and the emergence of the adult are processes, the time axis is the basis for ordering them (Hanson 1963a), and thus we can start at the beginning of development with the formation of the germ cells. Gamete formation has the constant feature of meiosis, but added to that are different modes of differentiating the haploid, functional gametes, especially the sperm. Sperm are more variable in form than eggs. This sometimes bizarre variety of spermatozoa often exceeds our present abilities to make rational, functional explanations of the observed form (Baccetti 1969). For example, we simply do not know why a *Drosophila* sperm needs to be a quarter the length of the body of the fly.

Fertilization can be analyzed in terms of chemical inhibitors and inducers of sperm action and egg response. The latter includes such events as appearance of a fertilization cone and fertilization membrane, of completion of meiotic events, and initiation of subsequent cleavages. All of these establish similarities and differences between species. Cleavages swiftly partition the zygote into many blastomeres. Patterns of cleavage and distribution of blastomeres are an important part of much phylogenetic speculation to date. Not so much because of a great variety of modes, but, on the contrary, because of great similarity of modes in otherwise widely different groups. The Spiralia

(Marcus 1958) are those metazoan phyla all showing spiral cleavage, and this includes all bilaterally symmetrical forms from Platyhelminthes (flatworms) up through unsegmented and segmented worms to the Mollusca and probably the Arthropoda. (For general accounts of the foregoing patterns of development and also of those mentioned below see Hyman (1940–1959), Kühn (1955), or Berrill (1961).)

Beyond these cleavage patterns, the next steps in development may be development of an internal cavity, the blastocoel; or the solid mass of blastomeres may go directly to formation of germ layers, with first the appearance of the outer layer or ectoderm and, then, the inner endoderm. The former will contribute to the adult epidermis, nervous system and some associated structures, and the latter will form the gastrodermis or lining of the adult gut cavity. A third layer, the mesoderm, appears and is the source of the remaining organ systems. Various processes such as inward folding or migration of cells, inductive interactions, selective adhesion of cells, and differential reproduction and death, all contribute in an orderly, coordinated manner to the progressive unfolding of events that finally result in the functional adult. This epigenetic process is capable of many deviations, modifications, additions, and deletions as one looks at one form and another and another. It is apparent, though, that there are general patterns and on these are layered various diversifications. Early patterns are basic patterns of development and are most inflexible. As one sees events and processes closer to the final molding of a given form, one also sees that these events and processes become more variable.

The general principle is that basic modes of development, occurring very early, have dependent on them all succeeding events—this is epigenesis. Changes in the early events usually result in massive changes from the normal pattern, and these are usually lethal or, at a minimum, result in reduced fitness of the product organism. Hence, the most probable changes are those that occur towards the end of a given ontogenetic sequence. This situation creates special opportunities and problems in evaluating developmental facts for evolutionary studies.

First, some comments on the problems. Changes coming at the end of a developmental sequence will affect only the final stages of formation of a structure and hence can be quite minor in their effects. As such, they may be hard to identify and may also represent specializations unique to one organism or a limited group of organisms and hence of minimal comparative value. Changes coming early in development will tend to be very conservative, that is, they will tend to introduce very little change, because of the possibility of causing excessively abnormal subsequent effects. Hence, this source of variability will probably be minor. Furthermore, there will be selection in favor of suppression of mutants, which may cause significant deviations from basic epigenetic patterns. Such suppression will build, in the long run, very stable and rigid patterns—the canalization (Waddington 1960) referred to earlier. This has the genetic advantage of allowing for storage of unexpressed genetic information, but the stability of the developmental pattern is a liability when a phylogenist is looking for diversity.

There are, also, the opportunities. Given the long periods of geologic time over which evolution has operated, the chance of preserving and accumulating long series of small changes is real, even if they are somewhat terminal in the developmental processes, and these will document considerable evolutionary history. Further, in any such pattern a great number of different changes can occur (Sewertzoff 1931; Hyman 1940) such as deletion of a step, or addition of one, or substitution, or emergence of a new one. And these can occur in all combinations. In addition, there are possible changes in rates of development of different parts of the same system—allometric phenomena— (Huxley and de Beer 1934), which introduces another variable into patterns of development. In extreme cases, such as pedogenesis, forms recognizably larval in anatomy are nonetheless sexually mature. The rich tapestry of events in embryonic development supplies intricacies of pattern that, when properly analyzed, are an invaluable addition to phylogenetic data. In fact, even the very conservatism that suppresses variability also aids by its tendency to preserve rigidly those changes that are selectively advantageous. Thus important differences in pattern do probably reflect very early evolutionary divergence of the groups concerned and encourage inferences regarding the evolutionary relations of the highest taxa.

Some day it may be possible to interpret developmental events directly in terms of the underlying genotype. But as was shown in the preceding pages, there is probably no one-to-one correspondence of gene and adult character except at the molecular level (Mayr 1963). The whole organism is vastly more complicated than the information stored in its DNA. Hence, degrees of differences in the development of one organism as compared to another are only indirectly referable to differences in the genotype, and even these estimates are beclouded by the phenomena of canalization and other devices of the gene pool, which result in the cohesion of all genetic information so as to achieve forms best adapted to a given niche.

FUNCTIONAL STRUCTURE OF POPULATIONS

Taking our cue from the various factors that change the stability predicted by the Hardy-Weinberg law, we can look at the organization of populations in terms of the genetic variability of their constituent demes (the result of selection), their breeding patterns and population size, migratory tendencies into and out of the population, and in terms of mutation rates. Next, the warning must be sounded that very few populations are well characterized in all of these regards. In some the demic structure is well known, in others breeding patterns are well studied, and in others there may exist only measures of the total population size or measures of genetic variability in terms of known lethals on a given chromosome. In fact, when we examine protozoan and lower metazoan populations in detail, we find no species in which all these factors are known in detail. Only Sonneborn's (1957) pioneering review of breeding systems in selected ciliates, with emphasis on *Paramecium,* gives us anything like the data that would be desired.

In brief, the description of the information content in natural populations is largely a formal exercise. We can see where we need to look for our data, but the reality is that, at present, the available data are minimally informative.

This is also true for any categories within or above populations. Here we have treated populations as species, i.e., as groups of interbreeding or potentially interbreeding individuals. Lewontin (1970) has broken groups of individuals into kin selection groups, demes, populations, species, and ecological communities. In that any of these units can have a variable phenotype that is liable to selection and is transmissible to the next generation, it is probably a unit of selection for evolution. (Lewontin also carries the analysis down to the molecular level.) But as his own discussion bears out, there is little systematic study of these groups of individuals over a significant number of instances and, therefore, the data do not allow meaningful comparative analysis at present.

CONCLUSIONS

This survey of the diversity and complexity of biological materials, ranging from genetic macromolecules to populations, makes it apparent that some aspects of organisms are much more informative than others in terms of materials potentially useful to phylogenetic inquiry. At the molecular level, DNA and proteins are the most promising. Their present usefulness depends on direct measures of similarities in their subunits. Such measurements by-pass approaches that would be dependent on as yet unavailable information on the detailed structure of these molecules, i.e., the full sequence of bases in DNA and conformational details in proteins. At the cellular level, the most complex and variable information comes from seeing cells as intricate systems of many organelles, and from viewing them as products of developmental events. Where, however, historical data are available, as with microfossils, even relatively simple systems can reveal evolutionary histories. In multicellular systems, anatomical details contain much highly specific and variable information. Embryonic development also provides a rich source of comparative studies, especially since, inherent in the epigenesis of development, there are identifiable conservative and innovative tendencies.

This total array of information will be even further augmented when we can reduce the separation between the information systems at the molecular level and those at the cell organelle level and above: the problem is our lack of knowledge concerning multimolecular assembly into organelle systems.

Looking back over this catalog of specific structural and functional patterns, a useful summarizing overview is Sonneborn's (1957) brilliantly insightful aphorism: The major tactic of evolution is repeat then vary. The starting point at the molecular level depends on whether the primordial living system or eobiont was a reflexively catalytic protein system or a naked gene (Allen 1957; Hanson 1966; Orgel 1968). In any case, the first cells possessing genetic systems had their molecular specificity determined by a sequence of polynucleotides whose functional expression depended on a sequence of amino acids in a polypeptide. As the polynucleotide lenghtened by duplication of internal segments (repetition), these segments then differentially mutated (variation) and in consequence so did the polypeptide determined by them. Such repetition and subsequent variation continues today (Lewis 1945) as the basis for genetic variation in gene pools and the consequent richness of enzymatic and structural proteins, which determine specific function and form. With the emergence of organelles, though these structures themselves show limited variability, they have been compounded (e.g., patterns of ciliation) and complexed (e.g., different congeries of organelles in different cells) to form highly distinctive and varied cell systems. And cells themselves become repetitive in simple cellular aggregates but differentiate through evolution and the appearance of developmental mechanisms to achieve complex multicellular systems, where initially they were simply compounded. Then populations, originally formed of a rather homogeneous, cohesive gene pool, will vary sufficiently to establish demes, and from these can come the progenitors for new species. Finally, from closely related species, the process runs to divergent species as a consequence of continued variation.

This pyramiding of organizational patterns through repetition and variation is another prerequisite for phylogeny. Added to our conclusion (Chapter 1) that the unitary origin of life and of living systems in the face of diverse niches is a necessary prerequisite for phylogeny, we now also see that the diversification of life while exploiting these niches was not so innovative as to entirely obliterate the historical record. At several levels of organization and in terms of well defined subunits of organization that show significant patterns of variation, we see that living systems are indeed rich historical documents. Next there is the critical question: How well can we read those documents?

Chapter 3
Phylogenetic Analysis

THERE has been only one evolutionary history on this earth. There have been many attempts—mythological, religious and scientific—to recount the course of that history. From a scientific point of view, they cannot all be correct; only one can be. How do we find it?

Without question, the first essential is a methodology of objective validity. Lacking it, another attempt at phylogenetic conclusions would be wasted effort. We need no more of the proverbial blindfolded wisemen who each gropes at his own portion of the elephant's anatomy and adds his subjective pronouncements on the nature of the beast to those of his equally ill-informed fellow sages. Ideally, an objective method removes the blindfolds of parochial authoritarianism and encourages a genuine scientific analysis of that magnificent beast we call organismic diversity. More realistically, since reliable analytical methods do not arise overnight, the first step is to draw up what best promises to be an objective method and then test it. Through testing there will eventually be determined its degree of general applicability. The purpose of this chapter is to formulate such a methodology. It will be applied in subsequent chapters to inaugurate there its usage as a critical phylogenetic tool.

Phylogeny and Natural Selection

In that phylogeny is an evolutionary study, its aims and procedures must be derived from and consistent with the known processes of evolutionary change through natural selection. From this premise there can be developed three procedural guidelines which will immediately help reduce the confusion inherent in much phylogenetic speculation today. These guidelines will be formulated first and then their dependence on evolutionary processes will be made explicit.

PROCEDURAL GUIDELINES

The first question to be asked is: What are the units of study in phylogeny? The answer is species. In that the flow of evolutionary history is embodied in a temporal stream of many different reproductively isolated populations, *species define the basis for interorganismic comparisons* (Rule I). (See

Tuomikoski (1967) for an explicit discussion of this point.) The most important consequence of this guideline is that, arguing from species, limitations are imposed on use of hypothetical forms. Such forms can include ancestral forms, primitive forms, stem forms, and the like. It will be necessary to refer to such forms because of the very facts of temporal sequence and evolutionary development, but if the reality of species is kept in mind it avoids abuse of these concepts. Such abuses arise from creating whole lineages of hypothetical forms to account for transitions between different groups, or from creating stem forms or ancestral types so thinly characterized as to be caricatures of living forms and regarding which it is essentially impossible to develop any critical argumentation on phylogenetic issues.

The next question is: What aspect of a species is to be studied phylogenetically? The functional biology of the species is most important for reasons given earlier, i.e., function determines survival. But function has its structural (and genetic) correlates, and these too can be examined and must be examined where functional information is lacking. More explicitly, *phylogenetic comparisons must examine the exploitive, homeostatic, and reproductive functions and/or their anatomical correlates whenever possible* (Rule II). One cannot understand the behavior and fate of a species in terms of a few characters, nor only in terms of the morphology of its members, nor only in terms of their exploitive or their homeostatic or reproductive characters. Ideally, the total biology of the organisms in a species must be considered in analyzing their fate in the face of natural selection. Practically, this would entail an impossible amount of data gathering, and so some choice is necessary for expediency; but choice must be guided by an attempt to sample all major functional and structural dimensions. (What "major" means will differ somewhat in different cases, as will be seen below.)

The special aim of this second rule is to avoid the bias inherent in trying to understand the course of evolution as it might appear from the isolated viewpoints of comparative biochemistry, or physiology, or selected aspects of development or morphology. Each of these areas is potentially useful to phylogenetic studies, but it is doubtful that any by itself can avoid the possible pitfalls of concentrating on atypically conservative or innovative characters, or misinterpreting convergences and parallelisms and reductions. Only a broad view of the organism can hope to inform us adequately of the life of the organism as a member of a given gene pool.

These rules (I and II) have the further consequence that at times we will have to conclude we have no representative, living or fossil, of a stem form and so can form no useful idea of such a species. In such a case, the conclusion is inevitable that the phylogenetic origins of the group descending from that stem form are obscure and will possibly ever be so. We must accept the realization that the whole puzzle of historical descent will not be solved in convincing fashion throughout all its detail.

Third, we ask: Looking at the biology of species, what kinds of evolutionary change or phylogenetic pattern might be expected? There is no specific answer in terms of a catalog of acceptable kinds of change. There is, however, the unavoidable criterion that no change which contravenes known processes of species formation can be accepted. More particularly, *every surviving innovative step must be selectively advantageous* (Rule III).

A major goal of this rule is to put an end to that type of morphological speculation which transforms one major body type into another by diagrams which purport to show how readily such changes could occur, e.g., obtaining the dorsal nerve cord of chordates from an annelid by dorso-ventral reversal, or deriving a polyp from a sponge by pulling out tentacles around the osculum of a single flagellated chamber. Presumably such games have passed as phylogenetic change, because evolution has been defined simply as change. It is obvious from the context of natural selection that a tentaculated flagellated chamber is neither a good sponge nor a good polyp and would be at a selective disadvantage against either form. It is not even a hopeful monster (Goldschmidt 1952). The only way of rendering such transformations plausible is to look for the accumulation of mutational events, in conjunction with favorable selection pressures, sufficient to carry out the changes in question.

NATURAL SELECTION: THE FOUNDATION OF EVOLUTIONARY THEORY

The foregoing rules, especially the last one, make explicit the procedural dependence of phylogenetic analysis on evolutionary processes. It is worth belaboring the point further so as to exorcise, once and for all, the greatest source of confusion in phylogeny, which is to conceive of evolution simply as change without the limitations imposed by the mechanism of natural selection acting on species. To make this argument as clear as possible it is worth reviewing the structure of Darwin's inquiry into the nature of organismic diversity.

The problem. The questions that initially aroused Darwin's awareness of the possibility of evolutionary change came, of course, from his observations of the wildlife in and near South America. In his own words:

"During the voyage of the *Beagle* I had been deeply impressed by discovering in the Pampean formation great fossil animals covered with armor like that on existing armadillos; secondly, by the manner in which closely allied animals replace one another in proceeding southwards over the Continent; and thirdly by the South American character of most of the productions of the Galapagos Archipelago, and more especially by the manner in which they differ slightly on each island of the group, none of the islands appearing to be very ancient in a geological sense.

"It was evident that such facts as these, as well as many others, could only be explained on the supposition that species gradually became modified; and the subject haunted me." from C. Darwin's *Autobiography* in F. Darwin (1887, p. 67).

These types of observations can be put in the general form of: If members of a group containing generally similar forms are found in differing environments, their differences are correlated with adaptive responses to their surroundings. Viewed logically, this formulation is a law, for it asserts a uniform correlation between different empirical phenomena. Such laws state ". . . that whenever and wherever conditions of a specified F occur, then so will, always and without exception, certain conditions of another kind G." (Hempel 1966, p. 54). Here F is *a group containing generally similar forms whose members are found in differing environments* and G is *their differences reflect adaptive responses to local surroundings*. This statement applies to species and their demes, to groups of species as seen in the Galapagos finches, and to related members of even higher taxonomic groupings.

The explanation. The next problem, as Darwin well understood, was to explain this law of adaptational variation, and he did so by the theory of natural selection. To understand the precise use of the term theory, we again look to specialists in logical analysis. In their necessarily rigorous terms, the framing of a theory requires the specification of *internal* and of *bridge principles* (Hempel 1966). It may be possible to state this more simply, but less succinctly, by saying a theory must refer to elements that need to be related either to each other or to other already accepted statements (Hanson 1963b) and there must also be those statements that do the relating. Leaving Darwin now and using contemporary studies, the elements of the internal principles are those associated with the species as a reproductively isolated population of organisms who exploit the environment in a unique fashion to maintain themselves and certain of their variations. These elements imply a parent-offspring relationship among successive generations in natural populations and the fact of variation among the members of the population. The bridge principles are those statements that relate the foregoing internal principles to visible and measurable facts of adaptedness in members of a given species. Actually, they contain essentially those statements that Darwin used to enunciate his theory of natural selection—that species can and do procreate in great numbers; that, nevertheless, the size of the reproductive population remains relatively constant; that there must, therefore, be a struggle for existence, which results in the elimination of many of the offspring; that heritable variations do occur; and that those variations of advantage in the context of the struggle for existence will, naturally, be preserved (Darwin 1859; Huxley 1943).

Testing the explanation. These bridge principles not only reflect Darwin's formulation of the theory of natural selection, they also contain references to antecedently available or pretheoretical terms (Hempel 1966), which terms are necessary for testing the theory. That is, using the concepts of reproductive potential, of mutability, and of relative advantage of mutants, we can make various predictions about the fate of a given population. These have been made, and the tests show, over and over again, the validity of natural selection in explaining the origins of adaptive differences among organisms (Dobzhansky 1965, 1970; Mayr 1963). This says that natural selection is the theoretical statement foundational to all understanding of evolutionary change. Recent attempts at more rigorous, deductive formulation of Darwin's thinking miss this essential emphasis that must be placed on natural selection. For example, Williams (1970) equates "Darwin's theory of evolution" to "the theory of anagenesis, or descent with adaptive modification." (p. 344) and she thereby by-passes natural selection as causal mechanism in favor of emphasizing species diversification.

This point has not been adequately understood as evidenced by the fact that the "theory of evolution" or simply "evolution" has been called a bad or weak or equivocal theory (Scriven 1959; Mayr 1961) because of its lack of predictive power. Only by hindsight, it is said, does evolution tell us anything important; never by foresight. It is clear from the context of such discussions that the writers have confused evolution as change with natural selection. They are not the same thing: change is the consequence of natural selection, not its equivalent. Change can be seen in successive generations of organisms and it is an adaptive change because of natural selection. Such successional changes are one definition of evolution and natural selection may be another. Unfortunately, because both are referred to as evolution, the false syllogism is too often unwittingly made, that change is, therefore, natural selection. Fisher (1930) put the problem clearly when he said, "Natural selection is not Evolution. Yet ever since the two words have been in common use, the theory of Natural Selection has been employed as a convenient abbreviation for the theory of Evolution by means of Natural Selection, put forward by Darwin and Wallace. This had the unfortunate consequence that the theory of Natural Selection itself has scarcely ever, if ever, received separate consideration." (from the Preface.)

The problem can be made explicit by looking at the function of the two concepts in the context of the structure of inquiry in natural science. Natural selection is a hypothesis or theory whose function is to explain evolutionary change. On the other hand, evolutionary change can be seen in Darwin's original sense of adaptational variation, of accounting for the similarities in fossil and contemporary armadillos, and the zoogeography of forms he observed along the South American continent and on the Galapagos Archipelago. The logical nature of this inference is hard to specify. It can be characterized as an intuitive induction or retroduction (N.R. Hanson 1958; Ghiselin 1969) in the sense that Kepler's laws of motion are not deducible consequences of his and Brahe's many observational data on planetary sightings, but rather are magnificantly succinct, mathematical summaries of patterns or regularities embodied in the observational data. Apart from the quantitative aspect, this same can be argued for Darwin's intuitive insight that his empirical information on fossils and on zoogeographical differences might best be encapsulated by the idea of change in living forms. In this context, the concept of change is a brilliant inductive generalization—far above the level of enumerative inductions such as robin's eggs are blue or cows have four legs—and genuinely deserving of comparison with Kepler's laws.

Darwin fully realized that his major problem was to formulate a convincing explanation as to how adapted changes occurred in nature's creatures. For example, in his notebook of 1837 he says, "With belief of transmutation [i.e., species change] and geographical grouping [i.e., adaptation], we are lead to endeavor to discover *causes* of change; . . . ." (F. Darwin 1887, p. 390.) And for this he produced the Theory of Natural Selection. It is this theory, and not the idea of change—neither as an intuitive induction nor as a predictable consequence of natural selection—that must be examined as to whether or not it carries predictive power. The contemporary answer is that natural selection is a very powerful theory, for there is overwhelming evidence today that when the variables inherent in natural selection are specified, i.e., selection coefficients, rates of mutation and migration, and population size and breeding patterns, then very precise predictions can be made *a priori*, which have led to subsequent validation by field or laboratory observation and have led to much further development of the theory. (See, for example, essays by Crow, Dobzhansky, Levins, MacArthur, Slobodkin, and Wallace in Lewontin 1968; Crow and Kimura 1970). Natural selection is the theoretical foundation for all studies of evolutionary change.

Homologous Relationships and Units of Phylogenetic Information

All phylogenetic conclusions come from comparing organisms from two or more species, and, therefore, we must know how to go about making comparisons, and just what it is that deserves comparison.

To develop the concepts necessary for answering the first question—How to make comparisons?—let us start by looking again at species diversification and transformation, this time to characterize the kinds of changes that occur there. We can use, for the present, a very simple set of categories. And also at present, where we are simply developing ideas, let us assume idealized situations such that in the case of the transformation of species A into B, where we have to compare A and B to determine what has gone on, that we have extensive information available on both the antecedent and the descendent groups. Similarly, in diversification, where species B has diverged from A, we will assume again the availability of extensive data.

In both situations it is clear that most characters will show no change. Of the few that are changed, they can be either transformations of already existing characters, or they may include a brand new feature—an innovation. Further comments are needed on both of these categories.

Transformations can include further development of a character already present either to (a) enlarge it essentially in its already existing form or function, (b) reduce it even to the point of loss, or (c) achieve a new structure or function derived from a preexisting one.

Innovations can be of two types. There may be added (a) another element like one that was already present, i.e., a character can be repeated, or (b) a new character is added not at the expense of an old one (a transformation), nor by repetition, but as an apparently totally new addition to the phenotype.

This brief classification of characters throws some interesting light on the classical approach to characters used in phylogenetic studies. Here we enter on the complicated problem of homology. Provisionally, it is enough to define a homology as a character which occurs in two or more different groups of organisms as a result of their descending from a common ancestor which possessed the trait in question. (Shortly we shall see the inadequacy of this definition.) Defined in this manner it is obvious that characters that do not change are homologous. Also, characters that transform are homologous. A repeated character is very close to a homology, as seen by the fact it was originally called serial homology by Owen (1848), the founder of the concept of homologous characters (see Boyden 1947; Remane 1956; Simpson 1961, for historical details), as distinct from a special homology, which includes homologies between two organisms. Only the brand new character is not homologous. This tells us that the classical concept of homology covers characters that show no change as well as those that can change in various ways, and it does not discriminate among them. This is enough to tell us that homology as we have just defined and applied it is a concept of limited use. This conclusion is not new, of course, and various attempts have been made to redefine and recast the term; the two most important ones are those by Remane (1956) and Hennig (1950, 1966). In that the concept has been central to much phylogenetic inquiry—phylogeny is based on homology (Remane 1955, 1956)—we must examine it in some detail.

HOMOLOGY

As Remane (1955, 1956) clearly pointed out, though we accept descent from a common ancestor as the explanation of homologies, we do not as a rule, and cannot in the case of taxa above the species level, use any sort of breeding analysis to test this genetic relationship. The test of homology, despite its explanation, is not genetic. Owen viewed the concept as a problem in comparative anatomy, whereas Darwin saw it as a problem of evolutionary descent. It would be well to restrict the term homology to Owen's approach, now considerably clarified by Remane, as we'll see, and refer to the Darwinian point of view as properly homogenetic (Simpson 1961), which refers to the explanation of homologies—of *why* they exist in the first place. Remane proposed that homologies be identified essentially on morphological grounds; he accepts serological and ethological data as providing additional lines of evidence. Most importantly, Remane has formulated operational criteria for identifying homologous structures. We must now examine with care Remane's concept of homology. Following that, there will be presented other views on homology leading to a broadened and revised view of the concept which will include a set of proposals to cover all dimensions of evolutionary stability and change in form and function with explicit recognition of what each can tell us in phylogenetic terms.

Remanian criteria. There are three major criteria, according to Remane (1956), for identifying homologies. They are the criterion of position ("Kriterium der Lage"), the criterion of the special quality of structures ("Kriterium der speciellen Qualität der Strukturen"), and the criterion of connections through intermediary forms (continuity criterion) ("Kriterium der Verknüpfung durch Zwischenformen [Stetigkeitskriterium]").

The positional criterion says that homologous structure lie in similar positions in comparable structural systems. To specify more sharply what is meant by a similar position, Remane elaborates three points. First, there can be a topographical similarity of position, which can be determined by the imposition of a system of coordinates. This allows superficially very different forms (Fig. 3–1) to be compared and parts homologized. Second, there can be similarity

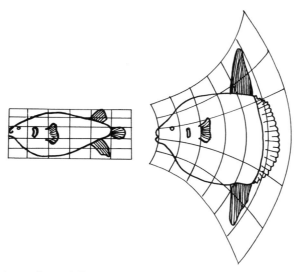

Fig. 3–1. The body outline of *Diodon*, porcupine fish, (left) is transformable into the outline of *Orthogoriscus*, ocean sunfish. The homologies between the parts of these fish establishes the coordinate system used to make the needed transformation. (After Thompson 1942.)

of position as a result of geometrically similar form. This is seen where one form can be significantly larger or smaller than another, but if the two agree in the geometry of their structure—as do the two figures in Fig. 3-2, though they are of different size—then their parts can be homologized. Third, there is a similarity of position as seen in the articulation of parts. Here homologies can be seen even though overall forms differ, providing the relations of component parts remain constant (Fig. 3-3). Thus, Remane states that prerequisites for applying the positional criteria are agreement in number of constituent parts and constancy of connections.

This brings us to the criterion of special quality of structures, or the compositional criterion. There may be such extensive displacement of a body part that the positional criterion is not a sufficient basis for determining homologies. In such cases homologies may be established by examining the subparts of the structure in question. This is illustrated schematically in Fig. 3-4A and more realistically by comparisons of various internal organs of flatworms (Fig. 3-4B), where the nature of tissues making up the organs allow one to identify homologous structures.

There can be instances where both the position and composition of a structure is so highly modified that it is apparently quite different from any other structure, but yet these dissimilar entities can be homologized if there are intermediate forms to bridge the gap between the extremes of the dissimilar forms. These intermediaries must be connected to their neighbors on either side, in this series, by fulfillment of both the positional and compositional criteria. If this can be done, then gradual changes can be so accommodated as to add up to the large differences in placement and nature of subparts that are seen in the extremes of the series. The intermediate forms can be other species, living or fossil, or can be developmental stages. The classi-

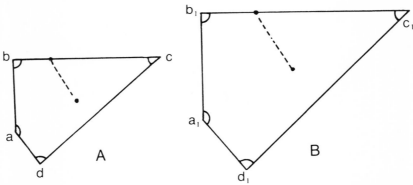

Fig. 3-2 Homology seen in similarity of geometric figures. The positional similarities are precisely defined by reference to point-to-point similarities in the positions of different points within each figure. (Redrawn, by permission, from Remane (1953) *Die Grundlage des naturlichen Systems der vergleichenden Anatomie und Phylogenetik.* 2 Aufl. Geest und Portig.)

cal example of such homologizing is seen in the homologous relationship of certain bones of the reptilian skull to bones of the middle ear in mammals (Fig. 3-5).

In addition to the major criteria, Remane also offers three subsidiary rules (Hilfskriterien or Hilfsprinzipien). It is to be expected that not all characters studied will be sufficiently complex so as to permit detailed correspondences and comparisons to be made, thus allowing unequivocal decisions in the light of the major criteria. Rather than discard such cases, they can be used as accessory sources of information as follows.

1) Structures in themselves simple can be understood to be homologous if they are found in a large number of the most similar forms (the probability

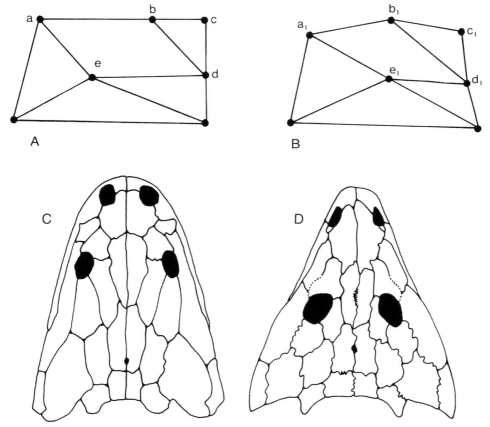

Fig. 3-3 Forms with somewhat different shape but similar internal articulations. A and B. Geometric forms illustrative of homology based most convincingly on similarity of internal connections. C and D. Skulls of two stegacephalons (C. *Metoposaurus*, D. *Lyrocephalus*), showing the same principle. (All redrawn, by permission, from Remane (1953) *Die Grundlage des naturlichen Systems der vergeleichenden Anatomie und Phylogenetik*. 2 Aufl. Geest und Portig.)

78 *Concepts and Methodology*

of homology increases with the number of related forms showing the character).

2) The probability of homology of simple structures increases with the presence of further similarities with the same distribution in very similar forms.

3) The probability of homology of characters decreases with the frequency of occurrence of the character in clearly unrelated forms. (Remane 1956, esp. Chapter 2.)

From the foregoing it is clear that Remane is operating from morphological principles in his attempt to define rules for identifying homologous structures. His attempt is not quite consistent, for he seems to take for granted that species are already identified and from that identification certain obvious relations apparently emerge such as "comparable structural systems" which is

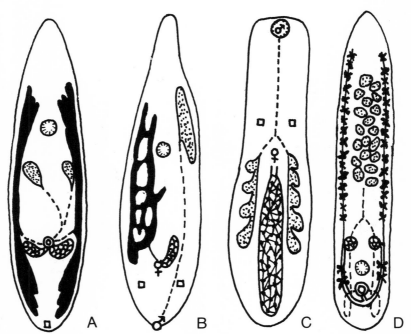

Fig. 3-4 Homologies determined by the composition of internal structures. The reproductive and excretory system of flatworms. A. *Macrorhynchus*. B. *Gyratrix*. C. *Prorhynchus*. D. *Monocelis*. Testes are stippled; ovaries are latticed; vitellaria (yolk glands) are solid black; and ovo-vitellaria are latticed with dots (e.g. c.) A square is the external opening of protonephridia; two concentric circles are the combined openings for both male and female gametes; ♀ is the site for a female opening and ♂ for a male opening; a circle with serrated border is the mouth. (Redrawn, by permission, from Remane (1953) *Die Grundlage des naturlichen Systems der vergleichenden Anatomie und Phylogenetik.* 2 Aufl. Geest und Portig.)

used as part of the positional criterion ("1. Homologie ergbit sich bei gleicher Lage in vergleichbaren Gefügesystemen" p. 58). The point is, how does one know they are comparable systems prior to establishing that fact through the criteria of homology? This raises the long standing criticism of the concept of homology, which goes back to R. Hertwig (1879), that it is part of a circular definition. This definition says, so the criticism goes, that related organisms are defined as those with homologous structures, and homologous structures are those occurring in related organisms. (See Hull (1967) for a cogent discussion of this and related problems.) This possibility of circularity is also seen in the subsidiary criteria, e.g., simple structures can be considered as homologous if they occur in a large number of the most similar forms ("nächstähnlicher Formen" p. 58). But one must ask again, how are these forms determined to be "most similar" apart from the use of homologies? Throughout his work Remane impresses one by his truly remarkable erudition as a morphologist and his perceptive insights into the nature of homol-

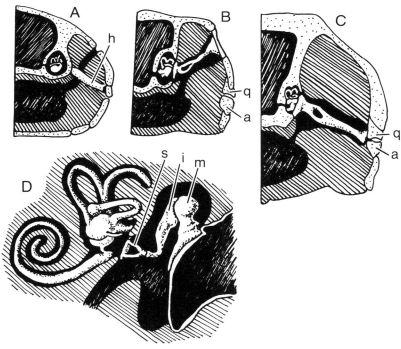

Fig. 3–5 Serial homology between the bones of the fish skull and mammalian middle ear. A. A fish. B. A primitive amphibian. C. A primitive reptile. D. A mammal. Note the changes in form and position of the articular (a), hyomandibular (h), and quadrate (q) bones; especially the hyomandibular as it becomes the stapes (s). The quadrate becomes the incus (i) and the articular becomes the malleus (m). (Redrawn, by permission, from Romer (1949) *The Vertebrate Body.* W. B. Saunders, Co.)

ogous structures *given the fact of species*. But nowhere is it made clear how one defines a species without using the concept of homology to bring together forms which have many homologous structures in common. To save the concept of homology, and any related concepts, as a phylogenetic tool, this circular reasoning must be broken. Some biologists, agreeing with O. Hertwig, believe it cannot be broken. (See Sokal and Sneath 1963.) But it can be, as follows.

The essential point is simple and obvious: species can be defined on the basis of interbreeding populations. That, in practical terms, the vast majority of species are defined on morphological grounds is logically irrelevant; that the only theoretically acceptable definition of species is that they are populations of actually or potentially interbreeding individuals, or reproductively isolated groups of individuals, is the unavoidable crux of the matter. The distinguished morphologist Naef (1919) was one of the earliest to recognize this. He says, in remarkably modern terms, that:

> "Nature presents us first of all with individuals (we speak only of multicellular forms) which occur in all degrees of similarity or difference. But among them there belong together natural groups of the specially similar ones that form a reproductive community, which is to say, that some of them can stand physically related to each other as ancestors and descendents and some as mates. Such a reproductive community, insofar as it is not produced by artificial means and it persists, that is, is viable through many generations, is called a species. There is no other possible basis for this concept, above all no 'morphological' one, and we can abandon at this point any further attempts in that direction." (Original in German. pp. 44–45.)

Given the concept of a species as a "reproductive community", we are aware of a biological entity that perpetuates itself isolated from other such communities and is composed of individuals varying to different degrees depending on their age, sex, and other variable characters. To compare such heterogeneous units (species) with each other the Remanian criteria can be used, though in modified form to avoid the logical inconsistencies cited above.

Reformulation of Remanian criteria. Presented herewith is a reformulation of the criteria of homology. There is one major criterion with two subcriteria, and also the extension of the major criterion.

1. Major criterion:
    a. Positional relationship or criterion: Homologous entities occupy similar positions in the systems of which they are a part.
    b. Compositional relationship or criterion: Homologous entities are composed of constituent parts similar in number, shape, or function performed, and in their interrelations to each other.

2. Extension of the major criterion:

   Serial relationship or criterion: Homologous entities can be arranged serially so that the least similar or extreme members of the series are joined by intermediate examples connected at each step by fulfillment of the positional and compositional relationships.

The subsidiary criteria can be restated as follows. These, it should be recalled, are for application to entities not sufficiently complex to allow application of the major criteria, especially the identification of compositional relationships.

3. Homologies are probable in entities apparently similar in position and composition that appear in species already known by the major criteria, to possess homologies. (The appearance of such structures and functions in species known not to possess homologies establishable by the major criteria, renders homologies improbable.)
4. Homologies are probable when similar entities occur in a large number of species.
5. The possibility of homology exists if sets of similar entities are similarly distributed in various species; this possibility increases with additions to such sets.

These formulations follow closely an earlier discussion of this problem (Hanson 1963a) and, as stated there, the present form of the criteria follow Remane's thinking quite closely. (The greatest difference is in the accessory criteria. The present criterion 3. corresponds to Remane's Hilfskriterien 6., here given in parentheses. The present order of 3, 4, and 5 represents a decreasing probability of identifying homologies.) Some further comments are also in order on this present formulation.

It is obvious that the operational basis for determining homologies is the determination of similarities. This deserves separate treatment as a procedural problem, and we will turn to that shortly. At present the point needs to be made absolutely explicit that when an investigator commences the search for homologies among members of two or more species, he is making interorganismic comparisons to establish degrees of similarity. Fundamentally, the criterion of compositional similarity is most basic, for, following Remane's analysis, even if the features being compared are variously placed in an organism, their similarity of subparts will assure us that a homologous relationship is present (Fig. 3-4). But there is an apparent paradox here, for the most interesting homologies are those between apparently dissimilar entities (Fig. 3-5). Here the serial relationship tells us that seemingly unlike features are homologous because, looking at intermediates between which differences are less pronounced and the positional and compositional relationships are realized, we find that further analysis uses gradual change in similarity to assure us of the homologous relationship. Also, the third subsidiary criterion is directly dependent on the similarities of location and makeup. The other two subsidiary criteria use similarity less rigorously, but are nontheless dependent on it. And we will further see, below, that the perennial problem of non-homologous similarities (e.g., convergence and parallelism) are resolved by rigorous adherence to observing positional and compositional relationships, especially the latter.

This now permits a definition of homology which is free of logical inconsistencies. *Homology is direct (positional and compositional relationships) or derived (serial relationship) similarity of structural and/or functional aspects of different organisms, especially as they are members of different species.* Homology defines, therefore, a relationship expressed in terms of degree of similarity. Again, the problem of how to determine similarities is apparent, and, again, we must say that problem will be treated later as a procedural question. There are still conceptual questions that need careful prior comments. The next one that needs consideration is the reason for using the word "entities" in the reformulated criteria rather than the phrase "structural and/or functional parts" as just used in the definition of homology.

As we mentioned earlier, Remane's emphasis in homological studies arises from his interest in morphology, and his criteria referred almost solely to structures as the source of data; elsewhere in his work he also recognizes the usefulness of behavioral patterns and immunology. In point of fact, the whole organism and the species to which it belongs lies in the purview of the phylogenist, and these possess functions as well as structures. It is obvious that ideally the concept of homology should apply to functions as well as structures (Waterman 1961; Hanson 1963a). Hence the equivocal term "entity" was inserted into the reformulated criteria to extend their sphere of applicability, and this extension is made explicit in the definition of homology given above. The question then arises of how to identify homologous positional and compositional relationships in physiological characters. The answer is to use the time axis inherent in all processes as the basic dimension for allowing comparisons. The temporal sequence of events—development is an especially good example—can be compared with that of another living system. In this way the positional criterion can be invoked. Further, to use the compositional criterion, any major process can be broken up into its subprocesses, and the way a given function operates can be compared in detail—number of constituent steps, their function, their connection with other steps—and decisions on relative similarity can be arrived at. It takes no further comment to show how serial relations can also be established with physiological characters (Hanson 1963a). Thus, the extension of the concept of homology readily encompasses functions as well as structure.

A next problem is to make explicit the significance of referring to comparisons between species. That homologies exist between members of the same species is both undeniable and trivial. Homologies and the phylogenetic studies derived from them become significant when interspecific studies are undertaken. This can be seen from the absurdities generated when pursuing homologies with disregard to the species concept, which is simply an exercise in comparing organisms. Comparison of two different adult males of the same species of the fruit fly *Drosophila* would reveal a completely homologous set of structures and functions, except possibly for certain molecular differences. If the adults were male and female, significant differences would be present and might well be considered nonhomologous in the case of the genitalia (compositional criterion). In fact, such differences between sexes of one species could be greater than between males of different species if one were looking at sibling species such as *D. pseudoobscura* and *D. persimilis*. And, finally, all adults would more resemble each other than they would resemble the larval form of any of their separate species. Hence, though pure homologizing might be fun—just comparing characters—any significant use of its data would necessitate disentangling in terms of the biology of species. Such knowledge tells us that two sexes and developmental forms are present in all sexually reproducing species, and that comparisons between species are most informative when they contrast equivalent examples from each group. When such care is not taken, and morphology alone reigns supreme, then, indeed, larvae are grouped together as "a natural group," as are distinct male and female forms, and subsequent research has the unenviable task of reexamining the uncritical and narrowly-focussed work of the earlier researchers. Historically, one can understand the not uncommon occurrence of such work. There is little or no excuse for it today. To avoid essentially aimless or misdirected comparative work, the concept of homology must respect the rest of our biological knowledge. In the present instance, this means that a species as the unit of evolution is also the unit of phylogenetic studies. (See also Rule I, above.)

A third problem in applying the concept of homologies is to comment further on the limits of its usefulness in defining the nature of the differences between two different species. Let us return to the idealized situation being discussed earlier, where we were examining what was happening to various characters when species A transformed into B or, in the other case, A gave rise to B as a diversification. Those characters that remained unchanged would be identified as homologies through their similar placement (positional criterion) and their similar makeup (compositional criterion). Probably no character would have changed so much as to be nonhomologous. If, however, we look at things a bit more realistically, in a situation where the historical relationship between two species, X and Y, is not obvious, then things are understandably more complicated. There can be characters essentially unchanged in both groups, and if they fulfill the rules of similar position and composition they can be considered homologous. There can be characters quite different in the two species, but if forms can be found to bridge the differences and establish a series of connecting intermediates (serial criterion), then these features, too, are homologous. This would apply to transformations or diversifications that enhance or reduce a character, or that change it to something new. And, finally, there can be quite new characters in one species or the other that show no evidence of homology. The concept, therefore, is important in identifying but does not terminologically discriminate between conservative and innovative characters—they are all called homologies. And it cannot deal with the class of completely innovative characters. This suggests the need for a more useful terminology. Before turning to that problem, we need to examine the role of these criteria for identifying homologous relationships in the context of nonhomogenetic resemblances and of reductions.

Homoplasy and reductions. There are five types of similarities or resemblances possible that are not due to inheritance from a common ancestry (Simpson 1961). These instances of homoplasy are convergence, parallelism, analogy, mimicry, and chance resemblance. Only the first two will concern us in detail, for the others offer no real opportunity for confusion with homologies but do deserve at least a brief examination. Chance similarities occur in separate taxa and are not related even indirectly, as by distant common ancestry or response to similar selection pressures, and so are quickly seen to not fulfill any of the criteria for homology. Mimicry is adaptive similarity of one edible species to another inedible one or similarity between two inedible ones, in relation to a predator, of course. The similarities are limited to gross resemblance in external anatomy, and the criteria of homology quickly dispel the possibility of true homology. Analogy, which is functional similarity without any relation to common ancestry or to specific adaptive pressures, e.g., wings in birds, bats, and butterflies, also shows only gross resemblances which are easily handled by the Remanian criteria, especially the compositional criterion.

Convergences and parallelisms are more difficult cases. The compositional and serial relationships are most important, but the accessory criteria are also useful. It is easiest to discuss these in the context of specific examples. A common example of convergence is the external form of sharks, ichthyosaurs, and dolphins (Fig. 3–6A). The placement of fins is similar, especially for the pectoral and the dorsal fins; they appear to fulfill the positional criterion. But the compositional criterion sets off the cartilaginous fin rays of the shark from the bony phalanges of the reptile and mammal. Regarding the pectoral fins, the search for intermediates to bridge the difference is, in effect, successful. One can connect differences in mammal and reptile to each other by various intermediates and arrive at a common reptilian ancestor, and from there proceed through the Amphibia and bony fishes to a common ancestor of teleosts and the chondrichthyian sharks. However, the dorsal fin is a different matter. In all three cases, the search for intermediates takes the phylogenist back to ancestors which lack such a structure. Hence, this structure is truly

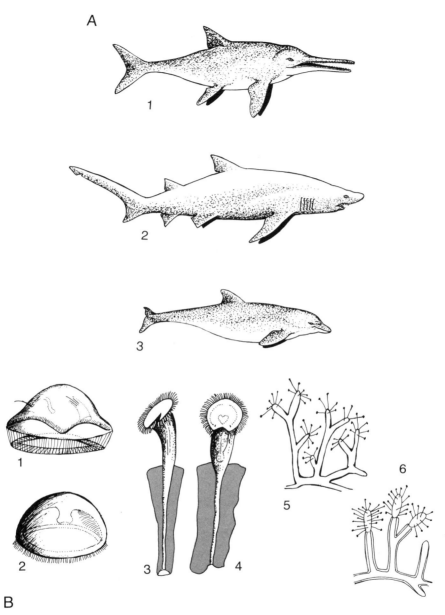

Fig. 3–6 Convergence in selected forms. A. In marine vertebrates similarities in external form are apparent between 1. icthyosaur (an extinct reptile), 2. sand shark (a cartilaginous fish), and 3. bottle-nosed dolphin (mammal). B. Among invertebrates, convergences are seen between pairs consisting of a protozoan and a member of another phylum. 1. The flagellate *Leptodiscus.* 2. The medusa *Homoeonema.* 3. The ciliate *Stentor.* 4. The rotifer *Ptygura.* 5. The suctorian *Dendrosoma.* 6. The polyp *Syncoryne.* (Redrawn, by permission, from Remane (1953) *Die Grundlage des naturlichen Systems der vergleichenden Anatomie and Phylogenetik.* 2 Aufl. Geest und Portig.)

convergent in all three lines. The same steps can be followed for convergences of the sessile protozoan, *Stentor,* and the rotifer *Ptygura* (Fig. 3–6B). Details of structures to be compared reveals the unicellular nature of the one organism and the multicellular nature of the other, and attempts to find intermediates only results in moving towards very dissimilar forms, at best. The criteria of homology can also apply to molecular convergence. Taking as an example the two major types of cytochrome c, which appear similar on the basis of their functional position (a terminal step in the respiratory chain—positional similarity), their amino acid residue sequences and the details of temperature and pH optima (compositional dissimilarity in structure and function) show, however, that the molecules are different, so different as to be considered nonhomologous (Margoliash and Smith 1965).

In the absence of compositional studies, it is possible to use the converse of the first of the accessory criteria to identify convergences. In the case of the hemoglobins, there would really be no homologies established by the major criteria that are held in common by the flatworms, annelids, insects, molluscs, echinoderms, and vertebrates along with presence of hemoglobin, except very general ones, which would be common to animals in general. That is to say, there would be very little evidence of descent of just these hemoglobin-carrying forms from a common ancestor.

The other accessory or subsidiary criteria also bear on the question of convergence. Convergences are the result of similar selection pressures acting on different species; their distribution is restricted to species with common functional needs, such as the rapid pursuit of smaller pelagic fish by most sharks, ichthyosaurs, and porpoises. It is, therefore, unlikely that many different species will share a common character unless it is by inheritance from a common ancestor (Criterion 4). Similarly, we would not expect there to be many sets of similar features all with the same distribution, unless, again, common ancestry were behind it (Criterion 5).

The instances of convergence discussed thus far are all examples of what we can call positional convergences, from the fact that at least they seem to fulfill the positional criterion of homology, i.e., they appear to lie in the same relative position in the organisms of which they are a part. Another category of convergences we can call compositional convergences. These are instances where the second Remanian criterion seems fulfilled but not the first. Examples are the cell organelles: unit membranes seem to be built the same way, regardless of their position in cells; flagella and cilia have the same fine structure regardless of whether they are a ciliary brush border of an epithelium or the two apical flagella of *Chlamydomonas,* or the tail of a sperm cell. These are all illustrations of the "limitations of organic design" (Pantin 1951; Grimstone 1959). They represent convergences resulting from a limited number of ways available to build, at the subcellular level, a structure that will optimally perform a given function. As we have indicated, such convergences are identified by apparent fulfillment of the compositional criterion, but lack of realization of the positional one.

The same lines of reasoning that are applied to detection of convergences apply equally to parallelisms, except that the serial criterion will bring us back to an ancestor lacking the character in question but otherwise possessing many features that are homologous with those in the descendant forms. The parallel evolution of African and South American porcupines illustrates this (Wood 1950; Simpson 1953b), in that their ancestor was a rodent of worldwide distribution. The independent evolution of the large woodpeckers in various parts of the world is another example of parallelism arising from a generalized ancestral type in which certain apparently similar characters appeared to be homologous but are really parallelisms (Bock 1963).

Reduction is the last of these problems, but it is quite different from the foregoing in its relation to questions of Remanian homology. Here there are two problems. The first is to determine whether a structure or function is relatively simple because it is reduced or because it is primitive. And the second, assuming the conclusion that the feature in question is reduced, is to determine its homologous relationship despite the loss of information entailed in its becoming progressively simpler.

Again we must think in selectionist terms to understand what causes reduction. It occurs when those selection pressures which built up the character in question in the first place are themselves reduced or removed. Mutation pressure, i.e., accumulation of random mutations, will now break the character down, rendering it vestigial and finally causing its total loss. There may also be selection against a structure that now performs a useless function (Prout 1964). Furthermore, reductions, like convergence, will affect only a few characters, at least initially. If the changes in the environment, which cause the changes in selection pressure, can be survived and exploited, then gradually more and more changes will occur, so that some existing characters will even evolve further—just the opposite of reduction—and others will be reduced completely—lost altogether.

Probably the classic examples of reduction are found in symbionts and sessile forms. In the former case—parasitic flatworms such as tapeworms are good examples—one finds loss of locomotory and nervous systems but highly developed integumentary structures, such as the scolex for attachment to the host, and a complex reproductive system for production of enormous numbers of progeny to insure infection of a new host. Reduction and further specialization proceed together.

Now, returning to the first problem, that of distinguishing primitive from reduced features, what can be said? The question becomes whether the organism with certain simple features is relatively simple *throughout*, as would be consistent with a primitive form, or is it a mixture of relatively complex and less complex characters, which is consistent with the specialization that goes with reductions. This means, in determining that a relatively simple feature is reduced, one must not look at that feature in isolation, but see it against the backdrop of the whole organism and the niche it is exploiting and, also, in comparison with other similar species. To return to the tapeworm: its attachment organelles and reproductive system are highly specialized and complex in comparison to free-swimming flatworms, but its locomotory and coordinatory capacities are relatively simple. Here the specializations and reductions are easy to spot. The same is true for the sessile, adult tunicates in comparison to their free-swimming larvae.

The second and remaining problem is to identify homologies where reduction is thought to have occurred. The answer is straightforward. If the feature in question still is sufficiently complex to allow point-to-point comparisons with other, apparently less reduced examples of that feature, then the serial criterion can be applied to bridge differences between the reduced and the less reduced states of the character in question. If, however, simplification has gone so far as to preclude application of the major criterion, the first auxiliary criterion can be used to see if other, known homologies coexist with this supposed homology. On this basis, useful information can be gained even as regards extreme cases of reduction.

In brief, though the problem of reduction can be a very complex one (Remane 1956), we can describe its essentials quite readily. Unlike convergences and parallelisms, which falsely suggest homologous relationships, reductions are concerned with loss of homology; but in that such loss involves only part of the organism, never the whole organism, the reduced parts can be identified and properly interpreted.

UNITS OF PHYLOGENETIC INFORMATION: SEMES

We now must deal with the following two problems. The single term homology expresses in a limited way the scope of biological phenomena it emcompasses. As we have seen, homologies depend in the final analysis on identifying certain kinds of similarities—positional and compositional—but they also reach far beyond these similarities through identification of what is called serial relationships. We need terms that will flexibly and precisely encompass the various stabilities and innovations that are presently crowded under the single terminological catchall of homology. That is the first problem. The second problem is the much-deferred procedural one of making explicit the nature of the operations needed to identify similarities of phylogenetic significance.

Both problems can be attacked by introducing the concept of a *seme* as a unit of phylogenetic information. The justification for new terminology was expressed in this manner by Johannsen (1911), who coined the term gene.

"It is a well-established fact that language is not only our servant, when we wish to express — or even conceal — our thoughts, but that it may also be our master, overpowering us by means of the notions attached to current words. This fact is the reason why it is desirable to create a new terminology in all cases where new or revised conceptions are being developed. Old terms are mostly compromised by their application in antiquated or erroneous theories and systems, from which they carry splinters of inadequate ideas, not always harmless to the developing insight." (p. 132)

A seme is an information-containing entity in an interbreeding population of organisms, but it will be most commonly used in reference to a structural or functional part of an organism, starting at the molecular level. In fact, the cognate term "semantide" was proposed by Zuckerkandl and Pauling (1965) for application to molecules. Similarly, Hennig (1950, 1966) has introduced the term "semaphoront" in reference to the unit characters used for comparisons by phylogenists and systematists. However, Zuckerkandl's and Pauling's term seems somewhat cumbersome, and its application was restricted to molecules. The concept of information containing entities includes much more than just molecules. Hennig's term is also cumbersome and has the further disadvantage of including all possible traits, characters, attributes, and conditions of an organism without regard to information content. For phylogenetic purposes, the key problem is information content: we can only work meaningfully with functional or structural features which are sufficiently complex and variable to allow meaningful comparisons. (The only exception is use of the subsidiary criteria, but these are not convincing in the total absence of information to which the major criteria can be applied.)

How do we determine that a trait or character is sufficiently complex to permit what we have called "meaningful comparisons," namely, that it is a seme?

Point-to-point correspondences. This is a complex problem whose resolution is not fully in sight. We can only illustrate the kinds of operations it entails. The search for homologies is commonly joined at a point where much prior information has become available. For example, most organisms have readily identifiable dorso-ventral, anterior-posterior, and right-left aspects, as well as various other more detailed anatomical parts. And this allows us, without much conscious effort, to go directly to comparisons of forelimbs or skulls or some other part of the anatomy. Despite an easy familiarity with various anatomical landmarks, we are, nonetheless, starting at a rather sophisticated level. The comparisons we make, as Woodger (1945) has clearly stated, are attempts at finding point-to-point correspondences. This is illustrated also by Remane's formal models seen in Figures 3–2, 3–3, and 3–4. These formal systems and examples provide little difficulty in satisfying us as regards "mean-

90 *Concepts and Methodology*

ingful comparisons." What these examples tell us is that we want to compare sets of identifiable subunits joined by identifiable connections between the subunits. But this is not yet explicit in telling us how many subunits and what relationship among them is optimally informative.

The basic issues here can be clarified by again looking at formalized systems (Fig. 3–7). Intuitively it is obvious that determination of the similarities between two round dots is a somewhat meaningless task; they can be superimposed one on the other in countless ways, thus obviating any but the most trivial point-to-point comparison, i.e., they are both round dots. But with two points, one round and one triangular, and joined by a line, we can see an axis of symmetry which must not be violated to achieve point-to-point superimposition. However, the figures can be rotated endlessly around that axis and still permit superimposition. Thus, though the argument for similarity is more compelling than when working with one dot, there is still room for con-

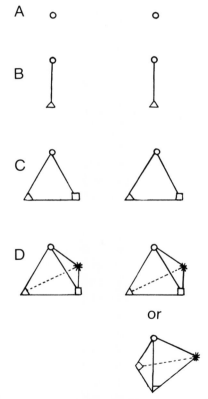

Fig. 3–7 Information content of various sets of subunits. A. Comparison of two single points. B. Two points or subunits in each figure, with the subunits separately identifiable. C. Two sets of three different subunits. D. Sets of four different subunits apiece. Figures are three dimensional. (See text for explanatory details.)

siderable ambiguity. Next, consider a three dot figure, composed of circular, triangular, and square dots joined to each other by lines. Here, two such figures can be identical, or, if one figure is flipped 180° (there is no upper or lower side specified here) it is the mirror image of the other. Thus we achieve only two possible outcomes from comparisons between three-point systems and ambiguity is reduced accordingly. And, finally, consider a four-dot system of circular, triangular, square, and asterisk dots forming a three-dimensional tetrahedon. When two such figures are compared they are either unambiguously identical or, if one point (the square in our figure) moves to the opposite side of the plane defined by the other three points, we now have a mirror image. The alternatives are reduced to identity *or* mirror image; in effect identity can be unambiguously specified.

The foregoing needs only a few further remarks. Five points add nothing to the rigor of the comparison, for with four points in three-dimensional space we can determine identity or lack of it, but extra points do aid in ease of identification and are useful in that case. This may be clearest when dealing with points lying only in two-dimensional space, e.g., the three point system just discussed, or Fig. 3–2, 3–3, and 3–4, for here added points add significantly to the validity of arguments. This also argues for the importance of making three-dimensional comparisons whenever possible. The reduction of three-dimensional structures to the two planes of a page of paper reduces information content if not taken into consideration. The need for many points is especially clear when we search for homologies between processes. In these cases the time axis is the only dimension available, and points of comparison are strung out as beads on a string. Obviously, then, the more beads we have that are similar and that lie in the same sequence in the instances being compared, the more similar are the strings of beads. In other words, the two sets of functions are alike to the degree they have similar sequences of similar subfunctions.

Also, we should make explicit the obvious fact that identification of similar points depends entirely on point-to-point comparisons. To argue that two circles are identical carries little conviction if they are not at least the same size. Likewise for triangles, squares, and asterisks. These are, however, easily distinguished from each other on the basis of point-to-point resemblances. In phylogeny we need only to be sure, for example, that it is two elbows we are comparing when we are looking at vertebrate forelimbs, and two wrists, and two shoulders. This is done by showing that bony articulations at these points correspond pointwise in one limb with articulations at only one other point on the other limb. Thus points which in themselves may not be very distinctive, can be at least distinguished from other points, and then the whole set of points, through definable interrelations, takes on a significant informational content.

Lastly, if we could specify that the probability of correspondence between any two paired points is 0.5, then the occurrence of correspondence (either on the basis of positional or compositional criteria) between five such pairs in succession would be $(0.5)^5$ or 1/32, or between seven such points it would be 1/128. These levels of improbability, assuming the null hypothesis that all correspondences, or all lack of them, are due to chance alone, are beyond the 5% and 1% levels respectively. It could let us argue that at least five to seven subunits (points of comparison) are needed to render a given trait sufficiently complex to be designated a seme. The essential problem is, however, that this computation depends on the initial probability of similarity being 0.5. There seems at present no way to determine what that value might actually be. In comparing two ciliated protozoans, if one chose to compare their nutritional requirements, the probability of homology would be close to 1.0, since both are animals. If one compared the contractile stalk on one with an adhesive area on the other, the chance of homology might be 0.0.

Therefore, at present, our response to what is a "meaningful comparison" must remain somewhat intuitive and can be stated as follows: A seme, to be used in the context of the positional and compositional criteria, should ideally contain four or more subunits in three-dimensional space or five or more subunits in two- and one-dimensional space, and these subunits must be identifiable as individuals and in relation to each other. When such semes are not present, the subsidiary criteria are the only ones applicable.

Types of semes. The neologisms to be proposed below are derived from Hennig's (1950, 1966) terminology as developed for his sophisticated argumentation of phylogenetic systematics. Hennig distinguishes, basically, between plesiomorphous and apomorphous characters. The former refers to characters initially present and the latter to derived characters. However, the suffix "-morphous" in Hennig's terminology is restrictive in its implication that only morphological characters are under consideration. We have seen that functions as well as structures carry the information studied by phylogenists. We therefore replace -morphous by -seme. Thus a *plesioseme* is an evolutionarily conservative trait and an *aposeme* is one showing innovation relative to a preexisting, plesiosemic condition. Mayr (1969) has referred to these as simply *ancestral* and *derived* characters or character states. In that we have tried to provide a fairly precise meaning to seme that is not coextensive with character—not all characters are semes, though all semes are characters—we will stick to the less familiar but more exact terminology being developed here. These terms and cognate ones, are placed in Table 3–1 along with other aspects of their use. The examination of this table will conclude our discussion and analysis of basic phylogenetic concepts.

In the first column of Table 3–1 is a summary of the possible fate of characters (semes) during species change. In the second column are the terms being proposed here. Plesiosemic entities will be the largest category when comparing species with a relatively recent common genetic heritage, though this may be unknown at the time of comparison. This is because of the very conservative nature of the evolutionary process. Aposemic features can be of four different kinds. When the change in character is simply enlarging on something already present, it will be termed *hypersemic*. And, conversely, when the change is reductive, the appropriate term is *hyposemic*. (It will be noted that we have omitted repeating apo- in each of these and the succeeding terms. This omission provides a less awkward word and its meaning is still clear, for, except as we refer to plesiosemes and neosemes, all other semes can be understood to be aposemes.) If a structure or function is repeated, either metamerically or as in some type of colony formation, the innovation is termed *polysemic*. When an old part or process is transformed into something new, this is designated as *episemic*. If there appears a brand new innovation, this is called *neosemic*. These neosemic changes will be least common in the evolution of one species into another. They and the episemic changes become quite apparent and may be very important in the longer view, which can reveal a significant difference between parental and much later, descendent forms. The episemic and neosemic changes can be referred to as innovations qualitatively different from the plesiosemic condition, whereas hyper-, hypo-, and polysemic changes are quantitatively different from the original condition. Another way of categorizing aposemic changes is to see them as homologous aposemes (hyper-, hypo-, poly- and episemes) in contrast to the nonhomologous neosemes.

Column III describes, for each case, the use of Remanian criteria in the context of each kind of change (including no change). Plesiosemes need only

Table 3–1 Terminology derived from the concept of the seme and its usage.

I Evolutionary fate of seme	II Terminology	III Identification of homology	IV Procedural usage
1. No change	plesioseme	positional and compositional relationships	establish evolutionary relationship
2. Change a. enlargement	aposeme hyperseme	positional, compositional, and serial relationships	evolutionary relationship and direction of change
b. reduction	hyposeme	″	″
c. repetition	polyseme	″	″
d. transformation	episeme	″	″
3. Newly emerged	neoseme	(not homologous)	direction of change

the positional and compositional criteria to identify their homologous status to each other. All the asposemic situations necessitate the serial criterion as well as the positional and compositional criteria to identify their homologous status to each other. The serial homologies or homonomies of polysemes show the curious situation that the positional criterion may be only partially fulfilled or even negative, but the compositional criterion will be adequately fulfilled. For true neosemes, neither of the first two criteria will be realized.

The last column (IV), brings us to procedural uses. Plesiosemes tell us the forms under comparison are genetically related and the greater the number of these semes the shorter the phyletic distance between the two groups. Aposemes also inform us of common descent but, in addition, describe the direction of evolutionary change. This is the basis of Hennig's use of aposemic characters and will be elaborated further in the next section of this chapter. Episemic characters or features tell us about patterns of relatedness and evolutionary direction. Neosomes can tell us about direction when used in conjunction with other characters, which establish the common characters of two groups. Against such a backdrop the neosemes show new directions of evolution.

These concepts, then, cover the full range of species change. How they are actually used to establish a pattern of phylogenetic relationships—the historical course of evolution—is the next major problem.

Phylogenetic Procedures

The aims of phylogeny are twofold: to reconstruct the course of evolutionary events, and to analyze those events so as to understand better the courses of evolutionary change. The task now is to come directly to grips with the first of those two goals. We will start by describing how the *phyletic distance* between any two species is determined. The logically subsequent step is to determine phyletic distances between more than two species and specify the evolutionary relations deducible from those conclusions. Then we will be ready to discuss the second goal of phylogeny.

COMPARISON OF TWO SPECIES

It is apparent now that the evolutionary relationship of two species is not precisely defined by their overall similarity to each other but rather by the nature of the similarities in certain traits. As Remane succinctly says "The *degree* of similarity plays only a subordinate role, the *kind* of similarity is central." (Remane 1956, p. 10.) Using the idea of semic characteristics, it is clear that if there are many plesiosemic traits and few aposemic ones held in common between two species, the phyletic distance between them is small. If, on the other hand, there are few plesiosemic characters and several aposemic characters in common, then the phyletic distance is greater than in the first case. Finally, an obvious third condition, if there are few plesiosemic and aposemic characters in common and many neosemic characters, the two species are only distantly related. (There might be no neosemic characters in the first case, a few in the second, and more in this last one. Convergent features are neosemes of similar appearance.) Therefore, as a first approximation, to establish phyletic distance through selected similarities, the relative number of plesiosemic entities as compared to aposemic ones is important.

Number of semes to be compared. In general the more semes compared the surer the conclusions they engender. There will be, naturally, limits to what is practicable. Also, there are some conceptual limits. The practical limits are circumscribed both by the data available, and, if that is not limiting, by the researcher's determination of when he has accumulated enough data to permit reasonably unequivocal conclusions with regard to the questions being asked.

The conceptual limits are best delimited by referring to Table 3-2, where there is listed the features of animal systems that were described earlier (Chapter 2) as showing that complexity and variability needed for phylogenetic studies. The reason for this approach is, in addition to the sampling problem, adherence to the phylogenetic rule (Rule II) that the total biology of the groups being compared should be taken into consideration. It will be apparent that the semes, as listed in the table, are not all equally useful. The list needs careful resorting in the light of the concept of semic entities. At present, the lack of sufficient comparative data on species structure and function (Chapter 2; Mayr 1963) excludes that level of organization from further consideration here. (The semic traits listed in the table are guesses as to what might eventually be important. As is obvious the guesses are guided by ideas from population genetics and implications from the Hardy-Weinberg law in particular. Recall also that selection can act on units below the species level, i.e., demes and kin groups, and perhaps, also, above the species level in ecological communities (Lewontin 1970).)

The multicellular and cellular features of living systems must be brought together as different aspects of one entity, the organism. This is necessary because subsequently in this work we will have to compare unicellular and multicellular forms, and a rational basis for comparison must be established. This is most readily achieved by using the ten categories of the organ systems and their functions, supplemented by two other aspects—general size and shape and symmetry, and body plan—which are common to all organisms and are sufficiently complex to be semic entities. The organ systems, in terms of their functions, represent elaborations, as was emphasized earlier, of functions present in all animal systems. (If our comparisons were between plants and animals we would have to rethink these features yet further.) For example,

Table 3-2 A listing of major animal semes.

1. Species semes
   a. Demic structure—genetic variability, etc.
   b. Breeding patterns
   c. Population size
   d. Migratory behavior
   e. Mutation rates
2. Semes found in organismic systems (multicellular and unicellular)
   a. Body organization
      1) Size, shape, and symmetry
      2) Special features—appendages, cavities, etc.
   b. Organ systems
      1) Integumentary
      2) Digestive (including nutritional needs)
      3) Circulatory
      4) Respiratory
      5) Excretory
      6) Muscular
      7) Skeletal
      8) Nervous
      9) Endocrine
      10) Reproductive
          a) Mating systems
          b) Development
3. Molecular semes
   a. DNA
   b. RNA
   c. Polypeptides and proteins
   d. Biochemical pathways and products

one can use the category of digestion to compare nutritional needs as well as ingestion and egestion. Also there are the structures associated with these functions. All of these can be treated as semes for comparative studies within and between protozoans and metazoans. This means, then, that there are potentially twelve general categories of comparisons that must be made in terms of the organism and its major parts. A little consideration will show that all twelve are not equally complex, especially in the protozoans, and therefore some of these, such as endocrine activities, will not permit meaningful comparisons; they are asemic traits. Furthermore, other aspects, such as reproduction, are very complex and can be considered to consist of more than one semic trait such as asexual and sexual processes and regeneration. However, this means that when two such semes are identified in one group of those being compared, they must be looked for in all others too.

Molecular semes must be treated separately from the organismic. They do not easily fit into the system-oriented categories and they have sufficient complexity to stand on their own in comparative analyses. Their categories are those of DNA, RNA, protein, and biochemical pathways with their special products. These plus the preceding ones indicate a possible number of 16 categories which will be increased if any category can be profitably subdivided and will be decreased if any category is too devoid of informational detail to be considered a seme.

Therefore, the answer to the question regarding how many semes should be used has a rough quantitative parameter of 10 to 20, which can be extended further depending on circumstances. To decrease it, however, has the obviously undesirable feature of narrowing the total area of point-to-point comparisons.

Semic comparisons. Now commence the actual comparisons between semes. Using general biological knowledge as to what functions and structures in one organism are to be compared with their counterparts in another organism—reproductive processes with reproductive processes, and so on—point-to-point comparisons are made to determine whether or not homologies exist. Where the positional and compositional criteria are fulfilled, it is arguable that the homologous relationship is plesiosemic. Where it appears that some intermediate form or forms are needed to make a convincing demonstration of homology, i.e., where the serial criterion is needed, then aposemy is indicated and established if the intermediates are indeed found. (This latter necessitates, of course, comparing more than the two organisms with which we have started.) There is apparent here a point of some ambiguity: how similar must semes be to be plesiosemic and when are differences sufficiently great to be designated aposemic? This is frankly a matter of judgment. Perhaps only between members of the same species will similarity reach the level of identity, but even here some intraspecific variation must be expected. By and large, in what follows, we have found that if 80% of the pairs of elements (points or subunits) in the two sets being compared correspond to each other, then the sets stand in a plesiosemic relationship. If there are elements present in both that can be compared (the pairs of points appear similar) but correspondence is less than 80%, and especially if insertion of an intermediate bridges the gaps in correspondence, then an aposemic relationship is clearly indicated. If, however, the elements in one system have no elements at all to correspond with in the other, the compared entities are neosemic or hyposemic (one seme is reduced). The choice between these two alternatives cannot be resolved when only two species are compared, for there is no point of reference from which to infer the direction of evolutionary change. We must, therefore, turn now to comparisons between more than two species.

COMPARISONS BETWEEN MORE THAN TWO SPECIES

It will be easiest to begin this discussion by examining the procedural steps necessary to analyze a small number of relatively similar species. Later we will discuss the problems of examining larger groups and heterogeneous groups.

A comparison of five selected species. In Table 3–3 are given five species (A through E) and a symbolic statement of the nature of their semes under ten categories (1 through 10). (Actually, species A through E represent certain ciliated protozoa and semes 1 to 10 are actual features of those species. It is clearer for present purposes to simply use letters and numbers so as to concentrate on the procedural steps.) The determination of whether two semes are alike or different is, of course, the determination of whether they are homologous or not. And as we have said above, if their homologous nature is settled simply by the positional and compositional criteria then they are plesiosemes. If the criterion of serial relationships also needs to be involved, then they are aposemes.

It is at this point that the term *plesiomorph* must be introduced. A *plesiomorph* is that species, or more precisely its neontological, paleontological, or embryological representative, which contains the largest number of plesiosemes in the group of species being studied. This term is derived from Hennig (1950, 1966) with the meaning changed somewhat. Hennig uses it to refer to characters, and in particular morphological ones. For us the suffix -morph, rather than denoting structure, refers to a genetic form which is part of the polymorphism of any species (Mayr 1969). The best alternative to plesiomorph would be plesiotype. However this is used to denote a species related to a specimen found in a different region or geological formation (Henderson, Henderson, and Kenneth 1963) or it refers to a specimen used subsequently to the original description of a species (Howell 1960). The plesiomorph is our way of handling the problem of a "primitive" or "stem" form or the "ancestral" type. However, the terms primitive, stem, and ancestral are used so many different ways and are, therefore, so ambiguous, that it seems best to avoid them entirely as concepts in a rigorous phylogenetic analysis. There will, of course, be occasion to refer to forms ancestral to the plesiomorph and, hence, ancestral to the group represented by the plesiomorph. To this hypothetical early form the term *eotype* (from the Greek *eos,* the dawn) shall be applied.

To determine the plesiomorph we must be able to read aposemic trends

Table 3–3 Five species, A through E, and their semes. A question mark indicates that the seme, though probably present, is undescribed. A zero indicates that the seme in question is absent.

Species	\multicolumn{10}{c}{Semes}
	1	2	3	4	5	6	7	8	9	10
A	1A	2A	3A	4A	5A	6A	7A	0	9A	10A
B	1A	2A	3A	4B	5A	6B	7A	0	9A	10A
C	1A	2C	3C	4C	5C	?	7C	0	?	10A
D	1D	2D	3D	4B	5A	6D	7D	8D	0	10D
E	1D	2D	3E	4A	5E	6B	7E	8E	0	10E

and determine which way they go, that is, decide which is the original form of the seme and which are the derived forms. Here we are dependent on Hennig's (1966) analysis of the problem. There are four criteria for determining the direction of an evolutionary transformation of a seme or, in the case of structures, a morphocline (Maslin 1952).

1. Geological character precedence. If, in the fossils of a given group of species, a certain character condition is found in the older fossils and another different but demonstrably homologous one is found in more recent fossils, then clearly the latter is the aposemic and the former the plesiosemic condition.

2. Chorological progression. (By this phrase Hennig is referring to use of distributional data. They are especially useful for certain terrestrial forms.) In certain cases it is possible to determine where a given group of species originated, and as they spread further, there occur various changes, the most extensive of which are correlated with the most extensive departures from the ecological niche and habitat of the original form. In such cases the direction of aposemic changes can be determined from distributional (and, therefore, ecological) studies.

3. Ontogenetic character precedence. This is the biogenetic law, in effect (see chapter 2). This general principle can be used in the sense that earlier stages in development are the more conservative ones, and if development shows one form of a seme appearing earlier in development than another, it is possible to argue that the one appearing earlier represents the plesiosemic condition. Great caution must be exercised here.

4. Transformation series. Here it is useful to go to certain "principles" developed by Maslin (1952). Of the ten principles analyzed, the principles of divergence are most useful and it is best to quote directly from Maslin's succinct formulations.

> "a. If one extreme of a morphocline resembles a condition found in the less modified members of related groups of the same rank, this extreme is primitive.
> b. If two or more morphoclines appear in different groups of organisms, and an extreme of each morphocline occurs in the same taxonomic unit, then these extremes are the primitive extremes.
> c. If two extremes of two clines are identical and are found in the same taxonomic unit, the identical extremes are primitive and the dissimilar extremes are derived." (p. 69)

Hennig also points out that if two or more character transformations occur with the same distribution and the direction of change is known for one of them—presumably on paleontological, chorological, or embryological grounds—then the others are determined in like manner.

The identification of the plesiomorph is important for the proper interpretation of the comparisons that must now be made between all possible

100 *Concepts and Methodology*

pairings of characters. The identification of plesiosemes poses no problem—a pair of highly similar traits stand in a plesiosemic relationship to each other. The identification of most aposemes, specifically hypersemes, polysemes, and episemes, also offer no difficulty. Here the serial criterion will be needed to make differences into a smooth continuum. The problem comes with hyposemes (reduction) and neosemes. In the absence of a plesiomorph, one cannot know which is the ancestral and which the derived condition. Therefore, complete reduction, that is, loss of a trait, could also be interpreted as a neoseme if one assumed the absence of the trait to be ancestral and then proceeded to a form possessing the trait. Hence, identification of the plesiomorph as a reference point is necessary. Its identification also makes clear the reading of direction in other aposemes, for, again, without a reference point, the direction of change in an aposemic situation is not determinable. With a plesiomorph at one end of such a series, the other end is the aposemic, or derived, extreme. In this way we see that the plesiomorph is representative of the ancestral form of a given group of species. It is our best representative, in terms of an actual species, of the phenotype from which a given monophyletic group arose.

Returning to the five species being examined in the present analysis, we conclude that species C is the plesiomorph. What we are doing is examining transformation series and, in particular, using the second principle quoted above from Maslin namely, that when the extremes of two clines lie in the same taxonomic unit, that unit contains the primitive extremes. This is illustrated in Table 3–4, where the variability in each seme has been arranged in terms of the clines present. In all semes except 6 through 9 we find that species C lies at one end of the cline in question. (In semes 1 and 10 species C shares the A condition of that seme. See Table 3–3.) In semes 6, 8, and 9 species C is asemic, so its position in those clines is undeterminable. In seme 7 it is not at the terminus of the cline, but as we will see later (Chapter 10) its

Table 3-4 Morphoclines in ten semes and their occurrence relative to the five species under discussion.

1	2	3	4	5	6	7	8	9	10
D	D	E	B	E	D	D	E	A	E
	D				A	E	D		D
								O	
	A	A	A	A	B	C			
							O		
A	C	C	C	C	?	A		?	A

position regarding this trait is probably due to special evolutionary development of this particular trait, i.e., this is one trait where the species is not plesiosemic and shows a special adaptation. We can therefore understand the special case of seme 7, and on the basis of the other semes conclude that C is the plesiomorph. The alternative would be to look at the other end of the clines, but this shows nowhere near the regularity shown by C, and examining each seme separately, as we did the exceptional seme 7 for C, does not result in plausible exceptions so that B, D, or E becomes a plausible plesiomorph.

We are now ready to make all possible pairs of comparisons between the semes of one species and another, starting with comparisons between C and all other species. In part A of Table 3–5 the inferred homologous relationships are given. Here species B and D are compared, for they include all possible semic relationships.

Next, phyletic distances can be calculated from the equation

$$R = \left(\frac{-p + (2a)^2 + (3n)^2}{t} \right) + 1$$

where p is number of cases of plesiosemy between two species, a the instances of aposemy, n the cases of neosemy, and t the total pairs of semes studied. These values are given in the right hand portion, Part B, of Table 3–5. It can be seen that this formulation accentuates differences between the species by, first, weighting the aposemes twice as much as the plesiosemes and weighting the

Table 3–5 The determination of the phyletic distance (R) between pairs of species. The species being compared are given at the left and the comparisons of semes as to whether the relation is plesiosemic (p), aposemic (a), neosemic (n), or indeterminate (?), is given for each of the ten semes studied.

Species Comparisons	1 2 3 4 5 6 7 8 9 10	Total semic pairs (t)	Plesio- semes (p)	Aposemes (a)	Neosomes (n)	Phyletic distance (R)
C and A	p a a a a ? a p ? p	8	3	5	0	13.1
C and B	p a a a a ? a p ? p	8	3	5	0	13.1
C and D	a a a a a ? a n ? a	8	0	7	1	26.6
C and E	a a a a a ? a n ? a	8	0	7	1	26.6
A and B	p p p a p a p p p p	10	8	2	0	1.8
A and D	a a a a p ? a n n a	9	1	6	2	20.0
A and E	a p a p a a a n n a	10	2	6	2	18.8
B and D	a a a p p ? a n a a	9	2	6	1	16.1
B and E	a p a a a p a n a a	10	2	7	1	21.3
D and E	p a a a a ? a a p a	9	2	7	0	22.6

neosemes three times as much as the plesiosemes. It further weights the aposemes and neosemes by the device of squaring their values. This magnifies the addition of semes to either of these categories. For example, two aposemic traits give the a term a value of 16 (i.e., $(2 \times 2)^2$) and three aposemic traits result in the value of 36. This magnification of differences has been found useful in the analyses which follow. Finally, in the extreme case of comparison of one species to itself to achieve a phyletic distance of zero, the value of 1 is added to the equation which cancels out the -1 resulting from $-p/t$, where $p = t$ and all the semes compared are plesiosemes.

The estimates of phyletic distance are used to construct the phylogenetic diagrams in Fig. 3–8. The first step is to place the species relative to each other in a manner that best reflects the phyletic distances between them (Fig. 3–8A).

There are two essential guidelines that must be observed in translating phyletic distances into a dendrogram. One is that *all pairs of comparisons must be made* and the second is that *the position of phyletic branchings must be checked against the original data*. The reasons for the first guideline are threefold. First, it will be noted from Fig. 3–8A (and from subsequent analyses) that the sum of two adjacent R values (e.g., D to B plus B to C, or E to A plus A to C) is less than the R value for the two extremes in the series

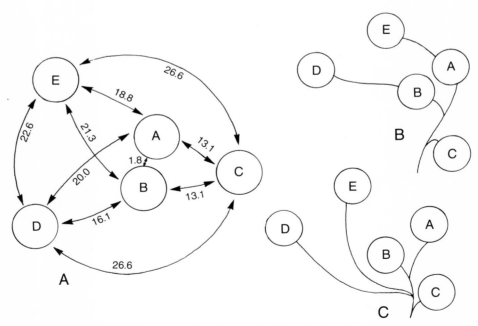

Fig. 3–8 Phylogenetic diagrams based on phyletic distance. A. Distance plots between five species, A through D. B. and C. Two possible dendrograms consistent with data in A.

(i.e., D to C, or E to C). This is because aposemes and neosemes are qualitative judgments and are not quantitative. Hence, aposemes exist between D and B and between B and C, and they are still identified simply as aposemes between D and C even though the change can be much greater than in the first two comparisons. This difference is lost in the present method of calculating R values. And the second fact which contributes to these discrepancies is the occurrence of asemic traits between certain species, and this means, unavoidably, different amounts of information are compared between different pairs of species. The foregoing warns the phylogenist that the two-dimensional summary of R values (Fig. 3–8A) is a limited representation of the evolutionary variation found in nature. This is further emphasized by the third point which is that one cannot assume because two species closely resemble a third and even a fourth species, that they will therefore be equally close to a fifth. (A and B are equidistant from C, almost equidistant from E, but quite different in phyletic distance from D.) All pairs of comparisons must be made, and the R values so obtained speak only to general phyletic relationships.

This brings us to our second guideline, which demands that the original data be consulted for determining the finer details of the dendrogram, in particular the locations of phyletic branching. There is no way of deciding on the basis of R values alone whether Fig. 3–8B or 3–8C is the more plausible phylogenetic scheme. The former shows a common starting point for C (the plesiomorph), for a line giving rise to A and B, and for two separate lines going to D and E, respectively. The latter scheme shows D and E coming off the line going to A and B. Only by going back to the original data can we justify dendrogram C over B. As was mentioned earlier, A, B, C, D, and E are ciliated protozoans. Corliss (1961) favors the view that the gymnostome ciliates, of which *Remanella multinucleatum*, species C in Table 3–3, is a member, is close to the original ciliate type (see chapter 10). *Paramecium aurelia* and *Tetrahymena pyriformis*, species A and B, respectively, are among the hymenostome ciliates lying about midway in the pattern of ciliate evolution, with *Tetrahymena* probably representing the less specialized form. And from it several lines evolve including the peritrichous forms, exemplified by *Vorticella campanula*, species D, and spirotrichous forms, represented by *Stentor coeruleus*, species E. The oral structures, for reasons to be reviewed in chapter 10, are especially important as an aposemic indicator of ciliate evolution, and probably oral structures like those found in gymnostomes give rise to structures like those in the hymenostomes, and from these there evolved two separate lines leading to the peritrichs and to the spirotrichs. This is not made clear in the first phylogenetic tree but is clear in the second one. In brief, we are saying that reference back to the actual forms being studied can tell us where one or another aposemic development is best thought to arise. Any time large R values are found, this is a clear indication that problems of interpretative judgment are at hand. The best solution is to find intermediate forms so that smaller R values are possible in all phylogenetic lineages. Such smaller values tell us that relatively complete series are being analyzed. Thus they provide an increased measure of confidence in the study which contains them.

Comparisons between many species. We can now turn to yet more numerous groups for comparison. In the absence of any experience with the species to be examined—let's say there are 200 species to be compared—one could simply proceed in the manner outlined in the preceding section and make all the necessary comparisons, the determination of the plesiomorph, and calculate all possible R values. These could then be used to construct a phylogenetic tree to complete that part of the study. But there is another approach, using accumulated experience with the forms being studied, which suggests we first separate them into groups of roughly similar form on the basis of a few characters. Such characters may come from the experience of others and be distilled into taxonomic keys or handbooks, or they may be somewhat educated guesses based simply on first hand experience. The reason for this initial sorting is practicality. It will tend to put more easily compared forms together, and comparisons among the groups can be carried out somewhat more expeditiously than otherwise. Eventually, however, there must be comparisons between the various groups of species, and these may be made on the basis of the same semic entities as were used within each group, or new semes may be used. Inevitably, as more diverse groups are being compared, what was plesiosemic within one group becomes aposemic when compared to another group.

This brings us to two problems in the comparison of many groups that need separate attention. The first we can call the maximum use of information, and the second is effective comparisons between quite different groups, groups which in systematic terms correspond to the highest taxa, e.g., phyla.

To use semic information to its fullest degree, we must separate the problem into comparisons being made between relatively homogeneous groups as compared to those being made between more heterogeneous forms. Let us take the former case first, for the latter will lead naturally into the problem of comparisons between higher taxa.

When relatively homogeneous groups, are compared, there will obviously be many characters held in common that show little variation and consequently tell us little about the evolutionary behavior of the forms in question. Attention is devoted to the variable or innovative features, the aposemic ones, as has been stressed by Hennig (1950, 1966) and others. When one studies the evolution of the birds, their beaks and feet are especially variable in the larger context of being feathered and winged vertebrates. Flight, a fringe benefit of being feathered, so to say, imposes very uniform structural restrictions on birds as a whole (Bock 1963). Within that adaptive type it is especially the bills and feet that show specializations documenting various evolutionary diversifications. Hence, within such groupings, the phylogenist ignores many if not the bulk of semic categories given earlier (Table 3-2) and concentrates on those which are of maximum information. It is clear that these categories of maximum information will differ for different groups. To compare birds with starfish would only be possible if we go back to the broad categories of the major semes, because starfish with their pentaradiate symmetry, specialization with regard to a sedentary mode of life, dependence on tube feet and the water vascular system for locomotion, etc., emphasize an aposemic diversification quite out of character with that found in birds. In brief, it is necessary, when elucidating evolutionary trends and divergences within a uniform group, to emphasize only a limited part of the biology of that group, but it must be kept in mind that these especially informative aposemes are only part of the total biology of the living systems being studied. This is a practical and understandable qualification to Rule II and is to be applied only *after* the rule has been first observed and then, subsequently, through search for homologies, the areas of aposemy have been identified.

To study heterogeneous groups, we must use broader categories, such as those given in Table 3-2, for only in this way can we be assured of systematic and thorough examination of the exploitive, homeostatic, and reproductive traits. It is also true that, because of the breadth of these semic categories, details within them may be quite variable and degrees of correspondence hard to determine with great precision. To minimize this clearly undesirable situation, various of the categories can be broken up into subcategories of separate subsemes and more detailed comparisons made. The danger here is the tendency to concentrate on those areas where detailed comparisons are possible and forget the other, less easily studied semes. Conclusions formulated in this biased manner will, in the long run, only impede phylogeny. Witness the endless debates, especially at the turn of the century, over phylogenetic interpretations of the biogenetic law where developmental semes were used to the exclusion of other data. Therefore, in comparing heterogeneous groups, the total biology of the forms being compared (Rule II) must be first examined; and second, where detailed comparisons are possible, they are to be made, but not in a manner that results in excessive reliance on them at the expense of the other less detailed semes. Examples will be given in later chapters of this kind of problem.

These broad, intergroup comparisons anticipate the extremely difficult question of how one compares organisms that in conventional systematic terms are members of different higher taxa. The practice of morphologists from Haeckel's (1874) time down through Naef (1919, 1931) to Remane (1954, 1955, 1956) and Hennig (1950, 1966) and others (Jägersten 1955; Marcus 1958; Ax 1961; Hadzi 1963) is to develop a hypothetical stem form or ancestral type that is thought to typify the group in question. For reasons given earlier (Rule I, the species or some actual representative of it is the unit of phylogenetic comparison), this procedure is not followed here. The logical candidates for these intergroup comparisons are the species, or their representatives, which are the plesiomorphs.

But one must realize that evolution need not be a sequence of plesiomorphs. That is, evolutionary development can occur within a group of species, and at some point above the plesiomorph an innovation or innovations can occur that lead to exploitation of a new adaptive zone with concomitant evolutionary development of a very significant sort. Hence, it is to be understood that when heterogeneous groups are subdivided, the plesiomorphs of these subgroupings of species are not to be compared solely with other plesiomorphs. To find the phyletic origin of a given monophyletic group, one looks for homologous relationships between its plesiomorph and any other species.

We must, however, keep one important difficulty with this approach clearly in mind. The plesiomorph may not represent closely the condition of the hypothetical stem form. Clear grounds for suspecting this would be the occurrence of some significant exception in the morphocline analysis, e.g., seme 7 in species C (Table 3-4), or some neosemic feature or features in addition to many plesiosemic ones. These traits may be understood as specializations not representative of the actual but unavailable stem form. This problem will of course have to be separately evaluated for each individual case of plesiomorphy.

PROCEDURAL SUMMARY

It will be useful now to summarize in outline the procedural steps needed for determining phylogenetic relationships, diagramming them as dendrograms, and evaluating the dendrograms.

1. Have available a summary of those features—structures and functions—which characterize as fully as possible the exploitive, homeostatic, and reproductive traits of the species in question. This can include 10 to 20 semes (Table 3-2).
2a. If the species are relatively homogeneous, treat them as a single group for intragroup comparisons, following steps 3 through 9.
2b. If the species are relatively heterogeneous, break them up into more homogeneous groups and then proceed for each group as in 3 through 9.
3. Determine what semes are present and how best to analyze them to show evolutionary relationships. This means minimize comparisons between invariable semes and maximize variable ones by breaking the latter up into various subsemes.
4. Determine whether the semes (or subsemes) are homologous or not by

use of Remanian criteria. This will provide a preliminary identification of plesiosemes and most aposemes.
5. Determine the plesiomorph.
6. Determine relationships between all semes by paired comparisons in terms of whether a pair shows plesiosemy, aposemy, neosemy, or is asemic.
7. Calculate all phyletic distances using an operationally defined procedure, such as

$$R = \left(\frac{-p + (2a)^2 + (3n)^2}{t} \right) + 1$$

8. Diagram the minimal possible phyletic distance between all pairs of species and convert this to a dendrogram.
9. Criticize plausibility of the dendrogram from information contained in the original data (contained in 1., above).
10. Compare plesiomorphs of various groups to other species to determine relationships among them, again using steps 3 through 9. In step 3, however, it may be necessary to use the very broadest of categories for the semes.

If one is dealing with very heterogeneous groups, it may be necessary to repeat the cycle of comparisons several times with appropriate adjustment of semes at each new cycle.

Evolutionary Innovation

Within the evolution of any group, a biologist will often find himself or herself concluding that this or that species (or collection of species) is an evolutionary "dead end," whereas another ensemble of forms is thought to have demonstrated a rich evolutionary potential. What exactly is meant by "dead end" and "rich evolutionary potential"? It is these problems we face as we come now to deal with the idea of evolutionary innovation.

Perhaps the two points which contribute most to confusing the issues here are these: (a) No living species is entirely devoid of evolutionary potential—it can still mutate and natural selection is acting on it—and none is, at the other extreme, bursting with all kinds of successful new changes—mutations are rare and among these natural selection finds only a few that further contribute to the fitness of a species. (b) There is marked disagreement among biologists as to the course of evolution, especially at the level of the highest taxa, and so there is little chance of clear documentation of forms whose history convincingly demonstrates one or the other of the two extremes of the spectrum of evolutionary innovation. In the absence of concrete examples, rigorous discussion becomes difficult. Particularly because of this last point, we can only introduce the discussion of this problem here and shall have to delay its further development until we can present our own documented examples of groups with limited and with significant innovative potential. In that our conclusions will arise from the phylogenetic analyses of Part II, it will only be in Part III that we complete our comments on this topic.

The key issue will be to understand how species can become highly adapted, i.e., specialize, and yet retain significant evolutionary plasticity. And by plasticity we must mean more than simply unexpressed genetic variability. By plasticity we mean the potential to not only invade a new niche but to also have the functional capabilities to exploit that niche and to even evolve into forms. This means there must be the evolutionary emergence of new functional patterns. Such evolution will entail major structural changes, too. We therefore want to know not only what was the sequence of species transformations and divergences that gave rise to a new organizational pattern, but why it came from a given starting point and not another. To be even more concrete, we are trying to find out why certain algae gave rise to protozoa, why these protozoan forms evolved the way they did, and why multicellular animals arose, as seems apparent, from only certain protozoans and not others. In this way we hope to understand better the evolutionary potential of living systems in general.

This precise ferreting out of the innovative steps in evolution is an ambitious goal, but it is a direct consequence of the methodology developed in the preceding pages. The plesiosemes are the conservative aspects of evolution and the apo- and neosemes are the innovative ones. That is, plesiosemes show us where selection pressures have remained largely unchanged, but aposemes and neosemes identify traits responding to new selection pressures. Therefore, by following the succession of aposemic and neosemic changes in a phyletic lineage we are describing the course of its evolutionary history. And to the degree that we can infer changes in function from these apo- and neosemic data, we will be able to pinpoint the course of that history in selectionist terms.

Other Procedures

The historical origins of organismic diversity have provoked speculation and serious study for centuries (Simpson 1961). Such study received its distinctive evolutionary aspect with Darwin, of course, and phylogeny took on its special role as elucidator of historical relationships with Haeckel (1866, 1874), who firmly established the dendrogram, at least symbolically, as the goal of such studies. At the turn of this century and through its first three decades, phylogenetic work continued to receive some attention but little new thought. With the emergence of "The New Systematics" (Huxley 1940) phylogeny received a major infusion of intellectual vigor. This came from recasting the species from a typological concept to a populational concept, and the recognition of the species as the unit of evolution (Fisher 1930; Haldane 1932; Wright 1949a, 1949b; Dobzhansky 1965, 1970; Mayr 1942, 1963). In that the unit of taxonomic organization, the species, was also the unit of evolution, evolutionary and taxonomic studies became closely associated, both in theory and in practice. So much so that, in 1961, Simpson was able to say, "With only the rarest exceptions, zoologists now agree that phylogeny can and should play its taxonomic role only in interpreting classifications based on other criteria." (p. 50) Simpson, it is to be acknowledged, is probably the major proponent of an evolutionary taxonomy, and at the time he wrote the cited passage, he gave reference to those few zoologists who disagreed with him. That criticism has swollen and probably has reached its peak in Sokal and Sneaths' *Principles of Numerical Taxonomy* (1963), where an alternative to evolutionary systematics has been attempted. But, nevertheless, this latter work goes on to talk of evolutionary implications of its mode of classifying organisms. Only the molecular evolutionists, of recent students of phylogeny, have approached the topic of evolutionary relationships without explicitly tying their findings to taxonomic inquiry.

We are faced, today, with at least four different outlooks on phylogenetic studies. One of these, i.e., the "New Systematics," conceptually integrates phylogeny and systematics. Another, e.g., numerical taxonomy, sees phylogenetic conclusions as a secondary and separate by-product of taxonomic investigation. A third viewpoint is that of those morphologists who continue the tradition known earlier as "idealistic morphology," a tradition stemming from Goethe (Naef 1931; Remane 1955). And fourth, there are those who pursue phylogeny purely as an evolutionary study, such as the molecular phylogenists. Such a viewpoint is also maintained in this volume, but with the important qualification that species as the basic units of evolution are also the basic units of phylogenetic study. Beyond that, taxonomic names and categories are used simply as indispensable conveniences for the accurate public discussion of various groups of organisms.

In the earlier pages of this chapter it was clear that, to some degree or another, we incorporated ideas from all of these major areas of contemporary work on phylogeny into our own phylogenetic procedures. To set in fair perspective not only our debt to these other procedures but also to show where and why our procedure differs from theirs, we will examine each point of view in terms of its contributions and limitations.

MORPHOLOGICAL PHYLOGENETICS

Adolf Remane (1955, 1956) is today the outstanding proponent of the morphological tradition in the study of evolution. As we have seen, it is to him that we owe that clarification of the concept of homology which distinguishes between homology as an explanation of similarity through descent and homology as a descriptive relationship that is operationally identified by application of specific criteria. This problem goes back to differences between Owen and Darwin (Simpson 1961), where the former, in terms of his special homologies, was concerned with direct identification of "... the same organ in different animals under every variety of form and function ..." (p. 80). (It is unfortunate that Owen said "same" rather than "similar", but his meaning is clear from his statement that the organ will vary in form and function.) And Darwin provided the theory that makes clear why they are similar. As Simpson points out, it might be best to refer to Owen's concept as homology and Darwin's as homogeny. In any case, Remane is clearly close to Owen's original attitude towards the concept and provides critical tools for making the needed identifications "under every variety of form and function," as Owen advocated.

A second point clearly established by Remane is that identification of homologies precedes phylogenetic interpretations. This follows from the homogenetic implications of homology. When we have found homologous relationships, then, and only then, can we infer the evolutionary descent that established them.

Our criticism of Remane's contributions touch on three topics. First, the concept of homology, as seen in his criteria, does still contain some logical inconsistencies, and also the criteria should be broadened explicitly to include more than anatomy. These points have been discussed in detail earlier and resulted in reformulation of the criteria. This reformulation does not detract from the debt owed Remane for his original insights into this problem.

Second, the concept of homology needed to be refined by recognizing different homologous relationships. This was done by following Hennig's concepts of plesio- and apomorphy, and in this way gaining terms that were more specific and reflected in more detail the variety of evolutionary trends inherent in homologous relationships. These trends are, unfortunately, nowhere treated adequately by Remane. The lack of integration of modern evolutionary theory to anatomical change is a major conceptual limitation of the morphological approach to homology and phylogeny.

Third, and finally, there is no adequate treatment of the problem of evolutionary potential. As we have argued, such studies are the second major goal of phylogenetic research. Though morphologists have always been concerned with the major trends in evolutionary history, their lack of appreciation of the dynamics of selection on natural populations has precluded any adequate analysis of these problems.

In brief, the morphological approach has been limited through inadequate attention to modern selectionist theory, but, to its everlasting credit, it has contributed operational criteria for defining homologies.

THE NEW SYSTEMATICS AND PHYLOGENY

Hennig's (1950, 1966) work provides the best transition from the morphological to the taxonomic approach to phylogeny. The latter, as we will see, can be divided into two groups, depending on their view as to how closely evolutionary and taxonomic concepts are related. Before getting further into this discussion some terminological clarifications are necessary. Simpson (1961) has defined systematics, taxonomy, classification, and nomenclature as follows.

"Systematics is the scientific study of the kinds and diversity of organisms and of any and all relationships among them." (p. 7)

"Taxonomy is the theoretical study of classification, including its bases, principles, procedures, and rules." (p. 11)

"Zoological classification is the ordering of animals into groups (or sets) on the basis of their relationships, that is, of associations by contiguity, similarity, or both." (p. 9)

"Zoological nomenclature is the application of distinctive names to each of the groups recognized in any zoological classification." (p. 9)

Clearly, systematics includes taxonomy, classification, and nomenclature, and the latter three show a logical dependence on each other, with classification being dependent on the "bases, principles, procedures, and rules" of taxonomy, and nomenclature being dependent on classification for "groups (or sets)" to be named. Other workers (Mayr, Linsley, and Usinger 1953; Mayr 1969) make less precise distinctions. Hennig (1950, 1966), in particular, sees systematics and taxonomy as synonymous, presumably feeling that taxonomy is the only kind of systematics that makes sense. Trivial relationships among organisms are not properly of scientific concern; hence, to define systematics as broadly as Simpson does is unnecessary.

Hennig's major work on phylogeny is *Phylogenetic Systematics* (1966) and he is emphatically opposed to the opinion that "Taxonomy is ordering without saying anything about the way in which this order is achieved." (Horn 1929). Hennig says, "We must disagree most strongly with this. In reality, nothing at all is achieved by a completely nontheoretical ordering of organisms." (p. 8) And he proceeds to develop the essential idea of "character phylogeny." (That apt term is attributable to Zimmerman (1959).) Hennig accepts Remane's analysis of homology and then, as we have seen, refines the idea further by referring to conservative homologies as plesiomorphic (identifiable by Remane's first two criteria) and to evolving homologies as apomorphic. With these concepts he has also gone on to make explicit the logic of arguments for determining the direction of evolutionary change, where he relies heavily on earlier workers, and for identifying evolutionary relationships, where he makes his own truly fundamental contribution to phylogenetic theory. His terminology and his handling of character phylogeny are important components of the procedures developed earlier in this chapter.

Because he is concerned with systematics, especially that of insects, Hennig never pushes this application of his procedures to taxa at the level of classes or higher. He does not, therefore, run into the most intriguing problems of the innovative potential of evolution. Coupled with this is a limited use of genetically oriented work on speciation, though he is very much more aware of this work than the morphologists. Consequently, Hennig's phylogenetic systematics does not arrive at a selectionist-oriented consideration of organismic potential, but it does advance the concept of homology and does develop incisive, logical patterns for its use.

Next, we consider what can be called the approach of the now somewhat aged "New Systematics." The first published formulation of this approach as the corporate viewpoint, so to speak, of a significant number of biologists is in the volume by that name edited by Huxley (1940). That viewpoint has been developed subsequently, and major works incorporating its opinions and conclusions are those of Mayr, Linsley, and Usinger (1953), Simpson (1961), Günther (1962), and Mayr (1969). As is well known, this point of view stems directly from contemporary evolutionary theory concerned with the formation, structure, and diversification of species. It possesses in abundance what the more structurally oriented views on phylogeny lacked, i.e., direct and deep conceptual dependence on the theory of natural selection. Our procedural system is also dependent on this point of view.

The limitations of the approach come, paradoxically, from close attention to the genetic aspect of species. First, the concept of homology is either ignored, mentioned only in passing, or defined in the homogenetic or Darwinian sense. For example, Günther (1962) makes no attempt to redefine homology. And Mayr, Linsley, and Usinger (1953) have only one reference each to "homologies" (p. 42) (also defined homogenetically) and to "homologous characters" (p. 123). With regard to the latter they say, "Without entering into the controversy on homology, it should be pointed out that classifications are based on homologous characters." This quotation indicates the authors' awareness of the unsatisfactory state of the concept of homology but also their awareness of its central, critical importance. Simpson's work proceeds with the homogenetic definition, as we have seen, and recognizes only Remane's 1956 work with two references (p. 93 and 120), and in the former refers to it as "... another exceptionally valuable book on taxonomy in general." Missed completely was the foundational analysis of homology. Mayr's (1969) more recent treatment of these problems refers more fully to Remane and the problem of homology but carries only "homologous" in the Glossary, and this adjective is defined as "A feature in two or more taxa which can be traced back to the same feature in the common ancestor of these taxa." This mixes a homogenetic interpretation with overtones of Remane's serial relationship.

Logically speaking, the idea of a common ancestor is not needed to identify a homologous relationship. It is important in explaining why the relationship exists. Because of irresolution of the problem of homology, Sokal and Sneath (1963) and others are completely correct in criticising homology when defined homogenetically as being a circular argument, and hence put their finger on the essential theoretical weakness of the New Systematics. It is nonetheless obvious that, though the concept of homology was never properly understood, even though basic to all their thinking, the New Systematist used the concept in the Remanian manner by making careful comparisons for similarity; and the result, guided by selectionist principles (our Rules I, II, and III), has resulted in dependable phylogenetic conclusions.

The second criticism of the New Systematics is the limited view taken of the problem of innovation. Certainly they have, especially from the paleontological side, treated it better than any of the others working in phylogeny. This is because of their base in selectionist theory. But looking at fossils results in limited use of homologies, that is, limited to hard parts or their preserved impressions, and thereby one deals only with limited information. One cannot speculate in a convincing fashion regarding the plesio-, apo-, and neosemic relations that existed between various forms when one can only observe a limited part of their morphology. Of necessity the paleontologist relies as much or more on stratigraphic data to determine ancestral and descendent forms than he does on inferences derived from the study of homologies. The result is a chronologically convincing sequence of forms, containing limited evidence, which speaks to the selection pressures behind the changes from an ancestral to a derived form. The neontologist who subscribes to the new systematics is largely interested in the dynamics of speciation. To overstate the case, he rarely sees the woods because of the trees; the larger view of evolutionary change does not interest him because of preoccupation with species formation. And in that changes at the species level are usually in terms of exploitive traits, he rarely concerns himself with the origin and establishment of the major evolutionary innovations which determine differences at the highest taxonomic levels—differences which concern new patterns of homeostatic control as seen in the emergence of respiratory, excretory, and circulatory innovations, or new patterns of development, or new body plans.

Thus the New Systematists have made major contributions to phylogeny in the elucidation of natural selection acting on species, but emphasis on that level of analysis has excluded treatment of other major problems in evolutionary history.

NUMERICAL TAXONOMY AND PHYLOGENY

Perhaps the best way to contrast this line of study with the one discussed is to use the terms currently in vogue, namely a phenetic versus a phyletic approach to systematics. The New Systematists and their sympathizers clearly espouse the goal of elucidating phyletic relations. The numerical taxonomists espouse the Adansonian view of strict attention to organismic phenotype as the source of critical data and as the basis for all categorization (see Heywood and McNeill 1964; Blackwelder 1967; Sokal and Sneath 1963, and earlier works cited therein.) In phenetic analyses, the taxonomic concepts are completely separated, by intent, from any evolutionary implications. Moreover, the basic unit for their taxonomy is not a species, at least not by design, but simply an operational taxonomic unit (OTU). This unit is defined by statistical techniques for measuring similarity. These techniques appear to be the essential contribution of this work. However, Sokal and Sneath are convinced that they have also clarified, following Adanson, the conceptual basis of taxonomy and, further, that they have found the OTU to be useful for phylogeny, too. In that they have presented a serious and detailed attempt to derive phylogenetic conclusions without the use of the species concept and without reference to a selectionist theory of evolution, and in that our attempt uses the species-selectionist view as its theoretical foundation, we must try to answer their arguments in detail.

We will first examine their foundational ideas, i.e., their definition of numerical taxonomy, their "Basic Position," and certain of their basic procedural concepts. Following this, there will be some discussion of the broader aims and goals of numerical taxonomy.

Regarding their idea of numerical taxonomy, Sokal and Sneath have this to say, "Before proceeding, it is necessary that we clearly define our use of the term 'numerical taxonomy.' We mean by it *the numerical evaluation of the affinity or similarity between taxonomic units and the ordering of these units into taxa on the basis of their affinities.*" (p. 48). The general sense of their meaning seems clear, but their logic is not. This definition implies circular reasoning, i.e., that affinities between taxonomic units are to be used to define taxonomic units between which affinities will be found. (It is ironic that just this criticism of circularity was directed by Sokal and Sneath (p. 61) at Mayr, Linsley, and Usinger's (1953) definition of a unit character.) What Sokal and Sneath are saying, because it becomes subsequently clear in later pages, is that the affinity or similarity between organisms is numerically evaluated and used as a basis for forming taxonomic units, and affinities among the organisms in these units are the basis for elaboration of yet further taxonomic units.

A more extended statement of their position is labelled Basic Points.

"Numerical taxonomy is based on the ideas first put forward by Adanson. They may be called Adansonian and are described concisely by the following axioms (modified from Sneath, 1958).
(1) The ideal taxonomy is that in which the taxa have the greatest content of information and which is based on as many characters as possible.

(2) A priori, every character is of equal weight in creating natural taxa.
(3) Overall similarity (or affinity) between any two entities is a function of the similarity of the many characters in which they are being compared.
(4) Distinct taxa can be constructed because of diverse character correlations in the groups under study.
(5) Taxonomy as conceived by us is therefore a strictly empirical science.
(6) Affinity is estimated independently of phylogenetic considerations." (p. 50)

All scientists, without exception, play the game of being incisive thinkers, and their work must first stand or fall on how well they formulate their problems. Sokal and Sneath call these six points "axioms," which is nonsense, if one defines axiom in its original sense, given by Euclid, of self-evident truths. In any case, it is not relevant to pursue further this unwitting attempt at logical rigor. To analyze the role of these points in numerical taxonomy, the following must be considered: First, point (1) is made of two parts—(a) taxa should have the greatest (possible) information content and (b) taxonomy should be based on as many characters as possible. Next, certain *procedural assumptions* are present in points (2), (1b) (as just defined), (3), and (4); namely, characters are of equal weight, there should be as many characters as possible, similarity (or affinity) is a function of character comparison, taxa reflect different character correlations. To these should be added another point, between (1b) and (3), which is that estimations of similarity can be made, or at least that characters can be scored as present or absent. There then remain points (1a), and (6) which can only be described as hoped-for outcomes of their procedures. What it seems the authors are saying, overall, is that they make certain assumptions, which may be formulated thus:

1. Organisms can be described by a catalog of their characters.
2. A given character can be variable between organisms (it shows degrees of similarity).
3. Degrees of similarity are objectively quantifiable.
4. These measurements can be used for taxonomic purposes.

Then, if the Adansonian view of taxonomy is accepted, their analytical goal is defined—straightforward, empirical determination of affinity. This statement of purposes allows guidelines for sampling of described characters so as to sufficiently characterize an organism. It results in taxa which are called "natural" because "... when the members of a group share many correlated attributes the 'implied information' or 'content of information' (Sneath 1957) is high; this amounts to Gilmour's (1940) dictum, that a system of classification is the more natural the more propositions there are that can be made regarding its constituent classes." (p. 19). Or again, "Indeed, we maintain that the elusive property of naturalness is simply the degree to which this principle obtains." i.e., "... that members of a group possess many attributes in common." (p. 18). It further allows "nested hierarchies"—taxa of lower level fitting into more inclusive, higher taxa—and in this way finally achieves "The prime purpose of a taxonomic system . . . economy of memory." (p. 170). This, in encapsulated form, would seem to be what numerical taxonomy is driving at.

Now let us turn to two problems that emerge from the procedures of numerical taxonomy. One is the idea of information content being maximal, in accord with one stated goal of their efforts. The other is simple clarity of procedure in resolving basic taxonomic problems.

The information content of a taxon, in Gilmour's sense (see preceding paragraph), is simply the sum of propositions that can be made regarding it. This point of view is reiterated later by the authors in a statement which may be as important as any on this point. In the context of an argument against weighting characters, the authors state that "It was argued that since natural taxa [Adansonian or numerical taxa] ideally contain the greatest possible content of implied information, this can only be measured in the number of statements which can be made about its members, which is independent of how important we may think any statement is." (p. 118) The first key phrase seems to be "content of implied information." One has to ask, What is "implied information"? If "implied" includes logical implication, then one can see that a taxon whose operational definition relates to certain bodies of theoretical knowledge immediately has a tremendous content of implied information, for its content ramifies the length and breadth of the related theory, in addition to any possible empirical statements (Hull 1968). Thus a taxonimic unit, such as the biologically defined species of interbreeding individuals, contains vastly more implied information than the empirical OTU. However, this, of course, is not what Sokal and Sneath wanted to argue. The other significant phrase here is ". . . independent of how important we may think any statement is." This reveals a philosophic position on the nature of scientific inquiry which is best discussed later.

The next point of specific commentary is one of several that could be raised about procedural matters. In discussing the selection of unit characters, Sokal and Sneath list certain inadmissible ones, among these are "invariant characters" (p. 67). (These are not characters always present, but rather, invariable characters.) It is of interest to see how rigorously the authors follow their own advice when they come to problems which specifically challenge the workability of their empirical approach. It can be recalled that Remane specifically inveighed against simple comparison as being adequate, for it would inevitably run up against developmental and sexual differences between members of the same species which were greater, in many instances, than differences between comparable members (e.g., adult males) of different species. Here is Sokal and Sneath's answer to that problem. "We think that sex differences will seldom prove troublesome in taxonomy. It seems certain that if one compared within one species all the features of a male with all the features of a female, the overall resemblance in most instances will be very high; the great majority of features will not be appreciably affected by sex." (p. 88). We agree with this statement, by itself, but do not see how it is consistent with the procedural rules which do not admit invariable characters. How can one compare all characters and still honor a list of inadmissable characters? Just omitting the invariable characters between males and females of the same species leaves pretty much nothing but differences. It is hard to see what taxonomically meaningful coefficient of resemblance comes from that. Of course Sokal and Sneath stated the comparison is between members of the same species. But use of the species concept would seem to be inadmissible to "a strictly empirical science."

The selected criticisms made thus far establish that Sokal and Sneath's exposition of the basic concepts and procedures of numerical taxonomy is too often ambiguous and logically inconsistent. A detailed critique would occupy more space than is appropriate here. Let us step around this problem and go to a general confrontation of what numerical taxonomy seeks to do; for, as already admitted, the general Adansonian point of view is essentially clear. Various authors have responded to the Adansonian approach, directly or indirectly (Mayr, Linsley, and Usinger 1953; Mayr 1969; Simpson 1961; Hull 1968), and reviewed Sokal and Sneath's book (e.g., Ross 1964; Ehrlich 1965). They have mainly commented on the application of numerical techniques to systematics, with limited criticisms from an evolutionary point of view. Here, our view is exclusively from the standpoint of phylogeny. Sokal and Sneath have argued against the possibility of a phylogeny based on anything other than fossil data (their Chap. 2) except through numerical taxonomy (their Chap. 8). The preceding sections of this chapter represent our answer to that point. The second question is the validity of the OTU as the conceptual base for phylogeny. There seem to be two arguments presented in its favor: (a) an OTU is a rigorously operational entity, and (b) it pretty much coincides with the biological species (p. 120–122) and so allows us to interpret affinities phylogenetically. Let us take these points in turn.

Being rigorously operational is not of itself an argument for using a concept theoretically (Hull 1968). Operational rigor is necessary for proper concept formation, but it is not a sufficient basis for guaranteeing, thereby, that the concept is also theoretically useful. The OTU is theoretically sterile except as it identifies with the species concept. Its sterility is guaranteed by the empirical approach to its formation—"In reality, nothing at all is achieved by a completely nontheoretical ordering of organisms." (Hennig 1966, p. 170)—and by its major stated goal, to aid in the "economy of memory." In that it is related to species, it immediately makes contact with one of biology's most powerful and rich concepts—the species as the unit of evolution. It is self-defeating to the essential aims of scientific inquiry to erect concepts that are by intention cut off from other concepts and theory; it is, in fact, the antithesis of the normal goals of science, which seek to unify knowledge. This same position is alluded to above in the author's phrase "... independent of how important we think any statement is." The simple fact of scientific life is that there are important statements and trivial ones. Only the former add up to significant scientific work, hence discrimination between the two is essential. Returning to the OTU and the biological species, phylogeny must build on the species concept, and the OTU is only useful to the extent it is a species and allows the theoretical interpretation of affinities as being homologous. But in that it does not weight affinities—does not, in Remane's terms, concern itself with kinds of similarities, that is, homologies of various sorts as opposed to similarities *per se*—it loses phylogenetic information and conceptual precision.

This is seen most clearly if we point out why phylogenies constructed from numerical taxonomy can say little or nothing on the origin and development of new evolutionary features. Let us start with the assumption that an OTU is another way, practically speaking, of defining a species. Numerical measures of affinities between species will more often measure homologous relationships than convergences. In fact, as Sokal and Sneath point out, convergences can be swamped by large samples of characters. All of this, thus far, argues cogently that numerical taxonomy (providing computers are available) can do phylogeny and perhaps do it more efficiently than going through the use of semes. That may be true up to the point of making the dendrogram. Beyond that, evaluating the dendrogram and inquiring into the actual paths of evolutionary change by following aposemic and neosemic developments, numerical taxonomy does not help. The OTU is fundamentally an inadequate substitute for the species in terms of phylogenetic studies, because it makes no contact with the realities of natural selection, as the species concept most emphatically does. If one says that one can always go back to the organisms after the dendrogram is made from numerical taxonomic procedures and make one's analysis in semes, or other comparable terms, then one is in effect admitting that one might as well have started with semes in the first place.

Numerical taxonomy does guarantee empirical success in rigorously classifying each and every organism we come across, but such success is not now and probably never will be essential to answering any major theoretical question in science. In its development of biometrical tools for measuring similarities between organisms, numerical taxonomy has surely made an important, useful contribution. Whether use of this technique will become essential to taxonomic procedures is yet to be seen. Its contributions to phylogeny are minimal.

MOLECULAR PHYLOGENY

Of the phylogenetic approaches examined thus far, none is more removed from taxonomic problems than is the molecular one. Its special contribution, to date, has been to establish that molecules do carry phylogenetic information; they are perfectly good semes. However, their interpretation in phylogenetic terms still poses serious problems. This is exemplified in the work of Fitch and Margoliash (1967). They find phyletic distance from mutational distance, that is, the numer of mutational steps necessary to pass from one amino acid sequence to another. When all possible distances were calculated for all the cytochromes c that they examined, they then determined the shortest distance possible between all these possibilities, using statistical assays, and utilized this as the basis for constructing their dendrogram. Fitch and Margoliash came to their method independently of Sokal and Sneath, whose method of measuring similarities through coefficients of resemblance is very similar.

The dendrogram produced by determining mutational distance has no internal information in it regarding its general plausibility except that it represents the simplest path, based on the genetic code, to go from the primary structure of one polypeptide to another. Whether or not the organisms represented by these molecules were parsimonious regarding their mutational changes, we have no way of knowing. In fact, King and Jukes (1969) have cogently argued that there must have been significant mutational change that had no dependence on natural selection, simply because ambiguity in the code would allow unexpressed mutations to occur. A lack of expression guarantees immunity from selection. Wisely, then, Fitch and Margoliash check their findings against phylogenetic conclusions derived independently from theirs, and they can do so because their data lie among the vertebrates whose phylogeny is well-known in classical (anatomical) terms. The fit between the two different approaches is good (Fig. 3–9) and is a most persuasive argument in favor of a molecular phylogeny.

However, the necessity for checking the results of a molecularly-based dendrogram against one constructed in the more familiar terms of homologies among organ systems also shows the present weakness of the molecular approach. It does not carry its own rigorous, internal, theoretical justification. Until one can discuss the changes between macromolecules in selectionist terms (see chapter 2) one can only argue molecular phylogenies in terms of Occam's razor. Happily for science, such a rule of thumb is very often helpful and even correct, but it is not a basis for critical argumentation.

Hopefully, molecular phylogeny will some day be of critical use. We can see one area where it promises to be the only source of information, and that is in instances of tachytelic evolution. Here, where there may well have been extensive anatomical change, especially if a selectively advantageous in-

120 Concepts and Methodology

novation lay in an exploitive seme, and many forms intermediate to the ancestral form and the present day survivors would be lost, thus obliterating the conventional anatomical data needed for determining homologies, it may turn out that molecular semes will be sufficiently conservative to provide conclusions otherwise unobtainable. Such conclusions will be soundly convincing only when molecular evolution has its own selectionist theory of molecular change. This, however, is not an argument against the use of

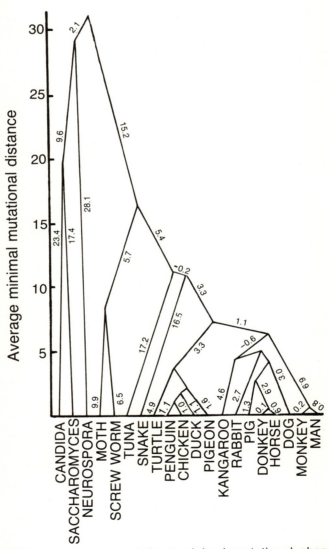

Fig. 3-9 Phyletic relations deduced from minimal mutational changes inferred to have occurred in the gene for cytochrome c. (Redrawn, by permission, from Fitch and Margoliash (1967). Copyright 1967 by the American Association for the Advancement of Science.)

molecular phylogeny. It simply cautions us to use it with appropriate care; for, to repeat, in the absence of independent evidence, one cannot tell where a molecular phylogeny is substantially correct and where it is patently wrong. That, thus far, it seems to be right much more often than wrong, is the basic argument in its favor. But that is established only by comparisons to more conventional studies.

A further criticism of molecular phylogeny is the one we have made for many of the other approaches, i.e., it can say very little about the innovative capacities of organisms. This is not only because, somewhat like numerical taxonomy, it is deficient in a consistent species-selectionist theoretical basis, but also because there is so much information above the molecular level that is evolutionarily significant but is incomprehensible in present molecular terms. It must be remembered that point-for-point transfer of information from DNA stops at polypeptides. There is no theory that allows us to predict from base sequences or from amino acid sequences precisely what the conformational pattern of a protein will be. Therefore many of the structures and their functions, such as cell organelles, tissues, and organ systems, whose origin and evolutionary development are of great interest to us, these evolutionary innovations will not be understood in the present terms of molecular phylogeny (Simpson 1964). In this sense molecular phylogeny is, like numerical taxonomy, mute regarding the second aim of phylogeny, the elucidation of evolutionary potential.

Conclusion

The foregoing has developed a consistent theoretical approach to phylogeny. Phylogeny is viewed as that aspect of evolutionary studies concerned with elucidating historical relationships between species and understanding the innovative steps underlying that history. The general relation to evolutionary ideas is evident in the guidelines that set limits to the allowable procedures and speculations of phylogeny, i.e., that the exploitive, homeostatic, and reproductive features of species be studied within the context of Darwinian natural selection.

Specific phylogenetic procedures all derive from the concept of homology. Criteria for identifying homologous relationships between comparable traits of two or more species have been developed. Application of these criteria depend on making point-to-point comparisons between the exploitive, homeostatic, or reproductive structures and functions to be compared. There is a basic level of complexity needed to allow for meaningful comparisons that has been explicated, and all traits showing such complexity are termed *semes*. Furthermore, semic relations can show direct similarity and therefore represent the universally conservative tendency of speciation; these are embodied in plesiosemes. Where similarity is established indirectly, through intermediates which bridge the apparent differences, the innovative tendency of evolution is manifest, and these are termed aposemes. Where quite new innovations occur and no convincing evidence of homology exists, then the new trait is termed a neoseme.

To determine the direction of evolutionary development in aposemic situations and to properly identify neosemes, it is necessary to establish a reference point. This is the plesiomorph, which is defined as that form (fossil or living, adult or developmental) that most clearly resembles the ancestral form of the species under study. Operations are described that aid in identification of the plesiomorph.

Phyletic distance, i.e., measured evolutionary divergence, is estimated from a weighted evaluation of plesio-, apo-, and neosemes. From these values dendrograms can be constructed. By examining the apo- and neosemic relations, in particular, among the species being studied, one can identify the factors underlying particular lines of evolutionary development.

And to close Part I. of our study, we note that the aspiration of elucidating phylogenetic relationships is dependent on the unitary origin of life with subsequent diversification into multifarious niches (Chapter 1). It is clear that much historical information is contained in the complexity and variability of the various traits of living things, ranging from molecules to species (Chapter 2). The procedure to be used in elucidating phylogenetic development depends on an analysis of homologous relationships (Chapter 3).

Part Two

Phylogeny of the Protozoa and Early Metazoa

Phylogeny of the Protozoa and Early Metazoa

LEEUWENHOEK'S draughtsman is not alone in his sentiment, "Dear God, what marvels there be in so small a creature!" (Dobell 1932, p. 147). Ever since the days when the great Dutchman began to bring to the rest of the world an awareness of the unicellular animals, there has been continuing wonder, curiosity, and serious study directed at the protozoa. Viewing them from the historical perspective of evolution, we see questions as to their origin, their evolutionary development as protozoa, and their further evolution into the metazoans. It is these three interrelated questions that are the concern of Part II.

Chapter 4
The Protozoa

LEEUWENHOEK simply called them animalcules. These tiny creatures collected from standing rain water looked and behaved like small animals—they ran about on legs, had short horns or ears, showed distress when their tails became entangled, and, in short, gave ample evidence that they were simply small editions of otherwise typical animal life. Some seventeen decades later, in 1845, von Siebold recognized the unicellular nature of these organisms and the term Protozoa took on a meaning it retains to this day (Hyman 1940).

In this chapter we wish to encompass, in brief form, the salient features of organismic and species organization in protozoans and to comment on protozoan origins. This provides the background necessary for the group-by-group analysis of protozoan phylogeny which follows in succeeding chapters.

Protozoan Systems

It is sufficient to define protozoans as unicellular animals, providing, of course, we are clear on what is implied by that short phrase with regard to the exploitive, homeostatic and reproductive capabilities of protozoans.

PROTOZOANS AS INGESTORS

Animals are those living things which satisfy their needs for energy and building materials by ingestion. Ingestion is easily identified in multicellular systems by the entrance of materials into the body through a mouth, the ingestive opening of a digestive system. In unicellular forms the ingestive system is seen to be food vacuole formation. This blends by degrees into the formation of yet smaller vacuoles, the pinocytotic vesicles, thus passing from cellular "eating" to cellular "drinking" (Chapman-Andresen 1962). Finally, at the finest level of uptake, assimilation occurs which refers to transportation or diffusion of molecules across the cell or plasma membrane. In that assimilation does not involve vacuole or vesicle formation, it would be convenient to define cellular ingestion as the uptake of nutrients by formation of membrane bounded vacuoles, but this would include pinocytosis and by convention, at least, that process does not strike most biologists as being ingestive. At the cellular level we can be satisfied with cellular "eating" or phagocytosis as our definition of ingestion (Hutner and Provasoli 1951) and see that it includes entry into the cell of microscopically visible, particulate, food materials by the device of vacuole or vesicle formation (Hall 1965). This is the essential exploitive function of these cells.

There must also be kept in mind certain of the corollaries of this type of exploitation of the environment. Particulate food can be and often is nonrandomly dispersed in the environment—it settles or it occurs in the form of other organisms only in selected localities—and it is often of a size necessitating the performance of work to capture and ingest it. For these reasons locomotion and particular devices for capture and ingestion are typically associated with the ingestatory or phagotrophic mode of life.

Locomotion. In all forms the basis for locomotion is transformation of chemical energy into mechanical energy through the operation of contractile proteins. Such proteins activate cilia and flagella or various types of pseudopodia. The significance of the locomotory habit in protozoa has been overemphasized, in that it provided earlier systematists with a naive basis for classification; and it also provided the excuse for most protozoan systematists to include the photosynthesizing flagellates among protozoan forms. As has been pointed out repeatedly (Lwoff 1951; Hutner and Provasoli 1951; Pringsheim 1963), the fundamental distinction between plants and animals is their mode of exploiting their environment—they are fundamentally producers or consumers, respectively. As we shall see below, these distinctions become blurred as they occur where transitions are occurring from phototrophy to phagotrophy. Pringsheim's detailed and authoritative critique of this problem includes this statement.

> "But the majority of authors have placed, quite early, the flagellates with chromatophores with the algae (e.g., B. Schnitz 1882, p. 12, note 2; Hansgirg 1886 and 1892; Lemmerman 1889, p. 102). If one goes at it that way, and today that is certainly best, it then becomes not a question of finding the boundary between algae and flagellates, but rather between those flagellates which one can place with the algae and those which do not go there." (Pringsheim 1963, p. 3–4. Original in German.)

There is no real problem in using the many characters found in photosynthetic flagellates to determine which of the colorless forms should be classified with them. These features include (a) the number, arrangement, elaborations, and mode of action of flagella; (b) structure and mode of division of the nucleus; (c) the type of cell division; (d) the nature of the cytoplasm; and (e) the nature of the reserve materials (starch, oil, paramylum), which are of special importance in algal taxonomy. To arbitrarily set off the leucoflagellates from the pigmented ones, or to set off all the flagellates from the rest of the algae, can only be justified on empirical grounds of simple categorization.

Where broader biological interests are to be served, such a classification has no justification whatsoever. By analogy, look at the case of the Tunicata, or sea squirts. When zoology was still in the midst of sorting out its diversity of organisms, the tunicates were understandably considered to be an invertebrate phylum. With Kowalevsky's (1866) celebrated discovery that the larval forms possess gills, notochord and dorsal nerve tube, the group was promptly and rationally placed in the chordates, where it has undisputably remained. With Pascher's (1931) recognition of the probability that unicellular, photosynthetic, flagellated organisms were the starting point of algal evolution, a great unifying conception was provided for the origin and early evolution of plants. To remove the photosynthetic flagellates from their systematic grouping with other algae is as ill-conceived, biologically, as to remove the Tunicata from among the chordates.

Coordination of locomotion and ingestion. A further corollary of the ingestive mode of life is that these microconsumers must coordinate locomotion and feeding. There is a selective advantage in being able to sense when food is in the environment and thus conserve energies more efficiently by using them when a return in terms of food capture is maximal. Coordinatory devices have, then, to integrate food sensing with orientation appropriate to bringing the consumer or predator into contact with the food, and then with triggering off the feeding or ingestive activity. The predaceous mode of life places a high selective advantage on the emergence of coordinatory devices. This is best exemplified by the pronounced trend toward cephalization in animals. That end of the organism which leads the way, so to speak, through the environment, is the end specially equipped with sensory organs and the nervous system for integrating the sensory inputs into motor commands. Also at the head or leading end is the mouth which needs to be apically or subapically located to ensure early contact with the prey.

The ideal form for a protozoan predator, it might seem, would be radial symmetry of a body streamlined for unhindered progress through the environment with an apically situated mouth. *Didinium* would be an example (Fig. 4-1). But as larger metazoan forms developed, with a permanent digestive system, a sensory-coordinatory system, and muscular-skeletal system, etc., simple radial symmetry of necessity would have to be modified. In any case, many forms—unicellular and multicellular—move against a solid substrate, which again makes radial symmetry unlikely except in sedentary forms. Bilateral symmetry is the common basis of organization in animal bodies. This allows anterior cephalization, ventral response to the substrate (or simply to the pull of gravity), and equal attention and mobility to right and left sides—in all, a functionally meaningful pattern of organization for a predator. However, with the complexity of niches available to the predaceous habit, and the

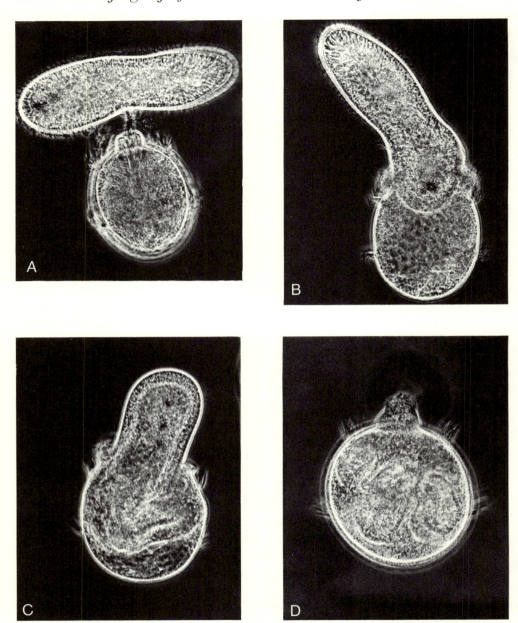

Fig. 4–1 The capture and ingestion of a paramecium by a didinium. A. The predator has contacted paramecium and is attached to it by discharged toxicysts and pexicysts. B. Ingestion has started. C. Ingestion is two-thirds completed. D. The ingested prey lies wholly within the didinium. (Photographs courtesy of Dr. Gregory Antipa.)

complexity of the organizational basis for the exploitive and homeostatic needs of predation, it is small wonder that perfect examples of bilateral symmetry, indeed, perfect symmetry of any type, are rare.

Loss of biosynthetic capacities. There is a final corollary deriving from the phagotrophic and predaceous mode of life. With successful exploitation of the niches available to consumers, animals were removed from the selective pressures that earlier gave rise to the photosynthetic producers. This meant that the extensive apparatus of photosynthesis would be broken down by mutation pressure unopposed by selection pressures, or could be summarily lost, as we'll see. The result is that very few organisms that are capable of photosynthesis are also efficient at some other mode of nutrition—either sapro- or osmotrophic (as in fungi) or phagotrophic. Where such forms do exist, they are among certain of the unicellular producers or algae as might be predicted, since it is from these forms that we think animals and fungi originally arose. The further point to be emphasized here is that dependence on other organisms as food resulted in loss of significant biosynthetic capabilities. As was mentioned in Chapter 2, Lwoff (1943), in his *L'Evolution Physiologique*, sees animal forms as dependent on preformed essential amino acids, certain vitamins, and other growth factors and on energy derived from the oxidation of reduced organic compounds; i.e., they are chemoheterotrophs, because these substances and energy sources are provided in their food and there is no longer a selective advantage to being able to synthesize them *de novo* or maintain the capabilities for photosynthesis. In fact, there is selection against maintaining those biosynthetic capabilities, since they represent luxuries. Thus, with emergence of phagotrophy, there also became established those limited and specialized biochemical capacities that have distinguished animal nutrition down to the present day, regardless of whether the forms were unicellular or multicellular.

UNICELLULARITY

The term unicellular is unambiguous when used in terms of structural detail. A cell, as we have seen, can be defined as a physical unit of nuclear material surrounded by cytoplasm and the whole bounded by a unit membrane. In this sense protozoa are predominantly single cells, for they are composed of more than one cellular unit only in rare cases of colony formation. However, the above tremendously simplified definition of a cell does carry the danger of oversimplifying the nature of the protozoa. Cells can be highly variable as a result of the numerous ways of assembling all their nuclear, cytoplasmic, and peripheral materials, and protozoan cells are unsurpassed in inventing highly complex and specialized variations on the organizational themes potentially available to cellular structure.

The alternative to describing protozoans as unicellular is to call them acellular (Dobell 1911; Hyman 1940; Corliss 1957), but this is valid only in the context of organismic organization. The protozoa, in this sense, are organisms whose body is not subdivided into cellular subparts, which subdivision is true for the Metazoa. There is little gained by this use of the term acellular; certainly it has led to no important conceptual insights, and it is not important to any line of experimental work. The term unicellular, however, immediately raises questions of historical origins and of the transition to systems composed of many cells. It implies, thereby, comparative problems of a functional and structural sort. Moreover, it is eminently sensible and useful to refer to the parts of the protozoan cell body in terms such as chromosome, mitochondrion, etc., which denote cellular organelles of universally established function. Such terms sound misplaced when used to described parts of something that is acellular, i.e., nominally devoid of cells. For these reasons we will retain here the use of unicellular in reference to the basic organizational pattern of the protozoa.

In all cases these cells are eukaryotic. The nuclei can be haploid, diploid, or polyploid and there can be one to many per cell. They are always bound by a nuclear membrane and double-ended chromosomes are, as a rule, present. The cell organelles are those typical of eukaryotes. Ingestion, even in parasitic forms (Rudzinska and Trager 1957, 1962), involves food vacuole formation. Locomotion is by pseudopodia, flagella, or cilia. Behavior of these cells shows various kinds and degrees of coordinated responses to the environment (Jennings 1923; Bullock and Horridge 1965), such that it is possible to infer the presence of some sort of sensory capabilities and their integration with motor abilities. The structural basis of these has not been established, not even in the ciliates where presumptive evidence for their occurrence is strongest (Jensen 1959; Bullock and Horridge 1965). Present evidence implicates the cell membrane (Naitoh and Eckert 1969; Eckert and Naitoh 1972).

The complex eukaryotic nature of these cells is an indication that they are not an early arrival on the evolutionary scene, even though they do stand at the beginning of animal evolution.

As was mentioned earlier (Chapter 2), the single protozoan cell, though lacking metazoan organ systems, nevertheless carries out all the functions associated with the more complex systems. Analogies between integument and cell membrane, between excretory system and contractile vacuoles make the parallels obvious. The most difficult parallel to establish relates to the endocrine system. Though there are chemically mediated effects between protozoan cells, as in the induction of mating in *Euplotes* (Kimball 1942), these are not of common occurrence. The point here, is however, that even though organized on a unicelluar basis, protozoan cells carry out a full array of homeostatic functions in utilizing the materials and energy sources taken in from their environment to maintain and reproduce themselves and their variants.

REPRODUCTION AND SEXUALITY

Because sexual reproduction is commonplace among higher plants and animals, biologists are accustomed to viewing the functions of sex and reproduction as if they are one, indissoluble process. Typically, the protozoa separate the two. Fission, of various sorts, achieves increase in number which is the function of reproduction. Fusion of gametic cells, conjugation, and certain unicellular sexual processes like pedogamy and autogamy can result in new genotypes, which is the basic role of sex (Muller 1932). The separation of sexuality and reproduction is not restricted to the protozoans; it occurs as two separable processes in most unicellular forms. On the other hand, as we have just said, the processes are typically combined in multicellular systems. This difference in the way sexuality and reproduction are integrated into the lives of uni- and multicellular forms has important corollaries in terms of chromosomal events, developmental patterns, and life cycles. And since these events characterize some of the major differences between uni- and multicellular forms, if we can understand how the transition occurs from one to the other type of organization, we are unravelling one of the major phylogenetic enigmas, i.e., the emergence of multicellular, diploid animals with segregated soma and germ and embryonic development.

Asexual reproduction. Asexual procreation is of various sorts: simple binary fission, as seen in ameboid forms (e.g., Fig. 7–3); multiple fission, as seen in certain multinucleate parasitic forms such as *Plasmodium* or *Grebnickiella* or in certain stages of the foraminiferan life cycle (Fig. 6–8); longitudinal or, better, symmetrigenic fission common to flagellated cells such as *Trypanosoma* (Fig. 5–2); transverse or homothetogenic fission, seen exclusively in ciliates such as *Paramecium* or *Tetrahymena* (Fig. 4–2); and in some species there is budding where a small portion of the parent cell is pinched off to give rise to a new one (Fig. 10–8).

There are various advantages to asexual reproduction. Foremost is its relative simplicity. This allows the process to be relatively rapid, and it necessitates only one parent. Given adequate food, one *Paramecium aurelia* can produce 64 paramecia in one 24-hour period. This is not as rapid as bacterial or viral proliferation, but is vastly more rapid than what is found among the metazoans. Also, asexual reproduction insures a certain stability of genotypes, it is a kind of brake on indiscriminate panmixia or outbreeding (Mayr 1963) and allows a well-adapted genotype, once established, to remain intact rather

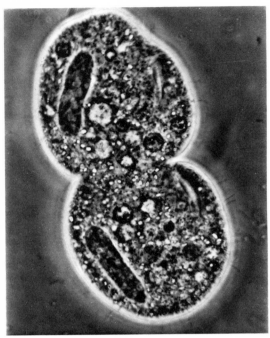

Fig. 4-2 Asexual reproduction in *Paramecium trichium*. The fission plane is, in effect, separating two complete sets of cellular organelles. In each presumptive fission product there is a macronucleus, a micronucleus (only one is in focus), a feeding organelle, two contractile vesicles, and other smaller structures such as mitochondria, ribosomes, etc. Other examples of asexual reproduction are given elsewhere: simple binary fission—Fig. 7-3, multiple fission—Fig. 6-8, longitudinal or symmetrigenic fission—Fig. 5-2, and budding—Fig. 10-8.

than being broken up by the chromosomal reshuffling that inevitably accompanies meiosis and fertilization.

The advantage of genetic stability becomes a liability when the environment changes. At such times organisms may need to draw on every possible source of variability they possess to find genotypes that will assure survival. At such times the only rare occurrence of sexuality can be a severe handicap. One partial offset to this limited genetic variability through sexual recombination is that the large number of individuals possible through asexual reproduction can provide the large populations needed to allow significant variation through spontaneous mutation. However, Sturtevant (1937) has argued that mutation rates are minimal in haploid or homozygous forms. In haploid forms, mutants are expressed almost immediately (there can be delay due to cytoplasmic lag, i.e., products of the normal phenotype may persist for a while—see Jinks (1964)) and so there is rapid weeding out of harmful versus helpful mutations going on continuously. Where diploidy or some higher degree of ploidy exists, the mutant can be hidden if it is recessive, as mutants are for the most part. Here, sexuality becomes an important device for experimenting with different combinations of the available variability. Unicellular organisms can do without the continual sexual recombination that occurs in sexual reproduction, but it seems unlikely that many of them can afford to forego it entirely. The protozoa, for the most part, exhibit a balance between asexual and sexual processes. Only in a few cases is there reason to think sexuality has been omitted entirely, and, because of the rarity of these cases, they seem to be cases of omission through loss, rather than original lack of innovation (see also Boyden 1953; Dougherty 1955; Stebbins 1960; Hawes 1963; Mayr 1963). This poses an interesting problem for the species concept. We will take that problem up after reviewing some aspects of protozoan sexual processes.

Sexual processes. It was O. Hertwig (1875–78) who first emphasized fusion of nuclei from different gametes as being the essence of fertilization. But the further question of why this is necessarily correlated with fertilization only has its answer in the evolutionary viewpoint of the creative aspect of sex, namely, the formation of new genotypes (Muller 1932). Thus sexuality can be defined as the bringing together of genetic material from two different sources to make new genotypes possible (Hanson 1967a).

In the protozoa, sexual processes involve either the total fusion of two cells, the temporary union of two cells during which nuclear material is exchanged, or the fusion of distinct nuclei within a single cell. The cells that fuse can be morphologically identical or can be differentiated to the extreme of one cell being smaller and motile with little in the way of metabolic reserves in the cell. When the fusing cells are morphologically alike, this is termed isogamy or isogamety; when they are so different that analogies to egg and sperm are pronounced, this is oogamy or oogamety (Wenrich 1954; Grell 1967). Copulation has been the term used by protozoologists (Kudo 1954) to cover these cases of gametic fusion. Conjugation refers to the temporary union of cells for the purpose of exchanging gametic material. It is limited to the ciliates. Uniparental sexual processes occur in certain species of paramecia, of which *P. aurelia* is the best example. A haploid nucleus undergoes mitosis and the products of this division fuse, resulting in homozygosity of the resultant diploid nucleus. This type of sexual process is the most perfect form of inbreeding possible (Sonneborn 1947, 1957). Other cases that approach this degree of inbreeding are known. One example involves the specialized ameboid form, *Actinophrys sol*. Here, a process known as pedogamy occurs, where the two daughter cells, formed following the meiotic divisions of a diploid parental cell, fuse, and thus there is sexual union between two sister cells (Bělař 1923).

Turning next to explicit discussion of the separation between sexual and reproductive processes in the protozoa, we can best illustrate the point by returning to *Paramecium aurelia*, which is endlessly cited in biology and genetics texts as exhibiting sexual reproduction. One paramecium makes more paramecia only by fission, and that is a reproductive process without any sexuality. Conjugation starts from the union of two physiologically reactive cells of complementary mating types and ends with there still being two cells. There is no reproduction but there can be new genotypes present in the exconjugants. Similary in autogamy, the process starts with one cell and ends with one cell; there is no reproduction, but there can be a new genotype. Both conjugation and autogamy are sexual processes; there is no reproduction (Sonneborn 1947, 1957) (Fig. 4–3).

Another example of sexuality without reproduction is seen in the flagellate *Trichonympha* (Cleveland 1949). Here there is anisogamy (Fig. 4–4). Sexuality results in the fusion of cells that, up to the time of differentiation into gametes, were cells typically capable of asexual reproduction. Sexual fusion reduces the number of cells by one half; it is the opposite of reproductive increase in cell number. But it can achieve new genotypes. Apparently genetic variability is so highly prized in selectionist terms that reduction in numbers is a price that will be paid to achieve it.

A final example of sexuality is that of the foraminiferan *Glabratella sulcata* (Grell 1958, 1967) (Fig. 4–5). This example brings us closest to sexual reproduction in the metazoan sense. For here two haploid, uninucleate cells come to lie together and then undergo multiple fission to form hundreds of haploid isogametes. It seems doubtful that these gametic cells can develop further unless they copulate with other similar cells. In that sense they are differentiated for the purpose of fertilization, and from the initial parent hundreds of gametes are formed which undergo fertilization; and hence there is an increase in number over the two original parent cells, and also new genotypes are achieved in the process. However, the parallel with metazoans is not exact. In metazoans, the parent organisms persist and survive the process of gamete formation, so that all progeny are clearly a gain in number over the two surviving parents. In *Glabratella* one can argue that actually the parents become many daughter cells through mitotic division, which is asexual reproduction, and then these hundreds of organisms reduce their number by sexual fusions. Thus, reproduction and sexuality are still separate and even opposed in their effects, as in *Trichonympha*. The semantic problem comes down to whether the unicells that initially "mate" are called the parents of the gametes that fuse, or whether the cells formed from the last mitotic division are called the parent cells. We need not make a ruling one way or the other at this time, for the essential point is not the semantic problem, but the realization that sexuality and reproduction in the protozoan are not the same as sexual reproduction in metazoans.

SPECIES STRUCTURE

The most extensive and insightful analysis of the breeding patterns, and also of the species problem, in protozoa is that of Sonneborn (1957). His study is based largely on the ciliates but includes comments on certain zooflagellates and amebas. This brilliantly lucid and provocative analysis deserves special attention on two counts. It is a pioneering study in the comparative

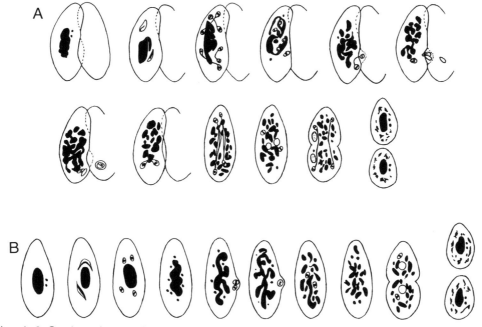

Fig. 4–3 Conjugation and autogamy as seen in cells of *Paramecium aurelia*. A. In conjugating *P. aurelia*, two cells, of complementary mating type, come to lie against each other with their ventral surfaces apposed. The same nuclear processes go on in both conjugants. Diagrammatic details are given here for only one of the two cells. The large macronucleus breaks down, ultimately being seen as only scattered fragments. The micronuclei (two per cell) undergo meiosis. Seven of these haploid products degenerate. The eighth, lying in the paroral cone, undergoes mitosis (fifth figure, first row). Reciprocal exchange of haploid nuclei occurs between the conjugants, and fertilization ensues from the fusion of one migratory nucleus with one stationary nucleus, restoring the diploid condition (first figure, second row). Thus, each conjugant is effectively hermaphroditic supplying both a "male" (migratory) and a "female" (stationary) nucleus or "gamete". The zygote nucleus divides twice, with the conjugants separating at the time of these divisions and concluding conjugation. Two nuclear products become macronuclei, which are segregated, one to each fission product, at the next fission, and two products become micronuclei, which divide mitotically at the next fission. B. In autogamy in *P. aurelia,* processes parallelling the foregoing occur, but in that only one cell is involved, there is no cross-fertilization. The haploid gamete nuclei in the paroral cone fuse with each other, i.e., there is selffertilization (sixth figure). (Redrawn from Sonneborn (1947), originally from Hertwig (1889).

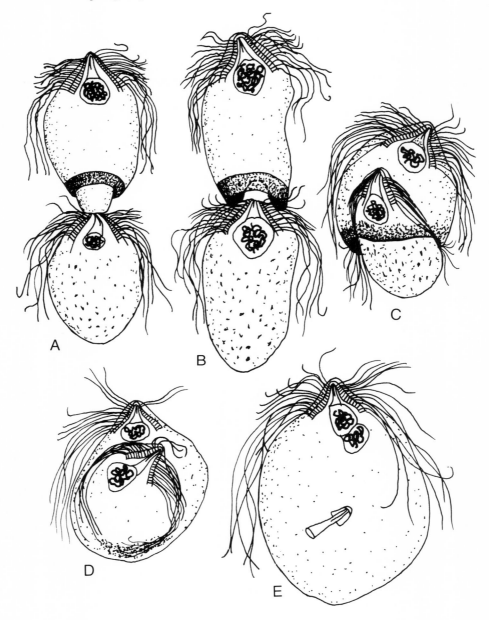

Fig. 4-4 Anisogamy in *Trichonympha*. A. Unlike cells begin to fuse, the anterior cell being distinguished by the ring of granules at its posterior end. B. and C. Entrance of posterior partner into anterior cell. D. Breakdown of engulfed cell. E. Fertilization. (Redrawn, by permission, from Cleveland (1949).)

structure of species and a sharp attack on the practical usefulness of the biological species concept.

Species in the ciliated protozoa. At the outset it will be helpful to outline the life cycle of *Paramecium aurelia*, the most intensively studied form, so as to familiarize ourselves with the general state of affairs within which the central importance of the breeding patterns take on their meaning. After that we

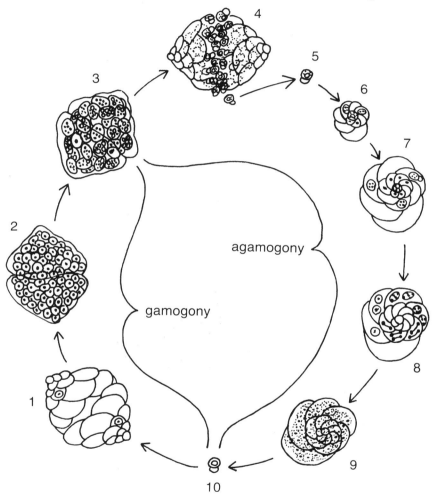

Fig. 4–5 The life cycle of *Glabratella sulcata*. 1. Pairs of cells undergoing gamogomy. 2. Multiple fission resulting in unicellular gametes. 3. Fusion of pairs of cells. 4. Release of zygotes. 5. –9. Development of multinucleate agamont. 10. Release of single gamont following multiple fission of agamont. (Redrawn, by permission, from Grell (1967) in *Research in Protozoology*, ed. T.-T. Chen, vol. 2, Pergamon Press, Inc.)

will turn to a specific example of inbreeding and follow that with an example of outbreeding. These will allow us to see, then, the general structure of breeding systems, their key features, and their evolutionary role. Finally, other protozoa will be mentioned and concluding comments made on the species concept in the protozoa, with special emphasis on the treatment of homogeneous groups of organisms that show no sexuality.

Let us somewhat arbitrarily start the description of a life cycle from the events of conjugation. Here, in *P. aurelia*, two cells of complementary mating type have come together and exchanged haploid nuclei (Fig. 4–3). Fertilization ensues, the cells separate and start to form their typical nuclear structures. Each of these two cells give rise, through subsequent asexual reproduction, to a clone of cells. Since the exconjugants are isogenic, this fact is recognized by referring to the two clones together as a synclone. In the smaller members of *P. aurelia*, each exconjugant produces two macronuclear anlagen which are segregated, one to each fission product, at the first division after conjugation. All the progeny of each of these cells now contains macronuclear products of one original macronucleus, and these cells are termed caryonides. Thus each synclone contains two clones, one from each exconjugant, and each clone contains two caryonides, one from each fission product of each exconjugant. The usefulness of these terms will become apparent when we come to discuss inbreeders and outbreeders in some detail.

Perhaps the next question would be to ask what happens to the descendants of each caryonide if they were given opportunity to reproduce asexually *ad infinitum*. The answer is that they would all eventually die. And that is the essential reason for talking of life cycles in *P. aurelia*; following sexual reorganization, such as conjugation, and in the absence of its recurrence in any cell, all the members of the synclone will die within about 300 fissions (Sonneborn 1954). Within these 300 fissions, various things can happen. If there is no sexual process, the cells show progressive weakening as seen in decreased fission rates, abnormal fission and abnormal nuclear division, among other symptoms of abnormality. However, if these cells are removed from the presence of excess food—which is the key condition to induce the ageing phenomenon—they can become sexually reactive and will mate with reactive cells of the complementary mating type; or, failing to find partners, they can undergo autogamy. In either case, completion of sexual processes resets the age clock to time zero, and they enter on a new life cycle. Such, in very abbreviated form, are the most critical aspects of the life cycle of *P. aurelia*.

The extreme inbreeders in *P. aurelia* are members of syngens 10 and 14. Syngen (Sonneborn 1957) refers to pairs of complementary mating types—syngen 10 contains mating types 19 and 20, syngen 14 contains mating types 27 and 28—and conjugation will occur between cells showing two such complementary mating types and not between these and any other mating types.

The only exceptions to this general rule are some instances where mating occurs but the exconjugant cells are, for the most part, nonviable. This means, in effect, each syngen is genetically isolated from all others; it conforms to the biological definition of a species. *P. aurelia*, containing sixteen syngens, is, therefore, made up of sixteen sibling species. (However, these sixteen include two which cover the larger form usually designated *P. multimicronucleatum*, which characteristically has four micronuclei as compared to the two typical of *P. aurelia*. Subsequently, Sonneborn (1970b) has referred to fourteen syngens as comprising the *P. aurelia* complex. In what follows here, we will follow the analysis as given in the 1957 paper.)

This introduces the species problem as it is found in the ciliates: in all cases carefully studied thus far, each morphological species is made up of several biological species. Sonneborn proposes that these biological species be call syngens rather than varieties and that term is found in the literature on *P. aurelia* following 1957. The term syngen is a valuable one, as we will see, though we will not follow Sonneborn's suggestion to use it wherever the biological species is referred to. Members of species or syngens 10 and 14 are termed inbreeders, because sexual processes are predominantly opportunities for autogamy or of mating with members of the same synclone. Conversely, exclusion of autogamy and restriction of mating to members of other synclones define outbreeders. How are these two conditions achieved?

Members of the inbreeding syngens can mate again as soon as they have completed conjugation. This means there is no period following conjugation during which no mating is possible, i.e., there is no immature period. Further, the mature period—the period during which mating can occur—lasts for only about 15 fissions and after that, decrease in food supply results in autogamy. How does this add up to inbreeding? One further point must be added and then the question can be decisively answered. Syngens 10 and 14 are both members of group B in the *P. aurelia* mating type systems. That is, the cytoplasm plays a conspicuous role in determining the mating type. This has the important consequence that there is only rarely a change of mating type in the two exconjugants from that which they initially showed. (This is, of course, an intriguing genetic problem—different phenotypes despite identical genotypes—but cannot be treated here. See Allen 1967 and Preer 1969 for recent reviews of the problem.) The presence of two complementary mating types immediately following conjugation and the absence of immaturity, means that the exconjugants can mate again with each other. Or, also, that members of the same synclone can mate again. That such matings are more probable than between members of different synclones comes from the facts of growth in nature. The paramecium most likely to be in the neighborhood of a paramecium that has just conjugated is, naturally, its partner. That they can remate is obviously a device permitting inbreeding. The short mature period means that distribution will not be very great before the ability to mate is excluded. Hence, the probability of swimming off far enough to achieve intersynclonal mating is present, but such matings are less likely to occur than intrasynclonal mating. And, if no mating occurs, the consequence is autogamy, the most extreme form of inbreeding. Sonneborn, very acutely, notes several other features of the life patterns of these forms that are also consistent with inbreeding: these paramecia are rare and highly localized in distribution, thus severely limiting the chances of outbreeding; the details of nuclear reorganization proceed relatively swiftly, thus allowing minimal opportunity for dispersal during the sexual processes; these forms occur in warm regions (southernmost parts of the United States) and the higher temperatures also speed up these processes; fission rates are among the highest known, which would lead to swift exploitation of environmental resources and consequent induction of sexual processes through starvation, and because these events would proceed rapidly, again time for dispersal is reduced. All in all, "The various characteristics of these varieties ... conspire to bring about mating between close relatives or within the population." (Sonneborn 1957, p. 207)

The strong outbreeders, such as syngens 15 and 16, contrast sharply with the inbreeders in each of the characteristics that have just been discussed. The outbreeders have immature periods lasting for months, or even a year when considering the low reproductive rate that probably prevails in nature. The period of maturity is apparently even longer than the immature period. In syngen 16 there are four complementary mating types, and a reactive cell of one type can mate with reactive cells of any of the other three types. The method of mating type inheritance is unknown for syngen 16, but syngen 15 belongs to the group A type. Here, new mating types are determined anew for each caryonide, hence the products of one clone can be of different types.

The long immature period insures that, after once mating, these cells cannot mate again for a considerable period of time, thus allowing a very significant opportunity for dispersal. After that they can mate and are mature for a long period, thus insuring the chance of meeting a stranger and, further, with the availability of three other mating types in syngen 16, it increases the chances of finding a suitable mate. Also, if mating is not successful, the sexual process that sets in to forestall ageing is not autogamy but is selfing, that is, some members of a given caryonide can change mating type and thus permit conjugation. This is a type of inbreeding, but not as intensive a one as the autogamy that occurs with the strong inbreeders. These outbreeding syngens are found widely dispersed from cold to warm regions of North America—they have not been looked for elsewhere—and apparently occur as many, small, local populations or demes. They also have lower fission rates than syngens 10 and 14. Here, too, there is seen, overall, to be a pattern of survival which points in one direction: extensive outbreeding.

The other syngens of *P. aurelia* lie variously between these extremes, usually favoring the inbreeding tendencies. None reaches the ultimate step of abolishing sexuality completely and thus holding the genotype stable through asexual reproduction, except for spontaneous mutation. And no syngen of this species of *Paramecium* is as extreme an outbreeder as *P. bursaria*. In syngen 1 of this latter species (the only intensively studied syngen), the immature period lasts three to five months under favorable conditions, and hence is much longer in the wild state; the mature period lasts about three years, and the period of declining vigor (senility) lasts about five years. During this latter period, conjugation can occur, as is consistent for an outbreeder. Other features also distinguishing an outbreeding habit are systems of multiple mating types (syngen 2 has eight mating types), the presence of synclonal uniformity of mating type (all caryonides have the same mating type after conjugation, which is a sure inhibitor of intrasynclonal mating), total absence of selfers, and autogamy (high mortality after crosses that are induced to achieve inbreeding). One result of this is wide distribution of common gene pools, and because of the relatively greater opportunities for gene flow, there is less differentiation of syngens in *P. bursaria* (three in the United States) as compared to *P. aurelia*, which as we noted, is predominantly inbreeding (14 or more syngens in the United States).

Breeding patterns can range, then, from obligate outbreeder, to facultative out- and inbreeder, to obligate inbreeder, to exclusive occurrence of asexual processes. This range of situations is made up of the following factors, the most important ones being listed first. (1) Immature period. The longer such a period lasts the greater the chance of dispersal beyond the area of close relatives; hence long immature periods characterize outbreeders and short ones are found with inbreeders. (2) Mature period. Again a long period marks an outbreeder and a short one an inbreeder. But immaturity and maturity should be considered together, as Sonneborn emphasizes. Short combined periods inhibit gene flow. In some cases a long immature period is followed by autogamy, which rejuvenates and avoids ageing but does not introduce a new immature period. Here the inbreeding resulting from autogamy is balanced by the increased opportunity for outbreeding. This example illustrates a point repeatedly made by Sonneborn, that the totality of the life processes must be analyzed to properly understand the breeding pattern of these ciliates. (3) Another characteristic of an inbreeder is its tendency to be relatively invariable; autogamy automatically results in homozygosity. Outbreeders will be relatively more variable. (4) In inbreeders, there would be expected a higher development of demic differences, and, among other aspects reflecting this, there would be differences in chromosome number and form—karyotypic differentiation would have occurred. Thus, in matings between demes, the F_1 would be quite fertile, but the F_2 would show varying levels of mortality related to chromosomal differences. These have been found in the highly inbreeding syngen 4 of *P. aurelia*. Breeding within a deme, however, would show no F_2 mortality. On the other hand, outbreeders would be expected to show mortality, when bred to members of their own deme or of their own synclone, as a result of accumulated lethal or detrimental genes becoming homozygous, these genes normally being suppressed by heterozygosity. We can interject at this point another aspect of karyotypic differentiation that could distinguish in- and outbreeders. In that the former are de-emphasizing genetic variability, devices to reduce intrachromosomal recombination would be of selective advantage. Normally, crossover suppression is thought to occur as a result of chromosomal inversions (Patterson and Stone 1952). This is useful where heterozygosity is the rule, and inversions would be pointless. The only alternative would seem to be small chromosomes, so small that pairing at a kinetochore region might mechanically inhibit the possibility of crossing-over for some distance on either side of it. If such is the case, we would expect small chromosomes with median kinetochores in inbreeding syngens of *P. aurelia* and relatively large chromosomes in *P. bursaria*. Certainly syngen 4 of *P. aurelia* has tiny chromosomes (Dippell 1954), about a micrometer in length at meiotic metaphase, in fact too small to see the location of the kinetochore. *P. bursaria* has larger chromosomes, about 10 micrometers long (Chen 1951). (5) Outbreeders would be expected to be widely distributed, and inbreeders will be relatively much more localized. (6) Rates of sexual processes and fission rates will be faster in inbreeders and relatively slow in outbreeders. This is best put in perspective by quoting a specially apt passage by Sonneborn on the relation of sexual processes and available food.

> "In *Paramecium*, and in Ciliates generally, the occurrence of sexual processes and genetic recombination is linked to decline in the food supply. This must be an inevitable and frequently recurring event in the lives of such organisms. The logarithmic increase in numbers resulting from repeated binary fission quickly yields populations that no conceivable food supply could support. Not even an amount of food equal to the bulk of the earth could support reproduction from a single individual for as much as 100 successive fissions. With due allowance for predation and other causes of accidental death, it is difficult to see how asexual reproduction could long continue without the onset of famine. The prevailing pattern of life must be to increase, multiply, and spread while the harvest lasts, forgetting about sex while enjoying food to the full, and to use hard times, when reproduction comes to a virtual standstill, as occasions for grasping opportunities to carry out sexual processes." (Sonneborn 1957, p. 203).

Those forms which reproduce rapidly will exhaust the food quickly and promptly mate before dispersive tendencies become pronounced. In such cases there will be inbreeding because mating will most likely occur between members of a clone and of a synclone. (7) Finally, there is the fate of these cells during senility. Outbreeders undergo selfing and inbreeders undergo autogamy. These have been commented on above in the context of discussing inbreeding and outbreeding syngens, and nothing more need be added here.

The foregoing study and its conclusions were developed largely from work on *P. aurelia* but are applicable, in detail, to the twelve other species of ciliates discussed by Sonneborn in 1957. There has been substantially nothing to change the analysis despite continued work on mating types in the ciliates (Allen 1967). Various papers have added to our knowledge of the occurrence of conjugation in nature (Siegel 1961), but this in nowise necessitates modification of Sonneborn's major conclusions.

Sexuality has not been studied in other protozoans in the detail that is found for the ciliates. Cleveland (1947, 1949, 1956b) has done extensive work in describing sexual phenomena in the zooflagellate symbionts of the wood roach *Cryptocercus*. His interpretations of these phenomena—many of them unusual, such as one-step meiosis—as newly-evolving manifestations of sexuality are surely wrong for reasons cogently argued by Sonneborn. Rather, they are better seen as the disappearance of sex as the symbionts respond to pressures for evolutionary stability within their hosts. Termites, which are regarded as evolving from the stock represented by the wood roaches, have symbionts similar to those in *Cryptocercus* but they show no sexual phenomena whatsoever, having arguably lost them completely. This final step past obligate inbreeding is not unexpected when organisms are living in restricted stable environments, which is certainly the case for the hindgut of termites, the hosts themselves living in environments engineered for significant stability and in colonies showing limited migration from one to another (Kirby 1949; Sonneborn 1957).

The species problem. There now remains the problem of clarifying the concept of species as it applies to the protozoa. Let us start by presenting Sonneborn's evaluation of the situation. The crux of the problem is seen in *P. aurelia*, where one morphologically defined species contains, today, 16 reproductively isolated groups, the syngens, each of which conforms to the definition of a biological species (Mayr 1963; Dobzhansky 1970; see Chapter 1). The morphological similarity of the different syngens means they would certainly be termed sibling species. The inclusion of these biological species within a single morphological species leads Sonneborn to propose three possible solutions, which are: (a) use the term species to cover both the morphologically unique and the reproductively isolated groups; (b) restrict species to biological species and find a new term for the morphologically defined groups; or (c) restrict species to the morphologically defined groups and find a new term for the forms sharing a common gene pool. To avoid the double standard of the first alternative, to avoid changing the widely accepted usage of a term and propose for it a more restricted application, and to avoid developing species names for all sibling species of ciliates, Sonneborn favors the third alternative. He proposes to refer to reproductive communities of organisms as "syngens" (Sonneborn 1957, p. 201). These are compelling logical and practical arguments.

However, these arguments have made absolutely no detectable headway against the entrenched position of the biological species concept. Sonneborn's prediction that "... there seems to be no possibility that biologists in general will ever agree to restrict the term species to carriers of a common pool of genes." (p. 294) is wrong. The harshest critics of the species concept, the numerical taxonomists, have agreed to the species defined as members of a common gene pool and define their phenetic unit as an OTU (operational taxonomic unit, Sokal and Sneath 1963), as we have seen. The battle has been won for the biological species concept as the sole possessor of the term species.

Now, if the foregoing arguments were solely semantic, we could desist from further discussion at this point. In fact, the whole discussion would not have gotten this far, for it could have been handled in a few lines referring to the reasons for proposing syngen and then concluding it remains only of local use by parameciologists. There are, however, two major ideas buried in this discussion of the species problem, of which the first is the general fact that the absence of sexuality today in a group of organisms is probably due to the loss of that process, and not that it never evolved (Boyden 1953; Dougherty 1955; Stebbins 1960; Hawes 1963; Mayr 1963), and the second is the idea of an ecospecies (Fisher 1930; Mayr 1963).

Let us start the discussion of the first point by another quotation from Sonneborn. "I for one certainly do not challenge the value of the biological species concept. It has great value. But its value is very narrowly limited in the first place to outbreeding organisms in principle and in the second place to the very small proportion of them that will in the foreseeable future be studied sufficiently." (1957, p. 200) Sonneborn does not adequately recognize the value of the biological species concept when he sees it as applicable only in principle to outbreeding organisms. (A similar position has been taken by Ehrlich (1961) but answered convincingly in favor of the biological species concept by Hull (1968).) That is a common and fundamental misapprehension of the application of the concept. In that all organisms living today either show sexuality as a regular part of their reproductive processes, or have it delimited in some way to part of their life cycle, or, if it is absent entirely, are descendants of forms that showed sexuality, must mean that sexual processes were established very early in the evolution of organisms. A reasonable guess would be that when organisms first evolved a nucleic acid based mechanism of information storage (Hanson 1966), then sexual processes for exploring its recombinatory potentials also evolved. The fact that viruses show genetic recombination (Hayes 1968) argues for the fact that sexuality of this sort could occur in other simple systems. This broad view of the early and universal establishment of sexual processes emphasizes again that present day restriction of sexuality is a secondary development and is the consequence of loss of selection pressures to maintain it or even of selection against it. It has been pointed that inbreeding can break down favorable local genotypes (Sonneborn 1957; Mayr 1963) by continually swamping them with genes more appropriate to other demes within the species. There are indeed circumstances where the variability of sexual recombination would be a detriment. These circumstances presumably are operative for inbreeding syngens of paramecia, but their exact nature in unknown. The situation seems clearer with the termite zooflagellates, where sexuality, as Sonneborn has clearly argued, has been wiped out entirely, and for the plausible reason that variability is a detriment to optimal survival in a very stable environment.

Rather than arguing that the biological species concept applies in principle only to outbreeders, we can argue that it applies in principle to all organisms back to the first emergence of sexuality in them. Loss of sexuality and its essential equivalent of obligate inbreeding are exceptions that test the rule from the very fact that these cases are those where outbreeding is arguably disadvantageous. The same is seen clearly in self-pollinating plants (Cleland 1964; Stebbins 1960). By implication then, the biological species concept is, in principle, applicable to all species. The fascinating problem is how that principle is used, even to the limit of eventually negating it in certain forms. Sonneborn's analysis of the breeding patterns in the protozoa stands as the first comprehensive comparative study which shows in one major group of organisms the full range of the application of the species concept.

And it raises one last question. In the absence of a common genetic resource, what keeps a groups of organisms homogeneous, at least sufficiently so in appearance so that they are usefully designated a species? The concept of the ecospecies is a possible answer. This can be understood by going to Fisher's study on natural selection, where he is discussing the structure of groups of organisms which show no sexuality.

> "In such an asexual group, systematic classification would not be impossible, for groups of related forms would exist which had arisen by divergence from a common ancestor. Species, properly speaking, we would scarcely expect to find, for each individual genotype would have an equal right to be regarded as specifically distinct, and no natural groups would exist bound together like species by a constant interchange of their germ plasm.
>
> The groups most nearly corresponding to species would be those adapted to fill so similar a place in nature that only one individual could replace another, or more explicitly, that an evolutionary improvement in any one individual threatens the existence of the descendants of all others. Within such a group the increase in numbers of the more favored types would be balanced by the continual extinction of lines less fitted to survive, so that, just as, looking backward, we could trace the ancestry of the whole group back to a single individual progenitor, so, looking forward, at any stage, we can foresee the time when the whole group then living will be the descendants of one particular individual of the existing population." (Fisher 1930, p. 121).

Fisher's neatly expressed system ignores the high probability of parallelism, but this does not negate his fundamental argument that only the possessors of a rather specific genotype could fill "a similar place in nature," i.e., a certain niche. Though Mayr (1963) states "It is too early for a definitive proposal concerning the application of the species concept to asexually or uniparentally reproducing organisms," (p. 27) he goes on to make a very provocative proposal for a point of view on these forms that takes Fisher's idea yet further.

> "For this phenomenon [of morphologically distinct asexual organisms] I have advanced the explanation (Mayr 1957) 'that the existing types are survivors among a great number of produced forms, that the surviving types are clustered around a limited number of adaptive peaks, and that ecological factors have given the former continuum a taxonomic structure.' Each adaptive peak is occupied by a different 'kind' of organism and if each 'kind' is sufficiently different from other kinds it will be legitimate to call such a cluster of genotypes a species . . ." (Mayr 1963, p. 27–28).

What this clearly implies is that in the absence of the cohesive force of sexual intercourse within a gene pool, there is still some factor or tendency which tends to drive organisms to certain functionally adaptive norms. Sonneborn clearly recognizes, as have all the systematists who have worked on these forms whether or not they knew propagation was only by asexual means, that perfectly distinct species are present here. The evolutionary unit is as distinct and isolated in asexual organisms as it is in sexual organisms, "no more and no less so" (Sonneborn 1957, p. 291), though it is not clear what exactly is meant by the "evolutionary unit" when referring to many distinct and presumably parallel lines of genetic descent. Finally, Dogiel, Poljanskij and Chejsin (1965, Chapter IX) present an extended discussion of asexual forms that nonetheless are found to occur as distinct species.

Does this further imply that the adaptive peak we identify through species occupation of them represent some concatenation of physical factors or resources in nature which occur independently of organisms? That is possible. It also is possible that organisms can only exploit nature in a discontinuous way; that not all sets of physical circumstances can be exploited. Therefore, superimposed on the discontinuities of nature are those of selected organismic capabilities. One obvious source of these latter is the problem of building functional proteins. Not all configurations of proteins are equally useful to survival. The evolutionary emergence of a protein or proteins that allow the organisms containing them to utilize a certain set of environmental circumstances would be a unifying principle underlying forms reproducing sexually, as well as those reproducing exclusively by asexual means. It would be easily overlooked in the former because of the cohesiveness of interbreeding; it only becomes visible in the latter when sexuality drops away. In this way the biological species is seen as a natural consequence of the occurrence of ecospecies. What better way to find that constellation of genetic information that is optimally appropriate for a set of environmental conditions than by the reshuffling and exploratory recombining provided by sexual processes? Thus certain forms will remain outbreeders as the environmental circumstances demand, and others will become inbreeders or even nonbreeders. Among protozoan species the whole range of possibilities is present.

Protozoan Origins

The preceding sections characterized the organismic nature of protozoans and surveyed the structure of certain of their species. We must now inquire how the first protozoan species arose. Detailed inquiry, as with the origins of specific groups of protozoa, will be discussed later.

Within the overarching general evolutionary trend of original primitive chemoheterotrophs evolving into photoautotrophs, followed by a resurgence of chemoheterotrophy as biotically formed precursors became abundant in the environment, the protozoa stand among the early ranks of the resurgent chemoheterotrophs. More particularly, the problem is to inquire into the general plausibility of an alga giving rise to a protozoan.

The loss of photosynthetic pigments by unicellular algae has received extensive study, and the particular problem of that loss resulting in viable chemoheterotrophic forms has been the object of detailed research, especially by Pringsheim (1963, and earlier works). The initial questions regarding this type of change are related to its known occurrence, the directions of the change (is it always pigmented forms giving rise to colorless ones?), and possible mechanisms which bring it about.

All major algal groups, except the brown algae (Phaeophyta) contain members that are colorless (Pringsheim 1963). Something of their variety is summarized in Table 4–1. Since all of these examples are naturally-occurring ones, it is obvious that loss of photosynthetic pigment does occur spontaneously, that it has occurred many times and in different groups, and that the colorless forms are capable of survival. Furthermore, there is no evidence of colorless forms ever giving rise to colored ones and, as will be clear from the mechanisms of loss of photosynthetic abilities, changes in that direction are highly improbable.

Pringsheim (1963) considers loss of photosynthetic pigments under four different categories. First, there is bleaching, which can be due to culture conditions, e.g., growth in the absence of light. Or simply continued growth in a medium rich in organic compounds will result in colorless forms appearing in the culture, as was first discovered by Zumstein (1900) working on *Euglena gracilis*. Restoring the cell to normal growth conditions can restore the normal pigments. Loss of pigment is reversible in these cases. Second, plastid size can be reduced. Pascher (1927), one of the early pioneers in analyzing

Table 4-1 Pigmented and colorless forms showing similar cellular morphology. (From Pringsheim (1963) *Farblose Algen* Fischer Verlag.) The first six categories of form are all found within the golden alga (Chrysophyta).

	PIGMENTED	COLORLESS
Single cells	Ochromonas Chromulina Mallomonas	Monas Oikomonas Mallomonas apochromatica
Spherical colonies	Synchromonas pallida	Anthophysa vegetans
Branching colonies with stalks	Chrysodendron ramosum	Dendromonas virgaria Monadodendron distans
Cells in tests	Epipyxis Stylopyxis Poteriochromonas	Stokesiella lepteca
Rhizopodial forms	Chrysarachnion insidians Rhizochrysis crassipes	Leukochrysis Amoeba stigmatica
Rhizopodial forms with tests	Eleuthropyxis Kybotion eremita Lagynion	Leukopyxis asymmetrica Platytheca micropora Heterolagynion
Bicoecaceae, cells with tests and one flagellum	Kephyrion Conradocystis dinobryonis	Donatomonas cylindrica Codonodendron ocellatum
Craspedomonadaceae, with collars and one flagellum	Stylochromonas minuta	Monosiga and others

the biological significance of colorless algae, obtained a series of unicellular green algae in *Chlamydomonas* with successively smaller plastids, until, finally, there were the extreme forms with no plastids but still showing pyrenoids—bodies associated with plastids which function as sites of starch accumulation—and then, finally, forms lacking even pyrenoids. Pascher thought of this series, as Pringsheim puts it, as analogous to the skeletal series showing evolution of the modern horse. But Pringsheim demurs, pointing out they do not make a linear series; at best they could have had some common ancestor from which they developed, each along its own lines. Pringsheim further points out that there was lacking in Pascher's series a very important form, one which showed plastids but lacked pigment, i.e., possessed leucoplasts. An example of such a form is known today in *Chlamydomonas*, which resulted from either a single gene mutation or a change in a nonchromosomal factor (Sager and Ryan 1961). Pringsheim concludes that differences in plastid size are the result of accumulated mutational steps, and that morphological series of different species, showing decreasing size, indicate different mutational situations in each species, rather than any actual phylogenetic sequence.

A third mode of pigment loss comes about through unequal distribution of plastids at fission. A fission product receiving no plastids is completely colorless and irreversibly so. If it can survive, it will produce a clone of colorless cells like itself. Original observations of loss of plastids through maldistribution (e.g. Scherfel 1901; Pascher 1918, reviewed in Pringsheim 1963) showed that the colorless condition was lethal. But more recently the successful culture of such colorless forms has been reported (Parke, Manton, and Clark 1955). Another aspect of this same problem is the situation where the cell divides faster than the plastids. Under such conditions the plastids are diluted out, and eventually divisions occur where one fission product receives none of the reduced number of plastids. This has been observed to occur in dark grown *Euglena mesnili* (Lwoff and Dusi 1935).

Lastly, loss of plastids can be achieved by selective destruction of the plastid, leaving a colorless, viable cell. Chemicals such as streptomycin can achieve this (Provasoli, Hutner, and Schatz 1948) as can ultraviolet irradiation (reviewed by Leedale 1967) and even visible red light (Leff and Krinsky 1967). All of these examples are taken from work with *Euglena gracilis*. The last example is very interesting, for, as the authors point out, there is reason to believe that the mutagenic effect is mediated by the photosynthetic pigments and acts on plastid DNA, not on nuclear DNA. Also, since the intensity of the light used is equivalent to normal sunlight it is apparent "... that in the case of microorganisms containing chlorophyll, sunlight may act as a natural mutagenic agent and thus serve as an important factor in the evolution of these organisms." (Leff and Krinsky 1967, p. 134.)

As is readily surmised, in all cases of complete loss of plastids, the colorless or apochlorotic condition is irreversible.

The fate of apochlorotic individuals is dependent on their abilities to use their environment as a source of energy and building blocks. Pringsheim points out that the transition from phototrophy to chemotrophy is not simply the loss of plastids. The cells must be able to function as chemoheterotrophs, that is, they must be able to use reduced carbon compounds as nutritional materials. This can be done by actively transporting substances dissolved in the surrounding medium across the cell membrane and then using them internally as needed. Such a capability is typical of the fungi and has been termed saprotrophy or, somewhat inaccurately, osmotrophy. Or, particulate matter found in the environment can be engulfed by a vacuole, digested and utilized. We have already discussed this as phagotrophy, or simply, as Pringsheim calls it, zootrophy. This means, actually, that the cells which successfully make the transition to zootrophy, which is our special interest here, are preadapted for such a mode of life. They are cells that, though capable of photosynthesis, are also capable of other modes of nutrition. A dramatic example of this is the chrysomonad *Ochromonas malhemensis* (Pringsheim 1952). These small flagellated cells can grow as phototrophs provided they have the necessary light, salts, and various growth factors. Under such conditions the chromatophores are conspicuous, and slow but steady growth occurs. Denied light, but with other factors kept constant, the cultures die out. Clearly photosynthesis is a necessary supplement to the various growth factors. Under other conditions, where the medium is rich in dissolved organic matter, *Ochromonas* can grow even in the dark. And when particulate matter is provided, including bacteria and fungal spores, these are phagocytized and broken down by the cell with ensuing growth. *Ochromonas* is an efficient saprotroph, phagotroph, and can on occasion act as a phototroph.

It has been pointed out (Pringsheim 1963) that if particulate food is present in the environment, zootrophy is more efficient than saprotrophy. Only a portion of the dissolved material in the surrounding environment can be selectively absorbed by a cell, and much diffuses away and is lost, whereas, with particulate material, its uptake and breakdown inside the cell assumes that most, if not all of the usable material, will be used by the cell. We can see why both fungal and animal modes probably arose from colorless algae able to pursue sapro- and zootrophy following the loss of pigment. For the cells which are potential phagotrophs, locomotion, either by flagella or pseudopodia, would be of selective advantage in allowing movement which would widen the area to be drawn on for food. For dissolved material, the ability to move would not be so important, though some dispersal mechanism at some time in the life cycle, especially to contact other cells for sexual fusion, would be advantageous.

The general picture we have, then, of the origins of animal-like unicells, is: (a) They came from unicellular algae. (b) These colorless algae were functionally preadapted to zootrophy in that they were capable of phagotrophy and locomotion before they lost the powers of photosynthesis. (c) The origins of zootrophy were multiple. That is, zootrophy or phagotrophy could have arisen from a multiplicity of causes and need not be confined to just one group of algae. In fact, it is found to occur naturally in many algal groups. As we shall see, and it should be no surprise in light of the foregoing, the protozoa have arisen from more than one algal line; they are polyphyletic. It is for that reason that we do not consider them as properly being a taxon and hence do not follow conventional nomenclatural procedure and refer to them as the Protozoa (Hanson 1967a, 1972). The term protozoa, has, nevertheless, practical use in collectively referring to ingestive or animal-like unicells. The term is used here as a common name for such organisms.

Phylogeny of the Protozoa

A few final comments are necessary now for outlining the pattern we shall follow in studying the early evolutionary history of animal systems.

Ideally, as was expounded in the chapter on phylogenetic procedure, phylogenetic inquiry proceeds from the analysis of small, uniform groups with identifiable plesiomorphs to larger, more diverse groups with their respective plesiomorphs. And, theoretically, the only limit to analyzing ever wider sets of organisms is the availability of the plesiomorphs themselves. The initial practical limits on this approach are the availability of plesiomorphs from small, uniform groups. In the protozoa, where there are 30,000–50,000 described species

(Mayr, Linsley, and Usinger 1953; Corliss 1967) it is physically impossible for one author to start at the level of uniform groups of species and their plesiomorphs and execute a phylogenetic analysis for the whole array of protozoans. But in doing otherwise one is perforce taking shortcuts and must, therefore, be aware of the pitfalls arising from the more abbreviated approach and should make clear why such an approach is to be attempted at all.

The pitfalls come from the fact that the definition of uniform groups and of their possible plesiomorphs will come from the work of others who may not follow the methodology being used in the present volume. Hence it is possible from the various ways in which workers define their groups—using one or a few characters, and only structural or developmental traits to the exclusion of others, and having any one of various ways of defining primitive or ancestral types—that one will be using groupings and forms that later analysis will reject. In at least partial answer to such reservations are these points: In various taxonomic groups, especially intensively studied groups, there is now fairly good taxonomic agreement at the ordinal level and above. These groups are usually the residuum of many decades of scrutiny, and, regardless of the systematic philosophy guiding the various authorities, these divisions have acquired a certain stability, which most often reflects a high probability of monophyletism in each such group. This is especially true if the workers espouse the philosophy of the New Systematics which lead them, as we have seen in a previous chapter, to an empirically correct though conceptually vague use of homology. Hence reliance on the data and conclusions of such workers, which is a necessity to fulfill the scope of the present survey, may not involve such serious methodological problems as might at first seem likely.

What results from such a procedure will, of course, lack the rigor and precision that come from a consistent application of consciously understood methodological principles. Nonetheless, certain positive gains will be apparent. These will include (a) a survey of the forms in question to identify semic traits, (b) an analysis of the semes in terms of establishing homologies, (c) preliminary attempts at defining plesiomorphs, (d) preliminary identification of plesio-, apo-, and neosemes, (e) tentative (and limited in number) dendrograms which can be compared to formulations of others, (f) discussions of the origins of major groups. This list represents, therefore, a serious first attempt at laying the groundwork for detailed phylogenetic analysis. If thought to be fruitful, specialists in the various groups of protozoans and lower metazoans can take this initial work further and expand, revise, or re-do it as the need arises.

This volume can only initiate the study of phylogeny by the method proposed here, but the goal is to make sufficient results apparent so that the reader will gain a critical feeling for the possibilities of this approach.

The procedure used is as follows. To reduce the heterogeneity of the protozoa and the lower metazoa to manageable proportions, there has been an attempt to deal only with certain major groups on a chapter by chapter basis. Within the chapter, the major groups are analyzed in terms of their subgroups. The major groups are the rhizoflagellates (zooflagellates and ameboid forms), heliozoans, radiolarians, acantharians, ciliates, sponges, cnidarians (coelenterates), and turbellarian flatworms. Omitted are symbiotic groups such as sporozoans, cnidosporidians, trematodes, and cestodes, and certain smaller groups of uncertain affinities, such as the slime molds, silicoflagellates, and coccolithophores. The symbiotic forms, though posing fascinating evolutionary problems, are arguably dead ends of evolutionary specialization and hence do not promise much help in following out larger evolutionary trends. Similarly the smaller specialized groups like the slime molds, etc., appear to be off the main lines of evolutionary development. Therefore these groups are not treated here. However, this does not say that the methodology developed here is not applicable to them. Just the opposite is true; the present method of phylogenetic analysis is, in principle, applicable to all species of organisms.

The foregoing groups do not neatly coincide with those of any of the standard taxonomic treatments of the protozoa (e.g., Grassé 1952a, 1953; Hall 1953; Kudo 1954; Dogiel, Poljanskij, and Chejsin 1965; Honigberg *et al.* 1964; Grell 1968). They do come close to the proposals of the French workers (Grassé 1952a, 1953) who have provided us with the most serious phylogenetic attempts to date. In this writer's opinion, their taxa are the closest to being monophyletic of any yet proposed, and for that obvious reason their proposals form the basis of our study here.

Given the foregoing groups as a starting point, the treatment of each group then falls under two broad headings: General Characteristics and Phylogeny. That is, the structural and functional biology of the group is summarized and then a phylogenetic analysis is initiated. Its depth varies with the author's experience with the group and with the general information available. Of necessity, much information comes from monographic and review sources. The original literature has also been examined. Because work on the protozoa covers all fields of biology, it was impossible to survey all possible journals. What was done, was to concentrate on journals specializing on protozoans, i.e. Acta Protozoologica, Archiv für Protistenkunde, Journal of Protozoology, and Protistologica, and to follow up on other journals as experience and other references dictated. Of necessity there will be omissions, but these are unintentional and inadvertent. As much as anything they reflect the limitations deriving from single authorship as already mentioned in the Preface.

Chapter 5
Rhizoflagellates: The Zooflagellates

THE deceptively simple appearance of the amebas led to the early view that they must be the simplest of animals, and, hence, representative of that primordial form from which all protozoans and metazoans subsequently evolved. This view has changed, and the colorless flagellates are now accorded the distinction of being the primordial animal. But there exist clear reasons for supposing a close relationship between the flagellates and amebas, and in view of this the French workers have done the obvious thing and created the taxon "Rhizoflagellés" (Grassé 1952c) which embraces both ameboid forms and the flagellates. This view has been perpetuated more recently in the Sarcomastigophora (Honigberg *et al.* 1964). Though these groups are the logical starting point for the phylogenetic study of the protozoa, it does not thereby follow that they can be successfully analyzed phylogenetically. We shall see reasons for the present situation of limited phylogenetic understanding of these forms. It will be simplest to recognize the flagellate-ameboid dichotomy to the degree that it exists and to first treat the animal flagellates as a group and then turn to the ameboid forms. Certain highly specialized ameboid forms—the Heliozoa, Radiolaria, and Acantharia—will be reserved for study in succeeding chapters.

Above all, the zooflagellates demonstrate the phenomenal organizational potential of the kinetosome and its surrounding cytoplasm. And they do so, in their most dramatic way, in the seemingly implausible location of the hindguts of wood roaches and termites. We will not be able to make clear the whole history of the evolution of the kinetosomal complex in these forms, but enough can be learned to document its real importance to this group and to sense its further importance to groups such as the ciliates.

General Characteristics

There is a broad, direct correlation of flagellar number with cell size. The smallest cells, a few micrometers long, have one or two flagella, and the largest cells, some hundreds of micrometers in their largest axis, can have thousands of flagella. In between these extremes are combinations of size and flagellar number, preserving the general correlation and showing, additionally, various patterns of arrangement of the flagella (Fig. 5-1). The shapes of the cells, generally ovoid, vary with life cycle stages and physiological state, and also show a variety of special contours. These reflect points of elaboration of

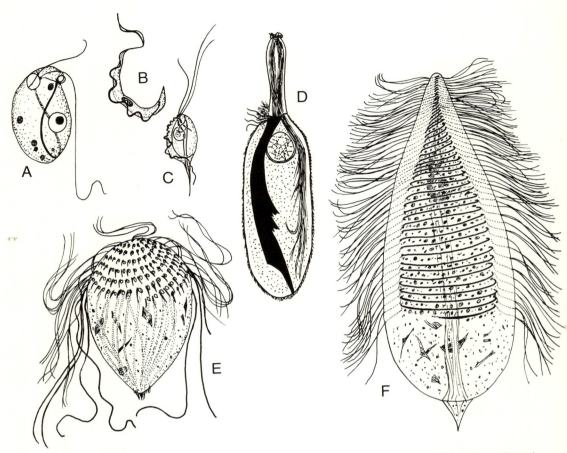

Fig. 5-1 Selected zooflagellates. A. *Bodo saltans* (ca. 8 μm), free living in polluted waters. B. *Trypanosoma gambiense* (15–30 μm), from the blood stream of humans. C. *Tritrichormonas batrachorum* (ca. 7–10 μm), from the urogenital tract of frogs. D. *Metacoronympha senta* (20–90 μm), from the hindgut of certain termites. E. *Oxymonas grandis* (several hundred μm), from the hindgut of certain termites. F. *Spirotrichonympha polygyra* (ca. 80 μm), from the hindgut of the wood roach (*Cryptocercus punctulata*). (A, B, D, and E redrawn, by permission from Grassé (ed.) (1952a) *Traité de Zoologie*, vol. I, part I. Masson et Cie, Paris. C. Redrawn, by permission, from Samuels (1957a). F. Redrawn, by permission, from Cleveland (1938).)

the flagellar apparatus, location of the terminus of rigid internal structures, or there can be convexities and concavities relating to various known and unknown functions. To present further necessary details the discussion will be divided into organizational and then functional aspects of these cell systems. Throughout we will be looking for structural and functional details sufficiently complex to be designated semes. And further, we must be alert for semes showing significant variability and, hence, possible evolutionary development.

In what follows it will be necessary to refer to various groups of zooflagellates in a categorical way. For example, the trypanosomes and forms similarly organized will be called the trypanosomatines. This follows from anglicizing the subordinal taxon Trypanosomatina (Honigberg et al. 1964). *Bodo* and similar forms will be the bodonines and these plus the trypanosomatines will be referred to as the kinetoplastids, i.e., kinetoplast-bearing members of the order Kinetoplastida. For the reader unfamiliar with protozoan taxonomy, the conventions of Honigberg et al. (1964) are such that orders terminate in -ida and suborders in -ina, hence, common names become -ids and -ines, respectively. However, in that family names end in -idae, e.g., Calonymphidae and Trichomonadidae, and they are conveniently contracted to the common names of calonymphid and trichomonadid, respectively, it will be clear that the common names will not preserve the full import of the proper taxonomic names. That will not, however, be a problem in what follows.

STRUCTURE

Kinetide. We will turn immediately to the kinetide, that assemblage of organelles associated with and including the flagellum and its basal body. This granule, previously termed the blepharoplast, is now called the kinetosome (Gibbons and Grimstone 1960; Pitelka 1963). This terminology recognizes the essential similarity in the fine structure of all basal bodies and uses the term applied to the ciliary basal body as the present generic term. The French have taken the logic of this terminology one step further and refer to the kinetosome and related structures as the "cinétide" (e.g., Chatton 1924; Joyon and Mignot 1969; Hollande and Carruette-Valentin 1971), or kinetide. This seems a sensible development, and we will follow their lead, remembering, however, that mastigont is the term favored in the current English literature (e.g., see papers by Honigberg and his associates, Mattern and Daniel). The major features of a flagellum and its kinetosome are seen in Fig. 2–4, given earlier. Beyond this essential pattern, there are various other structures, one or more of which can accompany the flagellum and its basal granule to make a kinetide; additionally, there may be many kinetides per cell. A listing of these accessory structures includes at least the following parts: undulating membrane, axostyle, parabasal body or apparatus, costa, flagellar rods, striated fibers, microtubules, and, possibly, mastigonemes. The first four structures are all visible in the light microscope. We will describe these briefly first, including both light and electron microscope observations. The remainder, seen with the electron microscope, will then be commented on.

The undulating membrane (Fig. 5–1, B and C) is always associated with a flagellum. The organization of this membranelle is variable. In forms such as the Calonymphidae, related to the Trichomonadidae which have undulating membranelles (Honigberg 1963), there is no undulating membranelle (Joyon *et al.* 1969). In the hypotrichomonads the membrane is not well developed (Mattern, Honiberg and Daniel, 1969). In the tritrichomonads the organelle is well developed and quite variable (Joyon *et al.* 1969). The variability that is most conspicuous is the variable contribution of the flagellum to the undulating membrane. In the trypanosomes (Vickerman 1969 and earlier workers), the outer membrane of the flagellum extends to fill the space between cell body and flagellum. In the tritrichomonads, *T. cricetis* shows the flagellar membrane extension but in *T. augusta* the cell membrane extends in a ridge to meet a moderate extension of the flagellar membrane (Joyon *et al* 1969; further details in Daniel, Mattern, and Honigberg 1970; Honigberg, Mattern, and Daniel 1971). In the trichomonads, the undulating membranelle is quite constant in structure and, as seen in *T. vaginalis* (Nielsen, Ludvik, and Nielson 1966), the extension of the cell membrane is called a "fin" against which the recurrent flagellum lies, a short distance below the outermost extension of this fin. A similar situation is seen in *T. gallinae* (Mattern, Honigberg, and Daniel 1967).

The axostyle is usually a long, rigid structure placed centrally in the cell and parallel to its long axis (Fig. 5–1 C, D, and E). Axostyles may be stout or delicate in their proportions, solid or grooved in their construction, and of one piece or multifibered in their composition. One end is invariably in close physical association with the kinetosome. This, the proximal end, can be swollen into a capitulum. The other end is often pointed and obtrudes from the cell surface. Some axostyles have rings around them towards their distal ends (Grassé 1952a pp. 704–861). Recent fine structure work reveals that the axostyle is formed of varying numbers of microtubules (Pitelka 1969; Honigberg, Mattern, and Daniel 1971, and earlier papers; and also Joyon and Mignot 1969). Usually this organelle is stiff but in some it is reported to be contractile (Grassé 1952a). In this latter group, the Oxymonadidae, the microtubules are organized into sheets (Grassé 1956a; Grimstone and Cleveland 1965).

The parabasal apparatus is highly variable, and there has been confusion over its identification, especially using light microscopy. In the first place, entities formerly called parabasals are now recognized as the Golgi apparatus (Grassé 1956b) and, in terms of their fine structure, are stacks of thin, flat vesicles, always lacking microsomal or ribosomal granules, and forming aggregates of various sizes and shapes in the cell (Pitelka 1963; Hollande and Valentin 1969). The classic definition of the parabasals (Grassé 1952b) as being composed of two substances, one chromophobic and the other chromophilic, and connected to the kinetosomes by fine threads, will probably all be resolvable into a typical Golgi apparatus through further work. Parabasals, as just defined, may exist as single or multiple units in a cell and may show a variety of shapes ranging from dots, rods, and threads, to bean-shaped, sausage-shaped, V-shaped, and branched entities. In certain groups, notably the

Joeniidae (Grassé 1952a) the parabasal is a complex, foliated structure; in other groups, e.g., the Trichomonadidae, it is usually rod shaped and curved (Honigberg 1963). There are also various patterns of filamentous connection between the parabasal bodies and the kinetosomes, i.e., the parabasal filament which is cross-striated in terms of its fine structure (Pitelka 1969). This variation is valuable in terms of diversity, but to use it to trace phylogenetic lineages is a difficult problem, as will be seen.

The costa appears in the light microscope as a rodlike structure, of varying length, curving posteriorly from one of the anteriorly placed kinetosomes. When a trailing or recurrent flagellum is present, it is the kinetosome of that flagellum that is the point of attachment of the costa. With electron microscopy, the costa is described as a striated rod with cross-bands repeating at about 60 mμ m intervals (Anderson and Beams 1959; Anderson 1967). However, Inoki and his group (Inoki, Nakanishi, and Nakabayashi 1959; Inoki *et al.* 1961) argue that the striated structure is a collagenlike fibril, and the costa is the point of attachment of the undulating membrane to the cell surface. In *T. gallinae*, Mattern, Honigberg, and Daniel (1967) find no evidence to support the interpretation of Inoki and his colleagues. In this trichomonad, the undulating membrane is an extension of a cell surface, and it contains a marginal lamella at its anterior portion, lying within the extension. There is no unique body at the base of this extension, and the structure designated a costa is cross-striated and lies internally. The costa is limited in its distribution to the family Trichomonadidae (Honigberg 1963) and, within this group, is a very uniform organelle whose function may be supportive or endoskeletal.

Striated fibrils include not only the costa but some of the fibrillar connections between kinetosomes and nuclei (rhizoplasts) and between kinetosomes and parabasals (parabasal filaments, mentioned above). Also, the so-called axial ribbons of certain flagellates show a striated fine structure (Pitelka 1963, 1969 — see especially summary on pp. 160–161, 1969). The diversity of these structures is perhaps matched by the variety of their distribution. They are found in various of the more complicated flagellates. Their function is not at all clear.

Microtubules, cylindrical structures about 18 to 25 mμ m in diameter, in addition to being an essential component of flagella, are not uncommon in the vicinity of the kinetosome and often appear to grow out from it. These structures also make up the fine structure of the spindles (Pitelka 1963, 1969; Hollande and Carruette-Valentin 1971; Honigberg, Mattern, and Daniel 1971) and are found as a subsurface structure, with a possible cytoskeletal function, in certain Trypanosomatidae (Angelopoulos 1970; Messier 1971).

Finally, associated with the flagellum itself, lying parallel to the microtubules of the flagellar axis and within the flagellar membrane, are microfilamentous entities in the trypanosomes (Fuge 1969). These electron-dense elements are found in the flagella of certain chrysomonads too. The mastigonemes—lateral projections of the flagellar membrane and characteristic of the unicellular algae—are absent from the zooflagellates, with the possible exception of the choanoflagellates (Petersen and Hansen 1954).

It should be emphasized that no kinetides exist which lack a kinetosome. The other structures are variable in their presence. Then, additionally, there may be several kinetosomes in a kinetide. Lastly, a given kinetide may itself become a unit of polymerization such that a cell (Fig. 5–1 D) may have many compound or complex kinetides contributing to its total organization, as seen in the calonymphids.

Other cytoplasmic organelles. There are yet to be mentioned other special cytoplasmic structures not usually considered part of the kinetide. The cresta lies at the anterior end of the cell in the trichomonads, and its narrow, bandlike appearance on the cell surface raises the possibility that it may be homologous to certain filamentous structures ("accessory filament") in other forms (Honigberg 1963). The pelta, a structure of enigmatic function found in the vicinity of the kinetide, is a crescentic area on the cell surface and associated with the capitulum of the axostyle. It is microtubular in composition, and its microtubules overlap those of the neighboring axostyle (Mattern, Honigberg, and Daniel 1967; Joyon *et al.* 1969). The pelta is carefully described in certain works which discuss zooflagellate anatomy (Kudo 1965; Kirby 1944; Honigberg 1963; for example) and is excluded from others (Grassé 1952a; Pitelka 1963). The aciculum has a somewhat similar treatment. This structure is described as a very slender internal rod in certain flagellates (Kudo 1965). It may be another form of the axostyle.

Also, various authors argue for the presence of a cytostome in certain of the animal flagellates. Pitelka (1963, Fig. 52) shows an invagination of the cell surface associated with fibrous structures in *Bodo saltans*, which result in an apparently permanent organelle that looks as if it might be of ingestive function. Honigberg (1963) reports that after patient and long search for a functional cytostome in *Trichomonas*, where such structures have been thought to be present, he can find no good evidence for such an organelle. However, work in the Trypanosomatidae has revealed convincing evidence, both structural and functional, for a cytostome. Steinert and Novikoff (1960) argued for such a structure in *Trypanosoma mega* with evidence of pinocytotic uptake of materials at its inner end. Work on *T. raiae* (Milder and Deane 1969; Preston 1969) has revealed details of the cytostome including a surrounding microtubular structure, evidence for uptake of ferritin at the "blind end" of the cytopharynx, and the formation of a second such structure in anticipation of fission when the two structures are segregated, one to each daughter cells.

The problem of the cytostome somewhat typifies the status of the organelles mentioned in the immediately preceding discussion. There is need for further careful work, best done by electron microscopy and with organisms cultured under specified growth conditions. This can clarify the status of the various structures thought to be present in these forms. Considerable physiological, developmental, and phylogenetic work is dependent on the prior, exact assessment of the morphology and function of these disputed organelles.

And there are, of course, certain cytoplasmic, nonkinetide organelles which have been critically studied and about which there is now good agreement as to their structure. Among these is the kinetoplast. This, as modern studies have shown (Steinert 1958, 1960), is that specialized part of the mitochondrion which contains DNA fibrils (Anderson 1967; Ozeki et al 1970, 1971). The Feulgen positive nature of the kinetoplast was known for years, but elucidation of its relation to mitochondria depended on careful electron microscopy.

The French protozoologists have successfully developed the concept of the atractophore, which is the site of formation of the extranuclear spindle (Hollande and Carruette-Valentin 1971, and earlier papers cited therein). These are not centrioles, because no basal body-like structures are present. Their function is to induce spindle formation and to bring spindle into contact with the nucleus. The atractophores are variable in composition, being microfibrillar, or having cross-striations, or being rods of electron dense material. Their formation is apparently determined by other organelles, particularly the special kinetosomes lying at the anterior end of the cell. These are called centrioles by Grimstone and Gibbons (1966), who describe them in *Trichonympha* and *Pseudotrichonympha*. The French refer to these centrioles as "cinétosomes privilegiés" and claim that they are active morphogenetically, determining the formation of the missing half-rostrum at fission in the trichonymphids. Hollande and Valentin (1967, 1969) claim that the atractophore is also present in the trichomonads, but here there are no special or privileged kinetosomes; presumably the existing kinetide somehow determines atractophore formation.

Similarly, the collar of the choanoflagellates was known for years, but only through careful fixation and observation under the electron microscope was its fine structure revealed. The presence of microtubules (Fjerdingstad 1961a, b) combined with observations (Hollande 1952c) on the pattern of flow of adherent materials—up one side and down the other, i.e., opposite directions on opposite sides of the collar—lead to the supposition that the collar is made up of fused pseudopodia and specifically of those pseudopodia called axopodia which have a set of microtubules in their center. Such axopodia have been found in a variety of forms, including certain unicellular algae, and are therefore of considerable phylogenetic interest. Cells provided with such collars form, as we shall see, a unique and quite homogeneous group.

Inevitably, an overview, such as the foregoing one, cannot cover all the details present in the literature; to balance the account, one can only refer selectively to certain details illustrating the enigmatic details which are generally omitted. Anderson (1967) reports large, spherical bodies lying next to the costa in *Trichomonas muris,* termed paracostal bodies. These are not described by others who have examined the trichomonads except that Daniel, Mattern, and Honigberg (1970) comment on the large number of conspicuous "spherical cytoplasmic inclusions" found in *Tritrichomonas muris.* Vickerman (1969) shows that acanthosomes, distinctively organized structures lying near the kinetosome and bearing phosphatase activity, are present in *Trypanosoma congolense.* And there is a puzzling "subpellicular organelle" reported from *T. congolense, T. Brucei,* and *T. vivax* (Taylor and Godfrey 1969). All of these are organelles whose three-dimensional complexity is sufficient to allow careful comparison with comparable structures in other forms but at present reports are limited to a few forms. Surely, much of the problem is still technical. Pitelka (1969) continually cautions against broad generalizing in the zooflagellates because of "inadequacy of sampling and variations in technique" (p. 304). The data from electron microscopy are still pouring in, and the flow of new observations shows no abatement. Much of this new work will continue to uncover structural complexities that will provide information for phylogenetic studies. But equally true, these data must be evaluated carefully and cautiously because of the admonitions voiced by experienced workers such as Pitelka.

In any case, carefully analyzed organelles such as kinetides and certain other cytoplasmic organelles, such as kinetoplasts, collars, and even cytostomes, are especially valuable for evolutionary studies, for they break the mold of ubiquity that commonly covers such universal organelles as flagella and kinetosomes, surface membranes, pseudopodia, and mitochondria. As was emphasized earlier, innovations which build on the standard organelles are the ones that are potentially of special use in the phylogeny of cell systems.

Nucleus. Before turning to cellular function, the nucleus must be mentioned. There is little electron microscopy that adds materially to the generalizations obtained from light microscopy except regarding details of chromosomal attachment to the nuclear membrane during division. This will be discussed below as part of reproduction. The present account is largely dependent on Grassé (1952b). In the zooflagellates there are one to many nuclei in a cell, and they are either haploid or diploid, depending on the stage of the life cycle. Diploidy, as a rule, is rare. As is true of protozoa in general, the nuclear membrane remains intact even during fission. Chromosomes are present in various sizes and numbers, as is to be expected of eukaryote cells. Nucleoli are also present (Grassé 1952b). Certain unusual aspects of chromosome behavior appear during fission, but these are best described as part of the functions of reproduction. A unique anatomical feature is the occasional association of the nucleus with the kinetide through the intermediary of a fibrous rhizoplast. The functional significance of this connection is not known. In some cells there is one nucleus for each kinetide, and where there is a polymerization of the kinetides, the nuclei likewise can be multiplied (Fig. 5-1 D). Such compounding of structure is a major organizational innovation of these cells (Dogiel, Poljanskij, and Chejsin 1965).

FUNCTIONS

Being animal cells, the zooflagellates are phagotrophic but, in that most of them have come to live within other organisms where foodstuffs are already available in solution, osmotrophy is another mode of feeding that occurs among these forms. Before proceeding further, it is necessary to define terms as they will be used throughout this volume in reference to different species of organisms living intimately together.

Symbiotic relationships. Symbiosis, in its literal sense of "life together," will be used as the general term which includes the following special cases: (a) *commensalism*, where one organism lives off another but without apparent harm or benefit to it; (b) *mutualism*, where both organisms benefit by the association; and (c) *parasitism*, where one lives off the other and the latter is demonstrably adversely affected. This usage of words follows Honigberg (1967) closely and allows application of the neutral term symbiosis when the nature of the relationship between two forms is not clear, which is often the case, and it provides categories for the more important relationships when they are known. Moreover, though this terminology does not jibe with common practice, which often uses parasitism as the all-inclusive term, it is essentially consistent with the proposals of such students of these problems as de Bary (1879), and, more recently, Caullery (1952) and Geiman (1964).

It is also appropriate to comment briefly on the postulated evolution of these various types of symbiosis. The selective advantage of symbiosis, in the long run, is to provide each partner in the relationship with advantages for survival that would otherwise not obtain. Viewed this way, Rogers (1962) has neatly summed up the problem by saying "The host-parasite [viz. symbiotic] relationship might be regarded as ultimately reaching a steady state in which two genotypes react to give a common phenotype on which selection might act." (p. 243) To reach this highly integrated steady state, Rogers envisages various steps leading up to it. In brief, they include initial contact between potential host and potential invader, penetration of host defenses, adjustment of host to invader, and vice versa, which means progressive mutual adaptations along biochemical, physiological, immunological and reproductive lines.

Such a viewpoint, which seems to be widely accepted, at least in its broader outlines, entails certain consequences in terms of the evolution of parasitism, commensalism, and mutualism, as defined above. Parasitism would appear to be an early stage of symbiosis. The wear and tear on the host is clearly of selective disadvantage, both to the host and to the parasite, since it damages or even destroys the essential basis for survival of them both. Hence, parasitism as an early stage is a relatively imperfect or inefficient aspect of symbosis. And commensalism, with an essentially neutral relation between commensal and host is the step that follows on parasitism when pathogenicity has been ameliorated by subsequent evolution. Mutualism, then, is the final and most efficient stage of this evolutionary sequence.

However, it would appear that there can be many exceptions to the foregoing scheme, which, as we said, may be true only in its broad outlines. For example, there may be no convincing illustrations of commensalism, where the host is quite unaffected by its symbiotic partner. Or, again, pathogenicity may be sufficiently successful where hosts are plentiful, as among bacteria and their viruses, so that rapid multiplication at the expense of the host, resulting in numerous offspring, is in fact as good a way to ensure survival of the parasite (which is all the parasite need be concerned with in selectionist terms) as a more cautious use of the host. Geiman (1964) feels the distinctions between the various types of symbiosis are not clear-cut—they really serve only man's didactic needs—and that symbionts will evolve those patterns of symbiosis to which they are driven by natural selection. This would make it necessary to tread cautiously when trying to follow postulated pathways of evolution from one symbiotic relationship to another.

Exploitive functions. The phagotrophy of the zooflagellates includes ingestion of bacteria both in free-living forms and symbiotic forms (feeding on the intestinal flora of their hosts, for example) and ingestion of nonliving particulate material, notably bits of wood, as in the case of the larger mutualistic symbionts of the wood roach and termites. In all cases the ingestion occurs by formation of food vacuoles and involves varying degrees of pseudopodial action to initiate the engulfment. In the small choanoflagellates there is a permanent pseudopodial structure at the base of the collar which is the site of formation of the food vacuoles. In the large termite flagellates, like *Trichonympha*, wood particles are ingested by the somewhat ameboid posterior end of the cell. Permanent feeding organelles, in the sense of a cytostome or gullet-like structure, are limited to only certain of the zooflagellates (see above), and sites of preferential uptake do occur, as with the choanflagellate feeding pseudopodium and the ameboid portions of trichonymphids. The occurrence of pinocytosis is probably widespread, if not universal, in the zooflagellates. Convincing electron micrographs of this process are seen, for example in *Joenia* (Hollande and Valentin 1969) and in the trypanosomatids (Steinert and Novikoff 1960), in conjunction with their small cytostome.

On the other hand, the occurrence of osmotrophy, postulated to occur among various of the symbiotic forms, implies movement of molecules across the cell membrane from the external environment into the cell. However, here too the possibility of pinocytosis is very real. It seems likely that, having evolved vacuole formation as a mode of uptake of nutrients, a cell would not be likely to lose it. When such cells inhabit some part of a host organism, which is a rich source of organic material variously concentrated—ranging from macromolecular aggregates to cellular debris to whole cells—the advantage of being able to take up material larger than simply molecular would be retained. The specialized blood flagellates seem, perhaps, the most likely candidates for osmotrophy, but in them, e.g., *Trypanosoma mega*, there is evidence of pinocytosis and a cytostomal type of phagotrophy.

Phagotrophic feeding is one important aspect of the exploitive function of animal cells, and locomotion is the other. As is obvious, the flagellum is the locomotory organelle of the zooflagellates. Though protozoan flagella show various patterns of motion (Jahn and Bovee 1965a, 1967), there seems to be no correlation, judging from the scanty data presently available, of these patterns with morphologically homogeneous groups, i.e., no phylogenetic significance to modes of locomotion. Superficially, there appear to be obvious differences between the helical motion of the trypanosomes, the rapid beating of the choanoflagellates, the somewhat jerky motions of the trichomonads, and the hula-skirt-like undulations of the trichonymphids. These, however, almost surely reflect a variety of mechanical problems, such as the undulating membrane of the trypanosomes, the single, free flagellum of *Codonosiga*, the compound kinetide of the trichomonads with cooperative action of various flagella, and the massive cell of *Trichonympha*, which could only be moved ponderously by the cooperative motion of many flagella moving rhythmically together. Hence, varieties of locomotory patterns are really as much a reflection of the total structure of the cell as of any peculiar flagellar motion itself. It is therefore easier and more accurate to compare the various zooflagellates in terms of their total anatomy than in terms of impressions of locomotory habits.

Homeostatic functions. The next general category of functions are those relating to homeostatics. There is little useful information here. The most conspicuous organelle in this regard is the contractile vacuole or pulsatile vesicle (Wigg, Bovee, and Jahn 1967). This organelle, which is largely responsible for osmoregulation (Kitching 1967), is found in the few freshwater species of zooflagellates and in certain of the marine forms too. It is absent in the symbiotic ones, which are in an osmotic balance with their hosts. These vesicles, because of their limited distribution and the scanty knowledge of their action, provide little evidence of information useful to phylogenetic studies.

There are, of course, other homeostatic functions of these cells—maintenance of their metabolic pools, transport mechanisms, internal pH, and so forth—but these would appear to be similar to cells in general and so provide no specific information of potential use to evolutionary studies. That comparative data in this area are steadily accumulating is very clear (Conner 1967), but much of it, at present, comes from work on the amebas and ciliates and is, therefore, of no use for study of the zooflagellates themselves.

An area of intensive work in the zooflagellates is the analysis of their nutrition and metabolism. Various reviews (Hall 1965; Dewey 1967; Honigberg 1967; Ryley 1967; Fulton 1969; Hutner et al. 1972) have surveyed this area, but, despite detailed descriptions, there has emerged surprisingly little that can be used from a phylogenetic viewpoint. It appears that the zooflagellates, and protozoa in general, share much in common regarding needs for amino acids, vitamins, and various purine and pyrimidine bases. Such a conclusion must be qualified by the realization that, though axenic culture of protozoans is common today, cultivation in defined media is limited to a few species, and therefore unequivocal comparative statements on nutritional requirements are quite limited. Further, general metabolic pathways common to many living systems are, not unexpectedly, present among the protozoans. Beyond this, there are various compounds whose utilization is restricted to one or a few species or a genus, but the occurrence of these phenomena is disparate and of little phylogenetic use, except for such obvious remarks as the need for hematin by the blood flagellates being an obvious adaptive response to their environment. The loss of certain enzyme systems—metabolic lesions—is a not uncommon feature of some of the symbiotic flagellates. These losses are apparently compensated for by materials provided by the host cells.

Certainly a large part of the problem here is the fact that much of the impetus for studying these forms comes from their relation to public health problems, especially regarding the trypanosomes and related forms and the trichomonads and their relations. Here the emphasis is understandably directed towards eventual chemotherapy, and for this reason these studies result in certain forms—the pathogenic ones—being intensively studied and other forms being slighted. Hence there is no systematic coverage of phylogenetically relevant forms and, of those studied, the analyses are not necessarily directed to evolutionary questions.

Reproductive functions. Binary fission invariably occurs as the longitudinal or symmetrigenic separation of the original cell into two daughter cells. One can immediately anticipate something of the complexity of such separations by recalling the possible complexity of the kinetide. However, it does not serve our purposes to review in detail the various ways in which these cells divide (Hanson 1967a). What is needed is to find patterns of development of the various organelles, for these are associated in various combinations to form the larger fission pattern of the whole cell. With regard to the kinetide, the formation of a new kinetosome or kinetosomes is the initial step in forming a new kinetidal complex in forms with relatively simple or few kinetides. Where there are many kinetides, as in *Trichonympha* (Cleveland *et al.* 1934), the division of the so-called centriole is the first step in preparation for new kinetide formation and cell division. Work reviewed by Hollande and Carruette-Valentin (1971) summarizes the brillant electron microscopy of these and other workers, which has shown Cleveland's centriole to be a rather variable organelle now called the atractophore. This organelle apparently induces formation of the spindle and leads certain spindle fibers, at least, into contact with the nucleus. We will return to behavior of the spindle and nucleus after completing the description of kinetide behavior during cell division.

In the zooflagellates the kinetosome and kinetoplast are apparently self-reproducing bodies along with the chromosomes. However, their mode of self-formation seems quite different from that employed by the chromosomes. Judging from kinetosome formation in the ciliates (Allen 1967; Dippell 1968), the formation of new kinetosomes is not a template-based mechanism but appears to be a progressive and sequential assembly of parts, of which the microtubular elements are most conspicuous. From new kinetosomes there arise new flagella, and, adjacent to the kinetosome, there arise a new axostyle, costa, and other parts of the kinetide. In relatively simple situations, such as *Trypanosoma* (Fig. 5–2 A), the new kinetosome develops a new flagellum and undulating membrane and these go to one fission product. The preexisting kinetide remains with the other cell. In trichomonad flagellates, with their complex kinetide, there is a more complicated distribution of organelles. In some cases there is a kind of biogenetic recapitulation. For example, *Hexamastix* has six flagella (5 directed anteriorly, one recurrent) instead of the sim-

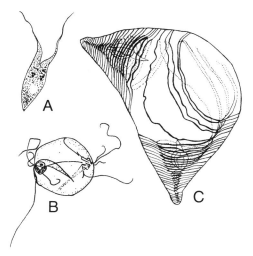

Fig. 5–2 Symmetrigenic fission in selected flagellates. A. *Trypanosoma cruzi* (Redrawn, by permission, from Noble, McRory and Beaver (1953).) B. *Tritrichormonas batrachorum* (Redrawn, by permission, from Samuels (1957b).) C. *Spirotrichonympha polygyra* (Redrawn, by permission, from Cleveland (1938).)

pler situation of four flagella, as seen in *Monocercomonas* or *Trichomonas* (three anterior and one recurrent). During fission in *Hexamastix*, cell separation can precede formation of the full complement of flagella, and where cells with three or four anterior flagella can be found, there will eventually, however, be developed the normal number of five (Honigberg 1963). And in forms with many kinetides, such as *Spirotrichonympha* (Fig. 5–2 C) yet another situation holds, where each daughter receives roughly half of the preexisting kinetides and then makes new ones to complete the normal complement. The fundamental control device here appears to be the formation of new kinetosomes, and the distribution of these determines the development of new kinetides in accordance with the genetic and developmental nature of the individual cell.

As to the kinetoplast, the DNA-containing fibrils are oriented parallel to the long axis of the cell; and then these are distributed, by constriction, into right and left portions, which go to each daughter cell (Burton and Dusanic 1968; Anderson and Hill 1969). This distribution is puzzling, as it does not seem to allow for an equivalent set of DNA in each fission product. When the DNA is labelled by H^3-thymidine, there is either no clear evidence of a distribution of the label to justify a semiconservative mode of replication (Burton and Dusanic 1968), or it is barely suggestive of that possibility (Anderson and Hill 1969). Further research is needed on this enigmatic situation.

The collar of the choanoflagellate, which always surrounds the single apical flagellum, behaves as if it were a compound structure, as is indeed indicated by its fine structure, for it separates into two parts for each daughter cell, and the missing parts are made anew in each cell. A unitary structure would be expected to go all to one cell, with the other cell making a new one, as occurs with the flagellar apparatus in these cells (Hollande 1952c).

The nucleus shows very distinctive behavior in fission in those flagellates possessing four or more flagella. Where the kinetides are simple, or if they are multiple but there is a nucleus with each simple kinetide, the kinetosome acts also as a centriole and determines the poles of the dividing nucleus. In the less complex forms, the spindle is intranuclear. Where, as in the complex forms, there is an atractophore separate from and in addition to the kinetides, the spindle fibers are extranuclear, and there are also aster fibers. The nuclear membrane, however, persists, and the chromosomes are seen to attach to the nuclear membrane (Cleveland 1935, 1949; Grassé 1952a; Hollande and Carruette-Valentin 1971). Such mitoses have been termed pleuromitoses and are characteristic of the complex zooflagellates. There are certain other special features too. In some forms a thread, the paradesmose, is present and runs from kinetide to kinetide. In several of the groups, especially the Spirotrichonymphida, the atractophore grows to astonishing lengths and produces a very flat spindle, which lies on the side of the nucleus. Features such as these allow the differentiation of nuclear behavior from one group to another, but these differences are of the nature of various discrete or discontinuous differences rather than of a series showing any sort of progressive development.

Finally, in terms of reproduction, zooflagellate life cycles deserve mention, and in this connection sexual processes can also be outlined. Life cycles uncomplicated by sexuality are found in the trypanosomes and their relatives. Sexuality has been most intensively studied in forms in the wood roach, *Cryptocercus,* and occur there only when the roach is molting (Cleveland 1956 and earlier). Cleveland (1965, 1966) also reports, rather briefly, some evidence of sexual activity in the more complex flagellates of termites. Sexuality has been suggested for the free-living form *Bodo*, but this has not been confirmed (Hollande 1952d).

The life cycles in the Trypanosomatina range from very simple patterns of development in a variety of single hosts to quite complex patterns involving two hosts. The enormous amount of work on these symbionts, some of which are parasitic in man, has achieved considerable clarification of their basic biology, along with control of their numbers, for reasons of public health and social economics. The life cycle of *Crithidia euryopthalmi*, a symbiont in the digestive tube of the bug *Euryophthalmus convivus*, is given in Fig. 5-3. Two points need emphasis. First, the form changes of the crithidial cell are correlated with their location in the host. The cycle starts with the small, ovoid, encysted form (A) with no external flagellum. The leishmanial, or amastigote, stage then passes into forms (F-H) which possess more than one

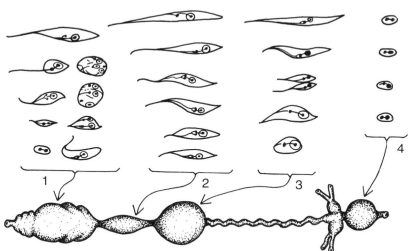

Fig. 5-3 Life cycle of *Crithidia euryophthalmi* in the digestive tube of the bedbug *Euryophthalmus convivus*. 1. Forms found in the crop. 2. Forms inhabiting the anterior midgut. 3. Forms lying in the posterior portion of the midgut. 4. Cysts in the rectum. Upon release and ingestion by another host, the cycle is reinitiated. (Redrawn, by permission, from Grassé (ed.) (1952a) *Traité de Zoologie*, Vol. I, part I. Masson et Cie.)

nucleus, kinetoplast, and kinetosome apiece. These are apparently reproductive stages, and from this there emerges the larger spindle-shaped cell, with a free terminal portion to the flagellum, which is termed the crithidial stage, or the choanomastigote form (Hoare 1967), in recognition of the emergence of the flagellum from a depression at the end of the cell (not shown in Fig. 5-3, but see Fig. 5-5). When the cells finally reach the lower end of the digestive tract, they encyst, again in the amastigote form, and are discharged from the host body in the feces. Ingestion of the fecal pellets by another bug results in infection and subsequent repetition of the developmental cycle.

The second point that needs emphasis is that the forms seen in this life cycle have their counterparts in the more complicated cycles of the trypanosomes. *T. grayi* (Fig. 5-4) shows the amastigote form (16) and the crithidial form, though here the flagellum emerges from its pocket on the side of the cell, as well as typical trypanosomal or trypomastigote forms (1, 2). There is here a sequence of forms in the insect host (stages 2–16) that again shows correlations with position in the host, though the amastigote form can apparently

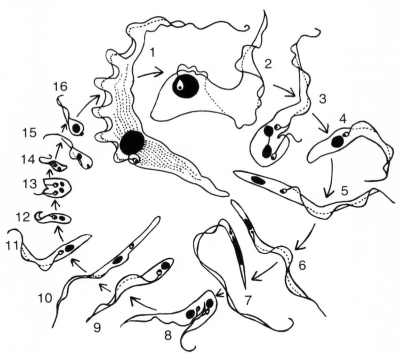

Fig. 5-4 Life cycle of *Trypanosoma grayi*. The form found in crocodile (*Crocodilus niloticas*) blood (1) is taken into the insect vector, *Glossina papilis,* where it undergoes a series of form changes (2–10) in the midgut. (Stage 8 is in fission.) Development in the posterior gut shows further changes (11–15) leading to the infective form (1). (Redrawn by permission, from Grassé, (ed.) (1952a) *Traité de Zoologie*, Vol. I, part I, Masson et Cie, Paris.)

be formed any time in the cycle. *T. grayi* not only shows more diversity of form than *C. euryophthalmi*, but it alternates between two hosts—the crocodile, where it remains in the blood (1), and the blood-sucking insect, *Glossina* (2-16), a species of tsetse fly.

Any further description of trypanosomatid life cycles would only provide slight variations on the themes illustrated by *C. euryophthalmi* and *T. grayi*. The various cell forms found in the group as a whole are given in Fig. 5-5. Note in all forms the constant relation of kinetoplast to mastigont. These forms are symbiotic in animals and certain plants. Where they complete their life cycle in a single host, as is true of members of the genera *Leptomonas* and *Crithidia*, the cell shape variation is restricted to amastigote forms or forms that have the flagellum emerge from the end of the cell body. Species of this type are found in other protozoa (Gillies and Hanson 1963; Freedman and Hanson 1970) or in rotifers, nematodes, molluscs, and insects (Grassé 1952d). Forms with two hosts—digenetic forms—include the genera *Leishmania* and *Trypanosoma*, and their hosts include a vertebrate (fish, amphibian, reptile, bird or mammal) and a leech if the vertebrate is aquatic, or an insect if the vertebrate is terrestrial. The forms found in certain plants with a milky sap also have an insect stage. In these latter the variations of the cell body are like the leptomonad species. But the other symbionts with two hosts

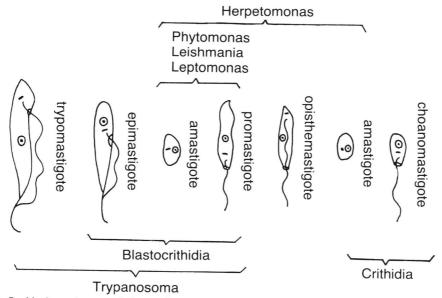

Fig. 5-5 Various forms of the cell body in trypanosomatids. Note, in particular, the relative placements of flagellum, kinetoplast (short, rod-like structure), and the nucleus (represented by a circle enclosing a dot). The occurrence of these various forms within different genera is indicated. (Redrawn, by permission, from Hoare (1967).)

show a wider range of variation, like that seen in *T. grayi*. As will be seen shortly, the distribution relative to hosts, the occurrence of monogenetic or digenetic life cycles, and the range of cell shapes in a life cycle are evidence for certain phylogenetic speculations.

The variation of the cell body is in all probability an example of phenotypic variation, i.e., the expression of different potentials of one and the same genotype. The significance of this sort of variation is hard to overestimate. A carefully studied example of phenotypic variation in the zooflagellates is the work of Inoki and his collaborators on the antigenic variation of *Trypanosoma gambiense* (Inoki *et al.* 1960 and earlier papers). From this work it is established that antigenic types change in the absence of any genetic change in a mutational or recombinational sense. The change is, in all probability, one of differential gene action; a gene or genes producing certain antigenic proteins are shut off, other genes are turned on. The transformation of the amastigote leishmanial form into the larger, flagellated leptomonad, *Leishmania donovani* (Rudzinska, d'Alesandro, and Trager 1964) is probably a similar case in point, as is also any drug resistance in this form (Hanson *et al.* 1963). Also, in *Trypanosoma mega*, the transformation into infective forms by addition of urea to the culture medium (Steinert 1960) is another example. Examination of the conditions of these changes leads to the conclusion that phenotypic change in the presence of a constant genotype must be the basis for this variation (Trager 1963; Hanson 1967a). Such a pattern of variation is of adaptive value for forms which have to adjust quickly to survive in the different environments provided by their host. The digestive tract of any animal varies in pH and other factors, along its length, and the digestive tract is quite a different environment from the blood stream. When the numbers of infecting organisms is low, clearly there are very few numbers within which mutation could occur to provide genotypes that in turn would provide phenotypes appropriate to the host environment. The mutation-selection mode of response to environmental diversity is appropriate to haploid organisms which exist in large numbers and is a mechanism of apparently great success among certain of them, e.g., the bacteria. The trypanosomes and probably many other protozoa—certainly, also, the ciliates as we'll see—have evolved another mechanism. Sonneborn has characterized it as follows:

> "Another and faster operating mechanism ... This mechanism involves (1) including in the common genotype of the species sets of loci for alternative and mutually exclusive phenotypes and (2) incorporating mechanisms for shifting expression from one to another on demand." (Sonneborn, 1957, p. 287)

The demand, in the case of the zooflagellates, would be stimuli from the host, very possibly of a chemical nature, as suggested by Steinert's work. Such triggers of differential gene action would endow these symbionts with the capacity for expressing those phenotypes which permit rapid, intimate adaptation to their host. Vickerman (1962) has suggested a mechanism involving the fine structure of the cell that would carry out these changes in the *Trypanosoma brucei* subgroup.

The life cycles of the termite and roach flagellate cannot be as yet resolved into as satisfying a picture of the genetic and selectionist basis of their adaptation. Actually, in one of the earlier analyses of the interrelations of these flagellates to their termite hosts, Kirby (1937) concludes that natural selection is inoperative. His problem is to account for the fact that the flagellates have apparently diversified within the seemingly constant niche of the host's hindgut. Further discussion of this problem is best postponed until further facts of the symbiont-host relation have been presented. Returning to the problem of host-symbiont interaction in determining symbiont phenotypes, it is perfectly clear (Cleveland *et al.* 1934, and other papers referred to in Cleveland 1956a, 1956b) that the insect moulting hormone can initiate the cycle of events associated with sexuality that occurs in certain of the zooflagellates.

In that the sexual process, apart from fission, which has already been outlined, is the chief developmental event in wood roach symbionts, their life cycles can be described in terms of the events which determine successive haploid and diploid phases in their life history. Details such as encystment and excystment before and after gamete formation, union of gametes, fate of kinetidal organelles during these processes will be omitted here, for they are in large part unique to the separate species. The larger aspects of timing and consequence of meiosis and fertilization are what will concern us now. Cleveland (1947) has summarized the major events. There are three ways to achieve the diploid condition from the haploid one, and only two of them are properly concerned with sexuality. The asexual process involves an endomitotic doubling of the chromosomes. This chromosomal duplication without cell division is followed by meiosis, and the haploid state is restored. Autogamy, the fusion of two separate nuclei of common origin from a single haploid parent nucleus, lies within the concept of sexuality we are using here; but it is clearly a case of extreme inbreeding, and genetically it is no different from the results of endomitosis. When diploidy results from fusion of gametic cells, the ensuing nuclear fusion guarantees a fertilization process that is potentially an outbreeding situation. What minimizes this aspect of the process is the fact that the fertilization occurs among the progeny of cells that originally infected the roach when in its first larval instar (Cleveland *et al.* 1934). Though these progeny do not represent a clone, for more than one representative of a given species was probably present in the first infection, the fact that roach colonies are themselves somewhat isolated implies that the flagellates in a colony represent a deme, and that most breeding occurs within the deme; moreover, it occurs between the asexual progeny of a limited part of that deme—that is, the inhabitants of one hindgut.

These various processes can be compared with the situations for permanent haploidy and diploidy, which are also known in the flagellates, and with the predominantly diploid condition found in ciliates, Metaphyta, and Metazoa. As discussed in the previous chapter, Cleveland (1947, 1956b) has interpreted his findings as illustrating the progressive emergence of sexual processes, but Sonneborn (1957), in a careful and convincing reinterpretation of these findings, concludes just the opposite, that sexuality is on the way out in these forms. The essential reasons for the latter view are twofold: first, the common occurrence of inbreeding tendencies is consistent with the relative constancy of the hindgut habitat of these protozoa, and, hence, variability due to chromosomal combination is quite plausibly disadvantageous; second, in the termites, which have evolved from the wood roaches or share a common ancestor with them (Cleveland *et al.* 1934), there is limited evidence of sexuality (Cleveland 1965, 1966), which is unexpected from Cleveland's point of view but readily understandable from Sonneborn's. The importance of these cycles is not so much for specific phylogenetic inquiry, for their scattered distribution and varied nature is hard to interpret, but rather they provide evidence for the continuing action of natural selection. These broader conclusions are among the arguments that can be brought against Kirby's view of the inoperation of natural selection.

Phylogeny

Briefly, then, in surveying the preceding account for those traits which are sufficiently complex to allow meaningful comparisons, namely, which are semic, the following emerges. Organizationally the kinetide is the most striking feature, for above the ubiquitous fine structural level of kinetosome and flagellum, we find other, associated organelles of characteristic fine structure and placement in the cell, and various compoundings of kinetosomes, flagella, and related structures. All are useful in point-to-point comparisons from one species to another.

In addition to the kinetide, there are other organelles showing readily-characterized features such as collars and kinetoplasts. These are of limited distribution, as are some other organelles. Altogether these form a category of important traits but of quite limited variability and distribution.

The nucleus is obviously eukaryotic and shows some peculiarities at fission, and certain cells are diploid and some multinucleate.

Functionally, the most important exploitive seme is the symbiotic relationship to hosts, which occurs with the majority of the zooflagellate species.

Regarding homeostatic functions, the nutritional adaptations of this group appear somewhat discontinuous in distibution and unique in nature, somewhat like that category of organelles that show limited variation and distribution.

Reproductively, the pleuromitotic nuclear division characterizes the more complex zooflagellates and is mediated through the spindle-forming organelle, the atractophore, which is fairly variable in structure. The sexual processes of certain of these cells show significant complexity and variability, but unfortunately these processes are carefully studied only in certain termite flagellates. Life cycles have been intensively studied and relate to the exploitive tendencies of this group. Details regarding correlations in cell shape and size with hosts and location in particular hosts provide significant material for the comparative studies related to phylogeny.

The next step is to define the homologous relationships within the zooflagellates and start locating plesiosemic and aposemic traits. And, as was exposited earlier, we are especially looking for those traits that characterize this group of organisms, that tell us what adaptive features they have evolved to exploit their environment and maintain and reproduce themselves, and that also show variability. We can omit consideration of such generalized features as membranes, mitochondria (though certain zooflagellates lack them entirely; a regressive loss related to symbiosis), and other basic organelles. These are common to animal cells in general and tell us little about zooflagellate evolution. The same is true for basic biochemical or biosynthetic activities.

HOMOLOGIES

The structural semes will be examined first and then the functional ones.

Kinetide. The organization of the microtubules of the flagellum and kinetosome shows no important differences from that found in other cells, and that ubiquity of organellar structure is of limited use to phylogenists. There are differences in the materials, especially filaments and other electron dense accumulations, that lie between the tubules. These were first described by Gibbons and Grimstone (1960) and seem to represent more the results of technical advances in fixing and resolving fine structure than real organizational differences (see Chapter 2). Pitelka, when commenting in general about the status of our knowledge on the zooflagellate flagellum, says this:

> "We face a similar lack of food for speculation as regards the structures seen occupying the center of the kinetosome in several species, as well as the fine fibrillar interconnections among kinetosomal fibrils and subfibrils illustrated by Gibbons and Grimstone. Their finding of elaborately different kinetosomal contents within the same species indicates that the matter must be of considerable importance to the kinetosome, but we cannot guess why or how." (Pitelka, 1963, p. 160.)

More important are various specializations associated with the basic flagellar-kinetosomal organelle. Mastigonemes, if finally shown to be present in the choanoflagellates—the only zooflagellate group where they might occur (Pitelka 1963)—would be an important characteristic. The complexity of their organization could provide an important semic trait for comparison with algal flagellates. Also, more details on the flagellar rods (Pitelka 1963; Fuge 1969)—the electron-dense microfilaments lying between the flagellar membrane and the tubules in forms such as the tryanosomatines and the bodonines—might permit homologizing of them to each other and to other forms, too. At present their simple organization is insufficiently complex to allow informative comparisons to be made. Pitelka (1963) notes the occurrence of such structures not only in the trypanosomatines and bodonines but also in the recurrent flagellum of some trichomonads, the adherent flagella of *Pyrsonympha* and in the phytoflagellates *Euglena* and *Peranema*.

The parabasals show a great array of shapes and sizes, and though there is no problem in establishing homologies among them, there is a question interpreting their variability. As was stated earlier, they are all presently understood as being identical with the Golgi apparatus, and their fine structure is homologous in all cases, being that of stacks of flattened vesicles bounded by unit membranes (Grassé 1956b). Regarding their diversity, it appears, thus far, there may be some trends in progressive thickening or thinning of parabasals as in the trichomonads, but, apart from this group, there appears little possibility of understanding this variation, especially in terms of trends; the various species often show their own peculiar organization of the parabasal. And various parallel developments, in terms of number or shape, may be present in other species.

A similar situation holds for the axostyle, which is composed of apparently homologous microtubules; but the problem of homologizing the various forms of the axostyle is very acute, for it can contain varying numbers of microtubules, and it can be hollow, delicate or stout, with or without a capitular expansion, and rigid or contractile. Comparative electron microscopy on a wider variety of forms is needed.

The undulating membrane, a relatively simple structure, shows variation, especially in the manner in which the space between the flagellar microtubules and the cell surface is filled (esp., Joyon and Mignot 1969). As noted previously, in some forms it is the flagellar membrane that extends to fill this space, in other cells the membrane of the cell surface forms an outward extension or ridge to meet or lie next to the flagellum. In these latter cases, the flagellum may or may not lie at the outer edge of this extension. Also, there can be an accessory filament lying at the anterior base of the membrane, as in the Trichomonadida (Honigberg 1963), or there can be a filament lying within the extension itself. These different ways of forming the undulating membrane and of juxtaposing its compositional elements renders homology doubtful as a general case. (The compositional criterion, in terms of articulation of component parts, is not fulfilled.) Within such otherwise relatively uniform groups as the Trypanosomatidae or the genus *Trichomonas*, there might well be homology. Beyond that, one must argue, from lack of fulfillment of the compositional criterion, that there has been more than one way to form an undulating membrane, and similar selection pressures have elicited several convergent responses.

The costa, the parabasal fibril, and other striated fibrils and bands show general positional similarities, but the variations in their striational patterns indicate compositional differences. That is, differences between costas and parabasal fibers and ciliary rootlets, but not so much differences within these categories. However, in comparing something like parabasal fibers, we are not looking at very complex traits: these structures lie in generally similar positions, are cross-striated, and fibrillar. The point-to-point comparisons are meager (here, we listed only three). It is probably best to treat them under the first subsidiary criterion and ask if they have a distribution that parallels that of other semes, known to be homologous, through application of the major criteria.

Thus far we have had limited success in establishing useful homologies for the various semic components of the kinetide. Either the structures are positionally and compositionally similar but show limited variation (kinetosomes and flagella), or show positional and compositional similarity but are highly and discontinously variable (parabasals or Golgi and axostyles), or are not complex enough to establish compositional similarity (striated components), or, finally, are possibly not homologous but are convergent (undulating membrane) when present. What can be done is to move to a higher level of organization and look at the kinetide as a whole—to compare it as a complex set of largely homologous suborganelles. With such an approach, it seems that corresponding numbers of flagella in the kinetides of two organisms, combined with other information about the correspondence or lack of it in terms of parabasals, axostyle, etc., would provide ample detail for treating the whole entity as a seme amenable to application of Remanian criteria of homologous relationships and would also provide that variability needed for determining aposemic relations, which are keys to analyzing phyletic development.

This, however, does not mean that every organism with a kinetide must have had a common ancestor with any other organism with a kinetide. As has been emphasized, since the components themselves cannot yet be completely homologized, one cannot argue as Pitelka (1963, p. 159) does in saying "At the same time, the repetition of similar patterns in very dissimilar flagellates emphasizes the underlying analogy—and surely homology—of the apparatus wherever it is found." If a similarity is not accompanied by other similarities of a similar distribution—and here the flagellates are admittedly dissimilar in traits other than the basic ones of flagellum and kinetosome—then the trait is unlikely to be homologous. This is the second accessory criterion of Remane, put in its negative form. It seems quite possible that kinetides could have arisen independently many times over. Or, starting from a minimal kinetide of kinetosome and flagellum, the compounding and complexing of parts could have arisen independently several different times. What is unlikely is that a given complex of kinetidal parts could have arisen independently many times over. More explicitly, if the kinetide is treated as a unit composed of various subunits, such as flagella of a certain number and type, presence of parabasals of roughly the same size and shape and probably of the same numbers, and similarly for axostyles and striated elements, then the compositional relationship of Remane can be applied with real force and, as we saw earlier, it is this criterion that is of first importance in distinguishing homologies from convergences. When compositional differences do arise, but homology is still suspected, then the serial relationship must be established to bridge the compositional differences, and this, *providing enough data are on hand*, can then decide the matter of homology conclusively.

Enough detail has been provided in the foregoing accounts to argue convincingly, at least regarding the zooflagellates with more than a minimal kinetide, that kinetides are indeed homologous. It might be that the basic kinetosome and flagellum of the choanoflagellates, the bicosocoecids, the kinetoplastids (bodonines and trypanosomatines) all descended separately from those of the other, more complex zooflagellate ancestry. Thus we would be talking about a distantly though fundamentally homologous structure when we consider a structure as basic to the kinetide as its kinetosome and flagellum. But this point of view does not allow us to establish direct and, therefore, close phyletic relations among the free-swimming collared flagellates and bodonine cells, nor among the remaining variously symbiotic forms. We will have to look to other features to provide that phylogenetic information. And apart from the symbiotic trypanosomes and their obvious relatives, the organelles added to the basic or minimal kinetide can provide that information because of the homologies between parabasal structures, axostyles, and other elements, and their physical relations to each other. The desired complexity of a structural seme is four distinguishable points; the elements just mentioned plus presence of anteriorly and posteriorly directed flagella provide just that complexity.

Parenthetically, we should note the following. Although the just-cited organizational or structural features of the kinetide are not exactly equivalent to four unique points in rigidly defined spatial relations to each other, they do represent a unique cluster or set of points, which can be compared to other such clusters or sets. This mode of point-to-point comparisons was not anticipated in the idealized situations given in Chapter 3 but is not inconsistent with the procedures developed there. The present situation is an example of the practical adjustments that inevitably follow on a formal, theoretical statment of analytic procedures. There will appear other similar examples in subsequent chapters, and these will be reviewed in Chapter 14, when the methodology proposed in Part I and applied in this part is finally appraised and evaluated.

Kinetoplasts, collars and other organelles. The kinetoplasts within the kinetoplastida and the collars of choanoflagellates show sufficient detail of placement and composition within each group so that they fulfill the Remanian criteria of homology. More explicitly, the collar is always positioned around the base of the single flagellum at the free end of the cell. Compositionally, it consists, in the forms thus far studied, of a circular palisade of microtubules whose relationships to each other and general similarity of construction argue for correspondences sufficient to validate the compositional criterion. As to the kinetoplast, its unique fine structure, i.e., DNA fibers localized within a mitochondrion, quite precisely define its composition in a very similar way in all forms studied thus far. If the variable position of the kinetoplast in the cell seems to invalidate the positional relationship, it should be recalled that, first, its position relative to the mitochondrion of the cell is constant in that it is part of the mitochondrion; second, its position relative to the kinetosome is constant; and, third, differences in its position relative to the nucleus and the cell as a whole can all be bridged by intermediate forms between the extremes of different placement (Fig. 5-5). However, the lack of variation in these structures themselves—the kinetoplast and collar—means they will be seen as plesiosemic homologies and not as aposemic ones.

Other organelles that have been mentioned previously, such as the pelta, the rhizoplast, the cytostome, and the like, all are limited in their distribution. They are best used as semes when more restricted groups within the zooflagellates are studied, otherwise they appear as isolated neosemes relative to some relatively more simple ancestral form.

Life cycles. A final category of homologies relates to exploitive functions and specifically to the host-symbiont relation of many of the zooflagellates. Clearly, the posterior portion of the digestive tract of the wood roach, *Cryptocercus*, and of the termites represents a highly similar mode of life, especially when it also involves digestion of wood ingested by the host and is, further, a mutualistic relation between host and symbionts. Such similar specific relationships argue for homology in terms of spatial position of the symbiotic partners and in terms of the detailed functional relations that derive from the positional relationship. Similar arguments for the trypanosomatines are complicated by the fact that there is a very broad spectrum of host-symbiont relations—from ciliated protozoans to higher land plants and to humans. However, it is just this range of hosts that permits of bridging the gaps between protozoan and vertebrate hosts.

In the trypanosomatines living in invertebrate hosts, the symbionts usually live in the gut of the host. It is thought that invasion of the blood of vertebrates is a consequence of being bitten by infected insects. Indeed, insect vectors are the explanation why a given species of symbiont can survive in such apparently divergent niches as invertebrate gut and vertebrate blood stream. It also explains why certain of the symbionts, e.g., *Phytomonas*, can also live in plants. It appears then, that members of this group of zooflagellates show a good deal of aposemic evolutionary development in terms of their exploitation of hosts.

It seems impossible to homologize the symbiotic habit of the trypanosomatines with those of the wood-consuming inhabitants of the roach or termite gut, especially the hypermastigid zooflagellates.

The trichomonadid flagellates are yet another story. Along with certain well defined representatives (see below) occurring in roach and termite hosts, they are also found in the digestive tract of invertebrates and vertebrates, in the urogenital tract of certain vertebrates, and on occasion certain species are found in the head and neck area (including the brain) and thoracic and abdominal viscera of vertebrates (Honigberg 1963). These zooflagellates come closest to being candidates for bridging the gaps in the distribution of the other zooflagellates. However, the chief difficulty is bridging the gaps in structure and organization between the various zooflagellates, such that even though similarities in host can exist between trypanosomatines and trichomonadids, and between the later and the hypermastigid forms, there are, in this series, some sharp discontinuities between the forms. The serial relationship is not convincingly fulfilled.

Furthermore, the similarities between hosts are probably superficial, for there is little reason to think that a trypanosomatine utilizes the same nutrients in the gut of an insect as a trichomonadid, and it is obviously clear that the latter do not ingest wood particles, as is true for many of the more complex zooflagellates. Hence, occurrence in a common host could be more fortuitous than of any phylogenetic import. But saying that does not, however, resolve the issues we face regarding the significance of host-symbiont relations in these forms. It is surely significant, and we shall have to return to this point later, that the vast majority of the zooflagellates are symbiotic. Either they did evolve by spreading from one host to another (but we've just commented on some of the difficulties of that view) or from free-living ancestors they invaded various hosts (with some subsequent further invasions), and the free-living ancestors have disappeared, leaving only the specialized symbiotic forms as remnants of the past history of this group. It is just this point we will return to later.

PHYLOGENETIC ANALYSIS

The next steps, which follow *seriatim*, are identification of a plesiomorph; determination of plesio-, apo-, and neosemic traits; and determination of phyletic relationships, as far as possible. To facilitate this phylogenetic study, we will first treat certain zooflagellate groups which show considerable intragroup homogeneity, i.e., the choanoflagellates or collared cells; the kinetoplastids, which contain the free-living bodonines and the symbiotic trypanosomatines (both with similar flagellar, pellicular, and cytostomal traits and, of course, containing a kinetoplast); the trichomonadids, with their complex kinetide; and then some remarks on the yet more complex hypermastigid forms.

Problem of the zooflagellate plesiomorph. The more prominent characteristics of selected zooflagellates are summarized in Table 5-1. The three representatives of the collared flagellates are seen to be very similar, with their most obvious differences relating to the extracellular gelatinous matrix, or lorica, which has long been recognized as a somewhat variable feature of these cells (Ellis 1929). There are certain differences in the collars, too. But there is no sure way—from fossils or from any sort of recapitulative development or from ecological distributions—to infer which is the earliest condition of these traits. From the generalized notion of structural simplicity one can rank *Codonosiga* as simpler than the colonial *Phalansterium* (Fig. 5-6). To make an evolutionary sequence of this series demands acceptance that evolution always goes from simpler to more complex and that reduction could not have occurred in these cells. Neither proposition is convincing among these cells. There is no way to identify, from data now on hand, a plesiomorphic species for the choanoflagellates. At best, one can describe certain traits that in all probability are present in the plesiomorph. These include small size (<15 μm) and ovoid body with some sort of extracellular matrix present. There would also be a single flagellum, apically located, and surrounded by a collar compounded of microtubules. There would be one parabasal body lying near the kinetosome, and there would be one nucleus. The cell might well be attached by a stalk to the substratum, and it probably would live in fresh water.

This description is detailed enough to permit us to seek homologies for its most distinctive traits among other forms, forms that would, therefore, be plausibly ancestral. de Saedeleer (1929a, 1929b) urged, in some detail, the derivation of the choanoflagellates from cyrtophorine chrysomonads, which can be represented by the form *Cyrtophora pedicellata* (Fig. 5-6D). The presence, in this form, of a stalk, axopodia surrounding the single flagellum, single nucleus, fresh-water habitat, and generally small size are all in favor of a general resemblance to a choanoflagellate. Hollande (1952c) reserves his judgment until cysts can be found for the collared cells and examined to see if they correspond to the distinctive cysts of the chrysomonads. If they do, Hol-

Table 5-1 The major features of zooflagellates as seen in species representing the various families recognized by Grassé and his colleagues (Grassé 1952a). The question marks in parentheses (left hand columns) indicate the absence of taxonomic designations. Numbers in parentheses following the species designate references, as follows: (1) Hollande 1952c; (2) Grassé and Deflandre 1952; (3) Grassé 1952d; (4) Hollande 1952d; (5) Grassé 1952a, p. 694–703; (6) Honigberg 1963; (7) Grassé 1952a,

Suborder	Family	Species
Choanoflagellata	Gymnoeraspedidae	*Codonosiga botrytis* (1)
	Salpingoecidae	*Salpingoeca gracilis* (1)
	Phalansteriidae	*Phalansterium consociatum* (1)
Bicoecidea	(?)	*Bicoeca lacustris* (2)
Trypanosomidea	Trypanosomidae	*Trypanosoma gambiense* (3)
Bodonidea	Bodonidae	*Bodo saltans* (4)
	Spiromonadidae	*Spiromonas augusta* (4)
	Bodomorphidae	*Bodomorpha minima* (4)
	Trypanophidae	*Trypanophis grobbeni* (4)
	Cercobodonidae	*Cercobodo neimi* (4)
	Thaumatomonadidae	*Thaumatomonas lauterhorni* (4)
Protermonadina	Proteromonadidae	*Proteromonas lacertae-viridis* (5)
	Karotomorphidae	*Karotomorpha bufonis* (5)
Trichomonadina	Monocercomonadidae	*Monocercomonas colubrorum* (6, 7)
	Trichomonadidae	*Trichomonas vaginalis* (6, 7)
	Devescovinidae	*Devescovina duboscqui* (7)
	Calonymphidae	*Coronympha octonaria* (7)
(?)	Polymastigidae	*Polymastix melonthae* (8)

p. 704–779; (8) Grasse 1952a, p. 780–788; (9) Grassé 1952a, p. 789–800; (10) Grassé 1952a, p. 801–823; (11) Grassé 1952a, p. 824–835; (12) Grassé 1952a, p. 836–847; (13) Grassé 1952a, p. 848–850; (14) Grassé 1952a, p. 851–857; (15) Grassé p. 858–861; (16) Grassé 1952a, p. 862–915; (17) Cleveland *et al.* 1934; (18) Grassé 1952a, p. 916–942; (19) Grassé 1952a, p. 965–982.

Cell body			Mastigont					
size µm	shape	special features	anterior flagella	recurrent flagella	axostyle	parabasal	number per cell	other features
<10	ovoid	gelatinous outer coat, stalk	1	–	–	1	1	–
<10	''	hyaline lorica stalk	1	–	–	1	1	–
10–15	''	gelatinous matrix for cells	1	–	–	1	1	–
<10	irregular ovoid	lorica	1	1	–	?	1	–
15–30	spindle		–	1	–	–	1	undulating membrane
<10	ovoid		1	1	–	–	2	–
<10	''		1	1	–	–	2	–
<10	''		1	1	–	1?	2	–
55–75	spindle		1	1	–	1?	2	undulating membrane
ca. 12	''		1	1	–	2	2	–
<10	ovoid	pseudopodia	1	1	–	1?	2	–
10–22	ovoid to spindle		2	–	–	1	2	–
12–16	spindle		4	–	–	1	2	–
<10	generally ovoid		3	1	1	1 rod	1	accessory fibril
<10	''		4	1	1	1 rod	1	'' undulating membrane
ca. 50	''	bacteria attached to cell	3	1	1	1 coiled	1	–
20–50	''	''	24	8	8	8	8	karyomastigont
5–22	ovoid		4	–	1 fibrous	?	1	–

Phylogeny of the Protozoa and Early Metazoa

Table 5-1. Major features of the zooflagellates, *continued*.

Suborder	Family	Species
Choanoflagellata	Gymnoeraspedidae	*Codonosiga botrytis* (1)
	Salpingoecidae	*Salpingoeca gracilis* (1)
	Phalansteriidae	*Phalansterium consociatum* (1)
Bicoecidea	(?)	*Bicoeca lacustris* (2)
Trypanosomidea	Trypanosomidae	*Trypanosoma gambiense* (3)
Bodonidea	Bodonidae	*Bodo saltans* (4)
	Spiromonadidae	*Spiromonas augusta* (4)
	Bodomorphidae	*Bodomorpha minima* (4)
	Trypanophidae	*Trypanophis grobbeni* (4)
	Cercobodonidae	*Cercobodo neimi* (4)
	Thaumatomonadidae	*Thaumatomonas lauterhorni* (4)
Protermonadina	Proteromonadidae	*Proteromonas lacertae-viridis* (5)
	Karotomorphidae	*Karotomorpha bufonis* (5)
Trichomonadina	Monocercomonadidae	*Monocercomonas colubrorum* (6, 7)
	Trichomonadidae	*Trichomonas vaginalis* (6, 7)
	Devescovinidae	*Devescovina duboscqui* (7)
	Calonymphidae	*Coronympha octonaria* (7)
(?)	Polymastigidae	*Polymastix melonthae* (8)

RHIZOFLAGELLATES: THE ZOOFLAGELLATES

Table 5–1. Major features of the zooflagellates, *continued.*

	Nucleus			Mode of life		
mitosis	spindle	centriole	Other structures	free-living	symbiotic	Sexual process
typical	?	?	collar	fresh-water		
″	?	?	″	″		
″	?	?	″ colonial	″?		
″	?	?		marine		?
″	?	?	kinetoplast		human and insect blood	
″	?	?	″	polluted water		yes?
″	?	?	″ cyst		coprophilic	″
″	?	?	kinetoplast?		″	
″	?	?	″		siphonophores chaetognaths	
″	?	?	cyst	fresh-water		
″	?	?	variable cell form	?		
″	?	present	kinetoplasts, rhizoplast, cyst		lizard rectum	
″	?	present	″		frog rectum	
″	present	centro-bleph.	pelta		squamate reptiles lower intestine	
″	″	″	costa, pelta, cyst		human urogenital tract	
″	″	″	cresta		termite lower intestine	
″	″ para-desmose	″			″	
″	?	?	rhizoplast		beetle rectum	

Table 5-1. Major features of the zooflagellates, *continued.*

Suborder	Family	Species
Pyrsonymphina	Pyrsonymphidae	*Pyrsonympha flagellata* (9)
	Streblomastigidae	*Streblomastix strix* (9)
Oxymonadina	Oxymonadidae	*Oxymonas grandis* (10)
Retortomonadina	Retortomonadidae	*Retortomonas gryllotalpae* (11)
Joeniidea	Joeniidae	*Joenia annectens* (12, 17)
(?)	Rhizonymphidae	*Rhizonympha jahieri* (13)
Lophomonadina	Lophomonadidae	*Lophomonas blattarum* (14)
(?)	Microjoenidae	*Microjoenia fallax* (15)
Trichonymphina	Trichonymphidae	*Trichonympha campanula* (16, 17)
	Eucomonymphidae	*Eucomonympha imla* (16, 17)
	Teratonymphidae	*Teratonympha mirabilis* (16)
	Hoplonymphidae	*Barbulanympha* sp. (17)
	Staurojoenidae	*Idionympha perissa* (16, 17)
Spirotrichonymphina	Spirotrichonymphidae	*Spirotrichonympha flagellata* (17, 18)
	Holomastigotidae	*Holomastigotes elongatum* (17, 18)
Distomatina	(?)	*Giardia intestinalis* (19)

Table 5-1. Major features of the zooflagellates, *continued*.

size μm	Cell body shape	special features	Mastigont anterior flagella	recurrent flagella	axostyle	parabasal	number per cell	other features
40–80 usually	pear-shaped	appears twisted	—	4–8	1 fibrous	1	1	—
15–530	slender spindle		4	—	—	—	1	—
30–>200	ovoid		4	—	1 fibrous	—	1	—
6–22	tear-shaped	depression on one side	1	1	—	1?	1	—
>200	ovoid		many in anterior rostrum	—	1 heavy	2 pinnate	?	—
ca. 200–400	,,	attachment disc	many	—	many	many	many	karyomastigont
ca. 50–110	,,	anterior tuft of flagella	many	—	many fused	many	many	—
<20	tear-shaped	apical circlet of flagella	4–10 pairs	—	1	several	several	—
ca. 200–400	often pear-shaped	anterior end covered with flagella	—	hundreds	—	many long rods	hundreds	—
100–165	,,	cell covered with flagella	—	,,	ca. 60	?	,,	—
90–275	slender ovoid	circular bands of flagella	—	,,	many fibrous	many thread-like	,,	—
ca. 200–400	acorn-shaped	two anterior areas with flagella	—	,,	,,	many thread and rod-like	,,	—
100–275	,,	four anterior areas with flagella	—	,,	,,	,,	,,	—
50–200	,,	spiral band of flagella	—	,,	1 tubular	along flagellar bands	,,	—
ca. 30	spindle	spiral bands of flagella	—	,,	—	in rows	,,	—
10–25	pear-shaped, flat on one side	bilateral symmetry	—	8	—	2 rods	2	—

188 *Phylogeny of the Protozoa and Early Metazoa*

Table 5–1.   Major features of the zooflagellates, *continued.*

Suborder	Family	Species
Pyrsonymphina	Pyrsonymphidae	*Pyrsonympha flagellata* (9)
	Streblomastigidae	*Streblomastix strix* (9)
Oxymonadina	Oxymonadidae	*Oxymonas grandis* (10)
Retortomonadina	Retortomonadidae	*Retortomonas gryllotalpae* (11)
Joeniidea	Joeniidae	*Joenia annectens* (12, 17)
(?)	Rhizonymphidae	*Rhizonympha jahieri* (13)
Lophomonadina	Lophomonadidae	*Lophomonas blattarum* (14)
(?)	Microjoenidae	*Microjoenia fallax* (15)
Trichonymphina	Trichonymphidae	*Trichonympha campanula* (16, 17)
	Eucomonymphidae	*Eucomonympha imla* (16, 17)
	Teratonymphidae	*Teratonympha mirabilis* (16)
	Hoplonymphidae	*Barbulanympha* sp. (17)
	Staurojoenidae	*Idionympha perissa* (16, 17)
Spirotrichonymphina	Spirotrichonymphidae	*Spirotrichonympha flagellata* (17, 18)
	Holomastigotidae	*Holomastigotes elongatum* (17, 18)
Distomatina	(?)	*Giardia intestinalis* (19)

Table 5-1. Major features of the zooflagellates, *continued*.

mitosis	Nucleus spindle	centriole	Other structures	Mode of life free-living	symbiotic	Sexual process
ʺ	?	present			termite gut	
ʺ	paradesmose	?	rostellum		ʺ	
ʺ	present	?	rostrum		ʺ	yes
?	?	centrobleph.	cyst		cricket gut	
pleuromitosis	extranuclear	present	rostrum		termite gut	
typical	?	ʺ	rhizoplast		ʺ	
ʺ	intranuclear paradesmose	ʺ	cyst		roach gut	
?	?	?	rhizoplast		termite gut	
pleuromitosis	extranuclear	large	complex rostrum, cyst		wood-roach gut	yes
ʺ	ʺ	ʺ	ʺ		ʺ	yes
ʺ	ʺ	ʺ	complex rostrum		ʺ	yes
ʺ	ʺ	ʺ	ʺ		ʺ	?
ʺ	ʺ	ʺ	ʺ		ʺ	yes
ʺ	ʺ	ʺ	modified rostrum		ʺ	?
?	?	?	no rostrum		termite gut	yes
present	intranuclear	present	rhizoplasts cysts		rat intestine	?

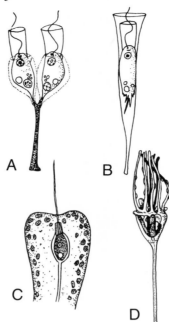

Fig. 5–6 Selected choanoflagellates and a chrysomonad. A. *Codonosiga botrytis*. B. *Salpingoeca gracilis*. C. *Phalansterium consociatum*. D. *Cyrtophora pedicellata*. (All redrawn, by permission, from Grassé (ed.) (1952a) *Traité de Zoologie,* Vol. I, part I. Masson et Cie.)

lande is of the opinion that this would be very strong evidence favoring this chrysomonad origin of the choanoflagellates.

From the point of view of homologies, and setting aside the unknown seme of the cyst for the moment, it seems clear that homologies are clearly possible, but where intermediate forms are critical—in bridging the gap between a circlet of separate axopodia in *Cyrtophora* and the fused or at least compacted ones for the collared cell—they are missing. In terms of the Remanian criteria, the compositional criterion is not really met. To bridge the gap, it is necessary to invoke the third criterion of serial relationships and use intermediate forms to make a gradual series of steps joining the rather different extremes. But such intermediates are not known, and the argument that cyrtophorine axopodia are homologous to the choanoflagellate collar is, at best, tenuous. And if this basic seme cannot be established as homologous in the forms under comparison, then other similarities, in themselves not sufficiently complex to be established as homologies by the positional, compositional, or serial relationships, also cannot arguably be homologous. This stems from use of the first subsidiary criterion. To be sure, inability to prove homology does not as a consequence disprove homology: it only says, on present evidence, we cannot be sure.

Next, we shall examine the Kinetoplastida. This taxonomic group contains not only the trypanosomatines and similar forms (Fig. 5–5) but also bodonines (Fig. 5–1A) and related forms. These are, at first appearance, somewhat different organisms: the former have only one flagellum per cell and the latter always have two; the former are always symbiotic and the latter contain many free-living forms. They are joined together by common presence of a kinetoplast, striated pellicular structure, presence of paraflagellar rods, and a so-called cytostomal tube in at least certain forms (Pitelka 1963). Nonetheless, because of the differences, it is easier to discuss the two subgroups separately, starting with the better known and more homogeneous group of the trypanosomatines.

Aposemic traits are needed for phylogeny and there are two such traits in the trypanosomatines: (a) variations in the form of the cell body including position of the kinetoplast and flagellum and (b) diversity of host relationships. The cell-shape variations, are, as we have seen, reflections of various stages in the life cycle of these forms (Fig. 5–5). They are, in effect, developmental stages. Hoare makes the obvious interpretation of them when he states, "The stages of development of trypanosomes in the insect vector can be regarded as a recapitulation of the forms of their monogenetic trypanosomatid ancestors, thus providing an example of Haeckel's biogenetic law among the Protozoa." (Hoare 1967, p. 57) Unfortunately, to be sure of recapitulative tendencies, we need to have a fixed starting point, which classically has been the zygote, and there is none in these zooflagellates. If one arbitrarily took the leptomonad or promastigote form as such a starting point, then the biogenetic law could be found to hold in fair degree because of the widespread occurrence of this form in most life cycles; also providing that the equally widespread amastigote (leishmanial) form was considered to be a specialized stage often associated with encystment. The only sort of objective evidence that this is the place to start the life cycle comes from interpretations of the host-symbiont relationships.

In that the trypanosomatines live in forms ranging from protozoa to higher vertebrates, it might be thought that this range of hosts is indicative of a time scale, the oldest symbionts being in the presumedly oldest invertebrates and so on up to the most recent symbionts in the recently evolved vertebrates. Such a loose supposition is attractive, really, only because it happens to work. The parasite of the ciliate *Paramecium trichium*, *Leptomonas karyophilus*, which localizes itself in the macronucleus of the ciliate (Gillies and Hanson 1963; Freedman and Hanson 1970), shows only leptomonad and leishmanial forms in its life cycle, and the same is true of the symbionts of other lower invertebrates. More complex life cycles appear only in conjunction with the vertebrates. The weakness in this correlation between simpler life cycles/simpler hosts and more complex life cycles/more complex hosts is seen when one asks why there is not an almost universal presence of trypanosomatine symbionts. For an underlying assumption of this correlation is that the symbionts are evolving along with the evolution of the Metazoa. It is true that there are probably many more species of trypanosomatids to be discovered. It is also true, at least for *L. karyophilus,* that it does not do well parasitizing other ciliates. In fact, that it is a parasite can be used as an argument that it is relatively a newcomer to the ciliate host. It is therefore also possible to propose that the distribution of the trypanosomatids reflects spotty invasion of various hosts starting from some as yet undetermined, probably aquatic, invertebrate as the ancestral host. The varieties of body form reflect the cell's best response to the host it is in, and since the cell probably does not have an unlimited range of viable, efficient forms it can take, starting from a given genotype, there will be some obvious repetition of form in the various hosts.

What it is that supplies the permissive cue for the various invasions is difficult to understand. Most, if not all, trypanosomatines can be cultured *in vitro* on a blood sugar medium, the famous NNN medium (Mackinnon and Hawes 1961), or some variation of it. This is true of the plant-dwelling forms and of the form living in the ciliate nucleus (Stuart and Hanson 1967) in addition to the typical blood-dwelling forms. This complex medium apparently supplies the range of needed metabolites found to be necessary in this group, a range which includes at least one purine derivative, a pteridine requirement (often referred to as "*Crithidia* factor"), and certain lipid and carbohydrate requirements (Guttman and Wallace 1964). Hence, it is hard to believe there is some common nutritional basis for the special evolution of host relationships evinced by the trypanosomatines.

Baker (1963) prefers the approach of infection of a common ancestor to account for the intestinal forms in insects and leeches. Such a common source for the segmented forms would suggest a very wide distribution in the arthropod and annelid phyla, or else very selective loss of the symbiont. There seems to be no detailed survey of large numbers of members of these phyla so as to check on the actual occurrence of infections, and hence there is no answer to Baker's speculations from that point of view. And of course it wouldn't answer, in any case, the prior question of the source of the initial infection of the primitive segmented host.

Another consideration regarding host-symbiont relations must be inserted here, for it bears on Hoare's views of evolution in the mammalian trypanosomes, views which differ somewhat from Baker's. This concerns the site of development of the symbiont in the invertebrate host, notably in the insect and leech. Hoare emphasizes a widespread and significant difference in sites of development in the invertebrate in the "anterior station," that is mouth and foregut, or in the "posterior station,' which is the hind gut and rectum (Hoare 1964, 1967). For Baker, development in the anterior station in the tsetse flies

(*Glossina*) is a carry-over from an ancestral situation found also in the leeches, where development occurs in the foregut. Such forms are grouped by Hoare (1964) as the Salivaria. Posterior station developers or Stercoraria (Hoare 1964) simply represent, in Baker's view, another trend in adaptation of symbiont to host. And where this latter type of development in insects has also been associated with infection of various vertebrates, the various trypanosomal forms in the vertebrates simply reflect those symbiont-host adaptations. Hence, in sum, as Lavier (1942–1943) has proposed, the phylogeny of the vertebrates is closely matched by the evolution of their trypanosomes.

Hoare's (1967) view argues that stercorarian forms represent highly adapted forms because (a) they are not pathogenic as the salivarian tsetse fly forms are, (b) they infect new hosts very readily (100% success in experimental infections as compared to 20% and 10% for certain salivarian forms), and (c) they do not easily lose their power to develop in the insect host (as is true of salivarian forms, especially after *in vitro* culture). The latter represent an evolutionary innovation in the sense that the original insect vectors, which are postulated to be stercorarian, are replaced by salivarian ones; that is, after the trypanosomes become endemic in wild ruminants, the insect vector changed. Why the change? In the stercorarians infection is contaminative: the feces of the insect are deposited on the mammal near the site of feeding and the infective forms in the feces then have to penetrate the host tissues. There is a certain element of chance in this mode of contamination. This, of course, was an advance over chance infection of one insect by the fecal matter of another in monogenetic cycles. But even more efficient than the use of the contaminative mode of infection of a second host was the injective mode of infection coming from salivarian systems. Here the bite, which allows the insect to feed, also inserts trypanosomes into the host, rather than simply letting them lie on the surface of the host. The salivarian system is now in the midst of evolving a more intimate association between the trypanosome and its insect host, and Hoare documents possible stages in this development.

Hoare's view, using the basis of selective pressures to support his arguments, is a less speculative approach than Baker's, and on that score is more acceptable. But Hoare, himself, recognizes the large component of speculation still remaining in his proposals, for there is as yet no way to establish that nonpathogenic forms (stercorarians) represent an older association than pathogenic forms (salivarians); nor are the presence of digenetic contaminative cycles necessarily of higher selective advantage than monogenetic ones—otherwise why do the monogenetic ones persist at all? Similarly, there is no objective demonstration of the advantage of injective over contaminative infection.

Much of this evolutionary thinking was earlier brought to a plausible focus in Grassé's proposal of the historical development of the trypanosomatines as a whole. He suggested:

"The leptomonads, perhaps, have originally been parasites of aquatic invertebrates, and among these the blood feeders (leeches) have transmitted them to fish, to amphibians, and to aquatic reptiles. Then the blood-feeding insects were contaminated in their turn by attacking the amphibians or the amphibious reptiles. Subsequently, they would have infected birds and mammals. At the same time, it is not absurd to suppose that these ancestors of the vertebrates have evolved while perpetuating their trypanosomes, which latter have been transmitted by means of terrestrial blood-feeding vectors (ticks or insects). For it is very possible that members of the genus *Leptomonas* were passed to terrestrial insects before the vertebrates had appeared on the surface of the earth." (Grassé 1952d, p. 662–663. Original in French.)

Grassé here anticipated the views worked out in more detail by both Baker and Hoare but is unable to choose between them. And there the problem must rest at present. So, to summarize, we can say that there are ways to define homologous relations between the trypanosomatines, but no way to rigorously determine the direction of the evolutionary trends of their two aposemic characters, cell shape variation (life cycle differences) and their various host-symbiont relations. Certain plausible schemes, especially Hoare's, which relies most heavily on selectionist principles, provide rather consistent, but not yet convincing, interpretations of our present knowledge. From the latter point of view, we can guess that the plesiomorph of these kinetoplastid forms would be a leptomonad organism.

Turning to the bodonines, the other kinetoplastid subgroup, we are faced with a situation much like that which we earlier examined in the collar flagellates. This is a somewhat variable group, especially in terms of cellular form, but with no way of determining any trends in the aposemy that is found. *Bodo saltans* is as plausible a plesiomorph as any form, and if we compare it to a leptomonad we find, besides the similarities listed earlier, differences in flagellar number (see Table 5-1) and in habitat. To find a common ancestor for these forms is difficult at best. Pitelka's speculations and conclusion are as pertinent as any in this context.

"Certain similarities of bodonids to euglenoid flagellates may be mentioned: the pellicular fiber system, the accessory rod within the flagellar membrane, as in *Peranema*, the reservoir (circumflagellar depression) with its system of fibrils arising from the kinetosomes. But it would be unwise at this stage to conclude that these resemblances are more than coincidental." (Pitelka, 1963, p. 143.)

Turning to particular emphasis on the kinetide as a tool in elucidating evolutionary relations, the Trichomonadida are of first interest. We will examine this group as it is defined by Kirby (1947) and Honigberg (1963). It includes four families—the Monocercomonadidae, the Devescovinidae, the Calonymphidae, and the Trichomonadidae—all of which possess kinetides with one recurrent and three to five anteriorly directed flagella. An undulating membrane and costa may or may not be present in association with the recurrent flagellum, and each possesses an axostyle and a parabasal body. Additionally, there is an extranuclear spindle with the kinetidal kinetosomes acting as centrioles; no sexuality is known; no cysts are known; these forms are all or nearly all symbiotic (Table 5–1).

In studying this assemblage of species, Honigberg has exploited fully the diversity of the kinetide, or, as he prefers, the mastigont. "In separating most of the general and all of the mastigont organelles, i.e., the recurrent and anterior flagella, the axostyle, the parabasal apparatus, and when present, the undulating membrane and costa, has been taken into account. . . . [The systematic] differentiation of the families, subfamilies, and genera has been based on the presence of several coordinate rather than of a single attribute." (Honigberg, 1963, p. 22.)

This means that the aposemy of the mastigont or kinetide has been the basis for the phylogenetic conclusions which arise from this systematic study. These conclusions are summarized in Fig. 5–7. It is apparent that the evolutionary approach of Kirby and Honigberg has been based on an intuitively correct and astute appreciation of the Remanian concept of homology. The only possible criticism is the basis of their determining the direction of aposemic change. Honigberg explicitly invokes recapitulation, where in fission there is a delay in development of the total flagellar complement and this delay reflects, as it always does, the flagellar constitution of the monocercomonad plesiomorph. This is a new application of the concept of recapitulation, since the starting point of development is not a sexually produced individual or stage. It is not, however, to be objected to on those grounds, for, theoretically, fission qualifies as a recapitulatory situation, and its starting point is obviously the single parent cell. This view of recapitulation is used, as we'll see, with great success in the ciliates. Another ordering principle used by Honigberg, but without a specific reference to it, is the criterion of correlation in transformation series (Hennig 1966), and the special cases of that criterion as defined by Maslin (1952). Specifically, where two morphoclines appear in different groups of organisms, and the extremes of both morphoclines are found in the same taxonomic unit, then this unit represents the primitive situation. In the case of the Trichomonadida, one morphocline is seen in the sequence of the monocercomonads → devescovinids → calonymphids. Here there is a progressive polymerization of the basic monocercomonad kinetide. The other morphocline goes through the monocercomonads to the trichomonads, where there is a tendency to increase the number of anterior flagella and, also, a development of a heavy, abruptly tapering axostyle. With members of the genus *Monocercomonas* at the meeting of these two morphoclines, we can argue for its primitive or, better, plesiomorphic nature.

Such argumentation is undoubtedly behind Honigberg's emphatic assertions, e.g., ". . . the primitive number of flagella among members of this order

196 *Phylogeny of the Protozoa and Early Metazoa*

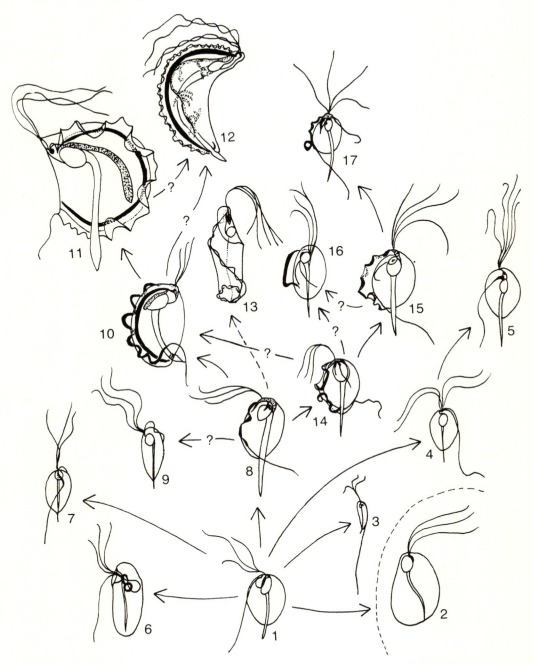

Fig. 5-7 Evolution of the Trichomonadida, with emphasis on the monocercomonad and trichomonad families, according to Honigberg. Forms 1-9 belong to the family Monocercomonadidae within which are three subfamilies. Subfamily Monocercomonodinae contains forms 1-5, and 7, which are, *Monocercomonas* (1), *Pro-*

is 4 (1 recurrent and 3 anterior) and . . . undoubtedly the primitive forms must have been like some of the present species of *Monocercomonas grassi.*" (p. 24) "The genus *Monocercomonas* undoubtedly represents the most primitive member of the order Trichomonodida." (p. 26)

As was pointed out in the chapter on phylogenetic procedure, the two surest criteria of evolutionary direction in aposemic traits are fossils and epigenetic recapitulation. (A special case of the latter is found here.) The special cases of correlated transformations are also useful but not as reliable, and therefore Honigberg's dependence on them must be used carefully. Let us look, therefore, at possible alternatives. The only other view would be, of course, to read the direction of change in the reverse of Honigberg's proposal and suggest that reduction has occurred. This is highly unlikely for reasons stemming from the nature of reduction (see Chapter 3). Reduction arises from specialized adaptation to a specific and often limited mode of life. It is, therefore, inconsistent with the conditions associated with reduction to find the monocercomonad family one of the most widely dispersed of this whole group and to find also its postulated ancestors—on this hypothesis, the devescovinids and calonymphids—restricted to the hind gut of termites. The reduction of trichomonads to monocercomonads is not so easily answered, for the trichomonads are themselves also widely dispersed. However, it can be pointed out that it should be exceptional to expect reduction to take the same path in all the trichomonad → monocercomonad lines. So that, really, here too, the possibility of reduction is minimal.

A last point is to inquire as to the selection pressures which drove those forms along the various evolutionary paths they followed. Unfortunately there is no real answer available. The fault lies in our lack of understanding of the functional details of the kinetide. What is the role of a recurrent flagellum as opposed to an anterior one? What is the specific function of an undulating membrane? of an axostyle? of different parabasals? of a costa? We really do not know. Until we can view these structures in functional terms, we will forever be limited in our phylogenetic understanding of these forms to just their morphological evolutionary relationships without the further and more satisfying knowledge of why just these relationships were the ones to be established.

---

trichomonas (2), *Tricercomitus* (3), *Tetratrichomastix* (4), *Hexamastix* (5), and *Monocercomonas moskowitzi* (7). *Chilomitus* (6) represents the subfamily Chilomitinae, and the third subfamily, Hypotrichomonadinae, contains *Hypotrichomonas* (8) and *Pseudotrichomonas* (9). The other family, Trichomonadidae also contains three subfamilies, as follows: subfamily Tritrichomonadinae is represented by *Tritrichomonas* (10), *Triomitopsis* (11), *Pseudotrypanosoma* (12); subfamily Pentatrichomonoidinae contains *Pentatrichomonoides* (13); and subfamily Trichomonadinae has *Trichomitus* (14), *Tetratrichomonas* (15), *Trichomonas* (16), and *Pentatrichomonas* (17). (Redrawn, by permission, from Honigberg (1963).)

Finally, comments are needed on the remaining zooflagellates, especially such complex ones as the hypermastigid forms which include large, polykinetidal cells such as *Lophomonas, Trichonympha, Spirotrichonympha* (Fig. 5-1F), and *Holomastigotes* (Table 5-1). Grassé's (1952c) distinctive taxonomic treatment of this group is to recognize seven or more separate orders. Honigberg *et al.* (1964) are slightly more conservative and recognize five orders. In both groups the nature of the kinetides is the dominant diagnostic feature. What the ordinal level separation among these groups says is that they are clearly separable taxa. One could go through them, as we have just done in the preceding discussion, and find the features which would be thought to be diagnostic of their respective, putative plesiomorphs. But these groups are more homogeneous than the trichomonadids and show less aposemy; they are more like the choanoflagellates and bodonines in their relative uniformity. They differ from these in one important respect: their plesiomorphs all show relatively complex kinetides, and it therefore seems probable that they arose from a zooflagellate (free-living?) ancestor rather than from an apochlorotic phytoflagellate.

What this comes down to is the following. In addition to the Choanoflagellata, Kinetoplastida, and Trichomonadida, there exist another five or more equally discrete groups of zooflagellates (Table 5-1). In them, plesiomorphs are identifiable with varying degrees of facility, most being somewhat conjectural. And then relations between these plesiomorphs are yet more conjectural, apart from certain broad conclusions, such as, that the choanoflagellates and bodonids arose from plant flagellates (to which one can also add the small, problematic group of the Bicosoecida) and that the symbiotic forms (all the remaining zooflagellates) came from some free-living ancestral forms or from further development of the symbiotic mode (probably both occurred). In any case, the symbiotic forms show such significant discontinuities between themselves that we are faced with a variety of kinetidal variants which are apparently homologous but equally divergent from one another. Aposemic trends are almost impossible to establish.

This becomes clearer when we attempt to determine phyletic distances and a dendrogram.

Phyletic relationships. From Table 5-1 we can see that four semes are available for phyletic analysis. Foremost in importance is the kinetide. The nature of its use in establishing homologies has already been fully discussed, and its essentials are that homologous subparts exist within the kinetide, and that variation among these parts and the compounding of whole kinetides shows aposemic variation but with important discontinuities. Second in importance is the symbiotic mode of life, which in the trypansomatines allows of extensive conjecture on aposemic trends, conjectures which may yet be resolved into a most informative pattern of evolutionary development. In other forms, the evolutionary aspect of host-symbiont relations is less obscure, especially so when we find many of the hypermastigid forms inhabiting either the wood roach or the termite hindgut and with essentially no variation in form between the examples in either host; in fact, they are treated as the same species.

Beyond these two semes there are general similarities in nuclear division and organization of the cell body. These are such broad resemblances that, if their distribution did not match those of the kinetidal and mode of life homologies, their homology would be suspect. (The first accessory criterion is being invoked here to argue in favor of homology.)

Lastly, the spotty occurrence of sexual phenomena and of various special organelles, namely, pelta, rostrum, cresta, cysts, etc., are of little use on the broad scale of zooflagellates as a whole, though they may be very useful within restricted groups. (For example, the structure of the pelta is an important feature in the trichomonadids (Honigberg 1963).)

Hence, we end up with four useful semes (kinetide, mode of life, nuclear behavior, gross cellular organization) and two traits (special organelles, sexual processes) of limited use. This is, at best, a meager basis for calculating phyletic distances, especially so since we argued earlier that the presence of about 10 or more semes is needed for sound conclusions. Add to this the difficulties we have had in finding a plesiomorph for the zooflagellates, and one can see that the phylogenetic analysis of this group is, at present, a very difficult problem. For these reasons, we will have to forego calculation of phyletic distances (R values) and the consequent formulation of a dendrogram. What remains is to understand why these limitations on zooflagellate phylogeny exist, and what, if anything, can be done to solve them.

PHYLOGENETIC CONCLUSIONS

If we look at Table 5–1 again we can see that our discussion of the collared flagellates, kinetoplastids, and trichomonadids has covered only 14 of the 34 families listed there. Of the others we are now asking: What of their evolution, and specifically why are the majority of them symbiotic and, indeed, mutualistically associated with termites and the wood roach? First, there is as yet no complete answer to this question. Grassé (1952a) suggests that his systematic ordering of the zooflagellates reflects separate evolutionary trends. His language vividly sketches the situation: "L'évolution des Zooflagellés symbiotes des Blattes et des Termites ... Elle a été buissonnante." (p. 955). This radiative explosion is suggestive of success in exploiting the symbiotic mode of life, but underlying it is the organizational potential of the kinetide and its kinetosome-centriolar control center. Both Grassé and Kirby (1937, 1949) find evidence that the flagellates have evolved significantly within their hosts. This relationship of host and symbiont has not resulted in the reductive specialization of the symbiont as so commonly happens, "... these higher zooflagellates have indulged in a profligate elaboration of structure" (Pitelka 1963, p. 143.) Why? What have been the selection pressures behind this? We do not know for sure. Grassé (1952a, p. 955, original in French) cryptically notes that "... the most archaic forms (*Foaina* for example) live side by side, even in the same hindgut, with the most highly evolved forms," and "... this fact is pregnant with theoretical implications."

From the vantage point of phylogenetic theory, the following three implications seem inherent in the zooflagellate state of affairs. First, there is a significant potential for evolutionary development within this group, which has its structural basis in that extraordinary complex of organelles called the kinetide or mastigont. Not only has a minimal kinetosome-plus-flagellum structure become polymerized, but it has become conjoined with a constellation of other structures (parabasal, axostyle, etc.), which perform as a morphogenetic unit and probably confer special, but as yet largely unspecifiable, physiological capabilities on the cells containing them. One such capability is specifiable, and that is the centriolar activity arising from the kinetosome itself or from the atractophore. In this way, the behavior of nucleus during division is related to the kinetide. And this leads to the karyomastigont, where a kinetide and nucleus form a unit of organization, function, and development. Cells such as *Metacoronympha senta* (Fig. 5-1 D) are seen with polymerized karyomastigonts, resulting in a multinucleate situation, which characterizes the Calonymphidae in general.

Second, the discontinuous nature of zooflagellate diversity argues for significant diversification prior to invasion of their various hosts. This is not so convincing for the trypanosomatines, where the current debate centers on paths of evolution within their hosts, but it seems to be the case for most of the others, which are located in the rectum of frogs, lizards, and reptiles, and also the urogenital tract of various warm-blooded forms and in the lower intestine of various insects, notably the wood roach and termites. If this conjecture is correct, then our chief problem is to learn what eliminated the free-living forms that presumably initiated development of the kinetide and were the natural reservoir from which the various hosts were invaded. A common cause of extinction in nature is competition; hence the question, What were the competitors of the free-living zooflagellates? There is a plausible answer to this when we come to examine the ciliated protozoa.

And third, we can provide an overview of the phylogeny of the zooflagellates in the following terms. Their origins are polyphyletic. Perhaps every free-living group, i.e., the choanoflagellates, the bicosoecids, and the bodonines, each had a separate origin—the collar cells from chrysomonads and the bodonids from euglenoids, with the bicosoecid origin still unresolved (Grassé and Deflandre 1952). And between these free-living forms and the symbiotic remainder of the zooflagellates there is a large gap, due to the suggested elimination of the free-living forms which presumably evolved from the simpler free-living ones. Thus the present day symbionts are seen as an array of a half-dozen or more relict groups, isolated in specialized habitats, and having undergone their own evolution starting from the different genetic heritages present in those of their free-living ancestors that first successfully invaded one or another multicellular host.

Chapter 6
Rhizoflagellates: The Ameboid Forms

FROM a phylogenetic point of view, there was something prophetic in the name Linnaeus (1758) gave to the first systematically treated ameba, *Chaos chaos*. Though the trained eye of a specialist distinguishes between amebas with convincing authority, the use of the regular classificatory traits is of little use phylogenetically. This can be made clear after reviewing the organizational and functional aspects of these organisms.

General Characteristics

As indicated earlier, certain ameboid forms—radiolarians, acantharians and heliozoans—will be treated separately from the rest of the amebas with which we are now concerned. Our primary interest lies in such forms as the amebo-flagellates, the very familiar naked amebas such as *Amoeba proteus*, the shelled amebas like *Difflugia* and also the Foraminiferida. Only slight attention will be given to highly specialized amebas such as the slime molds, the labyrinthulids, and the xenophyphorids.

Though the neophyte in ameboid taxonomy would tend promptly to agree with Dangeard's (1900) lament, that "Rien n'est plus difficile, en effet, que de déterminer une amibe," the trained eye does discriminate among a variety of consistent traits and thus differentiate those many morphological and physiological types designated as species. For example, Bovee (1970) provides a key to a suborder of amebas in which he used sixteen different traits to describe some thirteen new or poorly-known species. These traits include the following: cell size and shape, pseudopodial conformation, rate of locomotion, uroid conformation (form of the posterior end of a moving ameba), ectoplasmic markings, character of the endoplasm, nuclear structure, water expulsion vesicle (contractile vacuole), gas vesicle, "pseudovesicles," food vesicles, crystals, and granules, symbionts or parasites, cysts, division stages, and habitats. In that these amebas possessed no shells or tests, that trait was necessarily omitted, but for the amebas as a whole it is an additional, important trait. Also, life cycle patterns and sexual processes can be added to the list. But also, the variety of vesicles that Bovee has examined is not a consistently useful trait. However, overall, it is obvious that a significant number of complex traits is available not only for taxonomic studies but also for the purposes of phylogeny.

STRUCTURE

Pseudopodia are the most distinctive attributes of ameboid cells, and this survey of ameboid structure will start with them and include some comments on cell form and shape. Then nuclei and other intracellular organelles will be discussed with, finally, some comments on cysts.

Pseudopodia and cell shape. Some amebas, such as those of the limax type, are monopodial and the cell is a simple cylindrical form with generally rounded ends. Other amebas show multiple pseudopodia. Furthermore, pseudopodia can be broad and blunt or slender and pointed. Either type may branch, though bifurcations are rare in the broader pseudopodia. Formation of networks of pseudopodia due to anastomosis can occur; these are almost exclusively found in the slender pseudopodia. A heavy, blunt pseudopodium or pseudopod, is called a lobopodium. The slender ones are filopodia. Among the latter are reticulopodia, applied to anastomosing filopodia, and axopodia in which there is a relatively stiff unbranched axial filament composed of microtubules (Anderson and Beams 1960; Kitching 1964; Tilney 1968). The lobopodia may be composed of both granular endoplasm and clear, outer ectoplasm or may be formed exclusively of ectoplasm. The filopodia may be clear or may be granulated. The varieties of pseudopodia (Fig. 6-1) are described

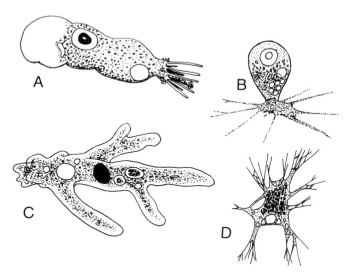

Fig. 6-1 Selected examples of pseudopodia. A. Monopodial lobopodium of *Vahlkampfia inornata*. (Redrawn, by permission, from Page (1967).) B. Filopodia of *Chlamydophrys* sp. (Redrawn, by permission, from MacKinnon and Hawes (1961).) C. Multipodial lobopodia of *Amoeba proteus*. (Redrawn, by permission, from MacKinnon and Hawes (1961).) D. Reticulopodia of *Penardia mutabilis*. (Redrawn, by permission, from Grasse, (1953) *Traité de Zoologie,* Vol. I, part II, Masson et Cie.)

in detail by Chatton (1953) and Deflandre (1953b) as seen by light microscopy, and they are relatively constant in a given form.

Beyond this basic description of pseudopodial types, one needs to know how variable the various types of pseudopodia may be. Do they blend into one another? Do different types of pseudopodia appear in the same cell? at the same time? at different times? What elicits their formation? The point of these questions is, of course, to learn the complexities of this trait, which promises to be a major source of semic information.

At this point there must be interjected comments on the physiology of pseudopodia, for their form is related to their mode of operation. The history of ameboid locomotion is a classic one in protozoology, from several points of view. It has engaged a distinguished roster of researchers both within and without the field of protozoan studies, for it represents a basic pattern of cellular motility (Allen and Kamiya 1964). It has engendered controversies that are still unresolved today, and in their wake there are more wrecks of scientific models than perhaps any other area of biology, with the possible exception of the related problem of muscle contraction. And, finally, ideas on ameboid locomotion are continually influencing other areas of protozoology, notably systematics. The best example of this is the total revision of all ameboid protozoa, i.e., the "Phylum Sarcodina" (Bovee 1970) that has come from Jahn's and Bovee's view (Jahn and Bovee 1965b, Bovee and Jahn 1965) that there are two fundamental mechanisms of ameboid motion, namely, the contraction-hydraulic and the shearing-flow systems. We will say more regarding these mechanisms when ameboid function is under review, below. This systematic revision is useful as a phenetic classification, though there are many who do not accept it as yet (Honigberg *et al.* 1964; Page 1968), but its implications for phylogeny are limited since it makes no attempt to recognize convergences, if they occur, and disregards whether or not other traits may show evolutionary variability. (As a phenetic classification, the phylogenetic concerns are, of course, not at all the issues being dealt with by Jahn and Bovee.) Hence, recognizing only two patterns of ameboid motion within the larger variety of pseudopodial form reduces the information content of pseudopodia from a phylogenetic point of view, which is counterproductive for our present efforts. We need therefore, to keep the varieties of pseudopodial form in front of us for the present.

The same species of ameba can show more than one type of pseudopodium, but does so in a very limited way. The small fresh-water amebas, for example, can exhibit a single lobopodium when moving along the substratum, but, when floating, they can be stellate with many extended, slender filopodia (Page 1967; Bovee 1970). Or, the lobopodial forms can show various types of lobopods: large blunt clear ones, which advance as a unit, or which show many small eruptive waves along their advancing edge, or the advancing edge can be parallel to the longer axis of the cell (the cell appears to move sideways), and so forth. But these differences usually characterize different species. Therefore, though a single species can show more than one type of pseudopodium, whatever type or types they show often evince a characteristic pattern within the given types expressed (Schaeffer 1926; Page 1967, 1968; Bovee 1970).

Pseudopods can be withdrawn for various reasons. Chemical or physical disturbances, onset of cell division, or encystment or transformation to a flagellated form are all instances where pseudopod formation is suppressed. It is, therefore, essential to consider those amebas whose behavioral repertoire has been extensively studied. Only from such studies does one learn the morphological variations available to a given morphologically defined species and thus provide sound and comprehensive taxonomic descriptions. And it is just such information on which the phylogenist is totally dependent.

The variations in pseudopodial form also correlate to a large extent with the varieties of cellular form. This is especially true regarding the uroid. This is the posterior end of an ameba, as defined by the direction of predominant pseudopodial flow. In stellate forms, or forms in which the cell body is largely enclosed in a test, the uroid is a meaningless concept. But in forms with one or a few pseudopodia and showing the contraction-hydraulic pattern of locomotion, there is a posterior area (Fig. 6–1 A and B) that can show characteristic protoplasmic formations ranging from a smooth rounded end, to a bumpy surface, to various types of spines and even pointed filaments. There may also be accumulations of particulate material on the surface of the uroid.

Nucleus. Next to the pseudopodia, the nucleus has received the most attention from morphologists and systematists concerned with ameboid forms. Page (1967), in a succinct historical survey, has divided systematists into those emphasizing details of nuclear structure and division and those emphasizing pseudopodial form and function. Both groups have met with success. Starting from Pénard (1902) with Volkonsky (1931), Singh (1951, 1952), and Page himself, among others, also using this approach, there has been considerable clarification of differences among amebas. But there have also been the usual problems coming from overdependence on fixed and stained material, e.g., limited observations on the total variability of the form, artifacts, and problems arising from misapplication of the technical limits (resolving power) of light microscopes. Page's work seems, on balance, to be a model of integration of the two schools of thought. His conclusion (Page 1967, 1968) that nuclear patterns correlate well with pseudopodial patterns allows Bovee (1970) to proceed confidently with his emphasis on pseudopodial analysis as a basis for species differentiation.

There can be one or many nuclei per cell; the larger cells, as a rule, are the multinucleate ones. Chatton (1953) distinguishes three categories of mitosis, as follows: (a) Promitosis—characterized chiefly by persistence of the intranuclear centrosome, anaphase contraction of the caryosomes, and persistence of the nuclear membrane, at least to anaphase. (b) Mesomitosis—differs from promitosis chiefly in loss of the carysome at early prophase. (c) Metamitosis—the centrosomal material (and centriole) are extranuclear. In all three types of mitosis the classic steps of pro-, meta-, ana-, and telophase are present. The caryosome, which does not take the Feulgen stain, seems to be a nucleolus, in modern terms. Within each of these three categories, many individual variations are found. In fact, acute observation will reveal something unique about every division, but variation between members of a species and the trivial nature of some of the differences renders them of limited value as stable traits. The more generally useful features are chromosome size, shape, and number as determined usually at metaphase; shape and structure of the spindle, of the centrosome and centriole; and behavior of the nuclear membrane and the nucleolus. The interphase nucleus can show various distinctive distributions of the chromotin, nucleoli, and other intranuclear bodies (esp., Bĕlař 1926), but again problems of individual variation as well as variation in culture conditions can affect these features, so that they are not always useful.

In fact, in some forms, the pattern of nuclear division is still unknown despite intensive work. For example, the giant, multinucleate ameba *Pelomyxa palustris* does not seem to undergo mitosis at all, but may undergo a "nuclear budding" (Daniels and Breyer 1967). Electron microscopy has not resolved the problem in this form; rather it has raised a variety of further problems in revealing that the nuclear membrane is lamellar and that there are several curious intranuclear structures there not typical of eukaryotic cells (Daniels and Breyer 1967; Andresen, Chapman-Andresen, and Nilsson 1968). Also, despite a general willingness to work with Chatton's categories of pro-, meso-, and metamitosis, workers are continually finding forms which do not fit comfortably into these groupings (Houssay and Prenant 1970; Page 1968). Page (1968, p. 24) states that, "Only general patterns of mitosis, rather than minute details, should be included in systematic criteria."

Other cytoplasmic organelles. Included here are various vesicles, crystals and symbiotic inclusions, and special structures such as chromidia. Again, such standard organelles as mitochondria, endoplasmic reticulum, ribosomes, and Golgi apparatus will not be considered because of their uniformity of structure and distribution. Flagella will be considered not as organelles in their own right, because of their uniformity, but as structures identifying stages in life cycles, i.e., flagellated zoospores or gametes, and will be mentioned later.

The cell membrane, or plasmalemma, of ameboid forms is potentially of importance for phylogenetic studies. More exactly, the mucoid layer, or glyocalyx, lying on the external surface of the plasmalemma, is the structure of special interest. But in that it appears to be secreted by the plasmalemma, the two entities can be considered together. In some forms the glycocalyx is absent (Daniels and Breyer 1967). In others, which is apparently true for most amebas (for example, Pitelka 1963; Fauré-Fremiet and André 1968), there is a significant surface glycocalyx, and it appears to play a significant role in food capture and ingestion (Chapman-Andresen 1972).

Vesicles, special inclusions, and unique structures are as yet of limited use for phylogeny. Either, as with vesicles, they are inadequately studied in a comparative way; or they are restricted in distribution, as with chromidial bodies in the genus *Arcella;* or else their occurrence, as with crystals or symbionts, may represent special physiological conditions or special pathogenic invasions, and in either case they seem of limited value in understanding the evolution of the cells in which they occur. Hence, though the presently available information on these traits may be providing useful characters for identifying given cells in a taxonomic sense, it cannot be usefully discussed further in terms of semic information.

This same conclusion applies to the general character of the cytoplasm, often used taxonomically, i.e., is it granular, opaque, refractile, and so forth.

Cysts and tests. Cysts are common among amebas, especially the parasitic forms. These coverings of the cell are laid down externally, i.e., secreted, and are transitory as compared to the relative permanence of the tests. Cysts are single chambered and without external opening until germination occurs. Their function can be fourfold: protection, especially against dessication; reproduction, as when fission occurs within the cyst; digestive, well-fed cells encyst in some forms while breaking down the contents of their food vacuoles; and sexual, in some cases it appears that fusion of cells and of their nuclei (i.e., fertilization) occurs within cysts. Cysts often show characteristic color, opacity, size and shape, contours, and thickness and texture of their walls—all sufficiently stable to be useful in many cases of identification of species. There seems to be no report, however, which has used the diversity of cysts to infer anything about the evolutionary history of cells capable of encystment.

Just the opposite is true of tests, and, in the case of Foraminiferida, the extensive phylogenetic speculations among these forms is based almost exclusively on the test. But speaking generally, first, regarding the shells or tests which are common among amebas, Deflandre (1953c) generalizes as follows. Tests can be of organic matter, often chitinous, secreted by the cells themselves, or there can be silicates or calcium carbonate in addition to the chitinous matter, or the tests can be made purely of inorganic minerals. All such tests formed by the cells themselves are termed endogenous or autogenous. In contrast, some tests are made of materials picked up from the environment, such as sand grains or discarded tests of other forms. These are exogenous or heterogenous tests, though they may contain an endogenous component,

Fig. 6–2 Phenotypes of *Difflugia corona* established by selection in the laboratory. (Redrawn, by permission, from Jennings (1916).)

namely, the organic matter in which the foreign materials are embedded. In addition to these compositional variations in ameboid shells, there are variations in shape, size, openings, ornamentation, and structure of the shell walls, and number, size and placement of chambers. In those with multiple chambers or locula, the number and shape are very constant, within a species.

The problem with the tests is not what they are, but what they do. Tests are easily collected and described, and from that point of view are a rich source of information. But to be more meaningful, in an evolutionary sense, that information must eventually be amenable to discussion in selectionist terms, and that raises the question of function, which will be discussed later on. There remains, though, the question of species specific pattern. The answer appears to be developmental and genetic. Thompson (1942) has argued that the apparent selectivity of *Difflugia* (Fig. 6–2) in aggregating sand grains of a particular size for its tests, has nothing to do with the esthetic impulses of the ameba, but rather is probably a consequence of surface energies of the ameba which allows only grains of a certain size to adhere, and they pack beside each other in terms of thermodynamic constraints, i.e, in patterns of minimal free energy. The same could be urged in terms of patterns of platelet orientation in the tests of *Euglypha* (Netzel 1972c) or the chamber formation of foraminiferans. Indeed, as we shall see when we come to reproductive functions, formation of tests does in fact appear to be a consequence of deposition around a cytoplasmic anlagen, or template. Test formation and structure is simply, then, a reflection of cellular organization to achieve certain functions as yet only dimly perceived by us.

FUNCTION

Examination of the functional aspect of amebas will start from a consideration of ameboid locomotion. This motion is a necessary prelude to understanding the exploitive abilities of the cell, for it is the basis of an almost continual search for food, and it is often at the pseudopodia that food capture and also pinocytosis occur—both are important in the uptake of nutrients. Then there will follow an examination of homeostatic and reproductive functions, with the latter including the complex subject of life cycles.

Ameboid locomotion. The puzzle of how they move has elicited more research on the ameba than any other aspect of their biology. It now appears that there are at least two major types of intracellular movement in the ameba. (Note that we are not now speaking of locomotion of the cells, but of movement of materials in their pseudopods.) Crudely put, these two types of motion are unidirectional and bidirectional flow. The first term needs immediately to be qualified: it refers to the general pattern of flow in a given area of the cytoplasm. If flow is determined by movement of granules, then, in a lobopod, some granules will flow faster than others, with some moving a little tangentially to the others, and at the tip of such a pseudopodium there can be observed lateral movement of granules. But the general impression is predominantly that of one major direction of movement. This is to be contrasted with bidirectional flow, where in one and the same pseudopod (and this is especially dramatic in thin filopodia only a few micrometers in diameter) equivalent numbers of granules are moving with equal speeds in opposite directions.

These two types of motion, in view of their proposed mode of operation, have been termed the contraction-hydraulic and the shearing-flow systems (Jahn and Bovee 1964, 1965a, 1965b, 1967; Bovee and Jahn 1965). The first is an updating of ideas stemming from Mast's (1926) gel-conversion theory, which in turn leads back to observations and ideas of workers in the 19th century. The constellation of major names associated with this problem (see review by Schaeffer 1920) gives some indication of the continuing deep fascination it holds for biologists, and it is a problem still not understood in many of its important details. The contraction-hydraulic point of view, briefly, conceives of amebas as capable of forming tubes of gel-like cytoplasm, and these tubes, which are open at one end and closed at the other, contract and force the fluid (sol state) cytoplasm within the tube out of the open end by hydraulic pressure. This view accounts for the unidirectional flow referred to above. Of course, to convert such intracellular movement into locomotion of the ameba, the role of the clear ectoplasm, the bounding membrane of the cell, and relations with the substrate must also be considered. And in the case of slime-mold plasmodia, it is proposed there are local sites of contraction ("origins"— Jahn and Bovee 1964, and earlier) which determine the direction of flow in the plasmodial channels.

Needless to say there have been numerous alternatives to the contraction hydraulic point of view. These include models wherein the force generated by the amebas is not hydraulic and is located at the anterior end of the pseudopod (Allen 1961), or models dependent on the action of the endoplasmic reticulum (Kavanau 1963). An extensive survey and discussion of these problems plus observations that fit no model can be found in Allen and Kamiya (1964). It is quite beyond the aims of this chapter to try to evaluate the various theories of the physiology of ameboid motion. We have commented on one such theory; there remains now the other major one of bidirectional motion.

Bidirectional motion is typical of filopodia and axopodia. The narrow diameter of these pseudopodia, and the fact that the moving granules appear to ride on the surface of these filopodia as visible bumps, argue against any mechanism dependent on a tubular organization of gelled cytoplasm. One view (Jahn and Rinaldi 1959) is that there are gel filaments flowing past each other in a shearing fashion. At the tip of a pseudopodium the filament turns tightly on itself so that there is the picture of a filament moving up one side of the filopodium, making a tight hairpin turn at the tip, and coming down the other side. The motion of both parts is dependent on their mutual interaction as they slip by each other ratchet fashion, and from the fact of their physical continuity. Such motion is used by cells to draw food particles along the filopodia into the cell. There are still debates on the nature of the surface of these filopodia, and, of course, on the mechanical and chemical basis of the moving filaments. This type of motion, the shearing-flow system, seems to function not so much for locomotion of the cell but predominantly to transport, into the cell, material stuck to the filopodia.

Exploitive function. Amebas feed by uptake of food particles into membrane bound vesicles or food vacuoles. This process of phagocytosis is undoubtedly complex (Chapman-Andresen 1967; Conner 1967) demanding energy, formation of membranes, and complex cell-medium-particle surface relationships that are both chemical and physical. A recent review of the problems (Chapman-Andresen 1972) makes clear that great progress has been made in understanding the general cycle of plasmalemma-to-vesicle (either endo- or pinocytic) to plasmalemma transformation. But there remain such further questions as the role of these transformations in locomotion, the role of uroid as a site of considerable endocytic activity, and the mechanism of replacement of the external mucous layer of glycocalyx. For no form has there developed a consistent set of statements providing a detailed explanation of these processes. When such ideas become available, there may well quickly develop some fascinating comparative studies to account for the selectivity of various cells regarding their food. Such food may be living matter ranging from bacteria through unicellular algae to other protozoa, such as other amebas, flagellates, and ciliates. It can also include nonliving material, such as starch grains or cellular debris. Once inside the food vacuole, the material is broken down enzymatically and the products taken into the cytoplasm, where they are metabolized further. Undigested materials are eliminated by eversion of the food vacuole at the cell surface. Though amebas can be cultivated on single food organisms (monaxenic culture, Dougherty 1953, 1963), they apparently show a wide range of food substances in nature. However, there appear to be no set patterns to their feeding habits, since their degree of fastidiousness varies from one species to another.

Dietary fastidiousness can also refer to growth factors. Some of the most narrow requirements for growth occur, not unexpectedly, in the parasitic forms (Johnson 1956). On the other hand, certain of the soil amebas, e.g., *Hartmanella rhysodes* (Band 1962), are reportedly grown on a mixture of seven amino acids, three vitamins, and salts, which is obviously a quite simple diet. Various other biochemical studies (carbohydrates: Ryley 1967; nitrogen metabolism: Kidder 1967; lipids: Dewey 1967; growth factors: Lilly 1967) reveal the scattered nature of our knowledge of the comparative biochemistry of even the best-studied forms, the ciliates and the amebas. No general patterns, studied in detail in a wide number of species, seem to be available for phylogenetic analysis.

Similarly, though the basic features of pinocytosis are clear, and though there may be differences in what initiates this "cellular drinking" and in what is taken up by this mechanism in different amebas, that information is not available now (Conner 1967). Hence this is not a trait of phylogenetic significance at present.

At this point the functional role of the test, in particular the foraminiferan test, deserves detailed consideration. Perhaps the first reaction is to assume the tests are protective, for they are also referred to as shells, and this term carries the connotation of protection. Then, too, they could be supportive, in the sense of being exoskeletons. Neither supposition can be validated or invalidated except through critical experimentation. There has been extensive experimentation on rhizopod tests, but this has been devoted to elucidating their morphology, development, genetics, evolution, and even ecology. The latter is the most helpful regarding their function. Our discussion must, therefore, start from ecological considerations, however meager, and then move to plausible inference when factual information is exhausted.

The best ecological work has been done on the foraminiferans. They are most generally described as marine, benthic forms, either crawling over the sea bottom itself, or on plants or other growth attached to the bottom. Only some two dozen species (Le Calvez 1953) are pelagic, floating forms. And, also, only a score or so are not marine. Of these, only a few species of *Allogromia* can be said to be a truly freshwater forms, for the others are in waters of a high mineral content, though lying a considerable distance inland (Loeblich and Tappan 1964). The typical marine, benthic forms are found at all depths in the ocean and at all latitudes. However, as would be expected, certain forms are characteristic of certain areas. This phenomenon of characteristic assemblages has been found for the Okhotsk Sea (Saidova 1960), for the northern Red Sea (Said 1950), for the North American Atlantic Coast and other places (see review by Myers and Cole 1957). In a study (Bandy 1960) largely based on ocean depth, it was also found that very clear correlations exist between shell type and bathymetric location. Distributions correlated with temperature are reviewed by Loeblich and Tappan (1964). Finally, there are interesting findings of large flat shells occurring on heavy mud bottoms. Further from the land the bottom muds are very fine — almost a water suspension — and here the foraminiferans are very small. These seem to be correlations of obvious adaptive significance (Myers 1943; Meyers and Cole 1957). Phleger's (1960) review of ecological and distributional data emphasizes the need for physiological data in understanding the biology of these amebas.

Somewhat similar conclusions are suggested for certain Thecamoebae, namely, amebas with single-celled chambers, often of endogenous organic matter, and living in fresh water. Characteristic biocoenoses, or species groupings of these forms, are found in specific localities in Eastern Europe (Moraczewski 1965) suggesting ecological constraints on their occurrence. That there is some physiological response to external conditions is seen in Heal's (1963) work, which shows that test size correlates with water content of the environment. (Small, flat tests occur in dry moss and large, round ones in water. These seem to be extremes of morphoclines from one species, at least in some cases.)

These considerations have told us little, if anything, regarding the functions of support and protection. In small cells, which are usually living in an aquatic environment, their buoyancy will prevent any collapse, if that is imminent, and hence an exoskeletal inference is extraneous and unconvincing. Protection is more difficult to dispose of. The first question is, protection against what? Predation from microcarnivores is one answer. Foraminiferan tests have been reported in the gut of decapod crustaceans (Le Calvez 1953). But tests will not prevent an ameba from being taken up by a small fish and swallowed, and digestive juices will not be held away from the living protoplasm when pores and openings exist in the test. Also, calcareous tests, as in foraminiferans, will readily dissolve in stomach acids. Protection, if there is any, must lie elsewhere. Food selectivity is shown by the feeding of microfauna. One can watch small crustaceans and annelids and turbellarians take up particles and reject some while retaining others. When disturbed, foraminiferans very rapidly withdraw their filopodia (and also the ectoplasmic covering of their test?), and in such a form might resemble an inorganic pellet to their predators. If this led to their rejection as food, it would be of selective advantage. The test would simply need to be strong enough to resist the initial uptake and subsequent rejection. If it is, then that appears to be the limits of protective function assignable to it.

There is one more possibility deserving mention, regarding test function, and that concerns a role as ballast. It can be seen best in this light by considering feeding in the testate forms. Specifically, regarding foraminiferans, it is clear that they are dependent on nutrient particles entrapped by their filopodia. These particles can vary from bacteria to small metazoans, and probably also include nonliving organic material, such as particles of detritus. The advantages of a test as ballast in this process might be twofold: it may counteract buoyancy, thus keeping the organisms on their substrate where they are maximally adapted to survive; and it may provide deadweight against which the filopodia can pull when they are contracting with relatively large entrapped prey. In this latter case, it would, at first glance, appear immaterial whether the prey were pulled to the cell body or vice versa. But the foraminifera are polypodial and have a considerable array of pseudopodia extended simultaneously; therefore, moving the cell body to food in one pseudopodium would necessitate lengthening all other pseudopodia proportionately. Hence it might be more efficient to leave the other food-gathering organelles undisturbed by collapsing just one, and this might be made more feasible by a central deadweight or ballast. That foraminiferans tend to attach to their substrate is seen not only in the way the filopodia pass along the substratum, but, also, workers comment on how the shape of the test "favors a firmer adherence to the substratum" (Berthold 1971), and there are certain forms, such as *Homotrema rubrum*, that attach the test directly to the substratum.

Homeostatic functions. Cysts are an obvious device for controlling the internal environment of cells, and many amebas, especially symbiotic and freshwater forms, use them extensively. There is not, however, much comparative information on cysts regarding details of encystment, excystment, and other functions related to these structures that would permit their use in terms of phylogenetic comparisons.

There is more information on contractile vacuoles or vesicles, also of obvious regulatory importance to the cell, but most of this is in terms of fine structure, and little that is of comparative physiological use. Again we must bypass a potentially important trait.

And similarly, the role of the cell membrane, or plasmalemma, will also receive only too brief mention. In the amebas this is an organelle of paramount importance for the maintenance of the life of these cells. There are many comparative studies in this area (Conner 1967; Chapman-Andresen 1972), but the focus is on using a variety of experimentally useful organisms to fill out a unitary view of membrane transport and of pino-, endo-, and phagocytosis, all based on membrane function, structure, and transformation to and from vesicles. Such a unified view is emerging, but its common features are stressed over its diversity, and rightly so. But ubiquitous traits, as has been emphasized before, are not of much use in phylogeny.

Metabolic pathways and nutritional patterns are either scantily known in amebas or reveal generalized patterns common to animal cells (Dewey 1967; Kidder 1967; Ryley 1967; Lilly 1967) and hence are not of phylogenetic significance.

Reproductive functions. Proliferation of ameboid cells include asexual reproduction through binary and multiple fission and sexual processes based on gamete formation, fertilization, and subsequent development. In the foraminiferans and slime molds, complex life cycles are known. Certain aspects of binary fission have already been discussed in the context of the nucleus. What has not been discussed are the cytoplasmic events. These become most interesting when we can view them as comparative expressions of the genetic potential of the cell. When we are dealing with the nonshelled, naked amebas, the cytoplasmic events are largely reduced to a rounding up of the cell except for small, blunt surface projections (reviewed by Hanson 1967a) and then subsequent cleavage. When shells or tests are present, then there is on view the formation of a new test. Such a case, "Externalizing the morphogenetic faculties of the cell . . . furnishes, in another way, matter for phylogenetic speculations more objective than that found in nonshelled protists." (Deflandre 1953c, p. 109. Original in French.) Such a view is in accord with Thompson's (1942) early argument that test patterns reflect the underlying physical properties and makeup of the cell and are thus revelatory of the fundamental architecture of cellular systems. The pioneering analyses of phenotypic variation are those carried out by Jennings (1916, 1923, 1941) in *Difflugia*. By visual selection of various phenotypes, Jennings established purebreeding differences in lines of *D. corona* in regard to size of the test, size and number of spines on the test, size of the opening of the test, and number of serrations, or "teeth", in the test opening (Fig. 6–2). Each line showed some variation around its characteristic mean. This illustrates a viewpoint long held by developmental geneticists and entirely consistent with Thompson's proposals, that the genes determine the potential of a system and the environmental context (experimental or natural) allows one or another dimension of that potential to come to expression. This is demonstrated especially clearly by the studies on the teeth in the shell aperture. To understand what is controlling tooth number, the facts of fission must be kept in mind. A new shell is formed for only one fission product, and it is formed in such a way that the aperture of the new shell is in close apposition to that of the pre-existing one. Exact details are not clear, because there is invariably a mass of detritus around fissioning cells (Jennings 1916). Clearer details have been reported for *D. oviformis* (Netzel 1972b). It seems plausible, however, to propose that the old aperture acts as a kind of template for the new one (Hanson 1967a). This would account for the high correlation in tooth number between old and new sister tests. It would also account for Jennings' experimental observations on tests with teeth removed (these teeth were not regenerated); here, to the degree the old test was normal, so was the new one, and to the extent the old one was abnormal, the new one also showed abnormalities but somewhat less so than the preexistent test. The trend to normality continued in successive generations until abnormalities were no longer present. This tendency towards a norm is attributed by Jennings to the genotype. In this sense the test is, indeed, an expression of the "morphogenetic faculties of the cell."

Similar results have been obtained with various species and strains of *Arcella* (Hegner 1920 and earlier; Jollos 1924; Deflandre 1928). In this genus, extensive field work by Hegner and Deflandre show the reality of variations studied in the laboratory. In neither *Difflugia* nor *Arcella* are any sexual processes known. These forms show norms of shape and size that are apparently genetically determined, as seen in their stability under various culture conditions, or their persistence after the selective forces which fostered the phenotype are removed, and, hence, can be termed species in a manner completely analogous to morphological norms of sexually reproducing forms. As we saw in Chapter 4, asexually reproducing forms have the same sort of variability that is associated with sexually reproducing forms (Sonneborn 1957; Dogiel, Poljanskij, and Chejsin 1965).

Recent studies on test formation in *Arcella* (Moraczewski 1965; Netzel 1971; Netzel and Heunert 1971) show clearly that this endogenous test is laid down on or close to the surface of a cytoplasmic extrusion provided by the cell body extending from the opening or pseudostome of the old test. A similar process occurs in *Centropyxis* (Netzel 1972a) and in the foraminiferans (Angell 1967; Berthold 1971). Differences in these processes depend on the nature of the test (organic or inorganic material), whether or not it is single- or multichambered, and on the role the pseudopodia may play in forming the extrusion, or template, which determines the form of the new test. With the advent of techniques for sectioning the test and new ways of viewing surface detail with high resolution (scanning electron microscopy), more and more detail will become available on the developmental processes of test formation. This has already been used extensively (e.g., Loeblich and Tappan 1964) regarding the taxonomy of the Foraminiferida and is useful for phylogenists, as will be seen below.

Multiple fission seems to differ very little from one ameboid form to another. The sequence of events is prior nuclear division to provide a multinucleate cell and then subsequent cytokinesis resulting in many daughter cells. However, the process is a distinctive one, and when it occurs in a given place in a life cycle, as is often the case, not only in ameboid forms but in the parasitic Sporozoa, too, it is a useful character in comparing various life cycles.

Life cycles are illustrated by the foraminiferan *Iridia lucida* (Fig. 6–3). Besides posing a fascinating variety of developmental problems in the transition from one cell type and state of nuclear organization to another, such a life

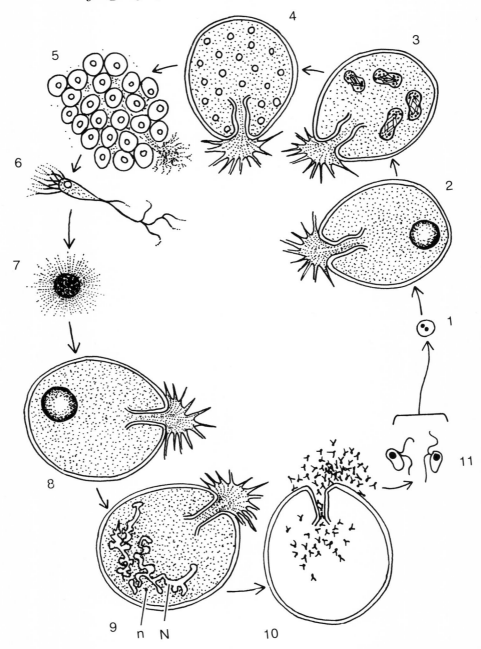

Fig. 6–3 Life cycles of *Iridia lucida*. 1. The zygote, with nuclei beginning to fuse. 2. Diploid schizont. 3. Mitotic nuclear divisions signalling the end of the vegetative period. 4. Mature schizont. 5. Following cytokinesis, uninucleate embryonic cells are released. Each one is reportedly haploid. These cells are first ameboid and bottom crawling (6) and then become spherically symmetical pelagic forms (7). 8. Transfor-

cycle provides certain items of phylogenetic information. In forms such as *Iridia*, it appears that the flagellated gametes only fuse if they come from different parent cells or gamonts. This ensures a certain degree of outbreeding, which is further enhanced by the extensive development intervening between fertilization and the next generation of gametes, which could allow dispersal of these forms. However, the foraminiferans are not very active forms, typically crawling slowly over their substrate in search of food. Dispersal is dependent, then, on that movement and such other as can occur in the flagellated gametes or the relatively active ameboid forms which appear between the agamont and gamont stages (Fig. 6–3). *Iridia* is unusual in having a spherical form with axopodia interjected into the life cycle (Fig. 6–3). In other foraminiferans, though the alternation of haploid gamont and diploid agamont is still present, various events of the sexual process are changed. In *Rotaliella heterocaryotica* (Grell 1956), there is no flagellated gamete, but, rather, gametes are formed by multiple fission of an initially uninucleate gamont. The fusion of gametes from the same haploid parent cell is a form of autogamy and exhibits, therefore, the strictest possible inbreeding. This has also been reported for *Allogramia laticollaris* (Lee and McEnery 1970). In some forms, the formation of flagellated gametes is suppressed, but the forms are nonetheless outbreeders. For example, in *Spirillina vivipara* two gamonts unite (plastogamy) and induce formation of ameboid gametes. Fusion of these gametes occurs only between products of the different gamonts (Myers 1936; Le Calvez 1953), thus assuring the opportunity for outbreeding. And in some of the Foraminiferida, going to the other extreme, the sexual phases of the life cycle are completely suppressed (Le Calvez 1953). This array of possibilities from outbreeding to autogamous inbreeding and to loss of all sexuality, is parallel to the array of possibilities found by Sonneborn in *Paramecium*. In the foraminiferans, however, the geographical distribution of the species and population sizes are not well known. Such information would be an interesting adjunct to the knowledge of breeding patterns.

As will be seen later, the slime molds present a variety of specializations, and it is easier to discuss them briefly as a group, apart from the generalizations we are examining here. It suffices for the moment to acknowledge their protozoan nature (Olive 1970) because of the phagotrophy of the amebas of the cellular slime molds and of the plasmodium of the noncellular forms. The parasitic nature of the Plasmodiophorales makes it difficult to decide whether or not they are phagotrophic. In these three groups, there is a de-

---

mation into the haploid gamont and return to a benthic habit. 9. Mature gamont with a macronucleus (N) and a micronucleus (n). 10. Dissolution of the macronucleus and multiple divisions of the micronucleus, followed by cytokinesis, result in release of flagellated gametes. 11. Two gametes (enlarged) and from different gamonts, prior to fusion. (Redrawn, by permission, from Grassé (ed.) (1953) *Traité de Zoologie,* Vol. I, part II, Masson et Cie.)

tailed life cycle involving spore formation in the fruiting bodies; the structure of the latter is the basis for the not uncommon classification of these forms among the fungi. Also there are other special groups that have been considered ameboid and grouped with the forms being studied here, namely the Xenophyophorida and Labyrinthulida, and the symbiotic *Piroplasmea* (Honigberg *et al.* 1964). And to these could be added the Stereomyxidae (Grell 1966). These we will touch on again later to indicate special problems relating to their phylogeny and will forego any discussion of their organization of functional capacities at this point.

SEMES

On present information there are seven potentially semic traits in the ameboid forms being discussed here. These include (1) pseudopodial form, (2) pseudopodial action, (3) nuclei and their division, (4) test morphology, (5) cysts, (6) life cycles, and (7) ecological data. Commenting briefly on each of these will bring us to certain other traits which also have the potential of being semic. A common problem with all of these traits is insufficiency of data.

Variety of pseudopodial form is seen in the categories of lobo-, filo-, and reticulopodia. These also determine, in large part, the nature of cellular form in these organisms. Altogether, then, one can consider the three-dimensional shape of the cell, the number of pseudopodia, their form, whether or not they branch, whether or not more than one pseudopodial type is present in a cell, position of the nucleus, and nature of the uroid as a constellation of data amenable to detailed interspecific comparisons. But, to anticipate a serious analytic problem, having semic complexity does not guarantee that the Remanian criteria for homology can be readily applied.

Patterns of ameboid motion can be followed by examining (a) pseudopod formation, which can include the initiation, extension, and formation of the advancing tip of a pseudopod, and also retraction; (b) flow of material within the pseudopod—directionality and distribution of moving cytoplasmic particles, rates of movement, transport of food or other larger materials in the pseudopod; and (c) pseudopodia and food capture, test formation, or other special activities. These three categories with their subpoints could well provide sequential processes with sufficient numbers of unique subprocesses to allow useful comparisons between species.

Nuclear structure is most informative when seen in terms of the details of nuclear fission. Again, dealing with a process, we need a temporal sequence of points. Here, the stages of mitosis are the obvious basic reference points, and then, in relation to pro-, meta-, ana-, and telophase, there can be determined the behavior of intra- or extranuclear centrioles and centrosomes, the dissolution of nucleoli (caryosomes), the behavior of the nuclear membrane, and the behavior of the chromosomes themselves. As mentioned above, the differences in these features are so distinctive that much use of these differences has already been made in the systematics of the ameboid forms, with pro-, meso-, and metamitosis being three categories of differences presently in use.

Tests need little comment, for their semic content is straightforward: their spatial structure can be compared through examining gross morphology, including internal chambers, if any, and by locating special points such as openings, pores, and various protrusions; their chemical makeup can also be compared.

Cyst morphology might well be considered on present information as semic. But its limited use in taxonomy, except in the symbiotic forms, suggest that at present this trait is of only special use in phylogeny, appearing as neosemic among a relatively few species of amebas. If this is too conservative a treatment of this trait, or if, in fact, two semes can be justified here, i.e., cyst morphology and cyst formation, then that is all to the good, for some more complex traits are needed if the recommended number of semes (about 10 to 20) is to be reached.

Life cycles are unfortunately incomplete. Loeblich and Tappan (1964) state that only twenty-some species of Foraminiferida have completely known cycles. But cycles are also known reasonably well in symbiotic forms, and in the remaining asexual, free-living amebas, the life-cycle is simple, consisting of vegetative growth and division and occasional encystment and excystment. This diversity of cycles, plus the fact that more work is accumulating in this area all the time, leads one to the optimistic inclusion of ameboid life cycles among the semic traits available here.

Finally, the diversity of niches exploited by ameboid forms allows the nature of freshwater ponds, of soil, and plant materials, of different hosts, and of various marine environments, to be used as a semic trait.

Also, test formation could eventually emerge as a seme. The present limited number of careful studies—not really more than half a dozen—show that the extruded cytoplasmic anlage, its relation to pseudopods, its fine structure, the patterns of secretion, deposition and joining of secreted substances, of relations to underlying cytoplasm—all of these can contribute to formulating a semic sequence of events.

Further work on the nature of intracellular symbionts, on vesicular structures, on metabolic pools and pathways, and especially on the plasmalemma and vesicle formation may well reveal semic information.

Eventually, we can believe that sufficient data will be available for phylogenetic analysis of the amebas, but at present the stockpile is meager and contains special difficulties, as we will see in the following phylogenetic analysis.

Phylogeny

The foregoing makes clear some of the heterogeneity that exists among the amebas and also brings out the limited areas wherein that diversity is found—pseudopodia, nuclei, tests, cysts, life cycles, and niches. The next problem is to see if, on the basis of these characters, we can find various homogeneous groups within which to start the search for homologies. As has been indicated earlier, a helpful context in which to search for homogeneous groups is the work of systematists.

TAXONOMY

Some decades ago Schaeffer (1920) introduced his discussion of the taxonomy of the amebas as follows:

> "A systematic study of the free-living amebas is beset with some peculiar and formidable difficulties. Among these may be mentioned the following. Amebas do not possess permanent differentiations of form such as are found in most other organisms. The life history of very few species is known satisfactorily; in fact it is doubtful whether there is a single species whose life history is known completely. Permanent mounts of many morphological characters can not be made; with the death of the ameba many characteristic structures disappear, and other interesting details of morphology can only be observed in the living, moving ameba. Many species are known only in the vegetative or trophic stage and probably pass through other stages only rarely. A large number of species do not occur abundantly in nature and can not be cultured artificially by methods now known, in sufficient number, to make studies of life cycles and nuclear events profitable. And finally the literature is burdened with a very large number of inadequately described species of which there are, of course, no "type" specimens in existence. While it is generally held that each group of organisms presents its own difficulties and requires methods peculiar to that group for the solution of its systematic problems, the amebas present a difficulty in the description of their morphological structure which is of a wholly different nature from that of any group of free-living organisms: viz, the shape of an ameba can not be shown by a single picture but only by a series of pictures . . ." (p. 8)

Though each of Schaeffer's concerns regarding an effective systematics of the amebas has had improvements rendered on it during the past decades, the tenor of his remarks accurately reflects the nature of the problems yet remaining with us.

Using three aspects of the morphology of the pseudopodia the French workers (Grassé 1953) and the Americans (Honigberg *et al.* 1964) have recognized three major groups of ameboid forms, but with certain important divergences in detail. The Lobosa (following the French nomenclature) include those forms with lobopods. Within this class are included the orders of the naked amebas (Amoebaea) and the testate forms (Testacealobosa), testifying to the taxonomic usefulness of presence or absence of tests. Within the next class, the Filosa, are forms with filopodia and again there are forms without (Aconchulina) and with tests (Testaceafilosa). In these organisms the filopodium is ectoplasmic, that is, clear and free of granules. By contrast, in the third and final class, the Granuloreticulosa, the filopodia are characteristically furnished with granules which run in currents throughout the filopodia and the reticulum, or network, formed by their extensive branching and fusing with one another. And within this class there are again orders recognized by the nature of the test; it is absent in the Athalamia, present in the Thalamia as very simple shells of chitin or heterogeneous, exogenous materials, and present in the Foraminiferida, a large group (perhaps numbering one half of the known protozoan species (Levine 1962)), usually with many chambered tests that are basically chitinoid but with calcareous, arenaceous, or siliceous materials also present.

At this point the French workers go to another superclass (the foregoing are all included in the superclass Rhizopodes) for the Actinopodes (combining the Radiolaria, Acantharia and Heliozoa), and state frankly their indecision as to whether the slime molds are fungi or protozoa. The American workers, on the other hand, group the typically ameboid forms under the class Rhizopodea, within the superclass Sarcodina, and then proceed to include the slime molds with the rhizopods. The Actinopodea are accorded a class status along with the Rhizopodea, and then there is added a third class, the parasitic Piroplasmea. These treatments of these forms are summarized in Table 6-1.

Another approach to the systematics of the amebas is based on the two modes of cytoplasmic motion mentioned earlier. According to this view (Jahn and Bovee 1965b; Bovee and Jahn 1965) there are two classes of ameboid forms: Hydraulea and Autotractea. The former includes forms with tubular or polytubular pseudopodia in which contraction of the gel tube provides a hydraulic force which is the basis of cellular motion and locomotion. The latter includes forms with slender filamentous pseudopodia (filopodia) in which two-way shearing flow is observable. Within each of these classes are groups reflecting further pseudopodial differences much along the lines of the French workers, in that groups such as the Lobida, Filida, and Granuloreticula are recognized. The actinopods and slime molds also find a place in this classification but are treated somewhat differently, in view of their diverse modes of cytoplasmic streaming, than the treatment accorded them by Honigberg *et al.* This third classification is also shown in Table 6-1.

What, now, can we say about phylogenetic relationships among these forms?

Table 6–1. Three systems for classifying ameboid forms.

from Grassé (1953)		from Honigberg et al (1964)		from Bovee and Jahn (1965)	
Superclasse	Rhizopodes	Superclass	Sarcodina	Superclass(?)	Sarcodina
Classe	Lobosa	Class	Rhizopodea	Class	Hydraulea
Order	Amoebaea	Subclass	Lobosia	Subclass	Cyclia
	Testacealobosa		Filosia	Order	Lobida
			Granuloreticulosia		Acrasida
Classe	Filosa		Mycetozoia		
Order	Aconchulina		Labyrinthulia	Subclass	Alternatia
	Testaceafilosa			Order	Mycetozoida
		Class	Actinopodea		Xenophyophorida
Classe	Granuloreticulosa				
Order	Athalamia	Class	Piroplasmea	Class	Autotractea
	Thalamia			Subclass	Filoreticulosia
	Foraminifères			Order	Granuloreticulida
Superclasse	Actinopodes				Filida
					Hyporadiolarida
					Radiolarida
				Subclass	Actinopodia

HOMOLOGIES

Despite earlier convictions (e.g., F. E. Schulze 1875) that the ameboid forms were, indeed, a homogeneous group, there seems to be no worker now who supports that view. Of course, one's position on this proposition depends on what is included under the phrase "ameboid forms." As we have seen, above, there is no detailed agreement on that problem. Let us put the problem in the form of a somewhat different question: Are there any characters that are demonstrably homologous, in the Remanian sense, among the ameboid forms or among any of their major subgroups?

First, let us direct these questions to the two modes of movement that supply the basis for defining the Hydraulea and the Autotractea. (It should be remembered that Jahn and Bovee have not claimed that the contraction-hydraulic are shearing-flow mechanisms are homologous in the various forms exhibiting them, but simply that they make very useful taxonomic criteria. We are asking the question regarding homology because of phylogenetic, rather than classificatory, interests in these forms.) The fact that the contraction-hydraulic mechanism is found so widely spread in nature (Jahn and Bovee 1967) — ranging from the protozoa into algal forms, being the possible basis for streaming in plant cells, and then being present in a variety of metazoan cells such as lymphocytes, leucocytes, and various embryonic cells — argues that this mode of cellular motion has arisen independently many times over. Its widespread occurrence is surely a convergent phenomenon in at least some of the cases where it occurs. Whether this is true among the ameboid protozoa is one of our present problems. It may be that there is a common ancestral group of ameboid forms, but equally likely is the possibility of multiple, separate origins of the contractile-hydraulic mechanism.

The shearing-flow system is not so widespread as the preceding system. It is reportedly present in the heliozoans, radiolarians, acantharians, proteomyxans, and the shelled amebas with filopodia (Jahn and Bovee 1965b; Bovee and Jahn 1965). It is also possibly present in the slender filopodia of the chrysomonads. However, as will be shown in the next chapter and already discussed by others (Trégouboff 1953b) the Heliozoa, Radiolaria and Acantharia probably have separate origins from each other, and, therefore the simultaneous occurrence of shearing-flow motion is then seen to be convergent. Certainly, the floating mode of life of these forms, especially the last two, is subject to selection pressures, making filopodia more advantageous than lobopodia; and, hence, a convergent solution to that problem would not be unexpected. The occurrence of filopodia in certain pelagic foraminiferans, e.g., *Orbulina universa* (Le Calvez 1953), or an apparently floating stage in their life cycle (Fig. 6–3), speaks to the fact of independent origin of this type of pseudopodium, presumably with shearing-flow cytoplasmic movement. Here, too, then, it is not possible at present to decide what forms with shearing-flow type of motion share it in common as a result of common ancestry, as compared to those that share it from responding to similar selection pressures.

The same quandary results when one looks at pseudopodia in terms of the Lobosa, the Filosa, and the Granuloreticulosa; there is no way to be sure on present evidence that the presence of lobopodia or clear or granular filopodia represents three instances of homologous structures. These traits are not sufficiently complex to permit convincing positive or negative criteria. This is a very important procedural point and needs some extended analysis to make clear what is involved here. To begin with, there is the paradox that pseudopodia are complex enough to be deemed semic traits, but yet are apparently not utilizable for the study of homology. Recalling that identification of homologies depends on similarities of position in the system of which they are a part and on similarity of composition, the analysis of these two categories of similarity is the crux of the problem here.

Establishing positional similarities in amebas will depend on, first, finding at least four unique points in three-dimensional space which in one species will match with another set of the same four points in another species. The logical tactic at this juncture is to transform ameboid form into a system of points. Let us try, referring to a relatively simple monopodial form at the outset. One unique point could be the advancing tip of the pseudopod. Another could be the uroid or posterior end. But what else is there? This is like the two point figure in Fig. 3–7 B, which specifies no unique similarities, for either of two organisms can be rotated, infinitely, around the axis of the two points and each change in position will still allow point-for-point superimpositions on the other organism. We need at least two more points in three-dimensional space. If the ameba's right and left sides could be defined relative to the axis and direction of movement, this would help. This now provides a form, given in Fig. 6–4, which provides four sets of points, A, P, L, and R, for comparison with other forms. (Needless to say, respecting the point made in the last lines of the quotation from Schaeffer, above, i.e., "... the shape of an ameba can not be shown by a single picture but only by a series of pictures," that, therefore, the determination of the placement of A, P, L, and R, will be determined as an average from many measurements from amebas of the species in question when cultured and observed under specified conditions.) However, this is not the end of the problem. First, the four points are in one plane, and one does not know whether one is looking at the ameba from

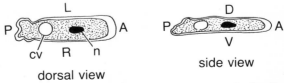

Fig. 6-4 Comparison of the form in monopodial amebas, indicating the difficulties in finding distinctive sites for point-to-point comparisons. A—anterior; D—dorsal; L—left; P—posterior; R—right; V—ventral; cv—contractile vesicle; n—nucleus.

above or below. The practical point, that microscopic observations are always from above (except for occasional inverted optical systems) does not lessen the logical concern that actually points L and R will be equivalent if the monopodial form in question is symmetrical about the A–P axis. This reduces, then, the strength of similarity. And it does not help to find other points along the A–P axis, such as position of contractile vesicle, nucleus, or endoplasmic boundary of the hyalin cap (Fig. 6–4). These do not change a two-dimensional figure into a three-dimensional one. What is really needed is to determine points on the dorsal (D) or ventral (V) surfaces of the cell. Then one would have the coordinates necessary for a three-dimensional comparison.

The second problem is more serious. All of the foregoing is passing by the initial question of defining positional relationships for it is comparing cells with cells, rather than taking parts of the cells to see if similar parts occupy similar positions in the systems in which they lie. The foregoing is really a study of the compositional relationship; it allows determination of whether one cell is composed of subparts (points A, P, L, R, D, and V) which are in the same relationship to each other in each cell. In effect, by comparing cell form, we inevitably compare whole systems and thus omit the first Remanian criterion of positional relationships, and concentrate only on the second—the compositional—relationship. It is as if we are trying to homologize whole organisms by point-to-point comparisons. And when that is done in terms of the distances between such nondistinctive points as anterior and posterior ends, left and right sides, and dorsal and ventral surfaces, we realize similarities can be trivial, because a large beetle could be seen as homologous to a shrew by that approach.

To avoid the simplistic errors seen in comparing beetle and shrew for homologies, we of course utilize prior information, which in the case of the amebas allows us to be sure we are comparing amebas and nothing else. That is, we are using compositional information that assures us we are dealing with unicellular, animal cells with pseudopodia. But these grosser features do not allow for point-to-point comparisons. First, unicellular is a common, invariable feature of most protozoa and so does not help in the phylogeny of amebas. "Animal" refers to mode of nutrition, but that too does not help. And we already have seen the difficulties in analyzing pseudopodia from the point of view of homologies.

The only possible further type of analysis might appear to be the following. Determine morphological similarity of composition and then, as the taxonomists do, introduce other traits to distinguish similarities and dissimilarities among the amebas, as by the presence or absence of tests or granular cytoplasm. This however, introduces a question of analysis whose resolution is not clear, at least to this writer. Namely, how can statements regarding composition be used in conjunction with a system of spatial points to resolve degrees of similarity? It would seem, intuitively, that such is possible, by reducing the statements to unique points simply enclosed within the spatial matrix—a phase space, as some mathematicians and physicists may refer to it. Thus there would have to be point-to-point similarity in space and congruence of nonspatial, descriptive points. This approach does not appeal to any mathematicians queried on the subject and will have to remain in limbo for the present.

What we come to is this: Combinations of morphological and descriptive detail provide sufficient information to classify amebas, but this is essentially a phenetic classification. It is not possible at present to use ameboid form in a rigorous application of the Remanian criteria for homology: in particular, the positional relationship cannot be applied. This does not mean that future work will not resolve the problems inherent in this approach to homology, but only that, at present, the question of homology is meaningless regarding the seme of pseudopodial and cellular form. If other semes are demonstrably homologous, we can then return to the pseudopodia and ask if specific pseudopodial forms are correlated with known homologies. In this way, using the subsidiary Remanian criteria, we can perhaps use the information available in ameboid form.

Turning now to the next semic trait, pseudopodial flow, we can deal with it briefly regarding the possibility of homologies within the various patterns that are possible here. Brevity is due to limited information. Though there is, thanks to Jahn and Bovee (Jahn and Bovee 1965b; Bovee and Jahn 1965; Jahn and Bovee 1967) considerable information on the distribution of the different types of ameboid motion, this is of limited use because of its reduction to two major types. Variations in it are known (Allen 1970) but not well understood, nor are the mechanisms of initiation and retraction of pseudopods nor many other desirable details. Furthermore, it will take an expert in this area, which this author is not, to resolve the various processes into patterns allowing of comparisons of similitude.

Nuclear structure and fission would seem to possess sufficient detail of the sort needed to analyze homologous relationships. However, only the processes of pro-, meso-, and metamitosis come close to being useful. Certainly the structure of these usually tiny nuclei and their ubiquitous placement internally in the cell offer nothing of special import. But the mitotic process, though well known in certain forms, is unknown to others, and disagreement as to how comprehensively these categories can be applied (Page 1967; Houssay and Prenant 1970) suggest that much more extensive work will be needed before one can use the fission process in a critical way.

The pattern of tests can be used for studies on homology, and have already been so used in the Foraminiferida. The location of the test on top of the cell body, amidst the radiating pseudopodia, and enclosing the nucleus and other organelles, provides a location relative to other cellular parts and organelles that is essentially similar from one testate cell to another. The composition of the tests can be quite variable until one reaches the chambered tests of the foraminifers and their three major types—calcareous, arenaceous, and porcellaneous—where differences are bridged, for the most part successfully, by intermediate forms. The availability of fossil tests is most important here. Outside the foraminifers, it may well be that such different tests as the accreted sand grains of *Difflugia* and the secreted scales of *Euglypha* represent independent origins of tests. Though the positional relationship is fulfilled, the compositional one is unfulfilled, and homology is doubtful if not absent. In that pseudopodia may show convergent evolution, the same could well be true of many of the tests outside of the Foraminiferida.

Cyst morphology and formation is known best among the symbiotic forms. We will forego further comments here, other than to reiterate what was remarked earlier. That is, the scattered occurrence of cysts in free-living forms may mean it is a neosemic character in the various species where it does occur. It could, therefore, represent another example of a convergent seme. Only more detailed knowledge of internal structure, chemical composition, and development will allow us to use the trait phylogenetically.

Life cycles are known for approximately twenty species of foraminiferans (Loeblich and Tappan 1964). This is not enough to be meaningful today for extensive phylogenetic analysis. But that which is known (Grell 1967) documents convincingly that the rich complexities of these life processes can be a gold mine of phylogenetic information. For the remaining amebas, again more information is needed. Perhaps, also, details on the patterns of DNA and protein synthesis would allow useful comparisons among those asexual forms with otherwise quite uneventful life cycles.

Ecological data, primarily directed to the way an ameba exploits its environment, i.e., food-getting, cues for cyst formation, and other life cycle transformations, are meager today. So again the semic information promises more than it offers.

In short, only the Foraminiferida offer any possibility of rigorous phylogenetic analysis.

Foraminiferida. D'Orbigny (1826) when 24 years old, in the first monograph on the foraminiferans, classified them as cephalopods, interpreting their array of delicate filopodia as microscopic equivalents of tentacles. Subsequent work, for example Dujardin's (1835) report to the Academy of Science in Paris, wherein he suggested a committee of that august body be appointed to corroborate his observations, helped remove these forms from the molluscs and place them closer to other unicellular forms. Interest in these forms continues, not only for purely biological reasons, but for paleontological ones too, fossil forms being found as early as the Cambrian (Loeblich and Tappan 1964). And in that these fossils are extremely valuable in the location of oil-bearing strata, there have also been understandable economic reasons for their study. In their magnificent review of the foraminiferans, Loeblich and Tappan (1964) report that as of 1958 there were 95 families (33 now extinct) which embrace 21,433 contemporary and fossil species. The foraminiferans are estimated to comprise about half of all the known species of protozoans (Levine 1962). These figures are in striking contrast to those presented for the preceding groups of ameboid forms and raise a variety of questions as to how to account for the success of the Foraminiferida. Another way of stating the problem is the phylogenetic one of inquiring into the evolutionary potential of this group. But, since such questions are answerable only after considerable prior analysis of homologies and their evolutionary development, we must start, then, with a survey of the biology of the Foraminiferida so as to discover and define those organizational and functional similarities that are of a homologous sort. The following account is drawn largely from Loeblich and Tappan (1964). Other general accounts are those of Galloway (1933) and Cushman (1948).

The foraminiferan cell system is composed of the following major parts: the cell body has pseudopodial extensions; the cell body, itself, has an outer clear ectoplasm in which the test lies, and within the test is the granular endoplasm; within the endoplasm are one or more nuclei; also, there are various inclusions consisting of food vacuoles, lipid droplets, bits of inorganic material, and pigmented droplets of various colors, depending on the species. Now let us examine each of these in a bit more detail, keeping the tests for the end, for they are the most important from a phylogenetic point of view. The pseudopodia are invariably filopodia of the granuloreticulate type. They are often extremely long, extending out over the substratum for distances many times greater than the diameter of the cell body. The clear ectoplasm surrounds the test and is continuous with the filopodial ectoplasm. The more differentiated endoplasm is obviously the center of much metabolic activity, judging by the inclusions. The nucleus or the nuclei, depending on the species and the stage of the life cycle, are situated more or less centrally. When one nucleus is present, it is often large, round, and with a conspicuous nuclear membrane. When many nuclei are present, as in the agamont stage, they are often somewhat smaller than the single nucleus characteristic of the gamont. In some cases the gamont can contain two different nuclei, as in *Iridia lucida* (Fig. 6–3): a micronucleus, which after many divisions supplies the nuclei needed for the uninucleate gametes, and a large macronucleus, which is somatic in the sense it does not pass on to successive generations but is lost with the test after gamogony occurs. In the endoplasm surrounding the nuclei are food vacuoles. Initially, digestion may occur in the bases of the pseudopodia, outside the test. When the contents of the vacuoles are broken down, the vacuoles enter the test, and digestion is completed in the endoplasm. In some cases, this prior digestion external to the test is necessary to reduce the vacuole contents to a consistency such that they can pass through the small openings of the test and reach the endoplasm. Much of the other inclusions seem to be consequences of the digestion of ingested material, since they consist of undigestible materials, subsequently to be egested externally, lipid droplets, and various pigment grains. Much of this material is eliminated from the cell body when the cell is preparing to go into gamogony; it is termed a "cytoplasmic purification" (Loeblich and Tappan 1964), seeming to be a protozoan prenuptial rite.

The test provides many complex aspects: chemical composition, disposition of minerals, variety of chambers, and variety of openings. Basically all the tests have a secreted, organic ground substance, usually called chitinous, within which are deposited various salts in various patterns. The nature of the chitinous matter is in some cases a mixture of proteinaceous and carbohydrate materials, a mucoprotein which is not a typical chitin. In other cases the distinction from regular chitin is not clear (Loeblich and Tappan 1964). It is simply easiest to follow convention for the present and refer to it as chitinous, realizing that this is in some cases a misleading use of the term. Where exogenous materials are accumulated to make the test, the term arenaceous is applied. Arenaceous tests are coarsely particulate in appearance. Endogenous minerals are deposited in the test, and they are often calcium carbonate, but magnesium carbonate, iron oxide, and silica are present, as well as traces of other materials. These can result in a granular or a clear hyaline appearance of the test, sometimes resulting in tests with a highly polished appearance (porcellaneous tests). The examination of the fine structure of the test (Wood 1948) reveals significant patterns of arrangement of the constituent crystals which, it turns out, are of great diagnostic use in defining taxa. In fact, Loeblich and Tappan (1964) consider test wall composition and microstructure as of ". . . primary importance for classifying the Foraminiferida, for these skeletal features are determined by the nature of the secreting protoplasm." (p. C153)

The structurally simplest shells, in terms of gross details, are those with single, spherical (except for apertures) chambers, termed unilocular tests. Beyond this, these tests can also be spindle-shaped, cylindrical, or with various evaginations (Fig. 6–5). Multilocular tests, with more than one chamber, can have their locula arranged in a great variety of ways. It is this variety, in large part, that determines the various foraminiferan species. The chambers show various types of openings between themselves and to the exterior. There are the relatively large and conspicuous apertures, or foramena (giving the group its name), which connect all chambers. In the last-formed chamber, the foramen provides a major opening to the outside. Also there are, in many shells, small perforations, or pores, in the walls. Also there are various canals and passages (stolons) within the walls. The chambers can be placed to result in coiling patterns, and the pitch and tightness of these coils are also characteristic patterns of the tests. The tests can show various external ornamentations,

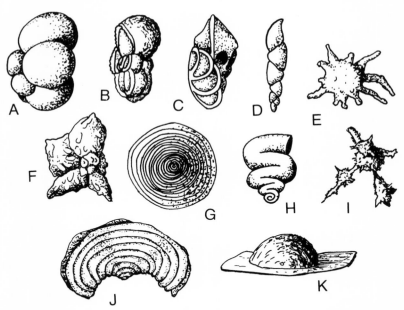

Fig. 6–5 Selected foraminiferan tests: A. globular, B. hemispherical, C. angular rhomboid, D. cuneate, E. radiate, F. radial elongate, G. circular, H. zigzag, I. irregular, J. semicircular, K. hemispherical. (Redrawn, by permission, from Loeblich and Tappan (1964) *Invertebrate Paleontology,* Part C. Geol. Soc. Am. and University of Kansas Press.)

such as spines. These are most highly developed on the planktonic forms. Lastly, there are differences in the tests in different stages of the life cycle, providing a special type of intraspecific dimorphism.

Wherever they are found, except for one exceptional parasitic form (Le Calvez 1953), the foraminiferans feed by capturing food with their filopodia. These organelles are extended out over the surrounding surface of the substratum and bacteria and unicellular algae—very often diatoms, other protozoa, and even small Crustacea, are caught. In some instances, it appears that the prey are immobolized upon contacting the filopod, presumably as a result of some chemical; but in other cases they seem simply to become entangled in the filopodia and are eventually engulfed and die, being subsequently digested. Most Foraminiferida crawl slowly over the substratum while feeding, but others are very sedentary, and others are very strongly attached to the substratum, as can be seen by anyone who tries to remove a form such as *Homotrema rubrum* from its place of attachment. In addition to serving as food-getting and locomotory and attaching organelles, the pseudopodia also are important in laying down new test walls and cysts (Le Calvez 1953; Angell 1967). Clearly the form of the underlying cytoplasm determines the shape of the new loculum, illustrating perfectly the statement of Loeblich and Tappan (1964) that the skeletal features are determined by the nature of the protoplasm.

Foraminiferans respond readily to changes in salinity, oxygen, pH, and temperature (Loeblich and Tappan 1964). Their form and feeding patterns can be changed and their life cycles affected. Presumably these are devices for reacting homeostatically to such changes, but within the confines of these cells, the cell membrane and general metabolic devices useful in making such responses are all that can be cited at present as the functional basis for such activities.

Reproductive functions are best described in terms of the generalized life cycle. The life cycle of *Iridia lucida* (Fig. 6–3) illustrates such a cycle of events. A schematic summary of foraminiferan life cycles is given in Fig. 6–6. The life cycle of *I. lucida* consists of the sequence of stages designated 1-2-3-4-1, a dimorphic cycle alternating between the haploid gamont and diploid agamont or schizont. In a form such as *Elphidium crispum* (Fig. 6–7), there are various sequences possible, such as 1-2-3-4-5-6-1, 1-2-3-4-1, and 5-6-5 (Sigal 1952). The first is trimorphic, the second dimorphic, and last is an apogamic cycle in which sexuality is suppressed. *Discorbis orbicularis* is reported to have exclusively an apogamic 5-6-5 cycle.

Stages d' and j' of *Elphidium*'s life cycle (Fig. 6–7) deserve further comment. The former, d', is termed a macrospheric test because of the large size of the central, first chamber or proloculum. And by the same token, j' is a microspheric test. The size of the proloculum depends on the size of the cell initiating it; a relatively large zoospore or a smaller zygote is especially important, for here the foraminiferans can display their versions of the biogenetic law in beautiful detail. The early stages of development, as laid down in the early chambers, can follow developmental patterns of ancestral forms, superimposed on which are the patterns of the contemporary form. This, combined with their fossil history, allows one authority to state that

> "The Foraminifera illustrate recapitulation probably more conveniently than other groups of animals. In many families the phylogeny as interpreted on the basis of recapitulation agrees exactly with the appearance of the various genera in geologic time." (Galloway 1933, p. 24.)

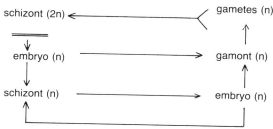

Fig. 6–6 Diagrammatic summary of foraminiferan life cycles. (Redrawn, by permission, from Sigal (1952) in *Traité de Paleontologie,* ed. J. Piveteau, Vol. I, Masson et Cie.)

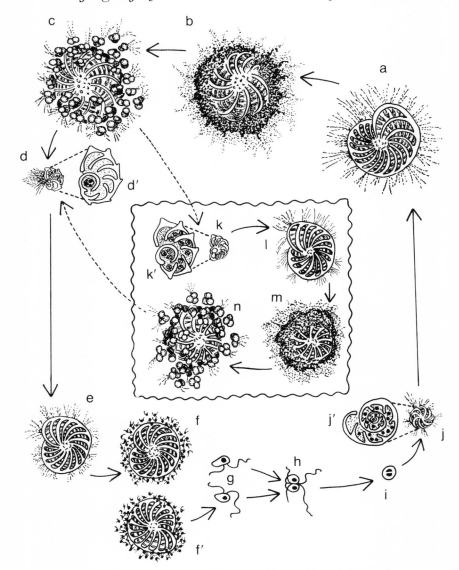

Fig. 6-7 The life cycle of *Elphidium crispum*. The schizont (a) undergoes multiple fission (schizogony, b, c) and then can follow either of two paths. Either the single, uninucleate cells develop into the gamont (d, e) which undergoes gamogony (f) producing flagellated gametes which fuse (g, h) and form a zygote (i), or, occasionally, there is a repetition of the schizont phase (k, m, n,) which then goes on to the gamont. Following fertilization, the schizont reappears (j). (Redrawn, by permission, from Grasse (ed.) (1953) *Traité de Zoologie,* Vol. I, part II, Masson et Cie.)

This can be illustrated by many cases. The Russian workers have taken the lead in applying Sewertzoff's (1931; and chapter 2) ideas of anaboly, deviation, and archallaxis to protozoan forms and to the Foraminiferida in particular.

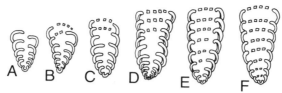

Fig. 6–8 Recapitulation of test development correlated with geologic age. A. *Textularia*. B. *Cribrostomum*. C. *Climacammina* (from the lower Mississippian). D. *Climacammina* (from the upper Mississippian). E. *Climacammina* (from the Pennsylvanian). F. *Cribrogenerina* (from the Permian). If specimens B, C, and D were unknown, one might propose evolutionary development as a kind of deviation from the original pattern. If A through E were unknown, then there is the impression of evolution through archallaxis, i.e., with the emergence of a genuinely new pattern. (Redrawn, by permission, from Pokorny (1958), *Grundzüge der Zoologischen Mikropäläntolgie*. Deutscher Verlag der Wissenschaft.)

Fig. 6–8 shows an example (Pokorny 1958) of progressive change in a test illustrating anabolic change and acceleration of the change, so that in its final form the new type of test appears very early. The intermediate steps (b–e) all give evidence of recapitulating the primitive stage (a) in their development. It should also be noted that these forms are found in the fossil record with stage (a) appearing among the earliest forms and (f) among the latest.

Here, then, we are in the extraordinarily fortunate position of having both fossil and ontogenetic evidence available for determining the direction of evolutionary trends.

The phylogeny of the Foraminiferida is derived almost exclusively from the homologies present in the nature of the tests. Though the complexities of the life cycle are almost ideal as another semic trait, the fact is that life cycles are completely known for only a score of species and partially known for another score or so (Loeblich and Tappan 1964). The details of nuclear structures and events parallel, roughly, the information known for life cycles. The pseudopodia and cytoplasmic inclusions show variations from species to species; in general, these differences are correlated with differences in test morphology and simply confirm what can be learned, usually in more detail, from studying the test.

The possibility of classifying the foraminiferans so as to reveal their "biological relations" (Brady 1884) was early appreciated, and, from the latter part of the 19th century down to the present, phylogenetic interests have played a major role in guiding the work of the systematists concerned with this group. (See Galloway 1933, for a short review of this history.) It is needless to emphasize the fact that many revisions followed the initial attempts and, of course, just because of continual revision the major trends of evolutionary development have become clearer. Embedded in these trends are essentially all the major processes known to occur in the evolution of other organisms. There is diversification and transformation of species, parallel developments

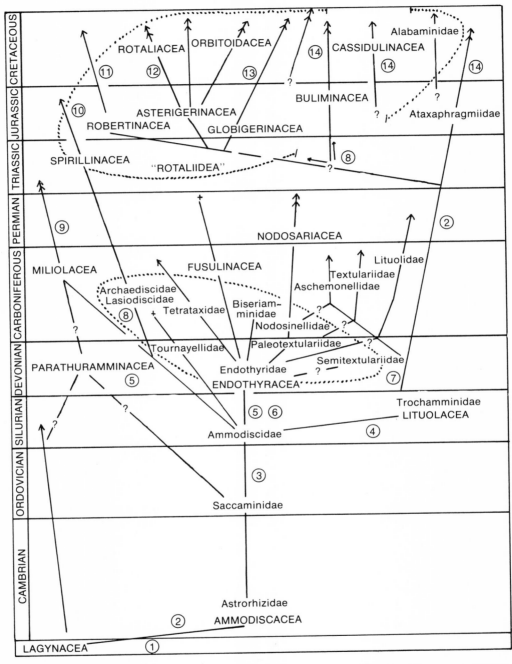

Fig. 6-9 Phylogenetic view of the *Foraminiferida* according to Glaessner (1963). Names of taxa (superfamilies recognized by Loeblich and Tappan (1961)) are placed in periods of their first-known occurrence. Double-headed arrows indicate intense

and divergent ones, reduction of traits and progressive complexity, extinction, variable rates of evolution, and biogenetic recapitulation of various sorts (Galloway 1933; Cushman 1949; Sigal 1952; Le Calvez 1953; Koenigswald et al. 1963). Patiently working their way through the ever-expanding diversity of known forms (many paleontologists, especially petrologists, were and are continually describing new forms without adequate reference to biological facts such as species variability or life cycle stages, nor reference to the previous literature, so that extensive duplication of forms plagues all critical systematic work here (Loeblich and Tappan 1964)), and developing ever more critical modes of comparison of test composition and microstructure, of chamber morphology and arrangement, of aperture and pore structure, and of the development of tests, the foraminiferan specialists have come to fairly good agreement on the general outlines of the major trends in the evolution of this group (Koenigswald et al. 1963). One summary diagram is that of Glaessner (1963) (Fig. 6-9). We can do no better than to quote directly the general conclusions that accompany this diagram in the original paper.

> "The Foraminifera as a highly differentiated group of Protozoa are not ancient but of a geologically relatively recent origin. Their differentiation occurred in two major bursts of evolution, one in Devonian and the other in Triassic-early Jurassic time. In each instance, it took about 50 million years before major taxonomic diversification was achieved. The first major evolutionary novelties were septation (periodic growth) and calcareous wall secretion. Both express changes in basic physiological processes which can be reasonably thought of as advantageous, though nonseptate, chitinous and agglutinated tests are still produced in living form. The novelties, however, provided the basis of intense further structural differentiation of the test and textural differentiation of the shell wall during *Late Paleozoic* times. While the Miliolacea, Nodosariacea and Lituolacea proved permanently successful, the structurally most advanced (and presumably most specialized) Fusulinidea failed to survive, and others were superseded by their descendants. The second great burst of evolution began in *Late Triassic-Early Jurassic* time and gathered momentum throughout the *Cretaceous*. Characteristically, it does not appear to have originated directly from or built on the results of the earlier differentiation but to have begun from a new starting point. It was achieved by a new departure in the direction of perfection of wall texture (as granulate or radiate [in terms of the disposition of crystalline material (Wood 1948)]), independently, as far as can be seen, of the earlier parallel trend to radiate wall texture in the Nodosariacea. It evolved now on the basis of

---

subsequent systematic development and single-headed arrows indicate continuing existence. Numbers refer to the nature of the tests, as follows: 1. chitinous, 2. agglutinoid, 3. coiling, 4. septate, 5. calcareous, 6. granular, 7. trochospiral, 8. radiate, 9. porcellaneous, 10. single crystal, 11. aragonite, 12. canals, 13. planktonic, 14. granulate. (Redrawn, by permission from Glaessner (1963) in *Evolutionary Trends in the Foraminifera*. Elsevier Publishing Company.)

trachospiral coiling [i.e., spirally coiled chambers, evolve on one side of the test and involute on the other]. Only rare departures to planispiral coiling [i.e., coiled in a single plane] occurred later. For reasons which are largely unknown, this combination of basic characters proved most successful. Among the new "achievements" were "bigger and better" (in bulk of the sarcode and strength and complexity of the test), larger Foraminifera, and, for the first time, successful adaptation to planktonic life in Foraminifera. Complications of the apertures reach extremes in the development of toothplates and other obstructions, for reasons which are unknown, as we know very little about the function of apertures. The building of the wall from aragonite, instead of the more common calcite, seems to have been persistent in a group which has decreased in numbers but increased in structural complexity since its first appearance. Another side branch builds single-crystal tests which remain entirely or largely nonseptate (Spirillinacea). The initiation of this trend is uncertain, as Late Paleozoic tests with similar structure have different textures. An important trend is towards increased morphological expression of the alternation of sexual and asexual generations, which is all we know of the evolution of sexuality in the Foraminifera. In other (apogamic) lines, the sexual generation may have been suppressed. A trend which manifests itself within the major lineages, often on the level of evolving species, is the increase in pore size. This again must be related to an unknown selective value in physiological efficiency."

And, the author concludes

"This review could be continued at length by enumerating many other trends in morphology, often well documented in phylomorphogenetic series [that is, successive evolutionary innovations]. This would go beyond its object, which is to review present knowledge (and lack of knowledge) in foraminiferal phylogeny and to show the main lines of their intense morphological and systematic differentiation during the last 400 million years. This fact, together with their abundance and the enormous time scale of stratigraphically well controlled sampling, would make them uniquely valuable objects in the study of evolution in the Protozoa. The value of further work on these lines will, however, depend on greater knowledge of not merely their morphology, but also their physiology and ecology." (Glaessner 1963), p. 20–21).

What remains to be added here are some comments deriving from our own phylogenetic procedures. Though it is clear that foraminiferan phylogeny has used the concept of homology, especially scrupulously applied to ever finer details of shell structure, this use was made with little explicit reference to that concept, as a review of the references already cited make clear. The reasons for this are twofold. First, with the evolutionary theory in hand, the wonderfully complete gradations in both fossil and developmental forms made obvious, without any methodological palaver, just what had to be done to work out the evolutionary relationships of this group. Second, classification was always as much an interest of the research work on these forms as were evolutionary studies. Individual workers naturally differed in their emphasis. Glaessner (1963) is critical of Loeblich and Tappan (1961) for their systematics, which does not adequately reflect phylogenetic conclusions, and in their later great review (1964), Loeblich and Tappan still give no explicit treatment to evolutionary considerations though admitting that such problems are appropriate questions in systematic work.

The questions we would raise now concern the evolutionary potential of the Foraminiferida. Why do these particular types of ameboid organisms represent such a successful group as seen in their long history and great diversity? And since the evolutionary diversity of the Foraminiferida, as presently described, is dependent on understanding the evolutionary trends in the development of the test, we must understand the biology of that product of the cell before we can talk about the reasons for its evolution. The summary by Glaessner, just quoted, explicitly refers to unknown selective advantages of various types of shell walls and apertures and ends with the plea for further enlightenment regarding the physiology and ecology of these forms.

Our best answers might be those suggested earlier, namely, a limited protective function plus a role as ballast. However, this, at first glance, only leads to further puzzles. A protective function explains the location of the test in surrounding the central metabolic and genetic machinery of these organisms but doesn't explain its variety. And the deadweight function explains neither location nor variety. The answers to these questions are better understood in terms of the way a cell body secretes a test. Test formation has a genetic basis, and it definitely appears to be a reflection of the morphogenetic capabilities of the cell. We must propose, then, that the patterns of walls and chambers and openings are a statement of how the cells are organized and how they grow. This is consistent with what little we do know of test formation. Initially, the power to secrete may have arisen from cyst formation, that ability being a preadaptation to test formation. Hence, conserving initial patterns of secretion would result in tests intimately dependent on the peripheral and surface organization of the cell for their form. Since ballast needs no particular form and protection is served by the cystlike origin of the test, the test would continue as a product of the peripheral parts of the cell and would be deposited in a manner continuing to reflect the organization of the cell. Where correlations occur between shell form and environment, these probably reflect correlation of shell with a cell body which is highly adapted to the environment. The remark (Loeblich and Tappan 1964, p. 88) that "Mostly features of test morphology are paralleled by similarly important differences of soft parts, cytoplasm, and nuclei, or by distinctive living habits and reproductive processes," is consistent with our thinking.

In conclusion, when future work clarifies the function and development of the foraminiferan test—whether along the lines suggested above or along

PHYLOGENETIC CONCLUSIONS

In a very real sense, our ability to reconstruct the evolutionary history of the amebas is not much advanced over that expressed about a century ago by Hertwig and Lesser (1874). Very cautious conclusions were put forth by these profoundly careful workers, and they insisted, most explicitly, on the consideration of many characters, each one of which would be clearly defined and well understood. However, though our phylogenetic reconstructions are not much beyond those of Hertwig and Lesser, we do have a clearer view of why the difficulties exist. The key issue is the polyphyletic nature of the amebas.

These rhizoflagellates are highly polyphyletic. There are certain forms, however, which can be brought together as monophyletic, and among these we would recognize the foraminiferans, and the three separate groups of slime molds—the cellular Acrasiales, the noncellular mycetozoans, and the parasitic Plasmodiophorales. The Xenophyophorida may also be monophyletic, as may certain families of the testate Lobosa and Filosa, but present information makes their identification a very unsure business. This lack of certainty seems to be an unavoidable consequence of their evolutionary history. This can be brought out by referring again to the work of systematists.

If one surveys the eleven families recognized by Chatton (1953) in the order Amoebaea (naked amebas) within the class Lobosa, the diagnostic features used include not only shape and nature of the lobopodia, but also general size and body contours (part of the lobopodial character); mode of life (freshwater, marine, soil dweller, parasite); cysts; life cycles; nuclear divisions (type of mitosis, spindle, centrioles); flagella, when present; and any special structures such as heavy pellicle or inclusion bodies. None of these features is used consistently throughout the whole of the Amoebaea to distinguish variously the different families. Rather, each family is a unique constellation of a variable number of the features just enumerated, so that though several families show mesomitotic divisions, they do not also all show the same ecology, cell shape, or general size. Hence, even such a limited group as the naked amebas with lobopodia offer no promise of being a homogeneous group; a group sharing at least one common homologous character and, hence, arguably monophyletic. This is emphasized over and over again by Deflandre (1953b, p. 92, p. 93; 1953c, p. 123, p. 139), where he candidly admits that classificatory similarity is not a guarantee of phylogenetic homogeneity. This problem and a possible evolutionary explanation for it are summed up in his comment

". . . one is led to think that all the amebas, even those which have no centrosome, are flagellates that have permanently lost that organelle. Since ameboid tendencies are found in many flagellates, even in autotrophic ones, one is led to suppose that the amebas are a polyphyletic group of forms which, originally flagellated and autotrophic, have lost their cell wall and developed the ameboid tendency which they are still using as an accessory mode of feeding (Pascher 1918)." (Deflandre 1953b, p. 36–37. Original in French.)

This postulated flagellate ancestry receives support from the occurrence of flagellated stages in the life cycle of ameboid forms. However, this argument must be used carefully, for just as one can argue for the multiple independent origin of pseudopodia, the same can be argued for flagella. Where there are selection pressures favoring dispersal, either to exploit new food resources or to find gametes of a complementary type for mating, flagella are a possible answer. And it does not resolve matters to point to the Mastigamoebidae (Chatton 1953) or Rhizomastigida (Honigberg et al. 1964), which are forms showing flagella and pseudopods at the same time, or which can go from ameboid to flagellate stages and back again depending on simple changes of the salt concentration in their culture medium. These organisms show that flagella and pseudopodia are not organelles segregated into separate parts of living nature, but that does not mean ". . . the concomitant occurrence of flagella and pseudopodia strikingly illustrates the basic affinities of generalized flagellates and amoebae . . ." (Honigberg et al. 1964, p. 11). In the absence of critical evidence that the flagella and pseudopodia of the mastigamebas are homologous to those organelles in other forms, we must recognize that the question of affinities is not answerable. The minimal view is that it is simply of selective advantage to have both flagella and pseudopodia in these forms, and the fact that they resemble those organelles in other forms may or may not, on present evidence, be of evolutionary importance in the sense of providing evidence of a transition from flagellated ancestors to ameboid descendants.

The foregoing remarks are applicable especially to the naked amebas but also are germane to the shelled forms with lobopodia and filopodia. The simple difference in pseudopodia, tests (shells), nuclear phenomena, life cycles and general character of the cytoplasm and body size and contour, offer no effective information for the mode of phylogenetic analysis we are using here. Are there other data that could be useful? Biochemical data? One attempt in this regard has been made in examining, comparatively, the protein composition of three different species of amebas. Unhappily for our purposes, the authors come to the following conclusion.

"In the light of our results it is doubtful that the electrophoretic techniques, which we initially thought might prove to be very helpful, will be useful for the taxonomic classification of asexual organisms—unless, of course, the taxonomic criteria change. If one extrapolates from the data presented here, it appears unlikely that *A. proteus* and *A. discoides* would be considered species on strictly biochemical grounds." (Kates and Goldstein 1964, p. 35.) More recent work underlines this uncertainty (Friz 1970). However, it can be borne in mind that techniques other than the comparison of the phsyical properties of the proteins, used in this study, may be studied, e.g., base sequence of nucleic acids or amino acid sequences of the proteins, and these may provide useful data.

It is possible that many of the amebas did evolve from other ameboid groups. However, the present lack of homologies allows no detailed specification of such possibilities. It is also possible that they arose from flagellated forms. Many workers favor the latter view, looking especially to the chrysomonad flagellates as especially good candidates for ancestral forms (see various chapters in Grassé 1953). The occurrence of ameboid tendencies—lobopodial and filopodial—in the chrysomonads, and of flagellated stages in many of the forms classed as amebas, is the basic reason for these speculations. But they are speculations; for, to repeat earlier comments, unless the pseudopodia and flagella can be shown to be homologous, convergence is another highly possible way to account for these similarities. There is, moreover, another argument quite strongly favorable to the view that the ameboid forms, or at least one such form (if we can then derive all other amebas from it), come from a flagellated ancestor. It is the following: photosynthetic eukaryotic cell systems in all probability preceded eukaryotic phagotrophic cells; in other words, algae were around before protozoa. The primitive algae, it seems universally agreed (Pascher 1918; Pringsheim 1963; among others), were flagellated and unicellular in most if not all cases, and hence, generally speaking, the flagellates were ancestral to the protozoan forms, be they flagellated, ameboid, or ciliated protozoa. Furthermore, no photosynthetic amebas are known. Hence, flagellated ancestry is the most probable one for the amebas. This could be from the apochlorotic Chrysomonadida, or from other colorless unicellular algae capable of phagocytosis, or from zooflagellates (which, of course, in all probability came from the colorless algae). Directly or indirectly, flagellated, unicellular algae are the most plausible ancestors of the ameboid forms, just as they are for the zooflagellates, too.

There remains one last general problem to be discussed, which is the correlation that exists between the presence of sexual processes in the larger, diverse groups of amebas and its apparent, and probably real, absence in the other groups. These other groups are relatively small and are characterized by the possession of a unique constellation of characters. They are, as has been emphasized, probably polyphyletic. Can we understand the occurrence of these sexless, specialized forms?

As has been argued in the previous chapter, essentially on the basis of

Sonneborn's (1957) analysis of protozoan breeding systems, lack of sexuality implies that there is no selective advantage to genotypic variability. Indeed, it may be of selective disadvantage, disrupting genetic information optimally adapted to a given niche. Such a niche must be very stable to place no premium on genetic diversity, but such seems to be the case in various protozoan forms. The result, in these instances among the ameboid forms, would be to expect highly specialized cells, and this in all probability would be the only way they could compete with the specially endowed testated forms. Shells are best developed in the ocean due to availability of salts; in fresh water, tests will have to be endogenous with inorganic materials picked up from the environment—sand grains, etc., as in *Difflugia*. The testate freshwater amebas seem to exhibit greater variability than the naked ones; they, too, however, are apparently highly specialized and sexless. We must argue that they, in their turn, are up against some sort of competition that can only be successfully met by extreme specialization. These competitors may be the freshwater ciliates.

In any case, the lack of sex and the presence of differing sets of unique characters to define the various small groups of these forms, is consistent with the idea that their extreme specialization resulted from modes of living that consist of very restricted adaptive peaks that are continuously available. Such specialization, also, would tend to wipe out less specialized ancestral traits, and so obscure possible homologies. This point of view, if correct, leaves us with the fascinating problem of determining the nature of the niches that are so constantly available as to permit the evolution of sexless ecospecies. And why should it happen so often to the amebas?

Chapter 7

Heliozoans

HAECKEL'S (1894) charmingly named "Sonnenthierchen", or sun animalcules, are a systematist's nightmare. It will be best to approach heliozoans through attempts at their classification, for this will inform us as to the heterogeneity of the group and indicate the difficulties we shall encounter when we ask phylogenetic questions about them.

Homogeneous Groupings

The creation of the taxon Heliozoa came from the slowly growing awareness that in the great group of the Radiolaria there were certain forms that, though showing spherical symmetry and fine pseudopodia, were, nonetheless, distinctly separable from the rest of the group. Such forms never showed a central capsule; their filopodia often were equipped with axial filaments and, hence, were axopodia; their skeletal material was often very simple; a typical calymma was absent; and they were most often found in fresh water. This set of somewhat discrete features was recognized by several zoologists, among them Hertwig (1879, among other works) and Bütschli (1887–1889), and the group came to have a distinct place in taxonomic zoology. But difficulties were at hand. Hertwig, speaking of the classification of ameboid forms in general (Rhizopodien, which included the Radiolaria and Heliozoa) said, in his discussion of their systematics,

"Here I would like to emphasize for a moment, that actually a splitting of the rhizopod groups is just as pointless as it is unnatural. For through the creation of many small, independent groups [taxa] the overall view of the classification is muddied, so that it is understandable that such a treatment has found no acceptance in most textbooks . . ." (Hertwig, R. 1879. p. 138. Original in German.)

Despite this plea, a recent treatment of the Heliozoa (Honigberg et al. 1964) has seen fit, and with good reason as we shall see, to put forms formerly classified as heliozoans into three different classes of protozoa, the original order of Heliozoa now occurring as the subclass Heliozoia. And even this last, much more selective group carries no assurance of evolutionary homogeneity in the sense of being monophyletic.

In Bütschli's (1887–1889) major review of the biology of the protozoa he recognized four heliozoan orders. These are based largely on the character of the skeleton as first used by Hertwig (1879) and Hertwig and Lesser (1874).

The Aphrothoraca contain heliozoan forms lacking a skeleton, only occasionally provided with mucilaginous or gelatinous material at the periphery of the cell, whose form is ameboid or can remain constantly spherical, and, of course, possessing numerous, radial, fine pseudopodia. Then, Bütschli adds, they may or may not possess contractile vacuoles and, curiously enough, nuclei. The Chlamydophora possess a typical heliozoan form, i.e., radial symmetry, with a delicate gelatinous envelope or one of tangled fibers or of platelets. The Chalarothoraca possess a peripheral skeleton of unconnected siliceous units. In contrast, the Desmothoraca have a one-piece spherical or nearly spherical skeleton pierced with many holes. These forms may or may not be stalked. This classification was perpetuated by Schaudinn (1896).

In 1904, Penard published his monograph on the freshwater heliozoans found in the environs of Geneva. He was critical of the systematic treatment then accorded the Heliozoa but conservatively went along with it. Two of his general comments became, however, the basis for later revisions. Penard emphasized the occurrence of true heliozoans as distinct from false ones, and among the former he argued for the occurrence of two distinct types of organization, one represented by *Actinophrys* and the other by *Acanthocystis* (Fig. 7–1).

Let us look briefly at these cell types before continuing with the discussion of heliozoan taxonomy. The best known form of *Actinophrys*, *A. sol*, is a spherical cell 40–50 micrometers in diameter with a clear difference between the outer ectoplasm, characterized chiefly by large vesicles of gelatinous material and a large contractile vacuole, and the highly granular endoplasm. The conspicuous nucleus lies centrally, always surrounded by a clear zone. The pseudopodia, which are rarely longer than the diameter of the cell, possess axonemes that abut internally on the nuclear membrane. *Acanthocystis aculeata* typifies the other cell type. It possesses a peripheral layer, or envelope, of solid, platelike spicules which form an almost solid cover to the cell, just inside the cell membrane. The ectoplasm contains various inclusions, sometimes zoochlorellae, and has one or several contractile vacuoles at its surface.

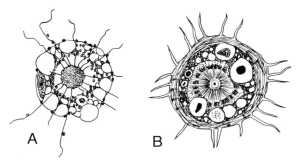

Fig. 7–1 Two heliozoans. A. *Actinophrys sol* (Redrawn from Bělař (1923).) B. *Acanthocystis aculeata* (Redrawn from Stern (1924).)

The endoplasm is eccentrically located and has a nucleus in it, located at the position where the peripheral ectoplasm is thinnest. In the center of the cell, but not in the center of the endoplasm, is the central granule. From it the axial filaments of the axopodia radiate out in all directions. Outside of the cell these filaments are covered by a thin layer of cytoplasm, thus forming the axopodia, whose length can be significantly greater than the diameter of the cell.

On the basis of these two cell types, Hartmann (1913) saw fit to designate two new heliozoan orders, the Actinophrydea and the Acanthocystidea. He also proposed a third order, the Clathrulinidea, which was essentially equivalent to the earlier Desmothoraca. These latter, however, were not part of Penard's "héliozoaires vrais." Kühn (1926) retained Hartmann's revisions but went further by including the Clathrulinidae in the Acanthocystidea now termed the Centrohelidia. This classification represents the extreme of the lumping tendency.

Valkonov (1940) accepted the Actinophrydia and Centrohelidia as the suborders of the Heliozoa and then went on to discuss five other groups which are included under Penard's title of "Pseudohéliozoaires." (Penard insisted the term had no taxonomic significance whatsoever, in his own mind.) These five groups include: (a) the Helioflagellata, forms with flagella which can transform into a heliozoan stage; (b) the Desmothoraka, forms with perforated shells but which lack the central grain of the Centrohelidia and lack also the axopodia of that group and of the Actinophrydia; (c) the Monadida, a group of two families whose representatives form protective cysts, some are parasitic, and some form flagellated zoospores (these had been earlier put into the provisional group Proteomylia (Doflein and Reichenow 1927–1929)); (d) the Proteomyxia, a collection of species with one feature in common, i.e., filopodia (this hardly distinctive character indicates the problems present in Valkanov's attempt to order the group); and (e) the Labyrinthuloidea, a very curious group which has often been placed within the slime molds but, because of long slender pseudopodia, was placed in the heliozoans as the least complicated of several modes of taxonomic treatment. Others, as we saw in the preceding chapter, suggest that they are a distinct ameboid group.

Trégouboff's (1953d) review of the heliozoans reflects many of the problems inherent in Valkonov's work but shows the stability now afforded to certain of the groups—groups whose names remain unchanged, whose representative genera are now also stable, but whose exact relation to the Heliozoa is difficult to determine in certain instances. Penard's distinction between true and false heliozoans gets further usage, taking on, implicitly, if not explicitly, the status of subclasses. The true heliozoans now seem stabilized as comprising the Actinophrydia and Centrohelidia. The false heliozoans, however, contain two *bona fide* orders, the Desmothoraca (essentially as defined earlier) and the Protomyxidea, now containing the helioflagellates and Valkonov's Monadida—the later as two separate families, the Zoosporidae and Azoosporidae. Then there is an appendix to the false heliozoans which contains the Labyrinthulidae and, finally, there are the Pseudo-Heliozoa "incertae cedis"—various genera of frankly unknown affinities but possibly related to the Heliozoa because of the presence of filopodia and roughly spherical symmetry.

The final attempt at the thankless task of classifying these forms that we shall mention here is that of Honigberg et al. (1964). Remaining in the subclass Heliozoia are the Actinophryida, the Centrohelida, and the Desmothoracida. There is no mention of true and false heliozoans; simply, three relatively homogeneous groups are included as sun animalcules. The Proteomyxidia, containing such forms as *Vampyrella* and the Zoosporidae and Azoosporidae, are raised to the level of a subclass. The helioflagellates are removed from them and placed among the Rhizomastigida—forms typically with blunt lobopodia but also alternating between flagellated and pseudopodial forms in their life cycle. However, because of the distinctly heliozoan character of the feeding stage of such forms as *Dimorpha floridanis* (Bovee 1960), it seems best to treat these forms, following Bovee, as heliozoans. And the Labyrinthulidae are also placed in the typically ameboid forms but, because of their unique nature, are given their own subclass.

The foregoing account will probably have been of interest only to veteran protozoologists. However, it contains lessons of general interest, especially to those interested in phylogeny at the cellular level. To see the heliozoans relatively clearly, one had first to look at their total biology, especially at their gross cytology, rather than just at their skeletal organization, which was overstressed by earlier workers. Second, one had to overcome the Adansonian tendencies of Hertwig to find simple, generally inclusive groups that were easily classified. Such an approach was, of course, rendered impossible, starting with Penard's analysis of the cell types of the heliozoans. Penard's approach led at first to a simplifying of the taxonomy in Kühn's hands, but the detailed work of Valkonov and Trégouboff restored the reality of heterogeneity to the group. We are left with the realization that there are six groups here each with some chance of homogeneity—the Actinophryida, Centrohelida, Desmothoracida, Proteomyxida, Helioflagellidae and the Labyrinthulida—as defined by Honigberg et al. (1964). Whether they are evolutionarily related to each other or, more exactly, whether we can determine the degree of evolutionary relationship, is yet another matter.

One last set of data must be furnished to finish the taxonomic picture. Penard's (1904) tally of the then described heliozoans totaled 52 freshwater and 11 marine species. In Trégouboff's (1953d) review of the group, there are about 20 genera of true heliozoans containing roughly 65 species, and 50–60 genera of pseudoheliozoans with 90–100 species. Many genera have only one species to represent it. For example, in the Azoosporidae there are about 10 genera and 30 species, but 20 of these species are in the one genus *Vampyrella*. This indicates the grab bag nature of the heliozoans. They represent a ragged assemblage of forms not classifiable, as we shall see in the next two chapters, in the other much more neatly circumscribed actinopodan taxa of the Radiolaria and Acantharia, and differing from most ameboid forms by the possession of delicate, radiating filo- or axopodia, rather than blunt lobopodia. The heroic taxonomic efforts just outlined give us today the groups listed above (still liable to further revision) with which to proceed in terms of phylogenetic questions.

The procedure which we shall follow can be quickly outlined. Each of the six key groups will be characterized as regards its distinctive organizational and functional features as far as is known. From this information we can proceed to look for homologous features within each group, and the rest of the phylogenetic analysis follows in order.

Organization and Function

The taxonomic terms used in this section will be those of Honigberg *et al.* (1964). The detailed descriptions of the various forms will, of course, be drawn from a variety of sources but largely from Trégouboff (1953d).

ACTINOPHRYIDA

This group contains four genera (Trégouboff (1953d), two of which—*Actinophrys* and *Actinosphaerium*—contain the best-studied species of heliozoans. *Actinosphaerium* contains only fresh water forms. *Actinophrys* contains one marine form, the rest are fresh-water. *Camptonema* contains one marine species, and *Vampyrellidium*, which is sometimes placed in the Proteomyxidia, contains one fresh-water species.

All members of the Actinophryida lack skeletal material, but all possess axopodia, which are variously arrayed in the cell body. In *Actinophrys* (Fig. 7–1A), the axial filaments terminate on the nuclear membrane. In *Actinosphaerium* they pass between the alveoli of the ectoplasm and stop at the edge of the endoplasm, which contains 20 to 500 nuclei. The axonemes of *Camptonema* each arise from the nuclear membrane of a single nucleus of these multinucleate cells. And in *Vampyrellidium* the relatively few axonemes stop, much as in *Actinosphaerium*, at the ecto-endoplasmic border. The axopodial filament, the axoneme (not to be confused with the more recently named central element of cilia), has been the object of various investigations with the light microscope (e.g., Raskin 1925; Rumjantzew and Wermel 1925), and there is agreement that it is a compound structure, composed of many small fibrils. Electron microscopy confirms this (Anderson and Beams 1960) and goes further, in being able to show the microtubule nature of the fibrils and their unique orientation, at least in *A. nucleofilum* (Tilney 1968)

(Fig. 7–2). In *Actinophrys* and *Actinosphaerium* the ectoplasm is largely composed of vacuoles or alveoli, whose contents are described simply as "Zellsaft" (Rumjantzew and Wermel 1925) or are frankly unknown (Anderson and Beams 1960). Also in this peripheral area of the cell are contractile vacuoles in the fresh-water forms. In *Vampyrellidium* the ectoplasm is simply a clear, hyaline border to the cell, distinct from the granular endoplasm. In *Camptonema* there is little difference between the endo- and ectoplasm. The nuclear situation has already been mentioned—*Actinophrys* is typically uninucleate, *Actinosphaerium* and *Camptonema* are both multinucleate. There is one nucleus in the heliozoan stage of *Vampyrellidium*, but occasional multinucleate stages are known (Ivanic 1934).

Though these forms have been thought of as floating micropredators (Beams and King 1941), like the Radiolaria and Acantharia, it seems from Penard's (1904) account that the Heliozoa are more often found attached by their axopodia to underwater plants than they are found freely floating in their environment. They can move over the bottom by rolling along on the ends of the axopodia. Penard recorded, in one instance, the feat of one moving 125 micrometers in thirty seconds by such a motion. Feeding is achieved by capturing of other small organisms that contact the axopodia. The immobilization of the prey leads to the supposition that some toxic substances are present on

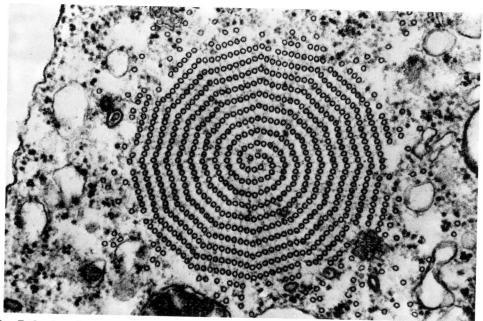

Fig. 7–2 An electron micrograph showing in cross-section the axoneme of *Echinosphaerium nucleofilum* and its constituent microtubules. (68,000X). (Original courtesy of L. E. Roth and Y. Shigenaka, Division of Biology, Kansas State University).

the axopodia (Penard 1904). The Actinophryida, especially *Actinophrys* and *Actinosphaerium*, are known to capture, ingest, and digest ciliates and small metazoans such as rotifers. In certain instances plastogamy occurs, that is, they form colonial aggregates, looking like clumps of burrs as Penard puts it, which seems to be an adaptation for the capture of large prey.

Reproductive functions include both asexual and sexual phenomena. Binary fission is the common mode of asexual formation of new cells. Details of the process are available for *Actinophrys sol* (Bělař 1923) and *Actinosphaerium eichhorni* (R. Hertwig 1899). Let us follow, in brief, the account for *A. sol*. Typical stages of mitosis are observable in the nucleus, the initiation of division being first signaled by the distinctly alveolar appearance of the nucleus in fixed and stained material. Chromosomes appear and at metaphase align themselves as short rods, parallel to the equatorial plane of the nucleus (Fig. 7–3). Anaphase movement of the separated chromosomal products also shows gradual fusion of the chromosomes into a deeply staining mass around which a new nuclear membrane develops. The old membrane has gradually disappeared, starting at the middle of the spindle and progressing polewards. The spindle is formed of parallel fibers whose ends terminate, not in a typical, pointed spindle apex, but as a truncated, somewhat concave surface lying against the polar cap material, which is functionally a centro-

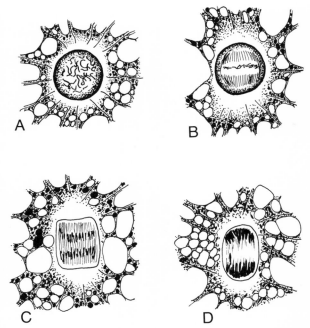

Fig. 7–3 Selected stages in the mitosis of *Actinophrys sol* during binary fission. A. Prophase. B. Metaphase. C. Anaphase. D. Telophase. In all cases the nucleus and only the immediately adjacent cytoplasm are shown. (Redrawn from Bělař (1923).)

some. This material appears in late prophase at opposite ends of the nucleus and gradually increases in amount until there are two conspicuous, clear, spherical areas at the ends of the nucleus, which is a rectangular body lying between the poles. Following telophase, when the fused chromosomes complete their poleward movement, the normal nuclear picture is restored. Bělař's account emphasizes the nuclear events, but his figures also make clear some of the details of the cytoplasmic events. There appears to be little change in the cytoplasm. The most obvious change is that the axonemes no longer touch the nuclear membrane. At the poles the axonemes stop at the periphery of the centrosomes, whereas at the sides of the nucleus they extend almost to the membrane. In *A. eichhorni* (R. Hertwig 1899), details of vegetative division are available on the nucleus, reflecting cytological interests in chromosomes, spindles, the nature of chromosomal movement, and questions of similarity or dissimilarity between protozoan and metazoan mitoses and meioses. Hertwig's account makes clear that the events in this form are very similar to those described later in *Actinophrys sol;* the now classical four stages of chromosomal behavior are definable, and special aspects such as anaphase fusion of chromosomes, presence of clear polar caps, persistence of the nuclear membrane around a rectangular nucleus, are all described. Hertwig does not explicitly describe the fate of the nuclear membrane but his careful drawings show its persistence to telophase and its apparent dissolution at the midregion of the spindle as cleavage advances. The origin of the membrane around each daughter nucleus seems to be nowhere discussed in his paper. He does explicitly propose that the polar caps be viewed as centrosomes, the same view that Bělař (1926) establishes so convincingly in his great review of protozoan nuclei. Discussing the similarity between the nuclear behavior of *Actinosphaerium eichhorni* and *Actinophrys sol*, Bělař says, "No one can, for example, doubt that the chromosomes of *Actinosphaerium* are homologous to those of *Actinophrys*, although we can only establish here division and orientation on the equatorial plate as the single shared characters." (Bělař 1926, p. 222. Original in German.)

Sexual processes are known in *Actinophrys sol* (Bělař 1923) and *Actinosphaerium eichhorni* (R. Hertwig 1899). In both cases fertilization occurs between haploid cells derived from diploid sister cells. This is, therefore, quite close inbreeding. The gametic cells which fuse, lie in a common cyst, hence, formation of flagellated gametes is functionally superfluous. However, there are some developmental changes that can be mentioned during the course of these events, collectively termed pedogamy. First, there is formation of the cyst wall, secreted by the cell, accompanied by considerable dedifferentiation, which includes loss of axopodia and dissolution of their axonemes. The vesicles of the ectoplasm also disappear, gradually. Then, pursuing the events in *A. sol*, the cell undergoes reduction divisions. All nuclei except one in each cell disintegrate, and then the two cells fuse, with fusion of their nuclei signalling completion of fertilization. Subsequently, on release from the cyst, the peripheral vesicles reappear, followed by differentiation of the axopodia and contractile vacuoles. Forty hours after emergence from the cyst, the cell can reinitiate pedogamy without an intervening cell division. This lack of an immature period is consistent with the inferred inbreeding propensities of these cells.

If we are to interpret the events of pedogamy in recapitulative terms, we would have to argue for a simple spherical, relatively undifferentiated cell as being the primitive form of these heliozoans. However, the highly inbreeding nature of the sexual process indicates that there has been considerable evolution away from the presumed original outbreeding situation. And the special occurrence of sexual processes within a cyst, rather than between free-swimming gametes, argues that pedogamy is a highly specialized process and probably no longer allows of a simple biogenetic interpretation of its events.

Homologies. We are now in a position to ask how many of the structural and functional specializations of the various actinophryidan forms can be homologized from one species to another. The semic features most promising in this regard are (1) the differentiation of ectoplasm and endoplasm, (2) the organization of the axopodia, (3) special cytoplasmic inclusions, (4) feeding and perhaps especially the habit of plastogamy, (5) details of fission, especially nuclear events, and (6) the sexual process of pedogamy. Less promising possibilities are (7) the spherical symmetry and general form of the cell body, and (8) details of nuclear organization in the vegetative cell, such as number and distribution of nuclei. These features will now be treated in some detail.

The character of the ectoplasm and endoplasm seems to be homologous in *Actinophrys* and *Actinosphaerium,* but it is doubtful that it can be convincingly homologized among all four genera of the Actinophryida. In *Actinophrys* and *Actinosphaerium* the conspicuous permanent vesicles of the ectoplasm and the occurrence of contractile vacuoles is a distinctive common feature. The clear separation of this layer from the inner more granular endoplasm also occurs in both forms. This separation is evident in *Vampyrellidium* but is inconspicuous in *Camptonema,* and in both of these genera there is no evidence of the alveolar ectoplasm.

Axopodia occur not only in these heliozoans but also in the Acantharia, the helioflagellates, certain chrysomonads, and certain Foraminiferida. Though there appear to be some differences in organization such that, for example, in the foraminiferans there is some branching of the axopodia and the axoneme is seen only under dark-field illumination (Valkonov 1940), the differences are relatively minor and hard to define in ways sufficient to see uniqueness, if it is there, for the axopodia of the different groups. An approach through electron microscopy seems to offer considerable promise, providing the technical problems, such as thin sectioning of cells with shells and spicules, are solved. There is, of course, the real possibility that there is a minimal variability of the axopodia even at the electron microscope level; that is, that these organelles will turn out to share a very similar fine structure and hence represent another example of convergence at the cellular organelle level.

Most workers who study the cytology of the heliozoans comment on the profusion of vacuoles and various inclusions in them (e.g., Rumjantzew and Wermel 1925). It appears that comparison of such elements might provide information on the possible homologizing of cells in these forms. Thus far it appears that no comparative work has been done. The individual reports are most difficult to compare because of the variety of nonspecific stains and various fixatives used. Their review has not been attempted here, and so it is not possible to say what the actual possibilities are with this material regarding homologies. An approach through electron microscopy is another obvious possibility. Anderson and Beams (1960) have presented a comprehensive, composite view of *Actinosphaerium nucleofilum* (Fig. 7-4), which could serve as a model for the comparative cytology of heliozoan cells, and other cells too.

Penard (1904) lists a great variety of forms, ranging from bacteria through unicellular and filamentous algae, various protozoa, to small metazoans, that are the food of the free-living heliozoans. However, it is not exactly clear what variety of forms is engulfed by any one species. No species seems to feed on everything in the range from bacteria to small multicellular animals, but, on the other hand, no species seems to specialize on one or a few closely related food organisms. It may be that there is too much overlap to allow any clear discrimination of feeding habits so as to adapt comparative data to the study of homologies. The mode of ingestion, of whatever food, may, however, be a more fruitful area. The study of immobilization of prey, of transport into the cell body (the axonemes are reported to liquefy (Trégouboff 1953d) and thus cause the collapse of the axopodium and placement of the prey on the cell surface, followed by engulfment into the cell inside a food vacuole), and then digestion, finally, in the endoplasm—all are areas open to comparative study. At present, however, these processes have been examined only superficially as general comments on feeding habits or have been inconsistently treated in isolated forms.

The fission process, especially details of mitotic events, seems to contain various distinctive features such as polar cap formation, fusion of anaphase chromosomes, and fate of the nuclear membrane, which would provide semic traits of great use to the study of homology. However, here, too, present information is adequate only in the two forms mentioned earlier, *Actinophrys sol* and *Actinosphaerium eichhorni*. In these it appears, as was mentioned earlier, that the two species show homologous processes.

The sexual processes in *Actinophrys* and *Actinosphaerium* appear to differ only in the initial multiple fissions that occur in *A. eichhorni* necessary to

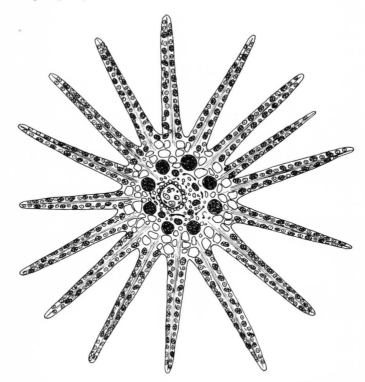

Fig. 7–4 Diagrammatic representation of *Echinosphaerium (Actinosphaerium) nucleofilum,* a composite view based on electron micrographs. Each radiating axopodium contains a central axis of axonemal material, ovoid mitochondria, peripheral vesicles with a central granule, and tubular endoplasmic reticulum (shown here as occasional rods). The cell body, proper, contains roundish nuclei (eight are shown), a centrally-located food vacuole surrounded by pinocytolic vesicles, ovoid mitochondria, and Golgi vesicles (shown as somewhat dumbbell-shaped structures internal to the nuclei). (Redrawn, by permission, from Andersen and Beams (1960).)

produce the uninculeate cells. These cells then undergo reduction divisions, fertilization, and restoration of the vegetative phenotype in a manner so similar to *Actinophrys sol* that pedogamy in these forms is homologous. Like processes are not known for *Vampyrellidium* and *Camptonema*.

As we indicated, general body form and details of nuclear organization might be semic. On present information, however, spherical symmetry is not unique to the Heliozoa, and there is no recognizable pattern to distribution of the axopodia which would transform that symmetry into a unique pattern allowing point-to-point comparisons with other patterns. However, such possibilities of unique cellular organization have not been exhaustively explored. Nuclear organization presents two possibilities: number and distribution of nuclei and also structure of the so-called resting, or interphase, nucleus. Data so far available offer no special clues on nuclear placement; if there is one nucleus per cell, then it is typically centrally located; if there are many, then they are scattered about the endoplasm and usually towards its periphery. These simple facts hardly add up to the complexity needed for semic traits. There are, of course, differences in the interphase nuclei of various cells. Bělař (1926 gives diagrams for *Actinophrys sol, Actinosphaerium eichhorni* and *Camptonema mutans*, in which the first has nucleolar material distributed as fine granules around the periphery of the nucleus, the second shows that material centrally located in an amorphous clump, and the third has eight nucleoli of varying sizes distributed throughout the nucleus. These details are not informative for comparative use concerning homologies.

In sum, we are left now with the only conclusion that *Actinosphaerium* and *Actinophrys* share certain homologous characters, i.e., organization of their ecto- and endoplasm, the mitotic behavior of their nuclei, and their pedogamous processes. This being so, using the first subsidiary criterion of Remane we can conclude that various of their other characters, such as the axopodia, feeding habits, and general cellular form, are also probably homologous. But homologies between these two genera and *Vampyrellidium* and *Camptonema* are not established. Hence, even though it is useful for purposes of classification to include these four genera in the order Actinophryida, it is doubtful on phylogenetic grounds whether any of them other than *Actinophrys* and *Actinosphaerium* can be thought to have a common ancestor. The proposal that *Vampyrellidium* is a primitive form in this group, representing the transition from ameboid ancestors to the more complex *Actinophrys* and *Actinosphaerium* (Ivanic 1934; Trégouboff 1953d), is based simply on the relative simplicity of *Vampyrellidium*. There is no way at present to convincingly establish the primitive nature of this form, as opposed to interpreting it as being secondarily simplified or coming from an ancestral form different from the other Actinophryida.

CENTROHELIDA

This group of heliozoans, represented by something like 15 genera and 60 or more species, have one special character in common. It is the possession of a central granule and its associated structures. This complex and conspicuous character is a seme providing significant unity to the group, and it determines in several important ways the general structure and function of the cells possessing it. Probably the most obvious consequences of the central granule and its surrounding parts—all simply referred to as a centroplast (Dobell 1917)—is the displacement of the nucleus and the so-called endoplasm to an eccentric position in the cell. Another point is that the axonemes all converge on the centroplast. In the case of *Wagnerella borealis* (Zuelzer 1909), the axonemes are reported to be continuous with the fine fibrils which radiate out through the centroplast from the clear, refractile central granule. Finally, it will be seen that the centroplast plays an important role in cell division, though Hovasse (1965) reports finding no centriolar fine structure to be present.

Trégouboff (1953d) divides the Centrohelida into three suborders on the basis of the skeletal material. These groups are essentially the first three orders of Bütschli's earlier classification, minus the forms removed to the Actinophryida. The first suborder, Aphrothoraca, has no skeletal material. The cells are uni- or multinucleate and can be sessile, with or without attaching stalks, or capable of freely floating in the water. The Chlamydophora have soft mucilaginous material as their outermost layer, to which can be stuck various materials of exogenous origin, as a sort or skeleton. Many of these forms are poorly studied. The presence of the centroplast in some cases is inferred from the eccentric position of the nucleus and endoplasm and has not been observed directly. The last suborder, the Chalarothoraca, also has an outer mucilaginous layer, in which are located either tangentially or radially or both, various isolated skeletal elements (Petersen and Hansen 1960). These are usually siliceous; rarely, they are chitinoid.

It is worth specially remarking on the organization of the ectoplasm and endoplasm in the light of some comments Penard (1904) makes regarding the conventional designation of these parts in these cells. Usually, the ectoplasm is thought to include everything from the cell surface into the relatively clear area which surrounds the nucleus and, usually, also the centroplast. Ectoplasm, so designated, consists of the outer mucilaginous layers and peripheral skeletal materials—endogenous or exogenous—and inner granular cytoplasm, in which food vacuoles and other inclusions occur. Endoplasm is limited to the clear area in the vicinity of the nucleus. Penard suggests that actually the ectoplasm should include all the inner granular area and its inclusions, including of course the centroplast and clear area around the nucleus. This suggestion makes especially good sense, Penard argues, because it puts digestion of food vacuoles in the endoplasm, where it occurs in all other protozoan cells. The clear area around the nucleus, he further urges, may be nothing more than the clear area that invariably surrounds the actinophryidan nucleus. In fact, he explicitly suggests homologizing the two. Penard's suggested reinterpretation of centrohelidan ecto- and endoplasm seems so sensible that we will follow it here, rather than keeping to conventional usage of these terms. *Acanthocystis aculeata* (Fig. 7–1B) illustrates the organization of the Centrohelida. However, to complicate the picture, it should be pointed out that not all authors see this separation of ectoplasm and endoplasm as clearly as Penard. Dobell (1917) for example, in describing *Oxnerella maritima* explicitly says such a differentiation is not visible in that form (Fig. 7–5). The axopodia, as was mentioned in connection with *Wagnerella*, radiate out from the centroplast. In a very important contribution to the fine structure of these cells, Hovasse (1965) shows that the microtublar components of the axial filaments do not lie in the striking pattern seen in *Actinosphaerium*. The microtubules in *Raphidiophrys* are simply aligned in parallel bundles that terminate at the edge of the centroplast. Their regular distribution is somewhat disturbed by the presence of the nucleus, and this latter structure, in its turn, is often wedge shaped, with the tip of the wedge directed towards the centroplast. This would seem to be a way of accommodating the presence of the nucleus to the simultaneous presence of axial filaments in its immediate environment.

Examining functional aspects of these cells, now, the axopodia are, of course, the means of capturing prey and of locomotion. In stalked or otherwise sessile forms, the axopodia are limited to their feeding role. The same variety of food that is ingested by the actinophryidans is also ingested by the centrohelidans.

Binary and multiple fission are two modes of asexual reproduction in the Centrohelida. Certain forms are also reported to form zoospores. The picture of fission is most readily presented by summarizing the highlights of the process in *Oxnerella maritima* (Dobell 1917) and *Acanthocystis aculeata* (Stern 1924). These are members of the Aphrothoraca and Chalarothoraca, respectively. These accounts center on the behavior of the centroplast, its central granule, and the nucleus. Apparently one of the first evidences of the onset of fission is the duplication of the central granule. Dobell figures clearly the dumbbell shape and subsequent separation into two granules that signifies self-duplication of this cytoplasmic organelle (Fig. 7–5). At this time the axopodia are being reduced, though their axonemes persist in the ecto- and endoplasm for some time thereafter, possibly as late as metaphase. The nucleus takes a position between the two centroplasts and proceeds to enter prophase by condensation of the chromatin into chromosomes. A spindle appears between the centroplasts, and the nuclear membrane starts to disappear (Stern 1924), first towards the poles. This is late prophase. It reappears in late anaphase. At metaphase the chromosomes are tightly aligned at the equatorial

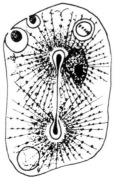

Fig. 7–5 *Oxnerella maritima* in an early division stage. The nucleus is to the right of the upper portion of the centrosome. (Redrawn from Dobell (1917).)

plate of the spindle and then start anaphase movement by a highly coordinated movement towards the two opposite spindle poles. The spindle fibers converge in typical fashion at each pole onto a clear area around each central granule, which in these cases appears to function as a centriole. As anaphase movement progresses, the chromosomes fuse and before they reach the ends of the spindle, movement ends. Telophase is apparent as the fused chromosomes, now included in a new nuclear membrane, start to form the chromatin of a new resting nucleus. The cell cleaves, sister cells separate, and from each centroplast new axonemes develop, and eventually axopodia reappear. This pattern of events differs only in relatively minor details between *Oxnerella* and *Acanthocystis*. Recall, however, that *Acanthocystis* is multinucleate, and in that each nucleus has its own spindle, Bělař (1916), with characteristic acuity, suggested that the centroplast could not really be a centriole for each of these nuclei. Hovasse (1965) was unable to find a centriolar structure in the centroplast of the uninucleate centrohelidan *Raphidiophrys elegans*, thus confirming Bělař's brilliant surmise.

Acanthocystis aculeata can also reproduce asexually through the formation of buds (Schaudinn 1896). The nuclear divisions associated with this process occur without the direct participation or even the division of the centroplast. Such divisions have also been reported for fissioning cells (Stern 1924). The buds are reported to be able to transform into biflagellated zoospores, which in turn form an ameboid stage, with lobopodia, and this finally develops into the heliozoan stage. The centroplast reappears from a granule which emerges from the nucleus of the cell. Unfortunately, these observations have never been confirmed by other workers and have never received the full, detailed account promised by Schaudinn, to fill in his somewhat preliminary report. We need not, however, go as far as Dobell's crusty comment made in reference to this report of zoospore formation, i.e., "As experience shows, Schaudinn's brief statements and incomplete descriptions are not always to be accepted as established facts; for the magnitude of his mistakes is beginning already to rival that of his success." (Dobell 1917, p. 537)

The only other member of the Centrohelida that is known to show budding is the remarkable stalked form *Wagnerella borealis*. This marine form, with a worldwide distribution, has a "head" with a centroplast, axopodia, and spicules. The nucleus, however, rests in the holdfast at the base of the stalk (Fig. 7–6). (The general organization of the organism and behavior of the nucleus is reminiscent of the much more familiar green alga, *Acetabularia*.) Zuelzer's (1909) detailed account of structure and reproduction of this form shows that, despite the stalked character and unusual location of the nucleus, the nature of the centroplast and its relation to the axopodia are typically centrohelidan. And, further, when the nucleus moves up to the head and division occurs, the centroplast acts like a centrosome, and the mitotic division figures resemble those described for *Oxnerella* and *Acanthocystis*. Also, budding can

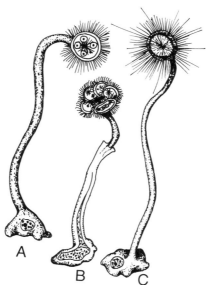

Fig. 7–6 *Wagnerella borealis*. A. Beginning of budding with multiple nuclei, each with a "karyosome", in the head region. Cytoplasm will cut off around each nucleus and the nucleated bud will escape from the parent organism. B. Multiple fission of the head-region. C. Vegetative plant with nucleus in the expanded hold-fast region. (Redrawn, by permission from Grassé (ed.) (1953) *Traité de Zoologie*, Vol. I, part II, Masson et Cie.)

occur in two different ways, depending on the behavior of the nucleus. The nucleus can lie in the head, divide several times, and then subsequent cell division will cut the head into individual uninucleate cells, each with a centroplast. This process differs from that reported for *Acanthocystis* in the concomitant division of nucleus and centroplast. Or, secondly, budding can occur using nuclear materials which were extruded from the nucleus in the holdfast and which migrate up the stalk to the head. Eventually the nucleus in the foot (Hauptnukleus) degenerates, but its products are the single nuclei of the cells formed by multiple cytokinesis of the head. This behavior of the so-called Hauptnukleus lead Hartmann (1911) to include it among his examples of polyenergid nuclei, i.e., nuclei which, like those in the Radiolaria, can subdivide many times, and yet each product carries a complete set of genetic information, as evidenced by the fact that they form the nuclear component of viable cells. In more modern terms, polyenergid nuclei are polyploid and capable of segregating out complete genomes.

Wagnerella borealis is not the only stalked form among the Centrohelida. Other examples are *Actinolophys pedunculata*, an aphrothoracan, and *Cienkowskya mereschkowski*, a chalarothoracan. As their classification indicates, the first has no spicules and the latter does. Work on the division of these forms (Villeneuve 1937) shows the similarity of their centroplasts and their binary fission to the forms we have already discussed. Especially important is the fact that in these two species the nucleus lies in the head in a typical centrohelidan fashion, suggesting, obviously, an organization intermediate to the stalkless centrohelidans and the condition in *Wagnerella*. No budding or other form of reproduction is known in these cells.

In fact, generally speaking, sexual processes are not known in the Centrohelida. Zuelzer (1909) suggests the possibility of gamete formation from some of her observations on *Wagnerella* but is unwilling to make a definite claim for it. In any case what she reports in nowise resembles the pedogamy of the Actinophryida.

Homologies. With the foregoing remarks in mind regarding the organization and feeding and reproduction of the Centrohelida, what can be said about homologous relationships among them? Foremost, structurally, is the uniqueness of the placement and composition of the centroplast. It fulfills the Remanian criteria of homology, and thus we can speak of that major item of evidence for the evolutionary affinity of those members of the Centrohelida in which such a structure is known to exist. It should be remembered that in some species placed in this group, the presence of a centroplast is simply inferred from the eccentric position of the nucleus. Beyond this key character no other structureal feature offers good evidence of being homologous, but the distribution of other characters with the same distribution as the centroplast (this is using Remane's first subsidiary criterion) renders their homologous nature probable. We refer here to presence of axopodia and possibly the organization of the cell body into ectoplasm and endoplasm in Penard's sense.

Functional homologies are found in the nature of binary fission, especially in the behavior of the centroplast, the chromosomes, and nuclear membrane. These provide a constellation of events that cannot be interpreted as convergent and hence also speak, along with centroplast structure, of genetic affinity. Budding and feeding and patterns of locomotion are too imprecisely known to allow of their homologizing by the major Remanian criteria. Their homologous nature can be inferred from the secondary criteria.

There is no fossil or developmental or other evidence that allows us to see the direction of evolution in those traits which are possibly aposemic. Two such traits are the skeletal elements and body form. The former has the possible sequence, reflected in Trégouboff's and earlier classifications, of no skeleton → mucilaginous ectoplasmic layer and exogenous skeleton → mucilaginous layer and endogenous skeleton. Body form has two major forms—stalked and unstalked. It is possible to think of development of skeletal materials being aposemic and with body form taking two lines of development, as follows. Stalked forms without skeletal materials, e.g., *Actinolophus*, gave rise to *Cienkowskya* and then *Wagnerella*, the latter two with spicules. In all three of these forms the fibrils extending from the centroplast include some which extend into the stalk; presumably the stalk is therefore a specialization of associated axonemes and their microtubules. As another development, there are the unstalked forms, going from *Oxnerella*, without any significant development of an ectoplasm, to *Acanthocystis* with its well-developed spicules. But what of the relation of the stalked to the unstalked form? Which, if either, is the plesiomorph? At present there are no criteria available—paleontogical, epigenetic, etc.—to solve this problem in an objective manner. The best that can be done is to speculate from the basic grounds of evolutionary progression going from less to more complex forms. This argument, as always, has no promise of certainty, only plausibility.

From such a point of view the Aphrothoraca, represented by *Oxnerella* and *Actinolophus*, are the simpler Centrohelida. As to which is a more plausible plesiomorph, this will depend on the chances of finding some possible ancestral form with a centroplastlike structure and centrohelidanlike fission. Also we would expect axopodia in this ancestor. If this form is stalked, our speculations would favor a stalked plesiomorph for the Centrohelida; if it were unstalked and spherical, we would guess that an *Oxnerella*-like form would be more plausible.

Because of the foregoing comments, it is more appropriate to turn next to the helioflagellidae rather than to finish the discussion of the heliozoans (as defined by Honigberg *et al.* 1964) by discussing the Desmothoraca.

HELIOFLAGELLIDAE

This group, containing 6 genera and 16 species in Trégouboff's (1953d) treatment, is best known through the freshwater genus *Dimorpha* and only this taxon is generally agreed on as being typically helioflagellate (Bovee 1960). The most notable feature of these cells is their ability to exist in two completely distinct forms, a heliozoan stage and a biflagellated stage. (Bovee and Wilson (1961) have described a third, ameboid stage, for *D. floridanis*.) The next feature of special interest is the occurrence of a central granule and centroplast, eccentric nucleus and axopodia in the heliozoan stage, and the transformation of the granule into basal granules for the flagella of the flagellated forms. Further, in terms of structure, *Dimorpha mutans* (Gruber 1881; Bělař 1926) is a stalked form, and *D. floridanis* (Bovee 1960) (Fig. 7–7A) is unstalked. *D. mutans* shows a flagellum arising from the central granule and passing into the stalk (Fig. 7–7B).

During division, both forms show nuclear behavior very like that of the Centrohelida. The greatest detail is found in the studies on *D. mutans* (Bělař 1926). The centroplast divides and acts like a centrosome; spindle fibers form between the two centrosomes; the nucleus migrates to a position on the spindle (prophase is now in progress); the nuclear membrane disappears;

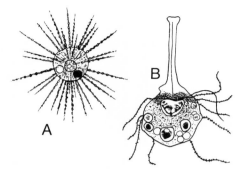

Fig. 7-7 Two examples of the Helioflagellidae. A. *Dimorpha floridanis*, showing the unstalked form with radiating axopodia and central granule. (Redrawn, by permission, from Bovee (1960).) B. *Dimorpha mutans*, showing the stalked form with its axopodia (granulated outline) and flagellum (smooth outline). (Redrawn, from Bělař (1926).)

metaphase shows chromosomes neatly aligned on the spindle's equatorial plate; anaphase and telophase follow, with dense clumping of the chromosomes in anaphase and new membranes appearing around the chromatin mass during this time. In *D. floridanis*, both heliozoan and flagellate stages are capable of binary fission. No sexual processes are reported.

These cells feed by means of the axopodia; *D. floridanis* feeds only in the heliozoan stage and captures small flagellates and ciliates.

Because of the presence of the distinct heliozoan stage, it seems unfortunate to have separated the Helioflagellidae off from the Heliozoa, as was done by Honigberg *et al.* (1964). Bovee's comment that the heliozoan phase is the more important, taxonomically, is supported by the biological fact of feeding occurring exclusively in that stage, as far as could be observed.

The time seems ripe for a specialist in these forms to undertake a species by species review of the helioflagellates to determine first the validity of the various species (for example, Bovee (1960) doubts the validity of *Dimorphella elegans*) and obtain detailed information on the centroplast structure and mode of fission. Then, a comparison with the aphrothoracan centrohelidans should be undertaken to determine homologies. Needless to say, the possibility of homologous relationships have already occurred to various authors (Stern 1924; Trégouboff 1953d). Especially useful would be electron microscopic studies of the whole centroplast in these forms.

Even assuming that subsequent work will confirm the speculations of an evolutionary affinity between the Centrohelida and Helioflagellidae, there is still no evidence to answer the question posed earlier, i.e., which is the plesiomorph, the stalked or the unstalked form? On morphological grounds, the structure of the heliozoan form of *D. mutans* appears less complex than that of *D. floridanis*. But, again, such arguments are never decisive. We must wait for the accumulation of more information before we can proceed any further with a careful phylogenetic analysis of these groups.

DESMOTHORACIDA

It can immediately be made clear why Penard and others since him do not consider the Desmothoracida to be typical heliozoans: they lack a centroplast and axopodia. The distinguishing features of this group are the presence of continuous perforated shells surrounding a more or less spherical cell body; the shells are chitinous but impregnated to varying degrees with silica, rarely being of pure silica; filopodia (never with axonemes) are present and may be branched; the cell body may or may not be attached to the substratum by means of a stalk; and reproduction by means of flagellated zoospores is known to occur as well as by binary and multiple fission. No sexual processes are reported.

This group contains six genera and twelve species (Trégouboff 1953d).

Except for *Clathrulina elegans*, members of this group are generally so fitfully described, usually only in terms of their gross morphology, that it is not possible to enter into the survey of their biology that is necessary for phylogenetic purposes.

The general body form of *C. elegans*—spherical head encased in a perforated shell from whose apertures filopodia extend, stalk with rootlike extensions at its base—is similar to that of the stalked Centrohelida. But internal details of the composition of the cell deny any evolutionary affinity. There is no special differentiation of ectoplasm and endoplasm, no centroplast, the nucleus is centrally located, and developing off one side of it is the stalk which Valkonov (1928) homologizes to a flagellum but without demonstrating a basal body or other essential details to make the homology convincing. Nuclear division (Bělař 1926) occurs in the apparent absence of centrosomal material.

The life cycle (Bardele 1972) proceeds from the stalked form through release of biflagellate swarmers to a naked "heliozoan," i.e., radially symmetrical with filopodia. This form settles on the bottom and forms a stalk and a perforated shell to restore the cycle to its starting point. Under unfavorable conditions, cysts can be formed. Excystation is by formation of swarmers, which transform, again, into naked "heliozoans" and then into stalked, shelled forms. Bardele's (1972) beautiful study also contains fine structural details on special organelles, possible storage of the plasma membrane, and details on the morphogenetic transformation of swarmer into filopodial ameboid form ("heliozoan"). Bardele also notes: "Although no rigid axial filament was observed in the filopods at light microscopical level, we were not surprised to find 10–20 *microtubules* in them. . . . The filopodial microtubules in *C. elegans* show no such precisely patterned array as it is known from "true" Heliozoa or Radiolaria. At best a somewhat irregular hexagonal packing may be disclosed in some micrographs. . . ." (p. 229)

In the absence of adequate information on the other members of this group, it is not possible to proceed to any meaningful phylogenetic inquiries within the Desmothoracida, though the possible relation of *Clathrulina* to other protozoa deserves some further comment. The presence of zoospores is suggestive of a flagellate ancestry, but it must be remembered that, for attached forms, zoospores represent a major means of dispersal. Hence the zoospore could be a new adaptation arising when attachment became a mode of life (the primitive state being, perhaps, some nonshelled ameboid form) and not any sort of biogenetic recapitulation. Valkonov (1928, 1940) has emphasized the character of the filopodia and the nature of the test in the Desmothoracida, as well as zoospore formation, to suggest a relationship with some of the shelled rhizopods. This remains a plausible point of view but needing much further imformation and analysis to test it adequately.

PROTEOMYXIDA

The two families, Zoosporidae and Azoosporidae, are the representatives of this group if we follow Honigberg *et al.* (1964) in removing the Helioflagellidae from this group. The Zoosporidae and Azoosporidae represent the Monadida of Valkonov (1940) and are rendered somewhat more homogeneous by the absence of helioflagellates. In Trégouboff's (1953d) handling of the Zoosporidae, 17 of the 18 genera he mentions contain one species apiece. They are largely parasitic on algae. In *Pseudospora,* the remaining genus, there are various species. All are parasites, living off of algae, diatoms, and various flagellates. The chief characteristic of this group is reproduction through zoospores, which are formed in a cyst. Zoospore formation, except for Schaudinn's (1896) unconfirmed report in *Acanthocystis aculeata,* and its occurrence in at least certain Desmothoracida, is unknown in the heliozoans. The Zoosporidae possess no axopodia; rather, in their ameboid stage they possess simple filopodia. No centroplast or elaboration of ectoplasm is apparent. In fact, the only heliozoan-like feature present is the very nonspecific character of spherical symmetry in the ameboid stage. Lack of detailed knowledge regarding the various members of the group, combined with specializations related to their parasitism, result in a very ragged group with little rigorous evidence of affinities among themselves and even less basis for comparisons with other groups.

Essentially the same picture holds for the Azoosporidae, whose eleven genera and approximately 30 species are dominated by *Vampyrella* with its 20 species. These forms, alternating between an ameboid and a so-called heliozoan stage, form cysts but never zoospores. The heliozoan forms are simply spherical with delicate unbranching filopodia. Most members of *Vampyrella* are parasites of algae. Phylogenetic analysis of the group must wait till the forms are better known.

LABYRINTHULIDAE

This is the last group to be mentioned under the rubric of the Helioza and the least helizoanlike of them all. A recent extensive review of the members of the single genus in this groups leads to this conclusion, among others: "The uniqueness of its gliding motility along slime tracks, lack of phagotrophy, and poorly understood congregation tendencies emphasize the present taxonomic isolation of the group." (Pokorny 1967, p. 697). Its former position in the Proteomyxida, along with helioflagellates, justified the comment that the Proteomyxida of earlier workers (Calkins 1926; Hall 1953) was "... a junk drawer of neglected creatures." (Pokorny 1967, p. 706). The most important fact for us is the lack of phagotrophy in this group. This, by definition (see chapter 4), can remove the group from the protozoa. Work on *Labyrinthula algeriensis* describes that species as having biflagellate spores and a leucoplast (Hollande and Enjumet 1955). On that basis, it was suggested that the genus *Labyrinthula* be put close to the colorless flagellates of the Chrysomonadida. This would seem to be a more plausible treatment of the group than retaining them in the protozoa. But, to further complicate the picture, Amon and Perkins (1968) report that the biflagellated zoospores are like those of Phycomycetes, hence they prefer a phylogenetic affinity to the fungi for these enigmatic organisms.

CONCLUSIONS

The heterogeneity of the heliozoans, despite their superficial similarity due to slender pseudopodia radiating from a spherical cell body, is the central issue in attempting to examine them phylogenetically. Only two genera of the Actinophryida give clear evidence of homologous relationships. The Centrohelida are the most homogeneous group, and to them the helioflagellates may well be attached. But the inability to find a plesimorph for these forms is the present limitation on the further analysis of evolutionary relationships among them. Some time ago Stern (1924) suggested that the Actinophryida, if they are a homogeneous (monophyletic) group, are quite distinct from the Centrohelida. In fact, the former probably came from ameboid forms (as indicated by their lack of a central granule homologous to a flagellar basal granule), whereas the Centrohelida, through forms like *Dimorpha*, very probably were derived from flagellates. Then, the further suggestion (Valkonov 1928) that the Desmothoracida are derivable from the testate or shelled amebas, suggests yet a third point of origin for heliozoan forms. Finally there are the proteomyxids, of quite unknown parentage, and the labyrinthinulids, which seem not to be protozoans, to complete the picture of obvious evolutionary diversity of the sun animalcules.

What does this diversity mean? It seems, speaking very broadly, that the correlated occurrence of spherical symmetry and filopodia was a common evolutionary development in various flagellate and ameboid lines. Spherical symmetry with slender pseudopodia apparently offers various, somewhat isolated possibilities for survival in various fresh water and occasional marine forms and hence results in convergent similarities starting from the various ways in which that body form can originate. Use of the criteria of homology, especially the compositional criterion, shows this convergence in the superficial similarities which provide the basis for the taxon Heliozoa. The Adansonian device of aiding memory by what seems the most "natural" way to classify these forms (Sokal and Sneath 1963) reveals little of the evolutionary diversity of these forms, even though the heliozoan concept effectively pulls together a messy assemblage of forms—a junk drawer—in a way that is admittedly useful for purposes of identification.

Chapter 8

Radiolarians

THE quite marvelous complexity of the radiolarians and the consequent difficulties in understanding their organization are typified by the fact that their most distinguished student, Ernst Haeckel, persisted for years in considering them multicellular. It was Richard Hertwig (1876, 1879) who finally sorted out the details of their organization, including the presence of symbiotic algae, to pronounce them, unequivocally, unicellular. These two great biologists laid the early foundations of our understanding of this group (Bütschli 1880–1882), but troubles plague us to this day in fully understanding the Radiolaria. For example, sexual processes, which are highly probable in this group (they form flagellated "swarmers"), have never been observed. The latest claims purporting to show their occurrence turned out to be release of flagellated algal symbionts from the host (Chatton 1920; Pätau 1937). Extended, detailed study of the radiolarians over the years (e.g. Haeckel 1862, 1887, 1888b; Haecker 1908; Popofsky 1909, 1913a, 1913b; Trégouboff 1953a, 1953c; Deflandre 1952, 1953d; Hollande and Cachon-Enjumet 1960) has elaborated further some of the details of the group, but the broad outlines of our understanding are still seen much in the terms Haeckel first developed.

General Characteristics

The essential diagnostic feature of the group is the presence of one or more perforated capsules lying centrally in the cell body. Added to this capsule are certain other special features that give the group a clearly recognizable homogeneity. The outermost parts of these organisms, that is, from the perforated capsule outwards, is termed the ectoplasm or calymma. Through this specialized area there extend the slender, branching, sometimes anastomosing, filopodia. Also in the calymma is found an often highly-developed gelatinous layer composed of numerous, homogeneous vacuoles. The spicules and shells of the so-called skeleton also lie in the calymma and are invariably composed of silica. Within the capsule is endoplasm and the nucleus, which appears to be invariably polyploid. All the radiolarians are marine forms.

STRUCTURE

Skeleton. Describing the foregoing features in more detail, we shall start with the skeletal elements, for these are among the most conspicuous and, certainly the most beautiful, parts of these organisms. They provided Haeckel the details needed for his extensive classification, which included some 4,000 species. In some cases—the radiolarian species are invariably defined morphologically—there are no skeletal elements. Such forms are scattered throughout the group. In certain other radiolarians the skeleton is exogenous, being an accumulation of shells and tests from other microfauna found in the environment. Where the test is endogenous, it is formed of silica. The simplest form is that of slender rods, which may be slightly curved. Or, again, they take the form of tetraxonic spicules with variations (Fig. 8–1). Furthermore, the siliceous material can ramify throughout the calymma resulting in a "spongy" structure (Trégouboff 1953c), or it can result in shells of various forms, the most remarkable of these being instances of concentric spheres, held apart by supports extending from one level to the next (Fig. 8–2). The shells may be completely enclosed except for fenestrations, or they may be open on one side, as well as being perforated in more or less regular patterns. The skeletons show various patterns of symmetry, from spherical through radial to bilateral.

Calymma or ectoplasm. The calymma can usually be seen to contain three layers. Outermost is a thin layer of cytoplasm with the cell membrane at its external surface. Next, there is a middle layer, whose most prominent feature is its numerous inclusions. These include food vacuoles and the globular alveoli of a substance identified, up to the present day, simply as gelatinous (Dogiel, Poljanskij, and Chejsin 1965). Innermost is a denser and less vacuolated area of cytoplasm immediately around the central capsule. Before leaving the calymma it is to be noted that in the living condition, radiolarians are

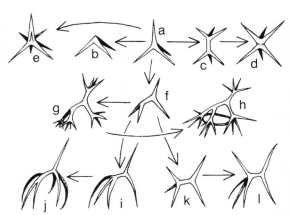

Fig. 8–1 Spicules of the nasellarian Radiolaria, also showing possible phylogenetic relations. The original type is thought to be the tetraxonic form (a). From it various other forms are derivable (b, c, e, and f), some of which, in turn give rise to yet other variations. (Original by Popofsky (1913b). Redrawn, by permission, from Grassé (ed.) (1953) *Traité de Zoologie* Vol. I, Part II, Masson et Cie.)

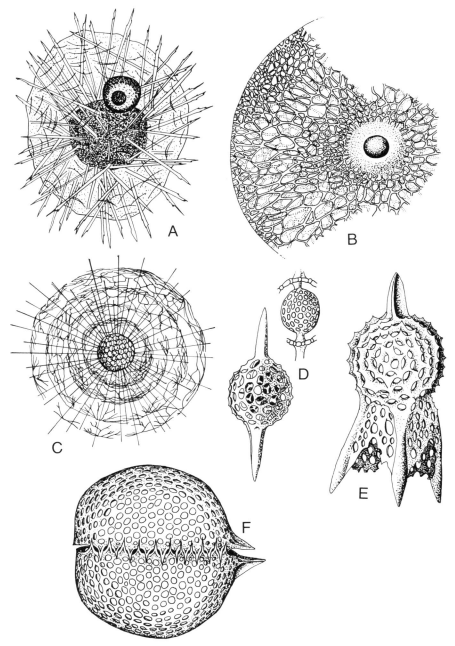

Fig. 8-2 Types of skeletal structure and organization. A. Radial spicules in *Aulacantha scolymantha*. B. Spongy skeleton from *Plegmosphaera exodictyon*. C. Skeleton with concentric layers in *Arachnosphaera myriacantha*. D. Skeleton with concentric shells from *Stylosphaera lithatractus*. E. Skeleton open on one side as in *Dictophimus sphaerocephalus*. F. Bivalved skeleton in *Conchidium argiope*. (Redrawn, by permission from Grasse (ed.) (1953) *Traite de Zoologie*, Vol. I, part II, Masson et Cie.)

often strikingly colored, being brown, red, green, yellow, blue, and various shades of these (Haeckel 1862, 1887; Trégouboff 1953c). These colors are sometimes diffused throughout the cell and are also due to colored inclusions, such as lipid droplets and photosynthetic symbionts.

Central capsule and endoplasm. The central capsule is bounded by a membrane, which is a double membrane in certain radiolarians, and this capsule is perforated according to three general patterns. In some forms there are many small perforations covering the whole surface of the capsule, or they may be concentrated into irregularly polygonal areas of perforation. This type of capsule is found in the group designated by Haeckel as the Spumellaria (Fig. 8–3 A), or the peripylar forms. A second type of perforation is the pres-

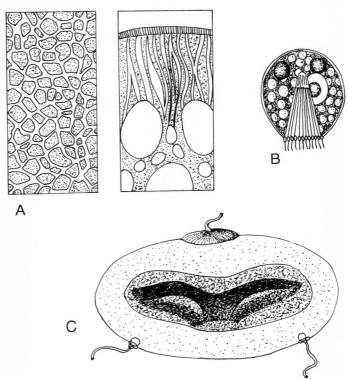

Fig. 8–3 Examples of three types of perforated capsules. A. On the left, a surface view of the capsule of *Thalassicola nucleata,* showing fine pores within irregularly polygonal fields. On the right, a transverse cut through the capsule reveals extensions into the endoplasm, which are continuous with the pores. B. The central capsule of *Tridyctiopus elegans* with pores at the base of the pyramidal podocone, which extends into the capsule. C. The capsule of *Tuscarora nationalis.* There are three pores, the large astropyle (top of capsule) and two smaller astropyles (lower left and right of capsule). The large internal structure is the nucleus. (All redrawn, by permission, from Grassé (ed.) (1953) *Traité de Zoologie,* Vol. I, part II, Masson et Cie.)

ence of a single field of pores at one localized site on the capsule. These pores are the external terminals of fine tubes, which extend into the capsule to form a conelike structure, the podocone (Fig. 8–3 B). This condition is found in the Nasellaria or monopylar forms. Finally, there is the capsule with specialized openings, usually three in number, found in the Phaeodaria, or tripylar radiolarians. These can be reduced to one large one, the astropyle, or there can be many smaller ones, the micropyles (Fig. 8–3 C). The nature of the capsule is only crudely specified by the fact it dissolves in hypochlorite solutions ("Eau de Javel"—Trégouboff 1953c) but is resistant to acetic acid. There may be one or, less commonly, several capsules per cell.

Fine structure studies on the capsule of various species of spumellarians lead Hollande, Cachon, and Cachon (1970) to the interpretation that the capsule is bound internally by the endoplasmic reticulum and externally by the plasmalemma, or cell membrane. This point of view has important consequences, which are spelled out by these researchers. The separation of the endo- and ectoplasm, which is demarcated by the capsule, also assures that mitochondria and Golgi bodies reside only in the endoplasm and that gelatinous materials, food vacuoles, and other inclusions and vesicles of the ectoplasm are always external to the capsule and enfolded in elaborations of the cell membrane. The pores or other passages through the membrane do not accommodate free flow of materials. The larger pores, or fusules, are for passage of axopodia from the endo- to the ectoplasm. Fissures occur between the plates, or plaques, which contain pores and which in aggregate compose the capsule, and these fissures appear to be sealed as regards any effective flow of material between the inner and outer cytoplasmic portions of the cell. Finally, the capsule itself is seen to be a "cortical differentiation of the cell and not an endogenous formation." (Hollande, Cachon, and Cachon 1970, p. 314). This would appear to say that the capsule is deposited on that side of the endoplasmic reticulum not enclosing the cytoplasmic contents of the endoplasm, much as if it were an intravesicular secretion. However, according to the authors' Fig. 1 (given here as Fig. 8–4), the capsule is continuous with the endogenous portions of the ectoplasm. This reversal of relations to the endoplasmic reticulum and to the ectoplasmic membranes (plasma membrane?) is hard to understand.

A more conservative interpretation of the electron micrographs and other data presented by Hollande, Cachon, and Cachon would simply state that the capsule is a secretion product of the cell and is bound on both sides by membranes, which is not unexpected for such intracellular products. Nothing need be said at present as to whether the inner membrane is endoplasmic reticulum and the outer one plasmalemma.

The function of the capsule can only be guessed at. That it is of major functional importance seems certain, because of its universal occurrence and its invariably well-developed condition. The pores could be a device, in con-

270 *Phylogeny of the Protozoa and Early Metazoa*

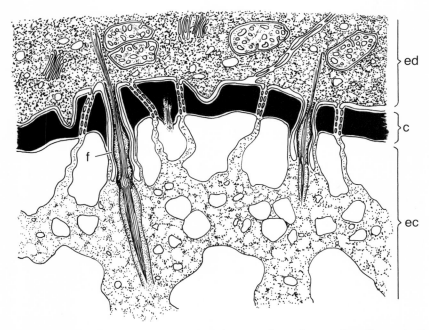

Fig. 8–4 Reconstruction, from electron micrographs, of a section through the capsule (c) of *Thalassophysa sanguinolenta*. Ectoderm (ec), endoderm (ed). Hollande, Cachon, and Cachon (1970), interpret the membrane on the endodermal surface of the capsule as being endoplasmic reticulum and that on the ectodermal side as being plasma or cell membrane. The axopodial microtubules pass through the capsule of specialized pores or fusules (f). (Redrawn, by permission, from Hollande, Cachon, and Cachon (1970).)

junction with the capsule membrane, for controlling movement of materials between the outer calymma, or ectoplasm, and the endoplasm within the capsule — a homeostatic device of some sort. There seem to be some lines of evidence suggestive of this in terms of endoplasmic differentiations aligned with the pores (Fig. 8–3 A), but this tells us nothing of the actual physiological events. Haeckel referred to the capsule as a reproductive organ in recognition of the fact that an organism reduced to just its capsule (with the contents, including the nucleus, intact) could regenerate a whole cell, and that swarmers were formed within it. Hertwig (1879), critically appraising this interpretation, points out that its specializations are not adjusted solely to reproduction and leans to Gegenbaur's view of a protective device, protective more in a physiological than a purely mechanical sense.

"... more correctly it would be designated a protective organ, for by means of it, part of the materials of the cell body, the nutritional reserves of the oil droplets, and especially the nucleus are withdrawn from environmental influences, which are only set off by the interposition of the central capsule and the intervening food-capturing and absorbing sarcode (cytoplasm)." (Hertwig 1879. p. 114. Original in German.) This speculation does make sense of the obvious correlation that exists, i.e., that the outer cytoplasm or calymma is highly specialized for food capture and digestion and for flotation and secretion of the skeletal elements, and that the material inside the capsule is much more homogeneous and has the nucleus as its most obvious organelle along with other typical cytoplasm organelles. Probably, the truth of the matter is a combination of Haeckel and Hertwig, with the central capsule being a structural element that aids and abets the specialized functioning of two major subdivisions of the radiolarian cell.

The distribution of cellular organelles described by Hollande, Cachon and Cachon (1970) certainly reinforces this speculation. But their observations of very limited access through pores or fissures also implies that passage of materials across the capsular boundary must be largely concerned with substances of molecular dimensions unless, of course, the fusules and the axopodial microtubules represent a special transportation device. Whatever the mechanism, it is clear that somehow products of digestion from the ectoplasmic food vacuoles must reach the endoplasm, and gene products and other results of biosynthetic activity must be available to the ectoplasm or calymma.

As we have said, the endoplasm within the capsule is freer of vacuoles than the calymma, though it may contain conspicuous numbers of fat or oil droplets. This part of the cell apparently never contacts the skeletal material. Deflandre (1953d) shows very convincing evidence of the capsule folding around the skeleton as it passes through the center of the cell. Figures given by others (Haeckel 1862, 1887, 1888b; Hertwig 1879; Trégouboff 1953c) would indicate the skeleton could lie in the capsule, though Hertwig, in characteristically thoughtful fashion, shows some cases where the skeleton actually lies in the grooves between papillae on the outer surface of the capsule (Hertwig 1879; p. 106, Plate V, Fig. 1b). Present evidence seems to favor Deflandre's view of this matter.

Axopodia. As recently as 1953, Deflandre and Grassé stated categorically that radiolarians are characterized by absence of axopodia. However, abundant recent research demonstrates the presences of a microtubular axis in the filopodia of these cells; axopodia are present in radiolarians. It appears, however, that no single fine-structure pattern is common to these cells. In certain sphaerellarians (a suborder of the peripylar Spumellaria) the axopodia are proximally inserted on the nuclear membrane and then pass through the endoplasm, through the capsule, and out through the endoplasmic calymma to the outside environment. In the very peculiar form *Sticholonche zanclea* (is it a radiolarian? an acantharian?) the axopods arise from the outer surface of the capsule.

In other radiolarians the axopodia originate from a distinctive area termed the axoplast. In many Sphaerellaria the nucleus lies next to or even wraps around the axoplast (periaxoplastid forms). In others the axoplast lies within the nucleus (centroaxoplastid forms) (Hollande and Cachon-Enjumet 1960; Hollande, Cachon, Cachon-Enjumet 1965). In other radiolarians, the monopylar nasellarians, it was seen that the extranuclear axoplast has certain relations to the developing skeleton and may well determine patterns of skeletal development.

The arrangement of axopodia seems determined not only by the emergence of microtubular patterns from the axoplast but also in their distribution relative to the capsule. They pass through the capsule at highly differentiated pores, called fusules (Hollande, Cachon, and Cachon 1970). It is not at all clear whether the axopodial microtubules determine the location and formation of these special pores, or the pores determine the location of the microtubules, or whether there is a cooperative interaction.

In any case, there are distinctive distributions of the axopodia in the radiolarians. Two patterns need some special comment, for they represent significant deviations from the spherical symmetry often thought to be common to these cells. In certain Spumellaria (Hollande, Cachon, and Cachon-Enjumet 1965) there is an axopod "priviligié," which is longer and stouter than others and is termed an axoflagellum, because it undulates, or at least oscillates, slowly. It lies in a special area of the cell, called with inflexible scientific logic, the axoflagellar cytoplasm. This structure is suggestive of the asymmetry found in many Nasellaria (monopylar radiolarians), where the axopodia extend from a central axoplast as a hollow cone (the podocone) out through one side of the capsule, emerging through a circlet of pores in that structure. This results in a cone of axopodia emerging on one side of the organism—a side usually designated as downwards or ventral—which implies an apparent geotropism on the part of these special cells. This seems to be the most specialized example of axopodial distribution to be found in the Radiolaria.

Thus far there have been described at least three different patterns of microtubular distribution within the axopodia (Hollande, Cachon, and Cachon-Enjumet 1965; Hollande, Cachon, and Valentin 1967; Cachon and Cachon 1971). We must wait to see what further research will reveal before generalizations will be possible regarding these patterns.

Nucleus. The radiolarian nucleus can contain up to 2700 chromosomes, or, more precisely, groups of linearly oriented chromosomes which are probably complete genomes (Grell 1953a; Grell and Ruthman 1964; Ruthman and Grell 1964). These highly polyploid nuclei are also termed polyenergid, from the further fact they can be subdivided into viable subparts; in the radiolarians these parts are complete genomes. Upon such division, the products of the polyenergid, primary nuclei, form the secondary nuclei which eventually come to lie in flagellated swarmers (Hertwig 1879; Huth 1913). Hertwig states that this swarmer formation is a property of all radiolarians. Apparently he has never been contradicted in this claim.

FUNCTIONS

Environmental exploitation. The radiolarians are floating predators. They are among those forms that wait for fortuitous contact (Fauré-Fremiet 1967a) with food particles in their environment, and they capture these particles by means of their fine pseudopodia. The particles are drawn into the cell body, enclosed in food vacuoles, and there digested. This food consists of a wide range of microscopic plankton—bacteria, algae, other protozoa, tiny invertebrate larvae, and the like. The amazingly large size of some of the radiolarians—they can be up to 30 mm in diameter (Trégouboff 1953c)—allows them to handle a considerable range of size in food materials.

Various devices act cooperatively to achieve a more than passive flotation of the radiolarian cell. That there must be some control over their buoyant density is apparent from the fact of their distribution (Haecker 1908; Trégouboff 1953c; Reschetnjak 1955). Certain forms are found characteristically at certain depths, and there is evidence of seasonal changes in depth. The skeletal elements, because of their high density, would be a deterrent to floating. Even hollowed ones, as in the Phaeodaria, would not be helpful unless gas filled, for which there is no evidence. It is true that the delicate and extensive elaboration of the spicules and shells provides them with forms that are adapted to generating hydrodynamic friction and thus minimizing sinkage in a fluid medium. Nevertheless, there must be something in the cells to overcome the negative buoyancy of the silica. In contrast to the Acantharia (next chapter) which contain a well-identified flotation apparatus, no such system is clearly identifiable here. It is obvious that the plentiful lipid materials are buoyant. It seems probable that the gelatin layer of the calymma might be a flotational adaptation, but how it would work is not known. At present, the extension and withdrawal of the pseudopodia seems also to be a device that could control the density of the cell, as well as being a food-gathering device. There are also reported to be vacuoles formed which are expanded and deflated (but they are not typical contractile vacuoles) in response to flotation needs (Trégouboff 1953c). In any case, these cells are able to maintain themselves as predators whose controlled vertical flotation through ocean waters in effect filters out prey by entrapping it in the pseudopodia. They may also feed in this manner on the carbon-rich aggregated particles that are formed spontaneously at the cell surface (Sutcliffe, Baylor, Menzel 1963; Baylor and Sutcliffe 1963).

Nutritionally, the radiolarians are phagotrophic chemoheterotrophs, i.e., perfectly good animals. Their symbiotic algae probably also supply some nutritional materials.

Homeostatic maintenance. Concerning homeostatic functions, we can say nothing, really, beyond expecting to find mechanisms typical for cells in general, i.e., there would be various devices for maintaining the internal pH, ion concentration, and metabolic pools, and so on. The central capsule, as we have mentioned, may, however, be a special homeostatic device in these cells.

Reproduction. Regarding reproductive functions, we can first repeat what was said earlier, which is that sexual processes are probable but unknown. Asexual reproduction is known and in some detail for a few species. There are two major types of asexual processes: binary fission and zoospore, or swarmer, formation. The latter results from multiple division of the initial primary nucleus into hundreds of small secondary nuclei. (See Huth (1913) for a detailed account. It is the presence of the secondary nuclei that lead Haeckel initially to consider the radiolarians as multicellular.) These nuclei become part of small, uninucleate, biflagellated cells. The subsequent fate of these cells is not known. They may meet with other similar cells, fuse, and form a zygote, which will develop into a new adult form. If such is the case, the asexual production of swarmers is the prelude to the sexual formation of new individuals. However, it seems that swarmer formation occurs in the colder months of the year, at least in the Mediterranean forms, when the cells are at lower depths (Trégouboff 1953c). If these conditions are necessary for swarmer development and subsequent fertilization, one can anticipate the great difficulties of reproducing such conditions in the laboratory.

In recent years, increasingly frequent reports by French workers have advanced greatly our knowledge of zoospore formation (Hollande and Cachon-Enjumet 1959; Hollande and Cachon-Enjumet 1960; Cachon-Enjumet 1964; Cachon, J. and M. 1969; Hollande, Cachon, and Cachon 1969). The major points which emerge are the following. All radiolarians apparently undergo vegetative growth with resulting increase in size of the cell and its nucleus, the latter apparently undergoing endomitosis to produce a polyploid organelle. Within this large, polyenergid nucleus, mitotic events, not clearly described, occur, and the nuclear membrane breaks down, with smaller nuclei appearing as discrete elements in the endoplasm. The capsule can also fragment and reform around the secondary nuclei. This results in the polycytic or collozoum stage. This apparently colonial aspect of the radiolarians had, for decades, resulted in the description of various species with multiple capsules. All of these descriptions, it would now appear, are spurious species; or more exactly, they are descriptions of the polycytic stage of the radiolarian life cycle. This stage can last for weeks or months and includes division of the polycytic organism into more such entities. Finally, nuclear divisions increase in frequency and daughter nuclei become smaller and more dense. Crystals appear within the capsules, and finally multiple cell divisions occur, partitioning the collozoum into many individual cells, each with a nucleus and a crystal. Differentiation of spores occurs, resulting in cells several micrometers in length and biflagellated (the flagella are of unequal length). The subsequent fate of these active cells is not known.

Additionally, Hollande, Cachon, and Cachon (1969) have described that in the course of division of the nuclei in the polycytic stage of *Thalanophysa* (a peripylar form), atractophores are present as the basis of chromosomal segregation. These structures, first described fully in the hypermastigote flagellates, are spindle-generating structures. In *Thalanophysa*, the spindles and their constituent microtubules are intranuclear. The atractophore itself is located just beneath the nuclear membrane. Just outside the membrane, opposite the atractophore, are two centrioles. During division, an atractophore separates into two parts, which, along with a centriole apiece, move apart 180° along the periphery of the nucleus. Centromeres are seen attached to the microtubules, and these move along the tubules and achieve chromosomal segregations. The authors refer to this mode of division as a special kind of pleuromitosis. (Typically, in pleuromitosis the centromeres lie on the nuclear membrane; here, apparently, the spindle fibers form the intermediary attachment between centromere and membrane, via the atractophore.)

Binary fission can be simple separation of one cell into two, with passive distribution of the skeletal elements. This takes place when the skeletal elements consist of separate spicules or are two separable valves of a single shell. In these cases the missing parts are formed for each fission product. (Such would be the case for *Aulacantha* (Fig. 8–2 A, and *Conchidium*, Fig. 8–2 F, for example.) Where solid shells are present, the preexisting cell presumably partitions its protoplasm so as to equip two complete protoplasmic systems, one of which will retain the old shell while the other one builds a new one, or else multiple fission occurs with spore formation. In binary fission, the nuclear behavior has been studied intensively in *Aulacantha scolymantha* (a tripylar form) by Borgert (1909), Grell and Ruthman (Grell 1953a; Grell and Ruthman 1964; Ruthman and Grell 1964) and Cachon-Enjumet (1961). Here the 1000 to over 2700 chromosomes become discrete bodies at prophase, and gather at the clear, homogeneous, central region of the nucleus at metaphase. (This region is designated the "Mutterplatte" by Grell and the archoplasm by the French workers.) The nuclear membrane has disappeared, and during anaphase the chromosomes move apart as two parallel masses of chromosomes. No centrioles or centromeres are apparent. There are microtubules among the chromosomes. Ruthman and Grell (1964) provide many fine structure details of this process, including details of the intracapsular endoplasm and its changes. The capsule itself splits in half, with one micropyle going to each daughter cell and the macropyle, which has been bisected by the fission plane, reforming its two missing halves (Cachon-Enjumet 1961). Grell has claimed and substantiated the view that each apparent chromosome in *Aulacantha* is actually a haploid genome. The constituent chromosomes are aligned end to end and form the unit of material which is segregated out to each zoospore (Grell 1953a).

There is, most unfortunately, no experimental information on the building of new tests. It has been proposed (Thompson 1942) that the pattern of the skeleton is determined by the character of the surrounding cytoplasm. This does not mean there is a simple relationship between cytoplasm and skeleton (Deflandre 1953c), but that in some cases there seems to be a crystallization process to produce the straight, needlelike or the textraxonal spicules (Fig. 8–1). In others, there is some interface difference in cytoplasmic layers to account for the shells (Fig. 8–2 C and D) or for the various ramifications of the spongy types (Fig. 8–2 B). The fenestrations of the shells could be "... spots where the delicate film (of deposited silica?) has given way and run into holes, and where surface tension has rounded off the broken edges and made rents into rounded apertures." (Thompson 1942, p. 723) Haecker (1908) has inferred from various separate observations on skeletal material in tripylar forms that there is first the formation of a basic needle ("Primitiv-nadel"). Then the gelatinous material precipitates around that, followed by deposition of the shell itself enclosing the needle, and the extent of the deposition seemingly follows available space between and around the cytoplasmic vacuoles. Haecker points out that the role of the gelatinous material may be analogous to spicule formation in sponges, which takes place in scleroblasts in the gelatinous mesoglea (Dreyer 1892). This view has also been endorsed by Dogiel, Poljanskij, and Chejsin (1965). There is, unfortunately, as yet, no experimental confirmation of these proposals. With the obvious possibility of significant evolutionary divergence among the skeletal parts in radiolarians, it might be most informative to learn, especially in the forms which have to build their skeletons anew following fission or zoospore formation, just how they do it. Is there a "seed" of spicular material which initiates development, and does this seed persist? Is the ubiquitous crystal of the zoospore such a seed? Haecker (1908) argues convincingly that there is a seed, as we have mentioned. In any case, is there evidence of processes that are understandable as being recapitulative of conservative formative processes? If there were, they might tell us the nature of the plesiosemic condition of the skeleton, a question we will come to later but will not be able to answer definitively.

Again, there is some contemporary research by French scholars that is suggesting new approaches in this field. Cachon and Cachon (1971) have analyzed the microtubule pattern in the axopods and they raise the point that "The axial symmetry which is conspicuous in the axopodial system has implied the axial symmetry of the whole cell particularly the siliceous skeleton." (p. 80) Prior to this Cachon, Cachon, and Ferru (1968) have proposed that the axoplast, the site of origin of axopodial microtubules, is the major determinative of cellular organization, especially in the nasellarian (monopylar) and spumellarian (peripylar) radiolarians. Thus far the work is based on careful speculation from descriptive data. When experimental testing of these hypotheses becomes feasible, we shall surely gain important insights into the mechanisms regulating cellular form.

In 1879, Hertwig said that the developmental history of the radiolarians is the most perplexing aspect of their biology. That is still true today.

Phylogeny

There is really only one aposemic feature in the radiolarians: their skeleton. The position of the axoplast may be another such feature. The calymma, the central capsule, the nucleus, and possibly the axopodia show specializations unique to the group but little variation within the group. They are symplesiosemes. Thus we are left with a very homogeneous set of unicellular creatures whose homogeneity is, of course, the chief difficulty in elucidating their evolution. Just how meaningful is a phylogeny based largely on the geometrical variations of glassy spines, spicules, and shells, especially when the mode of their formation is largely unknown? However, the fact that there are other specializations in these cells is of some help, and we shall first discuss the various aspects of structural variability in radiolarians so as to sort out certain major organizational patterns. From this we can make an informed guess as to the nature of the plesiomorph, and, from this characterization, infer at least the major evolutionary tendencies. Thus we can gain a perspective on the whole group, so as to judge its overall innovative potential. Finally, we can discuss the possibilities of determining the origins of the Radiolaria from nonradiolarian ancestors.

Homologies. If the generalization does hold, that all radiolarian nuclei are polyenergid, then the potentially useful seme of nuclear morphology is indeed simply a symplesioseme—a feature unique to the group but common to all its members. If, however, there are radiolarians which show simple haploid or diploid nuclei as their typical nuclear condition, then it may be possible to follow the aposemic development of the more complex polyploid or polyenergid condition from the simpler one. Even if such situations are not encountered, it is worth mentioning the other possibility of studying comparatively the transformation of the polyenergid primary nucleus into secondary nuclei and the further development of these into zoospore nuclei. The latter studies are, of course, possible, as our French colleagues are amply demonstrating. It seems just a question of time until there is accumulated sufficient data to allow comparative studies.

The calymma seems, *a priori*, to be a promising area for comparative work, because of the variety of gelatinous material, vacuoles, and other inclusions found in it, and because of the various functions it can perform.

However, it seems to be very difficult to preserve, for cytological studies and physiological studies apparently depend on the ability for extensive cultivation of radiolarians. Severe technical difficulties persist in both areas.

The role of the gelatin is obscure. It has been proposed as a flotational device, since, presumably, its specific gravity is less than that of water, and it would provide buoyancy. How the cell controls its gelatin content in response to depth is simply a matter for speculation; no one seems to have proposed even a tentative hypothesis. That the gelatin may somehow be a source of the silica needed for shell formation comes from Haecker's proposals on the mode of formation of the skeletal elements. His evidence is purely circumstantial.

The other special inclusions of the calymma include various pigments, fat and oil droplets, food vacuoles, and symbiotic algae. The function of the pigments is obscure. They could simply be by-products of the radiolarian metabolism. They might, if the cells are unpalatable to visually oriented predators, serve as a "don't-eat-me" warning, much in the manner that the much larger sponges and sea fans are highly colored in yellows, blues, browns and reds, or as unpalatable butterflies are brightly colored (Brower 1958). Certainly the free-floating radiolarian, hundreds or thousands of micrometers in diameter, would be a vulnerable morsel for planktonic fish larvae. Some protection against these marauders would seem essential. Chemicals plus warning coloration and the burrlike skeletons themselves might be of selective advantage as protection against would-be predators. Normally, lipid droplets are considered to be food reserves. This is, therefore, the most plausible suggestion for them in the radiolarian. They would also be buoyant, and their presence in the endoplasm has lead to the suggestion that the intracapsular material is the key to buoyancy (Haecker 1909). Whether such lipids might also act as repellents to aggressors is sheer speculation.

The symbionts, in all probability, provide nutritional materials to the cells. A study of their nature might be informative. Hennig (1966) discusses in detail Fahrenholz's rule, which concerns the theme that internal parasites can provide significant information on host evolution. Hennig is not optimistic about this source of information but does see some possibilities of useful information from it.

The axopodia have apparently never been analyzed in terms of the type of cytoplasmic streaming they posses. Such streaming is the basis of their new classification of ameboid forms (Jahn and Bovee 1965b; Bovee and Jahn 1965), and studies on the radiolarians would be of interest, if only from a comparative point of view.

The axopodia undoubtedly function in food capture; whether they also are a flotational device is more problematical. By their extension into the fluid medium around the cell, they certainly offer mechanical hindrance to sinking, and by increasing the surface of the cell, also offer more area for frictional resistance. Also, by their extension, they probably increase the volume of the cell without affecting its mass, and, thus, there is a decrease in density and an increase in buoyancy. Presumably then, withdrawal of pseudopods renders the cell maximally dense and achieves minimal friction to movement through the water, whereas extension of the filopodia would reverse these properties. There seems to have been no experimental work on this problem, even though it was first discussed in some detail over 60 years ago (Haecker 1909) and has been a question in biologists' minds since Haeckel's time. If there is a flotational role to the filopodia, it may have developed differently in forms whose horizontal distribution is restricted to different parts of the ocean (Haecker 1908; Reschetnjak 1955) and would show, therefore, variations from one group to another. This could be of phylogenetic interest.

Further work is needed on the potentially useful seme of axopodial fine structure. At present these structures show certain general features common to axopodia in general, namely, a microtubular axis and slender, tapering morphology. What could be of special concern to problems within the radiolarians is the role of the axoplast as a generative organelle for microtubules, the pattern of microtubule arrangement, the relation of axopodia to the capsular fusules, and the general distribution of axopodia in different cells.

The perforated capsule, the most distinctive feature of the radiolarians, presents us with three distinct categories of form. There are no intermediates between them, and each is rather constant in its own shape, especially the mono- and tripylar capsules. Our best guess as to the function of the capsule is that it is an intracellular differentiation that separates two functionally distinct parts of the cells and, by preventing free interchange of materials between the parts, permits each to specialize further than would otherwise be possible.

The foregoing comments emphasize that information of use to phylogenetic inquiry is very hard to come by in the radiolarians. Structures such as those described have unknown functions, admit of two or more plausible interpretations regarding possible functions, and are highly invariable in their structure so far as is presently known.

The skeletal parts certainly show aposemic diversity of form, but there is no agreement on their function. The skeleton is thought to have a supportive function (Dogiel, Poljanskij, and Chejsin 1965) to be of use in flotation (Trégouboff 1953c; Campbell 1954a), or, as we suggest below, it may be protective. However, the fact that not all Radiolaria are provided with spicules or shells shows that whatever function it performs, it is not an indispensable one. The variety of form in the skeletal material has only been hinted at by the sequence of complexity described earlier (Figs. 8-1, 8-2), starting from simple, straight spicules and tetraxonal ones, and then proceeding up through radially arranged ones and spongy structures to horizontal elaborations to form shells, and then shells of many layers and different geometrical arrangements.

If we are to see any evolutionary trends within the Radiolaria, using present knowledge, then it must be done on the basis of skeletal anatomy, for this is the only basis of significant trends within the group. Under these conditions it will not be profitable to go through the analysis that will provide estimates of phylogenetic distance between various representative radiolarians, for they can reflect only the aposemic development of skeletal material, for the most part, and this material has been examined many times for its evolutionary implications (Hertwig 1879; Haeckel 1888b; Popofsky 1913a, 1913b; Trégouboff 1953c; Campbell 1954a), with much the same conclusions being reached each time. We will summarize these later. What is of more interest is to try to understand what the limited and selected variability of the radiolarians means in terms of their evolution as a group. This we will try to do by answering the following two questions. What sort of evolutionary development did the Radiolaria undergo to fill the adaptive zone they presently occupy? And how did they come to invade this zone? In aswering the first question, we will have to examine various phylogenetic proposals and in the process gain an idea of the radiolarian plesiomorph. In answering the second question, we must face the difficult problem of ancestral origins.

PLESIOMORPH

The key to understanding radiolarian evolution is to see them as many, in fact, thousands of variations on the one complex theme of marine, unicellular, floating, axopodial predators. There are three basic body patterns, as determined by the three different types of capsules, which supply the basic organization for these forms. One might next raise the point that since the skeletal parts show such extensive development, why not examine trends in terms of them and let the capsules come where they may. But doing that would spread the three types of capsules throughout every range of skeletal type and would immediately suggest the contrary view, that the capsular types represent some basic organizational pattern within which there has been parallel but independent lines of evolution of the skeleton. Such, of course, is the classical Haeckelian view.

There is no independent source of information—paleontological, developmental, or distributional data, or correlated transformation series—which allows an objective argument that, in fact, the capsular types are indeed more conservative or plesiosemic than the skeletal parts. The argument for assigning them a basic position in the organizational patterns of radiolarians is that of Haeckel, who concluded that the invariable occurrence of the capsule indicated an early innovation basic to the radiolarian mode of life. The only possible aposemy in the capsules is that the peripylar condition preceded the uni- and tripylar. As Hertwig (1879) puts it, he finds it easier to think of a reduction of pore distribution to localized sites than to conceive of their emergence where they were initially absent. There is, at present, no rigorous phylogenetic argument which supports this view. Two possible future sources of further data are, first, that developmental studies, starting from zoospores, may possibly show us some day how the capsule develops and, by recapitulating conservative steps, tells us how these enigmatic structures evolved. Second, new fossil data may become available, now that techniques for recovering forms as small as bacteria and as old as 3×10^9 years have become available (Barghoorn and Schopf 1966). It is also true that earlier in the geological record spumellarian fossils (peripylar capsule) are more common than monopylar nassellarian ones, suggesting the earlier origin of the former (Deflandre 1953c).

There is no compelling reason to reject the Haeckelian view that the three capsule types represent three basic organizational patterns, and we will proceed on that basis, remembering, of course, the necessary qualifications on its validity just given. Within these three lines, we may find many striking examples of parallel trends in the skeletal parts. Forms whose skeleton is composed only of spicules are found in all three groups. Central fusion of the spicules is found in both the Spumellaria and the Phaeodaria. Other parallels are seen in the occurrence of spongy skeletons, of evolution of shells, of spherical and concentric shells, and so on, down to very fine details of structure (Haecker 1909; Dogiel, Poljanskij and Chejsin 1965, p. 530 esp.). Table 8-1 summarizes some of these features, including also the parallel occurrence of exogenous shells—inorganic material picked up from the environment—and also the lack of all skeletal parts. Such extensive similarities, arrived at independently, are not expected in a group exploiting the various possibilities of a limited organizational base. In other words, the three capsular types have apparently responded in similar ways to similar selection pressures, the results being that the spumellarians, nasellarians, and phaeodarians have each run through the gamut of skeletal possibilities from no spicules, to spicules, to "spongy" skeleton, and to solid skeletons of various sorts. They also have their own unique tendencies—the Nasellaria have a radially symmetrical shell, probably reflecting the radial symmetry of their monopylar capsule, and the Phaeodaria never show tetraxonic spicules, only monoaxonic ones—but the general impression is of an exploitable arena invisibly defining the radiolarian adaptive zone, and, within this, three cell (capsule) types have tried out all sorts of skeletal formations in exploiting the zone, and predictably, they have often come up with very similar solutions to their common problems of survival. One even wonders whether the three different capsules represent three independent origins of the radiolarian mode of life. Are the radiolarians monophyletic?

From the foregoing we can now understand the various phylogenetic proposals that have been made for the Radiolaria. All start with a plesiomorphic form, which is a marine, floating, filopodial predator, the nucleus is presum-

Table 8-1 Selected parallelisms in the Radiolaria. Each species is from a different family

Character	Major radiolarian group	Representative species
1. Absence of skeletal elements	Spumellaria	*Actissa* (= *Procyttarium*) *princeps*
	Nassellaria	*Cystidium princeps*
	Phaeodaria	*Phaeodina* sp.
2. Exogenous skeletal materials	Phaeodaria	*Cementella loricata*
		Miracella ovulum
3. Spicules	Nassellaria	*Triplagicantha abietina*
	Phaeodaria	*Atlanticetta craspedota*
4. Centrally fused spicules	Spumellaria	*Thallasothamnus pinetum*
	Phaeodaria	*Astracantha parodoxa*
5. "Spongy" skeleton	Spumellaria	*Arachnosphaera myriacantha*
	Phaeodaria	*Plegmosphaera exodictyon*
		Coelacantha ornata
		Coelodendrum gracillimum
6. Solid skeletons (i.e., shells)	Spumellaria	*Oroscena regalis*
		Carposphaera nodosa and many others
	Nassellaria	*Dictyophimus sphaerocephalus*
		Botryopyle dictyocephalus and many others
	Phaeodaria	*Castanella balfouri*
		Challengeron armata and many others

7. For parallelisms in details of shell formation see Haecker (1908) p. 573-576 and Dogiel, Poljanskij, and Chejsin (1965) p. 530.

ably polyenergid, and the capsule is thought to be peripylar. And let us repeat, these conclusions are not based on rigorous phylogenetic procedure, but simply on inference regarding the possible nature of a form that might be able to give rise to known radiolarian forms. For reasons already given—limited aposemic evolution, absence of critical paleontogical or development data—this is the only possible procedure at present.

The discussion of radiolarian phylogeny will first contrast Haeckel's and Hertwig's views on the subject. Then we will turn to proposals of Popofsky, whose views, representing an insightful analysis of skeletal evolution, have influenced more recent writers on the subject. From this we can then proceed to some comments on the radiolarian plesiomorph and its origin.

Haeckel (1888b), with his characteristic tendency to a unitary view of things, derived all the radiolaria from *Actissa* (= *Procyttarium*, Campbell 1954a) (Fig. 8-5), a genus of polypylar forms with a single capsule containing a polyenergid nucleus (he shows swarmers in his figures of *Actissa princeps*, Haeckel 1888b, Plate I, fig. la, b c) and no skeletal parts. From *Actissa*, he derives the rest of the Spumellaria directly, of course; he also passes to *Cys*-

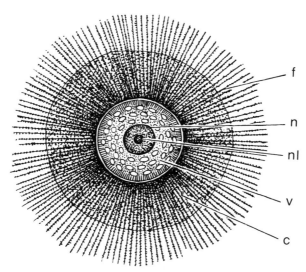

Fig. 8-5 A possible plesiomorphic radiolarian, *Actissa princeps*, first described by Haeckel (1887). (The genus *Actissa* has been replaced by *Procyttarium* (Campbell, 1954a).) c, central capsule; f, filopodium; n, nucleus; nl, nucleolus; v, intracapsular cytoplasm. (Redrawn, from Calkins (1901) *The Protozoa*. Macmillan Company; originally from Haeckel.)

tidium to derive the monopylar Nassellaria; and from *Actissa* he also goes to a hypothetical *Phaeometra*, from which there are derived four separate, actual phaeodarian forms to give rise to the tripylar radiolarians.

Hertwig's (1879) more cautious approach starts from basic forms, simply termed, "Urradiolaries," from which arise five different separate lines. Two are peripylar—the large forms with well-developed gelatinous material and reduced or no skeletons are in one group, and the ones with a less conspicuous calymma and significant skeletal development are in another. One is monopylar, and the remaining two are tripylar—again distinguishing conspicuous calymma and weak skeletal development from conspicuous calymma and strong skeletal development. This summary, Hertwig feels, best accords with the obvious, rampant, parallelisms in the group. If there is to be a single form representative of the primitive state, then he suggests that *Thalassolampe primordialis* be that "Grundform." This is a peripylar form, similar to Haeckel's *Actissa*.

Both authorities essentially agree, then, on the nature of the plesiomorph. (It is essentially an argument based on skeletal morphoclines plus capsular differences.) Their choice is a peripylar radiolarian lacking skeletal elements. They disagree on details of subsequent evolution. Hertwig's explicit recognition of the strong tendency towards parallelism seems more congruent with the facts, though we are not prepared to endorse all the details of his phylogeny, not so much because they are known to be incorrect, but because of the limited base of semic information on which such a phylogeny perforce rests.

Popofsky (1909, 1913a, 1913b) has built phylogenetic sequences simply on the basis of skeletal morphology and within the three separate lines of the capsular types. Popofsky starts in each case with the tetraxonic spicule and from this derives the major shell types of the Spumellaria and the Nasellaria. His phylogeny necessarily omits forms with no skeleton, or with exogenous skeletons, or with skeletons of straight spicules. He also does not treat the Phaeodaria. Hence, there are serious lacunae in his studies. But where he has made an intensive analysis, one is impressed, as Deflandre is when he says of Popofsky's work on the Nasellaria. "It is again to Popofsky that one owes the best phylogenetic essay on the Nasellaria, an essay of logical rigor throughout, resting on facts, and whose consistent application has lead to a natural classification of the group." (Deflandre 1953c, p. 405–406. Original in French.) However, Deflandre himself points out the difficulties of this work. Compare Fig. 8-6 and 8-7. The first figure is Popofsky's diagram of evolution within the Sphaerellaria—the heavily shelled peripylar forms. Fig. 8-7 is Deflandre's (1953c) modification of it. Deflandre clearly points out possibilities of parallel evolutionary development culminating in extensive similarities. This

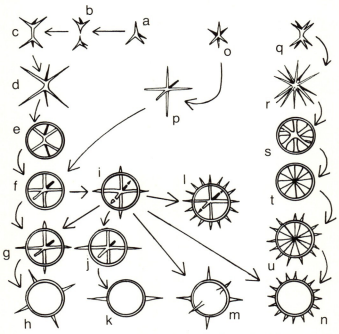

Fig. 8-6 Evolutionary relations among spumellarian skeletons according to Popofsky. Three independent origins of spicular materials are suggested (a, o, and q). These become surrounded by a spherical test, with significant adaptive radiation occurring from form f. Convergence is also apparent in forms l and n. (Redrawn, by permission, from Campbell (1954a) in *Treatise on Invertebrate Paleontology*. ed. P. C. Moore. Geological Society of American and University of Kansas Press.)

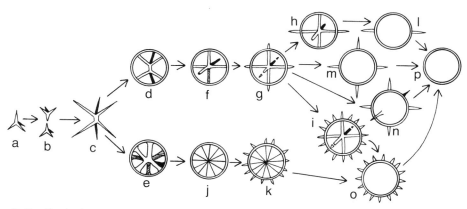

Fig. 8–7 Evolutionary relations among spumellarian skeletons according to Popofsky and modified by Deflandre (1953c). The tetraxonic spicule, through fusion, becomes dicentric (a–c) and the basis for the internal skeleton of spherical tests. Proceeding from d through g, there appears a form from which significant radiative evolution seems apparent (h, i, m, n and their subsequent evolution). Another line (e through o) shows a somewhat different evolutionary pattern (Redrawn, by permission, from Grassé (ed.) (1953) *Traité de Zoologie,* Vol. I, part II, Masson et Cie.)

states succinctly the problems of phylogeny when there are insufficient aposemic changes in a variety of characters so that they can check on each other. It seems impossible to develop an objectively definitive phylogeny of the Radiolaria on the sole basis of their skeletal elements. Popofsky's views are compelling (see the major review by Campbell 1954a, who apparently accepts them completely), but they are not convincing. Careful speculation by Cachon and Cachon (1969) proposes that the primitive spicules of the nasellarians is not a tetraxonic spicule but rather a rodlike one with terminal bifurcations, the latter lying in planes at right angles to each other. This too differs from Popofsky.

A final hopeful note is the continuing success of paleontological studies. Deflandre (1963) reports a lower Carboniferous fossil that appears to be a veritable "missing link" between certain peripylar forms (esp. the periaxoplastid sphaerellarians) and monopylar forms (esp. the podocone-containing cyrtellarians). This suggests an important bridge between the more spherical symmetry of the Spumellaria and the radial (and subsequent bilateral) symmetry of the Nassellaria.

ORIGINS

There is a consensus, then, that a skeletonless spumellarian form with one peripylar capsule is probably the best plesiomorph of the Radiolaria, e.g., *Actissa* or *Thalassolampe* or *Procyttarium* (Campbell 1954a). Such a form would, in addition to its capsular condition, possess a polyenergid nucleus, axopodia, and a flotation mechanism. The skeleton must, on this basis, be a later development of the radiolarians. How would such a form arise initially and why would it develop a skeleton in the majority of its forms?

To formulate a meaningful discussion of the origin of the radiolarians, we must consider the selection pressures that were being met by the postulated plesiomorph; we must consider these pressures in relation to the whole outer feeding-flotational calymma and the inner reproductive-genetic elements. The latter would need some buoyancy so as not to act as dead weight, and the presence of lipid droplets in the endoplasm could ameliorate that situation. These steps, if they occurred, probably occurred quite rapidly, giving rise to horotelic rates of evolution in the perfection of a marine, floating, filopodial predator with a perforated capsule around its polyenergid nucleus. In some such manner, the radiolarian plesiomorph might have emerged. What, now, is the reason for the appearance of the so called skeletons?

Probably, we can argue, the siliceous materials were initially protective. Not protective in the sense that clam shells are, but in the sense of sponge spicules (which are also supportive). That is, ingested spicules could irritate the internal tissues of the predator as powdered glass (silicates) will do, or as, in fact, sponge spicules will do. Anyone who has incautiously ripped apart a large hexactinellid sponge with his bare hands in the happy pursuit of the diverse fauna found in it, has perhaps paid for his enthusiasm the following day with swollen fingers—swollen because of the slivers of glassy spicules driven into them, but so fine that they were not felt entering. Analogously, ingestion of radiolarian spicules could cost the ingestor dearly. True, it would not save the ingested form, but it could protect the species by tending to eliminate those potential predators that would feed on the free-floating radiolarians. There would be selection favoring the forms that refrained from eating the radiolarians.

Where would these predators come from? Certainly by the Cambrian there were metazoans capable of consuming microscopic zooplankton. The free-floating radiolarians would be fair game for predators larger than themselves. Chemicals could come into play as protective devices, either to inhibit directly the feeding activities of predators, as would presumably be the case for those forms which never developed siliceous materials, or as a warning associated with the presence of the spicules and shells. When visual predation evolved, as with larval fish, warning coloring of the radiolarians could be of advantage.

The siliceous material might not only be protective but might be preadaptive in a skeletal role too. And before it, the gelatin which had a buoyant role might also have been subsequently used as the mechanism for depositing the silica.

From this we can see the origins of our present seeming confusion on assigning multiple functions to filopodia (feeding and flotation), gelatinous material (flotation and skeleton secretion), spicules and shells (support, flotation, and protection). They very probably do have multiple functions, and their multiple capacities may represent one of the reasons why the Radiolaria have been so successful in achieving such a complexity of functions within the bounds of a single cell.

A priori, there are three potential ancestral types for radiolarians; colorless algae with ameboid tendencies, ameboid protozoa (already evolved from colorless algae or zooflagellates), and degeneration of metazoans. The last suggestion is the most unorthodox, but when polyenergid nuclei are encountered, they suggest, at least to some biologists, that the compounding may be due to fusion of many nuclei. If cell membranes degenerated in a multicellular form, then a polyenergid nucleus is conceivable. Such a proposal has been made for the ciliates and their polyploid/polyenergid macronucleus. It would seem to be applicable here, though no published proposal of such a sort has been found. The refutation of such a suggestion depends on understanding regressive evolution (see Chapter 3), which is highly unlikely in a free-living predator, for it is difficult, if not impossible, to formulate the succession of forms that would lose the various adaptively significant specializations of multicellularity and yet compete successfully for a niche against other multicellular forms or against other unicellular forms. We will not pursue this point further here but turn to possible derivations of the Radiolaria from other unicellular forms.

That the radiolarian ancestor could have been another ameboid protozoan is a view put forth by Haeckel (1888b) in his proposal that the Heliozoia were ancestral to the radiolarians. This, however, is a double edged argument on morphological grounds, for one could argue that a nonskeletal form of the *Actissa-Thalassolampe* type could have been ancestral to the heliozoans, rather than the other way around.

There is another, more serious argument against Haeckel's proposal, and that is the implausibility of deriving marine forms from predominantly freshwater ones. The history of the freshwater forms clearly favors the probability of invasion from the sea, and except for well-documented cases in the mammals, such as cetaceans and pinnipeds, the traffic in the other direction is minimal. Furthermore, the well-established and successful radiolarians contrast sharply with the highly variable and heterogeneous group of heliozoans, such that the latter hardly seems a plausible source of the former. The only serious possibility would be a polyenergid, filopodial marine form with some sort of capsule. None is known at present.

In any case, we need to put such broad, speculative arguments aside and ask more specific questions. In particular, are there any nonradiolarian, unicellular forms which have a capsule like that of the radiolarian plesiomorph and also possess radiating filopodia? Since there are no such forms among the protozoans, except possibly for the acantharians, which we shall consider in the next chapter, we need to examine the only remaining possibility, i.e., the apochlorotic algae.

There are two algal groups which contain candidates for the role of radiolarian forerunner: the dinoflagellates (Pyrrophyta, or particularly the Dinophyceae), because of two forms *Gymnaster pentasterias* (Zimmerman 1930) and *Plectodinium nucleovolvatum* (Biecheler 1935) which possess perforated perinuclear capsules and siliceous spicules; and the chrysomonads (Chrysophyta, Chrysophyceae in particular), because of their extensive tendency to the apochlorotic condition and, among these colorless forms, the presence of ameboid forms some of which possess filopodia, such as *Leukochrysis* and certain other members of the Rhizochrysididae (Hollande 1952b; Pringsheim 1963). The problem with the postulated dinoflagellate ancestor, as we will see, is to understand the presence of the capsule and spicule in a context different from that in which they occur in the radiolarians. Neither *Gymnaster* nor *Plectodinium* have any reported tendencies to pseudopodial activity, much less being floating filopodial predators. How then can their perforated capsule be meaningful as, in the Radiolaria, a possible organizational-homeostatic device for maintaining the specializations of calymma and endoplasm? The answers seem to be (a) that it does not have a function comparable to that in the Radiolaria, and therefore it is a fortuitous convergence; or (b) that it does indeed have a different function in the two groups, and the dinoflagellate function (at present unknown) is a preadaptation for the radiolarian one. The same points of view apply to the spicules. Furthermore, regarding the spicules, if they are present in the ancestral form, this would render their presence highly likely in the radiolarian plesiomorph, and that would necessitate the assumption that absence of skeletal elements is not plesiosemic but hyposemic, a reduction.

Further study of this possible phylogenetic relationship between dinoflagellates and radiolarians by Hollande, Cachon, and Cachon (1970) results in the following points. The perinuclear capsule of *Gymnaster* and *Plectodinium* are not to be homologized with radiolarian capsules, for the perinuclear capsules have neither the location nor the structure of the radiolarian structure. That is, neither the positional nor compositional criteria for homology are fulfilled. Hollande and his colleagues go on to suggest that the theca, or external plates of dinoflagellates, such as those of the peridinians, have not only "the same chemical constitution and the same localization at the cell periphery, but their relations with the endoplasmic reticulum are very similar." (p. 316. Original in French.) This conclusion depends not only on the special interpretation of the membranes lying on either side of the capsule, but also on fine structure and chemical analysis of peridinian theca. The work of Kalley and Bisalputra (1970) is one cited as supporting these similarities. It seems to be a fair question whether the cellulose wall of dinoflagellates is the same as the glucoprotein of the radiolarian capsule, and whether the pores for trichocyst extrusion are homologous to those for axopods. Also, it needs to be determined whether or not the dinoflagellate cell wall is deposited outside or inside the cell membrane.

Hollande, Cachon, and Cachon (1970) also examine the possibility of a heliozoan ancestry for the radiolarians but reject it in favor of dinoflagellate origin, with the heliozoans representing a distinct but parallel evolutionary development relative to the Radiolaria.

The derivation of radiolarians from colorless, ameboid chrysomonads poses an altogether different set of problems. Here, the major reason for considering an evolutionary connection is the presence of filopodia in the chrysomonads. This is obviously a tenuous link; it is only one character and not an altogether complex one, so that rigorously determined comparisons are not possible between these and radiolarian filopodia. And, also, filopodia are not unique to these groups; they also occur in the Acantharia and Heliozoia. Hence, there is a large gap to be filled between the filopodial chrysomonads and radiolarians. This may reflect the necessary evolution to go from the substrate hugging *Leukochrysis* to the floating *Actissa* (*Procyttarium* or *Thalassolampe*) but it, nonetheless, makes it extraordinarily difficult to argue convincingly that such a large gap was actually bridged by evolutionary events.

CONCLUSIONS

The adaptive zone exploited by the Radiolaria is that of unicellular, floating micropredators, restricted to a marine environment and capturing their prey by means of filopodia. The group is organizationally a very homogeneous one; only the central capsule, showing three distinct types of structure, and the spicules and shells, showing great diversity in form, are known to show significant variation and, hence, display aposemic evolutionary development. The complexity of skeletal parts may have proceeded from emergence of spicules to "spongy" skeletal material, to various types of shells. There is, however, no critical evidence—paleontological, developmental, or other—which establishes the validity of the plesiomorph. Its nature is simply inferred in terms of a morphocline from a relatively simple form to more complex forms. The weakness of this argument must be kept in mind as a major limitation on our knowledge of radiolarian phylogeny. For example, it might also be that skeletal parts were present in the plesiomorph but have been lost independently in various groups. Another limitation in analyzing radiolarian phylogeny is the absence of sufficient numbers of aposemic traits so as to allow the probable delineation of evolution within the group. Skeletal parts show extensive development, but the possibility of widespread parallelism in the evolution of shells and spicules makes it impossible today to propose a definitive phylogeny on the basis of the skeletal parts alone.

Chapter 9

Acantharians

TOGETHER with the heliozoans and radiolarians, the acantharians were placed by Calkins (1926) in the taxon Actinopoda (rayed-foot). This is an example of Adansonian classification *par excellence.* Its usefulness has been attested to by decades of perpetuation (most recently by Honigberg *et al.* 1964), for indeed, spherical body form and radiating filopodia represent a distinctive mode of cellular organization, and classification is easy on that basis. Detailed study reveals that this homogeneity is superficial, and what is useful for Adansonian biology is not informative—in fact, is misleading—for Darwinian biology. The validity of this was most directly demonstrated by the Heliozoia and their heterogeneity. The Radiolaria were, by contrast, very homogeneous. This was achieved, however, by removing the Acantharia from them. Haeckel (1888a) had early recognized the singularity of the acantharians but had, nevertheless, continued to view them as a subtaxon within the Rhizopoda Radiariae (as he subtitled his monograph on the radiolarians, to perpetuate Johannes Müller's viewpoint on these forms). It was really with Schewiakoff's (1926) magistral monograph that biologists could finally see these forms as a distinct group in their own right (Trégouboff 1953b).

General Characteristics

Like the radiolarians, the acantharians are marine, floating predators. Their usually spherical, unicellular bodies, the presence of skeletal material, the presence of slender pseudopodia, and the not uncommon occurrence of a central capsule all contribute to the general resemblance between these two great groups of protozoan zooplankton. But the similarities are convergences, as we shall see from a detailed study of the composition of the separate features.

STRUCTURE

Cell body. The general contour of the cell body, apart from spicule protrusions and extended pseudopodia, is typically round, but sometimes ovoid. In forms with a highly developed flotational apparatus the surface layers are quite rigid and, hence, the body form is quite constant; forms with less differentiated superficial layers are more variable in contour. The cell body itself is visibly separable into an outer gelatinous layer and an inner portion designated by the German term "Weichkörper", which we will simply refer to as the inner cell body. This inner part is often visibly separable into an outer ectoplasm and an innermost endoplasm. Not only do these parts stain differently, indicating some chemical differences probably related to different physiologic states and functions, but, as might well be expected, the specialized organelles which they contain are quite different (Schewiakoff, 1926; Trégouboff, 1953b).

Endoplasm. Chief among the endoplasmic inclusions are the nuclei. Acantharia are typically multinucleate, the individual nuclei—there can be dozens of them—are aggregated in more or less concentric layers towards the periphery of the endoplasm. This applies to what can be called the adult state, where cells have reached their maximal size and are maximally differentiated. Very early stages show one primary nucleus, which divides into many, secondary nuclei. These seem to have various fates. Some persist to become the generative nuclei of the flagellated gametic cells, which are formed by multiple fission of the cell, involving especially the endoplasmic material. Others reputedly differentiate into myonemes, after first forming into deeply staining bodies called "trophochromidia" (Trégouboff 1953b). They are reported to be formed at the time of formation of generative nuclei from the secondary ones.

Food vacuoles are found in the endoplasm, and various droplets concerned with the breakdown of ingested food material and the storage of food reserves—notably lipids of different sorts—are also present.

Pigments, localized in droplets of varying size, are distributed throughout the endoplasm. As with the radiolarians, cells of a given species can be either yellow, brown, blue, or red or of various other shades. Usually the deepwater forms are colorless. Finally, there are "concretions" (Trégouboff 1953b), which are perhaps the residues of undigested material before their elimination from the cell. They are not found in young stages and can occur in a variety of shapes and sizes.

Ectoplasm. The ectoplasm, which is most distinct in the adult stages of the more complex acantharians, lies between the endoplasm and the outermost gelatinous layer. In younger stages and in less differentiated forms, the endo- and ectoplasm merge, indistinguishably, with each other. There are various inclusions in the ectoplasm that are located in vesicles: ingested food, lipid droplets, and pigmented droplets. Also present are symbiotic zooxanthellae. The major specializations of the ectoplasm are the pseudopodia and the capsule, when it is present. The skeleton also lies in the ectoplasm, but it is no more an ectoplasmic structure than it is endoplasmic or part of the gelatinous layer, since it lies in all three areas; we will treat it as a separate cellular specialization.

Filopodia. The pseudopodia are treated with the ectoplasm because they arise from it. They are of two types: slender, branching, and anastomosing filopodia, termed reticulopodia, and slender filopodia with a central axial filament or axoneme, i.e., axopodia. In certain, simpler acantharians, there is sometimes found a simple unbranching filopodium with no axial filament. This is called a flagelliform pseudopodium by Schewiakoff (1926), who felt it was homologous and, therefore, presumably derived from the flagella or flagellated cells. The reticulopodia and axopodia are especially abundant in the less complex cells, radiating profusely from all over the cell surface. In some forms, the reticulopodia appear gathered as tufts at the end of spicules, especially of very large spicules. The axopodia typically emerge from the cell surface in special apertures found lying within a band of circular elastic fibrils. Often there is one opening on each side of the polygon formed by these fibrils as they pass around a spicule tip, and, hence, the number of axopodia per spicule is reduced to a very consistent five or six, depending on the polygon with which they are associated.

Capsules. Capsules are found in about two-thirds of the Acantharia (Trégouboff 1953b), but these are all members of the one order Arthracantha, the largest of the four orders in Schewiakoff's classification, and invariably such cells are highly differentiated. This capsule differs from the radiolarian one in many important features. (a) It is never perforated except as it is traversed by axopodia and spicules. (b) It is of variable thickness in different forms—thickest in the most complex ones—and, as Trégouboff (1953b, p. 282) describes it, "It is a simple modification of the ectoplasm, of an order more physical than chemical, because its nature, as established by chemical reactions and staining, is exactly the same as that of the ectoplasm in the midst of which it differentiates, sometimes very late, after the Acantharian has become multinucleate." (Original in French.) (c) Its position, as just indicated, is not at the border of endo- and ectoplasm, but is within the ectoplasm, contrary to Haeckel's original conclusions. (d) It appears that the capsule, rather than having a homeostatic function as in the Radiolaria, is a structural device for steadying and holding fixed the position of the skeletal material.

Spicules and skeleton. The skeleton is composed of 10 diametral or 20 radial spicules, except in a few unusual forms whose relation to the Acantharia is somewhat uncertain. (See the actinelians, below.) These spicules are not formed of pure silica, as in the Radiolaria, but are composed of calcium-aluminum silicates, according to Schewiakoff's (1926) painstaking study. The common assertion that the acantharian skeleton is formed of strontium sulfate (celestine) comes from Bütschli's (1906, 1908) careful but preliminary analysis of the aberrant form *Podactinelius sessilis* and from Odum's (1941) preliminary analysis of one species of Acantharia. In this latter case the author, himself, reports that, because of the small amount of material available for analysis, he did not feel his results were definitive. Schewiakoff's work needs to be repeated, using the more refined techniques of Odum, for example. In any case, it clearly appears that radiolarian and acantharian skeletals are not composed of the same chemical compounds.

The arrangement of the spicules, in terms of the placement of their termini at the cell surface, follows a pattern first enunciated by Johannes Müller (1858) and since known as Müller's law (Fig. 9–1A). D'Arcy Thompson found a slightly different way of expressing this regularity (Fig. 9–1B). When the spicules are diametral, they traverse the whole cell, with their ends on opposite sides of the cell. The center of the cell is where these elements pass by each other, and the middle of each spicule sometimes shows irregularities in its otherwise quite regular structure. When the spicules are radial, they abut on each other at the cell's center. The manner of their meeting provided Schewiakoff (1926) with the characters needed for the major taxonomic subdivisions of his classification of the Acantharia. In some cases the ends are shaped to fit snugly against each other in a very rigid pattern, and in other cases the fit is more casual. Flutings along the axis of the spicules can converge on this central area to foliate the intersection. This provides the basis for the felicitous German term of Blätterkreuz (Fig. 9–2). At right angles to the long axis of the spicule there extend, in many species, elaborations of the spicules. These apophyses may be simple, short, needlelike protrusions or can be variously complex, forming grill-like or platelike extensions, parallel to the cell surface (Fig. 9–3). When the lateral extensions of one spicule meet those of another, there is then, in effect, a shell formed (Fig. 9–3G).

Flotational structures. There remains now the flotation or hydrostatic apparatus. In the simpler cells, this consists of a relatively homogeneous gelati-

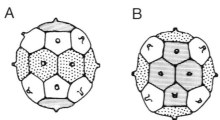

Fig. 9–1 The regular arrangement of acantharian spicules as seen in terms of their apophyseal plates (The spicules are represented as broken off, to facilitate viewing.) Polar spicules: horizontal shading of apophyseal plates. Equatorial spicules: stippled shading. Tropical spicules: unshaded. A. As viewed by D'Arcy Thompson. B. As viewed by Johannes Müller, illustrating Müller's Law. (Note both tropical bands are visible in A. and only one in B.)

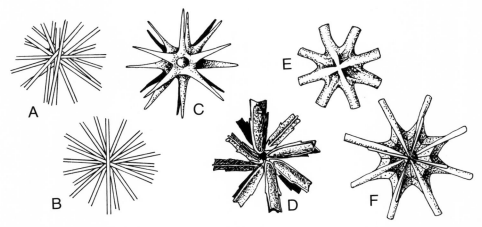

Fig. 9–2 The central intersection of spicules. A. Ten diametral spicules which simply traverse the cell. *Acanthochiasma rubescens.* B. Ten diametral spicules which cluster tightly in their mid-regions. *Acanthospira spiralis.* C. The spicules are apparently fused. *Astrolithium bulbiferum.* D. Twenty radial spicules with fitted internal ends. *Stauracon pallidus.* E. Twenty radial spicules with fitted and buttressed internal ends. *Acanthometra pellucida.* F. Twenty radial spicules with highly sculptured internal ends forming a Blätterkreuz. *Phyllostaurus siculus.* (Redrawn, by permission, from Grassé (ed.) (1953) *Traité de Zoologie,* Vol. I, part II, Masson et Cie. Originally from Schewiakoff (1926).)

nous substance lying as the outermost cell layer. In most cases, its hydrostatic function is not clear, apart from the fact that it contributes to the buoyancy of the cell by its presumably relatively low density. However, the relatively more complex cells have evolved a means of contracting of the gelatinous layer and thus controlling its density (Fig. 9–4). At least three sets of specializations interact to achieve controlled flotation: contractile elements called myonemes, the spicules, which supply the structure against which the myonemes can work to exert their force, and the gelatinous layer. In addition there are elastic fibers at the cell surface and so-called sensory fibrils, and both may also be important in flotational control. The latter fibrils may be more aptly termed coordinatory, as it would seem to be of advantage to coordinate the contraction of the various myonemes. However, their existence (Trégouboff 1953b) and function are problematical, and so we will examine them no further here. The myonemes (Fig. 9–4B) are attached to the distal ends of the spicules and, at their other end, to the inner surface of gelatinous layer, when it is a single layer. When it is a double layer, they attach to the inner boundary of the outer layer (Schewiakoff 1926). Contraction of the myoneme pulls the gelatinous layer (or at least its outer portion) against the cell surface. This surface is provided with radial (from each spicule tip) and circular (around each spicule tip, but some distance removed from it) elastic fibers

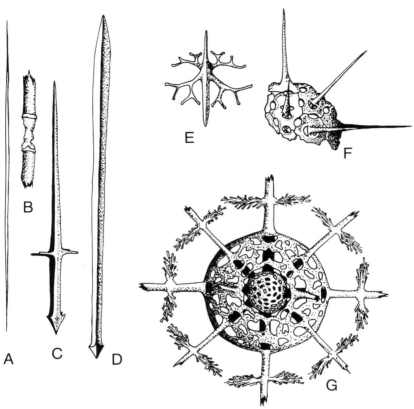

Fig. 9-3 Acantharian spicules and their apophyses. A. Simple diametral spicule from *Acanthochiasma fusiformis*. B. Midregion of a diametral spicule from *Acanthocyrta haeckeli*. C. Radial spicule with four simple apophyses from *Xiphacantha quadridentata*. D. Fluted radial spicule found in *Amphilonche elongata*. E. Radial spicule with four branching apophyses from *Staurapsis stauracantha*. F. Part of shell of *Dorataspis loricata*, showing contiguous apophyses and the spicules to which each is attached. G. The elaborate shell of *Phractopelta tessarapsis*. (Redrawn, by permission, from Grassé (ed.) (1953) *Traité de Zoologie*, Vol. I. part II, Masson et Cie. Originally from Schewiakoff (1926).)

(Fig. 9–4B). The circular ones, especially, appear designed to resist expansion and, therefore, designed to resist pressure from the underlying gelatin when it is forced up by the myonemes. Hence, the gelatin is effectively compressed. It now becomes obvious why the spicules are so neatly fitted to each other at their bases and why it is reasonable to propose that the capsule functions as a device to anchor the position of the spicules.

The foregoing is apparently a widely accepted picture. Major reinterpretation of the myonemes is proposed in fine structure studies by Febvre (1970, 1971). This worker suggests that the myonemes are composed of fibrils

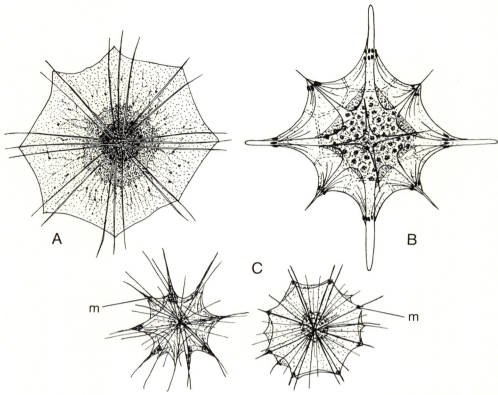

Fig. 9-4 Hydrostatic apparatus of acantharians. A. *Acanthochiasma rubescens*, showing clearly the outermost, broad, generally homogeneous layer of gelatin. B. *Acanthostaurus purpurascens*, showing the highly differentiated superficial and internal structures of the cell. The dark bodies encircling each axoneme near the junction of cell surface and the axoneme are contractile myonemes. (A and B redrawn, by permission, from Grasse (ed.) (1953) *Traité de Zoologie*, Vol. I, part II, Masson et Cie.) C. *Acanthophlegma krohni*, showing the relaxed (left figure) and contracted (right figure) states of the myonemes (m) and their respective effects on the shape of the cell. (Redrawn from Schewiakoff (1926).)

which are not striated; the apparent striations are due to patterns of superposition of the fibrils. Further, he challenged the conclusion that the fibrils are contractile. They may shorten the space they lie in by coiling on themselves but do not seem to change their actual length. Febvre questions whether they should even be called myonemes. That question is not answerable on present data. What remains still is that the fibrils demonstrably reduce the space they lie in and hence are analogs of contractile fibrils, regardless of the mechanism by which that reduction (coiling?) is achieved. At present we will continue to refer to these elements as myonemes and refer to their role in the cell as contractile, especially in reference to change in the density of the gelatinous layer.

FUNCTION

Environmental exploitation. The exploitative activities of acantharians are concerned with pseudopodial capture of prey encountered while freely floating in marine waters. In conjunction with describing the structure of the hydrostatic apparatus there was also outlined its function. Just a few further points need to be added here. The myonemes, first recognized as contractile by Hertwig (1879) and termed myophrisks by Haeckel (1888a), have been seen to contract to 1/5 or 1/9 of their extended length. Cross-striations are also reported (Schewiakoff 1926) but this has been challenged, as we have just seen, by Febvre (1971). How the myonemes are activated remains a mystery. Furthermore, it is not clear just why the Acantharia need such a highly developed apparatus. It may be that they follow the diurnal movement of phytoplankton (their prey, at least in part) and therefore somewhat rapid ascent and descent might be appropriate. If such is the case, light intensity could be a stimulus for activating or deactivating the hydrostatic mechanism. Also, it appears that Acantharia release their gametic cells at great depths. If this is done on one of their regular sojourns to the lower levels, then one does not need to invoke a special stimulator of the flotation mechanism relative to reproductive needs, which would otherwise be necessary. One hopes that the rich array of acantharian material studied so meticuously and brilliantly by Schewiakoff at Naples is still available to workers interested in further elucidating the biology of these cells in general, and of the hydrostatic device in particular.

The pseudopodia capture bacteria, diatoms, other unicellular algae, protozoa and, occasionally, small crustaceans such as copepods. The food is reported to be brought into the inner part of the cell body through the openings in the capsule that are provided for axopodia and spicules (Trégouboff 1953b). Undigested materials are removed by the same openings, and reticulopodia are often concerned with removal of these materials. (In this regard see, especially, Schewiakoff's observations (1926, p. 185) on elimination of an undigested rhizopod test from an acantharian.)

Homeostatic maintenance. Homeostatic devices in acantharians seem to be those of cells in general. The capsule does not seem to have the functions that were attributable to the radiolarian capsule.

Reproduction. Turning, lastly, to reproductive functions, we are in the position of being able to say a great deal more than was possible for the Radiolaria, but, of course, there is much more we would like to know. The major contributions here, as in so much of acantharian biology, come from Schewiakoff's great monograph of 1926. Asexual reproduction, by means of binary fission, is of limited occurrence, whereas sexual reproduction through swarmer formation, fertilization, and subsequent development is apparently the common mode of formation of new acantharians. Binary fission is restricted, in Schewiakoff's experience, to those forms with unfused diametral spicules, in particular, the family Acanthochiasmidae. The relative simplicity of body form allows a more or less even distribution of cell parts, including spicules, to the two daughter cells. The fission products then build the components needed to restore the normal condition, and, thus, binary fission is completed.

Swarmer formation and its subsequent events are vastly more complicated. Before entering into the details of the process, it is necessary to survey briefly the nature of the evidence which supplies these details for Schewiakoff's conclusion that he is studying sexual reproduction has been challenged by Hollande, Cachon, and Cachon-Enjumet (1965). Schewiakoff is certain of the widespread occurrence of swarmer formation, for he has seen the process occurring in some 20 different species, which are members of more than two-thirds of the different acantharian families. Furthermore, he has seen the other end of the process, i.e., developing young cells with primary nuclei and the emergence of secondary nuclei and later stages, in another wide range of species, some the same and some different from those in which swarmer formation was observed. Hence, *in toto*, he has seen the beginning, or end, or both stages of what he would term as sexual reproduction in 43 different species, representing 31 genera, a significant sample, certainly, of the 49 genera and 144 species of Acantharia recognized by Schewiakoff (1926). But, Schewiakoff also candidly points out, he has never actually seen fusion of the biflagellated cells, which he can only conclude are the result of sexual fusion. Also, he was never able to follow the transformation of the presumed zygote through to appearance of young cells with skeletons and so was never able to follow the entire process of swarmer formation, fusion, and subsequent development in any given form. There are good technical reasons for this. Sexual processes occur in most forms in the winter months and at the deeper levels—100 to 200 meters below the surface. Collecting cells in the process of swarmer formation and then hoping that the swarmers will be able to complete their development under laboratory conditions, which are in all probability quite, if not drastically, different from those prevailing at some distance below the surface of the sea, is clearly asking a great deal. One point in particular may offer special problems: If the absence of light is one of the necessary conditions for sexual reproduction and development, it could render observation of the process exceedingly difficult depending, of course, on the critical wave length. Schewiakoff's success comes from the fact of his patient pursuit of these forms for a period of over 25 years during many different months of the year. From this perserverance, he was able to accumulate the details of the story we have today.

Cells on the verge of swarmer formation (Schewiakoff uses the phrase gamete formation, but with the uncertainty that still lingers about the general fate of the flagellated cells it seems better to use the more neutral term of swarmer) can be readily identified by the fact that the ecto- and endoplasm appears quite homogeneous and rather opaque, the zooxanthellae are either eliminated from the cell or are sporulated at the very periphery of the ectoplasm or in the gelatinous layer, and the reticulopodia and axopodia are withdrawn, the axonemes of the latter being apparently broken down. The pigment of the cells is finely dispersed in tiny droplets. Multiple division of ecto- and endoplasm occurs, with the result that the whole cell body, apart from the skeletal and gelatinous materials and myonemes, is incorporated into the uninucleated swarmers. Schewiakoff concludes that the released swarmers go directly to fusion, if contact with appropriate partners is available. The phenotypic uniformity of these cells allows him to designate them as isogametes. Le Calvez (1938) argues that the swarmers undergo some further development more properly interpreted as gametocytes. No worker has confirmed the earlier interpretation of Moroff and Stiasny (1909) of *Acanthometron* being actually multicellular, i.e., the early stages of swarmer formation involve the presence of discrete cells (merozoites), each undergoing its own series of complicated nuclear transformations involving "macronuclear" and "micronuclear" phenomena. In fact, Schewiakoff is certain that Moroff and Stiasny were studying the symbionts of the Acantharia and not following the acantharian nuclei themselves.

Though, as was pointed out, Schewiakoff never observed transformation of a zygote into young cells with skeletons, the presumption that such occurs is a plausible one. The very small size of the cells with skeletons and their uninucleate condition renders their origin from zygotes very likely; the only other possibility being that a single swarmer metamorphoses directly into the miniature acantharian.

Hollande, Cachon, and Cachon-Enjumet (1965) propose that tetraflagellate forms are early stages in zoospore formation. (Hence, where le Calvez (1938) referred to gametocytes regarding these forms, Hollande and his coworkers would refer to them as sporocytes.) However, like Schewiakoff, they have never been able to follow the exact fate of these cells. Presumably they would suggest that tetraflagellate sporocytes divide into biflagellate zoospores, which then go on to develop into mature cells. They do, however, report one more step, and that is encystment prior to sporogenesis. This was observed in all species belonging to the Holacantha, Chaunacantha, and the Symphacantha (the lower orders) but not the Arthracantha. However, this encystment still throws no light on just how zoospores develop into young acantharians with filopodia and skeletons.

We return again, now, to Schewiakoff's observations. The development of the young cells into adults is seen most clearly in a form such as *Acanthostaurus purpurascens* (Fig. 9–5). The smallest stage with a skeleton is uninucleate but shows no capsule, no contractile elements, and the various in-

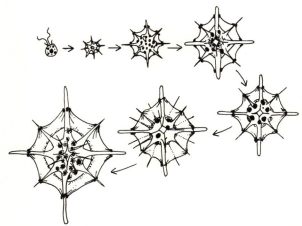

Fig. 9-5 Development of *Acanthostaurus purpurascens* starting from a biflagellated form, which Schewiakoff believed to be a zygote but may only be a zoospore. (Redrawn, by permission, from Grassé (ed.) (1953) *Traité de Zoologie* Vol. I, part II, Masson et Cie. Originally from Schewiakoff (1926).)

clusions of the adult are very meagerly represented. Pseudopodia would seem to be present, for they are necessary for feeding, and the fact that the cells grow clearly implies the availability of food from some source. However, Schewiakoff reports that the reticulopodia apparently first appear at the time of formation of the secondary, or generative, nuclei and the axopodia appear even later, when the capsule and circular elastic fibers are differentiated. The time of initiation of feeding is one of two major unresolved problems in an otherwise quite complete account of development from the young uninucleate form to the adult capable of forming swarmers. This epigenetic story seems to be very constant from one species to the next, so that a general summary of the sequence of events can be usefully applied to all the Acantharia.

From the very outset, the skeletal pattern defined by Müller's law is apparent. The spicules grow in length and breadth and the characteristic pattern of joining and fitting of the inner, basal ends of the radial spicules appears at the time of secondary nucleus formation. The apophyses do not appear until the spicules have reached their adult size, and then the lateral elaborations, characteristic for the species, begin to develop.

The nucleus has an interesting developmental history, especially in view of the claim that certain of its products transform directly into the contractile myonemes and, further, that the myonemes seem to multiply by longitudinal division. (This is the other unresolved developmental problem.) The originally compact single nucleus enlarges and becomes differentiated into central and peripheral parts. The latter are pinched off forming "trophochromidia" or "somatic" or "kinetic" nuclei. Schewiakoff uses these terms essentially interchangeably. In forms which develop myonemes, i.e., all the acantharians except the family Acanthochiasmidae, the "somatic nuclei" are reported to eventually become myonemes. Schewiakoff's report says that 40 such "nuclei" form. They segregate two per spicule, migrate distally along their respective spicules, and when they reach the position characteristic of the myoneme, they proceed to differentiate into such structures. In those forms where there are more than two myonemes associated with each spicule in the adult state, the two initial myonemes multiply by splitting longitudinally, repeatedly in some cases, to supply the number appropriate for a given form. Needless to say, this view of formation of the myonemes raises various important questions on gene action, organelle assembly, and differentiation. There seems to be no experimental work on this problem. One would be tempted to write it off as inaccurate interpretations of incomplete observations, except that competent observers of these forms have all made independent observations leading to the same conclusions (Moroff and Stiasny 1909; Schewiakoff 1926; Trégouboff 1953b). Hertwig (1879), who first recognized the contractile nature of the myonemes, says nothing in detail about their origin, nor does Febvre (1970, 1971).

The inner material of the primary nucleus divides to form the secondary nuclei, and these continue to divide, becoming the generative nuclei, which will eventually supply the single nucleus of each swarmer. There can be anywhere from 60 to 200 generative nuclei per cell, the number varying with the species. These nuclei lie towards the periphery of the endoplasm, when that portion of the inner cell body appears. They form concentric layers two or three nuclei deep.

At the time of differentiation of the primary nucleus into trophochromidia and secondary nuclei, many other differentiations occur in the cell. We have already mentioned the appearance of the specialized patterns of joining of the basal ends of the radial spicules. Also at this time the endoplasm and ectoplasm become clearly identifiable, and reticulopodia develop from the latter layer. The gelatin layer also develops further. In those forms in which two layers are present in the adult, the second layer now makes its appearance. The elastic fibers of the outermost edge of the gelatin layer begin to differentiate—first the radial fibers and then the so-called circular ones. The latter form five- or six-sided polygons around each spicule. The axopodial apertures at the cell surface, lying within the circular fiber sites, are now apparent, and well-developed axopodia are seen. At this time, the first invasion of the acantharians by zooxanthellae is seen to occur. (Is this the result of capture by axopodia?) The central capsule now is seen differentiating, and by this time the "somatic nuclei" have become clearly differentiated as myonemes.

The developing acantharian is now a multinucleated cell with differentiated endo- and ectoplasm, zooxanthellae, various inclusions (some of which are the direct consequence of feeding), a functional hydrostatic apparatus, and a complete skeleton except for some further apophysal development. It will grow little more and eventually will convert all of its inner cell body into some hundreds or thousands of swarmers. The skeleton, gelatin, and myonemes die, analogous to somatic tissue of higher forms, and the swarmers, the germ cells, initiate a new generation. Schewiakoff (1926, p. 723) has one remark which implies the Acantharia are outbreeders. He says the gametes ". . . copulate with those [other gametes], which are arisen from other individuals." (Original in German.) There are no experimental details given in support of this claim.

This picture of acantharian development is clearly an epigenetic one. In the earliest stages thus far seen, the essential features are presence of a nucleus, the skeleton and the two main parts of the cell body—gelatin layer and "Weichkörper." Then there is the further progressive development of all of these parts. This provides clear opportunity for biogenetic interpretations such that the early acantharian would be thought of as possessing a skeleton (showing the special acantharian configuration), a gelatin layer (but no contractile elements associated with it), and a relatively undifferentiated inner part within the gelatin layer, which contains the nuclear material.

Phylogeny

The task now is to identify the semic traits and the homologous relationships between them, and from there the variable aspects of acantharian biology will be examined to see if these are resolvable into aposemic traits. From what we have seen of the Acantharia thus far, it is fair to expect that not only will several aposemic trends be apparent, but that we will be able, on ontogenetic grounds, to establish the direction of their evolutionary development and, therefore, be able to move to a sound phylogenetic analysis of the group.

HOMOLOGIES

The wealth of systematically accumulated information on the various species of Acantharia allows a phylogenist to take several larger categories of traits and break them down into subtraits, each one sufficiently complex to be termed a seme. In particular, there are four categories of organizational characters and one of function which are readily subdivisible, overall, into twelve semes.

Semes. The functional category of reproduction can be subdivided into the semes of binary fission, swarmer formation, and postswarmer development. The first organizational category of inner cell body, or Weichkörper, comprises two semes: general shape and organization of the endo- and ectoplasm and the capsule. The flotation apparatus can be divided into gelatinous layer(s), myonemes, and elastic fibers. The skeleton is quite complex and consists of at least two semes: spicule organization, and apophyses and shells. Lastly, the filopodia can be reduced to the two semes of reticulopodia and axopodia.

Next, let us make more explicit the basis for designating the foregoing traits as semic. The essential argument is to find complexity sufficient to permit rigorous comparisons between species of the trait in question. We shall start with structural features.

The shape of the Weichkörper is delimited by the boundary it maintains with the surrounding gelatinous layer. Its form can be spherical or ameboid or stellate (as a result of extensions into the gelatin). Its composition can be determined by the array of the inclusions contained within it, each one constituting a point of comparison, as well as using details of endo- and ectoplasmic organization. The capsule does not separate endoplasm from ectoplasm, as in the Radiolaria, but encloses both, when it is present, and separates them from the more external gelatinous layer. The structure of the capsule in terms of thickness, numbers, and position of pores and other component features, all provide ample details for careful comparative studies.

The flotational mechanism is dominated, in terms of bulk, by the gelatinous layer. Its position relative to the capsule, in particular, and its single or double-layered structure, plus the presence of various inclusions, all allow detailed comparisons. The myonemes are complex structures, too. Their position, relative to spicules and gelatinous layer, is readily defined. Their number, size, and shape, and in some cases their fine structure, all allow of compositional comparisons. And the elastic fibers, too, can be carefully described in relation to their position in the cell and as regards their organizational patterns. In what follows, it will be seen that the variations in this seme are treated very conservatively; further detailed work on these fibers would, however, probably reveal significant variability allowing of much more interesting comparisons than those used here.

Spicules or skeletons show two obvious semic complexities. The first is to determine whether or not diametral spicules are present, and then there are the various patterns of central union of radial spicules, wherein the various geometrical shapes of the flutings and facets are readily compared on a point-to-point basis. The lateral extensions of the apophyses can be compared both as regards their position on a spicule and as regards the geometry of their elaborations. The same is true of shells, which are an extension of the apophyseal seme. The complexities are so rich here that fusion of apophyses, which constitute shells or tests, could be treated as a separate seme. Such a detailed study is better reserved for a subsequent, more precise analysis than is being attempted here.

The filopodia are of two types and allow an easy subdivision into axo-and reticulopodia. The composition of each is distinctive, and their distribution and number per cell can be used comparatively.

Finally, the reproductive semes can be briefly detailed. Fission involves a series of discrete steps involving nuclear division and segregation, segregation of the contents of the Weichkörper, the spicules, and other differentiations. This distinctive process is easily seen as being quite complex. And the same is true of swarmer formation. Postswarmer development will have to start from young uninucleate cells with small skeletons and the beginnings of Weichkörper development already present. But details of further differentiation potentially allow many detailed comparisons. Here, too, however, we will have to treat these very conservatively, because of very limited first-hand experience with these forms, and because in many cases Schewiakoff's data do not consist of much more than a rather cryptic comment that developmental stages were observed.

Homologous relationships and variability. The foregoing semes are readily homologized as one observes them in the various species where they occur. The very features which endow these traits with semic information allow for identification of those positional and compositional relationships which establish homology. Furthermore, many of the semes show significant variability.

It will not be necessary to review each seme in detail, but a quick review of the basis for homology in each is necessary, if only because an explicit awareness of each methodological step is necessary to allow those unfamiliar with the method to follow it with understanding. (Specifically, we are at step 4 of the procedure summarized at the end of Chapter 3.) To start, the most convincing homologies will be examined first, for they can then be used as the basis for validating the first subsidiary criterion, if the need should arise, as it will.

The spicules are always located as diametral or radial elements in the cell, and their number (10 diametral or 20 radial), their distribution (according to Müller's law), and their composition (probably calcium aluminum silicate) establish their homology in a convincing fashion. The transition from diametral to radial spicules comes about by observing that the former contain not only regular, uniform examples (Fig. 9–3A) but also those with special patterns in their mid-region (Fig. 9–3 B) where the spicules meet in the center of the cell. And then, among the radial spicules, there is a range—from those showing a casual contact in mid-cell, to those with highly sculptured and precisely fitted parts at their inner ends (Fig. 9–2). The spicules represent an aposemic trait.

Similarly the apophyses show homology in their location on spicules and in their composition and show aposemic development as one proceeds from simple lateral extensions of a spicule (Fig. 9–3D) to elaborations resulting in a shell (Fig. 9–3E).

The gelatinous layer can be homologized in terms of its peripheral position in the cell and in terms of its composition, as determined by its own texture and its inclusions. It shows some aposemic variability—thin or thick, and single or double layers—but none which denies the basic similarity of this seme in the various acantharian species.

The myonemes are a clear-cut example of an aposemic trait. Their basic homology can be seen in the way they are located relative to spicules and in their composition. However, some cells have none and others have up to 40–60 thread-like myonemes attached to each spicule. Between these extremes are cells with a few, stout myonemes, and others with a few to many myonemes and of varying sizes. The serial relationship, which is the basis for defining aposemy, is clearly in evidence here.

Similarly, the elastic fibers can be considered homologous. They show little variability at present, perhaps because they are not studied in detail. Treating them very conservatively, their variability can be reduced to their presence or absence, which in many cases is all the information supplied by Schewiakoff.

The inner cell body, or Weichkörper, can be considered, first, in terms of its general organization. Its position is readily determined and so is its general composition, since this includes such obvious features as endo- and ectoplasm, nuclei, zooxanthellae, fat bodies and oil droplets of various sizes, food vacuoles, concretions, and pigment granules. As we shall see shortly, the variability in differentiation of the endo- and ectoplasm correlate with the other aposemic trends (going from a slightly to well-differentiated border between these parts), and the other inclusions vary quite wildly in a rather rich profusion.

The capsule, absent in many forms, shows a similarity in location in all species and some similarity in composition (not so much chemically as in the location of pores and other apertures, especially for axopodia). It shows variation, but not to any striking degree, and not to an extent which casts doubts on the homology of the various examples of capsule formation.

The filopodia, to the degree they are known, present a special problem. The previous semes represent traits somewhat, if not completely, unique to the Acantharia. The reticulopodia and axopodia are not unique. We saw reticulopodia in the foraminiferans and axopodia in the radiolarians and heliozoans. We do not have the fine-structure detail needed for further discrimination. On present knowledge, these pseudopodia would appear to be as homologous within the Acantharia as between the Acantharia and other protozoans. It is difficult, therefore, to argue for a unique homology within this group. However, the fact that the acantharian reticulo- and axopodia have the same distribution as other known homologies (Weichkörper, flotation apparatus, spicules, etc.) allows us to argue, using the first subsidiary criterion of Remane, that these filopodial structures are, also, homologous. Furthermore, they show some variability in number and placement. Axopodia emerge from the whole cell surface or from special apertures in the capsule. Reticulopodia have various distributions relative to the spicules, especially.

It may well be that with added information, the reproductive cycle and the ecology of these forms will show variation sufficient to allow discrimination of aposemic changes among the different species. At present it seems that Schewiakoff's extensive findings on developmental events allow only the identification of one major aposemic (hyposemic) step, i.e., the loss of binary fission in the more complex forms. The facts are, as we noted above, that binary fission occurs only in the simplest forms, in the Acanthochiasmidae. All forms, it can be inferred, show swarmer formation and, possibly, sexual reproduction. At present the details of postswarmer development seem to be very uniform except for relatively minor differences due to production of somewhat different adult forms.

The first ontogenetic feature deserving comment is the presence of flagellated gametes. Such cells are often cited as evidence of the flagellated ancestry of various groups (e.g., Hyman 1940) but can be taken to ludicrous extremes, as when the structure of metazoan sperm and zooflagellates are compared (Alexieff 1924). The point is that at least one member of a potential pair of fusing gametes must be motile if there are no other means for bringing the cells together. And when, as in the case of the Acantharia, swarmers or gametes are released directly into the ocean, there is a quite evident premium on motility to insure a reasonable probability of effecting the unions necessary to perpetuate the species. Flagellation seems an obvious solution to such problems, and the widespread occurrence of such gametes could be a matter of convergence as much as descent from a common ancestor. The necessary analysis is to look at the fine structure of the gamete and the putative ancestral flagellate and decide the nature of the similarities on the basis of the compositional criterion of homology. In Fig. 9–6 is Alexieff's comparison of an idealized sperm to "... a *Trichomonas* with the parasome of *Devescovina*

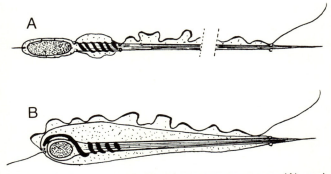

Fig. 9–6 Comparison between an idealized metazoan sperm (A) and a hypothetical zooflagellate (B). (Redrawn from Alexieff (1924).)

striata and a single anterior flagellum." (Alexieff 1924, p. 106) The attempted homologies between the spirally wound elements (p) are false—this is mitochondrial material in the sperm and Golgi material in the (nonexistent) flagellate. The acrosome of the sperm (A) contains the Golgi material of that cell, but yet it is homologized to a flagellum in the flagellate.

In the fine structure of the acantharian swarmer we find one flagellum or two flagella of unequal length (in one species they are of equal length) arising from a tiny cell body (2–6 μm in diameter) which is usually roundish in shape with a single nucleus and a few distinctive inclusions. There is very little specific information on which to develop a detailed picture of the swarmer. It would be especially important to know whether or not mastigonemes—the fine, lateral, hairlike projections of a flagellum—are present. This is apparently unknown. The most important positive bit of information we are left with is flagellar number and size.

The development of the inner cell body allows the interpretation that initially this part of the cell was quite changeable in form, there being various irregularities possible, at least in its contours. Also, there was initially little difference between the endoplasm and the ectoplasm.

The gelatinous layer seems to have been present very early in the evolutionary development of the Acantharia, judging from the fact of its very early presence in ontogeny. This is consistent with the assumption that to exploit the niche of a floating predator, a flotation device must of necessity have been one of the earliest innovations of the group. The progressive development of the flotation mechanism includes change from a relatively homogeneous single layer to a more differentiated layer of one or two parts and the development of (a) myonemes whose number increases somewhat and whose shape is variable, and (b) elastic fibers arranged radially and circularly around spicule apices at the cell surface.

The capsule arises late in development and we would therefore expect it only among the more complex Acantharia, that is, not in the plesiomorph or forms immediately derived from it.

The skeleton is present very early, showing the characteristic acantharian array of spicules. Its growth is essentially centrifugal; the spicules thicken and lengthen. When the spicules reach their adult size, apophyseal elaborations set in. Hence, the more complex shells are arguably limited to later stages of evolution. There is nothing in the developmental stages that provides a clue as to the ordering of the pattern of central union of spicules. That aspect will have to be treated as a correlated transformation series, namely, the order of the various patterns of union will have to be interpreted as they occur in conjunction with other aposemic changes whose direction is known.

The pseudopodia pose an interesting problem. The fact they are reported to appear after the skeleton and gelatin layer are well developed would, on a strict biogenetic interpretation, mean that they evolved after these structures.

But that simply does not make biological sense, for the pseudopodia are a necessary part of the predatory equipment of these organisms: floating has no point if it is not somehow associated with food getting, and all our present evidence indicates that the slender pseudopodia of the Acantharia are the essential means of feeding. (Photosynthetic symbionts could be of some help. But, note, again speaking biogenetically, they are in the cell after pseudopodial formation and hence could not be of help in the early stages of either ontogenetic or phylogenetic development.) We have already commented on the possibility that Schewiakoff's proposed time of appearance of pseudopodia is too late, for it does not account for the significant growth which precedes that time; growth that does not appear explicable solely in terms of food reserves in the form of lipid vesicles, as Schewiakoff proposes (1926, p. 717). It may be that the finely anastomosing reticulopodia are not present but that some other form of pseudopodia are present, such as lobopodia. These would, however, be most effective for a bottom-crawling form.

It may be pertinent at this point to speculate a little on that enigmatic stage, never found by anyone, which lies between the swarmer and the immature acantharian, with its incipient skeleton, gelatinous layer and inner body part. The flagella of the zygote (assuming there is a zygote) are nonfunctional according to Schewiakoff, and, judging from similar conditions in other cases of fusion of flagellated isogametes, the zygote rounds up into an immobile sphere. It probably now sinks further, as both Schewiakoff and Trégouboff speculate. How far down does it go before the appearance of a gel layer starts bringing it up? Could it go down to the ocean bottom and feed off of the detritus there as a small lobopodial ameba? Could it start its postzygotic growth there and continue to grow on that accumulated, ingested food? (The lipids carried over from swarmer formation might well be long depleted by now.) When it has differentiated a gelatinous layer, it might then gently lift off the bottom and start its quiet voyage back to surface waters. In the interim, there might be no feeding until this form reached the more plenteous surface waters, and hence Schewiakoff could then be correct regarding the time of formation of reticulopodia and axopodia but incorrect in his inference that the swarmers, zygotes, and postzygotic development are all nourished by the food reserves included initially in the swarmer. The need for further nutrition might supply the pressures needed to account for sexual reproduction occurring at great depths, the sinking of the zygote, and the delayed appearance of the characteristic acantharian pseudopodia.

The axopodia appear rather late in development and, therefore, would be expected in the more evolved forms. But both reticulopodia and axopodia are found in all Acantharia. Their late ontogenetic appearance would, therefore, appear to be some functional adaptation of developmental events to the feeding patterns of these cells. For the pseudopodia, then, a strict biogenetic interpretation of their development is inconsistent with our awareness that such feeding structures were probably present from the outset of the evolution of these forms. Their delayed ontogenetic appearance is, therefore, more likely to be a functional response to conditions in the life of these forms that we do not fully understand at present.

The probable presence of swarmer formation and subsequent epigenetic ontogeny in all the Acantharia argues that this type of reproduction and development has been present in the group from its beginning. Such a sophisticated mode of biological increase appears, at first glance, to be a puzzle, until we recall that there is evidence of a simpler mode, binary fission, which is present only in the simpler Acantharia. It appears, therefore, that sole reliance on a particular mode of sexual reproduction is something that evolved in addition to asexual reproduction and, presumably, finally rendered the latter superfluous. However, it is also highly likely that progressive sophistication of the cell itself made binary fission a progressively more complicated process, and this hastened its extinction when the alternative of multiple fission (and sexual reproduction) was at hand.

Nuclear behavior offers one final and obvious sequence of changes that deserve comment from the point of view of the biogenetic law. All adult forms are multinucleate, even the least complex ones. The neater view, biogenetically speaking, would have been to find a uninucleate condition in the simplest cells. However, it seems biological necessity has been interposed, and the rather large size of these cells (300 to several thousand micrometers in the adults) has necessitated more nuclear products than can be supplied by just one, presumably diploid, nucleus. (The actual time of occurrence of reduction divisions is not known. The suggestion of Le Calvez (1938), that the swarmers are gametocytes, implies that at that stage there occurs reduction of chromosomes to the haploid number.) In the radiolarians, certain Foraminiferida, and ciliates, the problem has been answered by polyploidy. But in the Acantharia and many zooflagellates, multinuclearity is the apparent answer. So, though surely the multinucleate condition arose from the uninucleate one, the functional advantages of the large cell imposed functional requirements on the nuclear condition very early in acantharian evolution, and, hence, multinuclearity is understandably the common adult condition, despite the ontogenetic sequence of events.

All of the foregoing results in a very consistent picture of the plesiomorph. (We set aside, of course, those traits common to unicellular eukaryotes with animal-like nutrition and concentrate on identifying the plesiomorphic condition of the aposemic traits we have been discussing.) The acantharian plesiomorph would appear as follows. The cell body has an outer homogeneous, single layer of gelatinous material that completely surrounds the inner cell body wherein lie the various cytoplasmic inclusions and nuclear materials characteristic of these forms. These nuclear features include a multinucleate condition, with the nuclei lying around the periphery of the more dense central area, the endoplasm, which is not sharply demarcated from the surrounding ectoplasm. From the latter arise large numbers of thin, ramifying reticulopodia, which traverse the gel layer and protrude from the cell surface. Also present are numerous axopodia, whose central axoneme seems to originate close to the endoplasm and then extends outward, through the ectoplasm and gel layer to provide long, delicate, unbranching pseudopodia radiating from all sides of the cell. They are especially abundant around the spicules. Also in this zone of ecto- and endoplasm are yellow, symbiotic algae and various vesicles containing lipids and pigmented materials. Also, there are food vacuoles and poorly understood, irregularly formed, dark granules termed concretions. The skeleton consists of 10 diametral spicules, each spicule a simple cylindrical needle (of calcium aluminium-silicate, probably) tapered to a point at both ends and arranged according to Müller's law. This cell can reproduce asexually by binary fission or can form swarmers, which probably carry out sexual reproduction. This plesiomorph is a marine organism and is exemplified by the species *Acanthochiasma rubescens* (Fig. 9–4A).

Schewiakoff gives details on the depth and time of year he encountered the various species in the Bay of Naples. He includes, also, data on general abundance and on abundance of forms developing swarmers. A cursory examination of the data reveals no specific patterns of occurrence and no correlations with a particular cellular form. However, the data need the expert attention of a marine ecologist for their proper interpretation. They may well contain considerable comparative information on the way the different species and groups of species exploit their environment, but at present these distributional data are inadequate for establishing the direction of aposemic change. The same is true for the fossil data. Apparently, acantharian skeletons are very poorly preserved. Their presence in the fossil record is meager and spotty and contributes very little to our knowledge of their history (Deflandre 1952; Campbell 1954a). We are left, therefore, with ontogenetic and morphological data and, as we have seen, this provides considerable information on the plesiomorphic condition.

PHYLOGENETIC ANALYSIS

We are now ready to turn to phyletic relationships. The specific forms we will study are the first species described by Schewiakoff (1926) in each of the 17 different families he recognizes in the Acantharia. Though this is a somewhat arbitrary way to sample the group, it is not as random as might appear, because of the fact that Schewiakoff arranged his taxa with his own phylogenetic conclusions in mind. The first species in any family is the structurally simpler member of the simpler genus in that family. In effect, the sampling device we are using is selecting for the plesiomorph of each family. However, since the comparisons we will be making are based on a methodology—

homologous semes, criteria for identifying the plesiomorph, phyletic distance, and the like—which is developed independently of Schewiakoff's evolutionary views on the Acantharia, the only significance to using members of these 17 families is that of a device for insuring a representative sample of the group as a whole. Some of these forms are illustrated in earlier figures (Fig. 9-4A and C); the other species are referred to in Table 9-1. For each species there is a detailed description of its cellular organization available and varying amounts of detail on its reproduction and ecology. From this has been culled the summary of the biological features of each species given in Table 9-1 and from these we have drawn a summary of intersemic homologies (Table 9-2).

Phyletic distance. We are now ready to determine the occurrence of plesiosemy, aposemy, and neosemy among the chosen species and from these to calculate phyletic distances (R values). It is first useful to make clear the basis for breaking down these 17 species into five subgroups: A,B; C,D,E; H,G,F; I,J,K,L,M,N; and O,P,Q. This was done on the basis of Schewiakoff's classification, which recognizes the mode of central union (seme 6) as of key importance. The practical gain in doing this is easily made clear. If we had made all possible comparisons between the 17 species and their 12 semes, there would have been 1632 such comparisons. By going to subgroups, and, of course, identifying a plesiomorph for each (which was done by identifying the form most like *Acanthochiasma rubescens*), and then comparing plesiomorphs and making any other comparisons that seemed appropriate, the total comparisons were reduced to 528 (Table 9-3), most of which are rapidly

Table 9-1 Summary of semic characteristics of seventeen species of Acantharia.
(Page references are to Schewiakoff (1926) and to figures published there.)
A. *Acanthochiasma rubescens* (p. 64-69; Pl. 1, fig. 1)   Inner cell-body rounded but showing variability of form; no sharp difference between ectoplasm and endoplasm; few symbionts; yellow and brown pigments; other inclusions including concretions; no capsule. Homogeneous, single gelatinous layer; no myonemes or elastic fibers. Skeleton with ten diametral, flexible, needlelike spicules; cylindrical; pointed tips. Many reticulopodia with broad bases; fibrillar; abundant axopodia. Binary fission. Swarmers with one flagellum.
B. *Acanthoplegma krohni* (p. 84-94; Pl. 1, fig. 3)   Inner cell-body rounded and ameboid; ecto- and endoplasm mingled; zooaxanthellae; yellow-gold pigments; concretions and other inclusions; no capsule. Two layers to gelatinous material; two myonemes per spicule; elastic fibers. Ten diametral spicules; thin, flexible, cylindrical; pointed tips; irregular midsection. Abundant, fine reticulopodia and axopodia. No binary fission in these or any of the succeeding forms. Uniflagellate swarmers.
C. *Conacon foliaceus* (p. 184-188; Pl. 18, fig. 1) "Weichkörper" rounded but with variable shape; colorless; endoplasm distinct from ectoplasm, but latter tends to mingle with gelatinous layer; few zooxanthellae; other typical inclusions present; no capsule. One homogeneous gelatinous layer; 4-6 myonemes per spicule (no mention of elastic fibers). 20 radial spicules, all of same length and shape;

Table 9-1 *(cont.)*

C. *Conacon foliaceus (cont.)*
rounded basal ends loosely opposed. Reticulopodia and axopodia. (Swarmers not observed in this species or family.) Postswarmer development observed.

D. *Gigarton fragilis* (p. 190–198; Pl. 18, fig. 2) Inner cell-body small, round and ameboid; separation of alveolar ectoplasm from yellow-red-brown endoplasm; no zooxanthellae seen; other typical inclusions present; no capsule. Two gelatinous layers with many vacuoles; 8–12 myonemes per spicule; a few elastic fibers. Many radial reticulopodia, cover spicules; axopodia. (Swarmers not observed.) Postswarmer development observed.

E. *Stauracon pallidus* (p. 224–228; Pl. 18, fig. 5) Ameboid inner cell-body (polygonal contours); few zooxanthellae; separation of endo- and ectoplasm; yellow-brown pigment; other typical inclusions; no capsule. Two gelatinous layers; 12–16 myonemes per spicule; elastic fibers. 20 radial spicules; 4 ribbed major ones, 3–4 times longer than other smooth ones; knobbed basal ends lightly apposed. Reticulopodia very numerous around spicules; 5–6 axopodia per spicule. (Swarmers not seen.) Postswarmer development observed in certain stages.

F. *Acantholithium stellatum* (p. 106–109; Pl. 8, fig. 1) Inner cell-body star shaped; clear separation of endo- and ectoplasm; few zooxanthellae; colorless or yellow gray; typical inclusions; no capsule. Two gelatinous layers; 8 myonemes per spicule; elastic fibers. 20 radial spicules; equal length and shape, with 4 ribs; meet in a homogeneous sphere. Abundant reticulopodia, highly branched; 4–5 axopodia per spicule. (Swarmers not seen.) Later developmental stages observed.

G. *Amphilithium concretum* (p. 133–135; Pl. 11, fig. 2) Weichkörper cylindrical or tube shaped; clear separation of endoplasm from ectoplasm; colorless; many zooxanthellae throughout endoplasm; other typical inclusions; no capsule. Two gelatinous layers; 12–14 myonemes per spicule; elastic fibers. 20 radial spicules; 2 very long ones, heavy and thick; others shorter, thin and needlelike; fused centrally in a homogenous sphere. Reticulopodia appear as tufts from spicule apices as well as radiating out from rest of cell surface; about 5 axopodia per spicule. Swarmers not seen in this genus; other members of this family show biflagellate swarmers with flagella of unequal length. Postswarmer development observed.

H. *Pseudolithium bifidum* (p. 154–160; Pl. 11, fig. 5) Rounded, ameboid Weichkörper; ectoplasm and endoplasm not clearly separated; pigment gray-yellow or brown or brown-green; zooxanthellae; other inclusions among which are distinctive concretions; no capsule. Two gelatinous layers; 16 myonemes per spicule; elastic fibers. 20 radial spicules, all of same length and shape; cylindrical, very thin; join in a small homogeneous sphere. Heavy network of reticulopodia; 6–8 axopodia per spicule. (No binary fission, contrary to Popofsky's claim.) Swarmers with two flagella of unequal length. Postswarmer development observed.

I. *Acanthometra pellucida* (p. 233–241; Pl. 22, fig. 1) Inner cell-body has regular polygonal shape except when certain spicules are longer than others; no pigment; zooxanthellae and other typical inclusions present; capsule present. Two gelatinous layers; 24–32–40 myonemes per spicule; elastic fibers. 20 radial spicules; usually of same length but 1, 2, or 4 can be longer than others; cylindrical or somewhat flattened in middle third; thin, flexible, pointed; pyramidal, 5–6 edged base. Many fine, branching reticulopodia; axopodia. Swarmers seen, 2 flagella of unequal length. Many postswarmer developmental stages seen.

J. *Lithopterida fenestrata* (p. 275–284; Pl. 28, fig. 1) Weichkörper expanded in equatorial plane, cross or star shaped; zooxanthellae; conspicuous fat droplets; other inclusions capsule present. One gelatinous layer, reaches to apophyses; 10–12–16 myonemes per spicule, threadlike; elastic fibers. 20 radial spicules; 4 cylindrical, major ones with conspicuous winglike grillwork apophyses; 16 needle-

Table 9-1 *(cont.)*

J. *Lithopterida fenestrata (cont.)*
like smaller ones; pyramidal bases. Reticulopodia; 5–6 axopodia per spicule. (Swarmers not observed.) Later developmental stages observed.

K. *Diploconus fasces* (p. 425–431; Pl. 36, fig. 1) Weichkörper hard to see because of shell; ellipsoidal; yellow-green; seems to be separation of ectoplasm and endoplasm; other inclusions, including zooxanthellae, present; capsule not definitely observed, thought to be present, is present in other members of family. One gelatinous layer; 8–12 threadlike myonemes per spicule; elastic fibers not observed for certain in any member of this family. 20 radial spicules; 2 long, strong, pointed; 18 shorter ones; shell; pyramidal bases meeting in a starburst configuration. Reticulopodia and axopodia present. (Swarmers not observed.) Later developmental stages observed.

L. *Pleuraspis costata* (p. 297–307; Pl. 26, fig. 5) Round inner cell body; many zooxanthellae; separate ecto- and endoplasm; various typical inclusions; capsule present. One gelatinous layer; 8 myonemes per spicule; elastic fibers. 20 radial spicules, same length, no ribs, somewhat flattened; strong apophyseal development forming a grill-like shell; cone-shaped basal ends tightly fitted to each other. Reticulopodia; 5–6 axopodia per spicule. Swarmers, biflagellate with equally long flagella. Later developmental stages seen.

M. *Coleaspis coronata* (p. 391–398; Pl. 37, fig. 1) Small, ellipsoidal Weichkörper; many zooxanthellae; endoplasm yellow-gray; many vacuoles and other inclusions; distinct ectoplasm; capsule present as a thin membrane. One gelatinous layer; 8–12 threadlike myonemes per spicule; no elastic fibers. 20 radial spicules; 2 large, strong, cylindrical; 18 shorter, two edged, all pointed; shell with thick walls; pyramidal bases. Reticulopodia and axopodia present. (Swarmers not seen.) Postswarmer development observed.

N. *Phractopelta dorataspis* (p. 443–448; Pl. 36, fig. 2) Rounded, inner cell body; clear separation of ecto- and endoplasm; yellow-brown pigment; many zooxanthellae; other typical inclusions present; thin capsule present. Gelatinous layer (single?), 6–8 myonemes per spicule; elastic fibers. 20 radial spicules, cylindrical, pointed; 2 concentric grillwork shells; 2 small pointed apopyses per spicule lying outside of shell; pyramidal bases. Many radial reticulopodia; axopodia. (No swarmers seen.) Young stages seen.

O. *Phyllostaurus siculus* (p. 460–466; Pl. 40, fig. 1) Inner cell body round to ellipsoidal; differences between endoplasm and ectoplasm; pigment dark brown-red to purple-red; also rose and yellow-green; rather few zoothanthellae; other inclusions also present; capsule. One gelatinous layer; 40–60 threadlike myonemes per spicule; elastic fibers. 20 radial spicules of same length and shape, cylindrical; pyramidal base meet in Blätterkreuz. Many tapering reticulopodia; 5–6 unbranching axopodia per spicule. Swarmers reported to be without flagella; later development seen.

P. *Xiphacantha quadridentata* (p. 503–507; Pl. 43, fig. 1) Round Weichkörper; opaque, red-brown; distinct endo- and ectoplasmic parts; zooxanthellae and other inclusions present; also, a capsule. One gelatinous layer; 40–60 threadlike myonemes; elastic fibers. 20 radial spicules; same length and shape; 4 ribs per spicule; pointed; 4 apophyses per spicule; Blätterkruz at base. Highly branched reticulopodia, axopodia. No swarmers observed. Young stages seen in *X. alata*.

Q. *Dictyacantha tetragonopa* (p. 531–534; Pl. 44, fig. 12) Round inner cell-body; separation of ectoplasm and endoplasm; yellow-brown pigment grains; zooxanthellae; other inclusions, too; capsule present. Thick gelatinous layer; about 50 myonemes per spicule; elastic fibers. 20 radial spicules, same length and shape, heavily built; 4 ribbed; grilled shell; large Blätterkreuz. Strongly branching pseudopodia; unbranching axopodia. Initiation of swarmer formation observed, but actual swarmers not seen. Later developmental stages not seen in this family.

Table 9-2 Intersemic homologies between seventeen species of Acantharia.

Species	Cellular morphology 1.	Flotation apparatus 2.		Skeleton			Pseudo- podia 8.		Reproduction		
			3.	4.	5.	6.	7.	9.	10.	11.	12.
A. *Acanthochiasma rubescens*	1A	2A	3A	4A	5A	6A	7A	8A	9A	10A	11A	12
B. *Acanthoplegma krohni*	1A	2A	3B	4B	5B	6B	7A	8A	9A	10B	11A	12
C. *Conacon foliaceus*	1C	2A	3A	4C	?	6C	7A	8C	9C	10B	?	12
D. *Gigarton fragilis*	1C	2A	3B	4D	5B	6C	7A	8C	9C	10B	?	12
E. *Stauracon pallidus*	1C	2A	3B	4E	5B	6C	7A	8C	9C	10B	?	12
F. *Acantholithium stellatum*	1F	2A	3B	4C	5B	6F	7A	8A	9C	10B	?	12
G. *Amphilithium concretum*	1F	2A	3B	4G	5B	6F	7A	8C	9C	10B	11G	12
H. *Pseudolithium bifidum*	1A	2A	3B	4E	5B	6F	7A	8C	9C	10B	11G	12
I. *Acanthometra pellucida*	1F	2I	3B	4I	5B	6I	7A	8C	9C	10B	11G	12
J. *Lithopterida fenestrata*	1F	2I	3J	4E	5B	6I	7J	8C	9C	10B	?	12
K. *Diploconus fasces*	1F	2I	3J	4D	?	6I	7K	8C	9C	10B	?	12
L. *Pleuraspis costata*	1F	2I	3J	4D	5B	6I	7L	8C	9C	10B	11L	12
M. *Coleaspis coronata*	1F	2I	3J	4D	5B	6I	7M	8C	9C	10B	?	12
N. *Phractopelta dorataspis*	1F	2I	3J	4N	5B	6I	7N	8C	9C	10B	?	12
O. *Phyllostaurus siculus*	1F	2I	3J	4O	5B	6O	7A	8C	9C	10B	11O	12
P. *Xiphacantha quadridentata*	1F	2I	3J	4O	5B	6O	7P	8C	9C	10B	11G	12
Q. *Dictyacantha tetragonopa*	1F	2I	3J	4O	5B	6O	7Q	8C	9C	10B	?	?

Key to Semes:

1. Cellular organization: 1A—rounded, ameboid, mingled ecto- and endoplasm; 1C—small, rounded, ameboid, separate ecto- and endoplasm; 1F—regular polygon, star shaped or cylindrical, separate ecto- and endoplasm.
2. Capsule: 2A—no capsule; 2I—capsule present.
3. Gelatinous layers: 3A—homogeneous; 3B—two layers; 3J—one complex layer.
4. Myonemes: 4A—no myonemes; 4B—two myonemes; 4C—four to six myonemes; 4D—eight to twelve myonemes; 4E—12–16 myonemes; 4G—12–24 myonemes; 4I—24–40 myonemes; 4N—six to eight myonemes; 4O—40–60 myonemes.
5. Elastic fibers: 5A—absent; 5B—present.
6. Spicule morphology: 6A—diametral; 6B—diametral plus marks at center of spicules; 6C—radial, bases loose; 6F—radial; bases fused into homogeneous structure; 6I—radial, pyramidal bases; 6O—radial bases showing Blatterkreuz pattern.
7. Spicule size and number, apophyses and shells: 7A—no apophyses, no shell; 7J—four large, 16 small, grillwork apophyses; 7K—two large, 18 small, shell; 7L—20 somewhat flattened, grill; 7M—two large, 18 shorter and double edged, thick shells; 7N—20, two concentric grilled shells; 7P—20, four with ribs, four with apophyses; 7Q—20, four with ribs, heavy, grilled shell.
8. Number of reticulopodia: 8A—abundant; 8C—less abundant.
9. Number of axopodia: 9A—abundant; 9C—less abundant.
10. Fission: 10A—fission present; 10B—fission absent.
11. Swarmers: 11A—one flagellum; 11G—two unequal flagella; 11L—two equal flagella; 11O—no flagella.
12. Postswarmer development: 12A—present.

ACANTHARIANS

Table 9-3 Homologous relationships between pairs of semes and phyletic distances among seventeen species of Acantharia.

Species comparisons	Homologies between semes 1 2 3 4 5 6 7 8 9 10 11 12	Semes compared t	Plesio-semes p	Ape-semes a	Neo-semes n	Phyletic distance R
A and B	p p a n n a p p p a p p	12	7	3	2	6.4
C	a p a n ? a p a a a ? p	10	3	6	1	25.2
H	p p a n n a p a a a a p	12	4	6	2	15.7
I	a a a n n a p a a a a p	12	2	8	2	23.2
O	a a a n n a p a a a a p	12	2	8	2	23.2
B and C	a p a a ? a p a a p ? p	10	4	6	0	15.0
H	p p p a p a p a a p a p	12	7	5	0	8.8
C and D	p p a a ? p p p p p ? p	10	8	2	0	1.8
E	p p a a ? p p p p p ? p	10	8	2	0	1.8
G	a p a a ? a a p p p ? p	10	5	5	0	10.5
H	a p a a ? a p p p p ? p	10	6	4	0	6.8
I	a p a a ? a p p p p ? p	10	6	4	0	6.8
O	a p a a ? a a p p p ? p	10	5	5	0	10.5
D and E	p p p a p p p p p p ? p	11	10	1	0	0.5
O	a a a a p a a p p p ? p	11	5	6	0	13.6
F and G	p p p a p p p a p p ? p	11	9	2	0	1.6
H	a p p a p p p a p p ? p	11	8	3	0	3.5
I	p a p a p a p a p p ? p	11	7	4	0	6.1
G and H	a p p a p p p p p p p p	12	10	2	0	1.5
I	p a p a p a p p p p p p	12	9	3	0	3.3
J	p a a a p a p p p p ? p	11	7	4	0	6.1
H and I	a a p a p a p p p p p p	12	8	4	0	5.7
O	a a a a p a a p p p a p	12	5	7	0	16.9
I and J	p p a a p p a p p p ? p	11	8	3	0	3.5
K	p p a a ? p a p p p ? p	10	7	3	0	3.9
L	p p a a p p a p p p a p	12	8	4	0	5.7
M	p p a a p p a p p p ? p	11	8	3	0	3.5
N	p p a a p p a p p p ? p	11	8	3	0	3.5
O	p p a a p a p p p a p	12	8	4	0	5.7
J and K	p p p a ? p a p p p ? p	10	8	2	0	1.8
L	p p p a p p a p p p ? p	11	9	2	0	1.6
M	p p p a p p a p p p ? p	11	9	2	0	1.6
N	p p p a p p a p p p ? p	11	9	2	0	1.6
O	p p p a p a a p p p ? p	11	8	3	0	3.5
K and L	p p p p ? p a p p p ? p	10	9	1	0	0.5
M	p p p p ? p a p p p ? p	10	9	1	0	0.5
N	p p p a ? p a p p p ? p	10	8	2	0	1.8
L and M	p p p p p a p p p ? p	11	10	1	0	0.5
	p p p a p a p p p ? p	11	9	2	0	1.6
M and N	p p p a p a p p p ? p	11	9	2	0	1.6
O and P	p p p p p a p p p a p	12	10	2	0	1.5
Q	p p p p p a p p ? ?	10	9	1	0	0.5
P and Q	p p p p p a p p p ? ?	10	9	1	0	0.5

made because of the relatively low variability in certain semes, e.g., 8 through 12. The validity of these groupings depends on whether or not the R values within subgroups are less than those between subgroups, i.e., whether or not the subgroups are indeed quite homogeneous. With only two exceptions—B is closer to H than to A and G is closer to I than to H—subgrouping homogeneity appears valid. Further comparisons of B to C and G to J and C support this contention.

To determine the course of evolutionary history among the maze of phyletic distances in Fig. 9-7 we need to use only the shortest phyletic distance between successive species. Starting with the plesiomorph, species A, *A. rubescens,* the next phylogenetic step would be species B. From this point a bifurcation is indicated, one arm going to C and the other to H. To do otherwise, i.e., to go from B to C to H, or from B to H to C is not only a longer phyletic distance but it also involves certain inconsistencies relative to ontogenetic events. C to H necessitates refusion of the central union of the spicules and refusion of ecto- and endoplasm. The reverse direction, H to C, necessitates reduction of myonemes and reduction of two gel layers to one and then back to two in D and E. These difficulties are avoided by assuming the somewhat parallel and separate evolution of C and H, with C representing the plesiomorph of the group with fused basal ends of the spicules, and H being the plesiomorph of the group with loosely joined basal ends.

From the H,G,F group, and specifically by way of G, i.e., *Amphilithium concretum,* there emerges *Acanthometra pellucida* (species I). However, there is an apparently very significant difference in form between *Amphilithium concretum* (species G) and *Acanthometra pellucida* (species I). The former species possesses one pair of enormous spicules that greatly elongate one diametral axis of the cell, including a proportional change in the dimensions of the inner cell body. However, it does not seem improbable that such a change in body shape is really due to a complex difference in cellular organization from the more typical spherical shape. Excessive deposition of skeletal material on the one key pair of spicules might result in correlated adjustments in the rest of the cell body, hence, the change in cell contour may be due to relatively simple genetic and developmental changes. The alternative solution would be to place *Acantholithium stellatum* in the place of *Amphilithium concretum,* since these two forms are very similar, except for *A. stellatum* possessing 8 myonemes and *A. concretum* possessing 12-24. In terms of these structures, the series of *Pseudolithium bifidum* (16 myonemes per spicule) → *A. concretum* (12-24 myonemes per spicule) → *Acanthometra pellucida* (24-32-40 myonemes per spicule) seems more plausible than if *A. stellatum* with its 8 myonemes per spicule were the intermediary form. Schewiakoff (see Fig. 9-8) prefers another sequence altogether: *A. stellatum,* he suggests, gave rise to both *Pseudolithium bifidum* and *Amphilithium con-*

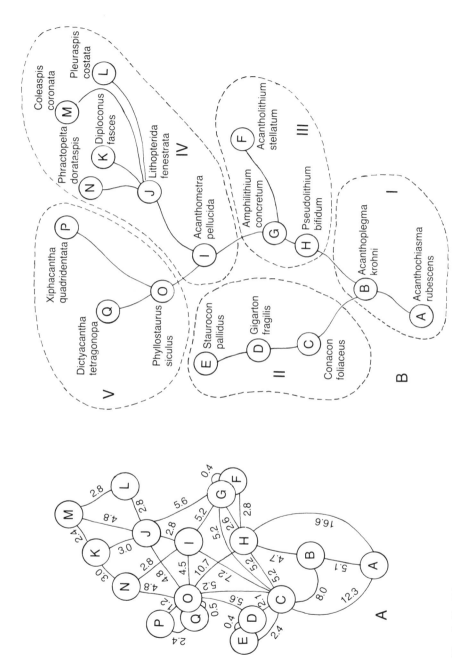

Fig. 9-7 Diagrams of phyletic relationships between seventeen species of Acantharia. A. Selected phyletic distances between the species in question (A through Q). B. Dendrogram interpreting the phyletic distances in terms of plausible evolutionary sequences. Numerals, I, II, III, and IV refer to the major subdivisions of the Acantharia as proposed by Schewiakoff, i.e., Holacantha, Chaunacantha, Symphacantha and Arthracantha, respectively. (See Fig. 9-8).

cretum. The reasons for not following his suggestion at present are given below, when we come to discussing his phylogenetic proposals.

Acanthometra pellucida is considered to be the plesiomorph of the major subgroup of the Acantharia represented by species I,J,K,L,M and N, all members of the Sphaenacantha in Schewiakoff's classification. From *Acanthometra pellucida* there is also derivable *Phyllostaurus siculus*, species O, and species P and Q, Schewiakoff's Phyllocantha. This makes both groups, which comprise the major taxon Arthracantha, all derivable from an acanthometrid ancestor. This is summarized in Fig. 9–7, B and in this diagram are also indicated the major taxa that have just been mentioned.

We can now turn to an explicit comparison of this dendrogram with the one proposed by Schewiakoff (Fig. 9–8). Overall, it would be expected that the two phylogenies would be in good agreement, for both are drawn from the same species descriptions and, more importantly, both utilize the same wide range of biological fact in coming to their conclusions. Differences might be expected, though, from the arbitrariness inherent in any classification that depends largely on one character, in this case Schewiakoff's use of the nature of central union of spicules. As we'll see, this is a minor difficulty.

Major agreement is found in the general relationships of the various species: the Holacantha give rise to the Chaunacantha and the Symphyacantha, and the latter give rise to the Arthracantha. We favor the view that the Phyllacantha arose from the acanthometrids whereas Schewiakoff suggests that the Phyllacantha may have arisen from the symphacanth family, Pseudolithidae. His diagram summarizes his indecision, explicitly expressed in his text, on this point.

Specific points of difference between the two phylogenies are as follows: Schewiakoff favors the Astrolithidae as representing the ancestral symphacanthan form (as we have just mentioned); we favor *Pseudolithium* as the plesiomorph. Our reason is the nature of the inner cell body. *P. bifidum* has an ameboid inner cell body with no clear difference between endoplasm and ectoplasm. In favor of Schewiakoff's view is the lower number of myonemes in *Acantholithium stellatum*. There is no way on present evidence to resolve this difference. Only further detailed study of the cellular structure and ontogeny of the forms in question would provide the particular data needed to resolve this point.

A second area of difference with Schewiakoff comes in our treatment of the lithopterids as a key group for radiative evolution of the rest of the Sphaenacantha. This view actually differs little from Schewiakoff who indicates the Lithopteridae as an early offshoot from the acanthometrids, who are the chief source of the subsequent radiative evolution. There seems little reason to sift out the details of these differences, which can really only be resolved by further intensive work on the forms in question.

Finally, there are the differences, earlier alluded to, as to the exact origin

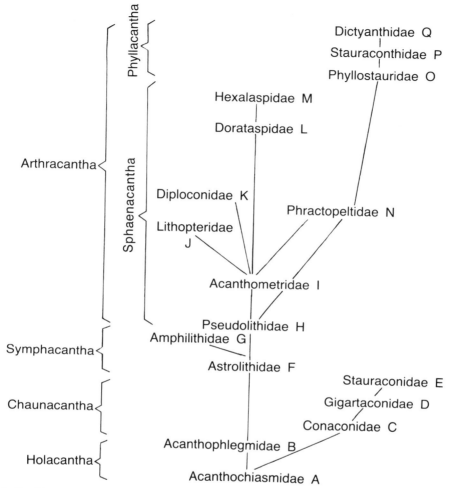

Fig. 9–8 Phylogeny of the Acantharia as proposed by Schewiakoff. Major taxonomic groups are given on the left and families lie within the dendrogram. Capital letters indicate the placement of the species presented in Fig. 9–7. (Redrawn from Schewiakoff (1926).)

of the Phyllacantha. Our proposal is simply more explicit than Schewiakoff's, and, again, detailed pursuit of the difference seems unrewarding for present purposes.

There are, of course, earlier phylogenetic proposals than Schewiakoff's, for example, Haeckel (1888b) and Hertwig (1879). But these were thoroughly reviewed by Schewiakoff, and, in any case, he added so much new material to our knowledge of the Acantharia in his monograph of 1926 that the earlier phylogenies were rendered quite obsolete. Perhaps the only further thing to be added is comment on the arbitrary—from an evolutionary point of view—

separation of certain groups in Schewiakoff's phylogeny. As we have already said, this comes from his attempt to classify as well as suggest evolutionary relationships. The Holacantha are not as homogeneous a group as the presence of diametral spicules would suggest and the Symphyacantha are quite close to the Acanthometridae of the sphaenacanthid Arthracantha. Distortions of phyletic distance in Schewiakoff's diagram are largely due to his desire to show the major taxa as a vertical axis on the left of his dendrogram. As has often been pointed out (e.g., Sokal and Sneath 1963), a phylogenetic diagram can not serve a variety of masters successfully. It should serve first the needs of showing evolutionary relations as phyletic distance. If one and the same diagram also tries to show taxonomic groupings, rates of change, numbers of species in a group, fossil versus recent forms, etc., it probably shows no one aspect effectively, and, therefore, all are done somewhat poorly.

Before turning to the problem of the origin of the Acantharia there needs to be some commentary on the Actinelia, for the actinelids are heterogeneous, and their relation to the acantharians is not at all clear. The major problem in discussing them is the lack of detailed knowledge. Except for *Podactinelius sessilis*, the most aberrant one of the lot from the point of view of the typical Acantharia, these forms have never been described in detail and, as Trégouboff (1953b) notes, few have been found again since their original discovery. The skeletal parts, as might be expected, have been best described. The spicules, all the same size and shape, are never arranged according to Müller's law and range in number from 18 to around 500. A capsule is present in *Podactinelius* and is reportedly present in other forms. In Fig. 9–9 are given some details on *P. sessilis*. This marine form, first described from specimens collected in Japan (Schröder 1908), was the form whose spicules were chemically analyzed by Bütschli (1908) and whose conclusion that strontium sulfate was probably present became, then, the sole predominant basis (until Odum's (1941) tentative study on *Acanthometra pellucida*) for concluding that the acantharian skeleton was, indeed, composed of strontium sulfate. One can only conclude that it was Bütschli's enormous personal prestige that fostered this conclusion, though Bütschli himself intended no such sweeping generalization. Regarding other features of this form, we can briefly make the following points. There is no gelatinous layer. Schröder believes certain peripherally located structures to be myonemes, but this seems highly doubtful, especially in view of no plausible function for them. The presence (or absence) of pseudopodia has not been determined. The capsule is perforated by the radial spicules which meet centrally and their basal ends are pyramidal in shape. Many nuclei are present, lying just within the capsule. The nature of the inclusion bodies is not known. There is a stalk which attaches this form to its substratum. Presumably, *P. sessilis* is in fact a sessile form, presumably animal-like in its feeding habits, which have never been reported on.

The most sensible treatment of the actinelids, for the present, is to refrain

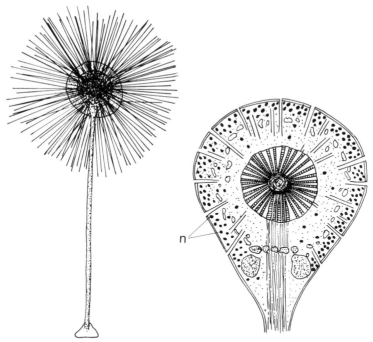

Fig. 9-9 *Podactinelius sessilis*. Left-hand figure showing vegetative form with radiative filopodia, stalk and hold-fast. At right, details of cell body. Nuclei (n) all peripheral. (Redrawn, by permission, from Grassé (ed.) (1953) *Traité de Zoologie*. Vol. 2, part II, Masson et Cie. Originally from Schröder (1908).)

from phylogenetic speculation regarding them. However, one feels compelled to draw attention to what may be an interesting parallel here with a possibly similar situation in the heliozoans, in the Centrohelida in particular. In that group, it was argued that the stalked and attached form *Dimorpha* might be an intermediate between the free-floating forms and a flagellate ancestor. The stalk of *Dimorpha* seems to arise from the central granule from which flagella and axopodia also radiate outwards. Is the stalk of *Podactinelius* a specialization of a flagellar structure? And, further, Schewiakoff has one brief observation (1926, p. 52) that he has seen flagellalike pseudopodia on rare occasions in the Acanthochiasmidae. The question is, therefore, did stalked forms represent evolutionary bridges from flagellates to various spherically symmetrical forms with axopodia?

Origins. As with the Radiolaria, there seem *a priori* to be three general possibilities for acantharian origins: colorless algae, other ameboid forms, and degenerate Metazoa. The latter has no plausible forms to support it, as was seen with a similar suggestion for the Radiolaria, and so we set it aside.

Among the possible ameboid ancestors of the Acantharia, one can consider the heliozoans, as did Haeckel, who bridged differences between these groups by inserting the actinelids as intermediate forms. But, as mentioned above, it is not clear that the actinelids have anything to do with the acantharians. Furthermore, to derive the marine acantharians from freshwater heliozoans is to move opposite to most evolutionary trends. (This point, also, was discussed in the previous chapter in considering a heliozoan ancestor for the radiolarians.) In general, most heliozoans seem to be specialized freshwater forms and unlikely candidates as ancestors to the Acantharia.

To derive the Acantharia from the Radiolaria, or vice versa, or to derive both from a common ancestor, flies in the face of their biological differences. There are numerous general similarities in the biology of these two great groups: spherical symmetry, slender radiating pseudopodia, skeletons, outer gel layers. But none of these is demonstrably homologous in the two groups; the detailed composition of their parts establishes the quite distinct and different organizational nature of the skeletons, pseudopodia, and cell structures in general. Both groups are floating micropredators, and their superficial similarities represent two independent solutions to similar selection pressures; they appear to be as good examples of convergence as found anywhere. To complete the argument for convergence, however, we should first complete the story of the origins of these groups. If they have a common ancestral form, they represent parallel rather than convergent evolution. All we have really established so far is the need to look for the ancestral radiolarian among the (a) dinoflagellates or (b) the colorless chrysomonads with ameboid tendencies, or (c) other ameboid forms (now excluding most heliozoans and acantharians). For the acantharians, the ameboid forms are still a possibility as are the colorless flagellates.

The acantharian plesiomorph, *Acanthochiasma rubescens* (Fig 9–4A), suggests that the ancestral form of the acantharians have pseudopodia, possibly of the axopodial type, and that the skeletal elements, if present, are simple spicules, possibly of calcium aluminium silicate, which are diametral in their placement in the cell. The characteristic gelatinous material could well have emerged in conjunction with invasion of the floating mode of life and probably was absent in the progenitor.

It might also be argued that the flagellated swarmers give biogenetic evidence of flagellated ancestry for this group. Such an interpretation must be followed with great caution. As argued earlier, flagellated gametes could be a convergent development in many forms. More intriguing would be the form that follows swarmer formation and that precedes the appearance of small developing acantharians—that mysterious, as yet never seen, form that we suggest might be a bottom-dwelling naked ameba. We badly need that ontogenetic information.

The chrysomonads contain the forms most similar to the putative starting point of the acantharians, but the resemblances are, at best, very general, being confined to the presence of ameboid forms with branching filopodia and very little else that is acantharian in character. For example, *Chrysamoeba pyrenoidifera* (Fig. 9–10) is provided with radiating filopodia, but it also possesses a single plastid and shows no skeletal material. Such a form indicates distant potentialities of the acantharian sort but no real trends in that direction.

It can, however, be expected that identification of the acantharian ancestor is going to be difficult. The most complex forms like *Xiphacantha quadridentata* and *Coleaspis coronata* still pursue the same mode of life as *Acanthachiasma rubescens* and with much the same cellular organization and organelles. Thus, with the emergence of the plesiomorphic type, there appeared a form showing the basic attributes needed to exploit the acantharian adaptive zone, and further evolution was largely limited to further variations on the themes established by selection acting on the acanthochiasmid form. Organisms ancestral to that form could well have been going through rapid, i.e., tachytelic evolution, and their numbers might have been small. Selection would have resulted in loss of the less efficient forms, which might have been the majority of them. Only with appearance of an efficient acantharian does evolution become horotelic, providing us a picture of predominantly cladogenetic development in this group as a whole, with limited evidence relating it to any ancestral form available today.

Before turning to concluding remarks it seems appropriate to enter here a few very ruminative remarks about acantharian evolution. The essential impression is one of some wonder at the rather uncomplicated manner in which acantharian organization and function fit a phylogenetic pattern of plesiomorph and aposemic development of certain traits and neosemy of certain others, e.g., the capsule. With the exception of the Actinelia, nothing is left out, all groups have their place, and quite wondrously the passage from the

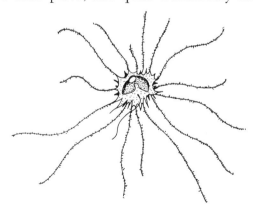

Fig. 9–10 *Chrysamoeba pyrenoidifera,* a chrysophytan with radiating, granular filopodia and one flagellum. (Redrawn, by permission, from Grassé (ed.) (1952) *Traité de Zoologie* Vol. 1, part I. Masson et Cie.)

plesiomorph to the most aposemic forms seems relatively unstrained and straightforward. If this group is as old as we tend to think the protozoa as being in general, where are the groups which show reductions and don't really fit in anywhere? Where are the parallelisms that, as in the Radiolaria, blur the distinctions between otherwise independent evolutionary trends? Does the biogenetic law really work as neatly as it seems to in the Acantharia? And so on. The Acantharia do seem, at least to this viewer, to be a freshly printed page from animal evolutionary history. How old are they? The oldest fossils are only from the Eocene (Campbell 1954a). Deflandre (1953d) discusses only radiolarian fossils and, in a footnote to Trégouboff's (1953b, p. 320) treatment of the acantharians, expresses doubt as to the existence of good fossil remnants of this group. Orlov (1962) goes further and says, "No absolutely authentic remains of Acantharia in the fossil state have been found." (p. 646) Does this mean the tests simply are not preserved from rocks more than 25 million years old? Radiolarian fossils are more than 20 times older, and why they should be preserved and not the acantharians is not at all clear. Perhaps the acantharians are, in fact, Coenozoic arrivals! Does their more elaborate flotation apparatus allow them to compete successfully with at least certain Radiolaria? Does the presence of the skeleton, even in the plesiomorph, mean they did evolve at a time when such protection was needed at the outset of their evolution, i.e., at a time when possible predators were already on the scene? Is part of their success due to an initial Batesian mimicry of the Radiolaria which now may be Müllerian? This would also help explain the close superficial resemblances between the groups. If this suggestion of a relatively recent origin of the Acantharia can be taken seriously, it poses a fascinating evolutionary situation. The Acantharia would seem to be successfully invading the adaptive zone preempted hundreds of millions of years earlier by the Radiolaria.

PHYLOGENETIC CONCLUSIONS

The Acantharia are all marine forms whose adaptive zone is apparently the same as that of the Radiolaria, that is, they are unicellular, floating micropredators that capture their prey by means of fine pseudopodia. The acantharians show variability in several important aspects of their organization and function. Hence, aposemic traits are present. These include the inner cell body, the flotation apparatus, the skeleton, the pseudopodia, and the reproductive system. The presence of development from a small, relatively undifferentiated stage allows interpretation of the direction of aposemic change by the sequence of ontogenetic events (biogenetic law) and confirms evidence derived from morphoclines. This allows identification of a plesiomorph, and from this the further steps of the phylogenetic analysis can be carried out. The conclusions reached by using this method agree closely with the earlier masterful work by Schewiakoff. The origin of the Acantharia is unclear. It may be from some ameboid form, perhaps a colorless chrysomonad showing development of filopodia. The distinctive form of the plesiomorph and the lack of suggestive homologies between it and other forms argues for considerable evolution leading to the appearance of the acantharian form. Once that form appeared, there seems to have been significant cladogenetic development.

Chapter 10

Ciliates

THE ciliated protozoa have many notable and distinctive features, and not the least of these is the united front they have consistently maintained in the face of some 300 years of human scrutiny. No other protozoans have shown the systematic stability of the ciliates; no other group is so neatly circumscribable as the ciliates. This unity and stability is not because this is a small, uniformly organized group with a discrete ecological distribution. It is, in fact, a large assemblage of forms, with over 6,000 described species showing a great diversity of size and form and exploiting a surprising range of habitats (Bütschli 1887–1889; Kahl 1930–1935, 1934; Corliss 1961, 1974a, 1975; Jankowski 1967, 1973). Ciliates reach a level of complexity that has been called pseudometazoan (Fauré-Fremiet 1952), and their impressively coordinated locomotor behavior, plus variously complex organelles, evoked from Leeuwenhoek the appropriate term "animalcule." He saw these animalcules first in 1674 (Dobell 1932), and a letter of 1675 contains a vivid and recognizable description of members of the genus *Vorticella*, a peritrichous ciliate.

"In the year 1675, about half-way through September (being busy with studying air, when I had much compressed it by means of water), I discovered living creatures in rain, which had stood but a few days in a new tub, that was painted blue within. This observation provoked me to investigate this water more narrowly; and especially because these little animals were, to my eye, more than ten thousand times smaller than the animalcule which Swammerdam has portrayed, and called by the name of Water-flea, or Water-louse, which you can see alive and moving with the bare eye.

"Of the first sort that I discovered in the said water, I saw, after divers observations, that the bodies consisted of 5, 6, 7, or 8 very clear globules, but without being able to discern any membrane or skin that held these globules together, or in which they were inclosed. When these animalcules bestirred 'emselves, they sometimes stuck out two little horns, which were continually moved, after the fashion of a horse's ears. The part between these little horns was flat, their body else being roundish, save only that it ran somewhat to a point at the hind end; at which pointed end it had a tail, near four times as long as the whole body, and looking as thick, when viewed through my microscope, as a spider's web. At the end of this tail there was a pellet, of the bigness of one of the globules of the body; and this tail I could perceive to be used by them for their movements in very clear water. These little animals were the most wretched creatures I have ever seen; for when, with the pellet, they did but hit on any particles or little filaments (of which there are many in water, especially if it hath but stood some days), they stuck intangled in them; and then pulled their body out into an oval, and did struggle, by strongly stretching themselves, to get their tail loose; whereby their whole body then sprang back towards the pellet of the tail, and their tails then coiled up serpent-wise, after the fashion of a copper or iron wire that, having been wound close about a round stick, and then taken off, kept all its windings. This motion, of stretching out and pulling together the tail, continued; and I have seen several hundred animalcules, caught fast by one another in a few filaments, lying within the compass of a coarse grain of sand." (From Dobell 1932, p. 117–119.)

This still fresh view of the vigorous activity of these ciliates contrasts sharply with the more limited and more random movements of the zooflagellates, the seemingly aimless cytoplasmic flow of ameboid forms, the poised flotation of the radiolarians, acantharians, and certain heliozoans, and the enclosed life of parasitic sporozoans and cnidosporidians. The dynamism of the ciliates has long provoked the curiosity of biologists, and the happy result is that today we have a massive accumulation of information about most aspects of their life. (A quick review of the major groups of the ciliates and their typical species is found in Fig. 10–23.)

General Characteristics

Both internally and at the cell surface, the ciliates show unique structural features. Internally, the duality of the nuclear apparatus is the distinctive feature. With only one exception, there is always one or more micronucleus and one or more macronucleus present. At the cell periphery, the surface ciliature and the feeding organelle are especially important. These features will be described first, along with other pertinent structures, and then the relation of structure to function will be presented.

STRUCTURE

There is about a 300-fold increase in size from the smallest ciliates, ten micrometers in length, to the largest ones; the usual size is 50 to 200 micrometers. The ciliate cell varies through different spherical and ovoid shapes to flattened forms and elongated, ribbonlike ones, and to various bell- and cone-shaped forms. Throughout, except for one group, which has apparently lost oral structures previously present, the location, size and form of the feeding structures adds to the contours of the cell body and actually strongly determines the body shape as an adaptation to feeding.

Ciliary units, kineties, and kinety patterns. The basal body, or kinetosome, is the essential visible component of the ciliature. This unit, sitting just below the cell surface, as it also does in the zooflagellates, gives rise to the external cilium and also has associated with it, internally, various fibrillar and microtubular elements (see especially the critical review by Grain 1969). This complex of basal body and associated elements has been termed a kinetide; it is the ciliate analog, and perhaps the homolog, of the zooflagellate mastigont. There is, however, a further development of this concept. The surface membranes associated with a kinetide are now construed as part of an integral structure termed the ciliary capsule (Ehret and Powers 1957; Ehret 1960), or the ciliary unit, and seen by Pitelka (1969, 1970), especially, as a basic constructional unit in these cells and termed a kinetosomal territory. (The terms ciliary unit or kinetosomal territory will be used interchangeably in these pages.)

The ciliary unit was first specified in terms of the cortical structures in *Paramecium* and subsequently expanded through work on *Tetrahymena* (Allen 1967, 1969, 1971) and many other forms (Grain 1969) (Fig. 10–1). As seen in these forms, it consists of the following parts: (1) basal body or kinetosome (there may be two in some units), (2) external cilium (or cilia when two kinetosomes are present), (3) a kinetodesma which extends to the right (the

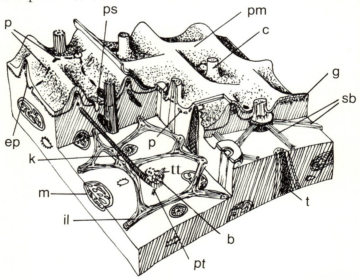

Fig. 10–1 Three-dimensional reconstruction from electron micrographs of the cortex of *Paramecium caudatum*. b—basal body, connected by a bridge to next anterior one; c—cilium shown truncated; ep—epiplasm; g—granulo-fibrillar meshwork; il—infraciliary lattice; k—kinetodesma; m—mitochondrion; p—perforations connecting adjacent alveoli; pm—plasma membrane; ps—parasomal sac; pt—posterior microtubules; sb—striated bands; t—trichocyst; tt—transverse microtubules. (Redrawn, by permission, from Allen (1971).)

cell's right) and anteriorly as a heavy fibril with cross striations, (4) transverse microtubules extending from the left anterior side of the kinetosome, (5) postciliary microtubules extending to the right posterior from the kinetosomes, (6) a parasomal sac which is to the right anterior of the kinetosome in *Paramecium* and directly anterior in *Tetrahymena*, (7) the surface unit membrane, and (8) the paired alveoli, bounded by unit membranes, which lie as two kidney-bean-shaped entities on either side of the kinetosome. (These terms are the English equivalents of the French "termes descriptifs" adopted at the Third International Congress of Protozoology (Puytorac 1970).)

The kinetosomal territories, thus described, show variation in all of their features except for the kinetosomes themselves. Cilia may or may not be present when a kinetosome is present, and when present they are sometimes club shaped or clavate (Wessenberg and Antipa 1968; Grain and Golinska 1969). The kinetodesma may be longer or shorter than those shown in Fig. 10–1; they can extend posteriorly rather than anteriorly and may even bend laterally; and they may or may not intertwine with other kinetodesmas. Grain (1968) has argued that the basic features of a kinetodesma are three: (a) periodic striations, (b) ectoplasmic location, i.e., they always lie close to the cell surface, and (c) the origin is always at the anterio-lateral side of the kinetosomal triplets. The two sets of microtubular bundles (transverse and postciliary) can contain a variable number of bundles, depending on the species; they can have various orientations relative to the kinetosome; and they can vary somewhat in their points of origin from the kinetosomal region. The parasomal sacs may or may not be present; if present, they are in the positions seen in paramecia or tetrahymena. The surface membrane is fairly constant but may be flat or depressed around the ciliary base. The alveoli are absent in many forms and, when present, may be collapsed or quite extended. Allen (1971) suggests that in *Paramecium caudatum* they are all interconnected.

The fine structure of kinetosome, cilium, unit membrane and microtubules have been discussed earlier (Chapter 2) and need no further comment. The makeup of the kinetodesmas is not clear. They may be compounded of microfibrils.

Kinetosomal territories can be combined into more complex structures. Chief among these is the kinety, typically seen as a meridional or longitudinal row of cilia and basal granules (Fig. 10–2). These meridional rows can be of various lengths and variously placed (Fig. 10–2A). Depending on such patterns of distribution, the ciliation can appear as horizontal girdles on the cell (Fig. 10–2B) or as localized patches (Fig. 10–2C). The cilia can be linearly combined into membranelles (Fig. 10–2D) or coalesced into cirri (Fig. 10–2E). The cilia in these specializations, or even individual cilia, can be of different lengths.

Cumulatively, the cilia of the cell body, apart from those lying in the feeding structures (see below), have been referred to as the somatic ciliature,

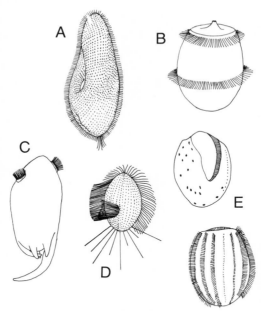

Fig. 10-2 Distribution of cilia. A. *Paramecium aurelia* showing complete ciliation of the body surface. The distribution of ciliary rows is seen as if the cell were treated by a modification of the Chatton-Lwoff silver impregnation technique. The dots represent the site of insertion of cilia, individual cilia being represented only along the contour of the cell. (Modified from Kaneda and Hanson, 1974.) B. *Didinium nasutum* with two girdles of cilia. Compare with Fig. 4-1.)Redrawn, by permission, from Jahn and Jahn (1949) *How to Know the Protozoa,* Wm. C. Brown Co.) C. *Epidinium parvicaudatum* showing two tufts of cilia. (Redrawn, by permission, from Corliss (1961) *The Ciliated Protozoa,* Pergamon Press. Originally from Kofoid and Christenson) D. *Pleuronema coronatum* has complete ciliation of the cell body with several, long caudal cilia and a highly developed paroral membranelle. (Redrawn, by permission, from Corliss (1961) *The Ciliated Protozoa.* Pergamon Press. Originally from Noland.) E. *Euplotes eurystomus,* showing the sharp difference in ciliation on the ventral (upper figure) and dorsal (lower figure) surfaces. The ventral surface is drawn as if the cell were treated by silver impregnation, hence only the sites of ciliary insertion are shown. On the cell's left margin is a single kinety; the banded structure is the adoral membranelle, and the other large dots are the cirri (clumps of cilia.) The dorsal surface shows seven ciliary rows, or kineties, and part of the adoral membranelle (anterior edge of the cell). (Redrawn with modifications, by permission, from Wise (1965).)

especially by the French workers (Fauré-Fremiet 1950a and perpetuated in Corliss 1961). This so-called somatic ciliature is to be distinguished from the buccal ciliature. However, "somatic" refers to the body (here the cell body) but the feeding parts are an integral aspect of that body; therefore, the terminological distinction is not a good one. Soma is also contrasted with germ, in biological thinking, and the feeding ciliature is not an analog of germ material, so again the term somatic is misleading. (All ciliature in the ciliates is somatic (Raabe 1967).) In what follows, we will refer to *surface ciliature,* instead of somatic ciliature, for the feeding ciliature is often in depressions or special fields. Or, in more precise terms, we see the ciliature as separated into two areas; one is directly related to feeding and is, therefore, the feeding or adoral ciliature; the other is not directly related to feeding and is the nonfeeding or aboral ciliature. (Aboral implies "without oral parts," rather than its other possible meaning of "away from oral parts.") This point will receive further attention when the feeding structures themselves are described and discussed below.

The kinety patterns of the surface, or aboral, ciliature are usually highly distinctive for a given species. This is especially true in a qualitative sense where one refers to a holotrich or a heterotrich pattern and even to much finer differences, such as between genera. Within a genus there is some difficulty in specifying a given species in terms of what Nanney (1966, 1967, 1968) has called its corticotype. Rather than finding a constancy in numbers of longitudinal kineties or ciliary meridians, Nanney has found in *Tetrahymena* that the angle between the feeding structures and the contractile vesicle pore, relative to the long axis of the cell, is the most constant feature. It is true that each interbreeding group, or sibling species (or syngen), of *T. pyriformis* has a mean number of ciliary meridians, but there is variation around these means and overlap from one group to the next. In *Glaucoma*, Cho (1971) has found the corticotypes even more variable than in *Tetrahymena*. It should also be noted that numbers of kineties can increase and decrease sharply within the cell cycle of one species. Roque and Puytorac (1967) describe an extraordinary increase from 43 to 300 kineties in *Ichthyophthirioides browni*.

Other cortical specializations. In addition to the ciliary units, there are other cortical specializations. These are too numerous to cover here in detail. We will only mention some of the better known ones, some of which are shown in Fig. 10–3. At the periphery of the cell body there are various organelles whose contents are readily extruded. The best known of these are trichocysts, as seen in paramecia, whose product is a long, striated, needlelike structure. The unextruded trichocysts alternate with the ciliary units, which make up a kinety.

Toxicysts release a fluid material that is capable of lysing prey, as in the case of *Dileptus*, or aiding in the capture of prey, as with *Didinium* (Wessenberg and Antipa 1968). In *Heliocoprorodon*, Puytorac and Kattar (1969) report three different kinds of toxicysts. According to their opinion, it is premature to attempt a comprehensive classification of toxicysts at present. Furthermore, this same organism also possesses mucocysts, which are smaller organelles but also thought to release materials to the exterior of the cell-body. Mucocysts have been reported from various ciliates including *Didinium* (Wessenberg and Antipa 1968).

Finally, there are haptocysts in the termini of suctorian tentacles (Bardele

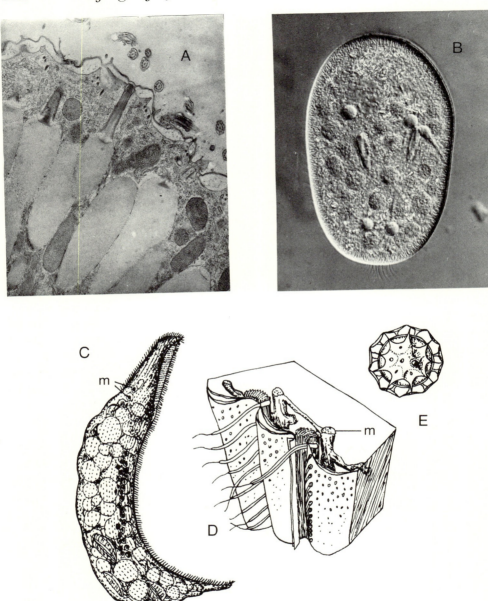

Fig. 10–3 Special organelles of ciliates. A. Trichocysts of *Paramecium aurelia*. (Courtesy of Jane Sibley.) B. Contractile vesicles (c) in dividing *Paramecium aurelia;* radial canals are especially prominent around the most posterior vesicle. C. Müller's vesicles (m) as seen in *Remanella multinucleata*. (Redrawn, by permission, from Raikov (1963a).) D. Reconstruction of the cortex of *Stentor*. The contractile elements or myonemes (m) lie subjacent to other surface structures. (Redrawn, by permission, from Tartar (1961) *The Biology of Stentor,* Pergamon Press) E. A cyst of *Bursaria truncatella*. (Redrawn, by permission, from Lund (1917).)

and Grell 1967). These are seen to be small bodies that play a role in attaching the tentacle tip to prey.

There are various stalks in certain attached forms, most of which arise from a specialized cortical structure, the scopula, which is derived from cilia (Fauré-Fremiet, Favard, and Carasso 1962; Randall and Hopkins 1962; Pitelka 1969). In addition, there are various thigmotactic surfaces in the ciliates. In some sand-dwelling forms, one surface of the cell is specialized for crawling among the sand grains and may be devoid of surface ciliature. In other forms, adhesive disks are formed, which may or may not involve ciliary functions. These are found in various ciliates with ectocommensal habits and, of course, show quite divergent forms. Tentacles are found in forms outside the well-known suctorians, but their occurrence is rare. In addition to tentacles, the suctoria show stalks, developed from a scopula, and some show external secretions (loricas, Fig. 10–17P) formed with variously different internal parts (discussed further below).

Lying just beneath the cell surface are the contractile vacuoles (Kitching 1967), or expulsive vesicles (Wigg, Bovee, and Jahn 1967), of the freshwater ciliates. They are connected to the external surface of the cell either quite directly by a pore or by ducts of different size and shape which connect vesicle and pore. These vesicles may also be provided with different kinds of subsidiary vesicles or canals. The cytoproct, or anal pore, is the site of elimination of undigested food vacuole contents. It is a constant feature, usually slitlike, of the cell surface, but lacking entirely in the suctorians (Rudzinska 1970).

Also worth noting are the statocystlike Müller's vesicles of certain ciliates, the skeletal plates of the complex symbionts of ruminants, various contractile fibrils (which permit extensive changes in bodily shape), and pigment granules of a variety of ciliates.

Electron microscopic studies are bringing to view important microfibril systems that are cortically or subcortically located. In addition to the contractile myofibrils of a few species, there are those referred to as cytoskeletons (Puytorac 1967; Grain and Golinska 1969) or as an infraciliary lattice (Pitelka 1965, 1969). More recently, in a brilliantly detailed study on *Paramecium caudatum*, Allen (1971) distinguishes not only an infraciliary lattice at the level of the proximal half of the kinetosomes and forming a polygonal network between them, but also two other microfibrillar systems—a granulo-fibrillar system running in the ectoplasmic ridges between the kineties, and striated fibrillar bundles radiating from basal bodies and lying just below the ciliary alveoli (Fig. 10–1). Also, just beneath many of the feeding parts, there are often found fibrillar reticula (Rouillier and Fauré-Fremiet 1957; Schneider 1964; Pitelka 1969; Wolfe 1970).

Feeding structures. The feeding organelle and its associated parts are such an important and highly differentiated aspect of the cell that they de-

Table 10-1 The feeding structures of *Paramecium*. The most internal part, i.e., site of food vacuole formation, lies at the top of the table. Horizontal rows represent terms applied to the same part.

from Kahl (1930-1935)	from von Gelei (1934)	from Wichterman (1953)	from Corliss (1961)	present proposal
Nahrungs-vakuole	Ösophagus mit der Empfangs-vakuole	food vacuole forming region	cytopharynx	cytopharynx
Schlund			cytostome	cytostome
Pharynx	Pharynx	esophagus pharynx or cytopharynx	buccal cavity	prostomial cavity
Mund	Mund	mouth	buccal overture	prostomial opening
Mundtrichter	Vestibulum	Vestibulum	Vestibule	adoral depression
Grube	Mulde	oral groove	oral groove	adoral groove

serve separate consideration. Immediately there is a terminological problem. This is illustrated by reviewing the many different labels which have been applied to the parts of the feeding organelle in paramecia (Table 10–1). The terminology proposed by Corliss (1961 and earlier, and derived from usage developed by Fauré-Fremiet) has done much to clarify this problem, and, more importantly, it attempts to use terms that indicate homologies among the different species of ciliates (Fig. 10–4). However, there are still certain difficulties, even with this widely accepted set of terms. One set of difficulties relates to the lack of logical clarity inherent in the terms themselves. For example, the buccal cavity lies external to the cytostome, but one would expect the reverse relationship, for "buccal" derives from the Latin word for cheek, and hence one would expect the mouth or cytostome to lie external to the cheek cavity. Further, the mixture of Greek (cytostome) and Latin cognates is unnecessary unless it is to be used for some special reason. No such reason is apparent here. Another point of criticism is that the terms proposed by Corliss have not succeeded in replacing many special usages, such as gullet (by paramecium workers), adoral zone, and peristome for various peritrichs. It seems that the proposed terminology has not been sufficiently useful to dominate this area of ciliate anatomy.

To attempt what may be futile, a new set of terms will be presented here in an attempt to rationalize further the description of ciliate feeding structures. These terms are developed with all of the ciliates in mind, are aimed at emphasizing homologous relations and the lack of them, and are based on the Greek stem *-stome* except where otherwise indicated (Table 10–1; Fig. 10–4). We start from the cytostome because of universal scientific references to the

CILIATES 335

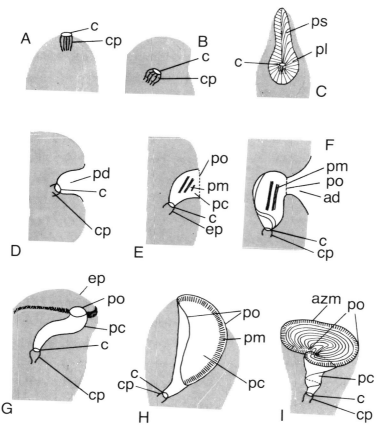

Fig. 10-4 The feeding organelle of ciliates. A. A rhabdophorine gymnostome, e.g. *Prorodon*. B. A cyrtophorine gymnostome. C. A different rhabdophorine, e.g., *Dileptus*. D. A trichostome, e.g., *Colpoda*. E. A tetrahymenine hymenostome. F. A peniculine hymenostome. G. A sessiline peritrich. H. A hypotrich. I. A heterotrich such as *Stentor*. ad—adoral depression, azm—adoral zone membranelles, c—cytostome, cp—cytopharynx, ep—epistomial disc, pc—prostomial cavity, pd—peristomial depression, pl—peristomial lips, pm—prostomial membranelles, po—prostomial opening, ps—proboscis. (A, B, C, D, E, F, G and H adapted from Corliss (1961); C adapted from Golinska (1972); and I adapted from Tuffrau (1967a).)

Greek word *kytos*, for vessel, now taken to mean cell. A cell mouth must, therefore, be a cytostome, and this accords with Corliss' starting point too.

From this we refer to all structures internal to the cytostome as poststomial. This will include the cytopharynx and various fiber systems and microtubules and their aggregations, as in the case of nemadesmas.

External to the cytostome there are two kinds of structures to be distinguished. Peristomial or prestomial structures will refer to those instances where the kineties or other surface structures relate to feeding. Thus we can recognize a peristomial depression, as in *Colpoda* (Fig. 10-4D), or peristomial lips with prestomial specializations, such as the ventral surface of the proboscis in *Dileptus* (Fig. 10-4C). The other kind of structure will be called prostomial and refers to a special cavity and its parts that is inserted between the cytostome and the surface or, better, the aboral structures. In particular, the kinety patterns of the latter are not continuous with the prostomium and its parts. The prostomium refers to the buccal cavity and associated structures, as designated by Corliss. Hence, the prostomium includes the prostomial walls, the prostomial membranelles, which are compound or polykineties and often referred to as peniculi, and it opens exteriorly by way of the prostomial opening (Fig. 10-4E, F, G, H, and I).

The prostomium can be greatly developed as in the peritrichous ciliates (Fig. 10-4G), and, in recognition of this, the term epistomial disc (Lom 1964) should remain in use as the appropriate cognate term. The other development of this structure, seen in the heterotrichs, is not so extreme (Fig. 10-4H), except in cases such as *Stentor* (Fig. 10-4I). In cases such as the hypotrichs (Fig. 10-4H), the feeding cavity should remain designated as a prostomial cavity, rather than the adoral, or buccal, cavity. Where this cavity, or depression, contains kineties that are from the cell surface (*Climacostomum* or *Stentor*), then the term adoral zone is appropriate.

The use of adoral zone becomes clearer if we go back to an example like *Paramecium* (Fig. 10-4F). Here the depression external to the prostomial opening (vestibulum and buccal overture, respectively, of Corliss) is a new development which, as will be shown subsequently, is *not* homologous to the peristomial depression of forms such as *Colpoda* (Fig. 10-4D). Hence, it deserves a different name, and a Latin cognate is appropriate. We propose the term adoral for this depression, and by extension the oral groove of paramecia is readily referred to as an adoral groove. Returning to spirotrichous ciliates such as *Climacostomum* or *Stentor*, which also incorporate surface ciliature into the feeding structures, we can refer to their organelle as containing an adoral zone. As we shall see, an adoral zone is not homologous to an adoral depression or groove, but they are convergent to an important degree. In both cases we are also able to conserve and rationalize the use of the term adoral, which is embedded in the current literature.

(Unfortunately, before achieving the preceding systematic review of the nomenclature of feeding parts in ciliates, Kaneda and Hanson (1974) proposed referring to the ciliature of *Paramecium* as composed of two zones, adoral and anoral. The term adoral zone is objectionable because it implies similarity to the adoral zone of other ciliates, such as *Stentor*. The similarity that exists is analogous, not homologous. We would now prefer to refer to the adoral area in paramecia as simply an adoral field. The phrase "anoral zone" is doubly objectionable. By parallel to adoral field it should now be an anoral field, rather than zone, which is an easy correction. But the word anoral (implying without relation to oral parts) is compounded of a Greek prefix on a Latin stem. Aboral would be better, wherein the Greek prefix ab- is taken to mean "without," rather than its more common meaning of "away" or "from," as was explained above. Thus, in paramecia, the ciliature is divided into two fields, one directly related to the feeding parts and called, therefore, an adoral field (which includes adoral depression and groove), and one not directly related to the feeding parts nor having a primary role in feeding and, therefore, termed an aboral field.)

Lastly, mention must be made of a crucial kinety which is either part of the adoral field or can also lie within the prostomial structures. This is the endoral kinety of paramecia (Yusa 1957), the undulating membrane of the tetrahymenines (Furgason 1940), pleuronematines, and others, and the paroral kinety of peritrichs (Noirot-Timothée and Lom 1965) and certain, at least of the heterotrichs (Tuffrau 1967a). Again, there is terminological confusion. We shall dispense with undulating membrane. (It is not a membrane, but composed of fused cilia. It is not as widely used a term as endoral or paroral kinety.) When this kinety lies outside the prostomial structures, as it does in *Paramecium* and other peniculine forms and in tetrahymenine forms, we shall call it a paroral kinety, for its position is typically along the right edge of the prostomial opening. When it lies partially or completely within the prostomial cavity, as it does in peritrichs and many heterotrichs, it shall be referred to as an endoral kinety. (Though this terminology will offend classicists, for it combines Greek prefixes with a Latin stem, it seems best to retain it because of its wide usage. Further, retention of the stem "-oral" identifies the evolutionary origin of this kinety as one lying originally, in the tetrahymenines, outside the prostomium.)

Extracellular structures. One last category of structures must be commented on before looking at the structures lying deeper in the ciliate cell body. Certain ciliates can lay down around themselves various secretion products, resulting in tubular or vaselike constructions. Some are clear, homogeneous structures; others are more heterogeneous, with adherent exogenous matter. Such tubes, or loricas, or thecas are quite distinct from the usually spherical cysts that the ciliates can also produce. The cysts often are characteristically sculptured and are not open at one end as are the loricas or thecas, though there may be a site on the cyst from which the cell typically emerges at excystment. As would be expected, electron microscopy is supplying many new details on these structures (e.g., Batisse 1969, and earlier papers; Holt and Chapman 1971).

Endoplasmic organelles. Turning to organelles lying below the cortex, these include the usual fine structure elements of mitochondria, endoplasmic reticulum, ribosomes, and food vacuoles (Pitelka 1963; Anderson 1967), including the presence of macromolecular clusters on the inner membrane of the mitochondrial cristae (Burton 1970). The Golgi apparatus, though often reported, is not conspicuously developed. Lysosomes have received limited attention in the ciliates (Estève 1970). In addition, there are also lipid droplets, various crystals and other metabolic products (Anderson 1967; Vivier, Legrand, and Petitprez 1969), and many examples of symbionts (Ball 1969).

Nuclei. It is the nuclei which show certain novel features. The micronucleus contains typical chromosomes, their number ranging from as few as four up to more than 40 (Kudo 1965). When the nucleus is not dividing, the chromatin can be coalesced into the center of the nucleus, leaving clear nucleoplasm around it—a vesicular nucleus—or the chromatin can be more or less loosely distributed throughout the nucleus. The micronuclei vary from one to a dozen or more per cell. Usually the number is not much above four. Their size is no more than a few micrometers in diameter. Rarely forms lacking micronuclei are found. No nucleoli are reported for the micronuclei. Though micronuclei are usually considered to be diploid, there have been reported variations in DNA level corresponding to 2n, 3n, 4n, 6n, and 8n micronuclei (Ovchinnikova 1970) and other variations are also known (Dupy-Blanc 1969). When dividing, the micronuclei can become very attenuated, running practically the whole length of the cell. The nuclear membrane does not break down during mitosis, though the central part of it, containing spindle matter in a crescentic form, can be separated from the two telophase nuclei and be broken down. The spindle is, therefore, intranuclear, and on it the chromosomes become oriented. No centrioles, centrosomes, or aster fibers are known in the ciliates.

The macronucleus (see review by Raikov 1969) is a polyploid, polyenergid accumulation of genetic material, essential to the survival of a ciliate. Only ciliates of the genus *Stephanopogon*, e.g., *S. mesnili* (Lwoff 1923, 1926) and *S. colpoda* (Dragesco 1963), are known to lack permanently a macronucleus. Though the macronucleus is formed from micronuclei, and these, as we have said, have distinct chromosomes, it is usually only under special circumstances that chromosomes can be distinguished in the macronucleus. Grell (1953a, 1953b) demonstrated chromosomes in the developing macronucleus of certain suctorians and provided direct evidence of their endomitosis. Since then, chromosomal structures have been seen in extended macronuclear fragments of *Paramecium bursaria* (Schwartz 1958), in dividing macronucli of *Nassula ornata* (Poljansky and Raikov 1961), and especially interesting are a series of reports of large polytene, banded chromosomes, very similar in appearance to dipteran salivary gland chromosomes, seen in the developing macronucleus of hypotrichous ciliates (Golikova 1964; Ammerman 1964, 1965; Péres-Silva and Alonso 1966).

The macronucleus is variously studded with nucleoli. Some are very small and appear as fine granules (Dippell 1963; Dippell and Sinton 1963), and others are large conspicuous structures (Raikov 1958, 1959, 1963b, 1969). There are also certain enigmatic entities: endosomes, areas of differential stainability (heteromeric macronuclei), and helical structures (e.g., Kaneda 1961; Dobrzhanska-Kaczanowska 1963; Radzikowski 1965). By contrast, nuclei which stain homogeneously are termed homomeric (Raikov 1969). During division, the nuclear membrane, which is double and provided with pores (Pitelka 1963), remains intact, and the content of the nucleus is apparently distributed amitotically and into two approximately equal halves (Morat 1970). Estimates of the ploidy of the nucleus give a wide range of figures. In *Paramecium aurelia*, the highest figure is 860 n (Woodard, Gelber, and Swift 1961), but in other forms the macronucleus may simply be diploid (Raikov, Cheissin, and Buse 1963). It is premature to attempt even a guess at what a typical ciliate ploidy level might be. There is usually one macronucleus per cell, but in an organism such as *Dileptus anser* there are 50 or more small macronuclei present. Though the macronucleus is commonly ovoid, it can also be a long slender rod or even beaded (Fig. 10–5).

There has been a long debate, still unresolved, as to whether the macronucleus contains subunits above the level of individual chromosomes. Nilsson (1970) presents evidence for such subnuclei in *Tetrahymena* in terms of cyclic changes in the distribution of intranuclear granules during the cell cycle. These granular aggregates could well represent genomic clusters of genetic material. However, there is no good evidence of membranes surrounding these units.

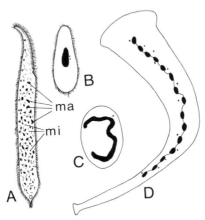

Fig. 10–5 Selected nuclear structures in ciliates. A. *Dileptus estuarinum* with many macronuclei (ma) and micronuclei (mi). (Redrawn, by permission, from Dragesco (1960).) B. One large macronucleus and two small micronuclei in *Paramecium aurelia*. C. *Euplotes* with one curved, rodlike macronucleus and one dotlike micronucleus. D. Beaded macronucleus and scattered, dotlike micronuclei in *Stentor coeruleus*. (B, C, and D, redrawn, by permission, from Corliss (1961) *The Ciliated Protozoa*, Pergamon Press)

Microtubules are seen in the macronucleus (Pitelka 1963, 1969; Bardele 1968; Mosevich 1968).

Endo- and ectoplasm. Lastly, the general organization of the cytoplasm into ecto- and endoplasm should be mentioned. These terms come from light microscope studies. The ectoplasm includes, in forms like *Paramecium*, the clear outermost layer of the cell, external to the granular endoplasm. It includes, therefore, most, if not all, of the cortex. The latter is, however, a very loose term, used as much for morphogenetic studies as for morphological ones. The granular endoplasm has a more solid or gellike outer layer and a fluid inner center. The food vacuoles circulate in the fluid inner endoplasm. The macronucleus lies centrally located but is rather fixed in position, clearly suggesting that it is somehow moored in place. Perhaps some of the fibrous material commonly found in the ciliate cytoplasm has, as one of its functions, the anchoring of the macronucleus in a constant location. In some forms, such as *Isotricha* (Noirot-Timothée 1958) and *Nyctotherus* (Albaret 1970), fibers form a baglike net or karyophore around the macronucleus.

Electron microscopy can also differentiate between ecto- and endoplasm, but the distinction is not an especially useful one, for the higher resolution of this form of viewing shows no distinct break between these two parts of the cell. Pitelka (1963) cited various workers, especially Puytorac, who apparently treat the level of the subcortical microfibril system, called the infraciliary lattice, as a visible demarcation of the ecto- and endoplasm boundary.

FUNCTIONS

A great deal is known regarding the function of many ciliate cells. This is especially true of reproductive functions. We will come to those after dealing first with environmental exploitation and homeostatic maintenance.

Environmental exploitation. The exploitive activities of ciliates go on in a great variety of habitats, and what is most remarkable is that often within well-defined taxonomic groups one finds this ecological diversity. There is no real correlation of form with ecological aspects except at the generic or species level (Fauré-Fremiet 1952); the ciliates have used each of their various body forms in many different ways. One finds freshwater and marine forms, forms feeding on bacteria or algae or other protozoa, free-swimming or sedentary or symbiotic forms, all within very similar cell types.

Feeding in ciliates can be either phagocytic or pinocytotic. The former is the vastly better described and understood process. It involves the cytostome and related parts, with ingestion leading to digestion of materials enclosed in food vacuoles and then egestion of undigested substances.

Ingestion can be by a filter-feeding process or by capture of relatively large prey. In the former, the adoral cilia, especially, set up currents which bring food into the prostomium, if there is one, and then to the cytostome. Such food must consist of materials small enough to be carried by ciliary currents. Bacteria are an obvious example. However, some ciliates, such as *Bursaria truncatella* are large enough to feed on fellow ciliates, like paramecium, by wafting the latter into a huge prostomium and them cramming the food, often doubled up, into food vacuoles. The ingestor here can be mobile, like *Bursaria* or various bacterial feeders such as *Paramecium*. Or the feeders can be sessile, like more typical filter feeders. *Stentor* and *Vorticella* are obvious examples.

The capture of large prey involves contacting and often attaching to cells as large as the captor. The captured cell can be engulfed intact, as is done by *Didinium* with regard to paramecia (Wessenburg and Antipa 1970) or by many of the herbivorous cyrtophorines, which ingest algal filaments. In some cases, e.g., *Dileptus*, there is lysis of the prey and ingestion of the remnant parts, or, in the case of the suctoria, the fluid contents of the prey pass down tentacles to the predator's cell body and finally the collapsed cortex of the prey is released from the end of the tentacle. These types of predation pose as yet largely unanswered problems with regard to the mechanics of capture and ingestion. The role of toxicysts in *Dileptus* (Grain and Golinska 1969) and especially in *Didinium* (Wessenberg and Antipa 1970) is clearly indicated as a means of immobilization and even of adhesive contact. The same is true for suctorian haptocysts (Bardele and Grell 1967). The uptake through the pharyngeal basket, or nassa, of the gymnostomes is also receiving detailed attention. Tucker (1968) has proposed mechanisms whereby the microtubules of the nassa can abet ingestion. With suctorians, Rudzinska (1970) and Bardele and Grell (1967) are also making suggestions involving microtubules and membrane formation within the tubules. Thus far, the analysis has described the chief elements engaged in capture and ingestion, i.e., toxicysts, haptocysts, etc., and microtubules and membrane formation but has understandably made little progress in terms of the molecular mechanisms involved in this obviously complex problem.

Digestion of the ingested materials takes place in food vacuoles and probably by the emptying of hydrolytic enzymes contained in lysosomes into the vacuoles (Estève 1970) so that food organisms are enzymatically degraded. Breakdown products are enclosed in pinocytotic vesicles, which come off the food vacuoles (Carasso, Favard, and Goldfischer 1964; Favard and Carasso 1964; Estève 1970). There is probably, also, active transport of molecules from inside the food vacuoles to the surrounding cytoplasm.

Egestion occurs at the cytoproct. The whole food vacuole is not eliminated; only its undigested contents are expelled. Thus there is a conservation of vacuole membranes. Are they recycled? This is not known; much less is it known how membranous materials or their precursors might move from the cytoproct to the cytostome for reuse.

Pinocytosis occurs at the level of food vacuoles and apparently can occur at the parasomal sacs. One report (Noirot-Timothée 1968) shows convincing evidence of the uptake of colloidal thorium oxide in the symbiont *Trichodinopsis paradoxa*. Whether this occurs in free-living forms and is a genuine source of nutrients remains to be seen.

It seems possible that eventually the role of environmental and physiological factors in controlling ciliate feeding will be shown to be of importance. Seshachar, Saxena, and Girgla (1971) show that degree of starvation is very important in determining the rate of prey capture and feeding in *Homalozoon*. Also, the numbers of the food organism in the medium, the susceptibility of the food organism to the predator's toxicysts, size of the prey, and the stimulation offered by the prey to grasping and swallowing all play a role in the feeding of this ciliate. Other such studies on the gymnostomes, especially, might be of great comparative interest.

Studies on the general ecology of the ciliates continue to reveal important new details on their responses to temperature, pH, salinity, the tension of various gases in their environment, and their many and often intricate symbiotic adaptations (Fauré-Fremiet 1967a; Noland and Gojdics 1967; Trager and Krassner 1967). These facts, too, can someday provide extensive comparative data.

Locomotion is predominantly by ciliary action. Only rarely are the peripherally located contractile fibrils used for such a purpose. Twisting and turning of the body changes the direction of swimming patterns, and, possibly, passage through an environment heavily entangled with debris of various sorts would be aided by contractibility of the cell body. But these are the generally less important roles of contractibility and are of minor importance for locomotion. Contraction, as seen in *Spirostomum* and *Stentor*, seems primarily to be a means of sudden withdrawal from various physical or chemical stimuli. In the gymnostomes, e.g., *Dileptus* or *Lacrymaria*, bending of the proboscis or contraction or extension of the "neck" are related to searching patterns for food.

Though much is known about the fine structure of the cilium, its functional aspects, especially in chemical terms, are only now being elucidated (Sleigh 1962; Child 1967). Thus far it seems that the mechanism of ciliary beat will turn out to be of universal occurrence, just as the structure is of ubiquitous distribution. For that reason, it is unnecessary to consider it further here for, important though the mechanism is biologically, its common occurrence and lack of variation in eukaryotic cells means it is of limited or of no use phylogenetically in this study.

It may be that the pattern of ciliary beat over the whole cell body can be of comparative interest. At present much effort has been concentrated on a few species of *Paramecium*. The coordination of waves of ciliary action is now relatively clear (Párducz 1967; Machemer 1969a, 1969b), and mechanisms underlying this coordination are being elucidated (Kinosita and Murakami 1967; Naitoh and Eckert 1969; Kuznicki 1970; Kung and Eckert 1972; Eckert and Naitoh 1972). However, at present extensive comparative information is not at hand.

Sensory capabilities have long been known and carefully described in ciliates (Jennings 1923). However, it is still largely unknown as to how these capabilities are mediated within a cell so as to result in effective locomotory or other responses. Recent fine-structure work is turning up modified, club-shaped, or clavate cilia, which are suspected of having a sensory function (Wessenburg and Antipa 1968; Grain and Golinska 1969; Grain 1970; Fauré-Fremiet and Garnier 1970). This, plus continuing work on membranes as transmitters of stimuli (Naitoh and Eckert 1969; Eckert and Naitoh 1972), may well start providing us with the long-sought basis for sensory-neuromotor coordination that many have postulated to occur in these animalcules of Leeuwenhoek.

Homeostatic maintenance. The visibly most obvious components of the homeostatic functions of the ciliates are the water-expulsion vesicles. Though variously formed and placed in the cells where they are found—usually in freshwater species—their function is consistently that of osmoregulation. A second function of excretion may or may not be important. There is no good agreement as to whether significant amounts of nitrogenous metabolic end products are accumulated in these vesicles and then are eliminated along with the water (Kitching 1967). Though it appears that the content of the expulsive vesicle is dependent of receiving the fluid accumulated in surrounding cytoplasmic tubules—nephridrial tubules (Schneider 1960; Organ, Bovee, and Jahn 1967)—this fluid may pass first to ampullae (radial canals in paramecia) before reaching the expulsive vesicle. Discharge of the vesicle is achieved by opening of an excretory canal with a pore at the cell surface. Details from electron micrographs reveal some differences in structure of this whole apparatus (Pitelka 1963), but the differences are not yet correlated with possible functional differences. In fact, the processes whereby the nephridial tubules are filled and are discharged into ampullae or the expulsive vesicles, and the control of discharge from the vesicles, these are all matters awaiting further clarification. The use of high speed cinematography (Wigg, Bovee, and Jahn 1967; Organ, Bovee, and Jahn 1967) has led to the proposal that the endoplasmic gel is active in expulsion of the vesicle, but in paramecium, the ampullae themselves show contractility, presumably generated by fibers in the ampulla wall. These proposals support the hypothesis suggested by Schneider (1960) on the basis of his electron microscope description of the expulsive apparatus in *P. caudatum*. Such approaches imply that in the future there will be significant comparative data on the morphology and physiology of the water expulsive systems in the ciliates, and in the protozoa as a whole.

To study effectively the comparative biochemistry of an organism it is desirable, if not necessary, to obtain it in a defined culture medium (Hall 1965), which is possible with *Tetrahymena* (Kidder and Dewey 1951; Kidder 1967) and *Paramecium* (Soldo and van Wagtendonk 1969; van Wagtendonk and Soldo 1970). The simplest summary of these results and others on protozoan nutrition, in general, is the following comparison between protozoan, metazoan, and bacterial nutrition. "Quite likely the various groups of protozoa differ from each other in metabolism as much as they do from metazoa. What *does* make protozoa valuable as guides to metazoan metabolism is that protozoa, as eukaryotes, share with metazoa a limited choice from among the myriad metabolic patterns in bacteria, . . ." (Hutner *et al.* 1972, p. 88) That is to say, there is a need for certain amino acids and purine and pyrimidine bases that are similar to the requirements of higher animals (Kidder 1967). Much the same pattern applies to growth requirements (Lilly 1967) and use of carbohydrates (Dewey 1967). Certain variations in metabolic pathways are being found (see reviews just cited), but it is not possible to use these for answering any significant evolutionary questions within the ciliates. The accumulation of data for a comprehensive picture of the metabolic activities of an organism is a long tedious process; the prospect of help from this quarter on phylogenetic problems is not bright.

Encystment and excystment can be added to the preceding list of regulatory processes that are readily observed in the ciliates. Though there have been rather detailed studies on the control of these processes, there is little understanding of how the controlling factors—pH, temperature, nutrients in the medium, etc.—actually work (van Wagtendonk 1955). For our present purposes, cysts are perhaps of more interest from a comparative structural viewpoint, than from a physiological one.

Other physiological processes, such as transport mechanisms (Conner 1967), are shared by the ciliates in common with other cell systems and therefore do not provide useful phylogenetic data.

Reproduction and sexual processes. Reproductive processes, especially fission, are quite varied, but they vary around the theme of cortical morphogenesis and stomatogenesis in cells whose fission plane typically cuts the cell in two at right angles to its longest axis. This transverse fission results in the tandem formation of daughter cells. The resulting distinctive alignment of cell organelles is the basis for the term *homothetogenic fission*, which contrasts with the longitudinal or symmetrigenic fission of the zooflagellates. The formation of new parts is solved by the ciliates in ways that offer important comparative developmental data, including apparent biogenetic recapitulations. Furthermore, the availability of Klein's (1926) silver impregnation technique, modified and improved by many workers (esp. Gelei 1934, 1935; Chatton and Lwoff 1936; Corliss 1953a), and, more recently, the use of protargol (Kirby 1950; Tuffrau 1967b) has allowed the accumulation of more than forty years of work on the patterns of the cortical surface and feeding structures in a large number of ciliates. We have now a rich reservoir of comparative information on this component of ciliate biology, the most important single component from the point of view of phylogenetic studies of these cells.

Homothetogenic fission is not the only asexual reproductive process in these cells, for there also occur budding and cyst formation. However, fission is the most important, and we will examine it in some detail before looking at the other ones. And then, following their examination, we will turn our attention to sexual processes and life cycles, to complete the discussion of reproductive function.

Fission, the formation of two cells where previously there was one, is the culmination of a complex series of events; it is best understood as the culmination of the growth, differentiation, and morphogenesis that regularly occurs within every cell cycle (Hanson 1967a). Though there is evidence that the events leading to a given fission start with the action of genes at the end of the previous cell cycle (Hanson and Kaneda 1968), it will serve our purposes to take the time from one fission to the next as the limits of one cell cycle. Within this period there are two broad categories of data that can be examined: (a) biochemical processes, as determined by measuring such parameters as DNA, RNA, and protein synthesis, rates of respiration, level of ATP, and so forth (e.g., Holz 1960; Woodard, Gelber, and Swift 1961; Kimball and Perdue 1962; Hanson 1967a), and (b) gross developmental patterns, as seen in changes in cellular and nuclear form and changing patterns of distribution of cortical and feeding structures (Lwoff 1950 and earlier references contained there; Tartar 1961, 1967; extensive references in Evans and Corliss 1964; Hanson 1967a). The biochemical work is concentrated in two genera—*Tetrahymena* and *Paramecium*—with other data, not so complete, found in *Stentor* (de Terra 1966, 1967) and *Euplotes* (Gall 1959; Prescott and Kimball 1961; Kimball and Prescott 1962). These data are insufficient for meaningful phylogenetic comparisons and are complicated by the fact that they represent different experimental approaches and are therefore somewhat difficult to compare. Actually, only the data on cortical and stomial morphogenesis are of extensive comparative use. Studies on changes in cell form (Jennings 1908; Kaneda and Hanson 1974) are more useful, at present, in terms of elucidating developmental control mechanisms rather than contributing to evolutionary studies. However, in that these more quantitative approaches to morphogenesis also contain qualitative accounts of the phenomena being studied, they are of present use when seen in that context.

The cortical morphogenetic events that precede the cleavage of a cell can be separated into three categories. Least useful is the development of new cytoprocts and new expulsive vesicles and their pores. More important is the fate of the kineties and modelling of the cell surface in conjunction with kinety development. And most important is stomatogenesis, the formation of new feeding structures. We will limit our account to stomatogenesis and development of surface kineties.

Fission: Development of the aboral surface. The logical place to start is the mode of formation of a new cortical unit; in particular, the formation of new kinetosomes. This question was first approached experimentally through the work of Chatton and Lwoff (summarized in Lwoff, 1950) and led to the exciting conclusions of direct genetic continuity of kinetosomes and of their pluripotency in development. Neither view is supported today (Sonneborn 1960, 1963, 1964; Grimstone 1961; Hanson 1967a; Fulton 1971). It does appear, however, that new kinetosomes are formed in a precise relation to preexisting ones (Dippell 1968; Allen 1969), but, as far as can be seen from electron micrographs, this relationship is not one of a template and its product, as seen in gene duplication. Rather, it seems possible that the preexisting kinetosome is responsible for determining an environment in which a new kinetosome is assembled, probably from molecular parts built elsewhere (Hanson 1967a). In any case, these studies on the details of kinetosome formation are still limited to only a few ciliates. It remains to be seen whether there will emerge a uniform pattern for all forms—which seems probable for so uniform a structure—or whether it will show some diversity.

At the next level of organization above the kinetosome, there is the problem of formation of the rest of the ciliary unit. Here again we are restrained from broad generalization because of lack of information in any significant number of species other than in *Paramecium* and *Tetrahymena*.

At this point we emerge from the realm of electron microscopy and can comment on data from the light microscope, largely derived from silver impregnation techniques. Such data, it must be remembered, come from the reduction of selected deposits of silver. In paramecia (Dippell 1962), it has been shown that the silver lies in the circumciliary depression at the base of the cilium and also penetrates into the base of the cilium. It apparently does *not* actually stain, so to speak, the kinetosome. Hence, earlier reports referring to the silver deposits as indicating the locus of kinetosomes are not precisely correct (see also Small 1968). Such deposits mark the base of a cilium and hence indicate the underlying position of a kinetosome. That silver impregnation and related techniques are not specific for the kinetosome is readily apparent in the observed silver-line systems. Here the silver delineates a linear cortical pattern quite distinct from any kinetosomes or cilia. Special difficulties are apparent when it is realized that the silver can be deposited in dotlike residues on trichocyst tips, parasomal sacs, and other organelles not part of a kinetide (Lwoff 1950; Dippell 1962; and others). The realization that silver is not specific for kinetosomes becomes crucial when nonciliated forms, such as suctorians, are examined (Chatton *et al.* 1929, Chatton, Lwoff, and Lwoff 1931; Lwoff 1950). In these forms, it was always assumed that dotlike deposits of reduced silver marked the location of kinetosomes. This has been vigorously challenged by Canella (1964). It will take careful electron microscopy to determine whether the silver deposits do or do not coincide with underlying kinetosomes. Much of the original thinking on the genetic continuity of kinetosomes rested on the suctorian work; it is most important to reexamine it critically, as Millecchia and Rudzinska (1972) have done with *Tokophrya*. The only foolproof assumption regarding the presence of kinetosomes when using the light microscope is that for each visible cilium there is a kinetosome. Therefore, any technique that deposits materials where cilia are present will also mark the location of kinetosomes. With appropriate care, then, the silver impregnation method can be used to delineate ciliary units and their aggregation into kineties or other compound ciliary structures. Recent studies are tending to compare the results of wet and dry silver impregnations along with protargol studies, with the latter technique being favored as probably the most reliable method (Small 1968). Even more reliable is, of course, to combine light and electron microscope studies (Puytorac 1968a; Antipa and Small 1971).

Using such approaches, it is possible to describe changes in ciliary patterns over the cell surface and in the oral structures during all phases of the cell cycle. Such changes include increase in number of cilia (or ciliary units) in a kinety, or appearance of new kineties, or fields of cilia adjacent to or quite separate from preexisting fields or kineties. It is this sort of information that forms the bulk of the data that have been used for phylogenetic purposes in the ciliates.

Increase in ciliary number along a kinety, on present evidence, always occurs through the appearance of a new cilium (and its kinetosome) just anterior to a preexisting cilium, and aligned in the axis of the kinety (Ehret and de Haller 1963; Sonneborn 1963, 1970a; Ehret 1967; Gillies and Hanson 1968; Dippell 1968; Allen 1969; Kaneda and Hanson 1974). In paramecia, this proliferation is largely limited to the middle third of the cell, i.e., that region to which cellular growth is largely confined, and in that area one, two, or three new kinetosomes can appear anterior to an old one (Gillies and Hanson 1968) (Fig. 10–6). Furthermore, this proliferation starts first at the right and left sides of the prostomial opening and then spreads around the equator of the cell. This pattern has been seen in *P. aurelia* (Sonneborn 1963, 1970a; Kaneda and Hanson 1974) and in *P. trichium* (Gillies and Hanson 1968). It has not been reported for other ciliates but should be easily observable in good silver preparations. Whether it will emerge as the usual pattern for ciliates, with areas of rather uniform surface ciliation, remains to be seen. In the hypo-

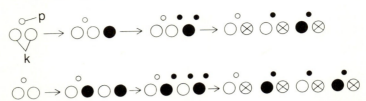

Fig. 10-6 The appearance of new cilia in a kinety, as in *Paramecium trichium*. (p, parasomal sac; k, kinetosome). The sequence of events in the top row represents the triplication of a kinetosomal territory composed of two kinetosomes and one parasomal sac. The large solid dark circle represents one new kinetosome. Following its appearance there are two new sacs (small solid circles) and finally three new kinetosomes appear (circles with crosses). The bottom row represents territory quadruplication. (Redrawn, by permission, from Gillies and Hanson (1968).)

trichs, the highly specialized ventral ciliature shows its own unique pattern of development. But on the dorsal surface, where regular ciliary meridians occur, the morphogenetic pattern is like the holotrichous situations. In *Euplotes* (Hammond 1937; Chatton 1942) there is an equatorial proliferation of new units in each kinety. In *Urostyla* (Jerka-Dziadosz 1964; 1972; Jerka-Dziadosz and Frankel 1969) there are two areas of proliferation, one-third and two-thirds of the way between the ends of the cell. Whether this proliferation always occurs according to Ehret's (1967) rule on antecorpy, i.e., new ones anterior to old ones, is not ascertainable by direct observation. Experimental excision of the posterior ends of *Paramecium aurelia* has revealed that ciliation covers the cut surface by posteriorly directed elongation of preexisting kineties; there is no proliferation of new kinetosomes into this naked area by appearance of the new organelles posterior to old ones (Chen-Shan 1970).

The appearance of new ciliary fields on the ventral surface of hypotrichs (Chatton and Séguéla 1940; Bonner 1954; Wise 1965) and also in cyrtophorine gymnostomes such as *Chilodon* (Chatton *et al* 1931) and *Chilodonella* (Kaczanowska and Kowalska 1969) shows that all ciliary and kinetosomal proliferation occurs at times of stomatogenesis to form the prostomial primordia (see below). Most striking is the apparent later proliferation that must occur in *Ichthyophthirioides* (Roque and Puytorac 1967) where, as mentioned earlier, the number of ciliary meridians increases from 43 to 300 at certain times in the life cycle. *Ichthyophthririus multifiliis* shows a comparable situation (Mugard 1948). The genetic control of cortical patterns has received extensive, careful analysis in *Euplotes minuta* (Frankel 1973). Despite measurable clonal variability, there is good evidence for a polygenic system determining the phenotypic potential of these cells.

The essential question to be asked here, in terms of evolutionary relationships, is whether the patterns of ciliation of progeny cells is the same or different between various species. And, if different, can the differences be bridged by intermediate forms. In brief, can developmental processes during homothetogenic fission be homologized between different ciliates? The answer is "Yes," as we shall see later in this chapter.

Fission: Stomatogenesis. This essential question of homology applies not only to kinety development on the cell surface but also to development of feeding structures. The preceding comments on the precautions necessary to proper interpretation of silver impregnation apply here too, for oral morphogenesis has been described in large part through the development of oral ciliature. Four patterns of oral morphogenesis have been described for ciliates (Evans and Corliss 1964). These are termed autonomous, semiautonomous, depending on stomatogenic kineties, and *de novo* formation. The autonomous pattern refers to cases where the new organelle develops in close conjunction with the preexisting one, and its kinetosomes seem to come from proliferation of a kinetosomal field more related to the old gullet than to the ciliature of the cell surface. *Paramecium* is cited as one of many examples here. However, experimental work (Hanson 1962; Hanson, Gillies, and Kaneda 1969; Hanson and Ungerleider 1973) renders this doubtful, as does descriptive work (Gillies and Hanson 1968; Kaneda and Hanson 1974), for in this species new kinetosomes arise in the adoral depression, beside the paroral kinety and not in the prostomium. Semiautonomous oral morphogenesis refers to derivation of new feeding structures from both the old stomial area and the surface ciliature. *Pseudomicrothorax* and *Pseudocohnilembus* are examples. Development of the new oral structures from a field associated with a given kinety is found in *Tetrahymena*, among other forms. However, the dependence on just one certain kinety is not fixed (Frankel 1960a). *De novo* formation refers to the apparent development of the primordian of the new feeding structures quite independently of any preexisting kinetosomal systems. *Euplotes* is quoted as an example. However, there is no way at present to rigorously define this independence—at the level or molecular precursors? of kinetosomes? of kinetides?—and no one has yet shown there is no migration of kinetosomes from a site near old ones to a new, apparently independent site (e.g., Wise, 1965, discusses this problem). That kinetosomes can migrate has been known for many years (Balamuth 1940, 1942); how widespread a phenomenon it is in the ciliates is not known.

These patterns of stomial development, judging from their reported distribution (Evans and Corliss 1964, p. 363–364), are not correlated with taxonomic or other major groupings. In fact, within the hymenostome order, three different modes prevail. This raises the question as to whether developmental events can be homologized when they appear to be so different; or, conversely, whether the classification of modes of stomatogenesis is at fault. The latter appears to be the case. There are two problems with this classification. First, it implies a basis in stomatogenic mechanisms—autonomous, semiautonomous, stomatogenic kineties, etc.—and this is misleading, for the mechanisms are not at all clear. And second, this classification looks only at what happens in the opisthe, whereas stomatogenesis is part of cellular morphogenesis and the events in *both* the proter and opisthe must be viewed together.

Jankowski (1967, 1973) has provided a somewhat different classification of stomatogenesis, in which emphasis is on the fate of kineties related to the opisthe feeding parts, and the series of four types shows progressively more and more independence from the surface ciliature. Briefly, the four modes, with Jankowski's examples, are: (1) *somatic:* somatic (i.e. surface) kineties around the cytostome give rise to the kineties with are related to the opisthe feeding structures, e.g., *Balantidium;* (2) *autonomous:* new feeding kineties arise in conjunction with surface kineties not associated with feeding structures, e.g., *Tetrahymena;* (3) *independent:* the new feeding kineties arise in conjunction with the undulating membranelle (UM), and, hence, the feeding kineties are thought to be "self-reproducing," e.g., *Proboveria;* and (4) *apokinetic:* the new feeding kineties are "not formed from kineties (either UM or somatic) but from a group of kinetosomes not carrying cilia and distributed without order over the body surface" (Jankowski 1967, p. 18), e.g., *Euplotes.* This scheme, like Corliss's, ignores the course of stomatogenesis in *both* proter and opisthe together; it implies mechanisms (autonomous, independent, etc.) which are problematical at best, and it focuses only on the ciliature while ignoring what happens to the morphogenesis of the feeding parts themselves. These are limitations that, under these modes, are of little use in phylogenetic studies, although Jankowski finds them very useful for his taxonomic work.

Tartar (1967) has offered another classification of stomatogenic patterns using ten different categories reflecting the participation of kineties, in various ways, or anlage formation, or stomatogenesis from fibers or preexisting mouth parts. This classification is eminently useful to a developmentalist but is confusing to one seeking phylogenetic patterns. Many of his categories include within them very diverse species of ciliates.

Returning to the point that stomatogenic events in both fission products must be kept in mind, another general classification of ciliate stomatogenesis has been developed (Hanson 1969). There are four modes; each will be briefly characterized and illustrated by examples.

(1) *Parafissional opisthe stomatogenesis* (Fig. 10-7A). This refers to situations where the proter retains its oral structures, with little or no morphological dedifferentiation, and the opisthe develops new oral structures in conjunction with development of the fission plane—both temporally and spatially. This is illustrated by the condition in *Dileptus anser* (Golinska and Doroszewski 1964) and is found in certain rhabdopharine gymnostomes and the similar trichostomes like *Pseudoprorodon* and *Coelosomides* (Fauré-Fremiet 1950b) and probably also in *Homalozoon* (Girgla 1971).

(2) *Prefissional opisthe stomatogenesis* (Fig. 10-7B). This mode occurs in the majority of ciliates and refers to formation of a new opisthe stomial area starting well before the cleavage plane is evident. The old area passes essentially intact to the proter. The opisthe oral area develops from kinetosomal fields located at various places in the cell, depending on the species, but al-

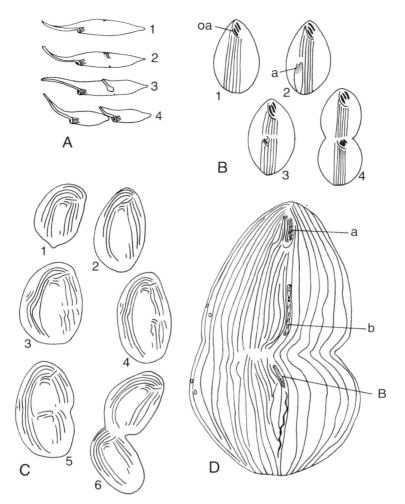

Fig. 10-7 Categories of ciliate stomatogenesis. A. Parafissional opisthe stomatogenesis in *Dileptus anser*. The time sequence is given by figures 1-4. Note that opisthe stomatogenesis is completed at the time the fission products are separating. (Redrawn, by permission, from Golinska and Doroszewski (1964).) B. Prefissional opisthe stomatogenesis in *Tetrahymena pyriformis*. The time sequence follows through figures 1-4. The old oral area (oa) lies at the anterior end of certain ventral kineties (the only ones shown, diagrammatically, by vertical lines). The new oral area (a) appears prior to fission and stomatogenesis is essentially completed by the time the fission furrow becomes apparent. (Redrawn, by permission, from Frankel (1962).) C. Prefissional proter and opisthe stomatogenesis in *Chilodonella uncinatus*. The time sequence is given by figures 1-6. The ventral kinety pattern of a nondividing cell is given in 1. and appears as two complete fields, prior to fission, in 6. The cytopharyngeal basket in 1. (not shown) dedifferentiates and two new baskets differentiate, one for each oral area, in 6. (Redrawn, by permission, from Dobrzanska-Kaczanowska (1963).) D. Postfissional proter and opisthe stomatogenesis in *Ichthyophthirius multifiliis*. The old oral area (a) disappears. The new areas ((b) proter, (B) opisthe) will differentiate after cytokinesis, now underway, is completed. (Redrawn, by permission, from Roque, Puytorac and Lom (1967).)

ways coming to lie posterior to the fission plane and being extensively developed when the fission plane becomes apparent. *Tetrahymena (Glaucoma)* (Frankel 1960a); *Paramecium* (Maupas 1889; Yusa 1957; Kaneda and Hanson 1974); *Stentor* (Schwartz 1935, Tartar 1961); *Euplotes* (Chatton and Séguéla 1940) all illustrate this condition. It is found in hymenostomes, peritrichs (Lom 1964), all spirotrichous forms (except *Bursaria*), and the entodiniomorphids, which are now removed from the spirotrichs (Wolska 1966b, 1971b; Jankowski 1967, 1973; Noirot-Timothee 1968; Corliss 1974a, 1974b).

(3) *Prefissional proter and opisthe stomatogenesis* (Fig. 10–7C). Here the old oral structures dedifferentiate and two new sets appear, one for each fission product, and are well formed by the time the cleavage furrow becomes obvious. This is illustrated by the condition in *Chilodon* (Chatton et al. 1931). In fact, it is typical of all cytophorine gymnostomes, being also found in certain of the rhabdophorine forms too, e.g., *Prorodon* (Puytorac and Savoie 1968), in the more complicated trichostomes like *Colpoda* and *Bresslaua*, and in the new order Scuticociliatida (Small 1968) as well as *Pseudomicrothorax* (Thompson and Corliss 1958), whose taxonomic position is debated (Corliss 1958a, 1958b; Fauré-Fremiet and André 1967). Also the Apostomatida show this type of stomatogenesis.

(4) *Postfissional proter and opisthe stomatogenesis* (Fig. 10–7D). This is described in the ophryoglenids (Roque, Puytorac, and Savoie 1965; Roque, Puytorac, and Lom 1967; Roque and Puytorac 1967) and two species of heterotrichs to date, i.e., *Bursaria truncatella* (Lund 1917) and *Neobursaridium gigas* (Dragesco and Tuffrau 1967). In these forms the preexisting feeding structures start dedifferentiating prior to fission and are almost nonexistent when fission occurs. Only after separation of daughter cells does stomatogenesis occur in both proter and opisthe. (These details are based on light microscopy; electron microscopy could give quite another picture regarding the degree of dedifferentiation and the timing of stomatogenesis.)

Forms omitted from this classification are the Chonotrichida, Suctorida, Astomatida, and those thigmotrichs excluded from the Scuticociliatida. The first group shows atypical fission, i.e., budding, and so the classification directed to stomatogenesis associated with fission understandably does not apply. The suctorians also bud and have atypical feeding structures; they are also reasonably set aside from the classification. Similarly for the mouthless astomes. The rhynchodine thigmotrichs will take further study; some reproduce by fission, but stomatogenic patterns are not clear, and others reproduce by budding, which again puts them in a special category (Raabe 1967, 1970a, 1970b).

Finally, it must be noted that there are variations within each of these categories. In the second category we have emphasized that the old prostomial structure remains intact and passes to the proter. This is true, but probably in all forms there is dedifferentiation of food vacuole-forming ability and, to varying degrees, loss of poststomial fibers. These lost elements reappear, of course. Also, the paroral (or endoral) kinety varies enormously in the role it plays. In the peritrichs, it seems to be a genuine stomatogenic kinety and is the site of kinetosome proliferation for the opisthe prostomium (Lom 1964). In certain hypotrichs, the paroral kinety, or undulating membranelle, participates in formation of the primordium of proter's cirral field (Jerka-Dziadosz 1964, 1972; Jerka-Dziadosz and Frankel 1969). In other hypotrichs, e.g., *Euplotes* (esp. Wise 1965), new feeding structures arise for the opisthe from an intracytoplasmic vesicle. The same is reported for the entodiniomorphids (Wolska 1966b; 1967a; Noirot-Timothee 1969). Furthermore, type 4 stomatogenesis recapitulates type 2, at least in the opisthe, and could be considered a subtype within type 2. In type 3, we have emphasized the fate of the nassa, but there are, in addition, observations on the ciliation around the cytostome. These behave differently in different forms, and in some ways appear to be a type 2 situation. These variants will become very important when we come to discuss possible homologies among these patterns of stomatogenesis.

Budding and cysts. We can turn next to other modes of asexual reproduction. The difference between budding and fission is largely a matter of degree: the bud is smaller than the parent and differs from it more than two fission products differ from each other. Nevertheless, budding involves problems of nuclear and cytoplasmic division, organization of new kineties, differentiation of other cell organelles, and molding of the cell body (Lwoff 1950; Dobrzanska-Kaczanowska 1963) (Fig. 10–8). One can treat it, therefore, in the same terms as fission.

The behavior of cells at encystment and excystment, and as regards changes within the cyst, are seen as problems of leaving and returning to the typical form of the cell. These processes involve the same sort of nuclear and cytoplasmic problems as fission and budding (Lund 1917; Tartar 1967). Additionally, one can pursue the factors which control encystment and excystment as physiological processes (van Wagtendonk 1955; Butzel and Horwitz 1965; Butzel and Bolten 1968). These studies do not allow, at present, for extensive interspecific comparisons.

Sexual processes. Sexual processes comprise the next set of functions we need to examine. Conjugation, cytogamy, and autogamy are the nuclear reorganizations that are sexual in the sense of our earlier definition, namely, providing the opportunity for new genotypes through fusion of nuclei. Other reorganizations, resembling conjugatory or autogamous events but lacking the essential feature of nuclear fusion, are endomixis and hemixis.

Conjugation has been described earlier, in conjunction with the discussion of breeding patterns in the protozoa (Fig. 4–3) and will only be briefly reviewed here for its major and usual features. Unusual aspects will be

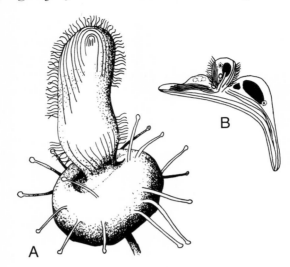

Fig. 10-8 Budding. A. The suctorian *Podophrya collini* producing a ciliated "larva", which, upon attaching to the substrate, will transform itself into a tentaculated adult. (Redrawn, by permission, from Guilcher (1951).) B. The thigmotrich *Sphenophrya dosiniae* reproducing by formation of a dorsal bud. (Kineties are diagrammed by solid lines; cilia are shown only on the bud.) (Redrawn, by permission, from Raabe (1970b).)

mentioned when we come to discuss homologies. In conjugation, two ciliates, of complementary mating type, come together with their ventral surfaces apposed. They remain together, paired, while certain precisely timed nuclear and cytoplasmic events transpire. Chief among the cytoplasmic events is dedifferentiation and eventual loss of the preexisting feeding structures, first described by Hertwig (1889). These are replaced immediately. Indeed, in *Paramecium aurelia,* the new structures are being formed while the old are being broken down (Hanson, Gillies, and Kaneda 1969). The pattern of reformation (Roque 1956a, 1956b; Porter 1960) is the same as formation of a new gullet at fission. In addition to loss of the mouth structures, trichocysts and cilia (but not their kinetosomes) are lost where the surfaces of the two conjugants are in contact (Porter 1960; Vivier and André 1961). As regards the macronucleus, it starts a series of changes which culminate in its degeneration through fragmentation, the fragments persisting, however, well beyond completion of conjugation. The micronuclei divide meiotically, and in each cell all haploid nuclei except one degenerate. The one that persists undergoes a mitotic division, producing two isogenic haploid nuclei. One of these passes into the partner cell and the other remains in the original cell and fuses with the gametic nucleus that comes its way from the partner. This reciprocal exchange of nuclei and subsequent fusion of haploid nuclei produces a diploid zygote nucleus in each cell, and these two nuclei are genetically identical (Sonneborn 1947). The zygote nucleus gives rise through division—the exact details vary somewhat from species to species—to new micronuclei and macronuclear anlagen. The latter rapidly differentiate into mature macronuclei. Thus the exconjugant cells can form new genotypes, and the germinal nature of the micronuclei is seen in contrast to the somatic nature of the macronuclei. Long ago Hertwig (1889) called the micronucleus the "Geschlechtskern" and the macronucleus the "Stoffwechselskern" in respect of their germinal and somatic (metabolic) functions.

Cytogamy is the same as the foregoing, except that there is no exchange of gamete nuclei, and self-fertilization occurs (Wichterman 1953 and earlier papers). Autogamy, as its name implies, involves self-fertilization and occurs without the pairing of cells that characterizes conjugation and cytogamy (Fig. 4-3).

The status of endomixis is complicated. It was first described (Woodruff and Erdmann 1914) as a nuclear reorganization in single cells of *Paramecium aurelia*, which, though it included eventual nuclear breakdown and division of micronuclei with their products reconstituting new micronuclei and the macronucleus, never included meiosis or fertilization. This report could never be confirmed, and only autogamy is known to occur as the typical sort of single cell nuclear reorganization in *P. aurelia* (Diller 1936; Sonneborn 1947). In other ciliates—*Loxodes rostrum*, *Remanella multinucleata* for example— the direct transformation of micro- into macronuclei is known to occur (Fauré-Fremiet 1954a; Raikov 1969 and a brilliant series of earlier papers) and has been termed an endomictic nuclear phenomenon. However, the absence of macronuclear breakdown makes these instances distinctly different from the original description of the process. If we define endomixis as it was originally, it then seems best to conclude that it does not occur. This means, then, the more recently described examples of endomixis must be some other process. This is clearly possible in view of Raikov's extensive work on these and related forms and his conclusion that they represent a primitive condition of the relation between micro- and macronucleus. This important point of view will be discussed more fully later.

Hemixis (Diller 1936), the last of the nuclear reorganizational processes, includes all types of reorganizations not identifiable as being any of the ones already discussed. It is, obviously, a catchall term and, by definition, includes a variety of phenomena, including abnormal events. The lack of regularity of hemictic events and their often peculiar nature, both argue for setting hemictic processes aside from consideration as processes of use in comparative phylogenetic studies, and we will do so now. Future work may, however, show this to be a mistaken decision.

Life cycles. Finally, to complete the general characterization of the Ciliophora, there are life cycles to be considered. The necessary comments can be kept brief, for there is no general pattern of cyclical change superimposed on cell cycles that includes large numbers of ciliates. The phenomenon of ageing (Woodruff and Baitsell 1911; Woodruff and Erdmann 1914; Sonneborn 1954) is found in a limited number of ciliates. It consists, basically, of the fact that in the absence of sexual reorganization of the nuclear material, the cells will eventually die out; in *Paramecium aurelia* the upper limit of their asexual propagation is about 300 fissions. In other ciliates, where sexual activities are excluded by the permanent absence of micronuclei, e.g., *Tetrahymena pyriformis* strain WH, there is no ageing effect evident. Another example of a life cycle is the sequence of events relating to mating activity (see Chapter 4). In *Paramecium bursaria*, for example, there is, following conjugation, a period of immaturity when for many fissions no conjugation can occur (Sonneborn 1957). Then one mating type can be expressed—the "adolescent period"—allowing certain matings, and then full expression of the mating type follows (Siegel and Cohen 1963). But this pattern is far from universal, and, as was recounted earlier, paramecia with a tendency to inbreed can express their mating potential immediately after completion of conjugation; the term life cycle has little application in such cases. In extreme cases (Heckmann 1967) the delay in expression of mating potentials can take a year or more. It seems best to view such variability in the expression of the genotype as questions of genetic expression relating to the biological peculiarities of given species, rather than trying to find in them an expression of some general phenomenon of progressive change common to all ciliates. Viewed in this way, expressions of mating potential are special adaptations of the various gene pools to various selection pressures. Other instances of variable gene expression, e.g., the antigenic types in *P. aurelia*, also probably represent special adaptations of the cells, but whose selective advantage is completely unknown at present (Sonneborn 1957).

A different view of ciliate life cycles is seen in the development of colonial forms, such as in the stalked peritrichous ciliates. Here, growth patterns follow a stable sequence of events, and certain cells are predictably differentiated in characteristic ways. The best-studied case is probably that of *Zoothamnium alternans* (Fauré-Fremiet 1930; Summers 1941). However, colonial forms are relatively rare in the ciliates as a whole, and no generally applicable principles occur from the study of their colonial development.

The clearest examples of life cycles are the sequence of changes found in the symbiotic apostomous ciliates (Chatton and Lwoff 1935; Lwoff 1950). These forms show changes in the length, form, and placement of their kineties that are correlated with their location in or out of their host, the availability of food and, therefore, are probably dependent on their various metabolic activities. As would be expected, such changes are highly adapted to the symbiotic mode of life of these organisms. What uniquely suits them is not found in other ciliates, and their life cycles, being exceptional rather than general, allow for phylogenetic speculations within the group but are of no use compared to other forms lacking such life cycles.

We repeat, therefore, that life cycles, even construing the term generously to include ageing and patterns of expression of certain phenotypic characters, i.e., any sequence of changes superimposed on successive cell cycles (Hanson 1967a), occur in the ciliates as a variable phenomenon and with an irregular occurrence, which renders difficult their use in comparative studies.

Phylogeny

The fact that the ciliates are a clearly definable group of organisms implies either that they are all descended from a common ancestor and still show traits recognizably relating them to each other, or there has been more than one origin of the group, but strong selection pressures have brought about convergent or parallel tendencies that give the group its apparent homogeneity. The latter point has never been vigorously espoused, though it has rightfully been raised as a possibility (Canella 1964); especially so in view of the fact that the characters treated as homologous (in particular Fauré-Fremiet 1950a; Corliss 1956, 1960, 1961) have never been shown to be such as the result of a critical, i.e., Remanian, analysis. However, as has been seen to be the case before, many workers, and especially those espousing the philosophy of the New Systematics, do implicitly employ thinking that is close to Remane's and develop phylogenetic conclusions that are only slightly modified by more rigorous analysis. A contribution of the more critical point of view, also as we have seen before, is to make explicit the rationale of the procedure, so as to clarify the data employed and render operational the steps used to arrive at stated conclusions regarding phyletic distances. Also, an explicit phylogenetic procedure goes further, proposing explanations for the various evolutionary innovations thought to be present, and leads to more precise formulation of proposals regarding the origin and trends within the group as a whole. The question of origin is a particularly vexing problem with the ciliates. Corliss says, referring to primitive ciliates, "Yet their origin (from zooflagellates?) still remains too much of a mystery even for speculation!" (1961, p. 36) Nonetheless, Corliss and others continue to speculate on the problem. An explicit methodology is essential to make speculation meaningful, and being equipped with such an approach, we will attempt our own speculations at the close of this chapter.

HOMOLOGOUS RELATIONSHIPS

There are fourteen readily defined semic traits in the ciliates, and most of these represent variable but homologizable characters that are uniquely informative about just this group of organisms. A rather complete phylogenetic analysis is possible with the ciliates.

Semes. The structural semes are ciliary units, kinety patterns, feeding parts, micronuclei, macronuclei, special organelles, and general cell-body organization. The functional semes are homothetogenic fission, budding, sto-

358 Phylogeny of the Protozoa and Early Metazoa

matogenesis, prezygotic or prefertilization sexual processes, postzygotic processes, nature of food, and ecological habit. These can be reviewed briefly to indicate the kind of point-to-point comparisons each allows and to indicate something of the variation found in each one.

Increasingly, detailed information on the fine structure of ciliary units is becoming available (Grain 1969). Fig. 10-9 illustrates the complex details available in four instances. In these cases, the relative position of the parts of each unit—kinetosomes, kinetodesma, transverse and postciliary microtubules, parasomal sac, and alveoli—all argue for homology. Unfortunately, this quality of detailed information is not yet available on a large number of ciliates. The first three examples in Fig. 10-9 are from rather similar-appearing forms (they are all hymenostomes), and this is reflected in the sim-

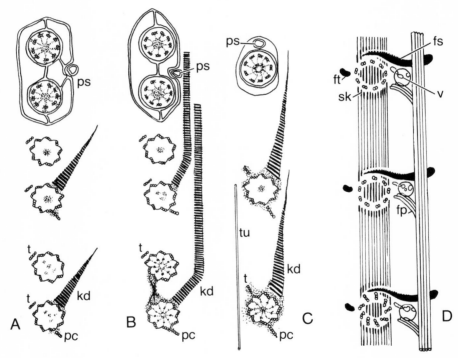

Fig. 10-9 Ciliary units found in selected ciliates seen as composite drawings from electron micrographs. All figures are presented with the cell's right on the viewer's right. In each set of figures (A, B, C, and D) structures are shown at different levels, the lowest part of each figure being the most internal level. A. *Frontonia invellatum*. B. *Paramecium aurelia*. C. *Tetrahymena pyriformis*. D. *Ignatocoma salbellarum*. fp—postciliary fiber (possibly homologous to postciliary microtubles), fs—striated fiber (possibly homologous to kinetodesma), ft—transverse fiber (possibly homologous to transverse microtubules), kd—kinetodesma, pc—postciliary microtubules, ps—parasomal sac, sk—subkinetal sheet, t—transverse microtubules, tu—basal microtubules, v—vesicle adjacent to parasomal sac. (A, B, and C redrawn, by permission, from Didier (1971). D redrawn, by permission, from Lom and Kozloff (1969).)

ilarity of their units. The fourth one is from *Ignotocoma sabellarum,* a rather specialized, thigmotrichous ciliate, and obvious differences exist between its ciliary units and the other three.

The two major groups of ciliary patterns are those seen in the four subclasses of ciliates (Honigberg *et al.* 1964). Here we are referring to the aboral surface ciliature, and in the holotrichous forms it is generally composed of uniform longitudinal rows of ciliary units (Fig. 10–2A and D). In the heterotrichous forms, this ciliature is also present in many species, but often there is a special ventral ciliature consisting of tufts of cilia or cirri (Fig. 10–2E). The peritrichous forms show little ciliation in the adult forms except for that associated with the feeding structures (Fig. 10–4G). And the suctorians show no ciliation in the adult form; their buds are, however, ciliated in the holotrichous manner (Fig. 10–8A). Within the first two groups—the holotrichs and heterotrichs—there is considerable variation; indeed, so much so that this feature plus feeding parts is the basis for most species descriptions. (Recall, however, that within these morphotypes there are, in the carefully studied instances, many gene pools, i.e., many biological species.) This variability in aboral kinety patterns will have to be used to bridge the differences among the four large categories of patterns seen in the systematics of these protozoans.

The major types of feeding parts and their major components have already been fairly thoroughly discussed (Fig. 10–4). In summary, the chief features of these semes are: universal presence of a cytosome; peri- and prestomial structures associated with a cytostome; a prostomium with its polykineties—three in number—and, on occasion, the presence, also, of an endoral kinety; adoral fields and, on occasion, a paroral kinety. These parts and their relations allow for detailed interspecific comparisons.

By special organelles we refer especially to contractile vesicles, cytoprocts, cortical and subcortical fiber systems, deposits of platelets at the surface, stalks and other attachment devices, the whole armamentarium of toxicysts, trichocysts, haptocysts, mucocysts, etc., tentacles and other extensions of the cell body, symbiotic algae and bacteria, special crystals and vacuoles and pigments, and also the array of external secretions—tests, loricas, and thecae. This constellation will have to be treated as the acantharian Weichkörper, that is, as an aggregate of discrete entities each of which can be treated as an item for comparison. In all probability, future work will allow the contractile vesicles, fiber systems, and extrusible organelles to be treated as separate semes. To do so now is premature.

Micronuclei are readily compared in terms of location, though this may not be critical since, depending on the stage of the cell cycle, they can be almost anywhere in the inner endoplasm. More importantly, their number, size, composition, chromosome number, and patterns of mitotic behavior can be carefully compared. (Meiotic behavior is reserved as a part of the sexual processes.)

Macronuclei show even more variability than the micronuclei. They differ in shape, size, number, composition, and behavior during fission. Accordingly, they allow even more detailed comparisons than their smaller partners. This includes some fine structure data that are usually unavailable for the micronuclei.

The last structural seme is general body organization. Here the symmetry of the cell body, its organization into ecto- and endoplasm, and the relative placement of feeding parts, contractile vesicles, cytoprocts, stalks, tentacles, etc., can all allow detailed statements regarding the degree of similarity between ciliate cell bodies.

The functional trait of asexual reproduction can be separated into the two semes of fission and budding. Fission in the ciliates is homothetogenic in nature, and within that category it varies depending on the ciliary patterns, general cell-body organization, and nature of feeding organelles that are formed for the two, tandemly oriented fission products. In that we shall treat formation of feeding organelles separately, the modes of fission with which we shall be concerned will depend on formation of aboral ciliation and gross organizational details. Regarding ciliation, there are certain general categories that first come to view. These are the ones exemplified by the holotrichous and the hypotrichous situations. In the former, which includes most of the heterotrichs, too, the basic pattern is a proliferation of units in the midregion of the body with a perkinetal (across the kineties) fission plane, followed by remodelling of the new ends of both daughter cells—the posterior end of the proter and the anterior end of the opisthe. Within this general picture, there are many variations, which include fission within cysts and dedifferentiations to various degrees prior to fission, e.g., untwisting of spiralled kineties, followed by redifferentiation. The hypotrich situation refers to a special ventral surface which is differentiated at fission from a special field of kinetosomes. This field gives rise to two ventral fields, one for each fission product, and the old field degenerates. There are, of course, variations within this general pattern, too.

It may be possible to some day find another seme within fission which relates to molecular preparation for fission, in contrast to the foregoing which refers exclusively to events at the organelle level. At present, however, the data on hand do not permit such a step.

Budding is of scattered occurrence in the ciliates but it is an important mode of asexual reproduction, and because it differs so strongly from fission it is treated as a separate seme. Where budding occurs, the bud is invariably smaller than the parent cell, and it undergoes developmental steps that do not appear in the parent, which remains relatively unchanged. The bud can be formed internally or externally. Its aboral ciliation may or may not reflect parental patterns, and there can be differentiation, almost *de novo* as it would appear, of various structures, including stomial parts, stalks, special aboral ciliation, etc. All of these differentiations allow for precise comparisons in bud formation.

Stomatogenesis has been categorized into four major patterns. The phylogenetic problem is to find intermediates between the various categories. To anticipate somewhat, the specific issues are going to be to bridge differences between parafissional opisthe stomatogenesis and proter and opisthe prefissional stomatogenesis, between the former and prefissional opisthe stomatogenesis, between proter and opisthe prefissional stomatogenesis and prefissional opisthe stomatogenesis, and between the latter and the final category of postfissional stomatogenesis in both proter and opisthe. The abundant details of structure and the clear sequence of known stomatogenic events in many species provide a rich set of comparative data.

The sexual processes, especially conjugation, offer such a precise and detailed set of data that they can be separated into two semes: (a) prezygotic or prefertilization events, and (b) postzygotic or postfertilization events. The prezygotic events can include the induction of reactivity among complementary mating types and the genetic basis of complementarity, as well as nuclear events (micronuclear meiosis and pregamic mitosis, macronuclear changes) and changes in cytoplasmic organelles such as ciliary and stomial dedifferentiation. Postzygotic events will include the separation of exconjugants and redifferentiation of cytoplasm structures, as well as reformation of the vegetative nuclear structures.

In time, with more information, autogamy could be treated as another seme. Its reported occurrence in hypotrichs (Luporini 1970; Jareño, Alonso, and Pérez-Silva 1970) in addition to *Paramecium* and other holotrichs, means that it is probably widespread in occurrence.

We can, at the present time, make comparative comments on the feeding habits of ciliates. They are all predators (except for certain symbionts) of one sort or another, and it is the nature of their predation that allows us to identify bacterial or algal feeders and feeders on other protozoa.

And finally, we can identify the habitats and, with some precision, the niches of various ciliates. They are readily identified as fresh- or salt-water dwellers and whether or not they are symbiotic. Within each of these categories, various further details are often possible with regard to various physical dimensions of the environment. As will be seen later, these data will be used conservatively, but as better data accumulate, it will surely be possible to make quite fine interspecific distinctions regarding ciliate ecology.

Homologies and variability. Though all members of the Ciliophora possess cilia in at least some phase of their life, this is not convincing evidence that ciliation is an homologous character in the group as a whole. The extraordinary uniformity of the ciliary fine structure and of its kinetosome raises the possibility of similarity through limitations of structural design. This is a case like the zooflagellate flagellum, where there is a real possibility of convergence.

Continuing further with the parallel to zooflagellates, we then ask, are the kinetides homologous, as we found mastigonts to be, or better, are the ciliary units homologous? The answer is: Yes, as far as can be determined at present. As noted earlier, where details are available, there are point-to-point similarities in the placement and composition of the parts of the kinetosomal territory. This argues convincingly for their homology (Fig. 10–9A, B, and C). There are, however, cases where significant differences appear, as between Fig. 10–9A, B, and C, on the one hand, and Fig. 10–9D on the other. It is here that the qualification, "as far as can be determined at present," becomes necessary, because for a rigorous identification of homology, we of course need intermediate forms to bridge the observed differences. Therefore, strictly speaking, the claim of homology between *all* ciliary units is somewhat premature at this point. If it is also possible to see that the ciliary units have a distribution which parallels other known homologies (first susidiary criterion), then they are also arguably homologous on that basis. Hence, let us continue to look for traits which are unconditionally homologous through fulfillment of the major criteria.

Kinety patterns of the surface or aboral ciliature are homologous throughout the ciliates. Monographs, such as Kahl's (1930–1935), document *in extenso* the positional and compositional similarity in the ciliary rows. Their cortical location and compositional dependence on the kinetosome and cilium is amply documented. Only two sources of apprehension remain. One is the homologous nature of composition in terms of the whole ciliary unit, for this, as was just discussed, is not a fully explored and resolved question as yet. The second is the question whether the great diversity of kinety patterns, especially the special ciliary patterns of peritrichs and suctorians, can be joined by intermediate forms to the predominant holo- and heterotrich patterns.

Before looking at these two extreme cases, it will be useful to illustrate the kinds of transitions possible in less extreme cases. The thigmotrichous ciliates are ectosymbionts—commensals, apparently—on mussels and, in some cases, holothurians (sea cucumbers). Looking at the family which contains the "exit form" (i.e., plesiomorph) of the whole order, Raabe (1970a) develops an orderly progression of forms which connects *Ancistrumina* to the more complex forms such *Boveria* and *Hemispeira* (Fig. 10–10). This analysis includes the prostomial (termed "adoral" by Raabe) ciliature as well as the surface or aboral kineties.

Three points are illustrated by this limited, but exemplary, set of data. (a) Transitions between differing sets of kinety patterns are readily possible. (b) The feeding ciliature are an important adjunct to the aboral patterns. (c) These transitions are within one family of ciliates and further, finer intermediate steps could be found if one surveyed larger numbers of species

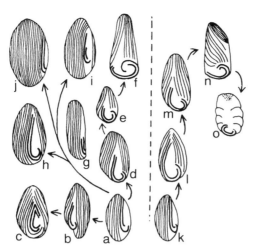

Fig. 10–10 Kinety patterns and possible evolutionary relationships among hemispeirid ciliates (Thigmotrichida). Stem forms are the *Ancistrina* (a) and *Protospira* (k). From the former there arise three lines (a–c; a–f; a–j, i). From the latter there is one evolutionary line (k–o). (Redrawn, by permission, from Raabe (1970a).)

within each genus. This is, of course, to be expected. The real problem lies in going in the other direction, to comparisons between families or between orders, as is the case for comparing holotrichs to peritrichs and to suctorians. We would expect greater difficulties in bridging the gaps between the higher taxa for reasons given in Part I of this study. The whole procedure for bridging gaps will not be followed in detail here; it involves, in its most rigorous form, the identification of plesiomorphs for genera, and then finding the plesiomorphs of the generic plesiomorphs, and so on up to the levels of orders and beyond (Chapter 3). As we stated earlier (Chapter 4) there is no room in this volume for such detailed argumentation, nor can one worker expect to know the 6,000 or more species of ciliates well enough to develop the whole phylogenetic argument in detail. Here, we are emphasizing the broad feature of the phylogenetic analysis to show its applicability and define some of the conclusions that arise from such a study. The problem at hand comes down to this: From surveying the literature, those forms that are judged to be illustrative of plausible transitions between significantly different species will be selected for discussion.

Thus, in the case of peritrich aboral surfaces, we can support the sequence already suggested by others (summarized in Corliss 1961) which starts within the holotrichs from the tetrahymenine hymenostomes and goes to the pleuronematous forms, and then to the arhynchodine thigmotrichs and finally to the peritrichs (Fig. 10–11).

The transition to the suctorians involves a different set of problems, for here there is loss of all ciliature in the adult. The comparison is then with the

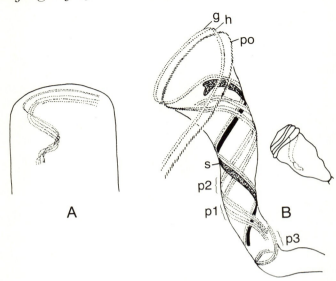

Fig. 10-11 Kinety patterns in feeding parts of holo- and peritrichous ciliates, illustrating homologous relations. A. A member of the family *Hysterocinetidae* (Thigmotrichida), showing the adoral and prostomial ciliation. (Redrawn, by permission, from Puytorac (1968a).) B. The right hand figure is the whole cell of *Campanella umbellaria* (except for the stalk and holdfast). The larger detailed figure represents the ciliation of the prostomial cavity. g—germinal kinety, homologous to paroral or endoral kineties of other ciliates, especially since the new stomatogenic field of kinetosomes proliferates next to it close to the anterior end of impregnable structure (s); h—haplokinety; po—polykinety which becomes the first band of penicular cilia (p1) in the prostomial cavity; p2—second band of penicular cilia; p3—third band of penicular cilia, possibly homologous to the quadrulus of holotrichous ciliates, as in *Paramecium*. (Redrawn, by permission, from Lom (1964).)

bud or "larva", which shows a definite holotrich pattern. But the question now becomes one of finding the intermediates that lead to budding and, equally important, to loss of cytostomal parts and their functional replacement by tentacles. The problem can be approached through an examination of this latter issue, which now raises the problem of homologies in feeding structures, that problem will be discussed in general terms before returning specifically to the suctorian kinety patterns.

Starting with Furgason's (1940) searching and important analysis of his newly-constituted genus *Tetrahymena*, there has been a tendency among many workers to see homologies among all varieties of oral structures in the ciliates. Some bases for these conclusions are seen in Fig. 10-4. Are these postulated homologies valid in the light of Remanian criteria?

The cytostome, we would argue, in support of Furgason's and Corliss and Fauré-Fremiet's views (latter summarized in Corliss 1961), is homologous throughout the ciliates. But it is such a variable structure that the arguments in support of homology continually depend on the fulfillment of the serial relationship. The only common feature of its position is that it is cortical or attached to the cortex by a peristomial depression or a prostomium. Other than that, it can lie at the anterior pole, towards the posterior pole, or anywhere on a lateral surface between these extremes. These latter positions are the intermediates needed to bridge the apical extremes of location. In terms of composition, the essential similarity is the presence of membrane-bound phagocytic vacuoles on the inner side of the cytostome. Hence, the major Remanian criteria of homology are fulfilled.

The remaining parts of the feeding apparatus introduce other complexities. When the cytostome is at or near the cell surface, then peristomial depressions and special organelles such as the nemadesma, toxicysts, and special prestomial structures can be analyzed for homologies through their position and composition. That they are homologous can be seen by the transitions that can be seen between extremes. The special form of *Teutophrys* is seen as a multiple body of *Dileptus* (Clément-Iftode and Versavel 1967). And *Dileptus*, with its round cytostome and toxicyst-laden prestomial parts and peristomial parts with nemadesma, could merge with the slit-like mouths of the other gymnostomes through transitional forms such as *Loxophyllum pseudosetigerum*, perhaps, or *L. kahli* (Dragesco 1960) (Fig. 10–12). There seems to be no problem, given the rich variety of gymnostome and trichostome forms, in finding easy transitions from one set of feeding parts to another in these two groups.

Among forms with a prostomium, we need to be able to homologize prostomial parts and to bridge the differences between forms with and without a

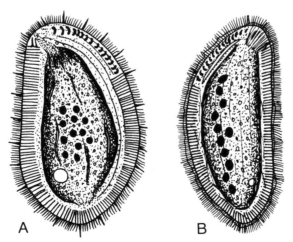

Fig. 10–12 Two gymnostomes with flattened bodies typical of sand-dwelling forms like *Stephanopogon mesnili* (Fig. 10–14A) and *Remanella multinucleata* (Fig. 10–3C), but which have roundish cytostomes rather than the slit-like structures of the two species just mentioned. A. *Loxophyllum pseudosetigerum*. B. *L. kahli*. (Redrawn, by permission, from Dragesco (1960).)

prostomium. Let us first examine homologies between different prostomial structures.

The presence of a paroral or endoral kinety and three prostomial or penicular polykineties in many prostomia is the logical starting point. (Recall that these are the undulating membrane (UM) and adoral zone membranes (AZM), respectively, of Furgason (1940).) In terms of the positional relationship, the paroral kinety consistently lies along the right posterior side of the prostomial opening, and the prostomial membranes are on the left wall of the cavity. When the paroral kinety assumes an endoral position, it still lies to the cell's right of the prostomial membranelles. And often there is a ribbed or ridged wall lying between these two sets of kineties. In terms of composition, the paroral and endoral kineties are often a haplokinety, and a prostomial membrane is a group of two to six kineties each, hence, a polykinety. Furthermore, differences among the paroral kineties and the prostomial membranes are often bridgeable by intermediate forms; thus the third or serial relationship again comes into force. This includes special coilings and expansions of the prostomium, as in certain heterotrichs (Tuffrau 1967a) and peritrichs (Fig. 10–11B).

However, it is not yet quite certain that all discontinuities are bridgeable. In some forms, like *Philaster* or *Pleuronema* (Small 1968), the homologizing of the ciliature anterior to the cytostome to prostomial membranelles still seems problematical. In particular, Small (1968) has explicitly denied homologous relationships between the prostomial structures of these and the tetrahymenine holotrichous ciliates. Small includes *Pleuronema* and related forms in a new ordinal taxon called the Scuticociliatida, because of a distinctive hook-shaped kinety lying to the right of the cytostome. This kinety and others which are part of the feeding structures, as well as the pattern of stomatogenesis, seem so distinctive to him that homologies are thought not to exist, and the creation of a new order of ciliates is justified. However, Small's definition of homology seems to depend solely on close similarity between the objects being compared. He does not recognize the serial relationship of Remane, which is the essential conceptual insight allowing workers to homologize seemingly disparate entities. Small's alpha, beta, gamma, and zeta membranoids are conceivably the homologs of prostomial (or peniculine) polykineties 1, 2, and 3, and the paroral kinety, respectively. Antipa and Small (1971) make a convincing argument for such homologizing in the case of *Conchopthirius*, and the same principle of looking beyond certain differences to larger similarities could also apply to the scuticociliates. In our view, the prostomial kineties of these forms are very probably the homologs of the prostomial kineties of the tetrahymenine forms.

A quite different problem emerges when one turns to bridging differences between forms with and without prostomial or other feeding parts. The circular nassa of the cytrophorine gymnostomes (Fig. 10–4A and B) seems to be quite distinct from the ciliated prostomial cavities (Fig. 10–4E) we have been discussing. However, Corliss (1958a, 1958b) has correctly seen *Pseudomicrothorax dubius* (Thompson and Corliss 1958) a suitable missing link Fig. 10–14E).

Finally we return to the tentacles of suctorians. They seem to be a new development, quite separate from typical mouths which have apparently been lost in these cells. But intermediates exist: *Actinobolina* possesses both a mouth and tentacles (Fig. 10–13) and Kahl's (1931) detailed study of tentacle structure led him to conclude that there are important similarities in the tentacular structure of suctorians and *Actinobolina* and other gymnostomes also equipped with these peculiar organelles. The tentacles of *Actinobolina* do not function directly in ingestion; they carry toxicysts at their termini, which are discharged upon contact with prey organisms. These tentacles are important in prey capture. Fine-structure studies reveal microtubule arrays in the tentacles of *Actinobolina* (Holt and Corliss 1973), and these same organelles are important in the structure and function of suctorian tentacles (Bardele and Grell 1967; Bardele 1972), which both capture and ingest prey.

The bridging of differences is achieved, then, in two ways. Either there are gradual transitions from one type to the other, involving changes in a given structure (or function), or there is the simultaneous presence of two different structures, e.g., the nassa and buccal ciliature in *Pseudomicrothorax* and tentacles and cytostome in *Actinobolina*, which in the extreme forms exist separately. In the case of *Pseudomicrothorax*, we see that the nassa is lost as we go to forms with further development of prostomial ciliature but which retain, nonetheless, a cytostome. Hence the cytostomes of *Nassula* and *Tetrahymena* are homologous, though *Tetrahymena* has nothing that is homologous to the nemadesma of the nassa, and *Nassula* has nothing homologous to the prostomium of *Tetrahymena*. The serial relationship, however, establishes the homology of the cytostome in these two forms. Similarly, the serial relationship makes clear the transition from forms with a cytostome but no tentacles to forms with tentacles but no cytostome. Here there is no common homology of cytostome to connect the two types of ingestatory apparatus. They are functionally analogous but not homologous.

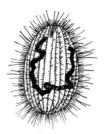

Fig. 10–13 *Actinobolina vorax* showing ciliation and slender tentacles (partially extended, longer than cilia), U-shaped macronucleus, contractile vesicle at posterior end, and cytostome at anterior end. (Redrawn from Wenrich (1929).)

The homologies that do persist into the suctorians are the kinety patterns. The sessile adult forms are not ciliated; the mobile larval forms are, and their ciliation is reminiscent of the regular longitudinal kineties found on *Actinobolina*.

A last seme, regarding surface structures is the general topography of such organelles as cytostome or prostomial opening, contractile vesicle pores, cytoproct, and suture lines of the kinety pattern. We can also include symmetry of the cell and general organization of ectoplasm and its relation to endoplasm. In brief, the general organization of the cell body. The composition of feeding structures, contractile vesicles, cytoprocts, ectoplasmic specializations, etc. provides quite specific points of reference in the three-dimensional space of the cell body, and therefore comparisons to homologize one cell-body's organization with another are readily possible. From what has already been said about transitions among various kinety parts and types of feeding structures, it can be inferred that transitions among the various kinds of cell-body topography can also be found and the various patterns homologized.

Nuclear dimorphism—the simultaneous presence of separate diploid and variously polyploid nuclei—is a homologous trait common to all except two species of ciliates. It is possible to focus further discussion on the micro- and macronuclei as two separate semes after first saying that these nuclei all lie in the inner endoplasm of their respective cells and, therefore, the positional relationship is essentially similar in all cases.

The composition of macronuclei is indeed alike in terms of their gross chemical and structural characteristics: they all contain DNA, RNA, and protein and are bound by a double unit membrane with pores (Raikov 1969). However, in terms of the structural organization of the DNA, RNA, and protein there appears some variation from one species to another, and, in terms of the degree of polyploidization present, there are also differences. The latter seems, however, to be only a matter of degree (Raikov 1958, 1959, 1963b, 1969; Raikov, Cheissin, Buse 1963; Woodard, Gelber, and Swift 1961) and hence intergrade in a manner not to cause difficulties for homologous relations from that point of view. The presence of endosomes and variously staining parts of macronuclei, the difference in size of various particles, in size and number of nucleoli, are all a more serious problem. When the function of these structures is more fully worked out, we will, of course, better understand the macronucleus and can then speak more meaningfully about similarities or degrees of similarity between nuclear structures. At present, most protozoologists would probably tend to the view that macronuclei are homologous and that the differences seen represent structural idiosyncrasies of various species that do not seriously raise the possibility of more than one origin of this distinctive ciliate organelle. The increasing number of reports of chromosomes in the macronucleus provide hope that its structural organization is now becoming more amenable to experimental study.

The differences in the visible form of the macronucleus can all be treated successfully through the presence of intermediate forms. In this regard, we are especially in debt to the work of Raikov (1969 and earlier papers). He describes in detail the process of differentiation of micronuclei into macronuclei and the subsequent polyploidization of the macronuclei at times other than during sexual reorganization of nuclear materials. This direct differentiation of micro- into macronuclei was reported earlier (Bütschli 1887–1889; Fauré-Fremiet 1954a), but it was Raikov's work on the gymnostomous forms that showed something of the distribution of the phenomenon and showed the various degrees of complexity of the differentiated macronuclei. What we now can see in the ciliates is a series going from *Stephanopogon mesnili* (Lwoff 1923, 1926) which lacks a macronucleus, through forms such as *Loxodes*, *Remanella*, and *Trachelocerca*, whose micronuclei produce more micronuclei by mitosis and macronuclei by direct differentiation but are then incapable of division, to forms such as *Dileptus anser* (Hayes 1938), whose many micro- and macronuclei both divide but asynchronously with fission, and finally to forms typical of the vast majority of ciliates, where both types of nuclei divide synchronously with the rest of the cell. In the last two cases macronuclear dependence on the micronucleus is only apparent at times of sexual reorganization.

The various shapes of the macronucleus—spherical, ovoid, cylindrical, beaded—vary without correlation with the taxonomic status of the ciliates. For example, in the genus *Stentor* there are ovoid (*S. auricula*), cylindrical (*S. roeseli*), and beaded (*S. ambiguum*) macronuclei (Kahl 1930–1935). The differences are probably correlated with size of the cell. In any case, they offer no real problems for homology, as there are no significant discontinuities between the various forms, and they can be considered as differences serially removed from the most common form of macronucleus, the ovoid one.

The micronucleus shows relatively little variation. Differences lie in chromosome number and the manner of dispersion of the chromosomes at interphase; in the vesicular nuclei they are collected centrally and in the dispersed nuclei they are scattered throughout the nucleoplasm. The persistence of the nuclear membrane, the absence of centrioles or centrosome, the presence of elongated intranuclear spindles, and their role as originators of macronuclei collectively distinguish the micronuclei from other diploid protozoan nuclei and render them homologous structures in the ciliates.

Other endoplasmic inclusions should be treated, as stated previously, somewhat as the Weichkörper of the Acantharia was treated. Namely, there are various discrete points for comparison, in the form of lipid inclusions, crystals, symbiotic algae, and bacteria, etc. But there is great difficulty in obtaining precise information for more than a few species, and some of these inclusions are highly variable in their occurrence. It would be unprofitable at present to attempt a systematic study of this seme.

We turn next to homologies among the various ciliate functions.

As we have said earlier, homothetogenic fission characterizes the fission process of ciliates. Though there are some differences in how the feeding structures are formed and in differentiation and morphogenesis of other aspects of the cortex, the ciliates show a basic pattern of separation into two cells that involves cleavage at right angles to the long axis of the cell and the development of similarly placed parts in the two daughter cells. The forms that come closest to being exceptions to this pattern are the oligotrichs, peritrichs and suctorians. We shall look at each of these groups.

In the oligotrichous ciliates, the fission pattern is termed enantiotropic (Fauré-Fremiet 1953). This refers to the fact that the polarity of the opisthe is the reverse of the proter's. In that the fission products are not mirror images of each other, the fission pattern is not symmetrigenic as in the zooflagellates. Following Fauré-Fremiet's interpretation, it is reasonable to view this enantiotropic mode simply as a variation of homothetogenic fission, for in all respects except the reversed polarity, fission is of the typical ciliate type. The functional basis for the changed polarity is not known.

In the peritrichs, which differ from the other ciliates in respect to their mode of conjugation too, fission at first appears to be symmetrigenic. However, such an interpretation depends on what are the ventral and dorsal surfaces of the cell. This question will receive more discussion later. For the present we will cite Fauré-Fremiet's (1950a) interpretation, which is that the stalk arises from the dorsal side of the cell, and the adoral field is really the ventral surface. Thus the fission plane divides the ventral and dorsal surfaces in two, as in other ciliates, though it does so by passing from the ventral to the dorsal side, rather than constricting equally on all sides. Furthermore, it must be noted that there are no intermediate stages between the typical peritrichous type of fission and that of other ciliates. Hence, the differences are not bridged by intermediate forms. To at least partially explain the absence of these forms, it is necessary to realize there are no forms that lie part way between the peritrichs and their probable ancestors, the thigmotrichs (Fauré-Fremiet 1950a). What this amounts to is that a cell must be a good thigmotrich or a good peritrich, something in between has little chance of competing with the specialized forms. And, therefore, intermediate forms are absent, which includes intermediate modes of fission.

Then there is budding, found in the suctorians (Fig. 10–8A) and in certain other ciliates, i.e. the chonotrichs and the rhynchodine thigmotrichs (Fig. 10–8B). There are reasons for suspecting the latter are highly evolved, symbiotic gymnostomous forms (Puytorac 1968b; Puytorac and Njiné 1970 and other references cited there) derived from an ancestor with a nassa. The same is true of the chonotrichs and the suctorians. In brief, there seem to have been three separate origins of these forms which show budding. Therefore, the budding process is not homologous but is a parallel evolutionary development in each group. This will be discussed again later.

Regarding stomatogenesis, we earlier identified four major patterns: parafissional and prefissional opisthe stomatogenesis, and prefissional and postfissional proter and opisthe stomatogenesis. The problem is to homologize these four patterns, and it can be done by viewing them as relating to each other (Table 10–5). *Dileptus* exemplifies the first type (Golinska and Doroszewski 1964), and *Prorodon* (Puytorac and Savoie 1968) illustrates the second mode, as do cyrtophorine gymnostomes (Deroux and Dragesco 1968; Kaczanowska and Kowalska 1969). There seems to be no good example of a bridge between these two patterns, but the fact that both are found in the same suborder, i.e., the rhabdophorine gymnostomes, suggests that the two are not highly dissimilar. *Pseudomicrothorax* is a possible transitional form between the second and third types, for here the nassa behaves as if it were a gymnostome structure with type 2 stomatogenesis, but the ciliature beside the cytostome remains intact in the proter and is newly formed for the opisthe, as in type 3 stomatogenesis. In fact, in *Chlamydodon mnemosyne*, Fauré-Fremiet (1950b) has found that the adoral ciliature for the opisthe is derived from three ventral kineties that extend posteriorly from the proter's cytostome, while in the proter the peristomial ciliature remains intact. This clearly anticipates the pattern in *Pseudomicrothorax dubius* and other type 3 forms. It should also be noted that structures that are in essence poststomial dedifferentiate in both type 2 and 3 (food vacuole formation poststomial fibers are lost in type 3 forms, the nassa is lost in type 2 species); whereas prostomial structures and the prestomial ciliature persist in proters and are formed anew for opisthes in both types of stomatogenesis.

And finally, it is the ophryoglenids that provide a transition from type 3 to

Table 10–5 Relations between four major types of ciliate stomatogenesis and genera exemplifying typical and transitional features.

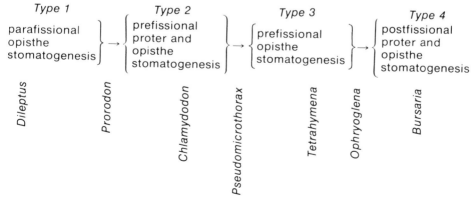

type 4. These forms show a very complex mode of formation of new feeding parts, involving especially the migration of a stomatogenic kinety to a position where it seems to determine the proliferation of new kinety fields for both proter and opisthe. These fields complete their morphogenesis after separation of the fission products (Roque, Puytorac, and Savoie 1965; Roque and Puytorac 1967; Roque, Puytorac, and Lom 1967). (Other examples of type 4 stomatogenesis are not common; we can think of only two, namely *Neobursaridium gigas* (Dragesco and Tuffrau 1967) and *Bursaria truncatella* (Lund 1917).) The reason for proposing the ophryoglenids as intermediate forms is because their stomatogenic kinety is also a paroral kinety, like that in the peniculine hymenostomes such as *Paramecium*. In both groups, proliferation of new kinetosomes for the new stomial fields occur beside this kinety. However, in *Paramecium* there is no migration of the kinety as in the ophryoglenids. And it should also be stated that the behavior of the paroral kinety is not known for certain in *Neobursaridium* or *Bursaria*. These large cells, with their massive dedifferentiation of the preexisting oral structures, do not readily stain with the silver technique. Perhaps Tuffrau's (1967b) protargol technique is needed here. He has certainly demonstrated astonishing details in the prostomial parts of nondividing *Bursaria* (Tuffrau 1967a).

Next, there is the complex process of conjugation. Let us look first at the prezygotic events. They are homologous in all known cases, with the possible exception of the peritrichous forms. The peritrichs are exceptional, in that they have different sized conjugants; the microconjugant fuses permanently to the macroconjugant, and only one gamete nucleus is formed from each conjugant (Finley 1952, 1969). It is obviously possible to see all three differences from normal conjugation as modifications of the normal process. Indeed, no one has suggested that peritrich conjugation is other than a somewhat modified version of the typical situation. However, there are no intermediates between the typical and modified version, e.g., forms with micro- and macroconjugants but without subsequent fusion and producing two, rather than one, gamete nucleus; and we have no apparent reasons, in selectionist terms, for the pattern of sexuality in these peritrichs. Despite these drawbacks, the general similarity to normal conjugation—union of two individuals, prezygotic reduction, division, and fertilization—all of these speak to the homologizing of events in conjugation of peritrichs and other ciliates. One hypotrichous form, *Urostyla hologama* (Heckmann 1965) also shows conjugant fusion, but this form lacks other characters that would relate it to peritrichous forms and provide a bridge to the other ciliates. Conjugant fusion in this species of *Urostyla* must be an independent, parallel appearance of this mode of conjugation.

There are certain other variants on the so-called normal prezygotic events of conjugation. Certain suctorians show anisogamous conjugants, fusion of conjugants, and formation of one gamete nucleus from each cell (Grell 1953a, 1967). But the suctorians also provide examples of more normal conjugation.

In them one does find a series of situations connecting the atypical to the typical sequence of events. One other variant is found in *Trachelocerca phoenicopterus,* where several gametic nuclei are formed, followed by multiple exchange of pronuclei and multiple synkaryon formation (Raikov 1958). This situation does not differ so much from the normal picture as to indicate a lack of homology.

The postzygotic events show much the same situation as the prezygotic events; that is, there is a very good general similarity in postzygotic mitoses, separation of conjugants, redifferentiation of dedifferentiated cytoplasmic organelles, differentiation with polyploidization of the macronucleus, and segregation of micro- and macronuclei to fission products to restore the normal nuclear complement. There are variations in the number of postzygotic fissions and in the fate of the nuclear products. Wichterman (1957) reports variations even with the one genus *Paramecium.* Ammerman (1965) has found an extraordinary loss of DNA in the period of macronuclear differentiation, and Reiff (1968) describes degeneration of certain nuclei in exconjugants of *Uronychia transfuga.*

The various other sexual processes—cytogamy and autogamy—seem to have a sporadic distribution among the Ciliophora. They can be viewed as modifications on conjugation evolved in response to selection pressures on the original outbreeding pattern to make it an inbreeding situation (Sonneborn 1957). Eventually sufficient information will be available regarding them to allow comparative remarks. They surely represent homologous events in the different ciliate species.

Another function that can be homologized among the ciliates is their phagotrophy. All of them, except certain symbiotic forms that are thought to have lost their ingestatory organelles, take up food in food vacuoles, commence digestion within the vacuoles, and release undigested materials through the cytoproct. The details of digestion and egestion are not so well known as to provide useful details for this study. But ingestion and something of the diversity of the foods utilized, i.e., bacterial feeders, carnivores, herbivores, symbionts of various types, are well enough known so as to permit comparisons in those terms.

And finally—the last seme—there is, of course, a good deal of information on the habitats and niches within the habitats that are exploited by the ciliates. We are thinking here of ecological features other than nutritional ones, which have just been mentioned. But even though the temperature, pH, salinity, oxygen tension, and other requirements can be spelled out in general terms (Fauré-Fremiet 1967a; Noland and Gojdics 1967), it is not possible, in many cases, to say much more than that certain forms are found in fresh water and others in salt water; or perhaps that some are found in stagnant waters and others in brackish water. Such broad descriptions are difficult, at best, to translate into the data items needed for point-to-point comparisons, and, necessarily, the discussion of homologies in terms of the niche preferences of ciliates is still an open question. Data collected with phylogenetic analyses in mind could be a rich source of information, but such analyses seem premature at the present.

In summary, the following are semes showing homologous relationships: ciliary units, aboral kinety patterns, feeding structures (including adoral ciliation), general body organization, macronuclei, micronuclei, fission, stomatogenesis, prezygotic events of conjugation, postzygotic events of conjugation, and feeding patterns. Many of these show considerable aposemic development, especially the general organization of the cell body, aboral kinety patterns, feeding structures, macronucleus, and stomatogenesis. Additionally, there are other features which are not homologizable but represent important evolutionary events, such as the occurrence of budding, which is a neosemic trait in each of three situations where it occurs. Similarly there are many special features of these cells, their stalks, loricas and vases, special ectoplasmic structures, such as myonemes and platelets and pigments, and special endoplasmic inclusions, which represent a great variety of apparently neosemic innovations; all of these can be used for defining phyletic relations among the various ciliate species.

To conclude this discussion of homologies, we can contrast our homologies within the ciliates with those of other workers who have addressed themselves to defining the salient or "fundamental" characters of ciliates. Fauré-Fremiet says:

> "Let us recall that the ciliates are characterized:
> (1) by the duality of their nuclear apparatus consisting of a micronucleus, which divides mitotically and undergoes, in the course of the sexual cycle, reduction divisions; and a megaploid macronucleus dividing amitotically, and reforming after degeneration (endomixis or conjugation), from a micronucleus, by an endomitotic process. . . .
> (2) by a ciliary apparatus, whose infraciliature includes, according to Chatton and Lwoff (1935), an assemblage of kineties [which are] independent and polarized (rule of desmodexy), and divided by transverse division during the course of fission." (Fauré-Fremiet, 1950a, p. 111–112, Original in French.)

Corliss (1961, p. 23–24) develops a longer list of six characters, which, though separately are "not entirely unique, . . . *in combination* are definitely distinctive of ciliates." In condensed form, this list contains the following points: (1) possession of cilia in at least one stage of the life cycle; (2) presence, without exception, of some sort of infraciliature, which "consists essentially of the basal granules or kinetosomes associated with cilia and ciliary organelles at their bases plus certain more or less interconnecting fibrils, the kinetodesmata . . ."; (3) two kinds of nuclei; (4) homothetogenic fission; (5) presence of sexual phenomena; (6) a functional mouth. Of these six traits, greatest emphasis is placed on the second one, the infraciliature. Corliss says, "The universality, stability, and persistence of this complex throughout the entire life cycle of all ciliates ('once a ciliate, always an infraciliature.' Corliss 1959) strongly suggests that it is a conservative, fundamental property of these protozoa."

And a similar listing of traits is to be found in the revision of protozoan systematics published by Honigberg *et al.* (1964).

The reason for presenting these listings of traits is to point out (a) that there is general agreement between characters deemed especially important by authorities on the ciliates and the conclusions derived from use of Remanian criteria for homologies, and (b) that of the characters most emphasized by these authorities, the infraciliature is predominant. If now we ask the obvious question: What is the basis for choosing these characters, since the Remanian criteria were not used, we come to something of an impasse. Fauré-Fremiet (1950a) and Corliss (1961), and Furgason (1940) before them, all speak of homologies and homologous structures, but they nowhere define their terms nor refer to a source which provides a definition. Corliss seems to prefer the notion of "fundamental character" as the basis for establishing the importance of the ciliature. His definition of this term seems to be that given in the quotation cited above—a trait that is universal, stable, and persists in all phases of the life cycle, also it is conservative. But such an approach also includes the cell membrane, mitochondria, endoplasmic reticulum, etc. The phrase "fundamental character" has no rigorously defined connections with any theoretical concept, and certainly it is not a logical consequence of evolutionary theory, which it must be if fundamental characters are to be used in phylogenetic inquiry. But yet this infraciliature is the trait, *par excellence*, for Fauré-Fremiet and Corliss in dealing with ciliate phylogeny. Hence, the impasse is that, though these workers have no clear exposition of their phylogenetic theory, they have worked with a trait that is homologous throughout the ciliates and, furthermore, it shows a good deal of aposemic variation. Therefore, the question is whether or not they have been able to read the aposemic trends correctly, in the absence of a critical understanding of the homologous nature and relationships of the infraciliature. As will be seen, the answer is both yes and no.

Plesiomorph and trends. In the absence of fossil or distributional evidence, only ontogenetic data and analysis of morphoclines allows determination of the direction of evolutionary trends. Such directionality must be determined to identify plesiomorphic forms. Let us start with ciliate ontogeny.

In that conjugation is apparently the basic sexual process, and in that fertilization, the common starting point for ontogenic development, does occur in conjugation, the cell with a synkaryon provides a glimpse of the primitive ciliate condition (Lwoff 1926). How informative is this glimpse? The fact that the oral structures, most notably, are dedifferentiated may tell us that at the time sexual processes had evolved in the ancestral form, functional oral structures were nonexistent. Then again, there may be a functional rather than an ontogenetic explanation for the loss of cytoplasmic organelles during conjugation: it may be that while the old macronucleus is degenerating, and before a new one if fully active, there are no gene products which will maintain the old feeding organelles. However, this is not altogether likely, for the new structures reappear before the new macronucleus is fully reformed, and reformation is not the restoration of lost parts but formation of a complete new gullet (Roque 1956a, 1956b; Porter 1960; Vivier and André 1961; Hanson, Gillies, and Kaneda 1969). This occurrence of a complete developmental process is quite consistent with an ontogenetic interpretation of the developmental events of conjugation. From this, three conclusions of a recapitulative sort suggest themselves: (a) the ancestral form of the ciliates lacked a feeding organelle, (b) the feeding structures so characteristic of ciliates evolved before their macronucleus did, and (c) the macronucleus arose from the micronucleus. And, further, regarding this last point, the ontogenetic data would argue that macronuclei came from the differentiation of mitotic products of diploid nuclei, and then these products became polyploid by endomitosis. Again, this agrees with other evidence and is in fact the actual evolutionary process proposed by Raikov (1969 and earlier). Whether the development of feeding structures at conjugation recapitulates the whole long history of their development remains to be seen. At present it seems to be untrue. For evidence at hand, both descriptive (Roque 1956a, 1956b) and experimental (Hanson, Gillies, and Kaneda 1969), on *Paramecium aurelia* says that prostomium development at conjugation is very like that at fission. This is a general rule (Tartar 1967; Hanson 1967a). But at fission we can distinguish at least four modes of stomatogenesis in the ciliate, and if they did evolve one from the other, as has already been suggested as being the case, there is no evidence of this in the ontogeny of the feeding apparatus at conjugation. Ontogenetic implications seem limited to the history of a given type of stomatogenesis. For example, in prefissional opisthe stomatogenesis (found in most ciliates), the new structures appear first as a field of new kinetosomes; then kineties differentiate, the prostomial cavity depresses into the cell and finally the cytostome and other structures appear. This would suggest that the cytostome arose, historically, in an area of special kinetosomal proliferation. There is, in this type of stomatogenesis, no suggestion of the nassa that accompanies the cytostome of the prefissional proter and opisthe stomatogenic type that, we will argue, precedes the appearance of a prostomium. (This is, in fact, another sound argument for arguing that the largely prostomial nassa is not homologous to the prostomium.) In conclusion, ontogenetic reasoning suggests long-range trends in macronuclear development, but in stomatogenesis the trends appear limited to a given type of stomatogenesis.

The other approach to analyzing trends is to look for the conjunction of the ends of morphoclines. One important morphocline, as we have already seen, consists of the various states of the macronucleus. One end of this series are the atypical homokaryotic forms of *Stephanopogon*, which lack macronuclear structures, and the other end is the typical condition of one large, polyploid macronucleus, capable of dividing synchronously with the cell.

Are there other trends with termini also lying in the homokaryotic cell? There may be three, but opinions will vary on how visible they are. Viewed from a vantage point that surveys the ciliates as a whole, the oral structures of *S. mesnili* appear relatively simple: the slitlike cytostome is at the cell surface (*S. mesnili* is classified as a rhabdophorine gymnostome, Corliss 1961); there is little evidence of special ciliature, though a paroral membranelle is claimed to be present, and some trichites or trichocysts are also reported (Lwoff 1936). In comparison, other ciliates are generally believed to have more complex, more highly evolved oral parts (except for secondary reduction as in the astomous, apparently parasitic forms) (Fauré-Fremiet 1950a; Puytorac, 1959a; Corliss 1959, 1961; Cheissin and Poljansky 1963; Raabe 1964a). However, a harder look at the oral parts of *S. mesnili* indicates what must be the obvious case, namely, the feeding structures of *S. mesnili* are specialized for the ingestion of diatoms, cryptomonads, and amebas, which this form feeds on as it moves over the substratum (Lwoff 1926). If there are trends, it is not so much from simpler to more complex oral structures, but from one kind of complexity to another.

The gymnostomous forms are a good example of the problems that face anyone looking for so-called primitive organs among living forms. Some traits will remain conservative and show little evolution; others will evolve rapidly. Especially, there will be heavy selection on traits relating to exploitation of the environment, for here there is always severe competition, and only by specialization can one form win out over others and minimize competition among remaining forms by finding a unique way of life. This has been put clearly by Puytorac and Njiné (1970, p. 443).

"But it seems to us, above all, necessary to view many of the ciliate species termed 'primitive' [and] belonging to gymnostomes and trichostomes of Fauré-Fremiet and Corliss, with extremely complex specializations of certain characters, not seen in other characters, as leading to the [evolutionary] development of numerous diverging lines, narrowly adapted to certain ecological conditions, but blocked thereby in their evolution and terminating in 'dead ends'." (Original in French.)

There have already been pointed out transitions from one kind of feeding structure to another—mouths to tentacles—in certain forms and also the transition from one kind of mouth to another—nassa to ciliated prostomial cavity. Within the gymnostomes, it may be possible to go from the slitlike mouths of Kahl's (1930–1935) pleurostomes to the round terminal mouths of his prostomes (both groups now included in the rhabdophorines) and to the round, ventral, nassa-type mouths of the cyrtophorine gymnostomes (Figs. 10–4, 10–12). However, others would probably propose starting with round apical mouths and branching to slit-like mouths (apically or laterally placed) and to round ventral mouths. There is no definitive treatment of the problem of evolutionary trends with the gymnostomous ciliates. The best we can say is that the slitlike mouth is obviously different from the rounded ones and may be joined to them by certain intermediate forms, thus a sort of trend is suggested. Whether *S. mesnili* sits at the extreme end of that morphocline remains to be seen from further work.

Another possible trend is seen in stomatogenesis (Table 10–5). Unfortunately, details of fission in *S. mesnili* are not published. If it shows parafissional opisthe stomatogenesis, then it would lie at one end of this spectrum of stomatogenic types. Apart from *Stephanopogon*, other ciliates do fall into one or another of the four categories, though with the following qualifications. (a) There seem to be few representative of the extreme categories. (b) Most of the gymnostomes and trichostomes and certain of the hymenostomes, i.e., the pleuronematous forms and certain thigmotrichs (or Small's scuticociliates) show prefissional proter and opisthe stomatogenesis. In the scuticociliates this may be an independent appearance of this type of stomatogenesis due to the posterior location of the cytostome. That is, prefissional opisthe stomatogenesis may have arisen from prefissional proter and opisthe stomatogenesis in the hymenostomes descended from a *Pseudomicrothorax* and their tetrahymenine ancestor, and then there may have been a reappearance of the older patterns when the posterior cytostome, which normally is retained by the proter in hymenstomes, moved to a posterior position. Its dedifferentiation followed by redifferentiation of two new fields, one for each fission product, would resolve the morphogenetic problem arising from the posterior location. (c) The rest of the hymenostomes (except for the ophryoglenids and *Neobursaridium*), the heterotrichs (except for *Bursaria*), and the peritrichs show prefissional opisthe stomatogenesis. (d) Budding arose independently in three different groups (chonotrichs, suctorians, and certain thigmotrichs). The distribution of these types of stomatogenesis is seen more clearly by reference to the phylogenetic tree given in Fig. 10–19.

The third and final trend pertains to the pattern of the surface ciliature. Again, the broad view suggests changes going from the relatively uniform ciliature of many gymnostomes to the more differentiated patterns of the peritrichs and spirotrichs. But what does a careful look at the gymnostomes themselves tell us? The cyrtophorine forms have quite complex ciliature (Chatton *et al* 1931; Kaczanowska and Kowalska 1969), certain of the rhabdophorines— the radially symmetrical ones—are uniformly ciliated, and other rhabdophorines, including *S. mesnili*, show ciliation on only one surface. In forms with only one surface ciliated, this seems an obvious adaptation to a bottom-crawling mode of existence (Puytorac and Njiné 1970). In any case, *S. mesnili* appears to have a somewhat specialized pattern. It is, therefore, very difficult at present to locate *S. mesnili* with respect to any specific trend in surface ciliary patterns.

In summary, *S. mesnili* is somewhere near the terminus of morphoclines of oral structures and surface ciliature when a broad view of ciliate diversity is taken, but the picture becomes much less clear when more precise details are looked for. There is, however, no denying its position relative to its nuclear apparatus. But because of the equivocal nature of our conclusions it is well to ask for other candidates for the ciliate plesiomorph and to see if a better case can be made for them.

Hanson (1958) proposed that *Remanella multinucleata*, a gymnostome with a slitlike mouth, be a possible representative of the primitive ciliate. His arguments were also largely based on the nuclear picture: *R. multinucleata* has many diploid micronuclei which give rise directly to the many small, diploid macronuclei (Fauré-Fremiet 1954a; Raikov 1963b). Unfortunately, at that time the author was unaware of Chatton's work on *Stephanopogon*, for it is the natural extension of the macronuclear picture that was being studied, as was subsequently made clear (Hanson 1963a). Hence, the original proposal of *R. multinucleata* was changed to *S. mesnili*, which simply brings us back to our current problem.

The only other candidate for the ciliate plesiomorph is *Holophrya* (Fauré-Fremiet 1950a; Corliss 1961) or *Prorodon* (Corliss 1959). *Holophrya* holds, however, the favored position today with these workers. "Such a genus as *Holophrya* ... seems a perfect ciliate 'prototype'; its simplicity in somatic and oral structures, its straightforward mode of perkinetal fission, its simple axis of symmetry, its generally unspecialized mode of life support the allegation." (Corliss 1961, p. 96). This citation illustrates the implicit assumption made by these workers regarding their basic phylogenetic working hypothesis: evolution proceeds from simple to complex forms. This explains their constant emphasis on "lower" and "higher" ciliates, an emphasis whose objective validity has always been debatable (Dobell 1911; Canella 1964; Finley 1969); and it explains why relative simplicity is used as the essential criterion in nominating *Holophrya* as the ciliate "prototype."

Though we can find no unexceptional justification for this approach from the phylogenetic theory being espoused here, it is nonetheless necessary to examine *Holophrya* to see if the right candidate has been put forward, even though we disagree with the reasons for the nomination. Probably the best reason for seriously considering *Holophrya* (neither Fauré-Fremiet nor Corliss propose a specific species; Kahl (1930–1935) lists 22 holophryid species) is its uniform surface ciliature. But is this really a more primitive condition than that found in *S. mesnili?* Or, more clearly expressed, perhaps uniform surface ciliature is just as highly adaptive to *Holophrya* as lack of ciliation on one surface is to *S. mesnili*. Holophryid species, according to Kahl (1930–1935) feed on red, green, and blue algae, also other small protists, infusorians, and flagellates, including *Synura* and dinoflagellates. Feeding occurs when the organisms, swimming freely through their aqueous environment, make contact with their food. For such swimming, ciliation over the whole surface is an advantage. For capture of prey, it is obviously advantageous to have the mouth at the apical end where food is first contacted, though in other ciliates other locations are also effective. (Again the comments, quoted above, of Puytorac and Njiné (1970) are pertinent in understanding the varieties of form in gymnostomes). But, in short, it is not possible to say that the organization of *Holophrya* is due to inheritance of an ancestral pattern rather than representing a form adapted to its mode of life. The so-called simplicity of this genus, including its radial symmetry, could well be the most efficient form for these ciliated cells and simplicity would, therefore, be adaptive rather than primitive. Of course, those favoring its role as prototype have done more than merely look at *Holophrya* itself, they have compared it to other ciliates and found it relatively simple. Except, however, for the macronucleus. In all the *Holophrya* listed by Kahl (1930–1935) there is always one macronucleus, larger than the micronuclei, and spherical, ovoid, or kidneyshaped in form. Analysis of these nuclei, their origin and ploidy, in the terms Raikov has carried out so effectively on other gymnostomes, is needed here. But so far there is no evidence of a homokaryote form or of forms showing diploid macronuclei which differentiate directly from micronuclei during fission. The primitive macronuclear picture seems to be absent from the genus *Holophrya*.

On present evidence, *Stephanopogon mesnili* seems the best example known today of the ciliate plesiomorph. However, endorsement of this candidate would be even more convincing if there were some evidence of trends in conjugational or divisional processes and to find that one extreme of these trends also occurs in *S. mesnili*. Neither conjugation nor fission has been described in detail for this species. Lwoff (1926) reports seeing what he believes were conjugating pairs, but details of the whole process were not obtained. Homothetogenic fission was observed many times, but the promised detailed account of cortical events has never appeared. Details of the nuclear picture have been published, and they provide one interesting point. Division is promitotic; that is, there is an intranuclear "caryosome" (nucleolus?) which persists through division. In other ciliates only mesomitotic divisions are known, i.e., there is no karyosome, or, if present, it disappears at prophase. Promitosis is known from ameboid and zooflagellate forms, but the same is true for mesomitosis (Grassé 1952c; Deflandre 1953a); hence, the significance of the occurrence of promitosis in *S. mesnili* is not really clear.

Before leaving the subject of the plesiomorph, two sets of final comments, stemming from Raikov's discussion of this problem, must be presented. First, Raikov presents four lines of evidence, apart from the nuclear picture, which he feels also argue for the primitive nature of the various ciliates with minimally differentiated nuclei. "(1) They exist only in the taxonomically lower representatives of the psammophilic fauna; higher organized interstitial forms (Heterotrichida, Hypotrichida) always have typical polyploid macronuclei. (2) Diploid macronuclei are met among Gymnostomatida as well as among the lower Trichostomatida (genus *Geleia*); thus they are not limited to any specialized taxonomic group. (3) Generally, the interstitial sand fauna retains many primitive forms (for example, turbellarians or crustaceans). (4) The macronuclei of diploid character are not an exclusive privilege of the psammophilic fauna—they are also met in typical freshwater pond species of the genus *Loxodes*." (Raikov 1963a, p. 255.) These arguments are of varying validity. The first one depends on the validity of "lower" in reference to taxonomic systems. If it refers solely to visible complexity, then this somewhat subjective judgment is further complicated by whether or not relative simplicity has any usable phylogenetic meaning. It does describe a general trend in all of evolution, undoubtedly, but the debates are endless when it comes to be applied to any given group, as we have already pointed out (Dobell 1911; Canella 1964). Next, Corliss (1961) classifies *Geleia* as a gymnostome rather than a trichostome, hence the broader distribution of these nuclei implied by Raikov is in doubt. Nevertheless his point here, and combined with point 4, is very important: there is no precise correlation of this special nuclear condition with the niches of the forms showing that condition. Such ciliates, though largely sand dwelling (Dragesco (1960) reports 60 species of this type), nevertheless show a great variety of body forms and considerable diversity of food, ranging from algae through protozoa to some bacterial feeders and including at least one freshwater genus. It is, as Raikov implies, quite improbable that similar selection pressures account for the nuclear picture. For this hypothesis of convergence to be true, it would necessitate an extraordinary case of multiple regressive evolution—not in exploitive characters, but solely with regard to the genetic apparatus. There is no basis for supposing reduction works in this manner (see Chapter 3). Raikov's third point is so vague as to be largely meaningless; also he cites, in point 1, important exceptions to this proposed generalization. (It will, however, be of interest to recall in Chapter 13 that certain turbellarians are found in the same habitat as the primitive ciliates.) In sum, these arguments regarding what is primitive in ciliates suffer from the weaknesses inherent in all such attempts which do not derive from selectionist theory. They are inductive generalizations which seize on patterns having little or no relation to evolutionary theory; hence their reference to "lower" forms (point 1), to arbitrary taxonomic authority (point 2), to other inductive generalizations (point 3), and the use of meaningless phrases such as "not an exclusive privilege of psammophilic fauna," all contribute to vitiating the point they wish to make. A point which may be valid but not for the reasons given.

The second and final, general set of comments pertaining to the plesiomorph is Raikov's (1963a, p. 254) important observation that the presence of homokaryote forms ... "provides a proof, that the nuclear dualism arose within the class Ciliata, i.e., later than the ciliates themselves did." Here is evidence in favor of the second recapitulative conclusion, that oral structures evolved before macronuclei did. And as Lwoff (1926) points out, the ciliature and nuclear apparatus evolve at different rates and in different ways in the Ciliophora. This was seen in our earlier discussion when we pointed out the diversity of macronuclear form within a genus, or how the same macronuclear form occurred in various genera. The same is true for various genera with species showing minimally developed macronuclei; they also contain species with well-developed nuclear components, e.g., *Trachelocerca*, *Remanella*, and *Loxodes*.

From this, it is clear that the pressures to evolve the typical macronuclear condition have not been so all-pervasive as have the pressures on the cortical structures. The nucleus has evolved in a more leisurely fashion, preserving various examples of its evolution, even the extreme one of its primordial absence. The cortex, which is directly concerned with the exploitive attributes of the cell, has been pushed much more rapidly in developing its possibilities. This emphasizes further the futility of looking for truly primitive conditions of the surface ciliature and the mouth parts. Such conditions have probably long disappeared. The problem of cortical evolution in known ciliates is one of diversification of already highly-adapted patterns of organization. (See also, Canella 1964.) The overall situation is a beautiful example of evolutionary mosaicism—a more conservative genetic apparatus, serving the reproductive and homeostatic needs of the cell, is encased in a highly specialized cortex whose response to the locomotory and feeding needs of these animal cells shows the results of much diverse innovation. That is one summary of the phylogenetic puzzle posed by the ciliates.

PHYLOGENETIC ANALYSIS

A phylogenetic study, it is to be remembered, has a twofold aim: to propose a set of historical-genetic relationships and to elucidate the basis for those relationships in evolutionary (selectionist) terms. The first goal has been attempted by various workers, most recently by Fauré-Fremiet (1950a), Corliss (1961, 1974a, 1974b, 1974c, 1974d), and Jankowski (1967, 1973), for the ciliates as a whole, and there have of course also been various special proposals for the different subgroups within the ciliates, many of which shall be referred to shortly. However, no worker has attempted the second, the explanatory part of the phylogenetic program. It is not a natural outgrowth of most methodologies: if one views evolution of the ciliates as proceeding generally from simpler to more complex forms, what more can be said after arranging the ciliates in a manner conforming to that sort of progression? A commitment, however, to phylogeny as an analysis of changes in homologous relations where species are the unit of comparison, naturally raises questions as to what forces account for the differences in given characters from one species to another.

Therefore, in what will now be discussed, we shall first go to the problem of elucidating the course of ciliate evolutionary history and then turn to the problem of the origins of the diversity within the group. The analysis of phyletic relations will be based on the orders and suborders recognized by Corliss (1961) and Honigberg et al. (1964), except for Small's (1968) recent revision. For almost each of these taxa we will propose a specific plesiomorph and then carry out the estimation of phyletic distances between them so as to construct a dendrogram for this sampling of ciliate diversity. The plesiomorphs will not, for the most part, coincide with the representative or typical genera Corliss provides for each order and suborder. The reason is obvious: the form presumed to represent the ancestral type of a given subgroup of ciliates is chosen by criteria other than those appropriate to recognizing a "typical" genus. Also, and unfortunately, there is no single source of data on the plesiomorphic species such as Schewiakoff provided so admirably for the Acantharia. Hence, data describing the various species is culled from many sources and is not altogether complete. Nonetheless, a phylogenetic analysis is made here to serve at least as an incitement to further work, and concrete proposals are the best way to see the limitations of a given analytical procedure.

Phyletic relationships. In Table 10-2 are given the species suggested as plesiomorphs for the various orders and suborders of the ciliates, plus a summary of the semic traits used to describe them and which serve also as the basis for determining their phyletic relationships.

First, some explanation must be given to make explicit the reasons for selecting the various plesiomorphic types. The methodology is that used for finding plesiomorphs in general, namely, use of recapitulative or trend data. *Stephanopogon mesnili* (Fig. 10-14A), the ciliate plesiomorph, will also serve as the plesiomorph of the first suborder we shall consider, the Rhabdophorina, in the order Gymnostomatida. The reasons for this choice need not be repeated; they are the same, of course, as given earlier. The other currently recognized gymnostome suborder is the Cyrtophorina. Here *Nassula ornata* (Fig. 10-14B), a form with quite uniform surface ciliature, is selected as the plesiomorph. This form suggests a position between the rhabdophorines and the cyrtophorines with a more complex ventral ciliature. Corliss (1961, p. 38) considers *Nassula* as . . . "probably at the base of the phylogenetic tree within the group [of the cyrtophorines] as it is known from present-day forms."

Heliochona scheuteni (Fig. 10-14C) is suggested as the plesiomorph of the Chonotrichida on the basis of analyses that show its development to contain stages reminiscent of those in cyrtophorine gymnostomes (Guilcher 1951;

Table 10-2 Semic traits of sixteen species of ciliates which are suggested as possible plesiomorphs of the various orders and suborders of the ciliates, except for the Astomatida.

A. *Stephanopogon mesnili* (Order Gymnostomatida, suborder Rhabdophorina) (Lwoff 1923, 1926) (Fig. 10-14A)
   1. Longitudinal ciliary rows on lower cell surface
   2. Terminal, slit-like cytostome with trichites
   3. Flattened ovoid cell with short anterior neck below cytostome
   4. Micronuclei usually two per cell, can be as many as 16
   5. No macronuclei
   6. Homothetogenic fission
   7. Stomatogenesis not described
   8. Conjugation probable; not observed in detail
   9. Postconjugational events not observed in detail
   10. Bottom feeder; food consists of diatoms, cryptomonads, and amebas
   11. No budding
   12. No special structures

B. *Nassula ornata* (Order Gymnostomatida, suborder Cyrtophorina) (Kahl 1930-1935) (Fig. 10-14B)
   1. Ventral surface with parallel kineties which curve anteriorly around the cytostome
   2. Cytostome at the surface surrounded by rods forming a nassa
   3. Flattened ovoid cell with slight point at left anterior margin; cytostome at anterior mid-ventral location
   4. Micronuclear condition not clear
   5. Single, large, round macronucleus; heteromeric
   6. Homothetogenic fission
   7. Prefission proter and opisthe stomatogenesis
   8. Conjugational events normal
   9. Postconjugational events normal
   10. Bottom-dwelling form, feeds on filamentous algae
   11. No budding
   12. No special structures

C. *Heliochona scheuteni* (Order Chonotrichida) (Guilcher 1951; Dobrzanska-Kaczanowska 1963) (Fig. 10-14C)
   1. No surface ciliature
   2. Oral area a complex, fan-shaped funnel with unique ciliation
   3. Vase-shaped cell with complex, fanlike oral area at one end
   4. Three micronuclei
   5. One ovoid, heteromeric macronucleus
   6. A highly modified homothetogenic fission, i.e., a kind of budding
   7. Stomatogenesis not classifiable into four categories being used here
   8. Conjugation is not reported but assumed to occur, since it is known in related forms.
   9. Postconjugational events not reported but probably of conventional sort except for stomatogenesis
   10. Marine form, commensal on crustaceans (gammarids)
   11. Budding occurs; ciliation of bud reminiscent of cyrtophorine pattern, e.g., *Allosphaerium paraconvexa*. (See left-hand portion of Fig. 10-14C).
   12. Adhesive disc present for attachment

D. *Coelosomides marina* (Order Trichostomatida) (Anigstein 1912; Kahl 1930-1935) (Fig. 10-14D)
   1. Surface ciliature in regular meridional rows

Table 10-2 *(cont.)*

D. *Coelosomides marina*
2. Cytostome at base of apically located, ciliated, peristomial depression
3. Ovoid cell-body, slightly flattened at anterior end.
4. One small micronucleus
5. Large, ovoid macronucleus located eccentrically in mid-region of cell
6. Homothetogenic fission
7. Stomatogenesis not observed but probably parafissional opisthe stomatogenesis as seen in related forms
8. Conjugation not reported in this species but assumed to occur
9. Postconjugational events not reported but assumed to follow basic features seen in other forms
10. Marine form feeding on filamentous blue-green algae
11. No budding
12. No special structures

E. *Pseudomicrothorax dubius* (Order Hymenostomatida, suborder Tetrahymenina) (Corliss 1958a, 1958b, 1961; Thompson and Corliss 1958) (Fig. 10-14E)
1. Uniform meridional kineties
2. Cytostome is an anteriorly placed nassa with four groups of membranes lying on the nearby cell surface, which are homologizable to the typical tetrahymenine prostomial ciliation
3. Ovoid cell body, which is flattened dorso-ventrally with dorsal surface slightly convex and ventral slightly concave
4. Single micronucleus
5. Single, centrally placed macronucleus
6. Homothetogenic fission
7. Stomatogenesis of the prefissional type in both proter and opisthe including formation of new oral membranelles
8. Conjugation not reported in this species but common and regular in other members of the suborder
9. Postconjugational events probably follow conventional pattern
10. Freshwater form feeding on algae and possibly bacteria
11. No budding
12. No special structures

F. *Frontonia marina* (Order Hymenostomatida, suborder Peniculina) (Kahl 1930-1935) (Fig. 10-14F)
1. Surface ciliation in regular meridional rows meeting along a conspicuous suture line
2. Cytostome at base of a prostomium containing tetrahymenine membranelles plus three polykineties on the right
3. Ovoid cell with slightly broadened and flattened anterior end with feeding structures on anterior mid-ventral side
4. Several micronuclei
5. One large, ovoid macronucleus
6. Homothetogenic fission
7. Stomatogenesis apparently a type of proter and opisthe prefission stomatogenesis; new feeding parts from field of kinetosomes appearing near polykineties on right of preexisting prostomial opening
8. Conjugation not reported but observed in related species
9. Postconjugational events assumed to follow basic pattern
10. Found in stagnant salt water; feeds on blue-green algae, diatoms, occasional rotifers
11. No budding
12. No special structures

Table 10-2 (cont.)

G. *Philaster digitiformis* (Order Scuticociliatida) (Mugard 1949; Puytorac, Roque, and Tuffrau 1966; Small 1968) (Fig. 10–14G)
   1. Uniform rows of meridional cilia; one long caudal cilium
   2. Heavily ciliated field anterior to prostomium and cytostome; membranelles to right of prostomium homologized with difficulty to tetrahymenine pattern
   3. Cell a somewhat slender ovoid, curved slightly to dorsal side; feeding parts with heavy ciliation lying on anterior, ventral third of cell body
   4. One conspicuous micronucleus
   5. Large, round, centrally placed macronucleus
   6. Homothetogenic fission
   7. Stomatogenesis a partial loss of old ciliation and formation of new proter and opisthe cytostomes prior to fission
   8. Conjugation not reported but probable
   9. Basic pattern of postconjugational events probable
   10. Marine forms; food not identified according to Kahl; probably feeds on unicellular algae
   11. No budding
   12. No special structures

H. *Glossatella* sp. (Order Peritrichida) (Kahl 1930-1935; Lom 1964) (Fig. 10–14H)
   1. Surface ciliature reduced to a girdle of diagonally directed, short kineties, lying between the epistomal region and site of contact with substratum
   2. Epistomial region bordered by an outer haplokinety and an inner polykinety which pass down into the prostomium where there is a germinal kinety and penicular ciliations
   3. Bell-shaped cell with the mouth of the bell being the epistomium
   4. One to a few micronuclei
   5. One roundish macronucleus
   6. Fission plane cuts through epistomium and separates cell body into superficially symmetrical halves
   7. Prefissional opisthe stomatogenesis
   8. Conjugation not reported for this species; inferring details from other species, conjugants of different sizes will fuse completely; there is only one gamete nucleus per conjugant
   9. Postconjugational details probably follow basic pattern
   10. Fresh-water form attached to the perch, *Perca fluviatilis,* and probably feeding on bacteria
   11. No budding
   12. Specialized scopula for attaching to host

I. *Parapodophrya soliformis* (Order Suctorida) (Kahl 1931, 1930-1935; Kudo 1965) (Fig. 10–14I)
   1. Uniform meridional ciliary rows on swarmer; none on sessile adult form
   2. No cytostome; ingestion by tentacles
   3. Spherical adult form with stalk and many radiating tentacles; swarmer is radially symmetrical, ovoid form with blunt anterior and carrying a pronounced bump
   4. Single micronucleus
   5. One round macronucleus
   6. Binary fission of sessile form appears symmetrigenic; also, there is transformation of sessile form into a swarmer
   7. No stomatogenesis
   8. Conjugation not reported; occurs in related forms
   9. Postconjugational events probably follow basic pattern

Table 10-2 *(cont.)*

I. *Parapodophrya soliformis (cont.)*
  10. Fresh-water form which captures holotrichous ciliates
  11. No budding in this species, but total transformation of sessile form into swarmer
  12. Tentacles, scopula, and stalk are specialized structures

J. *Spirophrya subparasitica* (Order Apostomatida) (Chatton and Lwoff 1935) (Fig. 10-14J)
  1. Spiral meridional rows of cilia; they despiral at fission and become longitudinal
  2. Rosette and short kineties make up feeding structures.
  3. Some variation depending on stage of life cycle; typically ovoid with pointed ends
  4. One micronucleus
  5. Single cylindrical macronucleus
  6. Homothetogenic fission with despiraling of surface kineties
  7. Prefissional proter and opisthe stomatogenesis
  8. Conjugation not reported but probable
  9. Postconjugational details probably follow basic pattern
  10. Marine symbiont on copepods and coelenterate polyps
  11. No budding
  12. No special structures

K. *Protocruzia adhaerens* (Order Heterotrichida) (Kahl 1930-1935; Jankowski 1964) (Fig. 10-14K)
  1. Uniform meridional rows of kineties
  2. Cytostome at base of a large prostomium with 4 to 8 membranelles along its left margin and an undulating membrane on its right, posterior margin
  3. Ovoid body with pointed anterior end; feeding structure lying in anterior third of the cell body
  4. One micronucleus
  5. One round, central macronucleus
  6. Homothetogenic fission
  7. Stomatogenesis not described
  8. Conjugation not reported but seen in related species
  9. Postconjugational details probably follow basic pattern
  10. Marine form probably feeding on bacteria
  11. No buds
  12. No special structures

L. *Kahliella acrobates* (Order Hypotrichida) (Kudo 1954; Tuffrau 1969) (Fig. 10-14L)
  1. Uniform meridional kineties; cilia organized as cirri on ventral surface
  2. Cytostome at base of well developed prostomium with membranelle band along left margin and undulating membrane at the right margin
  3. Dorso-ventrally flattened ovoid cell with prostomial cavity lying in anterior left quadrant of the ventral surface
  4. Two micronuclei, each adjacent to a macronucleus
  5. Two ovoid macronuclei
  6. Homothegenic fission
  7. Prefissional opisthe stomatogenesis
  8. Conjugation not reported but occurs in related forms
  9. Postconjugational details probably follow basic pattern
  10. Fresh-water form, probably feeds on small ciliates
  11. No budding
  12. No special structures

Table 10-2 *(cont.)*

M. *Raabena bella* (Order Entodiniomorphida) (Wolska 1967b, 1971b) (Fig. 10-14M)
   1. No surface ciliation except in certain tufts
   2. Cytostome in a prestomial depression
   3. Generally ovoid; flattened laterally; feeding structures at anterior end
   4. One micronucleus
   5. One large, elongate macronucleus
   6. Homothetogenic fission
   7. Stomatogenesis probably prefissional opisthe
   8. Conjugation not reported but seen in related forms
   9. Postconjugational details probably follow basic pattern
   10. Lives in caecum of Indian elephant (*Elephas maximus*)
   11. No budding
   12. Frontal lip at edge of prestomial depression

N. *Pelodinium reniforme* (Order Odontostomatida) (Kahl 1930-1935; Jankowski 1964; Kudo 1965) (Fig. 10-14N)
   1. Eight longitudinal kineties (4 on each side) but interrupted by cilia-free mid-regions
   2. Complex ciliature of prostomium—7 kineties and upper 2 form a membranellar structure
   3. Rounded cell body with anterior, short, trunklike structure curved ventrally over prostomial cavity
   4. Probably one micronucleus
   5. Usually two macronuclei
   6. Homothetogenic fission probable
   7. Opisthe prefission stomatogenesis probable
   8. Conjugation not reported but seen in related forms
   9. Postconjugational details probably follow basic pattern
   10. Fresh-water form feeding on decaying matter (sapropelic)
   11. No buds
   12. Special posterior indentation and spines

O. *Meseres cordiformis* (Order Oligotrichida) (Kahl 1930-1935) (Fig. 10-14O)
   1. Longitudinal kineties
   2. Large and complex developmental of membranes along border of prostomium
   3. Ovoid cell body, flattened anterior in region of feeding structures
   4. Single micronucleus
   5. Single macronucleus
   6. Enantiomorphic homothetogenic fission (Fauré-Fremiet 1969)
   7. Opisthe prefission stomatogenesis probable
   8. Conjugation not reported but seen in related forms
   9. Postconjugational details probably follow basic pattern
   10. Fresh-water form feeding on unicellular algae probably
   11. No budding
   12. No special structures

P. *Tintinnopsis nucula* (Order Tintinnida) (Campbell 1926 Fig. B, p. 189) (Fig. 10-14P)
   1. Longitudinal ciliary rows
   2. Strongly developed adoral zone of long membranelles and special tentaculoids; a prostomial cavity also present
   3. Vase-shaped cell about 70 μm in length with pedical at posterior end for attachment to lorica; cytoproct in posterior third of cell body
   4. Two solid micronuclei; one lying adjacent to anterior end of each macronucleus

Table 10–2 *(cont.)*
P. *Tintinnopsis nucula (cont.)*

  5. Two ovoid macronuclei
  6. Enantiomorphic homothetogenic fission (Fauré-Fremiet 1969)
  7. Prefission opisthe stomatogenesis
  8. Conjugation not reported in tintinnids but cellular attachments observed which are suggestive of sexual processes (personal communication from Dr. Kenneth Gold, Osborn Laboratories of Marine Sciences, Brooklyn, N.Y. 11224)
  9. Postconjugational events not observed
  10. Filter feeder principally ingesting "diatoms, dinoflagellates, radiolarians, small chrysomonads, and bacteria" (Campbell 1926, p. 184)
  11. No budding
  12. Lorica

Dobrzanska-Kaczanowska 1963). This line of thinking is especially important in bringing the chonotrichs into the Gymnostomatida as another suborder, as suggested by Dobrzanska-Kaczanowska. In defining the plesiomorph, though, we should compare the details of the recapitulation in *Heliochona* with those among other chonotrichs. However, such an analysis of these curious ectocommensal ciliates has not been done, and so the choice of *Heliochona* is an uneasy one. *Spirochona* (Guilcher 1951) is another possible candidate. Further work within the chonotrichs is the only way to decide this point.

The order Trichostomatida contains quite a variety of cell-body forms; (Fauré-Fremiet 1950c); there is need for careful reworking of this work. *Coelosomides marina* (Fig. 10–14D), because of its superficial resemblance to rhabdophorine forms, was often classified with that group (Kahl 1930–1935). However, the presence of a well-defined peristomial depression (or vestibulum (Corliss 1961)) demands its inclusion in the trichostomes. This possession of generalized rhabdophorine form as well as a trichostome feeding depression puts it in a position where it could represent the starting point of trichostomatous evolutionary development. Its general body organization lies at one end of a trend in trichostome body form whose other end is in the more complex feeding ciliature and somewhat twisted cell bodies of *Bresslaua*, *Colpoda*, and *Tillina*. Stout's (1960) analysis of the family Colpodidae uses the morphogenetic patterns of *Bresslaua*, which can be simple longitudinal meridians during fission and twisted meridians in the differentiated interfission cell, as a basis for deriving the phylogeny of that family. By use of Cartesian coordinates he shows the geometric plausibility of deriving the more complex trichostomatous Colpodidae from the rhabdophorine *Prorodon*. Others (Tuffrau 1952) have preferred *Pseudoprorodon*, but Stout shows that this involves more complex transformations than *Prorodon*. *Coelosomides* being very like *Prorodon* in its general form, but withal being a trichostome, is a logical candidate for plesiomorph of this order.

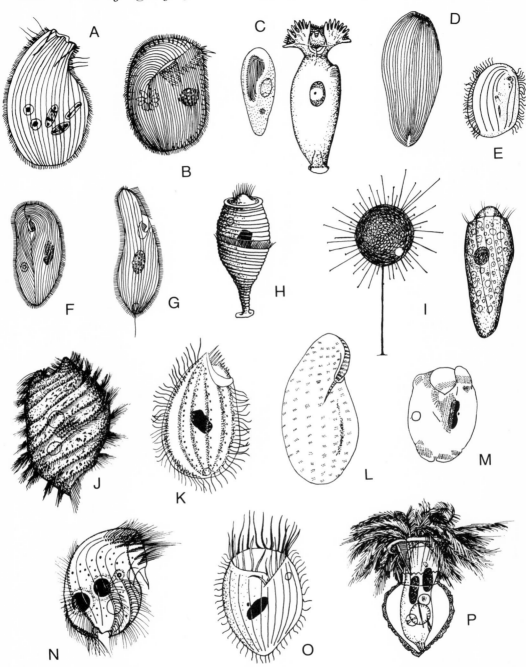

Fig. 10-14 Ciliates proposed as plesiomorphs of the various orders and suborders to which they belong. (See Table 10-2 for further details.) A. *Stephanopogon mesnili* (Redrawn, by permission, from Lwoff (1923).) B. *Nassula ornata* (Redrawn, from Kahl (1930–1935).) C. *Heliochona scheuteni* (Redrawn, by permission, from Dobrzanska-Kaczanowska (1963).) D. *Coelosomides marina* (Redrawn, by permission, from Kudo

The Hymenostomatida, containing the popular genera of *Tetrahymena* and *Paramecium*, might be represented plesiomorphically by *Pseudomicrothorax dubius*. There is a taxonomic debate, however, as to whether it is a hymenostome (Thompson and Corliss 1958; Corliss 1958a, 1958b) or a cyrtophorine gymnostome (Fauré-Fremiet and André 1967). (See Canella (1970) for a pungent discussion of the taxonomic confusion surrounding this species.) Furthermore, the fact that stomatogenesis in this species is of the prefissional type in both proter and opisthe (Thompson and Corliss 1958), as is true for cyrtophorines but not true for hymenostomes with their prefissional pattern only for the opisthe, also argues for its inclusion in the cyrtophorines. However, there seems to be no doubt about the "missing link" possibilities of this form, for its oral structures show both the nassa of cyrtophorines and the basic ciliature of the hymenostomes. For that reason we shall use it as a plesiomorph of the tetrahymenine hymenostomes though recognizing that future work will undoubtedly reveal a more distinctly hymenostome form as a better example. The peniculine suborder of the Hymenostomatida can be represented by *Frontonia marina* (Fig. 10–14F). This form is chosen because it is a marine organism (most of the forms chosen thus far are marine), and its relatively uniform ciliature and the surface location of its prostomial cavity are more like the gymnostome pattern than that of other peniculines.

The next order, the Scuticociliatida, represents a recent convulsive but justifiable upheaval in the midst of otherwise very stable taxonomic groups. In a very important paper, Small (1968) has seen the essential fact that the pleuronematine ciliates do not have a prostomial ciliature nor a stomatogenesis that justifies their inclusion any longer in the Hymenostomatida. Furthermore, the arhynchodine forms, formerly included in the thigmotrich order, also show similarities to the scuticociliate state of affairs, in particular they resemble the pleuronematine forms. Thus these two former suborders (Arhynchodina and Pleuronematina) of two different orders are now included in one new order, which takes its name from a characteristic aspect of stomatogenesis which shows a hooklike ciliary field in both new oral areas. The choice of *Philaster digitiformis* (Fig. 10–14G) is almost certainly a premature one for the plesiomorph of this group. The dust has not settled sufficiently

---

(1954) *Protozoology* 4th ed. Charles C. Thomas, Publisher.) E. *Pseudomicrothorax dubius* (Redrawn, by permission, from Corliss (1958a).) F. *Frontonia marina* (Redrawn, by permission, from Kudo (1954) *Protozoology* 4th ed. Charles C. Thomas, Publisher.) G. *Philaster digitiformis*. H. *Glossatella pisicola* (Redrawn from Kahl (1930–1935).) I. *Parapodophrya soliformis* (Redrawn, by permission, from Kahl (1931).) J. *Spirophrya subparasitica* (Redrawn, by permission, from Chatton and Lwoff (1935).) K. *Protocruzia adhaerans* (Redrawn from Kahl (1930–1935).) L. *Kahliella acrobates* (Redrawn, by permission, from Jankowski (1964).) M. *Raabena bella* (Redrawn, by permission from Wolska (1971a).) N. *Pelodinium reniforme* (Redrawn, by permission, from Jankowski (1964).) O. *Meseres cordiformis* (Redrawn from Kahl (1930–1935).) P. *Tintinnopsis nucula* (Redrawn, by permission, from Campbell (1926).)

around this taxonomic revision so as to see the group clearly. However, of the forms described as members of this taxon, and whose stomatogenesis is well-described, *Philaster* seems a plausible candidate as plesiomorph, standing at one end of a trend towards a more posterior placement of the cytostome and heavier development of the oral ciliature.

The next order to be discussed would have been the Thigmotrichida. But with removal of one suborder from them to the scuticociliates, this leaves only the suborder Rhynchodina in the former thigmotrichs, and this group may well have affinities with cyrtophorine gymnostomes (Puytorac 1968b; Lom and Kozloff 1969; Puytorac and Njiné 1970). This uncertain state of affairs suggests that it is best to leave, for the present, the rhynchodines aside from any phylogenetic discussions. That the thigmotrichs, as a whole, contain important evolutionary trends has been elegantly demonstrated by Raabe (1968, 1970a, 1970b). However, the trends were formulated in the context of a phylogeny only in his first paper (Fig. 10–10). Raabe's emphasis was on discussion of the evolutionary potential of certain cellular modes of organization rather than on actual phylogenetic relationships. His anaylsis provides rich material for someone familiar with this specialized group of forms, typically found in association with marine bivalve molluscs, to work out their evolution.

The next group, the Peritrichia, represents a highly specialized—some (Finley, McLaughlin, and Harrison 1959; Canella 1964) would say possibly the most specialized—group of ciliates. Our choice of *Glossatella* sp. as the plesiomorph comes from Lom's (1964) description of this form as having a buccal ciliature that puts it at one extreme—the less complex extreme, more like hymenostome forms—of the various types of feeding ciliature in the peritrichs (Fig. 10–11). This form has not been described in detail; we still await Lom's further work in this regard. In that the organism was collected from a perch (*Perca fluviatilis*), and in that forms with similar oral structures were described by Lom as *G. pisciola*, such a form (taken from Kahl 1930–1935) can be used to illustrate *Glossatella* sp. (Fig. 10–14H). It is to be noted that this species shows little development of the adhesive organelle, the scopula, which in other peritrichs can become a conspicuous stalk with contractile elements. In regard to this feature, too, *Glossatella* sits at the end, or near it, of a morphocline—another reason for treating it as a plesiomorph.

Finally, there are three orders of ciliates, the last that make up the subclass Holotrichia, which show quite sharp discontinuities with other ciliates. First, there are the Astomatida, symbiotic forms lacking any mouth. Today there seems general agreement that this is a polyphyletic group, that several different lines of ciliates independently gave rise to symbiotic forms that eventually lost their oral structures (see especially Puytorac 1959a, 1959b; Kaczanowski 1965). Eventually the members of this group will be placed in other ciliate orders in accordance with evidence on their various origins.

Hence, it is pointless to look for a plesiomorph of the astomatids as presently constituted. The Apostomatida are a small group of forms usually found in association with marine copepods and in some forms also utilizing a second host. Because they are highly adapted to their peculiar niche, they all show a rather uniform degree of specialization. The choice of *Spirophyra subparasitica* as the plesiomorph is too arbitrary to be satisfactory. The only objective reason for its selection is the relative simplicity of its life cycle. It lives on only one host, has one type of division, and returns to a rather simple ciliary pattern at fission. This puts it at the less complicated end of what might be seen as trends in these traits within the Aspostomatida (Chatton and Lwoff 1935) (Fig. 10–14J).

There remains now the Suctoria. This group is so different that it, like the Peritrichia, is placed by many workers (Honigberg et al. 1964) in its own subclass, along with the Holotrichia and the Spirotrichia. There have been many attempts to understand the evolutionary development of the Suctoria. Collins (1912), from his detailed acquaintance with the group, decided it must have arisen from the peritrichs. He cited the ciliary girdle, common to peritrichs and sometimes found on the ciliated suctorian swarmers, as one line of evidence. Also he found similarities in the oral structures (vestiges are present in the swarmers), the scopula for attaching the cells to the substratus, special fibrillar structures, and in certain aspects of asexual reproduction and conjugation. However, these similarities represent a mélange of similarities found in these groups taken as a whole and not found in specific forms, as must be the case when species are accepted as the units of evolution. Later, Kahl (1931, 1934) reviewed the problem again and derived the suctorians from gymnostome ciliates. His arguments make careful use of comparative data on tentacle formation—there are various tentaculated gymnostomes—and on a certain sequence in asexual reproduction. This sequence, according to Kahl (1931) proceeds from equal (homothetogenic) division in tentaculated rhabdophorines to forms with alternation between nonciliated, sessile, tentaculated cells and formation of ciliated swarmers; to sessile tentaculated forms capable of division (it is not clear from Kahl's account whether the ciliated stage is also capable of division); to forms like *Parapodophrya*, capable of division in the sessile state and transforming directly into the ciliated swarmer; and finally to the two further possibilities of internal and external budding of swarmers. Kahl is here clearly picking out a trend in tentacle formation and one in patterns of asexual reproduction, with the latter being filled out by hypothetical intermediate steps. Removing these intermediate steps, it becomes hard to read the direction of the trend; transformation into a swarmer is, indeed, different from giving rise to a swarmer by budding. But all suctorians have equally good tentacles, and a trend in their formation is also cut off by removal of hypothetical forms and the tentaculated rhabdophorines. It is, therefore, difficult to find objective evidence from Kahl's data (as distinct from his hypothetical phylogeny) for finding *Parapodophrya* as the most primitive suctorian (Kahl 1931). Other attempts to use trends to find the primitive suctorians have been used. Guilcher (1951), arguing that the swarmer is a stage regressed from the more typical ciliate ancestor, has proposed that the simplest swarmers, i.e., with no evidence of an oral anlage and very simple ciliature, are the most evolved. However, Kormos (1958) has cogently argued against such a straightforward interpretation of the morphological data, showing that in some cases the so-called oral anlage is a functional site of attachment and that, basically, structure must be interpreted functionally. On such a basis, he sees both simplification and complexities emerging in the swarmer. Furthermore, he also finds correlated trends, not in patterns of asexual reproduction, but in conjugation. In the form *Catharina florea*, he says conjugants are isogamous, whereas conjugants of other forms are anisogamous and can fuse (part of the evidence used by Collins for finding relationships with the peritrichs!). However, Kormos' brief discussion is distressingly incomplete on the rest of the biology of *C. florea* (the adult is not illustrated nor are there references to its original description—Kahl (1934), describes no such form). Being included in the family Discophryidae, *Catharina* (or *Coracatharina*, according to Corliss (1961)) may be stalkless but showing internal budding. That form of reproduction puts it in a more advanced evolutionary position, according to Kahl's thinking, with respect to other suctorians. There is, then, no agreement on the plesiomorph and no apparent way to generate such agreement. (See also Canella (1957), who also is skeptical regarding the establishment of phyletic relations between suctorians and other ciliates. The picture is further complicated by the suggestion of Paulin and Corliss (1969) that the internal budding and endosprits of "the taxonomically enigmatic ciliate *Cyathodinium*" suggests suctorian affinities.) For the present we shall follow Kahl's proposal, depending largely on his arguments for transformation into a swarmer as being evolutionarily antecedent to forming a swarmer by budding—though this is really an assumption and not by any means demonstrated fact. Also, Kahl could have argued for another trend: *Parapodophrya* (Fig. 10–14I) has a radial disposition of tentacles around the whole cell body, whereas other suctorians, including the Discophryidae, have less random distribution of the sucking appendages. Further study of the Suctoria, including macronuclear trends (round to lobulated structures?), is the only way to find the plesiomorph. This remains to be done in a convincing manner.

We are ready to look next at the spirotrichous ciliates. Here *Protocruzia adhaerans* has been chosen as the plesiomorph (Fig. 10–14K). This comes from Jankowski's (1964) study of these forms. This worker has worked out trends in body form and oral ciliature (Fig. 10–15). From this study it is clear that *P. adhaerans* is a satisfactory candidate as plesiomorph. It should be added that in some descriptions of this species, five anal cirri are reported to be present. (See Kahl 1935, p. 438, Fig. 72–30 and p. 532, Fig. 86–2.) If cirri

CILIATES

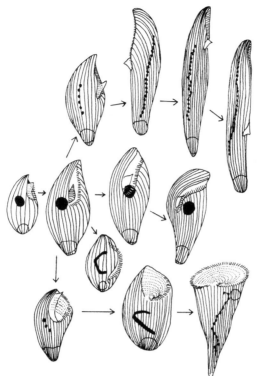

Fig. 10–15 Evolutionary trends in selected heterotrichous (spirotrichous) ciliates. (Redrawn, by permission, from Jankowski (1964).)

are present, then this form should be classified as a hypotrich. The plesiomorph of this latter group is well represented by *Kahliella acrobates* (Fig. 10–14L). Tuffrau (1969) has shown that this form has no special ventral field of cirri as in the more complex forms like *Euplotes* and *Uronychia*, but its ventral kineties are formed of compounded ciliary units and are meridional ciliary rows. There is also formation of new ventral fields of kinetosomes, which lead to the appropriate fields in the proter and opisthe. The opisthe obtains a whole new prostomial apparatus; the old prostomium persists for the proter.

For the Odontostomatida, again we follow Jankowski (1964) and propose *Pelodinium reniforme* (Fig. 10–14N) as the plesiomorph. The reasons are obvious from Fig. 10–16, which again illustrates trends in body form and ciliation.

For the Entodiniomorphida, Lubinsky (1957) initiates a phylogenetic analysis that locates a convincing plesiomorph for this group when this group is defined as containing only two families. Though his analysis is directed only at the family Ophryoscolecidae, it contains trends that are extended into the other family, the Cycloposthidae. The trends that Lubinsky uses are dia-

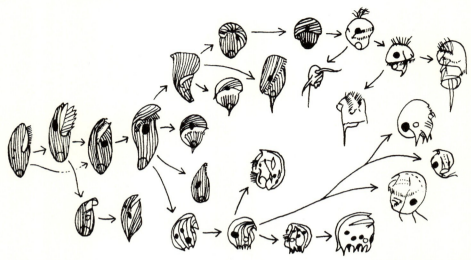

Fig. 10-16 Evolutionary trends in selected heterotrichous (odontostomatous) ciliates. (Redrawn, by permission, from Jankowski (1964).)

grammed in Fig. 10-17 and include appearance of caudal structures, ectoplasmic platelets, emergence of two ciliary tufts, enlargement and elongation of the macronucleus, and (not clear in the figure) extensive twisting of the body and rearranging of various internal and external parts. (The Cycloposthidae show even more extensive development of ciliary tufts, platelets, surface sculpturing, and skeletal platelets.) On this basis *Entodinium nanellum* is taken as a representative of the starting point of many of these evolutionary trends.

However, as Wolska (1971a, 1971b), in particular, has argued, there are reasons for removing the family Blepharocorythidae from the trichostomes and including them in the entodiniomorphids. (Grain (1966) and Noirot-Timothée (1969) are more cautious and withhold judgment as to the taxonomic position of this group.) If we follow Wolska's argument (also suggested by Noirot-Timothée (1969), *Raabena bella* (Fig. 10-14M) emerges as a possible plesiomorph of the entodiniomorphids, for she would derive this group from the Blepharocorythidae with *Raabena* (Wolska 1967b) as its more primitive representative, though she also suggests that the entodiniomorphids might share a common gymnostome ancestor, specifically *Didesmis*, with the blepharocorythids. Jankowski (1967, 1973) and Corliss (1974a, 1974b) tend to follow Wolska's arguments, and we will do so here. The reasons are that *Raabena* lies at the ends of morphoclines relating to placement, size, and number of ciliary zones and the form of the feeding parts. Therefore, if Wolska's phylogenesis is correct, the buccal cavity she describes is really a prestomial depression, deeply sculptured with a frontal lobe at one edge and its ciliation derived from the surface ciliation of its gymnostome ancestor, i.e., *Didesmis*

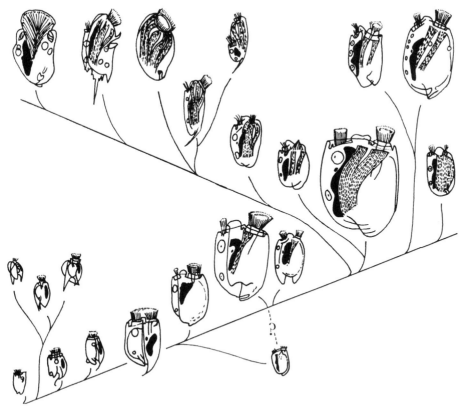

Fig. 10–17 Evolutionary trends in the ophryoscolecid entodiniomorphids. (Redrawn, by permission, from Lubinsky (1957).)

(compare Fig. 10–18 to Fig. 10–14M). This ancestry will be discussed further later.

The remaining two orders of spirotrichous ciliates, i.e., the Oligotrichida and the Tintinnida, pose certain difficulties in finding plausible plesiomorphs, for, like the problem in the Apostomatida, all show rather uniform levels of structural organization and trends are difficult to spot. By going through Kahl's compendium of species for the oligotrichs, *Meseres cordiformes* (Fig. 10–14O) was selected as appearing to be somewhere between *Protocruzia* and the vis-

Fig. 10–18 *Didesmis ovalis,* a gymnostome thought to be ancestral to *Raabena bella.* (Redrawn, by permission from Wolska (1971b).)

ibly more complex members of this order. Also, this species shows the opening of the fringe of prostomial membranelles which Fauré-Fremiet (1969a) points out as resembling certain heterotrichous forms.

For the predominantly marine, planktonic tintinnids, most monographs (Kofoid and Campbell 1929; Jörgensen 1924; Jörgensen and Kahl 1933; Campbell 1954) use an Adansonian taxonomy to separate out taxa on the basis of lorica morphology. For example, Kofoid and Campbell (1929) present 697 text figures for the 705 species they recognize, and only one diagrammatic figure (their Figure 14) represents the cell which manufactures the lorica; all the rest represent loricas which are the sole basis for the taxonomic groupings presented by these workers. Kofoid (1930) has attempted to identify evolutionary trends within the tintinnids, but this resolves itself into tracing serial relationships in various genera in relation to the size, elongation, multiplication of structural features (ribs, striae, and so on), and surface pattern. There is no evidence of a convincing stem form or plesiomorph in this discussion. The plesiomorph we chose here is arbitrarily selected so as to provide data for comparative studies. Such a well-studied form is *Tintinnopsis nucula* (Fig. 10–14P) (Campbell 1926). There are probably less complex tintinnids than this one but not so well known. Hence we will proceed with this unsatisfactory choice.

The nature of the foregoing sample of plesiomorphs is obviously now a mixed one; some we have been able to select through thoughtful use of phylogenetic principles, others have been selected more arbitrarily. They do allow us, however, to proceed with our phylogenetic study, though on a much more heuristic basis than is desirable. The following plesiomorphs are thought to have been derived on relatively sound grounds: *Stephanopogon mesnili, Nassula ornata, Coelosomides marina,* (possibly) *Pseudomicrothorax dubius, Protocruzia adhaerans, Kahliella acrobates, Raabena bella,* and *Pelodinium reniforme*. All the others are of doubtful validity.

From Table 10–2 we turn next to a summary of the intersemic comparisons expressed in terms of plesio-, apo-, and neosemic relations; in other words, comparisons expressed in terms of the evolutionary fate of the various semic traits (Table 10–3). Brief comments are needed to make explicit the way the traits listed across the top of the table were actually used. First, there is no mention of ciliary units, for this seme has not been studied in any of the forms listed in the table. The kinety patterns of the surface ciliature are quite variable, but the occurrence of uniform, meridional ciliation is a recurring pattern in several forms. We treat slight variations in this pattern conservatively by ignoring perturbations in it resulting from other variations which are major, such as lack of ciliation on one side of the cell body (*Stephanopogon*) or from the whole body surface (*Heliochona, Glossatella,* and *Parapodophrya*); these are treated as distinct differences.

The feeding structures are seen as unique in each species. This is certainly justified on the basis of their obvious differences. And where prostomial

Table 10-3 Intersemic homologies between sixteen species of ciliates.

	Semes			
Species	Cellular structures 1. 2. 3. 4. 5.	Reproduction and feeding 6. 7. 8. 9. 10.	Special features 11. 12.
A. *Stephanopogon mesnili*	1A 2A 3A 4A 5A	6A ? 8A 9A 10A	11A 12A
B. *Nassula ornata*	1B 2B 3B ? 5B	6A 7B 8A 9A 10B	11A 12A
C. *Heliochona scheuteni*	1C 2C 3C 4C 5B	6C 7C 8A 9A 10C	11C 12C
D. *Coelosomides marina*	1B 2D 3D 4D 5D	6A 7D 8A 9A 10A	11A 12A
E. *Pseudomicrothorax dubius*	1B 2E 3B 4D 5D	6A 7B 8A 9A 10B	11A 12A
F. *Frontonia marina*	1B 2F 3F 4C 5D	6A 7F 8A 9A 10F	11A 12A
G. *Philaster digitiformis*	1B 2G 3G 4D 5D	6A 7G 8A 9A 10F	11A 12A
H. *Glossatella sp.*	1C 2H 3H 4C 5D	6H 7H 8H 9A 10F	11A 12H
I. *Parapodophrya soliformis*	1C 2I 3I 4D 5D	6I 7I 8A 9A 10I	11I 12I
J. *Spirophrya subparasitica*	1B 2J 3J 4D 5J	6A 7B ? ? 10J	11A 12A
K. *Protocruzia adhaerans*	1B 2K 3G 4D 5D	6A ? 8A 9A 10F	11A 12A
L. *Kahliella acrobates*	1B 2L 3G 4L 5L	6A 7H 8A 9A 10F	11A 12A
M. *Raabena bella*	1M 2M 3D 4D 5D	6A 7M 8A 9A 10M	11A 12M
N. *Pelodinium reniforme*	1N 2N 3N 4D 5L	6A 7H 8A 9A 10C	11A 12N
O. *Meseres cordiformis*	1B 2O 3O 4D 5D	6O 7H 8A 9A 10F	11A 12A
P. *Tintinnopsis nucula*	1B 2P 3P 4L 5L	6O 7H 8A 9A 10C	11A 12P

Key to Semes:

1. Aboral kinety patterns: 1A—ciliation limited to one side of cell; 1B—uniform ciliation; 1C—no ciliation; 1M—tufts of cilia; 1N—longitudinal kineties interrupted by cilia-free midregions.
2. Feeding structures: 2A—slitlike cytostome; 2B—nassa present; 2C—fanshaped; 2D—cytostome and peristome present; 3E—nassa and stomial membranelles; 2F to 2H—and 2J to 2P—all with prostomial cavity but shaped differently; 2I—tentacles. (Emphasis on parts other than cytostome.)
3. General cellular organization: (For details of differences see Table 10-2).
4. Micronucleus: 4A—none; 4C—three or several; 4D—one; 4L—one per macronucleus.
5. Macronucleus: 5A—none; 5B—one, heteromeric; 5D—one, ovoid or round, 5J—one, cylindrical; 5L—two.
6. Fission: 6A—homothetogenic fission; 6C—budding; 6H—modified homothetogenic fission; 6I—budding (different from 6C); 6O—enantiomorphic homothetogenic fission.
7. Stomatogenesis: 7B—proter and opisthe prefissional; 7C—unique stomatogenesis; 7D—opisthe parafissional; 7F—modified proter and opisthe prefissional; 7G—modified porter and opisthe prefissional (different from 7F); 7H—opisthe prefissional; 7M—opisthe prefissional from a vesicle.
8. Prezygotic events of conjugation: 8A—characteristic pairing and meiosis; 8H—conjugants of different sizes.
9. Postzygotic events of conjugation: 9A—fertilization and restoration of normal nuclear situation.
10. Feeding: 10A—bottom feeder, largely unicellular algae; 10B—feeder, filamentous algae; 10C—commensal on crustaceans; 10F—feeds on a variety of algae and micrometazoa; 10I—holotrichous ciliates; 10J—symbiont on copepods and cnidarian polyps; 10M—in caecum of Indian elephant.
11. Budding: 11A—absent; 11C—budding; 11I—budding (different from 11C).
12. Special structures: 12A—none; 12C—adhesive disc; 12H—stalk; 12I—stalk (different from 12H); 12M—frontal lip; 12P—lorica.

parts appear for the first time, they are considered neosemic relative to cells lacking them.

General body organization also differs a great deal from one species to another, not so much in terms of the general outlines of the body, but more in terms of the relative placements of mouth parts, contractile vesicles, cytoprocts, and planes of flattening of the cell body. These are all considered aposemic trends.

The micronuclei are considered as essentially alike, except for the ones in *Stephanopogon mesnili*, which show promitotic division figures and do not ever give rise to macronuclei, and for differences in number.

For the macronucleus, the shape, number, and general nature of the internal structure (homomeric or heteromeric), were all considered.

Turning next to functions, fission was examined from the point of view of normal homothetogenic fission. Variants of the normal situation are seen in the enantiotropic pattern of the oligotrichs (*Meseres cordiformis*, not observed but assumed to be the case here) and the special pattern of the peritrich *Glossatella* sp.

Stomatogenesis was viewed as occurring within one of the four categories described previously. Where direct observations of stomatogenesis are not available, inferences were made where it seemed probable that the pattern was like that in closely similar species. Where even that inference could not be made, then no conjecture was supplied.

Conjugation was divided into two parts: first, the nature of the conjugants, their behavior, and prezygotic events and, second, postzygotic events. Most of the forms studied here have not been reported to have shown conjugation. However, its occurrence in many other members of the same group, except for the Apostomatida, make the occurrence of conjugation for the species in question a virtual certainty. And in cases such as the peritrichs, where the unique situation of macro- and microconjugants and their permanent fusion is known, it was inferred that that pattern held, though unobserved for *Glossatella*. Differences in postzygotic events centers largely on redifferentiation of the micro- and macronuclear complements. These details are lacking for all the species recorded here. That such redifferentiation occurs is certain, and hence all that can be done at present, due to lack of pertinent details, is to record this trait as a very plesiosemic one.

Feeding patterns were reduced to the broad categories of filter feeding, capture of small plants (algae) or small animals (other protozoa), or adaptations to a symbiotic mode.

Finally, there are two semes which recognize the occurrence of special traits. One is budding and the other is the presence of special organelles such as stalks, loricas, skeletal plates, and so on.

Table 10-4 summarizes the comparisons made between the various species and the phyletic distances (R values) that emerge from those com-

CILIATES

Table 10-4 Homologous relationships between pairs of semes and phyletic distances among sixteen species of ciliates.

													Semes com-pared t	Totals Plesio-semes p	Ape-semes a	Neo-semes n	Phyletic distance R
	Homologies between semes																
	1	2	3	4	5	6	7	8	9	10	11	12					
A and B	a	a	a	a	n	p	?	p	p	a	p	p	11	5	5	1	10.5
C	a	a	a	a	n	a	?	p	p	a	a	n	11	2	7	2	21.9
D	a	a	a	a	n	p	?	p	p	p	p	p	11	6	4	1	7.1
E	a	a	a	a	n	p	?	p	p	a	p	p	11	5	5	1	10.5
F	a	n	a	a	n	p	?	p	p	a	p	p	11	5	4	2	9.6
G	a	n	a	a	n	p	?	p	p	a	p	p	11	5	4	2	9.6
H	a	n	a	a	n	a	?	a	p	a	p	n	11	2	6	3	21.3
I	a	n	a	a	n	a	?	p	p	a	a	n	11	2	6	3	21.5
J	a	n	a	a	n	p	?	?	?	a	p	p	9	3	4	2	11.8
K	a	n	a	a	n	p	?	p	p	a	p	p	11	5	4	2	9.6
L	a	n	a	a	n	p	?	p	p	a	p	p	11	5	4	2	9.6
M	a	n	a	a	n	p	?	p	p	a	p	n	11	4	4	3	13.8
N	a	n	a	a	n	p	?	p	p	a	p	n	11	4	4	3	13.8
O	a	n	a	a	n	a	?	p	p	a	p	p	11	4	5	2	13.0
P	a	n	a	a	n	a	?	p	p	a	p	n	11	3	5	3	17.2
B and C	a	a	a	?	p	a	a	p	p	a	a	n	11	3	7	1	18.5
D	p	a	a	?	a	p	a	p	p	a	p	p	11	6	5	0	9.6
E	p	a	p	?	a	p	p	p	p	p	p	p	11	9	2	0	1.6
F	p	n	a	?	a	p	a	p	p	a	p	p	11	6	4	1	7.1
G	p	n	a	?	a	p	a	p	p	a	p	p	11	6	4	1	7.1
H	a	n	a	?	a	a	a	a	p	a	p	n	11	2	7	2	21.9
I	a	n	a	?	a	a	a	p	p	a	a	n	11	2	7	2	21.9
J	p	a	a	?	a	p	p	?	?	a	p	p	9	5	4	0	7.6
K	p	n	a	?	a	p	?	p	p	a	p	p	10	6	3	1	4.9
L	p	n	a	?	a	p	a	p	p	a	p	p	11	6	4	1	7.1
M	a	a	a	?	a	p	a	p	p	a	p	a	11	4	7	0	18.5
N	a	n	a	?	a	p	a	p	p	a	p	a	11	4	6	1	14.6
O	p	n	a	?	a	a	a	p	p	a	p	p	11	5	5	1	10.5
P	p	n	a	?	a	a	a	p	p	a	p	n	11	4	5	2	13.0
C and E	a	a	a	a	a	a	a	p	p	a	a	n	12	2	9	1	28.6
F	a	n	a	p	a	a	a	p	p	a	a	n	12	3	7	2	20.0
G	a	n	a	a	a	a	a	p	p	a	a	a	12	2	9	1	28.6
D and E	p	a	a	p	p	p	a	p	p	a	p	p	12	8	4	0	5.7
F	p	n	a	a	p	p	a	p	p	a	p	p	12	7	4	1	6.5

Table 10-4 (cont.)

	Homologies between semes 1 2 3 4 5 6 7 8 9 10 11 12	Semes compared t	Totals Plesiosemes p	Apesemes a	Neosemes n	Phyletic distance R
G	p n a p p p a p p a p p	12	8	3	1	4.1
J	p n a p a p a ? ? a p p	10	5	4	1	7.8
M	a a p p p p a p p a p a	12	7	5	0	8.8
E and F	p n a a p p a p p a p p	12	7	4	1	6.5
G	p n a p p p a p a a p p	12	7	4	1	6.5
H	a n a a p a a a p a p n	12	3	7	2	20.1
I	a n a p p a a p p a a n	12	4	6	2	15.7
J	p n a p a p p ? ? a p p	10	6	3	1	4.9
K	p n a p p a ? p p a p p	11	7	3	1	4.5
L	p n a a a p a p p a p p	12	6	5	1	9.6
M	a n a p p p a p p a p n	12	6	4	2	8.8
N	a n a p a p a p p a p n	12	5	5	2	11.9
O	p n a p p a a p p a p p	12	7	4	1	6.5
P	p n a a a a p p a p n	12	4	6	2	15.7
F and G	p a a a p p a p p p p p	12	8	4	0	5.7
H	a a a p p a a a p p p n	12	5	6	1	13.3
I	a n a a p a a p p a a n	12	3	7	2	20.1
J	p p a a a p a ? ? a p p	10	5	5	0	10.5
K	p a a a p p ? p p p p p	11	8	3	0	3.6
L	p a a a a p a p p p p p	12	7	5	0	8.8
M	a a a a p p a p p a p n	12	5	5	1	13.3
N	a a a a a p a p p a p n	12	4	7	1	17.8
O	p a a a p a a p p p p p	12	7	5	0	8.8
P	p a a a a a p p a p n	12	4	7	1	17.8
G and H	a a a a p a a a p p p n	12	4	7	1	17.8
I	a n a p p a a p p a a n	12	4	6	2	15.7
J	p a a p a p a ? ? a p p	10	5	5	0	10.5
K	p a p p p ? p p p p p	11	10	1	0	0.5
L	p a p a a p a p p p p p	12	8	4	0	5.7
M	a a a p p p a p p a p n	11	6	5	1	9.6
N	a a a p a p a p p a p n	12	5	6	1	13.3
O	p a a p p a a p p p p p	12	8	4	0	5.7
P	p a a a a a p p a p n	12	4	7	1	17.8
K and L	p a p a a p ? p p p p p	11	8	3	0	3.6
M	a a a p p p ? p p a p n	11	6	4	1	7.1

Table 10-4 (cont.)

		Homologies between semes 1 2 3 4 5 6 7 8 9 10 11 12	Semes com-pared t	Totals			Phyletic distance R
				Plesio-semes p	Ape-semes a	Neo-semes n	
	N	a a a p a p ? p p a p n	11	5	5	1	10.5
	O	p a a p p a ? p p a p n	11	6	4	1	7.1
	P	p a a p p a ? p p p p p	11	8	3	0	3.6
L and M		a a a a a p a p p a p n	12	4	7	1	17.8
	N	a a a a a p p p p a p n	12	5	6	1	13.3
	O	p a a a a a p p p p p p	12	7	5	0	8.8
	P	p a a p p a p p p a p n	12	7	4	1	6.5
H and I		p n a a p a a a p a a n	12	3	7	2	20.1
	J	a a a a a a a ? ? a p n	10	1	8	1	27.4
	K	a a a a p a ? a p p p n	11	4	6	1	14.6
	L	a a a a a a p a p p p n	12	4	7	1	17.8
O and M		a a a p p a a p p a p n	12	5	6	1	13.3
	N	a a a p a a p p p a p n	12	5	6	1	13.3
	P	p a a a a p p p p a p n	12	6	5	1	9.6

parisons. These values are converted into a dendrogram in Fig. 10–19. To discuss the dendogram is largely to raise the question of the origins of the various groups.

Origins. Starting from the plesiomorph *Stephanopogon mesnili*, the two forms discussed here which are closest to it in terms of phyletic distance are *Coelosomides marina* and *Nassula ornata*. (From here on we shall usually refer to these and the other forms being analyzed by their generic names rather than follow the convention of using the specific name, e.g., *S. mesnili*. The reason is that we are arguing that evolution proceeded from a form like *S. mesnili*, a plesiomorph, to forms like those represented by other plesiomorphs. Using the generic names catches the feel of this procedure better than using the specific names.) But *Coelosomides* is closer to *Stephanopogon* than is *Nassula* and the distance between *Coelosomides* and *Nassula* is less than between either of them and *Stephanopogon*. From this either of two things can be concluded: (a) evolution proceeded from *Stephanopogon* to *Coelosomides* and then to *Nassula*, or (b) from *Stephanopogon* two lines diverged, one to Coelosomides and the other to *Nassula*. The absence of trichitelike entities in the cytostome of *Coelosomides* and their presence in *Stephanopogon* and other forms interposable as intermediates between the latter and *Nassula* make the derivation of *Nassula* from *Stephanopogon* the more likely line of evolution. (In this discussion, it must be borne in mind that phyletic distances are not statements of direct evolutionary

404 *Phylogeny of the Protozoa and Early Metazoa*

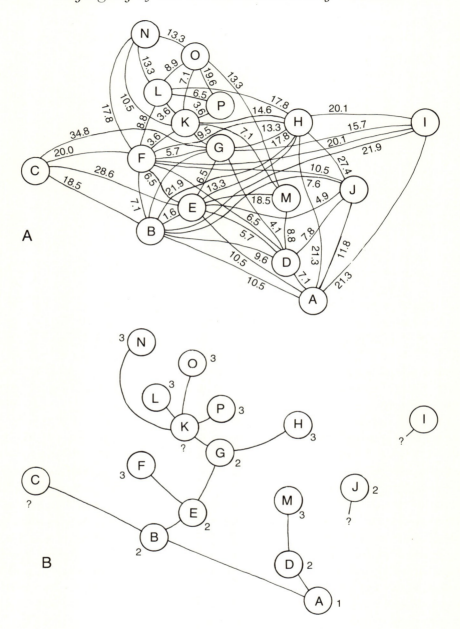

Fig. 10–19 Phylogenetic relations among selected ciliates. A. Certain phyletic distances. B. Dendrogram based largely on phyletic distances. The distribution of the four modes of stomatogenesis are also given. (1—parafissional opisthe, 2—prefissional proter and opisthe, 3—prefissional opisthe, 4—postfissional proter and opisthe, not represented here; see Table 10–5 and Fig. 10–7.)

relations. We cannot, from the data examined here, argue that *Coelosomides* arose from *Stephanopogon*. All we are saying is that *Coelosomides* more resembles, in terms of homologous relationships, *Stephanopogon* than any other ciliate presently under consideration. Therefore, it more likely arose from a form like, or represented by, *Stephanopogon* than any of the other forms under discussion.)

At this point emerges the vexing problem of derivation of the entodiniomorphids. As already mentioned, Wolska, Noirot-Timothée, Jankowski, and Corliss are ready to accept a trichostome origin for these endocommensals of mammalian digestive tracts. The proposed origin (Wolska 1971a, 1971b) is not from a free-swimming form like *Coelosomides*, but from the mammalian endocommensal *Didesmis ovalis*, for example (Wolska 1966a). A measure of phyletic distance between this latter form and *Raabena bella* would provide an R value of 1.5, there being only two aposemes among the traits analyzed in Table 10–3. Feeding structures differ in ciliation and stomatogenesis is of the parafissional opisthe mode in *Didesmis* (Wolska, 1966a), whereas *Raabena* seems to show a prefissional opisthe mode (Wolska, 1967b). Otherwise the other traits appear to be plesiosemes. (Compare Fig. 10–14M and Fig. 10–18.) The one troublesome point in this analysis is the transition between the two modes of stomatogenesis. With semes other than those relating to structure and development of feeding structures apparently homologous, it would appear homology is likely here too. Wolska is aware of the problem. She comments, "It could be postulated that a leap in development occurred—perhaps in connection with the atrophy of nonciliated kinetosomes on the body which exist in *Didesmis*—and caused that morphogenesis assumed [and caused morphogenesis to assume] the feature of restoring the ciliature entirely de novo without transmission of any rudiments of the parental ciliature to the opisthe". (Wolska 1967b, p. 289) The comment on *de novo* formation of the ciliature refers to a "vesicle", which is the anlagen or primordium of the opisthe buccal parts. This description of stomatogenesis, quite incomplete still, suggests that the prefissional opisthe stomatogenesis of *Raabena* utilizes a mechanism different from that seen in anarchic field development in the hymenostomes. In any case, if the foregoing phylogenesis is accepted—and there is much in its favor—it represents an independent origin of prefissional opisthe stomatogenesis from that in the other lines of ciliates. We have already touched on this problem regarding the scuticociliates and their descendants, and will discuss it again shortly.

Continuing on, estimates of phyletic distance show that the chonotrich *Heliochona scheuteni* is best derived from cyrtophorine forms such as *Nassula* as the work on its ciliary patterns had already established (Guilcher 1951; Dobrzanska-Kaczanowska 1963). Also arising from *Nassula* would be a form like *Pseudomicrothorax*, and from this there could be two lines of further evolution. One would go to *Frontonia*, probably by way of some tetrahymenine form, and signal the emergence of prostomium and opisthe prefissional stomatogenesis; the other would go to *Philaster*, with appearance of a heavy ciliature homologizable with the prostomial ciliation of *Frontonia* but lacking a prostomial cavity and retaining proter and opisthe prefissional stomatogenesis. (However, this pattern of stomatogenesis in the scuticoid forms may be an independently developed example of proter and opisthe prefissional stomatogenesis, as suggested earlier. It could be derived from a tetrahymenine form with opisthe prefissional stomatogenesis, but because of the posterior placement of the cytostome, it evolved a pattern paralleling that seen in the cyrtophorines. Further work is needed to resolve this dilemma.)

Small (personal communication) says that stomatogenesis in *Protocruzia* is of the same general pattern as in the scuticociliates, i.e., it is proter and opisthe prefissional stomatogenesis. This observation deserves publication because of its importance in determining the origin of the heterotrichous ciliates and the rest of the spirotrichous forms from them. Among these latter forms, prefission opisthe stomatogenesis is the rule (except for *Bursaria*), and one would expect that pattern to hold for *Protocruzia*, too. At present we shall make no decision (see Table 10-2 and 10-3) on this important point. It could be decisive in the derivation of the heterotrichs—from the scuticociliates or from tetrahymenine hymenostomes.

In terms of its feeding ciliature and morphogenesis at fission, the relationship of *Glossatella* to hymenostones is not implausible. The evolution of the sessile habit in hymenostomes has been suggested in forms like *Urocentrum* (Bradbury 1965), but this intermediate is not as convincing as others derived by way of pleuronematines (in Small's scuticoid forms) and the arhynchodine thigmotrichous ciliates (also scuticoid) (Fauré-Fremiet 1950a, 1954c; Corliss 1956, 1959, 1961; Raabe 1964b; see Fig. 10–11). Here similarities in form, development of adhesive structures, and development of large prostomial or epistomial fields all are suggestive of evolution towards peritrichs. Again the pattern of stomatogenesis in the scuticoid forms poses a dilemma. If they intervene in the evolution of the peritrichs, then the opisthe prefissional stomatogenesis of the latter group, which is also found in the hymenostomes and entodiniomorphids, is of independent origin from its occurrence in those two groups. This would be yet another possible example of parallel evolutionary development of stomatogemic patterns. (Another one, just mentioned, was the possible independent occurrence of prefission proter and opisthe stomatogenesis in cyrtophorines and the scuticociliates.) All that seems clear for the moment is the large value of R for peritrichs relative to other forms being studied here. This means there has been significant evolutionary development of the peritrichs, and intermediates must be sought to provide a convincing transition to them from other ciliates.

Even more so than with the peritrichs, the apostomatids and suctorians are distinct from the other ciliates, as well as distant from each other. (The desirable three-dimensional spacing cannot be represented accurately on the two-dimensional surface of Fig. 10–19B.)

The apostomatids, represented by *Spirophrya subparasitica*, are closest to *Coelosomides* and *Pseudomicrothorax*. This reflects its regular ciliation of the cell surface and the curious rosette structure associated with the feeding area (a degenerate nassalike cytostome?) but, especially in the process of stomatogenesis, two new oral areas are formed and the old one is lost. However, other characteristics, largely ecological but including loss of sexuality, remain to form a very significant gap betweeen *Spirophrya* and even its more similar free-living form. Because of this distance, it is not possible to state with any assurance the probable origin of the apostomatids beyond invoking an ancestor with a uniformly ciliated body and possibly lacking a prostomial cavity (hence, a gymnostomous ciliate, as others have proposed (Corliss 1958b, 1961)).

Equally, the evidence of great phyletic separation between the suctorians and the other ciliates renders it very difficult to make meaningful comments on their origins. All we can do at present is to point to some of the more obvious approaches to these problems. The suctorian problem was discussed in some detail, above, in connection with attempts to identify the plesiomorph of the group. It is sufficient to add here that the presence of trends in asexual reproductive patterns, in surface ciliation of the swarmer, in macronuclear organization, in number and distribution of tentacles, and in the nature of a scopula and stalk and a lorica (Collins 1912; Kahl 1931, 1934, Kormos 1958 and earlier papers; Paulin and Corliss 1969), all speak realistically to the future possibility of a meaningful discussion of the phylogenetic origin of these forms.

If Small's unpublished observations on stomatogenesis in *Protocruzia adhaerans* are correct, the heterotrichous ciliates appear very closely related to philasterid-type ciliates, i.e., the scuticociliates. Traditionally, the heterotrichs are derived from tetrahymenine hymenostomes (Furgason 1940; Corliss 1961) because of postulated homologies between the prostomial membranelles — by increasing their number one can convert a *Tetrahymena*-like prostomium into a *Protocruzia*-like one. The problem cannot be resolved here.

The step from the hetero- to the hypotrich necessitates compounding of cilia into cirri on the ventral surface and evolving their mode of formation at fission through the formation of new cirral fields. Such a sequence of changes seems plausible and would appear to involve selection pressures for further development of the hymenostome buccal or prostomial cavity and ciliature, along with special ventral locomotory ciliature.

This emphasis on exploiting a heavily ciliated stomial area and developing further the cell surface ciliature can be seen in the further fate of heterotrichous forms as specializations become distinctive in the three orders of the oligotrichs, odontostomes, and tintinnids. It appears the best interpretation of the phyletic distances is to derive each group separately from the heterotrichs. Jankowski (1964) finds it possible to originate the odontostome form, as seen in *Pelodinium reniforme*, from *Metopus*-like heterotrichs. The origins of the plesiomorphs of the other two orders are still obscure, though Corliss (1961) (Fig. 10–20) suggests that the tintinnids were derived from the oligotrichs, and recently (Corliss 1974a, 1974b) subsumes them as a suborder within the order Oligotrichida. Grain (1972) offers some fine-structure data in support of this.

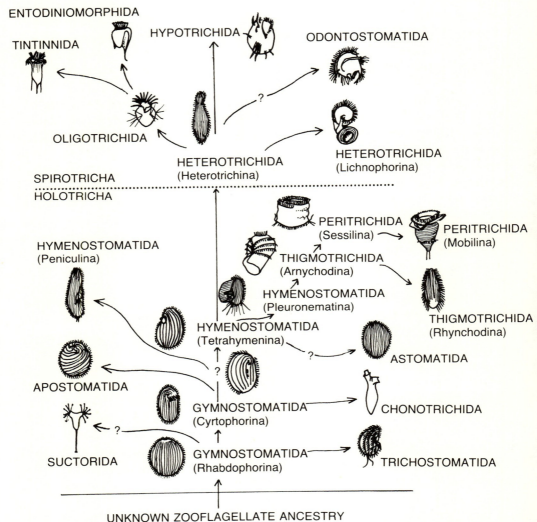

Fig. 10–20 A proposed phylogeny of the ciliates. (Redrawn, by permission, from Corliss (1961) *The Ciliated Protozoa,* Pergamon Press.)

Before going on to the larger phylogenetic conclusions that can be developed from the foregoing analysis, certain comments need to be made regarding certain phylogenetic details and regarding the weighting used for apo-and neosemes. In Chapter 3, there was emphasized the need to return to the original data so as to check on the validity of the dendrogram. The present study confirms the importance of that procedural guideline. For example, according to the R values, *Glossatella*, a peritrich, is phylogenetically closer to *Frontonia* than to *Philaster* (R = 13.3 and 17.8 respectively). For various reasons already referred to (Fig. 10–11), this is not a defensible conclusion. Reexamining Table 10–4 and the comparisons between species F and H and G and H, we find the difference resolves into one seme that is plesiosemic between F and H and aposemic between G and H. That is, micronuclei of *Glossatella* and *Frontonia* are alike, but are significantly different between *Glossatella* and *Philaster*. This one trait is not sufficient ground to derive *Glossatella* from the hymenostomes rather than from the scuticoid forms.

Regarding the problem of weighting, we see it expressed in several places. One example is the difference in the R values for species A and C and for A and H (Table 10–4). In the former comparison there are two plesio-, seven apo-, and three neosemes. The latter pairing should have been more divergent than the former, but yet they have the smaller R value. The reason lies in the effect of squaring larger numbers of aposemes; specifically, $(2 \times 7)^2 = 196$ whereas $(2 \times 6)^2 = 144$. The difference between 196 and 144 is not made up by changing the neosemes from two to three, i.e., $(3 \times 2)^2 = 36$ and $(3 \times 3)^2 = 81$. Hence, since $196 + 36 > 144 + 81$, the A and C comparison will have a larger R value than the A and H value, which is inconsistent with the observable evolutionary divergence between these species. The methodological problems will be discussed again in Chapter 14. They cause no serious problem now, for they are easily detected and appropriate allowances can be made for them.

PHYLOGENETIC CONCLUSIONS

It can be seen that the phylogenetic conclusions we have come to here are very close to those of Fauré-Fremiet (1950a) and Corliss (1956, 1961). For reasons made clear earlier, this was to be expected. It is also close to the ideas of Jankowski (1973) and Corliss (1974a, 1974b). Further discussion now will bear on differences between our proposal and those of these other workers and on a general appraisal of the validity of the present scheme (Fig. 10–19B) and how it can be significantly improved.

Other phylogenetic conclusions. Corliss' (1961) work is the most detailed exposition of the views that directly reflect Fauré-Fremiet's original proposals for ciliate phylogeny (Fig. 10–20). This view of ciliate evolution dominates the field today. The selection of a different plesiomorph from the ciliate prototype of Corliss has already been discussed fully. Our hesitancy to draw such precise origins for the apostomatids and suctorians as Corliss (though he explicitly labels his proposal with a question mark) lies in our awareness of their great divergence from other ciliate forms. We do agree with Corliss as to the most likely locus of origin of these forms.

Difference in treatment of the hymenostomes and forms derived from them result largely from appearance of Small's (1968) scuticociliates, a revision, of course, unavailable to Corliss, and our new classification of ciliate stomatogenesis. It suffices to point out the relative isolation of the peritrichs from other forms in our scheme is the result of omitting the many possible intermediates that other workers have suggested (Fig. 10–11). This is largely a limitation of building a phylogeny solely from plesiomorphs and illustrates most forcefully the major limitation of our phylogenetic conclusions. More on this shortly. Corliss' phylogeny presents several of the intermediates thought to lie on the possible path to peritrich evolutionary differentiation.

The origin of heterotrichous ciliates from scuticociliates or from a tetrahymenine-cyrtophorine is a major difference between our proposal and the Fauré-Fremiet and Corliss viewpoint. Here again the difficulties arise from examining only a limited number of forms and from the still open question as to the mode of stomatogenesis employed by *Protocruzia*. If that mode is like that found in the scuticociliates, then intermediates should be sought along those lines. If stomatogenesis is like that in the tetrahymenines, then intermediate forms should be sought in that area.

We find it difficult to derive the tintinnids from the oligotrichs rather than more directly from the heterotrichs. However, there is no decisive evidence, to date, for or against one or the other view.

One thing our diagram (Fig. 10–19B) does emphasize, and which is not so clear in Corliss' view (Fig. 10–20), is the degree of divergence among the various groups. This comes from the semiquantitative nature of the R values. They indicate that from *Stephanopogon* to *Protocruzia* the differences are hardly greater than from *Stephanopogon* to the suctorians or peritrichs. And from *Nassula*, representing the cyrtophorine gymnostomes, the distance to heterotrich forms is less than to the suctoria, peritrichs, and apostomatids. These observations offer support to those protozoologists who wish to recognize the diversity of the ciliates and keep the peritrichs and perhaps also the suctorians as separate subclasses (Finley, McLaughlin, and Harrison 1959; Hyman 1959; Cheissin and Poljansky 1963; Raabe 1964b; Canella 1964). However, these degrees of divergence are exaggerated by our exclusive attention to plesiomorphs.

Turning next to Corliss' reformulation of phylogenetic relations (Corliss 1974a, 1974b) which follow closely Jankowski's (1973) proposal, except for treatment of the scuticociliates (ignored by Jankowski), we see four points which need comment. (Compare Fig. 10-19B and Fig. 10-21.) (1) The especially important work of the French protozoologists is largely responsible for our increased awareness of the complexity and diversity of the gymnostome forms. A great deal of work is needed to understand better the diverse evolutionary lines present here and to discover which ones have led to further evolutionary developments. Our own analysis (Fig. 10-19B) does not do justice to this area. (2) The removal of the Entodiniomorphida from the spirotrichs is a bold step, for which there is sound evidence. But the puzzling gap between the patterns of stomatogenesis in these forms and their suggested parental form, *Didesmis*, should be commented on again and the search for intermediates strongly urged. (3) The suctorians are provisionally derived by Corliss, not from the gymnostomes, as we have argued following Kahl (1931), but from the Hypostomata—a taxon containing the cyrtophorine gymnostomes such as *Nassula*, the rhynchodines (thigmotrichs), apostomatids, and chonotrichs. Jankowski is much more noncommittal. (4) Corliss seems to have modified his earlier stout defense of a tetrahymenine form as being pivotal in ciliate evolution. Our own analysis emphasizing the scuticociliates as being as important as the hymenostomes is recognized in Corliss' new scheme. Jan-

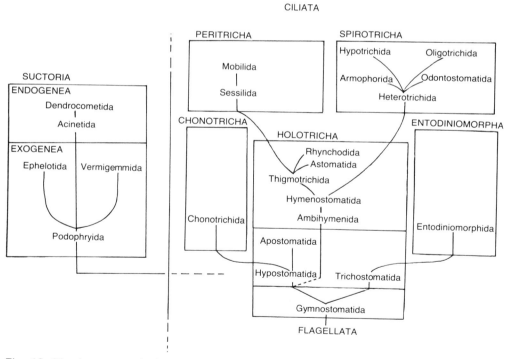

Fig. 10-21 A proposed phylogeny of the ciliates (Redrawn, by permission, from Jankowski (1973).)

kowski's lack of treatment of the scuticociliates, to date, avoids this problem area thus far. This point deserves further discussion, but in a broader context than that being considered here.

Ciliates can show evolution, in a model fashion, as successive branchings from a continually growing stem-line of development. Such a line is seen in the sequence: *Stephanopogon, Nassula, Philaster, Protocruzia, Kahliella*, and *Meseres. Philaster* is the only form not on our previous list of more reliable plesiomorphs. In any case, what has just been alluded to as the stem-line of ciliate evolution is weakest at its most critical juncture, which is the evolutionary radiation that comes with the origin of the ciliated prostomial cavity. Corliss' earlier emphasis on tetrahymenine forms seems now to have been premature. First, *Pseudomicrothorax* does not have a tetrahymenine pattern of stomatogenesis, nor do the scuticociliates, nor does *Protocruzia*, according to Small. It appears that the cyrtophorine pattern may persist well into the time when ciliated prostomial cavities are present. The same morphogenetic mode builds quite different feeding structures—the nassa of cyrtophorines and ciliated oral areas too. The typical hymenostome pattern of stomatogenesis—prefissional opisthe stomatogenesis—may have arisen independently four different times: in the hymenostomes, the peritrichs, the more complex heterotrichs, and in the entodiniomorphids. Prefissional proter and opisthe stomatogenesis seems to have been the pivotal mode of stomatogenesis, with the possible exception of the transition to the entodiniomorphids.

We need to comment here on emergence of postfissional proter and opisthe stomatogenesis. This is seen to occur in the ophryoglenids (Roque, Puytorac, and Savoie, 1965; Roque, Puytorac, and Lom 1967; Roque and Puytorac 1967), *Neobursaridium* (Dragesco and Tuffrau 1967), and *Bursaria* (Lund 1917). The ophryoglenids and *Neobursaridium* are safely ensconsed in the peniculine hymenostomes, but *Bursaria* is treated as a heterotrich (as was *Neobursaridium* initially). Tuffrau (1967a) has shown that *Bursaria* has a complex paroral kinety, and one can only wonder if this genus is not also better seen as a peniculine hymenostome, which takes the evolutionary tendencies of *Neobursaridium* yet further. If so, postfissional proter and opisthe stomatogenesis would appear as a special development restricted to the most complex peniculines.

Returning now to the phylogenetic problems, we continue with comments on the general trends of ciliate evolution. At several points along the major line or stem-line, there have been various degrees of adaptive, cladogenetic radiation, with perhaps the most varied burst coming at the beginning, in the gymnostomes, and at the end, in the various spirotrichous forms diverging from the heterotrichous plesiomorph of that whole class. As the phylogeny of each group of ciliates becomes better known through critical analysis, it will be possible to find homologies not just between plesiomorphs of the various groups but between plesiomorphs of the derived group and some special ancestral form lying above the plesiomorph of the parental group, as is already argued for the derivation of the entodiniomorphids from a didesmidlike form in the Trichostomatida (see Fig. 10–21 and Fig. 10–17M). Thus the peritrichs are not, as we have already stated, reliably derived from any of the forms discussed here, but more probably come from some series of forms originally derived from the hymenostomes or the scuticociliates. The discovery of these intermediates is the task of future phylogenetic work.

A definitive phylogeny needs the cooperative effort of thoughtful specialists in every group of ciliates. Only in this way can the crude approximation of ciliate evolutionary history, previously limited by inadequate theory and limited, as in these pages, to comparisons of plesiomorphs, achieve the full potential available in this magnificent material. And by using operations based on sound selectionist theory and carefully defined concepts of homology, there can be achieved general agreement on the evolutionary behavior of these forms. If these procedures can satisfy the criticisms made over the years by Canella (1957, 1958, 1964, 1970), who has rightly denounced the validity of "lower" and "higher" forms and criteria of so-called "fundamental" characters, then we have indeed achieved a useful advance in phylogenetic thinking.

Major evolutionary problems. From present work we are in a position to see the outlines of the larger evolutionary tendencies in the ciliates. Structures and processes concerned with the physiological maintenance of the cell and the reproduction of the cell, plus sexual processes and the nuclear apparatus, are all quite conservative. However, these appear to have been bestowed on the ciliates in a highly developed form, judging from their almost universal occurrence as complex semes showing limited aposemic development. Conjugation, especially, is little modified throughout the group, and the same for homothetogenic fission; the nuclear apparatus displays its distinctive organization in the vast majority of species. On the other hand, as has been repeatedly emphasized, it is the cortical structures, concerned with locomotion or adhesion and with feeding, that are most variable. Of the innovative features, stomatogenesis shows an especially interesting pattern (Fig. 10–19B). There is a profound problem, combining genetics, development, and evolution, awaiting solution as to how the ciliates have gone from one stomatogenic mode to another, and, with just four modes and their variants, have managed to build some thousands of different oral structures. The permanent phagocytic cytostome—the essential exploitive innovation of the ciliates, rarely if ever present in other protozoa—must of functional necessity lie in the cortex. And since that cortex carried the extraordinary morphogenetic properties of the kinetosomes (Fauré-Fremiet 1950b; 1954b, Sonneborn 1963, 1970a; Tartar 1967; Hanson 1967a), here was a special combination of events which has been exploited to remarkable degrees of structural complexity by the ciliates and all within the organizational confines of a unicellular system.

"... one is tempted to admit that these organisms have realized all the possible combinations implied by their basic ciliate apparatus. This would explain that their diversification has been completed, one might say, within a closed circumference; that which can yet arise within this diversification of structural innovations based on cytoplasmic differentiations is limited; the tentacle of the suctorians, for example, seems without a future." (Fauré-Fremiet 1952, p. 281. Original in French.)

If the future of the ciliates is limited, what of their past? Where did they come from? The absence of fossils or of biogenetic evidence incorporated in the life of *Stephanopogon* exclude any direct inferences. But we do have a good plesiomorph and from this can get a quite clear picture of the primitive or eotypic ciliate. The further question is to then ask whether any of its essential features are yet apparent in other organisms and to then evaluate the plausibility of evolutionary connections.

The ciliate eotype, or ancestral form, would be an ovoid or possibly somewhat flattened cell, about 40 to 80 micrometers long. Its surface would carry quite uniform, longitudinal rows of cilia. There would be a slitlike, permanent mouth towards the anterior end of the cell. (Anterior would be defined by the most common direction of locomotion, this being largely parallel to the kineties, since motion would be dependent on coordinated ciliary beat.) The mouth or its immediate area might well be provided with organelles, such as toxicysts to aid in feeding. This would also probably be a marine form, spending much of its time as a bottom feeder, ingesting unicellular algae, small flagellates, occasional detritus which, in the absence of animal forms larger than itself, would be the decaying parts of plants. Bacteria would probably be ingested with the detritus. Internally, the cell would have various specializations associated with the ciliary basal bodies, and further internally, there would be the endoplasm, probably quite firm peripherally and more fluid internally, where food vacuoles would come to lie following ingestion. There would be two or more diploid nuclei, possibly promitotic, but dividing in synchrony with the cell at division. Cell divisions would be perkinetal, i.e., cutting across the kineties and at right angles to the longer axis of the cell. A new cytostome would develop at the time of fission just posterior to the fission plane to provide the opisthe with a cytostome. These cells would also be capable of sexual activities in the form of conjugation. Morphologically isogamous, but probably physiologically discrete, the conjugant cells would unite with ventral surfaces apposed; reduction divisions of the nuclei would supply haploid gametic nuclei; some of these, or maybe only one per cell, would be mutually exchanged and fertilization would ensue, with restoration of the typical nuclear situation in each cell.

Are there any nonciliate organisms showing one or more of the foregoing features which are plausibly homologous to them and which, therefore, supply evidence as to the origin of the ciliates? There are two places to look—the opalinids and the zooflagellates. We have not previously mentioned the former organisms. They are ciliated symbionts of cold-blooded vertebrates, being especially abundant in the rectum of frogs. Furthermore, they possess many diploid nuclei, and fission can be interkinetal (parallel to the kineties) or perkinetal (across the kineties). They lack a mouth and do not conjugate. (General descriptions in Grassé 1952 and Wessenberg 1961.) For many years these organisms were classified as ciliates. Then the French workers looked past the external ciliation, and largely on the basis of a single type of nucleus, absence of a ciliate life cycle (especially the lack of conjugation), and fission that can be longitudinal or symmetrigenic (interkinetal), they included the opalinids among the zooflagellates. The taxonomic question is still not fully resolved (Honigberg et al. 1964). A neutral position is to leave the opalinids in limbo—neither ciliates nor flagellates.

Our problem is not a taxonomic one but a phylogenetic one. Are there homologous relationships between the opalinids and *Stephanopogon mesnili?* The answer is inconclusive (as it will also be when looking for homologies between the ciliate plesiomorph and the zooflagellates proper). In particular, it is difficult to see how vertebrate symbionts could be ancestral to ciliates, which evolved long before vertebrates appeared on earth. The only possible use of opalinids in the present context is to suggest they evolved from some ancestral form that might also have been ancestral to the ciliates. That form could have arisen from the zooflagellates. So let us now examine further the possibility of a zooflagellate ancestry.

These are the forms usually proffered as having some chance of being ancestral to the ciliates (Hyman 1940; Grassé 1952b; Corliss 1956, 1961; Pitelka 1963; Kudo 1965 and earlier editions of his text). This suggestion is based on the common occurrence in ciliates, and in the more complex flagellates especially, of extensive development of the kinetosome and its associated structures. And then the subject is usually not taken much further. In 1912, Minchin put the situation in words which Corliss (1956) found pertinent to the contemporary view of the problem. Minchin said, "From these various considerations [largely morphological] it seems highly probable that the Ciliata are descended from flagellate ancestors; but it is not possible at present to indicate with any approach to exactness the line of descent." (Minchin 1912, p. 455)

However, it has been suggested (Hanson 1958) that various features of certain zooflagellates resemble key ciliate features. For example, multiple nuclei are found in *Calonympha*, a heavily flagellated member of the Trichomonadida. The multiflagellated *Urinympha* and *Notila* are diploid. Various forms show sexuality, though it invariably involves fusion of the cells. But the problem with these similarities is that they are scattered among various zooflagellates and that such zooflagellates are all symbiotic, either in the roach or termite gut. It is impossible to make a sensible argument for ciliate origins when the necessary traits must be borrowed from different organisms that are adapted to highly specific niches involving symbiosis.

The only sure way around that difficulty is to find a free-living, preferably marine, zooflagellate exhibiting all the key traits. As was seen in Chapter 5, there are no such flagellates known, and the chance of their being discovered is really nil. However, that such forms once existed is not improbable. Referring to our earlier discussion of zooflagellate evolution, it will be recalled that there are compelling reasons for looking at the present-day distribution of zooflagellates as being the result of survival of only those capable of invading very specialized niches. Something, it was stated earlier, had very possibly wiped out the vast majority of free-living zooflagellates, leaving behind only specialized relicts of a once possibly abundant and quite certainly highly varied array of these cells. It is not unreasonable to suggest (Hanson 1958) that the ciliates were at least one agent in the extinction of free-living zooflagellates, and subsequent evolution of micrometazoans could have completed the process.

Indeed, in comparing a ciliate—even the eotypic ciliate—with flagellates, one can imagine that a form provided with a permanent ingestatory structure would be a more efficient consumer than a form dependent on ameboid ingestion of food, as would presumably have been true of large, complex, free-living zooflagellates. And if one supposed that a multinucleated (diploid nuclei), heavily flagellated protozoan evolved a permanent cytostome, this could have been the founder of the line giving rise to the ciliate eotype and finally to the known ciliates of today. Pygmalion-like, this creation of the zooflagellates came to dominate its creator.

Such prose, unfortunately, proves nothing. We must really agree with the pessimists and admit that there is no convincing evidence regarding the zooflagellate origin of the ciliates. What we can do, however, is point out what the nature of the transition must have been, if it did occur, and thus see the problem as clearly as possible. And secondly, we can point out why we should expect never to have a clean answer. The latter point first: The proposed key innovation—development of a true cytostome—is an exploitive feature, and use of it to compete successfully with flagellates, or even to consume them directly, as well as the significant evolution that must have occurred to develop the mouth eventually seen in *Stephanopogon*, clearly implies that forms ancestral to the eotypic mouth would be casualties of its evolution. The appearance of a cytostome-equipped protozoan doomed closely related forms not so equipped. A hiatus between survivors related to the ancestral form and the descendants would be inevitable.

What might have been the changes present in the intermediate forms that would bridge ancestral and descendant types? Conjugation would have to change from fusion of cells to temporary union of cells with exchange of gamete nuclei. This is a big demand. However, the first problem in such questions of the evolution of complex traits is to know if the outcome would be of selective advantage over the prior condition. If the answer is positive, no excess of time need be spent looking for intermediates; selection pressure is its own best explanation. In the present instance, if conjugation can produce two new genotypes, different from parental ones, where gametic fusion formerly produced only one, the advantage is on the side of conjugation because of increased variability, and its development from a fusion process is not improbable.

Symmetrigenic fission would have to change to homothetogenic fission. Also, since fission in the zooflagellates typically involves a centriole and a large spindle whose orientation may well be correlated with the fission plane, we must also postulate loss of an extranuclear centriole and spindle. Again, are such losses of selective advantage to a ciliate? This is very difficult to answer. It may be that localization of all morphogenesis in the cortex, including formation of a constrictive girdle at times of cell cleavage, is advantageous, if only because it frees the cortex even more from ties to the nuclear machinery and allows both sets of structures to evolve independently of each other, which is precisely the case in the ciliates. This independence can be seen as providing a degree of developmental versatility not attained by the zooflagellates. Why the cleavage plane should move from interkinetal to perkinetal fission is not clear. It may be related to location of the cytostome. The differentiated mouth at one end of the cell may have inhibited, or at least somewhat complicated, the formation of a fission plane. Since an anterior mouth is the best place for that structure, it was the fission plane that was moved. Such conjecture is all we have to go on at present on this important problem.

The development of macronuclei can be seen to have occurred after the ciliates were well differentiated (Raikov 1969). Hence the nuclear picture presents only the question of whether diploid nuclei were present in the ancestral zooflagellate. They quite plausibly were there.

The surface ciliature seems to be derivable from rows of mastigonts. However, actual homologies between mastigonts and kinetides can only be established by fulfilling the serial relationship, and we have no hope of doing so. Hence this transition must remain simply as a possibility. However, functional, morphogenetic, and certain fine structure similarities argue favorably for this possibility.

The permanent cytostome would be a selectively advantageous advance over feeding by ameboid ingestion, and that is the essential argument in favor of the development of ciliate feeding structures. We should not expect ever to find living intermediates between mouthless zooflagellates and the typical ciliate condition, except in instances of developmental recapitulation. Conjugation, as we saw, contains an almost mouthless stage in it. And with that we run out the thread of plausible speculation to its end.

From a selectionist point of view, therefore, the origin of the ciliates from zooflagellates is a paradoxical problem: selective advantages of ciliate organization as compared to the zooflagellates make the transition not unreasonable; the action of selection is what has removed the evidence needed to substantiate the argument. This is, of course, not an uncommon situation (Hanson 1972); it represents the logical limit of phylogenetic analysis in such situations. At first glance, it seems to be a kind of uncertainty principle analogous to Bohr's (1933) suggestions regarding the study of life itself, wherein the method of analysis (reduction to nonviable parts) destroys the problem we wish to analyze (viability). Evolution, however, puts its own twist on the problem, for it is not that experimental manipulations, as such, set the limits to inquiry, but that the theoretical explanations we are using, i.e., the evolutionary process, is seen to necessitate the difficulties in elucidating evolutionary history. Selectionist theory says we must expect breaks in the continuity of phylogenetic evidence.

The ciliates are a magnificent product of evolution; within the group there are scores of problems awaiting clarification of just how evolution proceeded in bringing to expression the extraordinary potentials of these cell systems. They have reached a level that, the greatest student of these forms says, raises them ". . . to the rank of pseudometazoans." (Fauré-Fremiet 1952) Such a comment now poses the inevitable question: How did the real multicellular animals, the eumetazoans, actually arise?

Chapter 11

Sponges

A DELIGHTFULLY naive thumb-nail sketch of the sponges, reflecting then current beliefs in spontaneous generation, is found in Gerard's Herball, printed in 1636: "There is found growing upon the rockes near unto the sea a certaine matter wrought together of the foame or froth of the sea which we call Spunges . . ." (Quoted in I.B.J. Sollas, 1906, p. 167.) However, except for the origin of the first sponges, it is only sponges that make more sponges, and we must look far beyond the generative powers of sea foam to understand the origin and organization of this group.

Sponges are essentially cavitated systems for filtering water. Their lives are so intimately adapted to this function that many workers subscribe to a view such as this: "Among the Metazoa, the sponges constitute a phylum characterized by an exceptional anatomical structure having no homolog in the animal kingdom." (Lévi 1956, p. 5.) Though no other animal (nor plant, either) is comparably organized in its gross anatomy, there are, of course, details of cellular organization, of reproductive function, and other traits that can be examined in an effort to understand the origin and evolution of the sponges—to understand the historical development of animal cells acting cooperatively to form a water channel system for recovering particulate food from an aqueous environment.

In what follows we shall have occasion to refer to calcareous sponges, demosponges and hyalosponges. This classification, used for many decades now, refers to skeletal elements. Calcareous sponges, calcisponges, or Calcarea have only calcium carbonate spicules or lithistid skeletons made of fused skeletal elements called desmas. The other two classes all have siliceous spicules or skeletons, with or without proteinaceous spongin, and some have only spongin for supporting material. The hyalosponges, or Hexactinellida, have only siliceous skeletal parts, and these consist of six-rayed spicules or their fusion products as skeletons, or of some spicule type derived from the six-rayed condition. The demosponges (Demospongiae) contain forms with monaxonic or tetraxonic spicules, with or without spongin, and the horny sponges possess only spongin. Hartman and Goreau (1970) have urged recognition of a fourth sponge class, the Sclerospongiae or coralline sponges. No detailed treatment of this group has been published thus far, and, therefore we cannot treat it in detail.

Subdivisions within these three great classes vary in stability (Table 11-1). The instability seems to arise from the fact that earlier classifications were based simply on the geometry of the spicules and their distribution among various groups. As further information, notably embryological, has accumulated, it has been seen that spicule geometry, though useful for Adansonian systematics, does not reflect possible phylogenetic relations. This situation was to be expected. Adansonian classifications are bound to face revision when more complex traits are analyzed and more searching analyses of evolutionary trends become available. Lévi's (1956) revision of the Demospongieae is a perfect case in point. However, the largest subdivisions of the sponges—the three (or four) classes—are not threatened by revision, since apparently the spicule differences are so deep that they do reflect important evolutionary distinctions. Other traits also are found to separate these classes.

Table 11-1 Three classifications of sponges. It should be noted in Brien's scheme that no precise taxonomic level is given for the *Calcarea homocoela* and the *Calcarea heterocoela*. Further, Brien includes the Homosclerophorida under the Tetractinellida, a treatment differing from that of Lévi.

	Hyman (1940)	Brien (1968) (after Lévi 1956)		Hartman (1971)
Class	Calcarea or Calcispongiae	Calcarea	Class	Calcarea
			Subclass	Calcaronea
Order	Homocoela	*Calcarea homocoela*	Order	Leucosolenida
				Sycettida
	Heterocoela	*Calcarea heterocoela*	Subclass	Calcinea
			Order	Clathrinida
				Leucettida
				Pharetronida
Class	Demospongiae	Demospongiae	Class	Demospongiae
Subclass	Monaxonida	Ceractinomorphae	Subclass	Ceractinomorpha
Order	Halichondrina	Halichondridae	Order	Halichondrida
	Hadromerina	Haploscleridae		Haplosclerida
	Poecilosclerina	Poecilosclerida		Poecilosclerida
Subclass	Keratosa	Dendroceratidae		Dendroceratida
		Dictyoceratidae		Dictyoceratida
Subclass	Tetractinellida	Tetractinomorphae	Subclass	Tetractinomorpha
Order	Myxospongida		Order	Homosclerophori
	Carnosa	Clavaxinellida		Clavexinellida
	Choristida	Tetractinellida		Choristida
Class	Hexactinellida	Hexactinellida	Class	Hexactinellida
			Subclass	Hexasterophora
Order	Hexasterophora	Hexasterophores	Order	Hexactinosa
				Lynchnisacosa
				Lyssacinosa
				Reticulosa
			Subclass	Amphidiscophora
Order	Amphidiscophora	Amphidiscophores	Order	Amphidiscosa
				Hemidiscosa

General Characteristics

The polymorphism and general indeterminancy of body form in the Porifera are implied by the 17th century quotation cited above, as well as an awareness of the marine littoral habitat where sponges are common. Sponges are best known morphologically, there being some 10,000 species descriptions, although half of these are thought to be synonyms (Laubenfells 1955), due undoubtedly in large part to the polymorphism just alluded to. For reasons of constancy of form and ease of preservation, spicular hard parts form an important source of information about sponges and have been of major use to sponge systematists. Furthermore, sexual reproduction is of common, if not universal, occurrence, and such a process is a most important dimension in analyzing evolutionary relationships within the sponges. However, the reproductive behavior of sponges is nowhere near so easily studied as their morphology. Not all sponges, by any means, are shallow water forms, and our ignorance of much of the biology of deeper water forms is profound. This is especially unfortunate, because it seems, from studies such as Lévi's (1956) classic one on the development and phylogeny of the demosponges, that development is the key trait in understanding sponge evolution.

With these problems in mind, we turn to a survey of what is known about sponge structure and function. This account is based on Sollas (1906), Henstchel (1923–1925), Hyman (1940), a symposium volume of the Zoological Society of London (Fry 1970), Tuzet (1973) and others as indicated.

STRUCTURES

Body form. Though radial symmetry is the commonest general description of sponge organization, it is a term honored more in the breach of the generalization than in its realization. The fact that tubular channels exist in all sponges for the passage of water is the basis for the notion of a radially symmetrical organization, but in that sponges compound these channels in such multifarious ways and show such irregular growth patterns, the notion of radiality is not rigorously applicable. Sponges are more often asymmetric than symmetric and show a polarization in terms of being attached to the substratum with the rest of the body extending into the aqueous environment and having a one-way flow of water through their bodies (only rarely reversible). The result is forms that are generally somewhat amorphously cylindrical vase- or urn-shaped with one, but usually many, water channels and the body attached, in various ways, directly to the substratum, which can be bottom muds or sand but is commonly a more solid base of rocks, corals, or the like (Fig. 11–1).

The flagellated chamber is the constant feature of the water channel system. Tubes leading to and away from these chambers show almost endless variability

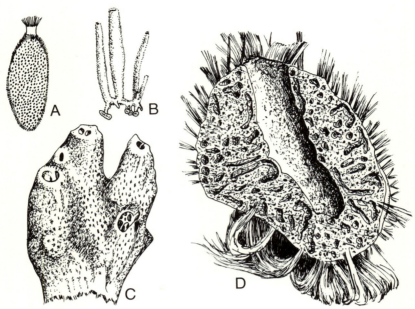

Fig. 11-1 Selected sponges illustrating variations in radial symmetry. A. *Sycon*. B. *Leucosolenia* showing a cluster of ascon-type tubes. C. *Tedania* showing various excurrent openings or oscula towards the apices of the sponge body. D. *Pheronema*, in longitudinal section, revealing the central spongocoel. Hairlike structures at the base of the figure are root spicules, used for attachment to the substratum. (All redrawn, by permission, from Hyman (1940) *The Invertebrates. Protozoa through Ctenophora*. McGraw-Hill Book Company, Inc.)

in their ramifications, even to the point, in a few forms, of being essentially nonexistent, but a well-formed chamber is, itself, a *sine qua non* of sponge existence. Water leaves the chamber through one opening but may enter through one or many openings. If there is one incurrent opening, it lies at the opposite end of the chamber from the excurrent one. If there are many incurrent openings, they are so small as to be designated pores and are found all over the sides of the chamber, but with the excurrent pore again situated at one end of the chamber. The chambers themselves are ovoid or cylindrical in shape, and their lining is composed of collar cells or choanocytes (Fig. 11-2). It is the flagellar activity of these cells that generates the water currents in the sponges. Outside the choanocytes, the chamber wall is made up of various types of amebocytes, often lying in a jellylike mesoglea, usually with, but occasionally without, spicules and bounded externally in most cases (the hyalosponges are the conspicous exception) with an epidermal layer of much flattened pinacocytes.

In no case is a sponge composed of a single flagellated chamber, but forms such as *Leucosolenia* and *Clathrina* are composed of several tubes, standing alone and upright and joined at their bases, and each tube is in effect a single flagellated chamber (Fig. 11-1B). In this situation water enters

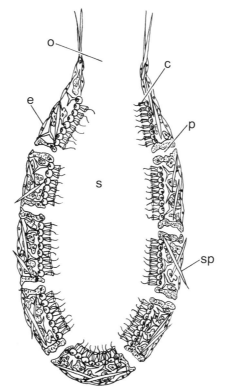

Fig. 11-2 Diagrammatic longitudinal section of an ascon-type flagellated chamber. c—layer of choanocytes, e—epidermal cell (the mesenchymal cells lie between the choanocytes and epidermis), o—osculum, p—pore through a porocyte, s—spongocoel, sp—spicule. (Redrawn, by permission, from Hyman (1940) *The Invertebrates. Protozoa through Ctenophora.* McGraw-Hill Book Company, Inc.)

through small pores in special cells, the porocytes, and goes to the spongocoel (Hyman 1940), or cavity of the chamber, where food particles are removed by entrapment on the collar of the choanocytes. Water leaves the spongocoel by the relatively large excurrent pore, or osculum. This simple asconal type of organization is not common.

More complicated is the sycon, wherein the flagellated chambers are found to lie around the periphery of a common chamber into which their excurrent water flows and out of which the water then passes through a single large osculum. This condition is found in the genus *Sycon* and is represented, diagrammatically, in Fig. 11-3B. A yet more complicated syconoid condition is seen where the wall containing the many flagellated chambers develops a thickened outer, cortical layer. The result is that water must now first come through dermal pores in the outermost layer, into subdermal spaces which are continuous with incurrent canals lying between flagellated chambers, out of the chambers into the large common spongocoel, and then out the osculum

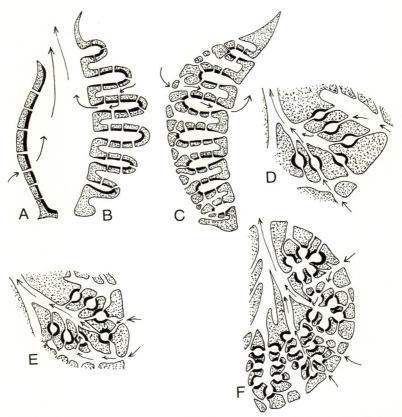

Fig. 11-3 Schematic view of sponge walls. A. Asconoid type. B. Syconoid, early stage without cortex. C. Syconoid stage with cortex. D.–F. Various leuconoid conditions. D. Diplodal. E. Aphodal. F. Without special incurrent and excurrent channels for each chamber, the eurypylous condition. (Arrows indicate the direction of water flow.) (Redrawn, by permission, from Hyman (1940) *The Invertebrates. Protozoa through Ctenophora.* McGraw-Hill Book Company, Inc.)

(Fig. 11-3B). The thickening of the wall in the syconoid forms is accompanied by the arrangement of spicules, in special patterns in the cortex and also lying in the mesenchyme, so as to achieve a supportive function.

Leuconoid forms, which are the great majority in the Porifera, show yet further development of the water passages. Here the chambers may each have their own excurrent and incurrent tubes (the diplodal condition). The latter tubes come off of incurrent channels in turn connected to subdermal spaces and dermal pores, and the former lead to common excurrent channels terminating eventually in oscula. Or the chambers may be fed directly by a large common incurrent channel and the only channels are at the excurrent side (the aphodal condition). Or yet again, and this is similar to the syconoid condition, the chambers open off of common channels and feed into common channels (the eurypylous condition). The difference from the syconoid mode is that the excurrent channels may combine several times to form a quite large channel leading to the osculum. In this way the spongocoel of the sycon is replaced by a series of channels. (Fig. 11–3C, D, E). Where there is significant development of the cortex to the extent of showing cell and spicule types and spicule arrangements that are unique to the superficial parts of the sponge, then this outer cortical layer is designated the ectosome, in distinction to the choanosome, which contains the flagellated chambers and excurrent channels.

A final type of chamber arrangement is termed the rhagon. This is found as the simplest form of the demosponges (Fig. 11–4). It consists of pyramidal spongocoels, surmounted by an osculum. More or less eurypylous chambers empty into common excurrent chambers, which feed into the common spongocoel. The rhagon type is probably best seen in young forms (Fig. 11–4). Some authors (e.g., Laubenfells 1955) do not recognize the leucon and simply refer to the most complicated sponge forms as rhagons. Others see the rhagon as a simpler stage of the leucon. The implication is that the leucon arose from a rhagonoid form as well as from asconoid or syconoid ones. The indeterminacy of the final or adult leuconoid form precludes, however, seeing in it any characteristics that always identify it as being derived from one source or another. Hence leuconoid is applied to all the vast array of forms not clearly asconoid, syconoid, or rhagonoid. From a phylogenetic point of view, it is a grab-bag term, but the reason for it seems clear. Hyman puts the whole problem succinctly when she says, "The vast majority of sponges are constructed on the leuconoid plan, which exhibits innumerable variations and has undoubtedly arisen independently over and over again in different groups of sponges. No doubt much the greater frequency of the leuconoid than the other canal systems may be attributed to the large size permitted by this type

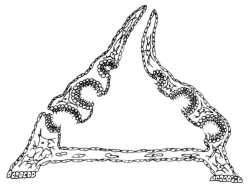

Fig. 11–4 Larval stage of rhagonoid organization. (Redrawn, by permission, from Hyman (1940) *The Invertebrates. Protozoa through Ctenophora.* McGraw-Hill Book Company, Inc.)

426 *Phylogeny of the Protozoa and Early Metazoa*

of structure and its great efficiency in producing a water current." (Hyman 1940, p. 293).

Histology. The choanocyte is the most distinctive cell type in the sponges. Some authors speak of a generalized type of this cell (Tuzet 1963, Fig. 11–5A) and others mention its various forms (Lévi 1956, Fig. 11–5B). In any case, the constant presence of certain features is widely agreed upon. There is never more than one flagellum, and this arises at the unattached end of the cell and extends into the cavity of the flagellated chamber. Surrounding the flagellum is a collar which, as electron microscopy shows, is composed of one row of microtubules of an average diameter of about 0.10–0.15 micrometers (Rasmont *et al.* 1957). The flagellum terminates proximally on a basal body

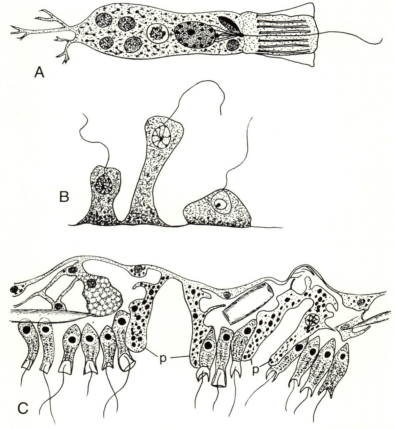

Fig. 11–5 Cell types in sponges. A. Choanocyte. (Redrawn, by permission, from Tuzet (1963).) B. Variability of form among choanocytes. (Redrawn, by permission, from Lévi (1956).) C. Porocytes (p) as seen in *Leucosolenia,* an asconoid member of the Calcarea. (Redrawn, by permission, from Hyman (1940) *The Invertebrates. Protozoa through Ctenophora.* McGraw-Hill Book Company, Inc.)

(kinetosome), and to this is connected, according to light microscopic investigations, a parabasal body and various fibrils, among which is a paradesmose whose other end terminates on a granule (centriole) lying next to the nuclear membrane (Tuzet 1963; Fig. 11–5A). This pattern is not universal. In the Calcaronea of the calcisponges, the choanocyte nucleus is apical and the flagellum seems to arise from it (Hartman 1971), and in the Calcinea there is no connection between the basal nucleus and the apical flagellum (Hartman 1971). This placement of the nucleus has been used as a taxonomic character by various workers (e.g., Sollas 1906). The cytoplasm of the cell contains the usual organelles—mitochondria, endoplasmic reticulum, Golgi apparatus, etc.—as well as food vacuoles. The base of the cell may show ameboid extensions and the overall shape of the cell may be columnar, ovoid, or pear shaped with the broad expansion basal.

The epidermal cell or pinacocyte is not found in the hexactinellids or glass sponges (hyalosponges) but is present in the two much larger groups of the calcareous sponges, or calcisponges, and the demosponges. It is usually a very flat cell, of polygonal shape when viewed on its external surface, and provides an outer covering for the sponges, interrupted only by pore cells, dermal pores, or extrusion of spicules. Fine structure details are given by Bagby (1970).

The external layer of pinacocytes and the internal one of choanocytes are the closest sponges come to true tissue formation. But the looseness of organization of the cells in these two instances leads most workers to disclaim that typical epithelial tissues are present (Hyman 1940; Laubenfels 1955). And certainly the loose, seemingly randomly organized ameboid cells and mesoglea of the rest of the sponge body are not to be characterized as tissue. The sponges do appear, then, to be an aggregation of different cell types not yet at the tissue level of organization.

The ameboid cells, just mentioned, represent the last major category of sponge cells. These amebocytes have a great variety of shapes and sizes and an equal variety of functions, it seems. A quick listing of these cells includes the following: archaeocytes—large amebocytes that may be the sole source of gametes and important in regeneration; chromocytes—pigmented amebocytes; desmocytes—fiber cells, common in the demosponges; gland cells—amebocytes specialized for secretion; gametocytes—amebocytes committed to development into eggs or sperm and rapidly losing their ameboid form; myocytes—muscle cells (see Bagby 1966 for fine-structure details), which are important as sphincters around pores or oscula; scleroblasts—cells which lay down the various types of skeletal material; thesocytes—cells containing food reserves. The foregoing list, adapted from Hyman (1940), is not meant to imply that all amebocyte types are present in all sponges, for they are not. It does point up, however, some of the various functions played by the amebocytes and certainly poses acute problems in cellular differentiation.

It is useful to see a different listing of this variety and a different view, also, of sponge organization. Lévi (1956) views the sponge as being formed of choanocytes and ectomesenchyme. The latter contains three types of amebocytes, macrocytes, fibrils, spherical cells, and dermal cells as well as pinacocytes. Also, the jellylike mesoglea is understood to be present in greater or lesser abundance, depending on the species. The emphasis of this view is on the cells of the flagellated chamber as compared to all other cells. The latter are polytypic as compared to the monotypic collar cells. It is the polytypic ectomesenchymal cells that carry the burden of sponge variety.

Using electron microscopy, Borojevic and Lévi (1964) have described eight different cell types in the demosponge *Ophlitospongia seriata*. That listing includes the following types: archeocytes (amebocytes), pinacocytes, and collencytes; gray cells ("cellules grises"); cells with globules ("cellules globifères"); cells with small granules ("cellules microgranulaires"); cells with rhabditic material ("cellules rhabdifères"); choanocytes; and reproductive and skeletogenic cells. (The latter two being lumped, for unclear reasons, into one category.) And Cowden (1970), examining the "connective tissue" in six different species of demosponges, reports considerable variability in his findings.

In recent years Tuzet and others (Tuzet and Pavans de Cecatty 1952, 1953; Pavans de Cecatty 1955; Roskin 1957; Tuzet 1963, 1973) have claimed that nerve cells are present. Careful reviews of this work (Jones 1962; Bullock and Horridge 1965) do not confirm nor definitely deny the claim, but the weight of the evidence seems to be on the negative side at present.

It is clear that there is as yet no widely-accepted view on the details of amebocyte diversity in the sponges. The safest general conclusion would seem to be that diversity of ameboid cells is present and is a reflection of different activities. Whether this diversity is the result of one ameboid type differentiating in various directions in response to varying stimuli (determined by functional states of the sponge body?) or whether there are permanently differentiated strains of ameboid cells is yet to be determined. Lévi (1970) favors deriving all cell types, except perhaps the choanocytes, from archaeocytes.

Skeleton. Sponge spicules are inorganic though apparently laid down in all cases around an organic core. Sponge fibers, the other skeletal element, are of a horny proteinaceous material, a keratin. Looking first at the spicules, they are classifiable, as we have seen, into those formed of calcium carbonate and those that are siliceous; they are also classifiable in terms of their size and general location in the sponge and in terms of the number of axes and spines present. The latter classifications are more detailed and discriminatory, hence of greater use from a phylogenetic viewpoint.

In terms of size, spicules are seen as megascleres or microscleres. Most megascleres have shafts greater than three micrometers in diameter and an overall length greater than 100 micrometers (Laubenfels 1955). Giant examples occur, such as the anchoring spicule of *Monoraphis*, which can be one centimeter thick and two to three meters long (Hyman 1940). Microscleres are usually less than one micrometer thick in shaft or spine diameter and 10 to 100 micrometers long (Laubenfels 1955). Megascleres are typically seen in rather regular alignments at the surface, along radial canals or in the choanosome (the area of flagellated chambers) in general, and at the borders of the spongocoel. Their array is clearly suggestive of a supportive function internally and a protective function at the external surface (Fig. 11-6). The microscleres are commonly described as lying in the "fleshy" part of the sponge, that is, in the mesenchyme. Their somewhat random distribution and small size renders their function enigmatic.

In terms of numbers of axes and spines, the following abbreviated outline of spicule diversity can be used (Hyman 1940).

(1) Monaxons. These are spicules with one axis, and are, hence, needle or pinlike. When both ends of the axis are similar in form—pointed, rounded, knobbed—this implies growth has occurred at both ends and the monaxon is called diactinate or is a rhabd. When the ends are different, one being blunt and the other one variously shaped and implying unilateral growth of the spicule, it is called monactinate or a style. In the sponges with calcareous spicules, monaxons are common and represent one of the two chief patterns of spicule formation, the other being the tetraxon. In the sponges with siliceous spicules, it seems monaxons are simplified triaxons where two of the three intersecting axes are much reduced leaving only one. This supposition is supported by the occurrence of knobs on the siliceous monaxons which suggests the reduction or, better, lack of further development of other spines. Examples of monaxons are seen in Fig. 11-6A, B, C, and E.

(2) Tetraxons. Here the rays point towards the apices of a tetrahedon. When the rays are of equal length, the tetraxon receives the name calthrops. When the rays are unequal, usually with one large one and three short ones, the spicule is a triaene, with a large rhabdome and a cladome of three smaller rays. In the calcareous sponges, the tetraxon is a common form, but more common is the triatine form, where the rhabdome has apparently been lost, leaving three rays, which, when resting on their tips, outline three edges of a flattened tetrahedon—the rays, therefore, are not lying in one plane. The tetraxons of siliceous sponges seem to be derived from the triaxonic condition by loss of one axis (two rays) resulting in a tetraxon all of whose rays lie in one plane. These would more properly be called diaxons or tetractinate (four spined) forms. These terms seem to be unused by sponge morphologists (Fig. 11-6A, B, D, E).

(3) Triaxons. These are hexactinate spicules found only in the hyalosponges. The form is the result of three axes lying at right angles to each other and all passing through one point of intersection to give a six-rayed figure (Fig. 11-6C).

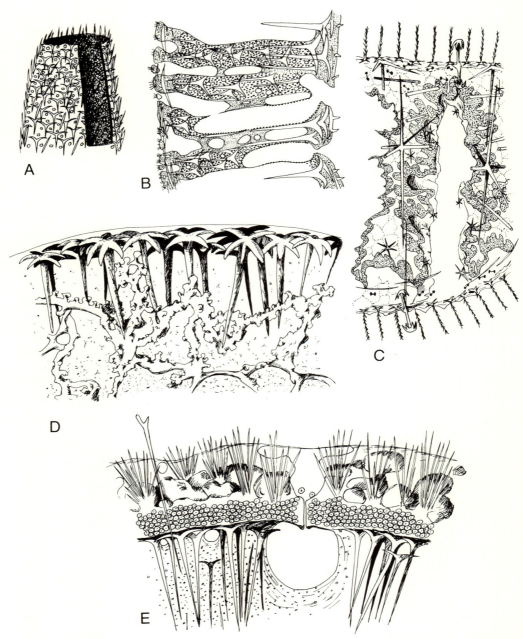

Fig. 11–6 Distribution of megascleres in selected sponges. A. Oscular end of *Leucosolenia* with a section removed. B. Section through *Grantia*. C. Section through *Hyalonoema*. D. Section from *Corallistes*. E. Section through *Geodia*. (All redrawn, by permission, from Hyman (1940) *The Invertebrates. Protozoa through Ctenophora.* McGraw-Hill Book Company, Inc.)

(4) Polyaxons. These are spicules with more than three axes and more than six rays. They are commonly microscleric forms, often of the multi-rayed type appropriately called asters.

(5) Spheres. As the name implies, these are rounded spicules with no obvious axes or rays.

(6) Desmas. This is a special type of siliceous megasclere, found only in certain sponges, and consists of ordinary, usually small, monaxonic, triaene, or tetraxonic spicules upon which more silica has been deposited in an irregular fashion. This added material distorts the original shape of the first spicules so that they may no longer be identified. (Fig. 11-6D)

Desmas also illustrate a complexity in spicule formation that is found in various forms. This is the fusion of spicules into quite elaborate skeletal structures. (However, the term "skeletal structures" commonly refers to both fused and unfused spicules.) This is seen in the pharetronid calcareous sponges, in the lithistid forms of the siliceous demosponges (Fig. 11-6D), and in siliceous hyalosponges where the hexactine spicules form a beautifully regular lattice, a dictyonine skeleton (Fig. 11-7).

The foregoing comments do not come close to revealing the variety that can be found in spicules, in that there has been no mention of the curving or spiralling of rays nor of the terminal elaborations that can be present. What has been described are the basic patterns. These patterns seem to be minimally reducible to three: the monaxons and tetraxons of the calcarea (all megascleres), and the triaxon of the siliceous sponges with its variations through loss of rays. This simplification further assumes that polyaxonic forms are some sort of polymerization of the simpler patterns and that spheres are

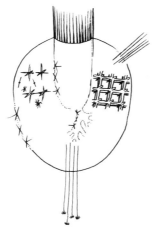

Fig. 11-7 Spicules of a hyalosponge schematically placed, and enlarged, in the sponge body whose osculum is at the top of the figure. (Redrawn from Hentschel (1923-1925). Original from F. E. Schulze.)

another variation wherein concentric deposition has replaced rayed deposition of material. It is too early to judge the validity of these generalizations. What is known of spicule formation attributes formation of each ray to a cell at the distal end of a ray, with the initial geometry somehow controlled by the mechanisms of initial deposition. As in the formation of radiolarian and acantharian tests, there is surely a conjunction of factors here, involving at least the tendencies of calcium carbonate and silica to form crystalline forms and the geometry of the cells which deposit this inorganic material. How these interact to determine the variety of spicules observed in the sponges, and what role natural selection plays in culling out the more efficient productions from the less efficient ones, remains to be solved.

The proteinaceous skeleton, formed of keratin and called spongin, is found in conjunction with siliceous spicules or by itself. It consists, typically, of an anastomosing network of keratin fibers. Occasionally, polyaxonic keratin spicules are found. The keratin fibrous network may have, in addition to endogenously produced siliceous spicules, exogenously produced sand grains. The keratin may also be found as a tough, leathery outer covering to certain sponges, such as *Hircinia* (Fig. 11-8). In the demosponges the keratin can function as a kind of cement to hold monaxonic spicules together to form nets or to bind them together into spiculofibers. In the nets, the keratin can occur only at the interstices of the net, or it can follow along all parts of the net, to-

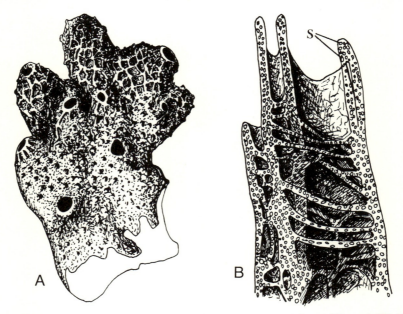

Fig. 11-8 The horny sponge *Hircinia*, showing its general form (A) and (B) the keratinous skeleton with embedded sand grains (s). (Redrawn, by permission, from Hyman (1940) *The Invertebrates. Protozoa through Ctenophora.* McGraw-Hill Book Company, Inc.)

tally surrounding the inorganic spicules (Fig. 11-9). Keratin is laid down by special cells, amebocytes, termed spongioblasts. What the morphogenetic movements are that place the secreted inorganic spicules in their special places with the keratin is apparently quite unknown.

Those few sponges lacking spicules and the spongin skeleton are not common; they are small and rely on a rather stiff mesoglea to support their bodies. An important phylogenetic question is whether the absence of all spicules and spongin is an original or a derived situation. This will be discussed later in the context of evolutionary trends.

FUNCTIONS

Environmental exploitation. Sponges meet their nutritional needs by filtering out minute food particles from the water that passes through their bodies. This activity is graphically illustrated by the following statistics (Parker 1910). In *Leucandra*, a small leuconoid, calcareous sponge, about 22.5 liters of

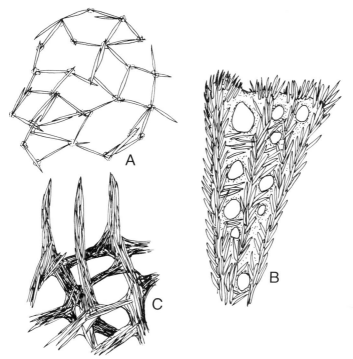

Fig. 11-9 Skeletons of keratin combined with monaxonic spicules. A. Spicules with keratin at their termini. B. Spicules embedded in a continuous keratin matrix. C. Spicules in keratin and forming rows of spiculofibers. (All redrawn, by permission, from Hyman (1940) *The Invertebrates. Protozoa through Ctenophora.* McGraw-Hill Book Company, Inc.)

water are filtered in one day. The experimental specimen was about 10 centimeters high and one centimeter in diameter and was estimated to contain 2,250,000 flagellated chambers. Obviously larger forms could pass 100 or more liters of water through their bodies in one 24 hour period.

The water currents are set up by the beating of flagella of the choanocytes. In the absence of any evidence of a nervous-coordinating system it seems that each cell must act separately. The effective beat of the flagellum depends on a screwlike activity, which forces water away from the cell. This sets up currents, which would sweep food particles up against the outside of the collar where they can be entrapped. As each cell sets up a current away from itself, the hydrodynamic flow so generated could resolve itself into an excurrent situation at the largest opening in the chamber. Excurrent pores tend to be larger than incurrent ones, and it is also reported that choanocytes seem to be tipped towards the excurrent pore. These simple mechanical factors apparently are sufficient to generate one-way currents. It can be seen that small chambers are more efficient than large ones, because the latter would tend to have a stagnant area in the center of the chamber. Thus it is not surprising to find the prevalence of leuconoid structure in sponges and to find syconoid and asconoid forms as rare and small.

There seems to be no doubt that food particles are initially trapped by the choanocytes and ingested by them. The details of the process are not well studied. Food materials also appear in amebocytes, and it is presumed that food substances are transferred to them from the choanocytes. The amebocytes conceivable function, therefore, as a kind of circulatory system, bringing nutrients to the nonfeeding cells of the sponge body. Whether reports of extensive intercellular communications are important in nutrient transfer as well as in other types of communication (Pavans de Cecatty, Thiney, and Garronne 1970) remains to be seen.

Digestion apparently occurs in both choanocytes and amebocytes. In any case, there is no evidence of extracellular digestion, and older references to the spongocoel as a gastral cavity are quite misleading. All digestion is intracellular and apparently occurs by use of typical animal enzymes, some of which have been found in sponge preparations (Rasmont 1968; Hammen and Florkin 1968).

Homeostatic mechanisms. Those simple generalizations that can be made about digestion and distribution of food products by amebocytes have been made above in connection with feeding.

That sponges lack specific respiratory, circulatory, excretory, and probably also nervous systems precludes any special comment on those topics. Functions that are related to transport of materials and riddance of wastes occur by means of devices common to all cells—diffusion, active transport, etc.—and supply no information of phylogenetic significance.

Reproduction. The complexities of sexual reproduction in sponges are still being unravelled; we can only summarize the present state of affairs. It seems that gametogenesis and fertilization may follow a rather stable pattern throughout this phylum, but cleavage and subsequent larval development are more variable, centering, however, around the formation and later development of two larval forms—the amphiblastula and the parenchymella.

Gametes appear to arise from enlarged amebocytes, but there are reports that choanocytes can also form gametes. What is referred to as the "classical account" (Hyman 1940) states that large amebocytes, classifiable as archaeocytes, are the only source of gametes. The ovocyte comes from an archaeocyte with a large nucleus and nucleolus. It continues to grow, presumably at the expense of nearby nurse cells (trophocytes). Eventually it undergoes meiotic divisions and a mature egg is formed. Sperm are also thought to arise only from archeocytes. These cells may become enclosed by flattened covering cells to form a spermatocyst. Inside the spermatocyst, reduction divisions occur, and subsequent spermiogenesis results in sperm with round heads, midpieces and long tails—a common sperm type in the metazoa.

Sperm enter other sponges by way of the incurrent flow of water and achieve fertilization in a unique fashion, first described by Gatenby (1920, 1927) and confirmed by Duboscq and Tuzet (1937) and Tuzet and Pavans de Cecatty (1958). The sperm is captured or simply enters a choanocyte, which now acts as a "carrier cell" (Tuzet 1963) and transports the cell to the site of a mature ovum. There, after the carrier cell and ovum are in intimate contact, the sperm head is transferred to the egg and fertilization ensues. The zygote starts development, either by equal or unequal holoblastic cleavage, and when it reaches the stage of an externally ciliated blastula, it migrates to the excurrent channels of the parent and is swept out to continue its development, swimming free for several hours and then settling down to metamorphose into a typical sponge.

The period from initiation of cleavage to attaching of the flagellated larva will be described by first commenting on the parenchymella larva and then the amphiblastula (Fig. 11-10). In Lévi's (1956, 1970) view, there are two kinds of each of these larvae. Generally speaking the parenchymella is a blastula stage, more correctly a diploblastula, in which outer flagellated cells almost completely surround an inner solid mass of cells, which will eventually form the ectomesenchyme of the adult sponge. According to Lévi (1963), in sponges, the simplest development pattern from a zygote is the development of the parenchymella of the demosponge *Clathrina*. Here morular delamination, i.e., the cutting off of the inner ends of very early blastomeres, results in the two layers of the cells of the parenchymellar diploblastula. Another pattern, much more common in demosponges and exemplified by *Halisarca*, is the ingression of cells from many sides of a solid blastula. Somewhat similar to this last pattern, but nonetheless distinct in important particulars, is pa-

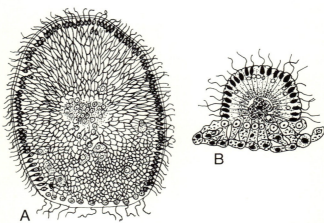

Fig. 11-10 Sponge larvae. A. Parenchymella of *Halisarca dujardini*. (Redrawn, by permission, from Brien (1973).) B. Amphiblastula of *Leucosolenia botryoides*. (Redrawn, by permission, from Tuzet (1973).)

renchymella formation in certain calcareous sponges, where unipolar or multipolar ingression in a hollow blastula results in the parenchymella.

The amphiblastula, as the name suggests, is a mass of cells differentiated into two types. But rather than have an internal-external polarization as in the parenchymella, the amphiblastula shows an anterior-posterior polarization. Anteriorly there are the flagellated cells, which will later form the choanosome—these are the choanoblasts—and posteriorly there are the future ectomesenchymal cells, or the ectoblasts. These are often separated by "enigmatic cells" (Lévi 1963) not distinctly choanoblastic or ectoblastic. In the demosponges, where amphiblastulae are not common, this larva arises from early blastomeres—the morula stage—with rather similar-appearing cells simply separated by the "enigmatic cells." In the calcareous sponges, above the asconoid type, where amphiblastulae are the rule, the amphiblastula can be identified very early. At the sixteen cell stage, eight cells (those lying towards the maternal choanocytes) are ectoblastic and the other eight are choanoblastic. The latter soon become flagellated (but lack collars) and face inward and are partially enclosed by the ectoblastic cells. An opening in the center of the choanoblasts is a larval mouth, whereby the diploblastula feeds on nearby parental cells. Duboscq and Tuzet (1937) refer to this stage as a stomoblastula. By inversion, the inner flagellated layer becomes external, and the ectoblasts are simultaneously and partially exposed at the posterior end of the larva. The larva now swims free of the parent and settles down.

In all these forms it is seen that an early differentiation of cell types has occurred, into flagellated cells, which are choanoblastic and are either externally or anteriorly located on the diploblastula, and into ectoblastic cells, which are internal or posteriorly located. From the point of view of classical metazoan embryology it may be asked why these double layered or bipartite larvae are not termed gastrulae. Lévi's (1963) answer is that gastrulation is properly defined as the process wherein future endodermal cells move inwardly. Furthermore, cytodifferentiation is most conspicuous, at least in the coelomate metazoans, following gastrulation. The situation in the early sponge larvae differs from the above. Inward movement of the so-called endoderm, i.e., the presumptive choanocytic cells, does not occur until after the larvae have settled down—a stage we have not yet discussed in the present account—and by that time the essential determination of cells is completed.

From this it becomes very evident, according to Lévi's (1956, 1963) careful analysis, that sponge development has many unique aspects. He emphasizes the early cytodifferentiation and unique diploblastula development prior to gastrulation.

Now let us look at gastrulation. In both types of diploblastulae, the flagellated cells migrate internally, in effect displacing the ectoblastic cells to an external position. The attached larvae continue to develop with internal spaces appearing, which are now lined with recognizable choanocytes; water channels appear, and the ectosome becomes organized into an adult aspect. Some calcisponges remain at the asconoid stage, if this is the adult form for the species. Others go on to the syconoid stage. The leuconoid condition may arise from further development of a sycon or from an ascon, directly. In demosponges, gastrulation leads to a rhagon from which the leuconoid condition subsequently appears.

Thus, it can be seen that sponge gastrulation amounts to morphogenetic movement of predetermined cells to sites where their potencies come to expression and achieve the adult phenotype. Here, too, however, differences with the classic pattern of animal development are apparent—differences long ago made clear by W.J. Sollas and Bütschli and confirmed by others (I.B.J. Sollas 1906; Hentschel 1923-25; Hyman 1940; Hartman 1971). The essential point is that the outer flagellated layer, which in other metazoan larvae is considered to be ectoderm (giving rise to epidermis), is endoderm in the sponges or, better, choanoblastic (giving rise to the feeding cells, the choanocytes, of the flagellated chamber). And the inner cells of the sponge larvae, normally thought of as endoderm, are in fact ectoderm or ectoblastic (giving rise to the ectomesenchyme). There is, therefore, a clean reversal of the fates of primary double layers of the early embryo of the sponge as compared to the classic metazoan situation. This is, of course, of phylogenetic importance and will be referred to later.

To summarize: Sponge development, starting from gametes, includes release of the sperm to the outer environment, followed by a unique pattern of internal fertilization utilizing a choanocyte modified to a carrier cell. Larval development includes the unique diploblastula, seen either as an amphiblastula or a parenchymella, both having two different modes of origin, and in which cytodifferentiation into ectoblastic and choanoblastic cells has occurred early. The free-swimming larva of this stage settles down and gastrulates, in the sense of showing inward movement of future feeding cells, and morphogenetic movements of the cells continue so as to produce a functional sponge.

It can be seen, as a last remark, that inversion as a phylogenetic clue is limited to only certain calcareous sponges and represents the most complicated of all the developmental patterns. Duboscq and Tuzet (1937) regard this as a palingenetic or primitive trait. Lévi (1963) has thoughtfully suggested that inwardly directed flagella may be the consequence of initial cleavage patterns and that inversion of the stomoblastula is brought about to achieve external placement of flagella for the free-swimming larva. And then, return of the flagella internally, as in all other sponges, occurs when locomotion is not at a premium but formation of a flagellated chamber is. On Lévi's view, inversion represents a response of the sponges to a functional need not met by the initial organization of the larva, a situation not seen in other sponges, where the initial flagellation is always external.

Asexual reproduction in sponges can occur by budding or by formation of reduction bodies but is more commonly brought about by gemmule formation. Budding seems to be found only in certain sponges where regularly, or because of adverse conditions, terminal parts are dropped off—"constrict off" (Hyman 1940)—and round up to eventually regenerate a new sponge. Reduction bodies are the products of massive degeneration of sponges, apparently because of adverse conditions, with the result that a spherical mass remains covered by epidermal cells with internal amebocytes. This cystlike structure can regenerate a new sponge.

All freshwater and many marine sponges can reproduce by gemmules. In freshwater forms, the gemmule is a rather elaborate structure formed through the cooperative action of various cell types. The following is Hyman's admirable summary of this process.

"The gemmules of freshwater sponges begin by the gathering of amoebocytes, probably archaeocytes, into groups. Through the cooperation of special nurse cells, or trophocytes, these amoebocytes become filled with food reserves in the form of oval bodies composed of glycoprotein or lipoprotein. Other amoebocytes join the group and arrange in a columnar layer encircling the rounded mass of amoebocytes. The columnar layer secretes a thick hard inner membrane and later a thin outer one. Meantime scleroblasts throughout the sponge have been secreting amphidisk spicules, and they carry these amphidisks into the columnar layer, where they place them radially between the inner and outer membranes. When the gemmule is finished, columnar cells, trophocytes, and scleroblasts depart, and the completed gemmule is a small hard ball consisting of a mass of food-laden archaeocytes enclosed in a wall composed of two membranes with a layer of amphidisks (lacking in *Spongilla*) between them and pierced by an outlet, the *micropyle*. The freshwater sponges usually form large numbers of gemmules in the autumn and then disintegrate; the gemmules remain in the remnants of the sponge body or fall to the bottom. They can withstand freezing and a certain amount of drying and hatch, under favorable conditions, in nature usually in spring, when temperatures rise." (Hyman 1940, p. 307 and 309.)

Upon hatching, the contents stream out and cell division ensues, various cell types emerge in the mass of cells, and ultimately there appears a small sponge.

In marine sponges, gemmule formation also starts with archaeocytes, which become enclosed in a membranous covering. In this case, the covering is formed from other cells themselves. These enclosed cells eventually become columnar and flagellated, except at the posterior pole, and the final result is a larva very much like the amphiblastula.

Finally, in connection with reproduction, the phenomenal powers of regeneration among sponges can be mentioned. Following the pioneer of H. V. Wilson (1907) who separated sponge cells by forcing sponges through fine silk cloth, other workers have used the same technique to follow the fate of the isolated cells. It is apparent that single cells by themselves do not regenerate a whole sponge, but small aggregates of cells can do so. The mechanisms of this aggregation implies selective mechanisms of surface recognition (Humphreys 1970; Curtis 1970), which result in a somewhat orderly pattern to the aggregates so formed. These clumps then have the potential for developing into tiny sponges, which, upon feeding, return finally to the parental form. The sponges are alone among the Metazoa in displaying these phenomenal regenerative powers. The only other groups approaching them are the Cnidarians and certain flatworms.

Phylogeny

Sponge phylogeny, in its broadest sense, aims at elucidating the diversity that has evolved within multicellular organisms depending on a water channel system with flagellated chambers for food procurement. Our first problem is to find semes. From there we proceed to identification of homologies and of the sponge plesiomorph. The phylogenetic relationships will be only broadly reviewed, for the details of sponge evolution rely heavily on a highly specialized knowledge of spicules and skeletons (fossil and recent), and this is beyond the competence of this reviewer. Furthermore, in what follows there will be no discussion of the new sclerosponge class recently erected by Hartman and Goreau (1970) because of the paucity of its treatment to date.

SEMES AND HOMOLOGOUS RELATIONSHIPS

There are many sponge traits that can be identified on present evidence as sufficiently complex to be termed semes, and therefore they are useful in defining homologies. Cell types, flagellated chambers, asexual reproduction, and feeding and digestion are conservative traits showing little variability. On the other hand, body form, spicules and skeletal parts, and larval forms show significant variability and will, therefore, be especially important in outlining phylogenetic trends.

Cell types. The choanocyte makes the flagellated chamber possible, and this in turn makes sponges possible. Are the choanocytes all homologous in the various species of sponge? Or are there differences which might suggest convergence and multiple origin of choanocytes? And what is the condition with other cell types in the sponges?

From the similarity of the choanocytes in the various species of sponges, it is concluded that they are all homologous. They are located only in flagellated chambers; thus the positional relationship is apparent. They all possess one flagellum with a surrounding collar and one nucleus. The collar in all cases thus far studied (Rasmont 1959; Fjerdingstad 1961a, 1961b) is made up of vertical tubules standing side by side. There is always a parabasal structure and centriolar body. Thus compositionally the choanocytes appear related. Those differences which are observed are in relation to size and to placement of the nuclei (apical or basal) and to presence or absence of a paradesmose—the thread joining basal body to nucleus. Neither of these seems sufficiently important to deny the reality of the compositional relationship.

Looking at the other cell types, that is, the various amebocytes and dermal cells—pinacocytes and porocytes—one finds many general similarities and some differences. Certainly amebocytes are limited to the area between the flagellated chambers and the pinacocytes, and the latter are always at the sponge surface, whether it be the outer surface or the lining of water channels. Positional similarities are obvious. Compositionally, the pinacocytes appear very similar as mononucleated, polygonal, flattened cells. Amebocytes vary in composition, but they represent a problem compounded by their variety of functions and by our ignorance of how they arise and perhaps even change their type on occasion. If one compared archaeocytes in all sponges, there would be obvious structural similarities, and the same for thesocytes, chromocytes too, possibly, and others of the important subtypes. Hence, critical comparisons among the amebocytes must be made in terms of the subtypes. But then those variations that do occur—for example, not all sponges have fiber cells or myocytes, some have spongioblasts for laying down spongin (Vacelet 1971)—raise certain problems. Do they represent aposemic or even neosemic innovations? On the whole, however, it may be said that the category of amebocytes is probably homologous in the sponges; and looked at as a broad generalization, the presence of this gross cell type is better treated for now as a plesiosemic trait, even though specialists will probably be able to exploit the aposemy apparently present here.

Another major plesiosemic trait is the flagellated chamber. It is hard to argue for fulfillment of the positional relationship in strictly anatomical terms because of the variety of water channel systems. It is nonetheless obvious that the chambers are always located between incurrent and excurrent openings, accepting the further point that incurrent openings can consist of many pores, as in asconoid forms, or one or two prosopyles, as in other forms. Compositionally, all flagellated chambers are related by the common possession of a choanocytic lining.

Looking next to functions, there are three plesiosemic characteristics in sponges, i.e., gametogenesis and fertilization, asexual reproduction, and feeding and digestion.

Since there is still uncertainty as to whether gametes differentiate from only archaeocytes, or from choanocytes, too, it is not possible to say that the start of the reproductive function is always located at the same point. However, subsequent steps—formation of ovocytes and spermatocysts, reduction divisions, and formation of the gametes—all seem to be composed of similar steps, as far as is now known, in the sponges. The almost total absence of knowledge of these processes in the deep-sea sponges, including the whole group of the hyalosponges, means generalizations must still be seen as tentative.

The same is true for fertilization. If the use of a carrier is indeed universal in the sponges, this remarkable feature is then, because of its complexity and uniqueness, a very significant homologous relationship, tying the calci- and demosponges together. Nothing is known in this regard for the hyalosponges.

Gemmules constitute a possible common feature in asexual reproduction, but the extent of their distribution is not adequately known, and it is true that the gemmule of freshwater sponges differs from that of marine forms. It is probably premature to classify this trait as plesiosemic, recognizing the aposemy of the freshwater forms and our limited knowledge of other forms. That being so, not much weight need be afforded this character for the present, and its conservative treatment as plesiosemic seems safest for the present. Similar comments might be made for the rarer process of reduction-body formation.

The modes of food capture, ingestion, and digestion are understandably plesiosemic in view of the plesiosemy of their organizational basis, i.e., the flagellated chamber. The various steps in these processes, however, could vary in size or nature of particles captured, in mode of ingestion, and pattern of digestion, especially, in the latter case, with regard to transfer of food to amebocytes. It may well be that further studies will turn up such differences.

On present evidence, though, it seems unlikely that such differences, when they appear, will be sufficiently great so as to render the nutritional processes nonhomologous in the various poriferan species.

The noticeably variable aspects of sponge biology are the organization of the general body form, the spicules and skeletal parts, and, lastly, larval forms.

Body form. The leuconoid body is the essential problem here, and the key issue can be highlighted by asking what sort of classification of sponge diversity would ensue if body form, rather than skeletal parts, were used as the sole informational base of systematic efforts. It might be that asconoid and syconoid forms would be separated off as small minor groups and all the leuconoids lumped together. There might be distinctions then made among the latter on the basis of such simple features as branched or unbranched, stalked or unstalked, flattened or rounded, and so forth and in various combinations of these traits. In any case, it is doubtful if taxonomic boundaries would coincide with those of the present day Calcarea, Hyalospongiae, and Demospongiae, or the Sclerospongiae.

What does such a conclusion mean? Most simply it says that, in terms of the leuconoid body alone, trends or large general patterns of organization are not present that provide an obvious basis for taxonomy (regardless of one's philosophy—Adansonian, New Systematics, or whatever). We seem to have, to use the metaphor of the evolutionary tree, leaves (species) which show a set of characters—stalked-unstalked, rounded-flattened, etc.—that are found in all possible combinations, and it is not at all clear why one trait rather than another should be used as the basis for taxonomic separations. These traits certainly make possible dichotomous keys, for identification; they provide a possible approach for Adansonian taxonomy if an investigator gives way to his intuition as regards what is "natural"—sponges are "naturally" flattened; perhaps, rather than upright, or unstalked rather than stalked; but they provide no clear picture of trends, which is what a phylogenist wants of an aposemic character.

The best we can do at present includes the following two points. (a) Evidence of trends in body form are limited to the classical examples of ascon-sycon-leucon in the Calcarea, and if we also look at embryonic forms, there is a rhagon-leucon sequence in the Demospongiae. (b) The variety of leuconoid forms will be treated like the "Weichkörper" of the Acantharia; namely, it is a variable entity with enough details in common to argue for a general compositional similarity, rendering homology likely. But since serial relationships are not clear for tying together all the multifarious different combinations of traits, homology here will have to rest on the use of subsidiary criteria, i.e., leuconoid forms have the same distribution as homologies clearly established by the positional and compositional relationships with regard to such semes as the flagellated chamber, etc.

There still remains, though, the problem of understanding what has occurred, evolutionarily speaking, to result in this seemingly jumbled variation on leuconoid forms. We shall return to that problem later.

Skeletal parts. The supportive (and protective) parts of sponges, as we have seen, are classifiable into three groups on the basis of their composition and form. Simply on the basis of composition, the Calcarea are to be separated from forms with siliceous spicules, and sponges with exclusively horny skeletal material can be separated into a third group. But these last forms blend into the second group because of forms with both siliceous spicules and keratinized skeletons. Essentially there are two groups then, in terms of chemistry of the skeletons—the calcareous sponges and the noncalcareous ones with siliceous and/or spongin skeletal parts. There is no known transition between these two groups, and since the compositional relationship cannot be fulfilled, we must conclude that the two different types of spicules are not homologous, but represent parallel evolution. We say parallel rather than convergent evolution because the common occurrence of flagellated chambers clearly argues for a common ancestry and, therefore, the spiculization of sponges arose after the appearance of the flagellated chamber.

The next problem is to look at the nature of the aposemy within these two groups of calcareous and silico-keratin sponges. Beginning with the calcareous forms, there are the facts that monaxons and tetraxons are the basic spicule forms and that they are always megascleric. These two basic types show many variants including monactinate and diactinate monaxons of curved or straight form, and tetractinate and the very common triactinate variants of the tetraxon. Also, there is fusion of these elements by a calcareous cement to form the continuous skeletons of the pharetronid sponges, which have a fossil history going back to the Permian (Laubenfels 1955; Vacelet 1970). The discontinuity between spicular types, especially the clear discontinuity between monaxons and tetraxons, is due to different patterns of deposition of the spicules, hence the question of homology does not demand intermediates to join these differences. The more important differences are the variations within the monaxons and within the tetraxons and the use of these, by cementing, to form the continuous skeletons. In that these variations seem continuous— many intermediates between extreme forms—we can accept that all calcium carbonate spicules and skeletons are homologous.

This variety of spiculation occurs, however, as part of the skeletal equipment of essentially all the Calcarea. That is, all have both monaxons and tetraxons and their derivatives. The former are often peripheral in their location and the latter often associated with the water channels and chamber. Even the pharetronid forms have some free spicules—triaenes of a tuning-fork shape. Spicular and skeletal variety in the Calcarea is expressed, then, as a common property of the species in this group. It is not possible to see real evidence of increasing complexity except for the few forms with fused spicules. Judging from the fossil record, these forms were once more abundant than today, and, therefore, it seems the innovation of fusion was of limited evolutionary advantage.

In the species included in the silico-keratin group, the variety of skeletal parts is considerably greater than in the previous group. It ranges from forms with no spicular or skeletal material whatever, to a variety of spicule forms— tetractinate and hexactinate, micro- and megascleres—to combinations of inorganic and organic substances. But as in the Calcarea, the spicules themselves, when they are present, show a variety of the several forms possible rather than showing a progressive complexity from one form to another. It is as if each species has chosen from the abundance of possible spicule variations those that are functionally effective for a given body form, and therefore, as with the body forms themselves, there is a bewildering array of combinations of various mega- and microscleres and various types of spined and spherical spicules.

Some semblance of progressive change can be seen in the overall change from forms with no spicules to forms with spicules and then to forms with spicules fused into skeletons. Within these last two steps, however, confusing variation occurs with the realization that forms with skeletal parts can have only siliceous ones, or only keratin ones, or both. And when both occur, the keratin can bind the inorganic spicules into continuous nets or spiculofibers. There is no simple, clear sequence to the variations found in the skeletal parts of these forms. This applies to both the Demospongiae and the Hyalospongiae. As with the leuconoid body form, we are left with the problem of accounting for this kind of variation.

Larvae. Lack of information regarding the Hyalospongiae means that our discussion will concentrate on the diploblastulae of the Calcarea and Demospongiae. In the account, above, of sexual reproduction in the Porifera— largely based on Lévi (1956, 1963) and augmented by Hartman (1971)—we saw that the amphiblastula and parenchymella occur in both the calcareous and the demosponges. There are differences in the origins of the two larval types in each class. In the Calcarea, the parenchymella arises from a simple hollow blastula and amphiblastula from a stomoblastula. In the Demospongiae, a solid morula, by delamination of the outer layer of cells of variable size, provides the inner cells to form a solid blastula, which then forms a parenchymella. The rare demosponge amphiblastulae come from morulae with somewhat similar cells in its two parts which are separated by enigmatic cells. These differing origins of the larvae suggest that they are not homologous, though superficially similar in appearance, and they represent independent but parallel evolutionary developments. This lack of homology is discernible in some other details of organization. The position of the nuclei in the flagellated cells of *Leucosolenia*, a calcisponge, remain apical, whereas in *Clathrina*, a demosponge, the nuclei move from an apical to a basal position. The parallel nature is emphasized by several points of general similarity (Lévi 1963) that, however, apply to all diploblastulae: they show axial symmetry; they swim along a spiral course; the anterior end is choanoblastic and the posterior is ectoblastic; cytological differentiation is comparable to what will appear in the adult. Despite these similarities, and for reasons still unknown in functional terms, the sponges have inserted two types of diploblastulae into two different lines of development.

The question of homology is whether the amphiblastula and parenchymella within each class can be homologized. Homology is probable in that the parenchymella can be thought of as an amphiblastula with extensive ectoblastic proliferation, the products of such proliferation being located within a cup of choanoblastic cells. Or, alternatively, the amphiblastula can be derived from a form with parenchymellar potencies by inhibition of ectoblast proliferation. Both views have been proposed by earlier workers but Lévi (1963) thinks the former is more probable. In that the demosponges develop from a similar morular stage, the homologizing of the two larvae seems possible, since it needs only a change in pattern of ectoblast proliferation to derive one type from the other. The case is more complicated for the calcisponges, for the stomoblastula starts with inwardly directed flagella and undergoes inversion to form the amphiblastula, whereas the parenchymella comes directly from a blastula with externally directed flagella. In this group differences in ectoblast proliferation and in flagellar orientation must be considered before a final decision is reached on deriving one larval form from the other.

The aposemy of larval forms, on present evidence, offers the view of two nonhomologous sets of diploblastulae; within each set, homology may be present between the two differing larval types.

The foregoing aposemic traits offer us a curious condition. In regard to larval forms and the skeletal materials, we have concluded that two distinct lines of evolutionary development exist. They both show parallel evolution, and, very importantly, they coincide, in that the division into two groups always places the calcisponges on one side and the demosponges on the other. In regard to spicules, the hyalosponges join the demosponges. Looking again at body types, we see the ascon-sycon-leucon pattern of the calcareous forms as opposed to the rhagon-leucon of the demosponges. (Here we can only refer to the hyalosponges as leucons.) These comments lead to the generalization that a deep separation exists between the calcisponges and all the others. But, nevertheless, the groups are united by their plesiosemes, and by the fact that parallel evolution has occurred in the nonhomologous traits. The aposemes offer no direct further support to the monophyletic nature of the sponges—none is needed, of course—because all the aposemic traits are bifurcated into two groups of characters showing homology within each group but only parallel evolution between the sets of traits.

Monophyly. The argument for the monophyletic nature of the sponges is good but has its own peculiarities, especially with regard to the aposemic characters. Let us start with the plesiomorphic traits: These are certain cell types, the flagellated chamber, the feeding pattern, and special aspects of gamete formation. These features do not occur in any other animal or plant forms and are therefore synaposemic (Hennig 1966) relative to these other organisms. Or as Lévi (1956) puts it, "Among the metazoans, the sponges constitute a phylum characterized by an exceptional anatomy without homology in [the rest of] the animal kingdom." (p. 5. Original in French.) The more variable traits of the sponges—their aposemic characters—show limited homology. The spicules, as such, are not unique to these animals, but the fused skeletons are, and to that extent are synaposemes. The body form composed of various patterns of water channels and the diploblastic larvae are unique to sponges and therefore synaposemic. There is no doubt of the monophyletic nature of the sponges, but there is also strong evidence that much of the later development of sponges has been on two major, independent but parallel lines. This argues strongly for much parallel development within each of these lines.

Plesiomorph. This final section will concentrate on the problems of the sponge plesiomorph and on its origin. The discussion of evolution within the phylum will receive only limited attention for reasons to be made clear after discussing the plesiomorph.

The two possible sources of what we are calling direct evidence for the nature of the plesiomorph, are, in the sponges, fossils and ontogeny. Distributional evidence is presently of no use.

Though the sponges have a fossil history going back to the Cambrian, this tells us little more than siliceous and calcium carbonate spicules were present in the earliest fossils. The fossil information does document the development of certain sponge families (Finks 1970) and allows careful speculation on evolution within the class Demospongia (Reid 1970), but it tells us nothing about the form of the simplest sponges nor about the relations between the three sponge classes (Laubenfels 1955). Ziegler and Rietschel (1970) claim that only leuconoid forms are seen in the fossil record.

Developmental knowledge is more informative. The first functional sponge, judging from the embryology of the group, was an ascon in the calcisponges and a rhagon in the demosponges. In the latter, though there is appearance of spicules as early as the gastrula stage in some forms, in others they appear later, and, of course, in the few simpler forms lacking spicules entirely, they never appear. It is difficult, therefore, on ontogenetic grounds alone, to know if spicules were present in the early rhagon or not. For reasons adduced below, we can conclude that spicules were probably not present. In the calcisponges, the forms that later develop into syconoid or leuconoid forms show early formation of spicules, but the permanently asconoid forms show a late development of spicules. This suggests that spicule formation arose after the asconoid form evolved, but as sponge form increased in size, and supportive elements became more important, they appeared earlier and earlier in development, perhaps to insure adequate numbers being present to meet the needs of the growing, larger bodies.

It might be argued that the earlier larval forms of the diploblastula type represent the sponge plesiomorph. This is doubtful from the very fact that they are not functional sponges; they lack flagellated chambers, choanocytes, and are free-swimming rather than attached. They are indeed larval forms, forms on the way to a functional adult type and possibly showing reminiscences of a presponge organization, but also showing adaptations to being an effective developmental stage for the Porifera. This last is seen in their free-swimming habit which provides the otherwise immobile sponges with a chance for dissemination, a major function of larval forms. It can further be argued that absence of the collar in the flagellated larvae is an adaptation to free-swimming. When that function has been carried out, and the sponge settles down, then the flagellar apparatus is moved internally and adapted to the needs of the flagellated chamber, which includes now the appearance of the collar so essential to feeding. It would be most interesting to have a comparative study of flagellar fine structure and the mechanics of flagellar motion in larvae and choanocytes of the same species of sponge. There seems to be no such work published thus far.

This discussion leads us, at most, to glimpses of two separate plesiomorphs—one for the calci- and one for the demosponges. What further comments are possible when we look at the morphocline data? There are three sets of aposemes to be examined as morphoclines: body types, skeletal parts, and larval forms.

There are two body-type sequences which have been familiar to invertebrate zoologists for years, i.e., ascon- sycon- leucon, and rhagon (larval form, to be sure)—leucon. On the classical and never-certain argument that evolution proceeds from simpler to complex forms, the ascon and rhagon were thereby considered as primitive. However, the surer way of determining morphocline direction is to look for conjunction of two morphoclines. (Also, as is immediately obvious, the body type morphoclines are in agreement with the ontogenetic data, and this consistency helps both arguments.)

The calcareous skeletal parts have only one general morphocline, and that is between free spicules and fused ones forming a skeleton. The asconoid and syconoid forms never have the fused skeletons that are found only in the pharetronid leuconoid forms. This morphocline does not pinpoint the ascon form as one terminus of a morphocline; it simply includes it there along with the majority of other calcisponges. This, therefore, is of little help in making a specific argument for the ascons in support of the body form morphocline, but it doesn't hurt that argument.

In the demosponges there are more morphoclines discernible, and they make a somewhat complicated picture. The picture might be sorted into component parts as follows: at one terminus are (a) forms with no spicules, and at the other are forms variously supplied with spicules and/or spongin fibers of keratin. These include forms with only siliceous spicules and these can terminate in (b) forms with fused siliceous spicules. Also there are forms with siliceous and keratin elements, and a terminus of their organization can be (c) the skeletons composed of both spongin and siliceous spicules. And finally there are (d) forms with only spongin fibers organized as extensive skeletons. It is not clear whether these last forms represent a morphocline coming directly from spicule-free forms or from forms with keratin and siliceous spicules from which the latter were later lost. Either possibility forms a perfectly good morphocline in the formal sense. From such a point of view, the convergence of the morphoclines on the forms lacking the spicules argues for their being the primitive end of the morphocline. Forms such as *Oscarella* or *Halisarca* represent possible plesiomorphs (Lévi 1956; Hartman 1971). These two are, respectively, the simpler forms of the two subclasses of the demosponges, the Céractinomorphes and Tétractinomorphes (according to Lévi's (1956) classification (Table 11–1)). It is not possible here to decide which of these is the better single representative of the demosponge plesiomorph, (Brien (1968) favors *Oscarella* and Borojevic (1970) favors *Halisarca*.)

Larval types provide us with no useful morphocline data today, because at present we are not sure whether the amphiblastula arose from the parenchymella, or vice versa. It seems that the latter may be more probable. Further studies are the only way to resolve this problem.

In conclusion, with regard to the poriferan plesiomorph, we come to no simple position. An ascon-type sponge is the best candidate for the calcisponges, but since there is no perfect ascon form known, we take the closest known form which is *Leucosolenia,* made up of a group of ascons joined only at their bases. *Leucosolenia* is well provided with spicules as an adult but not in its earliest ascon stage of development. For the demosponges, developmental evidence, body type, but especially spicule evidence point to a nonspiculated plesiomorph, and at present we have two equally probable candidates—*Oscarella* and *Halisarca*. As regards the hyalosponges, we have said nothing because of (a) lack of ontogenetic data, (b) the prevalence of a leuconoid body form, and (c) skeletal formation and spicules are widespread and offer no convincing directionality to a morphocline. We have, therefore, no proposal for a hyalosponge plesiomorph. On present evidence it seems justified to take the conclusions at face value and say that we can effectively go no further than to propose one plesiomorph, *Leucosolenia*, for the calcisponges and two for the demosponges, *Halisarca* and *Oscarella*. (In no case are actual species being proposed, only genera, emphasizing further the uncertainty of these conclusions.) Such conclusions differ from the general consensus on the nature of the primitive sponge. However, the approach through a more rigorous phylogenetic theory provides a more explicit evaluation of how one arrives at the stated conclusions and their validity.

A different view of the plesiomorph has been stated by those whose approach to phylogeny is primarily morphological. A type of stereogastrula, to look at Jägersten's (1955, 1959) arguments, is proposed as the starting point of sponge evolution. This view depends largely on interpreting the diploblastula as a recapitulation of ancestral forms. As was pointed out, though, the sponge larvae are certainly adapted to the needs of sponge biology and may be quite modified from ancestral forms.

The only further generalizations we would make on early sponge forms do not involve the larva but rather speak to what we can call the sponge eotype, a hypothetical form ancestral to all the poriferan plesiomorphs. Because of the monophyletic nature of the sponges we can argue for a single eotype. The two different patterns of spiculation suggest, as stated earlier, that spicules arose after the sponge *Bauplan*, seen in the eotype, appeared. The eotype would then possibly be an asconoid form without spicules.

PHYLOGENETIC ANALYSIS

The absence of any clear case for a single sponge plesiomorph renders it difficult and almost fruitless to attempt calculations of phyletic distance except within each sponge class, or even subclass, in the demosponges. The major difficulty for a nonspecialist in this group, which includes the writer, is the interpretation of spicule morphology. On that basis, largely, Reid (1970) has proposed the demosponge phylogeny shown in Fig. 11–11A. In contrast Lévi (1956) presents his phylogenetic views (Fig. 11–11B) based chiefly on embryological data. It is very difficult to cross connect these two sets of conclusions, because Reid refers to many taxa which are known as fossils, and Lévi uses a French taxonomy based on living forms. The sponge specialists must communicate much better among themselves, before a relative outsider can hope to enter their discussions to offer alternative points of view for their consideration.

What can be usefully mentioned here is an overall view of sponge evolution and, particularly, of the patterns of variation seen in the aposemic traits. We shall start with the latter. They key to this problem is the conclusion that

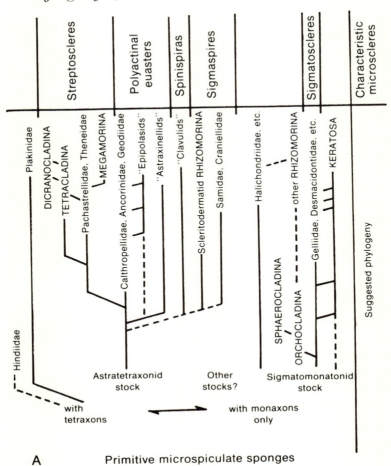
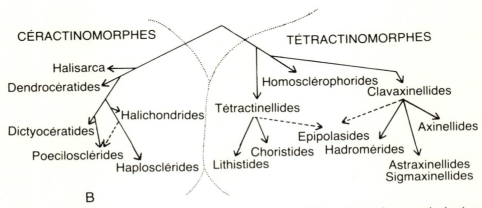

Fig. 11-11 Demosponge phylogeny. A. After Reid (1970, redrawn by permission) and based largely on spicular criteria. B. After Lévi (1956, redrawn by permission) and reflecting embryological information in large part.

the leuconoid form is apparently the most efficient of all the sponge body types. There is developmental evidence that it can arise from the asconoid, the syconoid, and the rhagonoid forms. Evolutionarily, then, it is eminently plausible to conclude, as many sponge specialists already have (Hyman 1940; Lévi 1956; and others) that the leucon has arisen many times, independently of other such origins. In short, there are compelling reasons to see in the sponges much parallel evolution (Borojevic 1970). The foregoing is based on ontogenetic and general morphological considerations.

The same conclusion comes from a selectionist interpretation of phylogenetic theory, as follows. Sponge biology centers around the use of the flagellated chamber as the essential device for exploiting the environment for food. This filter-feeding device is developed evolutionarily in conjunction with spicules and skeletons, which support and protect ever larger and more complex water channels, which lead into and drain the flagellated chambers. It is not surprising, under these conditions, that parallel evolution commonly occurs, for the selection pressures are all in favor of combining and recombining the possible permutations of efficient chambers, channel systems, and skeletal parts.

And all of this has one further important consequence. In that evolutionary innovation has depended on an exploitive trait, any advantage achieved by innovation will tend to wipe out near competitors. Continued evolution will result in continued loss of ancestral forms with only successful innovators surviving. Hence, present-day sponges are the product of forces that have wiped out most evidence of ancestral forms and whose collective appearance is a crazy-quilt amalgam of those various modes of growth patterns, of channel, chamber, and spicule formation and fusion, that are possible through parallel evolution. The difficulties in finding the sponge plesiomorph and in adequately describing sponge variability now become much more understandable.

Origins. The critical traits for determining the origin of the Porifera are the flagellated chamber and its choanocytes. In fact, we can go one step further and say the choanocyte will hold the solution to the problem. The setting aside of the flagellated chamber is done for selectionist reasons. It is the innovation which makes sponges possible; other forms have choanocytes or choanocyte-like cells, but none have them organized into feeding chambers. And for the reasons we have used to argue for loss of the majority of simpler sponges relative to the leuconoids, these same reasons would argue all sponges were eliminated until one reaches a forebear which is so unsponge-like as not to be in competition with the sponges. This brings us to the limit of our phylogenetic analysis, namely, the identification of that trait which makes invasion of a given adaptive zone possible, but which at the same time is the cause of the hiatus, through competition, between known inhabitants of the new adaptive zone and the plausible ancestral forms. In the case of the sponges, the hiatus in terms of general body form is very clear, but the nature of the choanocyte remains.

The choanoflagellates, zooflagellate protozoans, are the classical locus of forms having cells like the sponge collar cells. The critical comparisons of de Saedeleer (1929a, 1929b) are still useful today (Table 11-2). These are a convincing set of similarities. The most important question that remains is to determine how close choanoflagellate organization comes to the putative sponge plesiomorphs or to the sponge eotype. Here the case of *Proterospongia* must be examined.

Proterospongia haeckeli was first described by Kent (1880–1882) and then not reported again except sporadically (see Tuzet 1963) and from specimens of somewhat doubtful validity. Indeed, it was finally argued by some workers (Dubosq and Tuzet 1937; Hollande 1952a) that Kent's form was simply a sponge fragment. However, in more recent years *Proterospongia* has been rediscovered (Ertl 1968). These reports, though, concern *freshwater* forms. In that all sponges except for the demosponge family Spongillidae are marine, it makes it difficult to accept that the freshwater choanoflagellates are ancestral to sponges. It may be that freshwater is one area where colonial, protozoan choanoflagellates are free from competition from the more efficiently organized sponges. This is simply conjecture. It is still quite probable that marine *Proterospongia* still exist and, when found, hopefully electron microscope studies can be made of their fine structure.

Proterospongia is certainly the form closest to modern sponges, not only because of its choanocyte but also because of its colonial nature, the mesogleal mass which holds the cells together, and the presence of amebocytes in

Table 11-2 Comparison of sponge and choanoflagellate collar cells. Based on de Saedeler (1929a) and Hollande (1952c).

	Choanoflagellate	*Sponge*
Flagellum:	two bends; bursts of activity; plane of beating an inverted cone, rarely also helical	two bends; bursts of activity; beats in a single plane, asymmetrically.
Ameboid tendencies:	pathological; rarely normal (?).	pathological and common
Attaching pseudopodia:	number: two position: base of the collar. functional activity: rhythmic; in a very few species, irregular used for capture of prey with few exceptions.	number: indefinite position: base of the collar. functional activity: irregular not used for capture of prey
Parabasal body:	dictyosome (Golgi apparatus) in contact with blepharoplost (kinetosome)	(Same)

the mesoglea. Nothing is known as to whether or not sexual reproduction occurs in *Proterospongia*. From this line of evidence, the sponges are seen to have achieved their multicellularity by integration of a colonial protozoan. The special pattern this integration took on, to achieve the eotypic asconoid sponge and finally the complex leuconoid forms, with their variously developed skeletons, has resulted in macroscopic animals still dependent on microscopic food particles. The techniques for this filtration set the sponges apart from all other animals. It seems most probable that extreme specialization has set the sponges aside as a dead end in animal evolution (Hyman 1940), albeit a most fascinating one.

Chapter 12
Cnidarians

THE Cnidaria, in contrast to the sponges but in common with all other multicellular animals, are termed Eumetazoa (Hyman 1940). They have definite form and symmetry, the individual organisms are easily defined, and their cells are aggregated into tissues. The fact that adult cnidarians occur in either of two body forms, the polyp or the medusa, understandably caused difficulties in the early attempts to classify this group. Even recently, the polypoid and medusoid forms of one and the same species have been referred to by different generic names, e.g., *Stephanocyphus* and *Nausithae* were applied to the polypoid scyphistoma and medusa, respectively, of one scyphozoan species (Hyman 1940). Furthermore, the radial symmetry of these forms initially led to an Adansonian type of classification, wherein all radiate animals were grouped together in the phylum Radiata. Intensive work led to finer insights, with Leuckart, in 1847, separating the Coelenterata from the Echinodermata. But the Coelenterata contained sponges, cnidarians, and combjellies (ctenophores). Hatschek (1888), sensing the diversity of these groups, recognized three separate phyla—the Spongiaria, Cnidaria, and Ctenophora, a situation with us today. The Cnidaria, as presently constituted, contains three classes, and these were recognized as such by the end of the nineteenth century. The Hydrozoa contains polyps such as *Hydra* and certain medusoid jellyfish. Some life cycles contain only the polypoid or hydroid forms, others have only the medusa as the adult stage, and other life cycles contain both body forms. The Scyphozoa are predominantly medusae, e.g., the common *Aurelia*, but attached forms exist in developmental stages and in the adult form of certain scyphozoans. The Anthozoa are exclusively polypoid, containing the familiar sea anemones, corals, and gorgonians. Thus stabilized in taxonomic terms, the cnidarians appear as a relatively uniform group, showing internal variations on the theme of attached polyps or free-swimming medusae. Both body types, wherever found, are at least superficially radially symmetrical and typically furnished with tentacles surrounding a mouth, which connects to a gastrovascular cavity. The tissues of these simple bodies consist mainly of an outer epidermal epithelium and an inner gastrodermal one, with varying amounts of mesoglea in between.

As to the role of the Cnidaria in metazoan evolution, there has probably never been so much disagreement as there is today. Earlier, and this view is the commoner one today, the relatively simple structure of the cnidarians and their radial symmetry were thought to be obvious evidence of their primitive nature, and hence they were often assumed to be the starting point of eumetozoan evolutionary development. However, there are other simple eumetazoans, and none of the other multicellular animals (except the ctenophores and echinoderms) show radial symmetry, nor is it at all clear why such symmetry must be presumed to be primitive. Other workers, therefore, see the Cnidaria from a different perspective, as derived, for example, from flatworms and representing a specialized dead end in evolution, having little or nothing to do with the origin of other eumetazoans.

This disagreement about the general evolutionary position of the cnidarians carries over into attempts to understand the history of the forms within the group. Almost all possible juxtapositions of the three classes have been proposed, namely, that hydrozoans are most like ancestral forms and from them scyphozoans and then anthozoans were evolved, or, some reversal of that sequence is proposed, or, that none is more ancestral than the other and the diversity of the phylum is due to equally intensive adaptive radiation within all three classes. The views of evolutionary direction within the Cnidaria are too often colored by presuppositions derived from some larger scheme of evolutionary development. This is unacceptable phylogenetic procedure, for we must first let the intragroup homologies tell us all they can about evolution within the group, before we go to questions of the origin of the group or to the problem of deriving other groups from this first group. The first problem is, therefore, to elucidate as far as possible the evolutionary history within and between the hydrozoans, scyphozoans, and anthozoans.

General Characteristics

The predominantly radially symmetrical polypoid and medusoid body types function as sessile or free-swimming organisms, which show coordinated activities for food getting and food utilization and which reproduce both sexually and asexually. In elaborating on this sketch of the cnidarian forms and their functions, materials are drawn largely from Hyman's (1940, 1959) treatment of the Cnidaria, from Rees (1966a), and from *Chemical Zoology* Vol. II (Florkin and Scheer 1968). Other sources will be acknowledged where appropriate.

STRUCTURES

Body form. The size of polyps and jellyfish range from minute adult forms about a millimeter long to other forms whose length, including that of the long tentacles, may be more than 13 meters. As specific examples, the tiny sand-dwelling *Halammohydra* (Remane 1927; Swedmark and Teissier 1966; Bush and Zinn 1970) is about 1 mm long and the jellyfish *Physalia pelagica* has been reported to have individuals 13 meters long (MacGinnitie and Mac-

Ginnitie 1949). In between these extremes a common size is that of one to several centimeters, which includes single individuals as well as colonial aggregates of individuals.

The symmetry of single cnidarians is radial or biradial in the majority of cases, but certain of the anthozoan polyps show a bilateral symmetry which is most conspicuous in terms of internal structure. These three types of symmetry blend into each other, in the sense that forms with slightly modified radial symmetry exist as intermediates between forms with distinct radial and distinct biradial symmetry. Intermediates between biradial and bilateral forms are also found. There is good reason, therefore, to think of symmetry as showing continuous aposemic differences, rather than being separated discontinuously into three distinct types. This view is reinforced by the fact that a tetramerous type of organization is commonly present in polyps and medusae. It is seen in external morphology by the common occurrence of four or eight tentacles or of some other, yet higher, multiple of four. However, on occasion, there are six to twelve tentacles and even some examples where tentacle number seems determined simply by the space available. Then, too, there is a certain natural variation, such that species characterized by four tentacles, for example, may have rare individuals with three or five. Another external structure that can also mirror this tetramerous pattern are the lobes of the mouth in the medusa; this is especially characteristic of certain scyphomedusae. Internally, the evidence for tetramerous organization is seen in the distribution of radial canals in the hydromedusae and in the occurrence of septa in scyphomedusae and anthozoan polyps. However, here a not uncommon tendency to a hexamerous pattern in the polyps, general variability, as well as some polymerization of the tetramerous condition, all tend to obscure or even deny the simpler pattern.

The predominance of the tetramerous condition has given rise to a set of terms that are useful in describing the placement of the radial canals or the position of oral lobes. These are called perradii. Equidistant between the perradii are the interradii. And equidistant between each perradius and interradius are the adradii. Fig. 12–1 illustrates this terminology and provides diagrammatic examples of patterns of cnidarian symmetry.

It is next necessary to describe polyps and medusae in more detail. In the polyps there is always a base for attachment to the substratum, a more or less columnar stalk or stem extending up from the base, and a terminal portion with tentacles surrounding the mouth. These three parts are variously differentiated when one considers hydrozoan as opposed to anthozoan polyps (Fig. 12–2). The hydrozoan terminal portion, the hydranth, is often elongated, with tentacles and even reproductive structures variously disturbed on it; and the mouth lies at the very apex of the hydranth often on a conelike projection called hypostome. The anthozoan polyp, on the other hand, terminates in a flattened oral disk, again with the mouth centrally located in among the ten-

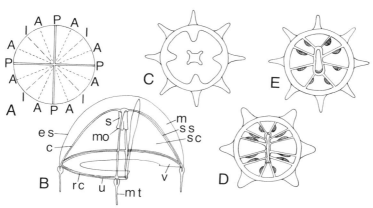

Fig. 12–1 Cnidarian symmetry. A. Diagram illustrating the location of radii in the medusa. A—adradius, I—interradius, P—perradius. B. Diagram of a hydrozoan medusa with most of one quadrant (between two perradii) removed. Tetramerous radial symmetry. c—radial canal, es—exumbrellar surface, m—mesoglea, mo—mouth, mt—marginal tentacle, rc—ring canal, s—stomach, sc—subumbrellar cavity, ss—subumbrellar surface, u—umbrellar margin, v—velum. (A and B redrawn, by permission, from Russell (1953) *The Medusae of the British Isles.* Cambridge University Press.) C. Diagram of a tetramerous scyphomedusa with eight tentacles. D. Diagrammatic cross section of a hexamerous anthozoan (sea anemone) showing biradial symmetry. E. Diagrammatic cross section of an octomerous anthozoan (alcyonarian polyp) showing bilateral symmetry. (C, D, and E redrawn, by permission, from Hyman (1940) *The Invertebrates. Protozoa through Ctenophora.* McGraw-Hill Book Company, Inc.)

tacles. The stalk of the anthozoan is proportionally shorter and broader than its hydrozoan counterpart and encloses a gastrovascular cavity much subdivided by septa (Fig. 12–1 D,E). Such septa are absent in the hydrozoan polyps. The base of the hydrozoan form may be simply an adhesive disk in solitary forms such as the well-known *Hydra*, or it may be part of a tubelike coenosarc, which can connect many polyps into a colonial organization. There is usually a single gastrodermal tube within the coenosarc, but in many species of *Tubularia* and similar forms, there are multiple tubes. The coenosarc is usually covered by a chitinous secretion, the perisarc, which may or may not extend as a cup around the hydranths. The coenosarc can occur as various parts of a colony: as the stem of delicate terminal branches, as the main stem of the colony—the hydrocaulus, or as rootlike stolons running along the substratum. In some forms the hydrocaulus may result from the compounding of several coenosarc tubes. Also, the pattern of outgrowth from the coenosarc (hydrocaulus) determines the form of the often fragile, feathery-appearing, and delicately beautiful hydrozoan colonies. In the anthozoan polyps, the base is an adhesive disk in single forms, or it is part of a continuous extension of living material that ties individual polyps together into colonies, somewhat as in the Hydrozoa. In the Anthozoa, gastrodermal tubes, or solenia, run through stolons uniting the colony. Or, very often, interconnecting solenia are

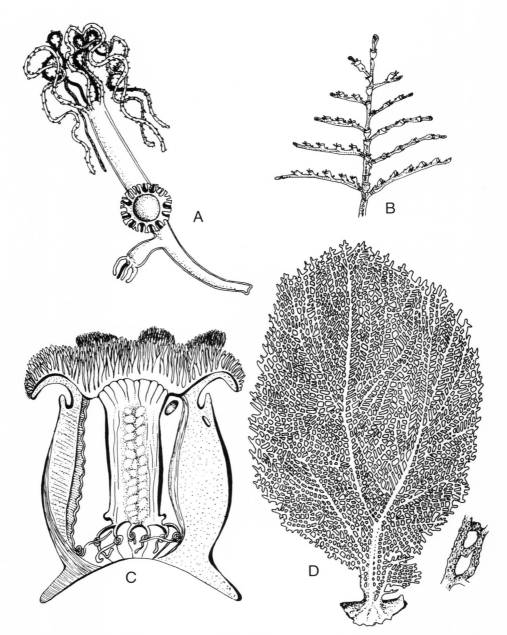

Fig. 12-2 Representative polyps and colonial growth. A. *Hydra littoralis,* a solitary freshwater form seen here reproducing asexually by budding another polyp and sexually by means of an egg, which is shown fertilized and surrounded by an embryonic theca. B. *Plumularia,* a colonial hydrozoan showing vertical hydrocaulus and side branches, each with a growing tip. Bellshaped structures along the branches are perisarc covering of hydranths; at the base of these are specialized stinging polyps, the dactylozooids. C. Cutaway diagram of a sea anemone, based on *Metridium.* D. *Gorgonia,* a sea fan, with a detail from the colony shown at the right (the dark holes mark the sites of polyp extrusion). (All redrawn, by permission, from Hyman (1940) *The Invertebrates. Protozoa through Ctenophora.* McGraw-Hill Book Company, Inc.)

embedded in mesoglea and the whole structure enclosed in epidermis. From this fleshy coenenchyme the oral ends of the polyps project outwards.

Many of the anthozoans produce skeletal material that is either calcareous or a tough, horny protein, or it may be a combination of the two. Skeletal material may be deposited internally as spicules but is best known through the external secretions produced by the stony corals. The sessile nature of the polyps and the plantlike aspect of many of the colonies contributed to the early inclusion of these forms among the zoophyta, forms thought, many decades ago, to be transitional between plants and animals (Hyman 1940). Indeed, an underwater seascape of coral heads, rising twenty to thirty feet or more from a white sand bottom, bedecked with the tracery of sea fans and sea whips jutting out from the geometric jumble of hemispheric, many-hued brain corals, rose corals, and their relatives, unavoidably suggest the overused simile to terrestrial gardens. But polyps are carnivores and only green due to symbiotic algae. And their other colors (Goodwin 1968) have, of course, nothing to do with problems of pollination, though the array of yellows, purples, drab or russet browns, flesh pinks, magenta, and olive tones vie in their own way with the petalled brilliance of land plants.

Before leaving the polyps and turning to the equally fascinating medusae, the polymorphism of the hydrozoan polyp must be mentioned. Three different categories of such zooids are recognized. The feeding polyp, or gastrozooid, has been the center of the discussion thus far. In addition, there are polyps incapable of feeding and specialized only for defense and food capture. These dactylozooids are heavily armed with stinging cells, characteristic of the Cnidaria, accumulated as a terminal knob or simply distributed along the dactylozooid which is in the form of a long, single tentacular structure called a tentaculozooid. Lastly, there are the polyps modified for asexual reproduction; these gonozooids bud off medusae. Medusa buds, can, however, also arise from the surface of gastrozooids. It should be obvious that this polymorphism is only of functional significance when the various types of polyps are associated cooperatively in a highly integrated colony, so that the specialized functions become complementary.

Medusae can be considered as the most extreme form of cnidarian polymorphism, for here the organizational pattern is highly modified to function as an unattached, free-swimming form in most cases. Pigmentation is less prominent here than in the polyps, the heavily gelatinous bells being often quite transparent or a soft blue tone with only special tentacular, feeding, or sensory structures showing conspicuous coloration. However, some medusae—often the attached ones or rare sessile ones, and occasionally free-swimming ones—show conspicuous color markings.

The hydromedusae are distinguished by the presence of a craspedon, or velum, which is a shelf of tissue, reported to be entirely epidermal by some authors (Russell 1953) but considered by others to have also some mesoglea.

This velum extends into the space beneath the medusa bell, or umbrella (Fig. 12–1B). The umbrella can be of various shapes (Fig. 12–3), but is almost always sufficiently curved so that a convex exumbrellar surface is obvious. The velum projects into the subumbrellar space and defines the size of the subumbrellar opening. The mouth is located in the center of the subumbrellar surface and lies at the end of a projection extending from that surface for variable lengths into the subumbrellar space. In some cases it passes out through the opening demarcated by the velum. This structure bearing the mouth is the manubrium and surrounds those parts of the digestive cavity directly connected to the mouth. From the digestive cavity there extend the radial canals, which go to the edge of the umbrella and there join the ring canal. The latter runs around the circumference of the umbrella and, like the radial canals, is lined with gastrodermis. Usually there are four radial canals, lying along the perradii, but in some forms there can be eight or sixteen more of these canals, which are typically unbranched; but, when there are many of them branching is not uncommonly found. In these craspedote medusae, the gonads are located in the epidermis along the radial canals or on the sides of the manubrium. Along the margin of the umbrella are located the tentacles and the sense organs. The latter will be more fully discussed later. Tentacles are usually located perradially, and thus there are commonly four on a medusa. They can also be located adradially and interradially, and even in some cases crowded together to fill up the entire margin. At the other extreme, for instance in *Hybocodon*, there is only one tentacle sitting at the end of one perradius; in *Amphinema* there are two such tentacles (Russell 1953). Often the tentacles arise from a basal swelling, or bulb. Tentacles are usually unbranched, i.e., entire. They are always provided with stinging cells and their enclosed nematocysts and, in some cases such as *Gonionemus*, also carry adhesive pods. Sometimes, interspersed between the typical marginal tentacles, are less conspicuous tentaclelike extensions termed cirri, and one species has very small solid tentacles (the marginal ones are hollow in this form) termed tentaculoids (Russell 1953).

Before considering the scyphomedusae, a few words must be used to de-

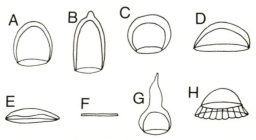

Fig. 12–3 Diagrams of various hydromedusan umbrellas. A. *Sarsia*. B. *Aglantha*. C. *Bougainvillia*. D. *Phialella*. E. *Aequorea*. F. *Obelia*. G. *Amphinema*. H. A narcomedusa. (Redrawn, by permission, from Russell (1953) *The Medusae of the British Isles*. Cambridge University Press.)

scribe the special colonial structure of the siphonophore hydrozoans (Fig. 12–4). In these forms, specialization of the component parts and their mutual interdependence has gone so far that the colony is properly considered an in-

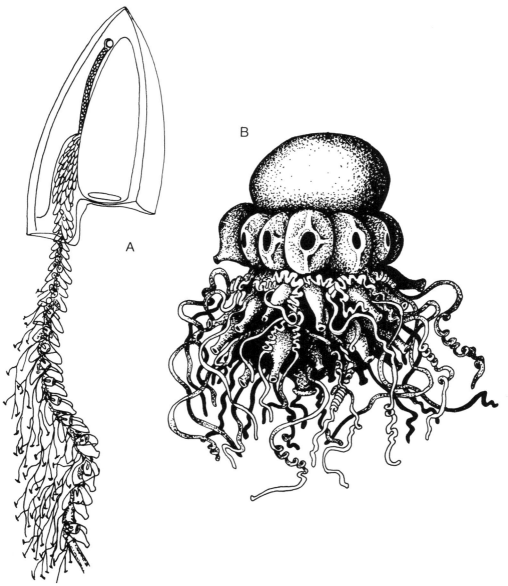

Fig. 12–4 Examples of siphonophore organization. A. *Muggiaea,* showing a single swimming bell and a long stem bearing cormidia, i.e., bells, each one bearing a gastrozooid and gonophore. B. *Stephalia,* with an apical bell float, a circlet of swimming bells, and a pendant mass of gastrozooids and dactylozooids. (Both redrawn, by permission, from Hyman (1940) *The Invertebrates. Protozoa through Ctenophora.* McGraw-Hill Book Company, Inc.)

dividual in its own right. It appears that, as if to evolve further complexity, these cnidarians have eschewed further modifying of cells and epithelial layers but have taken as their unit of organization the polypoid or medusoid states. Superimposed on these is the higher level of individuality derived from the associated functions of these complex parts. In the more complex siphonophores there are, in fact, three different polypoid types and four medusoid ones. The medusoid parts include swimming bells; bracts, which are bell-like enclosures of various polypoid parts; gonophores, on which the gonads are borne; and pneumatophores, or medusoid forms, modified into gas-filled floats. The polypoid forms form feeding parts or gastrozooids, dactylozooids for protection and food capture, and gonozooids, which develop the gonophores. These polypoid and medusoid persons, as they are termed, develop into highly distinctive versions of gastrozooids, gonozooids, bracts, and gonophores, such that they receive their own special labels. This, in effect, emphasizes that the siphonophores have developed their own distinctive direction of evolution not only within the Hydrozoa but also within the Cnidaria.

Scyphomedusae are acraspedote, i.e., lacking a velum. They also differ from the craspedote hydromedusae in having gastric pouches arranged on a tetramerous plan. These pouches are formed by four interradial septa. Characteristically, the subumbrellar surface has four interradial pits, or funnels, which indent that surface and sink into the septa. As in the hydromedusae, the bell or umbrella is variable in shape, showing in some forms a boxlike form—hence the name Cubomedusae for that group—and in others showing a circular groove or coronal groove around the exumbrellar surface—these are the Coronatae. The margin of the bell is often scalloped into lappets. Sensory organs and tentacles occur on the margin, usually in a definite number and inserted in the identations between lappets. Usually, the tentacles are hollow, highly contractile, and bear nematocysts throughout or only on clearly defined surfaces. Capitate tentacles, with nematocyst armed heads, are characteristic of the Stauromedusae which contain sessile forms with trumpet-shaped medusoids and are commonly attached to the substratum at the apex of their exumbrellar surface. The Rhizostomeae have no marginal tentacles.

The manubrium is often rather short and is squarish. At its four corners, typically, there are lobes; commonly these appear simply as long, frilled, perradial expansions of the manubrium (Fig. 12–5). In the rhizostomous forms, which lack marginal tentacles, the strongly-developed oral lobes, armed with nematocysts, become the essential device for capturing food. The gastrovascular cavity lies internally behind the mouth and is connected to the four gastric pouches formed by ingrowth of interradial septa into the digestive cavity. In some forms, slender radial canals develop peripherally from the gastric pouches and reach the umbrellar margin, where a circular ring canal may or may not be present. The radial canals are often numerous, do not follow precisely the major radii in their distribution, and may be branched.

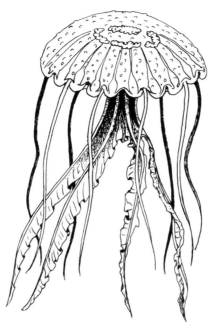

Fig. 12–5 *Pelagea*, a scyphomedusan, showing extensive development of four oral lobes. (Redrawn, by permission, from Hyman (1940) *The Invertebrates. Protozoa through Ctenophora.* McGraw-Hill Book Company, Inc.)

Histology. The polypoid and medusoid bodies just described all show an outer epidermal and an inner gastrodermal epithelium, between which is the mesoglea (Hyman 1940; Bouillon 1968). The principle cell types of each of these layers will be described, and then the various organ systems will be briefly presented. The epidermis is occasionally syncytial but usually cellular and consists of a variety of cells functionally specialized for the most part. The most common types of cells are the so-called supporting cells, which are cuboidal or columnar in most cases. They may, however, be tall and slender in anemones or very flat on the exumbrellar surface of medusae. These cells may contain granules at their outer ends and may be ciliated or flagellated. (The difference between the latter two conditions seems to be dependent on the number of vibratile elements on each supporting cell and somewhat on their length. A single such element seems to be rather consistently called a flagellum, and shorter, numerous elements are termed cilia.) The bases of these cells extend to the mesoglea and may be extended, parallel to the epidermal-mesogleal interface, in slender, tapering processes containing contractile fibers. In such cases, the cells are called epithelio-muscular cells. The contractile processes, when present, are usually oriented longitudinally. In some cases, the contractile fibers lie within their own spindle-shaped cells, quite apart from epithelial cells. In such cases they are typically embedded in the mesoglea.

Gland cells are common in the epidermis of tentacles, oral regions, and the pedal disk of polyps. These cells may produce a mucous type of secretion and contain large granules or may be finely granulated. The latter cells are much less common than the former.

The cnidoblasts with their suddenly extrusible component, the nematocyst, are so characteristic of the Cnidaria that their discussion will be postponed until the gastrodermal cells have been discussed. Then there will be a general discussion of different nematocyst types and their distribution.

Sensory cells are common in tentacles and the oral area and may be concentrated in patches in the surface epithelium to form, in effect, local areas of sensory epithelium. These cells are elongated and terminate in a point or a bulb or in sensory bristles. The nucleus is located basally. From the base there are one or more fine fibrils, often branched, which pass to the nerve cells.

The nerve cells, called ganglion cells, are usually unpolarized (Bullock and Horridge 1965), with two or three neurites extending from them. Electron microscopy of the synapses in *Gonionemus* implies, however, that there is one-way transmission of impulses (Westfall 1971). These ganglion cells lie near the mesoglea, forming a network among the basal ends of the other epithelial cells.

Interstitial cells are small, relatively undifferentiated cells occurring singly or in clumps near the base of the epidermis. They are important in the development of the Cnidaria, for they give rise to cnidoblasts, differentiate into gametes and other cell types, and are important in regeneration and asexual processes such as budding.

A general picture of the epidermis is conveyed by the following sentences:

"The covering layer of coelenterates is thus not a single epithelium but is composed of three strata: an outer wide stratum composing the main portions of the supporting cells interspersed with gland and sensory cells, a nervous stratum on a level with the bases of these cells, and an innermost muscle stratum, next to or embedded in the mesoglea." (Hyman 1940, p. 379).

In general, the types of cells found in the epidermis are present in the gastrodermis, too. The more conspicuous cells, often columnar or cuboidal, are called nutritive cells. In areas of active digestive activity, the cells are typically columnar, and the gastrodermal border is thrown into folds. These cells often contain food vacuoles in their distal halves and often are flagellated or ciliated. Bilbaut and Pavans de Cecatty (1971) have described a special configuration at the base of the flagellum in the endodermal cells in the octocorallarian *Veretillum cynomorium*. The flagellum is buttressed by slender ridges, which arise from the cell surface and radiate out from the base of the flagellum. These authors find this organization to resemble the circlet of so-called microvilli that make up the collar of choanocytes and to resemble the microvilli in enteropneusts (Norrevang 1964). These resemblances will be mentioned again later.

Similar to the muscular fibers in the base of the epidermal supporting cells, contractile fibers can be found in the nutritive cells. Hence, these are properly called nutritive-muscular cells. These processes usually run in a circular direction close to the mesoglea. Where, as in anthozoans, the musculature is well developed, the nutritive-muscular cells may have their myonemes running both in circular and longitudinal directions. The fine structure of muscle in the scyphozoan jellyfish *Chrysaora* reveals thick and thin filaments and, in general, a structure like that of metazoan muscle including actomyosin and ATPase activity (Perkins, Ramsey, and Street 1971).

Gland cells are common in the gastrodermis. Those with a presumed mucous secretion are especially to be found adjacent to the mouth. Their products are probably specially important to ingestion of food. Other types of gland cells are found lining the digestive cavity proper and are probably the source of enzymes used to break down ingested food. Some of these cells, which are thought to be gland cells, may well be storage cells holding food reserves.

Sensory cells like those seen in the epidermis are also present in the gastrodermis.

Also, there are interstitial cells, comparable to the same type of cells in the epidermis but considerably less common.

Nematocysts are found in both the epidermis and gastrodermis, often localized in patches. They are the product of special cells, the cnidoblasts or nematocytes. Nematocytes are common, epidermally, on tentacles or near the mouth. Westfall (1971) reports an interesting organization of nematocytes, epitheliomuscular and nerve cells in *Gonionemus*. Gastrodermally, nematocytes and their internal nematocysts are concentrated on the septal edges facing the digestive cavity and on the gastral filaments, which extend into the cavity. Nematocysts themselves are small spherical, ovoid, or pyriform organelles containing a coiled tube, which is everted upon appropriate stimulation. In the Hydrozoa only (Hand 1961) there is a cnidocil, or trigger, of the nematocyte, which apparently must be stimulated to induce release of the nematocyst thread.

There are two kinds of nematocysts: spirocysts and nematocysts proper. The former are single walled, often stain with acid dyes, are permeable to water, and have a simple, unarmed, reversible tube. The nematocysts proper have double-walled capsules, stain more readily with basic dyes, are impermeable to water (except after ejection), and carry a tube armed at least partially by spines. Spirocysts are limited to the Zoantharia (anthozoans), but nematocysts of various shapes, especially with regard to the everted tube, are found throughout the Cnidaria.

The fundamental study, leading to a widely accepted classification of nematocytes, is that of Weill (1934). Weill recognizes seventeen kinds of nematocysts, which are differentiated on the basis of combinations of certain characters, e.g., tubes open or closed at their distal end after eversion, tube tapering or of uniform diameter, presence or absence of a butt (*hampe*, in French), or, lastly, butt of uniform diameter or swollen at base or at summit (Fig. 12–6). It is not necessary for our purpose to go further into Weill's classification, which has been extended by subsequent work to recognize some twenty different nematocysts (Hand 1961). (And new types are still being added (Lacassagne 1968; Calder 1971).) It is sufficient to point out that there is a good correlation between nematocyst distribution and the systematics of

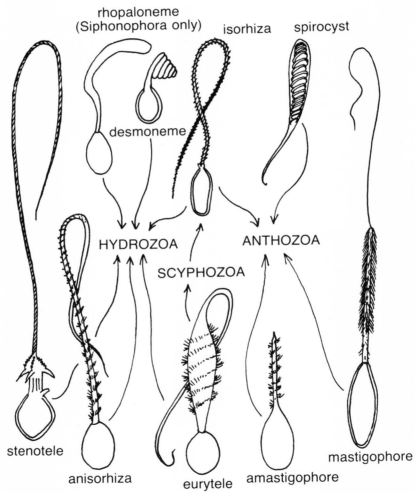

Fig. 12–6 Selected nematocysts and their distribution among the major groups of Cnidaria. (Redrawn, by permission, from Hill and Wells (1956) in *Treatise on Invertebrates Paleontology* (F) *Coelenterata* Geol. Soc. Am., University of Kansas Press.)

the cnidarians. This is seen not so much in finding a given nematocyst to occur solely in a given taxon, but rather that the taxon is characterized by a certain array of nematocysts. This becomes more understandable when it is seen that nematocysts function in differing ways. Some enter the foreign body against which they are ejected, others coil around available protrusions (Fig. 12–7). Of those that pierce, they are variously armed not only with spines but also with toxins. The latter are of differing effectiveness, depending somewhat on the victim. Also some cnidoblasts are triggered by one cue and others by another. Hence, it is seen that the differences among nematocysts reflect different functional capabilities. Their occurrence is then, in all probability, largely determined by selective forces. Certain constellations of nematocyst types represent a given set of capabilities, and a different set occurs with other nematocysts. The result would be that given nematocyst armatures are characteristic of different cnidarian taxa (Calder 1971), and the correlations that occur reflect the food-getting habits and defensive needs of the various cnidarian species. More detailed work is needed to fully elucidate the details of nematocyst operation, especially in terms of stimuli for discharge and the nature of the toxins and their effects.

The mesoglea is the third layer of the cnidarian body. Though it is variable in amount, cellular nature, and included structures, such as organic fibers and inorganic spicules and crystals, it seems best to simply refer to it collectively as the mesoglea (G. Chapman 1966), even though others have argued otherwise (Hyman 1940) i.e., that its variety is deserving of several more specific terms. In the Hydrozoa, the polypoid forms show very little mesoglea; it is at most a delicate, gelatinous cement for the two epithelia. In the hydromedusae, as in medusae generally, there is an abundance of meso-

Fig. 12–7 Two different types of nematocyst seen on the tail bristle of the freshwater crustacean *Cyclops* after pulling away from *Hydra*. Desmonemes are coiled bristle hairs and two stenoteles are seen puncturing the bristle itself. (Redrawn, by permission, from Hyman (1940) *The Invertebrates. Protozoa through Ctenophora.* McGraw-Hill Book Company, Inc.)

gleal material. It is, however, acellular, consisting of fibers, gelatinous matrix, inorganic materials and bound water. In scyphozoans, cells are present along with the noncellular material, but this is not invariable. Cases are known where, in two species of *Cyanea,* one has a cellular mesoglea and in the other the mesoglea is acellular; or in *Chrysaora,* the scyphistomal (larval) mesoglea is cellular, but the medusoid (adult) mesoglea is not (D.M. Chapman 1966). Anthozoan mesoglea is always cellular, containing ameboid and more typical connective tissue cells along with the non-cellular material.

The noncellular material is made up largely of water and contains about 3% of its bulk as various salts and 1% of its bulk as organic material. The latter consists largely of collagen fibers and a matrix of mucoprotein. These latter substances provide the cnidarian with the necessary property of binding together epithelial layers but also can function as a mechanical support (due to the viscous nature of the mucoprotein reinforced by fibers) and may also be used as a storage depot and for flotation (G. Chapman 1966).

Muscular system. The term "system" as commonly applied to the array of contractile fibers in cnidarians is somewhat misleading, for definite muscle tissue is only clearly seen in the septal muscle bands of anthozoans, where muscle cells are completely separated from epithelial supporting cells and have sunk into the mesoglea to form bundles of contractile tissues. Otherwise, the common condition is for muscle fibers to be distributed as myonemes, which are part of the basal portion of epithelial cells and, therefore, are properly regarded as epithelial rather than being a system in their own right.

In hydroid polyps, the epithelial myonemes form an outer (epidermal) longitudinal layer of fibers and an inner (gastrodermal) circular layer. This theme of longitudinal and circular fibers is found with various modifications throughout the Cnidaria. Hydromedusae show little gastrodermal musculature, and the epidermal aspect is largely confined to circular myonemes in the exumbrellar surface and in the velum. In the scyphomedusae the picture is much the same except for absence of the velum and the presence in many forms of mesogleal muscle bundles. In the Anthozoa, where muscle bundles are not uncommon, the epidermal musculature is best seen in the tentacles and oral disc, and the regular gastrodermal musculature is well developed and concentrated in well-defined muscle bands in many forms.

Sensory and nervous system. Here too the term "system" connotes more than is actually present. The sensory system is composed either of sensory cells, described above, or of certain localized sensory structures—not prominent on medusae—which are innervated by sensory cells. Processes from the sensory cells make synaptic contact with neurites from the nerve cells, and these latter are synaptically interconnected to form a loose network in the base of the epidermis and the gastrodermis. These two networks can be connected through the mesoglea.

The nervous apparatus is more highly developed in the free-swimming medusae than in the sessile polyps. Polyps show epidermal and gastrodermal nerve nets, the former more strongly developed and concentrated around the mouth and in the pedal disk. Sensory cells are also more common in the epidermis but especially on the tentacles and hypostome. In the medusae it is best to consider hydro- and schyphomedusae separately. Hydromedusae show a well-developed epidermal net in the subumbrella with concentrations on the manubrium, along the radial canals, and at the margin of the bell. This marginal nerve net forms two rings; one receives sensory processes from tentacles, patches of sensory cells, and special structures like tentacle bulbs and ocelli, the other is connected to organelles like statocysts and the velar and subumbrellar musculature. The sensory equipment, in addition to sensory cells, light sensitive ocelli, and statocysts of varying types, includes special sense clubs and tactile combs which may be modified tentacles. Scyphomedusae, as far as known, have a nerve net much like that in the hydromedusae except for lack of marginal nerve rings in most forms (the Cubomedusae have a nerve ring). Passano and Passano (1971) describe bipolar and tripolar nerve cells with unbranched neurites as composing a synaptic net in the scyphozoan endoderm. It is similar to the ectodermal net. The sensory parts include sensory cells and sense organs called rhopalioids in certain forms (Stauromedusae) and rhopalia in the remaining ones. These lie in the indentations of the umbrellar marginal and are organs of balance; in effect, statocysts. The depression in which they lie is often lined with sense cells and forms a sensory pit. Ocelli can also be present in the clublike rhopalial structures. The rhopalia often have collections of nerve cells (rhopalial ganglia) associated with them as do nematocyst batteries in some cases.

The nervous structures of the Cnidaria seem to express degrees of organization clearly correlated with the functional activity of the organisms. The less active and often stationary polyps are minimally equipped, being able to respond especially to stimuli on the tentacles and mouth region. Such stimuli, following on capture of prey by nematocyst armed tentacles, can lead to coordinated shortening of tentacles toward the mouth, which in turn can be stimulated to open, and then the prey is ingested by coordinated muscular activities of the hypostome and body wall. Needless to add, such relatively simple behavioral responses in conjunction with the relatively uncomplicated nervous system have led many many workers to study this "primitive" sensory-nervous system and its motor pathway (e.g. Parker 1919; Pantin 1956; Passano 1963), and various ideas on the nature of such systems are available. However, what is relevant to our present needs are comparative data within the Cnidaria to compare the function and structure among various polypoid and medusoid species. The studies just referred to have concentrated on a few well-known forms, especially *Hydra*, and similar studies on other forms are rare (Bullock and Horridge 1965). The structure of medusoid systems is complicated by the need for coordinating the swimming movements of the umbrella and for maintaining an effective swimming position (statocysts and ocelli). Comparisons with polyps reflect more immediate adaptive needs rather than evolutionary trends. As we will see, such a conclusion is a common one throughout these forms.

Digestive system. This system has already been largely described in terms of the digestive cavity and its septa and canals, its gastrodermal lining, and the mouth. What can be added here is to describe more fully the most complicated condition of the digestive system, which is found in the Anthozoa. Here the lining of the mouth extends down into the digestive cavity, forming a kind of pharynx. This lining is derived from the epidermis and, hence, is a stomodeum. It is differentiated into a slitlike structure with a ciliate groove or siphonoglyph at one or both ends. The siphonoglyph is considered to be a specialization for bringing water currents into the digestive cavity to aid in circulation of the contents of the cavity for both digestive and respiratory purposes. The cavity itself is subdivided by septa extending from the outer body wall across to the pharynx. The septation follows an octomerous and a hexamerous pattern in the Alcyonaria and Zoantharia, respectively.

The octomerous pattern (Fig. 12–1E) shown by the septa in the alcyonarian suborder is repeated in the eight pinnate or branched tentacles that are always present. The septa are evenly spaced around the gastrovascular cavity with longitudinal retractor muscles for the oral disk lying on one side of the septa. The septa closest to the siphonoglyph have septal filaments, heavily supplied with gland cells, along their inner edges below the pharynx. The two filaments farthest from the siphonoglyph have a band of flagellated, epidermal cells on their innermost edge which are used to direct a current of water out of the digestive cavity, thus completing the circulation initiated at the siphonoglyph.

The septal pattern of the hexamerous zoantharians is more complex. Septa may all be complete, namely extend from body wall to pharynx, or some may be incomplete. Details will be minimized here, except to point out that the septa of adult anemones are always paired, and there are never less than six pairs. Differing patterns of incomplete septa intercalated between complete septa are to be found. However, a pattern of bilateral symmetry is retained, with the axis of symmetry dividing the mouth with its one or two siphonoglyphs into mirror halves. Again the inner, free edges of the septa typically bear septal filaments. These are armed with glands and nematocysts and may terminate basally in threads or acontia. This overall bilateral organization of the Anthozoan has been the source of lively phylogenetic controversy. Is it evidence of a bilateral ancestor of the Cnidaria? Is it evidence of an evolutionary trend away from a biradial or radial ancestor of the Cnidaria? Is it simply a further result of the functional bilaterality imposed by development of siphonoglyphs? These points will be discussed again later.

Reproductive system. The nature of the gonads will be briefly described and then all further discussion of reproduction, sexual and asexual, will be relegated to the discussion of reproductive processes.

Regarding the structural organization of the gonads, Hyman (1940, p. 431) says, relative to the hydrozoans, "The gonads can scarcely be designated organs, being merely accumulations of sex cells in definite sites." This comment can equally well apply to the other two classes of the Cnidaria. In general, organs as such are essentially absent from these animals, the only possible exceptions being the sensory devices, such as statocysts and rhopalia, of the umbrellar margin of medusae.

Gonads are located in the epidermis of hydrozoans and the gastrodermis of scyphozoans and anthozoans. More precisely, the hydrozoan polyps can develop accumulations of sperm or ova in the epidermis of the upper trunk. In hydromedusae, gonads can lie along the radial canals or on the sides of the manubrium. In the scyphomedusae, gonads are on the sides of the gastric septa, and in the anthozoans they lie on the edge of the septa, behind the septal filaments.

Usually sexes are separate, but some forms are hermaphroditic.

Skeletal materials. Materials used for internal or external support of the cnidarians can be organic or inorganic or both. The organic materials can take the form of proteinaceous secretions laid down within a colonial aggregation and they can support the upright posture of forms such as sea fans and sea pens—members of the alcyonarian Anthozoa. Or the organic material can be a chitinous exoskeleton, such as the perisarc of hydrozoan colonies. Mixtures of horny, proteinaceous material and calcareous spicules are used as the endoskeleton in certain anthozoans, and endo- or exoskeletons of calcareous material are characteristic of many other anthozoans and of certain hydrozoans (Milliporinae). It has also been argued that water is used as a hydraulic skeleton by single anemones. Closure of the mouth, thus holding the digestive cavity with fluid contents intact, provides a noncompressible volume against which the muscles of the body wall can act, especially those in the foot to achieve pedal locomotion (Batham and Pantin 1950; Clark 1964).

This concludes our discussion of organized structures at the cell or tissue level that are related to specific bodily functions. It is to be noted, therefore, that in the Cnidaria there are no cellular or tissue organizations specifically concerned with excretion, respiration, or circulation. All of these are taken care of by diffusion. Such is possible, of course, in organisms composed primarily of two epithelial layers. The bulk of that mesoglea which lies in between the epithelia is largely aqueous and the "tiny dash of organic material" (G. Chapman 1966) associated with it apparently has its metabolic needs also taken care of by diffusion.

FUNCTIONS

Environmental exploitation. Except for the rare symbiotic forms, the Cnidaria feed by ingestion of prey captured and killed by nematocysts located in tentacles, in most cases, but also around the mouth and within the digestive cavity. Prey, for the Cnidaria, is other animals of a size and sensitivity to nematocysts such that they can be captured and ingested. Such food ranges from microscopic copepods to fish. There has been fairly intensive work on the mechanisms of feeding behavior. Earlier work (reviewed by Lenhoff 1968) showed that glutathione acts as a stimulator of the feeding response. More recently it has been shown that asparagine controls the contraction and bending of tentacles and that glutathione controls ingestion once the food has contacted the mouth (Linstedt 1971). Other factors, such as pH are also implicated in these processes.

Both polyps and medusae are passive predators, waiting for prey to contact them and giving no evidence of seeking out their victims. Solitary polyps are capable of locomotion—*Hydra* was seen long ago to execute a cartwheeling type of motion, alternately having tentacles and then basal disk attached to the substratum, and sea anemones are capable of quite vigorous pedal locomotion, using wavelike contractions of their basal disk to move across the substratum. Medusae move by expulsion of water from the subumbrellar surface due to contraction of the margin of the bell. Apart from the fact that sense organs orient the bell, there seems to be no directional basis to locomotory patterns. The simplicity of the nerve net is consistent with this simple behavior.

Homeostatic functions. Digestion is both extra- and intracellular in the cnidarians. Ingested materials are first partially broken down by enzymes released from gastrodermal cells. The resulting broth is then taken up in food vacuoles by the nutritive cells and converted into food stuffs utilizable by the rest of the body. Some of these products of digestion may go into reserves held in gastrodermal cells. As far as is known, such a pattern is common to the free-living members of this phylum.

Under the broad interpretation we are using for homeostatic functions, we can also mention behavioral responses of cnidarians to external stimuli, excluding the feeding stimuli mentioned in the preceding section. Where studied, such responses appear relatively nonspecific and most effective at the point of stimulus. That is, responses to mechanical or chemical stimuli are generally similar, and the reaction involves a local contraction of a tentacle or other part of the body. Further stimulation or stronger stimulation can result in a general contraction of the body, but, again, this can occur for any strong stimulus. Variations in response can be correlated with the body part stimulated and have reference to the degree of development of the sensory apparatus, nerve net, and musculature at that point. The limited nature of the behavioral responses is not unexpected in view of the relatively uncomplicated nature of the organism. Studies on the nervous system of cnidarians have been summarized above, and though a good deal is known anatomically, little is known from a functional point of view except for a few well-studied species, such as *Hydra*. This limited comparative information on the cnidarians, as a whole, makes it premature to try to develop significant comparative data regarding sensory-coordinatory functions in this group of animals. Authorities such as Bullock and Horridge (1965) present an overview that seems to emphasize the general similarities in function among the various cnidarian groups rather than finding significant differences.

The absence of excretory, respiratory, or circulatory systems precludes specific discussion of these typically homeostatic functions. Diffusion suffices, apparently, to take care of the processes included under these functions, and in that context there are no special capabilities here that are distinctive and useful for comparative studies.

Reproduction. Sexual reproduction is known throughout the Cnidaria, but it is somewhat variable in terms of the various forms that comprise a given life cycle. Development starts from the fertilized egg. Sperm, typically with rounded heads, short midpiece, and a long tail, are released externally and fertilize the ova either still in the parent or released into the surrounding water. It seems chemotactic responses by sperm can determine union with eggs. In the hydroid *Gonothyrea*, free-swimming sperm contact tentacles of the female gonophores and then proceed along the tentacles to the internally placed ovum where fertilization occurs (Miller 1970). Subsequent development is either internal or external, depending on the site of fertilization.

Cleavage patterns vary somewhat. In the Hydrozoa, cleavage of the zygote is indeterminate, holoblastic, adequal, and radial. In the Scyphozoa, two orders (Stauromedusae and Cubomedusae) show holoblastic and equal cleavage, whereas the remaining forms (Coronatae, Semaeostomae, and Rhizostomae) show holoblastic and somewhat unequal division (Berrill 1949; Thiel 1966). In the Anthozoa, cleavage is complete and equal. What results from cleavage is a ciliated coeloblastula. This hollow sphere of cells becomes a solid stereogastrula by single and multiple inwandering of cells, by delamination, and rarely, as in the Coronatae, by invagination. This stereogastrula, or planula, is ciliated in all forms except the scyphozoan Stauromedusae, where it is a small wormlike form. The planula is often ovoid, but can also be pear shaped, moving with the blunt end foremost. The fate of the planula is variable (Fig. 12-8). In the hydrozoans, it can change into an actinula, which in turn may become a polyp in the Actinulida (Swedmark and Teissier 1966), or into a medusa. Or the hydrozoan planula can turn into a polyp, which will become sexually mature in its own right or will bud off medusae. Where medusae occur, these form gonads and so recommence the life cycle through gamete formation. In the scyphozoans, the planula develops into a polypoid medusa in certain forms; in the others, it forms a special polyp, the scyphistoma, which buds off medusae by successive transverse fissions. In certain forms, these medusae pass through a larval medusoid stage, the ephyra, before achieving the distinctive adult form. Finally, in the anthozoans, where there is no medusa, the planula develops directly into the polyp.

At this point, there can be pointed out a very interesting feature common to all planulae; there is always one cilium per cell (Widersten 1968). This curious situation is not found in other eumetazoan larvae, nor is it always the case where vibratile elements are present in adult cells; these may have more than one cilium per cell. One convention is to refer to conditions such as this in the planula as flagellation rather than ciliation. This reflects ideas stemming from the protozoa, where many flagellated cells carry only one flagellum, but ciliates have many cilia. Such a terminology may also carry a phylogenetic implication. Since phylogeny is not profitably pursued by such simple generalizations as flagellar or ciliary number by itself, we can set aside any such complications here. It will be simplest for our present study to re-

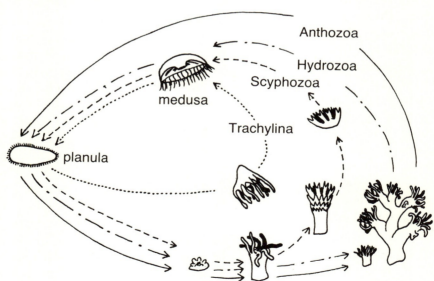

Fig. 12-8 Various life cycles in the Cnidaria. (Redrawn, by permission, from Hill and Wells (1956) in *Treatise on Invertebrate Paleontology*. (F) *Coelenterata*. Geol. Soc. Am., University of Kansas Press.)

tain the commonly used term, cilia, for the locomotory elements of planula, but not forgetting the special conditions of their occurrence here.

The transformation of the solid planula into the polyp or medusa form, carrying a hollow digestive cavity, occurs always in generally the same manner. The planula settles down on its anterior pole, which now becomes the aboral pole of the next stage of development. The inner mass of cells splits to form a cavity lined with endoderm (the future gastrodermis) and an opening breaks through the unattached posterior end of the planula to form the mouth. This is now the oral pole of the organism. Then tentacles appear and other structures are differentiated in detail as is appropriate to the various species.

Beyond the preceding purely descriptive account of development, there are now appearing more analytically oriented studies. For example, Müller (1969a, 1969b) and Brändle (1971) have developed detailed models to account for cnidarian morphogenesis and differentiation in terms of gene action, activators, gradients, and inducers. As in the ciliates, some forms are intensively studied, but the majority have received no treatment in experimental terms. Hence, this work, rich though it is in terms of complex processes, has insufficient breadth to allow of significant interspecific comparisons.

The same situation holds for the experimental analysis of regeneration. Here, steps have been taken to capitalize on the dramatic regenerative capacities of cnidarians. Burnett's group has shown that complete organisms can be regenerated from pieces of pure gastrodermis or pure epidermis (Lowell and Burnett 1969), and there has been determined the source of secretory cells during regeneration by means of radioactive labelling (Rose and Burnett 1970). These and other such studies will surely eventually provide us with useful data on regenerative patterns that can be used phylogenetically when a larger sample of cnidarians has been studied.

Asexual reproduction can be viewed as a reproductive step that helps complete a life cycle or as a process which simply multiplies the forms in a given stage of the life cycle. That is, a polyp can give rise to a medusa, which is the adult form of a given series of developmental phases, or it can give rise to more polyps by budding. A medusa can bud off more medusae; however, medusae are never known to form polyps. In addition to budding, frustule formation is also known in the hydrozoans. This is a kind of budding wherein planulalike forms are pinched off. Also, fissioning of polyps is known. New polyps can also be formed from detached parts of polyps by what is essentially a regenerative process. This is seen most dramatically in the phenomenon of pedal laceration seen in anthozoans. Here anemones actually pull away from parts of the basal disk which remains attached to the substratum. This can be done repeatedly to leave behind a trail of detached parts, and each one of these can regenerate into a new polyp. The anthozoan polyps also show isolated examples of longitudinal and transverse fission, as well as budding. In the Scyphozoa, the planula larva can transform into a stolonous larva, and from this various upright, polypoid parts can be developed which become scyphistomae. The scyphistoma, by transverse constrictions or strobilization can form medusae, but can also form new scyphistomae. It can also put out stolonlike buds which will form new scyphistomae, or it can bud off planulalike forms, which will then differentiate into new scyphistomae.

There seems to be no special pattern to the known occurrence of these asexual processes. On present evidence it seems that the various buddings, fission processes, etc., have arisen independently, probably several different times, and reflect the procreative needs of individual species. On this view, there is apparently some functional capability of the Cnidaria that renders asexual processes as readily evolvable. Certainly the relative simplicity of their structure and the widespread presence of undifferentiated interstitial cells are a necessary part of the story, but these cannot be considered as sufficient to explain all the asexual tendencies. A complete explanation must wait for further understanding of developmental phenomena.

This rather complicated account of the reproductive capabilities of the Cnidaria is a fitting way to conclude the statements on general characteristics of the group. It points up the fact that there is much diversity within these animals but a diversity that is not readily, if at all, resolvable into trends.

Phylogeny

There are various informative traits that clearly set off the Cnidaria from other organisms and allow one to make a clear argument for the monophyletic nature of the group. Within the phylum, however, the variability of many traits, though initially suggestive of important opportunities for phylogenetic analysis, can be seen to represent many different but sometimes apparently repetitive variations on a few simple themes. Thus the intragroup variation is more a picture of independent experiments in adaptive radiation than an example of sustained development of definable trends. This will add serious difficulties to finding the plesiomorph of the Cnidaria and pose problems in trying to grapple with relationships of groups within the Cnidaria. The easiest place to start our examination of these problems is with the least variable characters. From there we can move to more complex situations and see more clearly what are the semic traits that give the Cnidaria their special character and are the basis of their diversity.

HOMOLOGOUS SEMES

Body layers. The three body layers are perhaps the classic example of a conservative homologous trait in the Cnidaria. This is not to say there is no variation in the epidermis, gastrodermis and mesoglea of the various jellyfish and polyps. There is, of course, but the essential fact remains that this type of organization is found in all groups. The variations, except for the mesoglea, are best seen as variations in one component or another of the epithelial layers—the musculature, the sensory and nervous cells, the nematocysts—and will be so treated. What is of immediate concern is that the relative locations of the three layers can be seen to be similar, and their component parts are generally similar everywhere—sensory cells, supporting-muscular cells, nutritive-muscular cells, collagen fibers, and so on. Hence, the positional and compositional relations are fulfilled in terms of homologizing the body layers.

Body types. Another recognizable homology is the Bauplan, or body type, of the cnidarians. The fact that polyps and medusae both share the typical body plan does not really detract from its validity as a common organizational pattern. Numerous transitions, both morphological and developmental, bridge that gap. It is hard to speak precisely of positional relationships when one is discussing the whole organism for, by definition, an organism is not part of a network (Gefügesystem) or another such system. One can only say whole organisms are being compared and then pass to the compositional relationships to see if the major parts (mouth, tentacles, gastro-vascular cavity and body walls—umbrellar or pertaining to the body column) all bear similar relationships to each other, which they do in the Cnidaria as a whole. Further, their symmetrical arrangements are relatively conservative. In the hydrozoans and scyphozoans a radial symmetry is apparent that is most often tetramerous. In the sycphozoans this feature is also seen in their septa. The anthozoans show a superficial radial symmetry, but internal specialization—siphonoglyph, septa and retractor muscles—reveal a bilaterality which may well have been imposed on an ancestral radial pattern of a tetramerous (now octomerous) and a hexamerous nature. Finally, there must be recognized differences in body proportions and numbers of parts (e.g., tentacles, septa, etc.). These, however, are better treated in terms of specific body forms within the larger context of the Bauplan, just as musculature can be treated separately from the epithelia in which it is commonly found.

There are certain functional traits in the Cnidaria which are plesiosemic. The mode of *capturing and ingesting prey*, the patterns of *digestion, locomotion* (allowing for differences in medusae, as compared to polyps), and many *behavioral responses* all play equivalent roles in the overall functioning of these animals and are composed of similar subprocesses.

Development up to the planula. Reproductive functions contain both plesiosemic and aposemic components. Sexual reproduction is quite plesiosemic in terms of gamete formation, fertilization and appearance of the planula, though the cleavage patterns between fertilization and stereogastrula formation are somewhat variable. From the planula on, development goes along many different paths. These include insertion and subsequent deletion or extensive modification of certain larval stages and the occurrence of various types of asexual reproduction. Events from the planula on will be considered aposemic, of necessity, and events up to and including planula formation can be interpreted as largely plesiosemic. We turn next to aposemic features of the coelenterates.

Body forms. These are naturally aposemic in that the arrangements of parts reflect adaptation to specific niches. Such specificity will, of course, modify underlying plans of organization in a variety of ways. Here it will be necessary to use the serial relationship to demonstrate homologous relationships between the variety of tentacles, mouth parts, umbrellas, and polypoid stalks. There seem to be no significant discontinuities in the various cnidarian forms that would necessitate arguing for separate origins of the various kinds of polypoid or medusoid bodies.

Nematocysts. The twenty-odd different types of nematocysts are clearly aposemic. The essential characters of an intracellular location and capsular construction with the pleated, inverted thread allow us to argue for the homology of all nematocysts. (Electron microscopy of the spirocyst is needed before they can be included as homologs of the nematocyst.) The different placements of spines and of swellings and so on, all make the nematocysts appear highly variable. The overview of nematocyst variability has the further dimension of more than one type of nematocyst often occurring in a given species. It is, however, the characteristic association of nematocysts that is so useful taxonomically (Calder 1971). The aposemy in these microscopic structures takes on a superficial appearance, at times, of random variation, since trends in the variants are hard to find, and since the function of the nematocysts is so poorly known. The latter precludes systematized conclusions regarding the distribution of these organelles. It does not, on the other hand, justify the statement that "... the possibility exists that nematocysts may be another example of variation without functional significance." (Hand 1961, p. 195) Hand, himself, in the article quoted from, points out that the function of only three nematocyst types is clear. This should be clear warning against premature guesses as to the reasons for nematocyst variation. At present, the fact of nematocyst variation is with us, superimposed on certain similar positional and compositional details that establish the homologous nature of this organelle.

Sensory-nervous system. Variations in the sensory-nervous system are of two types. In the nerve net itself are seen variations in the concentration and location of the ganglionic cells. This, as was said before, is correlated with the functions of various parts of the organism such that tentacles and mouth parts, which must respond to feeding stimuli and act cooperatively to achieve ingestion, are relatively well provided with a nerve net. The same is true for the umbrellar margin. Secondly, different organisms have different responses to these needs. Hence, for instance, there is variation in the distribution of sensory cells and sense organs along the umbrellar margin. These differences, however, do not obscure the common location of the nerve net at the base of the epithelia nor change the fact of the common makeup of sensory and ganglion cells. The sensory organelles may, however, have shown separate origins (as of statocysts in the hydro- as compared to scyphomedusae), and so there can be some neosemy in these traits. It seems wisest to treat the sensory-nervous system as largely aposemic.

Musculature. The same discussion that was applied to the preceding system can be applied to the musculature. It varies most in placement and degree of development, with some forms showing special developments, such as the septal retractor muscles of the Anthozoa. Compositionally, there is little variation except between fibers lying in the base of epithelial cells or lying free as distinct muscle cells.

Gonads. The gonads might be considered compositionally as a plesiosemic trait, but their variety of external form and of position, especially in the hydrozoans, suggests that they be treated as an aposeme. The positional variations and compositional conservatism renders them, too, a trait like the nervous and muscular systems.

Larval forms. The final aposemic character to be mentioned is the variety of larval forms, or, in broader terms, development after the planula stage. Here, the variety is in terms of the various ways the planula reaches the sexually mature adult stage (Fig. 12-8). The planula can go to a polyp, which in certain hydrozoans and in the anthozoans is the mature adult, or it can go from the polyp to the medusa, or, finally, it can go almost directly to the medusa. The last case is seen in the trachyline hydrozoans, where the planula goes to the actinular larva, which resembles a simple medusa (Fig. 12-9) and then to the medusa proper. The transition from the polyp to the medusa has several variations, including budding of medusae off of a feeding hydranth, development of a reproductive hydranth, development of reproductive hydranths specialized for medusa formation by budding (all of these in the Hydrozoa), and transverse constriction of special polypoid forms or scyphistomes to give larval medusae (ephyrae), which then develop further into typical medusae (found in the Scyphozoa).

These cases show some special larval forms, i.e., the actinula, ephyra, and scyphistoma, but these are protomedusae or polypoid in nature, and represent variations on the basic body-plan theme of the Cnidaria. The aposemy in postplanula development lies not so much, therefore, in new products of developmental differentiation but in modifying, adding to, or losing preexisting

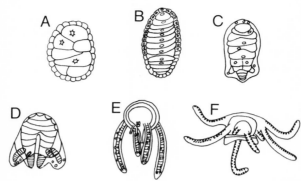

Fig. 12–9 Development of *Aglaura hemistoma*. A. Early planula. B. Late planula. C. and D. Transformation into medusa underway, by way of actinular larva. E. Actinular larva. F. Young medusa. (Redrawn, by permission, from Rees (1966b).)

ones. The composition of the process, so to speak, is not changed so much as the position of a given stage in the developmental sequence. Throughout, two rules are maintained: a medusa forms a planula via gametes and fertilization, and a medusa never directly forms a polyp. Apart from that, most other possibilities all appear one way or another, especially when the various asexual processes are also considered: a planula can give rise to other planulas, or to polyps, or to (larval) medusae; polyps can give rise to planuloid forms (frustules), other polyps, or medusae; and medusae can bud off medusae or form a planulae only by sexual means. The greatest variety of these steps occurs in the Hydrozoa, whereas the development of the Scyphozoa and Anthozoa is more stable.

Also, in terms of aposemic developmental patterns, there should be mentioned the extensive occurrence of colonial forms. These are seen in a variety of growth patterns in the hydrozoans and anthozoans. It is quite probable that such tendencies arose several times independently, as evidenced by the fact that different skeletal elements—perisarc, horny proteinaceous materials, and calcium carbonates—have been used to support the colonies. And if, as seems probable (see below) the plesiomorph of the phylum was a solitary form, then colony building is really a neosemic trait.

From the preceding discussion of homologies we can make the following statements. Medusoid and polyploid body plan and associated symmetry; two epithelial layers with intervening mesoglea; similarity of feeding, digestion, locomotion and behavioral patterns; and similarity of sexual processes and development to the planula stage—these are all quite stable and therefore plesiosemic traits unique to the Cnidaria. Aposemic traits include certain body forms, nematocysts, the nerve net and various sensory structures, the musculature, gonads, and postplanula development. These aposemic and plesiosemic traits together define a group of complex characters that are shared by cnidarians and no other group of organisms: this is the synaposemy of the Cnidaria relative to other living things that allow us to argue for the monophyletic nature of these distinctive organisms. It has also been suggested (D. Chapman 1966) that the cnidarians have two other unique features, i.e., desmocytes and a certain basal body apparatus. Details supporting this claim have yet to appear.

It is, however, especially important to note that with the sole exception of the nematocysts, the aposemic characters lie in traits that relate to variations on the polypoid or medusoid theme of organization. The variations in form, nervous and sensory tissue, musculature, and gonads, and development beyond the planula, are variations not so much in composition of structural materials but in rearrangement of constituent parts, reflecting variations in how polyps and medusae are put together or develop. This is seen also in colony formation, which, since the phylum seems determined to keep the polyp-medusa theme inviolate, is the only possible further evolutionary step; namely, to use the basic body plan as a unit for a higher level of organization. This is best seen in the Siphonophora, where the cooperating polyps and medusae are so modified, due to subserving the needs of the colony as a whole, that they are hardly recognizable still as polyps or medusae, and the colony has its own individuality; it is really an organism in its own right (Mackie 1965). Much of the specialization in these forms comes from loss of features such that a multipurpose medusa is reduced to the nectophore, whose single purpose is swimming (Mackie 1965); it lacks mouth, manubrium, tentacles, and sense organs.

The next step is to find the cnidarian plesiomorph. There are paleontological and ontogenetic data to look at as well as a set of aposemic traits for possible morphoclines.

PLESIOMORPH

Fossils. The earliest possible fossil cnidarian is a Pre-Cambrian scyphomedusa (Hill and Wells 1956). Hydrozoans, scyphomedusans, and possible anthozoans are known sporadically from Cambrian rocks (Fig. 12–10). From then on, the record becomes better, as would be expected, with certain forms becoming extinct, but with many representatives of the three major classes existing continuously down to the present day. This picture, when examined in closer detail (Moore 1956 and various contributors therein), tells us that the representatives of the major cnidarian groups were well evolved by the time the earliest fossils were being formed, and subsequent fossilization is a record of trends within these groups but tells us little or nothing about the interrelations between the Hydrozoa, Scyphozoa, and Anthozoa and certainly provides no direct evidence on the common plesiomorphic form.

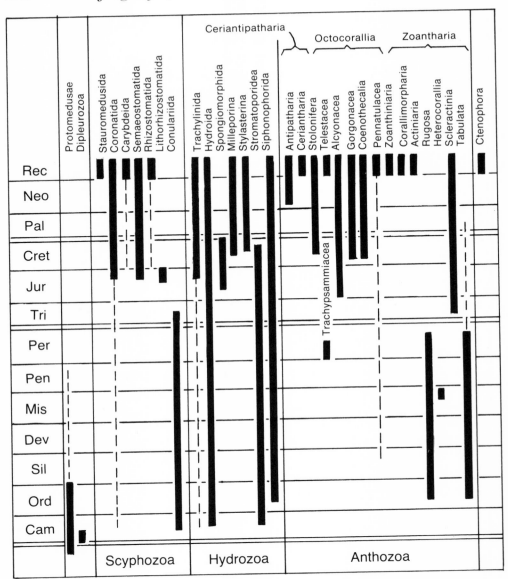

Fig. 12-10 Geological occurrence of the Cnidaria. (Redrawn, by permission, from Moore (1956) in *Treatise on Invertebrate Paleontology.* (F.) *Coelenterata.* Geol. Soc. Am., University of Kansas Press.)

Ontogeny. The ontogeny of the Cnidaria is difficult to interpret in biogenetic terms. The basic pattern seems to be that a solid, ciliated (one cilium/cell) stereogastrula or planula arises from the zygote and forms a medusa. Between the planula and medusa, a polypoid stage is often inserted, and if the medusa is lost, then the polyp can be the source of new gametes. The basis for postulating that the medusa is the original adult form is the following: (a) if a medusa is present in the life cycle, it, and not the polyp, produces gametes, and (b) polyps can directly form medusae, but never the reverse. This suggests, then, that the medusa, probably tetraradiate and like the actinuloid medusa (since here there is no polyp stage in development), is the primitive, functional cnidarian. This idea goes back to Böhm (1878), Brooks (1886) and more recently Hyman (1940), Rees (1966b) and Swedmark and Teissier (1966). The planula itself, sometimes cited as the starting point of the Cnidaria, can only be, at best, a reminiscence of the protocnidarian. Its inability to feed, its solid endoderm, its lack of tentacles and mouth, all argue that it does not represent the Cnidaria as such. Just what it does tell us as a protocnidarian will be discussed later, when the highly speculative problem of cnidarian origins will be examined.

It is to be repeated that the cnidarian ontogenetic pattern is highly variable from the planula on. Nevertheless, for the reasons given, the medusoid form is considered the original adult form of the Cnidaria and the polyp a larval stage in its origin, though latterly an adult stage in its own right in many forms. However, on ontogenetic evidence alone, the foregoing proposals can be neither affirmed nor negated, for it is possible to suppose that the polyp was originally the adult and through prolongation of the life cycle by adding a medusa, the latter assumed adult functions, notably gametic formation.

These two different interpretations of cnidarian ontogeny are part of the arguments for proposing that the hydrozoans, with their actinula larva, contain the most primitive forms, and from them the Scyphozoa and Anthozoa are derived; or for proposing the reverse sequence, that anthozoans, with their exclusive dependence on polyps as the adult form, are representative of the primitive state in this phylum and the Scyphozoa and Hydrozoa are products of further evolution. There is only one other source of information that can help us in this dilemma, and that is to look at evolutionary trends or morphoclines.

Morphoclines. Two obvious gross trends are changes in symmetry and in septation. The hydrozoans and scyphozoans show a tetramerous radial symmetry, and the anthozoans show a hexamerous bilateral symmetry. Here one can broadly describe the trend as follows:

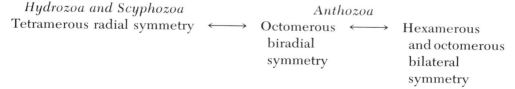

Hydrozoa and Scyphozoa	*Anthozoa*	
Tetramerous radial symmetry ⟷	Octomerous biradial symmetry ⟷	Hexamerous and octomerous bilateral symmetry

The double-headed arrows indicate that on this evidence alone the direction of the trends is not ascertainable.

In terms of septation, the morphoclines run somewhat differently. The hydrozoans show no separation, but both scyphozoans and anthozoans do. In the latter, despite the probably homologous nature of the septa, they do have certain obvious differences, described earlier and summarized by the terms septal pouches in the Scyphozoa and septal walls in the Anthozoa. There are two ways to express diagrammatically the morphoclines to be seen here. One indicates that lack of separation in the hydrozoan is one terminus of a morphocline and the septate condition itself the other terminus but having two different conditions of expression—scyphozoan and anthozoan. This is consistent with the interpretation of the septa as homologous in both classes. If the septa are not homologous, then the morphocline can have the two septate conditions as termini with the hydrozoans in the middle. The least complicated interpretation of either scheme is to derive the septate from the nonseptate condition, i.e., the scyphozoan and anthozoan condition would arise from a common septate ancestor that came from the hydrozoans, for example,

or both conditions would be derived independently and in parallel from nonseptate forms, thus,

$$\begin{array}{l}\text{nonseptate} \longrightarrow \text{septate scyphozoans} \\ \text{hydrozoans} \longrightarrow \text{septate anthozoans.}\end{array}$$

This argument for deriving septate forms from nonseptate forms has the weakness of any probability statement, i.e., it is not certain and other alternatives are also possible. What does improve its plausibility is that it agrees with what is thought to be the more likely of the interpretations of the ontogenetic data, i.e., that hydrozoans are probably ancestral, because they retain examples of planula-to-medusa development, via the actinulid protomedusa.

Let us turn next to other aposemic characters to see if they contain morphoclines that add to or detract from the above tentative conclusions.

The distribution of nematocysts provides little help in defining morphoclines. One summary of their occurrence in cnidarian classes (Hill and Wells 1956) is given in Fig. 12–6. It is clear from this figure that certain arrays of nematocysts can characterize certain taxa, as stated previously, but beyond that there is no good evidence of trends. Though formerly it was maintained that the trachyline hydrozoans, which include the forms with actinulid larvae having only one kind of nematocyst, Lacassagne (1968) has shown three kinds of nematocysts to be present in *Halammohydra*. This makes it difficult to press the case of the trachylines in terms of nematocyst type, for there are other cnidaria which are monotypic with regard to nematocysts. There is no way at present to decide which is a primitive and which a derived type of nematocyst, i.e., there is no convincing nematocyst phylogeny. Hence we can say nothing that is really useful regarding the use of nematocysts to define morphoclinal trends.

The sensory-nervous system shows various degrees of development, but these are closely correlated with the functional activities of the forms possessing them. Hence, polyps on the whole are less well equipped with sensory organelles than the more active medusae. The location of nerve-net concentrations reflects usage of body parts—tentacles, the mouth, and umbrellar margin have significant development of ganglion cell nets. Trends, if there are any that merit the term, are correlated with special activities of this or that form rather than showing how a given ancestral condition is progressively elaborated or reduced, or both. Explicitly, it appears that an ancestral condition of nerve-net sensory cells and the potential for differentiating sensory organelles were initially present in the cnidarian eotype, and that this system developed in various ways, probably repeating itself as independent, parallel developments as different polypoid or medusoid forms responded independently to similar selection pressures. Trends are largely impossible to define in such a mosaic of variation.

And, to be brief, the same situation appears to hold with the musculature, the gonads, and the variety of larval forms. It leads to the conclusion that morphoclines relevant to the Cnidaria as a whole are for most characters not reliably defined. (Within smaller groups they can be found, as in distribution of tentacles in the hydrozoans (Kühn 1914; Hyman 1940) or in the fossil trends of certain corals (Schindewolf 1942).) It means, further, that the search for the plesiomorph stops with the data derived from ontogeny and from symmetry and septation morphoclines, and these data, as we have just seen, do not permit much precision in identifying the plesiomorph and, hence, also, allow little confidence in our conclusions.

The study of Swedmark and Teissier (1966) of the Actinulida seems to be as direct a statement as is possible today on the nature of the cnidarian plesiomorph. These authors conclude that the various species of *Halammohydra* and *Otohydra* represent forms that are neither sessile polyps nor free-swimming medusae, but, rather, are actinulae adapted to a sandy habitat. They are specialized in terms of their psammophilic niche but are primitive in terms of other less specialized characters—". . . at one and the same time as archaic and specialized types of organization . . ." (p. 131). This is a recent formulation of the actinula theory of Böhm (1878) and Brooks (1886) with refinements added by later workers, notably Hyman (1940), Garstang (1946), Tatton (1954) and Rees (1966b). All of these workers favor this view largely because of the ontogenetic data summarized earlier.

To suggest a specific form as the plesiomorph of this group, we will propose *Otohydra vagans* (Swedmark and Teissier 1966). This species seems somewhat less specialized than the members of *Halammohydra*, in that it lacks an adhesive organ and nerve ring, though it still possesses statocysts, like the other genus. (The statocyst is an organelle not uncommon in sand-dwelling forms. See Chapter 13.) The evolutionary development of these sensory organelles from the vibration receptor function attributed to sterocilia has been outlined by Horridge (1969). The earlier interpretation of these sand-dwelling hydrozoans as regressed medusae (Remane 1927) has been shown to be an incorrect interpretation of their morphology (Swedmark and Teissier 1966, and earlier papers of these authors).

Otohydra vagans can be briefly described as follows (Fig. 12–11). It is a small form, about 1 mm. long, entirely covered with cilia, which are used for locomotion. There are 12 to 16 tentacles and statocysts around the hypostome. Stenoteles are the only type of nematocyst found thus far in the *Otohydra*. Unlike *Halammohydra*, *Otohydra* is hermaphroditic. This may be another specialization in this form, since most cnidarians are dioecious. After fertilization, the gastrula is formed by total, equal cleavages leading to a morula whose blastocoel becomes filled by delamination. This development occurs within a brood pouch in *Otohydra*. There is no freeswimming planula stage. The gastrula transforms directly into the young actinula, first with four tentacle rudiments. Then four more appear and, also, the first four statocysts. The mouth is formed when the larva leaves the parent as an actinula. This then differentiates into the mature *Otohydra*.

Before leaving the topic of the plesiomorph, it is necessary to comment on other views on the subject. These are predominantly those that prefer an anthozoan as the primitive cnidarian. Unlike those who favor a hydrozoan and come to close agreement on the actinulid forms as being the best candidate, largely for ontogenetic reasons, a variety of reasons are given for the anthozoans being primitive, and only very vague suggestions are found as to which specific form or forms is best seen as a concrete example of the viewpoint espoused.

A quick summary of the major points is as follows: Hadzi (1963 and earlier) has long held the view that cnidarian evolution is best understood by proceeding from rhabdocoel flatworms to bilaterally symmetrical cnidarians such as the anthozoans. However, Carter (1954) and Pantin (1966) have both made explicit the errors in Hadzi's interpretations of structures thought to be sufficiently similar in rhabdocoels and anthozoans as to justify his phylogenetic views. These and other criticisms (Remane 1963a) have seriously weakened, if not rendered highly improbable, Hadzi's argument.

Jägersten (1955, 1959) favors the anthozoans as primitive because of his interpretation of the hypothetical ancestor as a bilaterally symmetrical gas-

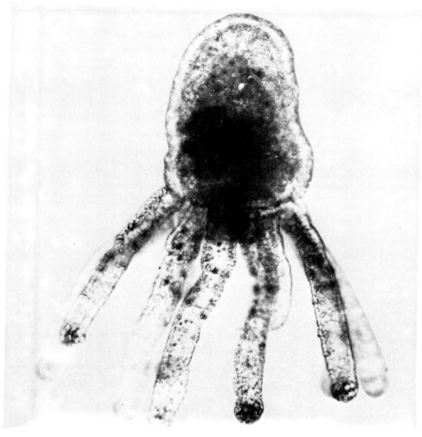

Fig. 12-11 *Otohydra vagans*. (Courtesy of B. Swedmark.)

traea—a bilaterogastraea—and because actinian (sea anemone) larvae show features similar to the postulated bilaterogastraea. The essential criticism to be raised against this hypothesis is its methodological inadequacy. Plesiomorphs are not defined by finding forms which resemble hypothetical ancestors. Indeed, this is the reverse of the method that must be followed: hypothetical ancestors are inferred from those actual forms which by paleontogical, chorological, embryological, and morphoclinal data are thought likely to be most primitive. If one tries to argue for an anthozoan plesiomorph on ontogenetic data alone, which is part of Jägersten's argument, then one must face what seem to be the more cogent arguments of those favoring the hydrozoan point of view, i.e., that the polyp is a secondarily derived form and that the medusoid forms are primitive.

Remane's (1963b and other papers) views can be criticized in the same vein as Jägersten's. His conclusions are the result of a theoretical bias and not of analysis of the group itself in a rigorous phylogenetic framework. Remane's theoretical bias, however, is not the bilaterogastraea theory but the enterocoel theory, which demands that the primitive cnidarians have a pouched gastrovascular cavity. In that the anthozoans provide such a condition, they are interpreted thereby as being primitive. The inadequacies of the enterocoel theory itself have been presented by Hartman (1963), in addition to the unacceptable nature of this approach to phylogenetic problems.

Pantin (1966, see discussion), on purely morphological grounds, favors the actiniarian Anthozoa as the primitive cnidarians. Hand (1966) has argued that the Actiniaria are better derived from the madreporine corals, and if he is correct, these forms would be the cnidarian plesiomorph. However, Hand is arguing only about the evolution of the Actiniaria and favors the trachyline Hydrozoa as primitive cnidarians (Hand 1959, 1963). It is not at all clear that Pantin accepts Hand's proposed derivation of the actiniarians from the corals. Pantin finds that ". . . the Anthozoa do not show that extreme functional differentiation of nematocysts such as you see in *Hydra*; on the other hand, they show a wide spectrum of curious forms such as spirocysts. But to me, most important, is the essential simplicity of the nervous system compared with other Cnidaria. So far as we know at present, it is a comparatively simple nerve sheet. In the Scyphozoa, as Dr. Horridge has shown, we find two intraepithelium nerve nets at least. When you get into the Hydrozoa, you have got to have greater degrees of complication." (p. 17) This argument based on relative simplicity is a very uncertain one, as we have seen, and only becomes relatively convincing when (a) the long run of evolution is being considered (demanding an overview of several phyla usually; certain large phyla such as Mollusca and Chordata are obvious exceptions) and (b) where adaptive responses can be ruled out as an alternative explanation of varying degrees of visible complexity. The former case does not hold here, of course, and the latter point—adaptation—is particularly difficult to eliminate in the Cnidaria. In fact, from earlier discussions in this chapter, it would seem that differences in nervous organization would be expected to have adaptive correlations, and that the relatively sessile actiniarians would need only a minimal nervous system. Pantin's proposal, based on relative simplicity, is not convincing, but as is also true with such arguments, cannot be dismissed.

In conclusion, we return to the view that *Otohydra vagans* is the best possible candidate today for the cnidarian plesiomorph.

PHYLOGENETIC ANALYSIS

We shall forego an analysis of phyletic distances and relationships in the Cnidaria. First, because of the tenuous nature of the evidence that suggests *Otohydra vagans* as the plesiomorph, including differences of opinion as to whether it is primitively or secondarily simple, it seems prudent to wait for further evaluation of this and related forms. Second, and more important, the enormous plasticity of the Cnidaria in their form and internal structures makes it extremely difficult for a nonspecialist in this group to have any real confidence in forms he might choose to gain a representative sample of the various major taxa within the phylum. This is reinforced by the still unsettled nature of the taxonomy in many parts of the phylum; e.g., Hand (1966) proposes raising two of the zoantharian orders to subclass status to improve the probability of homogeneity (and monophyly) in the remaining zoantharians, and a comparison of cnidarian systematics in Hyman (1940) and Hyman (1959) shows other evidence of significant taxonomic change. It seems best to forego what must be quite premature attempts at explicit phylogenetic conclusions until the phylogeny of smaller groups has been worked out more fully and to see, thereby, what forms emerge as plesiomorphs of their particular groups. These, then, can be used for a more meaningful phylogenetic analysis than is now possible.

What can be considered, to some extent at least, is the origin of the presently-proposed cnidarian plesiomorph, *Otohydra vagans*. This discussion will also provide the opportunity to comment on the question raised earlier, and thus far not properly answered, namely, the curious pattern of variation seen in the Cnidaria.

Origins. To look for some noncnidarian form with traits homologous to at least some of those found in *Otohydra vagans* is the necessary approach to elucidating the origin of the Cnidaria. What features of *Otohydra* seem likely to have been conserved from an ancestral condition? Several traits can be suggested: (1) tetraradiate symmetry, (2) a ciliated body surface, but this may simply reflect locomotory needs (however the nature of the locomotory structure could be of interest), (3) habitat and food-getting habits, (4) ontogeny and reproductive habits. On the other hand, distinctive cnidarian traits that seem less conservative are presence of statocysts, specialization of tentacles, and presence of nematocysts. Regarding these latter, in particular, Pickens and Skaer (1966) conclude that, on the basis of current electron microscopy, the cnidarian nematocysts are peculiar to the Cnidaria. The explosively extrusible trichocysts, and toxicysts of ciliates, ejectisomes of cryptomonads, cnidocysts of certain dinoflagellates, polar capsules of Cnidosporidia, colloblasts of ctenophores, and sagittocysts of acoel Turbellaria, all of these are unconvincing as homologs of nematocysts.

Starting with the locomotor organelles, the fact that there is one cilium per cell leads one to wonder where such a condition might be originally present. It is unlike anything in the ciliates, obviously, and is closest to a group of flagellated cells. The absence of cell walls and any evidence of phototrophic nutrition suggest some colonial zooflagellate form. This is reinforced by the ameboid mode of formation of food vacuoles by the cnidarian gastrodermis—an ingestive, or animal, mode of food intake. The problem with this suggestion is one that has been apparent ever since Haeckel (1866) started his speculations on the origin of "diploblastic" animals from an invaginated, colorless volvocine colony: there are no living forms that adequately represent the suggested zooflagellate colony. The volvocine coenobia, so commonly and uncritically used in textbooks in a loose neo-Haeckelian sense, are always photosynthetic forms, biflagellated, with cell walls and other features that are not homologous with the postulated zooflagellate colony and certainly not related to the ancestral cnidarian (Hanson 1958). Sponges, as we have seen, can be derived from colonial choanoflagellates, but those colonies with their distinctive collar cells are not a suitable candidate for an ancestral cnidarian. In the zooflagellates proper, described in Chapter 5, it was seen that not only are free-living forms rare, they are always single, never colonial.

The absence of any plausible zooflagellate form suggests one should look elsewhere for the cnidarian ancestor. But where? Other flagellate forms are largely photosynthetic when colonial forms are examined. One is forced back to unicellular zooflagellates if one stays with animallike nutrition, which is obviously reasonable for the ancestral form. And it is an indeed forlorn hope to find any plausible evidence of homology between such a form and *Otohydra*.

The answer to this dilemma, as proposed by many workers, again starting with Haeckel and exemplified recently by Jägersten (1955, 1959), is to propose hypothetical intermediates. Actually, all that is convincing in the use of these forms is to reiterate how different presently known Cnidaria are from the other living forms, and in particular, how much evolutionary history probably intervened between the zooflagellates (if they are ancestral) and the earliest cnidarians.

The fact of this gap becomes understandable when we realize that the Cnidaria are predominantly a cladogenetic group; they show evolutionary development around the theme of nematocyst-bearing tentacles surrounding a mouth which leads to a gastrovascular cavity. Their anagenetic development is limited almost entirely to colonial forms, the Siphonophora showing the greatest development in this regard. We are, then, viewing a group that has rung an impressive variety of changes on its quite simple ancestral organizational plan of medusoid body. What are the consequences of that evolutionary history?

The Cnidaria are an old group, and if they did first evolve, as speculation and *Otohydra* indicate, as a bottom-dwelling colony of zooflagellates feeding by ameboid ingestion of food particles (later these cells specialized into the lining of a digestive cavity), then they were tiny and probably living in a world where unicellular forms and bacterial detritus were their food. This was very possibly an efficient mode of life until competition from mouthed forms, e.g., the ciliates and bilaterally symmetrical metazoan worms became a fact of life in the early seas. Escape from predation might have been the further development of secretions used for food capture that were also repellent to enemies. Continued selection pressure along this line could have resulted in primitive nematocysts. Such organelles might also have been a preadaptation for food capture. With this was evolved the cnidarian eotype. This form was already far removed from the zooflagellate ancestor, which was either consumed by enemies or had lost out competitively to the nematocyst armed forms. As all animal forms continued to evolve, the food available to the Cnidaria also changed, notably getting larger. In this situation the cnidarians evolved further, but still retaining the limitations presumably inherent in their origin from simple colonial forms, but increasing their size in single forms, or by colony formation, with help from mesogleal development and from endo- and exoskeletons. The result would be a great array of attached and free-swimming forms reflecting the diversity and size of the food the cnidarians had available in the Cambrian seas and subsequently.

On this view the original exploitive features of the eotype were pushed to the extremes seen in the polyp-medusa form, only escaping to go further through colony formation. And because of the dual pressure first from other ingestors (both as competitors and as predators), and secondly from expansion of their food resources, the Cnidaria known today are far removed from their ancestral forms. Also, their innovations were in terms of the exploitive characters of food capture, hence their history is one of variation on that theme and would be as varied as the various modes of efficient exploitation would permit. As in the sponges, parallel evolution would be of common occurrence. (An Adansonian or numerical taxonomic treatment of their variety may well be the best way to handle these two phyla. Preliminary work along this line in the Cnidaria appears successful (Powers 1970).) This point of view presents a way of understanding the phylogenetic isolation of the Cnidaria and their curiously complex aposemic variability. It also says we can never expect to be confident of the derivation of this group from any other, at least in terms of the kind of data presently available. Perhaps, and hopefully, for it is our last resort, molecular data in the form of conservative polypeptide sequences can provide information not otherwise obtainable.

Finally, there is the question as to why the Cnidaria have clung so tenaciously to their simple organization and low level of tissue differentiation. It seems the pressures were surely present for evolving a more complex nervous and muscular system and for more complex orienting organs, as is seen in other eumetazoans. Why, then, were these organs not developed? The answer may lie in the limited potential of the postulated ancestral zooflagellate cells. Such cells, presumably supplied with one flagellum apiece, would not have had the great organizational and developmental potential of the higher zooflagellates and ciliates. Starting from such limited origins, as did the sponges from another even more limited group of flagellates, the Cnidaria were doomed to limited evolutionary development, which not even colony formation could overcome.

Chapter 13
Flatworms: Turbellarians

THUS far, this study of the origins and early evolution of animals has not encountered the organization that characterizes most animals, namely, multicellularity with tissue differentiations in a bilaterally symmetrical body. The Platyhelminthes or flatworms do exemplify that mode. There are differences of opinion as to where these worms stand in the course of evolution of the bilateral animals—at the bottom of the Bilateria or somewhere further up—and the question will be dealt with in due course. However, that the flatworms are properly included (Hyman 1951; Beauchamp 1961a) in the Bilateria is not challenged, and thereby our phylogenetic study now reaches the level of typical animal organization.

As presently constituted, this phylum includes the predominantly free-living turbellarians and two large groups of symbiotic forms, the flukes (trematodes) and tapeworms (cestodes). These symbiotic forms and the curious symbiotic Temnocephalida, will not be examined in this chapter. Such specialized forms are in all probability evolutionary deadends; they apparently arose from free-living turbellarians and specialized along unique lines.

General Characteristics

Except for other references as indicated, the following account is based largely on Bresslau (1933), Hyman (1951, 1959) and Beauchamp (1961b).

STRUCTURES

External morphology. Turbellarian worms vary in size from adult individuals no more than a few hundred micrometers long to those 50 cm. or more in length (the land planarians). Most are, however, less than 5 cm. long (Hyman 1951). Typically the worms are much longer than they are wide, and the body itself is often dorso-ventrally flattened, though many forms are quite round. Commonly, the ventral crawling surface is flat and the dorsal surface is slightly to distinctly convex. Anteriorly, the body terminates in a head, which can be distinguished externally in some forms by the presence of lobes, tentacles, eyespots or other specializations, often, but not always, of a sensory nature. The posterior end is usually regularly tapered but may show lobes or, rarely, thin tail-like appendages. In general, the contours of the turbellarian body are quite regular; the most notable exception probably is the body of the polyclads when showing active locomotion for, at such times, the margins of the body can show extensive undulations and foldings. Selected body forms of free-living flatworms are given in Fig. 13-1.

Finer features of the external morphology include the following. Cilia commonly cover the entire surface of these worms. In some cases ciliation is limited to the ventral surface and in some cases it is absent from areas where sensory organs are present. Bristles are present in certain forms and probably serve a sensory role. Where the bristles are located posteriorly, their function is quite problematical. There is always a mouth in the free-living Turbellaria, and it is found in the midline of the ventral surface, usually some place in the middle third of the body. It is usually a small circular opening but can also be slitlike. The genital orifices also lie on the ventral surface. There is at least a male opening, but in addition there can be female openings, and in some species these are combined into a single orifice. And in extreme cases there may be three genital orifices, where copulatory and egg-laying functions have separate body openings (Fig. 13-2). The sensory structures, alluded to briefly above, may involve bristles, but also tentacles, lobes, and sensory pits in certain forms. Pigmented eyespots are sometimes located on cephalic lobes but are often internally located. However, because of their pigmentation, they are often a conspicuous component of the general details one observes in examining the gross morphology of these worms. The same is true for the statocyst, except that its refractile nature, rather than pigmentation, accounts for its conspicuousness.

Pigmentation is obviously present in the flatworms. Colors may be vivid—reds, oranges, violets, often contrasting with dark stripes or bands. More often the colors are drab browns, blacks and olive greens. The pigment itself can be in special pigment cells or dispersed throughout other cells, in granules or apparently even in solution. The pigment can lie in the superficial layer of the body, but more commonly it is found in the deeper parts. Symbiotic algae are not uncommon.

Body layers. Turning next to internal structures, the identification of three body layers is found in all the turbellarians. These layers are the outermost epidermis (often completely ciliated), the innermost gastrodermis lining the gastrovascular cavity or intestine—only in the acoels is this digestive epithelium lacking—and an intermediate layer of solid mesenchyme or parenchyma. The careful work of electron microscopists (Pedersen 1961a, 1961b, 1964, 1966; Skaer 1961; Dorey 1965) has shown that all tissues of the flatworms studied thus far are cellular. The only possible exception is the central parenchyma lying next to the mouth in acoels. If the limited number of forms studied thus far is a representative sample—it includes a half dozen species

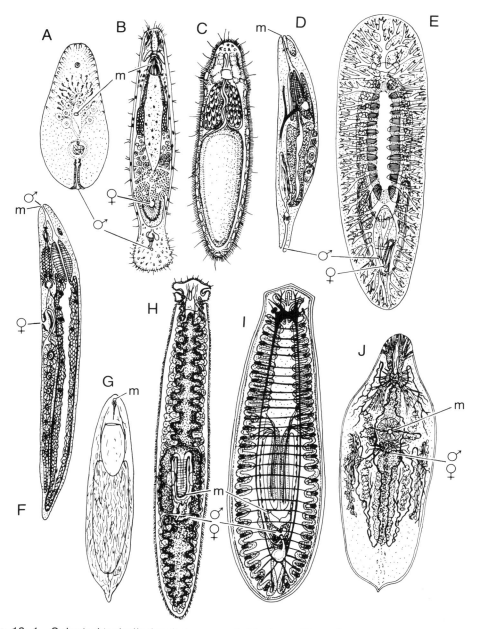

Fig. 13–1 Selected turbellarian flatworms. A. *Haploposthia rubra*. B. *Macrostomum appendiculatum*. C. *Stenostomum ventronephridium*. D. *Proporoplana jenseni*. E. *Notoplana alcinoi*. F. *Plagiostomum album*. G. *Prorhynchus* sp. H. *Bothrioplana semperi*. I. *Procerodes lobata*. J. *Mesostoma ehrenbergi*. m—mouth, ♀—female gonopore, ♂—male gonopore. (A. Redrawn, by permission, from Westblad (1945). C. Redrawn, by permission, from Marcus (1945). F. Redrawn, by permission, from Hyman (1951) *The Invertebrates—Platyhelminthes and Rhynchocoela.* McGraw-Hill Book Company, Inc. B, D, E, G, H, I, and J. Redrawn, by permission, from Beauchamp (1961b) in *Traité de Zoologie.* Vol. IV, Masson et Cie.)

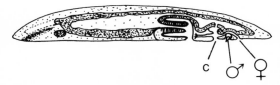

Fig. 13-2 Sagittal section of *Monocelis* illustrating, especially, the three genital orifices. (c). copulatory opening, ♂ male opening with penis papilla, ♀ female opening for release of fertilized eggs. (Redrawn, by permission, from Beauchamp (1962) in *Traité de Zoologie,* Vol IV, Masson et Cie).

from quite different taxonomic groups—and generalizations from them are in order, it appears that the long-standing claims of syncytial tissues in the Turbellaria must now be set aside. Thus far, all fine structure studies agree on the cellular nature of these worms. This is especially important for the acoelous forms, where the supposed syncytial nature was an important part of certain phylogenetic arguments (Steinböck 1958; Hanson 1958, 1963a). Certain specialists on these worms, notably Reisinger (1959) and Ax (1961, 1963), had long argued for the presence of extensive cellularization. The difficulties in establishing their claims by direct light microscope observation have been made clear by electron microscopy. The turbellarian tissue cells very commonly show irregular shapes with extensive cytoplasmic projections. This interdigitation of cells is essentially impossible to unravel by the light microscope (Pedersen 1961a, 1961b; 1964) and, hence, the understandable conclusion that many of the tissues, especially the parenchyma, are syncytial.

The epidermis is an epithelium one cell-layer deep. Its external border is usually ciliated, and it often rests internally on a basement membrane. This is a highly developed mass of fibers in the polyclads, is less conspicuous in the planarians, and absent in the acoels. The filaments making up subepidermal membranes are found in the intercellular spaces in the planarians and polyclads.

It would be of great interest to have extensive comparative details on ciliary organization. Dorey (1965) has provided interesting details of the cilia and their subsurface components, i.e., whole kinetides, in the Acoela. Also, he has worked out the surface pattern and interconnections of these kinetides (Fig. 13-3). Interestingly, the kinetides lie in ordered longitudinal rows; various illustrations of this can be found (Fig. 13-4A). However, the details of the pattern, as seen in *Convoluta roscoffensis* (Dorey 1965) (Fig. 13-4B), show that there is not a direct connection between neighboring kinetides of the same kinety. The connection, presumably via "ciliary rootlets" or microtubules, is to the two kinetides of the adjacent right and left rows, and then from them through their rootlets to the next anterior kinetide. It will be of great interest to know how generally this pattern is maintained across cell boundaries. Dorey's plates show regularity of ciliary distribution from one epidermal cell to the next, but whether the rootlets maintain their pattern too is not yet answered. If the function of the microtubules is to space and anchor the cilia and kinetosomes, then one expects continuity of the rootlet system regardless of cell boundaries, though it then becomes difficult to understand how the separate cells cooperate in developing shared rootlet structures. Dorey's observations on regeneration of host epidermal cells through replacement cells formed in the parenchyma, suggest an experimental situation where rootlet formation and ciliary orientation might be profitably analyzed. Unfortunately, there is no other work on the turbellarian epidermis comparable in detail to Dorey's on selected acoels; and more information is needed on the epidermis before extensive comparisons can be made on it as a tissue.

Pedersen's (1961a, 1961b, 1964, 1966) detailed studies on the parenchyma of acoels, planarians, and polyclads shows that they contain similar neoblastic cells (their identification is not entirely clear in the acoels) and fixed parenchymal cells. The most significant parenchymal difference is the significant amount of intercellular material present in the polyclads. This is minimal or absent in the other two groups. Pedersen feels this difference may relate to the size of the organisms. "The acoels are very small, while the polyclads range as the largest turbellarians, and those triclads previously examined come in between. Perhaps it is more economical for the animals to use extracellular products instead of cells in building up their connective tissue." (Pedersen 1966, p. 114).

Studies on the gastrodermis are more in the state of the epidermal work, namely, its basic features are known, but detailed work on several quite different species is not available. It is to be emphasized, though, that if there is any syncytial tissue in the acoels, it is in the central or digestive parenchyma, just inside the mouth. Pedersen (1964) is hesitant to make any definite claims for or against a syncytial interpretation of this material. Nonetheless, it is clear that acoels lack a gastrodermis such as is found in the other turbellarians. However, in these latter forms, at times of heavy feeding, the demarcation of the differences between gastrodermis and parenchyma becomes obscure, and the digestive cavity itself seems to disappear. The gastrodermal cells with enclosed food vacuoles fill the central body area of the worms (Beauchamp 1961b). Beauchamp, especially, thinks that absence of a hollow digestive sac in the acoels is not a sharp discontinuity between them and the other Turbellaria, since in these latter a digestive sac can be transformed to a condition not unlike that seen in the Acoela.

Secretory elements. These structures are epidermal or subepidermal glands and their products, such as rhabdoids. Rhabdoids are of various sorts (Hyman 1951), the most common being rhabdites, which are secreted by either epidermal or mesenchymal gland cells and are short rodlike structures located in the epidermis at right angles to the cell surface. Other rhabdoid en-

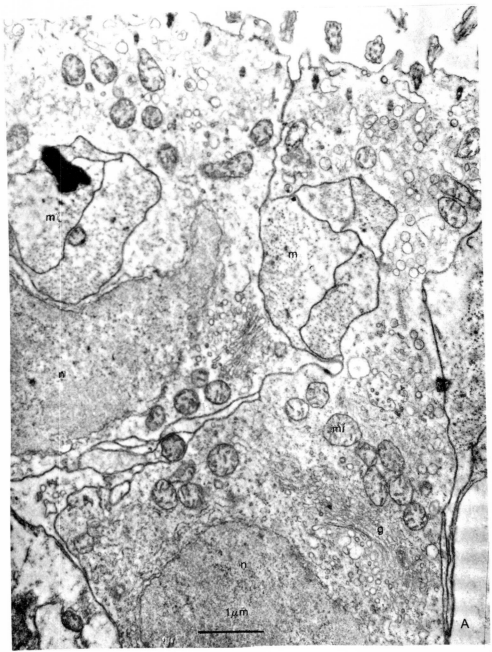

Fig. 13-3 Fine structure of *Convoluta roscoffensis*. A. Transverse section through the dorsal surface. Note the outline of the epidermal cell whose nucleus (n) lies at the left central position in the micrograph. There are many cilia associated with each cell. Subepidermal muscle cells (m) with their microfilaments are apparent. mi—mitochondrion, g—Golgi apparatus. B. A nearly tangential section of the dorsal epidermis. Note

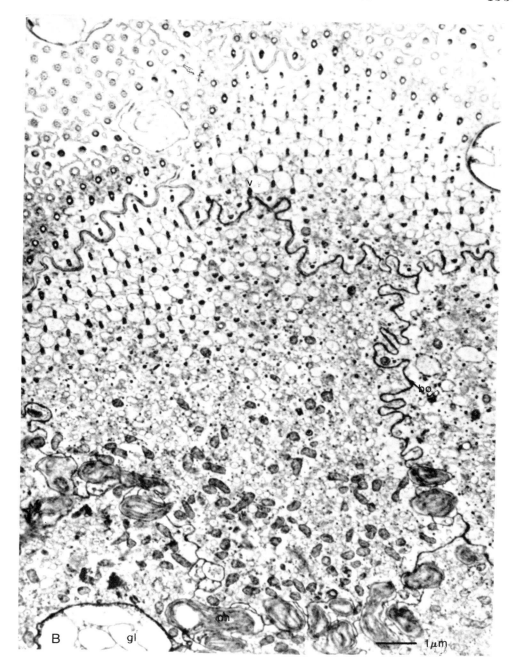

the positioning of the cilia. (See also Fig. 13-4 B.) bo—undulating boundary between epidermal cells, ch—chloroplast of endosymbiont, f—fused microvilli between cilia, gl—gland, v—superficial vacuole. (Originals courtesy of Dr. A. E. Dorey, published in Dorey (1965).)

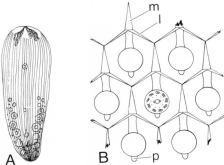

Fig. 13-4 Ciliary rows and kinetides from acoels. A. "Typical acoel, from life, California", according to Hyman. Note the longitudinal ciliary rows. (Redrawn, by permission, from Hyman (1951) *The Invertebrates. Platyhelminthes and Rhynchocoela.* McGraw-Hill Book Company, Inc.) B. Details of ciliation, reconstructed from electron micrographs, in *Convoluta roscoffensis.* 1—lateral rootlet, m—main rootlet, p—posterior rootlet. (Redrawn, by permission, from Dorey (1965).)

tities are the pseudorhabdites of polyclads and alleocoelous forms, which apparently produce mucous or slimy secretions, and the sagittocysts of acoels, which contain needlelike bodies. The exact function of the rhabdoids is not clear: they appear to produce materials related to defense, or to entanglement of prey, or to adhesion, for even the rodlike rhabdites are reported to dissolve upon extrusion and provide a mucous material which remains on the surface (Beauchamp 1961b).

Returning to the gland cells themselves, the useful and simple classification of Hyman (1951)—eosinophilous and cyanophilous—appears to be the most effective descriptive approach to these structures. The majority of them are subepidermal, and uptake of blue color from hematoxylin designates them cyanophilous, and uptake of red from eosin, fuchsin, and other such dyes identifies them as eosinophilous or erythrophilous. (The term eosinophilous seems to be most commonly used.) These glands are commonly unicellular but may on occasion, as in the case of the frontal gland of many acoels and certain other, larger forms, be aggregated and release their products at a common, localized part of the cell surface.

The cyanophilous glands are especially abundant, occurring on all parts of the body surface but most commonly on the anterior ventral surface. Eosinophilous glands are less common and occur sparingly over the entire body surface. Regarding their function, Hyman (1951, p. 72) says " . . . it is inferred with reasonable probability that the cyanophilous secretion is of slimy or mucouslike nature and acts to lubricate the ventral surface and to form a slime trail on which the cilia get a purchase; and the eosinophilous secretion is of a sticky or adhesive nature, used for adhesion, capture of prey, cementing eggs to a substratum, etc. There are a number of kinds of glands in each category, and each of these presumably plays a specific role in the activities of the animal."

The exact nature of the secretions of these glands is not known. Typically, the eosinophilous glands contain a granular secretion product, but this says nothing as regards its chemical nature.

Musculature. The musculature of the flatworms is best divided into three categories. What we shall discuss now is the musculature associated with the body wall and that which traverses the parenchyma—the general musculature. What will be set aside for treatment under two other headings is the musculature of the pharynx and of the copulatory apparatus. General body musculature is situated, for the most part, just below the epidermis and its basement membrane. In certain smaller acoels and small and less complex forms of other groups, it reportedly lies in the base of the epidermis itself. The outermost fibers are circular with respect to the long axis of the body, and the inner fibers are longitudinal, that is parallel to the body axis. In larger forms, diagonally placed fibers can lie between the outer and inner muscle layers. In fact, one can see a tendency to more muscle fibers and more layers as one proceeds from the relatively delicate development of these structures in the acoels up to the large and complex polyclads and land planarians. These latter show very complex development of parenchymal muscles, which are present in the acoels as more or less occasional dorso-ventral fibers.

As to the composition of the fibers, they were originally thought to be long slender strands, sometimes striated, lying in the surrounding tissue as isolated fibers. The critical silver preparations of Gelei (1930 and earlier) showed, however, that the fibers are enclosed by a membrane which also included a nucleus, set to one side of the fibrous material. This view of the cellular nature of the general body musculature has been perfectly corroborated by electron microscopy (Pedersen 1961a, 1961b, 1964, 1966; Dorey 1965), although this later work has found no evidence of striations (Fig. 13–5).

Nervous system. Discussion of the nervous system is dependent on light microscope work, for electron microscopists have thus far provided no detailed data on this important tissue. As with the musculature, there are forms in which the nervous system appears to be epidermal. In these forms—certain acoels and other relatively simple forms like *Hofstenia* (Hyman 1951; Bullock and Horridge 1965)—the nervous structures consist largely of longitudinal strands, usually five pairs, spaced more or less regularly around the periphery of the body. The material designated as a brain is either lateral thickenings of those cords at their anterior end or is a layer of material around the statocyst. The situation is not clear. In addition to longitudinal cords and so-called brain, there are various lateral extensions of the cord which interconnect the cords and thus make a plexus of the whole system. In all other forms the ner-

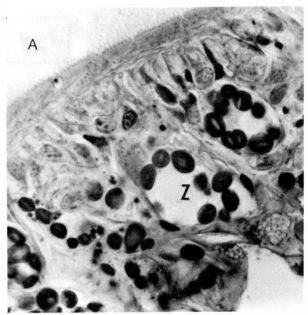

Fig. 13-5 Muscle cells of turbellarians A. Light microscopy of *Convoluta convoluta* (Acoela). The ciliated epidermis is seen at the upper left, and immediately subjacent is the subepidermal musculature. The rest of the tissue is peripheral parenchyma, with zoochlorellae conspicuous as round, dark bodies (Z). (Courtesy of Professor K. Pedersen, originally in Pedersen (1964).) B. Electron micrograph of *Discocelides langi* (Polycladida). A parenchymal muscle cell (MU) is shown with microfilaments in cross section (M—mitochondrion). (Courtesy of Professor K. Pedersen originally in Pedersen (1966).)

vous material lies below (internal to) the subepidermal musculature. The submuscular plexus, to use Hyman's (1951) term, is the locus of the most extensive development of the nervous material. In the polyclads the netlike nature of the acoel system is also found to be present but heavily developed, especially on the ventral side and in regard to a conspicuous brain located in the parenchyma. Other forms show emphasis on the longitudinal cords. In forms like the kalyptorhynchids and dalyelloids, the tendency is to heavy development of the ventral pair of cords and presence of the others simply as short cords, or they may even be absent. In the triclads, represented by planarians, the nervous system is represented by a brain and ventral nerve cords with heavy cross connectives or commissures. In the land planarians, the whole structure becomes an almost solid ventral plate whose anterior specializations are somewhat arbitrarily designated as brains. In their review of the longitudinal cords and commissures of turbellarians, Bullock and Horridge list nine major organizational patterns. Additionally, the brain, surrounded by parenchyma, is seen as a localized thickening of the cords or as a distinctly organized mass of nervous tissue, which, in more complex forms, shows an outer

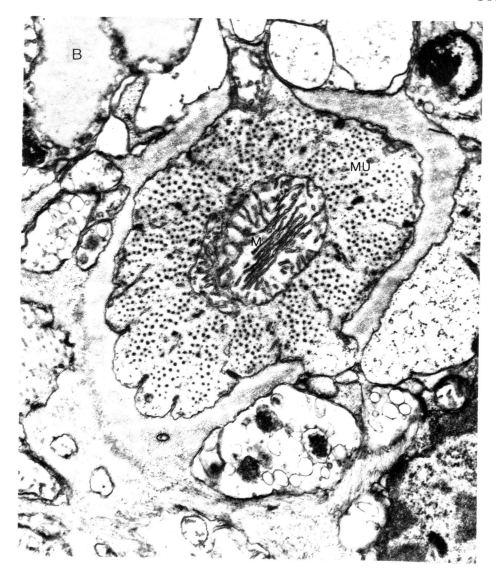

layer of cells and an inner core of fibers. This is a major organizational change from the simpler metazoans, i.e., sponges and cnidarians, and presages a typical feature of higher invertebrates (Bullock and Horridge 1965).

Another aspect of the nervous material in the Turbellaria is the sensory paraphernalia, including statocyst, eyespots, various other types of receptor cells (tango-, chemo-, and rheoreceptors), and the cells that innervate all of them. The statocyst and eyespots lie in the parenchyma in close association with the brain, or they may rise from the brain (Westblad 1937). Though there is nervous tissue around the statocyst, actual sensory endings have not been traced into the statocyst (Hyman 1951). The structure of the statocyst seems to be quite constant. It consists of a double layered vesicle with the inner space containing a statolith. Except for the land planarians, the eyespot is a cup of pigment cells with the club-shaped, distal ends of the retinal cells projecting into the cavity of the cup. In the land planarians studied thus far, the retinal clubs pass between the pigment cells and the clubs of some of them are reported to face the cup opening. Such eyes are of the converse type; the others are the inverse type (Hyman 1951). Eyespots are common throughout the Turbellaria being typical for all except the simplest forms, such as most acoels. On the other hand, statocysts are common in the simpler forms—acoels, catenulids, monocelids, *Hofstenia*. In the acoels both types of sensory apparatus are present simultaneously (Fig. 13–6).

The receptors for touch, chemicals, and water currents have their distal ends at the cell surface, and the rest of the cell extends internally into the parenchyma and joins synaptically with the central nervous system. Bristles

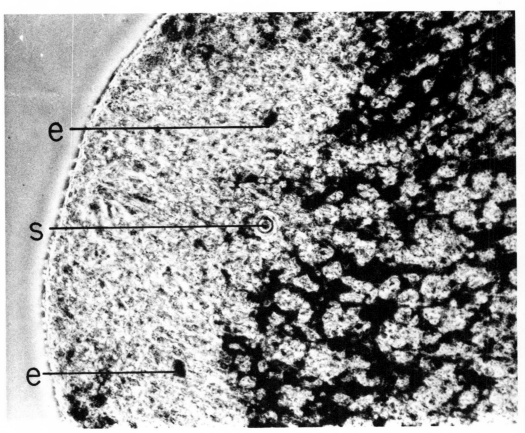

Fig. 13–6 The anterior end of *Convoluta sutcliffei* showing statocyst (s) and eyespots (e).

mark the surface location of tangoreceptors, and the turbellarians are richly provided with them. Gelei (1930) found in *Mesostoma* that every epidermal cell is pierced by three to five such bristles. Chemoreceptors are apparently often localized in sensory grooves or pits. These are variously distributed in turbellarians. However, the absence of sensory pits and grooves does not mean chemoreceptors are absent. The almost universal ability of these worms to locate the source of food juices in their environment indicates the widespread and probably highly developed ability to orient in response to chemical stimuli.

Rheoreceptive cells have been identified in relatively few forms (Hyman 1951); there is little to be said in a comparative sense other than they appear similar in location and composition.

Excretory system. This system is a protonephridium; namely, a tube opening exteriorly through an excretory pore and terminating internally in a blind ending. Such a system is absent in the acoels and shows various complications in the forms that do possess it. The excretory opening or nephridiopore can consist of one or more pores located usually on the ventral side but found also on the dorsal side. The blind ending is furnished with a ciliated flame bulb. One cell can apparently form more than one bulb (Reisinger 1923). The tubule connecting flame bulb and nephridiopore is an epithelium one cell thick. Cilia can be present in the lumen of the tubule (Pedersen 1961b). Whereas catenulid forms have a single tubule, there may be paired tubules in other forms. In others there can be two, three, or four tubules on each side of the body. In triclads these tubules may have multiple nephridiopores (up to several hundred). The tubules are located in the parenchyma close to the epidermis.

Digestive system. This aspect of the turbellarian worms is best discussed in two parts: the pharynx and the intestine. Actually, a third part, the mouth, can also be distinguished, but this is defined most simply as the external opening of the digestive system, and as such is a relatively uniform structure throughout the group. Where the mouth parts may show interesting complications, these are best discussed in conjunction with the pharynx. The acoels possess a mouth and simple ciliated pharynx but lack a gastrovascular cavity. The pharynx can be classified into three separate categories—simple, bulbous, and plicate—and there is a plethora of intermediate forms. Hence, extremes of location at anterior or posterior ends are joined by more centrally located ones, including those lying in the common midventral position. In composition, too, varying degrees of complexity in musculature, innervation, supply of gland cells and degree of distinctness from the parenchyma are all present (Fig. 13-7). However, one must be careful in identifying trends here. It is one thing to find a graded series of a given character and quite possibly

506 *Phylogeny of the Protozoa and Early Metazoa*

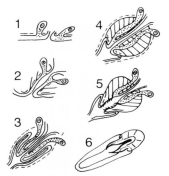

Fig. 13-7 Selected pharyngeal structures. 1. Found in Acoela. 2. Common to most archoophorans, excluding the polyclads. 3. Occurs in the polyclads, triclads, Seriata and some Prolecithophora. 4. Found in certain Prolecithophora and neorhabdocoels. 5 and 6. These are least common and occur in limited groups; 5 is found in certain mesostomes and 6 in the gnathostomulids. (Redrawn, by permission, from Karling (1963).)

another thing to try to argue that this is an actual evolutionary progression. It might be that the plicate pharynx and the bulbous pharynx are two separate developments of pharyngeal muscle around the simple pharynx, rather than that the bulbous is derived from the plicate type, as is suggested by Fig. 13-7 and as is proposed by several workers (Ax 1961, 1963; Karling 1963).

In a newly described species, *Hofstenia pardii*, Papi (1957) finds a condition intermediate to the acoels and other turbellarians with regard to the gastrovascular cavity or intestine, which is completely absent in the Acoela. In *H. pardii* the simple pharynx leads dorsally and then posteriorly to the esophagus and then terminates in the digestive parenchyma. An intestinal epithelium is completely lacking. The intestine of many of the forms with a plicate pharynx, e.g., the triclads and polyclads, is complex in its subdivision into anterior and posterior trunks, each with extensive diverticula. In *Bothrioplana*, a seriatid (Table 13-2), a less complicated digestive system is present, and in various neorhabdocoels more or less regular tubes are present. When the pharynx is anterior or posterior, the intestine is, respectively, a posterior or anterior extension of it. In all cases the intestine lies in the central part of the body surrounded by parenchyma and always connected to the pharynx. The gastrodermis is a one-celled epithelium. There is some variety of cells present, but they usually consist of larger cells capable of phagocytosis and smaller cells with granular contents. The latter presumably release their contents to initiate digestion in the intestinal cavity (Jennings 1968). There is a variable degree of musculature developed beneath the gastrodermis in certain forms, and certain forms show cilia lining not only the pharynx but also the intestine.

Reproductive system. The final structural system to be examined is the most complex of all, the hermaphroditic reproductive system. A minimally complex acoel, such as *Afronta* (Fig. 13-8A), shows gametes developing in the parenchyma, without any surrounding gonadal tissue, and with only male copulatory structures present. The latter exist as a relatively simple canal with a weakly differentiated seminal vesicle at its inner end and its gonopore opening at the posterior tip of the organism. In contrast we can look at the complex genitalia of the larger acoels (Fig. 13-8B-F). Here, even though the sex cells still differentiate in direct contact with the parenchyma, there is extensive development of the male and female copulatory apparatus. In other turbellarians one finds an almost endless set of variations on the pattern seen in the complex acoels. And within rather narrowly defined groups, such as the Monocelidae, there can be extensive differences in the arrangement of similar parts (Fig. 13-9). New features do arise in various turbellarian forms, especially notable being the appearance of yolk glands. Also, gonadal tissue becomes apparent.

There are two specific areas within reproductive structures that need further comment. These are the sperm and chromosomal studies. Major advances in our knowledge of turbellarian sperm are due largely to Hendelberg (1969,

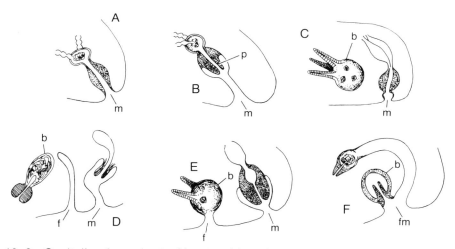

Fig. 13-8 Genitalia of acoels. A. *Afronta* with male canal opening externally through a male gonopore (m). B. *Proporus,* showing general features like *Afronta* but with a penis papilla (p) in the male canal. C. *Palmenia,* similar to the foregoing but the seminal bursa (b) unconnected to the male canal. D. *Convoluta,* with a vaginal canal leading to, but unconnected with, the seminal bursa. The female gonopore (f) is the external opening of the female canal or vagina. E. *Amphiscolops,* with the vagina connected to the seminal bursa. F. *Otocelis* has a common gonopore (fm) leading internal to both the bursa and male genitalia. (Redrawn, by permission, from Hyman (1951) *The Invertebrates Platyhelminthes and Rhynchocoela.* McGraw-Hill Book Co.)

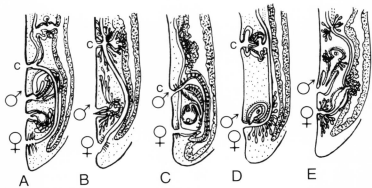

Fig. 13-9 Genitalia of selected Monocelidae. A. *Minona evelinae* B. *Monocelis fusca* C. *Archilopsis unipunctata* D. *Peraclistus oophagus* E. *Automalos hamatus*. (c) common gonopore, ♀ female gonopore, ♂ male gonopore. (Redrawn, by permission, from Beauchamp (1961b) in *Traite de Zoologie*. Vol. IV, Masson et Cie. Original compiled from various sources.)

1970). His survey of forms from most of the major turbellarian taxa justifies the generalization that all sperm develop from spherical spermatids provided with two flagella. These flagella become incorporated into the mature sperm during spermiogenesis or may hang free (the Nebengeisseln of the older German literature). In some groups, e.g., the rhabdocoels, prolecithophorans, and macrostomids, there are species with no flagella in the mature sperm and hence there is a loss or reduction of flagella during spermiogenesis in these cases. In cases where the flagella persist and are incorporated into the sperm, they often form the borders of the two undulating membranes characterizing the spermatozoan. This is seen in the acoels. These sperm are called thread shaped or filiform because of their long slender appearance. The nucleus is a slender rod located towards one end of the sperm cell. The ultrastructure of these sperm has shown some variability in the typical flagellar fine structure, but in the acoels the expected 9 + 2 structure is present.

Hendelberg concludes that the biflagellate condition is the original or primitive one in the Turbellaria, and it can be evolutionarily reduced to one or none.

Benazzi (1960, 1963) has pioneered on chromosomal studies on Turbellaria and his work on germ cells of certain triclads (planarians of the genera *Dugesia* and *Polycelis*) has laid bare some of the microevolutionary development of those species. This includes triploid and polyploid strains and unusual chromosomal behavior such as pseudogamy, asynapsis, and chromosomal elimination. Unfortunately for the purposes of this analysis, the studies do not yet encompass a wide array of turbellarians. More recent work is still analyzing *Dugesia* (Benazzi, Giannini, and Puccinelli 1970) or another triclad genus *Bdellocephala* (Umylina 1971).

As for the nature of mitosis itself, there is an extraordinary imbalance of information. The best work on adult forms still appears to be an early paper by Stern (1925), where careful drawings of epidermal mitotic figures in *Stenostomum* reveals a persistence of nuclear membranes, absence of aster fibers, and possibly an intranuclear centriole. These details are confirmed in a relatively short report on the division of parenchymal cells in the acoels *Amphiscolops langerhansi* and *Convoluta sutcliffei* by Hanson (1961) except for the intranuclear centriole, which was nowhere in evidence. There is considerable detail on mitosis during cleavage. In fact Costello (1960, 1961) has developed a rather detailed proposal for the role of centrioles in determining cleavage planes from his studies on early embryos of the acoel *Polychoerus carmelensis*. In this material the nuclear membrane does break down, and there are extranuclear centrioles and aster fibers. Most figures of cleavage mitoses in the Turbellaria show the same picture (Hyman 1951; Beauchamp 1961b).

FUNCTION

There now remain to be examined the functional traits of Turbellaria. Regarding the exploitive functions of feeding, locomotion, and general behavioral response to the environment, there are either only broad, general observations or else detailed work but limited to only a few forms. The same is generally true of the homeostatic functions of digestion and secretion. However, regarding developmental processes, we will be able to speak in terms explicitly more useful to our phylogenetic concerns.

Environmental exploitation. All of the free-living forms are ingestors, utilizing their mouths and pharynges to take food into their bodies. In forms with a simple or a bulbous pharynx, food is taken in whole. The size of the prey is roughly correlated with the size of the predator; hence smaller forms capture and ingest smaller food, such as unicellular algae, protozoa, and other small worms; larger forms ingest larger prey, including small crustaceans. The secretions that entangle the prey also may act as a paralyzing agent. As many workers have noted, it is astonishing to watch the almost incredible expansion of the usually insignificant mouth to engulf prey almost as big as the predator. This provides a spectacle of voraciousness not commonly appreciated as being present here. The plicate pharynx, though expansible, seems to serve largely as a tube for sucking up food, such as large chunks of nutritive material or the more fluid parts or juices of captured material. Such pharynges are extrusible through the mouth and become, in effect, the functional mouth.

Locomotion is basically dependent on ciliary mechanisms, but the undulatory swimming of polyclads and the variations in body form found in many Turbellaria speak to the importance of the musculature as an increasingly important adjunct to cilia. The turbellarians have apparently not yet reached that large size that cannot be moved by ciliary action, though the cilia do need the help of adhesive secretions to give them added traction on the substratum. But in the larger forms there are the beginnings of those muscular developments that presage the advent of locomotion dependent on contractile systems (Clark 1963, 1964).

The extensive behavioral repertoire of the flatworms is found in other Bilateria, but it is not found so extensively developed in the lower radiate animals nor in the protozoa. It can be argued that the extensive development of nervous tissue, which is not seen in unicellular animals, sponges, or cnidarians, is the basis for the flatworms' various and special sensitivities to touch, chemicals, water currents, and so forth.

As to the functioning of the nervous system itself, understanding is at a level characterized by Bullock and Horridge (1965) as "... still very primitive, and the detailed, basic studies of functional capacities of the peripheral plexuses, cords, and brain are yet to be done." (p. 356) Intriguing discoveries lie ahead as suggested by the work of Koopowitz (1970) on the brain. He reports decerebration as inhibiting the timing of the patterns used to capture prey but as having no effect on the ability to feed.

Homeostatic mechanisms. The beginning of extensive comparative work in digestion are now on hand (Willier, Hyman, and Rifenburgh 1925; Jennings 1957, 1968). In many turbellarians, much if not all the digestion occurs after the food is taken into vacuoles by the phagocytic cells. This is clearly the case in certain acoels, neorhabdocoels, and triclads. In the acotylean polyclads some digestion occurs in the lumen of the digestive cavity and the rest in phagocytic cells, and in cotylean forms it seems that digestion is almost wholly carried out in the digestive cavity. Food reserves are fat, glycogen, and protein. Thus the pattern of digestive activities is largely intracellular but shows a change to extracellular breakdown of the food in certain forms. Jennings (1968) reviews the present understanding of breakdown of food. His own earlier work (1957, 1963, and other papers) has shown that proteolytic enzymes (endopeptidases) are released by both the pharynx and the gastrodermis.

Studies on intermediary metabolism (Read 1968) are not extensive on the free-living flatworms. Respiratory studies suggest the not surprising conclusion that cytochrome oxidase is present. There are some interesting pigments present, such as hemoglobin, but their function is not known. Food reserves are only characterized in general terms and are reported to fluctuate seasonally and relative to population density (Baddington and Mettrick 1971). Little seems to be known regarding the actual mechanism of excretion via protonephridia in the Turbellaria. The presence of the protonephridia tells us that these organisms have reached an organizational level where specializations for ridding the body of wastes are important. Apparently larger forms and marine forms have the greatest need for such a system, judging by the complexity of their protonephridia. Paucity of significant data on the excretory process itself denies use of this trait in phylogenetic studies.

There are no respiratory or circulatory systems in turbellarians.

Reproduction. Turbellaria reproduce by sexual and asexual means, the former being of universal occurrence and the latter being found only in certain forms. In sexual reproduction the pattern of mating which leads to fertilization is similar in almost all forms, in that two worms typically place their ventral surfaces together, exchange sperm, and fertilization occurs internally. The chief variation on this theme is the relatively rare case of sperm injection through the epidermis of the partner, mostly by means of special penis structures, after which the sperm wander through the parenchyma of the impregnated worm until they fertilize an egg (this occurs in a few acoels and a few polyclads (Hyman 1951)).

Cleavage is of the spiral sort where entolecithal eggs are present, notably in the Acoela and Polycladida. Where ectolecithal eggs are present, as in the Tricladida and neorhabdocoels or eulecithophores, the cleavage pattern is highly indeterminate. Ax (1961) cites certain authorities who claim to see evidence of spiral cleavage here, but Beauchamp (1961b) states categorically that all evidence of spiral patterns is completely lost. Where spiral patterns do occur, they are of two types. In the polyclads, cleavage is highly reminiscent of what is seen in the other spiralian phyla, especially the annelids and molluscs (Fig. 13–10A). It also appears to be determinate. By contrast, the Acoela show spiral cleavage of a unique sort. It is into duets rather than quartets (Fig. 13–10B), and the cleavage products, at least in *Childia*, are able to regulate through the third cleavage (Boyer 1971). Seilern-Aspang (1957) and Beauchamp (1961b) report that in *Macrostomum* the cleavage pattern lies between the acoel and the polyclad situation. The early stages seem to show a duet pattern, but as cleavage proceeds the quartet distribution of blastomeres be-

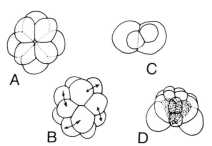

Fig. 13–10 Spiral cleavage in polyclads and acoels. A and B. Quartet formation in *Planocera inquiena*. C and D. Duet formation in *Convoluta roscoffensis*. (Redrawn, by permission, from Beauchamp (1961b) in *Traite de Zoologie*. Vol. IV, Masson et Cie.)

comes apparent. Development beyond early cleavage in the acoels leads first to a solid gastrula (stereogastrula), with subsequent hatching of an immature worm. Feeding and subsequent growth leads also to differentiation of the internal genitalia and the adult worm capable of sexual reproduction. Apelt (1969) has studied three species of *Convoluta* and the ecology of their reproduction. Similar studies are lacking on other forms. In the polyclads also, further cleavage leads to a stereogastrula, and before hatching the pharynx and gastrovascular cavity are differentiated. The pharynx first appears as a simple ciliated tube running internally, and subsequently it develops the special features distinctive of a given species. Larval forms (Götte's and Müller's larvae) can appear at this stage, followed by further development into the adult. Kato (1940) has reported that Müller's larvae may develop within the egg capsule before being liberated.

Ectolecithal development involves a membrane reputedly formed of a single layer of blastomeres within which are yolk cells and other blastomeres. The latter cleave further and come to lie along the membrane at one point. They then differentiate a simple pharynx, which engulfs all the yolk cells, thus placing them within the developing intestine. This first pharynx is apparently provisional, for it dedifferentiates. Another, permanent pharynx appears later when the mesenchymal cells develop further. They come to lie as eccentric masses in the embryo, and from them other organ systems, except the genitalia, appear in due course. The ectoderm, in many cases, appears as a layer of cells lining the inner surface of the original outer membrane. A different mode of ectoderm formation is seen in *Hydrolimax* (Newton 1970). Here the original epidermal material forms within the center of embryo and then moves to the surface and spreads over the yolk and mesenchymal cells. After hatching, the genitalia differentiate and the worms mature into adults. In *Stenostomum sthenum* it is reported that under certain conditions the testes can also produce eggs, i.e., the male gonads are potentially bisexual (Borkott 1970).

The chemistry of reproduction is just beginning to be studied in turbellarians. For example, eggs of *Dugesia* contain high level of RNA (Gremigni and Benazzi 1969) but its functional significance is not elucidated.

Overall, then, we find significant diversity in turbellarian development. In polyclads and acoels there is spiral cleavage with certain real differences between them. The former also show larval forms not seen in other groups. In the triclads, there is formed a solid stereogastrula within which a pharynx and intestine appear, but which later degenerate and are lost. This is not known in other forms. On the other hand, certain regularities are present: the invagination for the mouth is the first part of the digestive system to appear in all forms; the pharynx appears to develop from a simple pharynx to a plicate one or to a bulbous one; the intestine is a solid mass of gastrodermal cells before it hollows out the intestinal cavity; young worms, except for a few cases of neoteny (Heath (1928) reports fully developed gonads in the Müller's larva of *Graffizoon lobata*) are immature, and genitalia appear to be the last organs to develop.

Asexual reproduction is well known in the planarians and in the catenulid and stenostomid forms, where a pattern of transverse fission is to be found. It seems some races of *Dugesia dorotocephala* are almost permanently committed to asexual reproduction, whereas in others sexual reproduction predominates (Jenkins 1970). Whether or not there is a genetic basis for this or whether it has an adaptive significance is not known. Architomy (Marcus and MacNae 1954) occurs in a South African acoel. In this process there is no preformed fission plane, and the organism breaks up readily from external forces—in this case probably wave action. This seems to be a very special situation. In another acoel (Ax and Schulz 1959) asexual reproduction by paratomy has been reported; that is, division anticipated by development of new structures. A more typical fission-like process has been reported for *Amphiscolops langerhansi* (Hanson 1961). The foregoing cases involving the Acoela are all found in the larger forms. It seems that they are separately originated, as is also the situation in the planarians. There are no useful phylogenetic conclusions to be drawn from them.

Regeneration has been extensively studied in planarians (Wolff 1962; and other reviews, e.g., Goss 1969; Rose 1970). They show quite extraordinary abilities to replace lost parts. This is also true for at least certain (Steinböck 1954, 1963b) and perhaps all acoels, with the exception of the statocyst, which seems never to be regenerated (Hanson 1967b). An extremely interesting report on regeneration of the nervous system is that of Koidl (1970). In *Bothrioplana* the normal orthogonal nervous system (two dorsal and two ventral cords with regularly spaced rings of commissures) is not replaced directly during the regenerative process. Rather, an irregular nerve net appears, which only later becomes orthogonal. Koidl rightly wonders if this is a recapitulative phenomenon. Detailed studies regarding the nature of the regeneration blastema and the source of its component parts has been limited to the planarians, and, hence, there is little comparative information within the Turbellaria. Gabriel (1971 and earlier papers) is using various chemical inhibitors and radioactively labelled compounds to elucidate the molecular patterns of regeneration. This process of regeneration may turn out to be of phylogenetic importance, but at present this is not the case.

Phylogeny

HOMOLOGIES

We need to review briefly the traits sufficiently complex to be treated as semes, and then homologies must be established among them, including some consideration of the plesio- or aposemic nature of the homologies. This is prerequisite to defining the plesiomorph and to moving to a more detailed phylogenetic analysis.

Semes. From the foregoing discussion of general characteristics there emerge eight structural and three, possibly five, functional traits that are semic. The structural semes are, in the order discussed above, external morphology, body layers, secretory elements, musculature, and nervous, excretory, digestive, and reproductive systems. Functionally, patterns of ingestion and digestion and of sexual reproduction are semic, with the latter being separable into two semes, one being gamete formation and mating and the other embryonic development. Digestion may eventually be treated as a separate seme, and regeneration might be considered by some also to be semic.

The points needed for comparison in external morphology refer to the bilateral symmetry of the worms—their right and left, anterior and posterior, and dorsal and ventral sides. Then added to this are the placement of eyespots, statocysts, and other sensory devices, such as bristles, and the location of openings, such as the oral and genital ones. All of these permit specific spatial relationships to be compared among various species.

The body layers are first to be compared in terms of their position in the whole worm and then in terms of parts of the worm body that allow various comparisons to be made between the body layers. In this way positional and compositional comparisons can be established. For example, the placement of what is thought to be mesenchymal material or parenchyma must lie between epi- and gastrodermal tissues to be homologous. And within this parenchyma comparisons can be made for the presence of muscle tissue, reproductive and excretory organs, etc., so as to define the compositional details of the mesenchymal layer. The epidermis and gastrodermis can be treated in a similar fashion.

Secretory and extrusible elements must be compared on a cellular basis, for many of these structures are represented by single cells, such as the rhabdites and saggitocysts. Frontal glands are compounded of many cells and can be studied in that perspective. In all of these, location in the body is important as is actual compostion of secretory cells and their products. There is limited electron microscopy on these elements, but more is sure to be reported. In its absence, staining and light microscopy offer us the best information on the composition of these materials.

The musculature and the other systems of the turbellarians, i.e., nervous, excretory, digestive, and reproductive systems, all present the same kind of problem as regards their classification as semes appropriate to phylogenetic studies. They are all located in specifiable parts of the turbellarian organism, especially in reference to body layers, and they all have unique details of composition that allow for point-by-point comparison as one proceeds from one species to another.

Feeding and digestion must be examined as a process separable into various component parts. These would include the location and capture of prey and the nature of the prey and means of immobilizing it. Also, modes of ingestion and actions of the mouth and pharynx are significant. Finally, details of digestion can be considered, such as vacuole formation, release of digestive enzymes, and uptake of materials into the gastrodermis. In time, it may well be possible to recognize two semes here, by separating feeding and digestion. At present, due to somewhat limited information, the two are joined together.

Gamete formation and fertilization provide a rich seme in which location of differentiating oocytes and spermatocytes, their relation to other parts of the genitalia, the mating process with exchange of sperm and subsequent fertilization, all provide many points for detailed comparisons of these processes.

Similarly, embryogenesis, with its familiar sequence of cleavage, gastrulation, and differentiation of the many details characterizing adult worms, is another important source of semic information.

Plesiosemic and aposemic homologies. When one passes rapidly over the foregoing list of semes, it is relatively straightforward to make a case for homologies. The variations in body form and other features of external morphology are not so far-ranging nor discontinuous as to exclude positional similarities in the location of sensory structures and body openings. The compositional similarity of these elements is clear, either from their inherent structure, e.g., statocysts, or from their relations to other distinctive structures, e.g., the small gonopore connecting to the male genitalia.

The body layers are especially straightforward in defining their positional and compositional similarities. They show relatively little variation except as one studies differences in their constituent organ systems. The gastrodermis perhaps shows the greatest variability, in that it is absent as a tissue in the acoels, for they lack an intestine. The central parenchyma of the acoels, in that it can contain food vacuoles, is homologizable with the gastrodermis of those forms with a gastrodermal or gastrovascular cavity or intestine (Pedersen 1961a, 1961b). Gastrodermal tissue is capable of both pinocytosis and phagocytosis (Jennings 1957, 1968) and at times of heavy feeding seems to break down into a vacuolated mass, apparently to facilitate digestion, thus creating a situation intermediate to that seen in acoels and that exemplifying a more common gastrodermal organization, i.e., a continuous epithelium of the intestine. As cited earlier, Papi's (1957) new species, *Hofstenia pardii*, is a significant intermediate between acoels and other Turbellaria.

In terms of homologies, the glandular materials of flatworms pose special problems. As we have seen, they are located in essentially equivalent positions wherever they are found. Also their composition is similar in the sense that they are usually unicellular, but the uncertainties regarding the chemical nature of the secretion means that we cannot entirely satisfy outselves regarding the compositional relationship. Furthermore, what similarities in position and composition we do see are the result of a broad overview of these structures. Many variations are contained within the two broad categories of classification that are popularly used, i.e., eosinophilous and cyanophilous. In terms of the proposed functions of these glands, which include such diverse things as food capture, locomotion, adhesion, protection, attachment of eggs, and also cyst formation, we could well expect a diversity of glandular form and function. It appears wiser for the present to conclude simply that homologies are likely in the secretory cells and their allied structures (rhabdoids), and only when more information on the nature of the secretions and on the variety of the glands is on hand will we be able to say anything meaningful about trends that may be present in the variety of secretory elements present.

The general body musculature of the Turbellaria is clearly homologous in all forms and shows definite trends towards increased complexity. This is expressed by increased numbers of subepidermal muscle layers and of parenchymal muscle. There may have also been an initial change in position from epidermal to subepidermal. The direction of this change depends on whether the simpler acoels can be thought to resemble the turbellarian plesiomorph or to be a case of regressive evolution.

The nervous tissue is so similar in position and composition as to justify concluding that it represents development from some common ancestral source. As with the body musculature, there is a question as to whether the original location was epidermal and then became subepidermal, or whether the former position is some kind of unique specialization. In terms of compositional variation several trends seem apparent. The extensive development of the nerve plexus in polyclads is one. Retention of a few cords is another, and that retention, especially of the ventral cords, with development of many lateral commissures as in the planarians, is yet another trend. In all lines there has been a development of special tissue which can be called a brain.

Homologies in the sensory equipment of the Turbellaria must distinguish between categories of structures. Where they occur, statocysts seem to be clearly homologous and show little variation. The same is generally true for eyespots, but they show variation in number, in position in the anterior end of the organism, and in innervation of the pigment cup. These variations do not show sufficient discontinuities so as to argue against homologous relationships among them. Trends may be identified in change of number of eyespots and in type of pigment cup (inverse or converse). Regarding rheo-, chemo-, and tangoreceptors, there seems to be insufficient data to find trends, but, generally speaking, there are probably homologous relationships between these types of receptors in the various Turbellaria.

The protonephridial system of the Turbellaria is probably homologous through the group. (Such a system is also found in many of the acoelomate and pseudocoelomate animals. Hence, there is the possibility that it arose independently several times. Its absence in certain Turbellaria suggest that it arose within this group, or was lost from certain of its members; again the vexing question of how one interprets the condition in the acoels.) There has been some change in protonephridial structure within the Turbellaria as seen in the varying complexity of the tubule and its nephridiopores.

Overall, the digestive system seems homologous in all forms. The transition between the absence of an intestine in the acoels and the gastrodermis in other forms is seen in stages where the gastrodermal cavity disappears and cells replete with food vacuoles, much like the phagocytic parenchymal cells of acoels, are seen to fill the center of the body as in the case of Papi's (1957) *Hofstenia pardii*. From these forms, the gastrodermal structures ramify into the extensive systems of the triclads and polyclads. All lie within the parenchyma, have a gastrodermal epithelium, and are attached to the pharynx. The pharynx, too, shows all stages of gradation between simpler and more complex forms. Hence, pharyngeal trends are also present and represent material to be used for analyzing evolutionary developments.

That the genitalia are all part of homologous structures seems quite plausible from the general point of view that they are all located in roughly the same part of the body—in the parenchyma with gonopores commonly ventral—and they are composed of the same parts—ova, ovaries, yolk glands, ducts, seminal receptacles and bursa, sperm, testes, etc. Indeed the occurrence of a similar complexity of parts is really the best argument here for homology. However, it is a fact that critical search for detailed fulfillment of the positional and compositional relationships becomes a nightmare. Recalling Remane's insistence on similarity of contextual relationships, we find in one form that the opening of the male ducts is through the pharyngeal cavity (*Prorhynchus*), in another it combines with the oviduct (Fig. 13–9), and in others it is a separate gonopore. And the seminal bursa may be connected to the intestine (Fig. 13–9). How can this diversity be understood and still honor our initial intuition that the turbellarian genitalia are indeed homologous entities?

Karling (1963 and see other papers) has proposed the most useful point of view. He stresses that in pharyngeal and copulatory structures (Fig. 13–7, 13–11) similar organizational patterns occur. These can also be found in adhe-

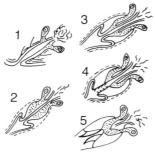

Fig. 13–11 Selected copulatory structures. 1. Found in the acoels. 2. Common to various forms. 3. Also found in various forms. 4. Occurs in *Paracicerina maristoi*. 5. Found in *Carcharodopharynx arcanus*. (Redrawn, by permission, from Karling (1963).)

sive organs. Such parallelisms suggest that both similar functional needs and similar developmental potentials have been operative. The former includes use of adhesive secretions and muscular contractions to hold and ingest food (pharynx), to adhere to a partner and insert a penis for impregnation (male copulatory structures), and to attach to and hold onto a substratum or prey (adhesive organ). The developmental potentials include the property of epithelial cells to form "... depressed glands and to produce cuticular muscle fibers, giving rise to pits and annular folds such as the pharynx plicatus, etc." (Karling 1963, p. 230). Also, the underlying parenchyma can contribute to muscular developments that include protractors and retractors and also develop the septa separating muscles from parenchyma. Hence, not only are there homologies evident in the male copulatory apparatus but also in the parallel development of the pharyngeal structures, as we saw earlier. And since it is found also in certain adhesive structures, we have here an example of serial similarities on one organism, or homonomy (Remane 1956).

Regarding the female parts, there is no general pattern that emerges to account for its development. Karling readily recognizes a diversity of opinion. However, he continues to use his twin guidelines of evolutionarily determined function—selection pressure to successfully inseminate and then store gametes—and developmental potential. He points out that prior presence of oviducts provides a structure that, upon expansion, will serve as a seminal receptacle or even seminal bursa. He implies that the invaginative properties of the epidermis can also be exploited to build new copulatory ducts and vesicles. Also the presence of the digestive tube allows another preformed exit for germinal products. Hence, the peculiar needs of various species are met in various ways depending on copulatory functions, preexisting internal structures, and the general developmental resources of turbellarian tissues. Thus, the diversity of the genitalia.

If the question of trends is raised regarding this diversity, it can be seen that it poses problems that are probably answerable only in the general terms presented by Karling. We can see overall differences in degree of complexity; we have at present no detailed account describing the species-by-species progression of events in any given group, e.g., the Monocelida (Fig. 13–9). Perhaps, we never will; our best hope may be intensive study of the ontogeny of these worms. We must, then, limit ourselves to the general trends. In such terms we have already commented on the extensive changes apparent between the simpler acoels and other forms and how, within the acoels themselves, there has appeared a high level of complexity.

To continue these speculations somewhat further, it can be suggested that germ cells were the first reproductive materials; then there appeared tissues for conducting them externally. These first appeared in conjunction with male parts because of the necessity of safeguarding sperm exchange during copulation. Egg laying led to formation of female ducts. These were utilized to further the efficiency of copulation through formation of vaginal structures and sperm storage vesicles. There appears then a picture of the progressive perfecting of the organs associated with sexual reproduction. It remains to be seen what evidence we will be able to adduce for or against this scheme.

Next, we turn to the category of functional semes and examine them for homologous relationships. Feeding and digestion appear homologous throughout the Turbellaria. Though there are variations in the food organisms preyed upon by different species of worms, the mode of capture and migration through the mouth and pharynx consists of comparably similar steps, and differences that are present appear bridgeable by intermediate steps. For example, the active predation of large acoels on microcrustaceans and the slow-moving scavenging of planarians is bridgeable by intermediate modes of hunting and ingestion. Also, digestion is comparably similar. There is some extracellular digestion where an intestine is present and then there is much evidence of pino- and phagocytosis of food materials which leads to intracellular digestion.

Patterns of sexual reproduction are somewhat unusual. The Turbellaria seem to be very conservative regarding their method of mating, but in an exploratory mood regarding blastomere formation, and conservative again in the later stages of ontogeny. The comparisons possible in gamete formation, the various details of copulation, and exchange of gametes with internal fertilization show a similarity that argues for homology. There is an unresolved difference of opinion regarding whether or not ectolecithal development is related to entolecithal development. Ax (1961) says it is, but Beauchamp (1961b) is apparently unconvinced. However, Beauchamp finds evidence in *Macrostomum* of late cleavage and early organogenesis patterns that are similar to those in ectolecithal eggs, namely, a provisional outer membrane and the presence of cells like the yolk or vitelline cells of the triclads. Further work, inevitably, is needed to clear up this point. The sequence of events in the formation of the digestive system and of the genitalia, especially, argue for homologies in the organogenesis of the Turbellaria, though here again much more information can be gathered and would be most valuable.

One might remark parenthetically at this point that systematic work on the Turbellaria is far from being thoroughly developed. The taxonomy is still unsettled (e.g., Reynoldson and Bellamy 1970; Schilke 1970), and there are continual reports of many new species (Marcus 1954 and earlier papers; Kozloff 1965; Lanfranchi 1969; Kenk 1970; Ax 1971). Hopefully among these new forms will be ones supplying intermediate stages to bridge better such variational extremes as those found in the cleavage of ento- and ectolecithal eggs.

In summary, we can draw up a list of homologous traits which tend to be more conservative (plesiosemic) and another list of those which show significant innovation (aposemic). The conservative traits are general organization and perhaps also the secretory and sensory structures. The latter two are doubtfully treated as plesiosemic because the comparative data are very hard to evaluate. They are treated here provisionally as plesiosemic because secretory and sensory organelles appear to be of the same general sort throughout the Turbellaria, even though it is recognized that some forms are more richly provided with such elements than are other ones. Generally speaking, too, the feeding and ingestatory habits are quite similar throughout these flatworms and, therefore, can be termed plesiosemic. Also the copulatory process is similar throughout these worms, and the sequence of events in organogenesis also appears conservative. The more aposemic traits are those concerning the general body form with its variously placed apertures, the musculature, the nervous, excretory, and digestive systems, to mention first the structural semes. In the second and third there are changes in position from epidermal to subepidermal as well as significant further development of structures, development of new structures (new muscle layers, new nerve plexuses and layers), and reduction in some cases (loss of dorsal and lateral nerve cords). The excretory and digestive system are absent in certain forms, and in the others they show extensive degrees of elaboration, as in the digestive system especially (pharyngeal variations and intestinal diverticula). The reproductive system is highly aposemic, showing a relatively few examples of simpler structures and many complex ones and, among the latter, exhibiting many variations. Finally, cleavage patterns either show highly aposemic development, or are not homologous at all with the ectolecithal pattern being possibly neosemic. However, with homology present in mating patterns and organogenesis, namely at either end of the cleavage processes, it appears that cleavage too will some day be homologized, with forms like *Macrostomum* serving as one of the needed intermediates.

Variable characters showing trends and sexual reproduction with its ontogenetic development offer significant opportunities for identifying a turbellarian plesiomorph.

Plesiomorph. If we assume that the embryology of the flatworms is recapitulative, it then suggests that the acoel type of organization is primitive. In both types of development, from ento- and from ectolecithal eggs, the endoderm (presumptive gastrodermis) is solid before it hollows out into an intestine. Also, the inpocketing of the mouth is apparent at the time the endoderm is solid. The later development of tissues in the parenchyma allow one to suppose that the earlier stages of development, which are more like acoeloid forms than that of any other turbellarian, are, in fact, reminiscences of an ancestral form resembling the present Acoela.

An examination of morphoclines also points to the Acoela as being plesiomorphic. In our earlier discussion of the various traits of the Turbellaria, we saw time and again, even in predominantly conservative characters where variation is minimal, that the Acoela stand at one extreme of the changes that were apparent in a given structural or functional part of these organisms. This is seen in terms of form, where the Acoela contain the smallest Turbellaria; it is seen in terms of their parenchymal organization; in terms of the presence of statocysts and rare presence of eyespots, which characterize most other Turbellarians; in terms of their relatively simple muscular and nervous systems; in terms of the absence of intestine and protonephridia, which occur elsewhere in these forms; in terms of the extreme simplicity of the genitalia in some forms; in the reduction of flagella in sperm; and in terms of the duet pattern of spiral cleavage which sets the Acoela apart from the other Turbellaria. The morphocline evidence overwhelmingly nominates a member of the acoels as a plesiomorph. The actual species that can be used here will be that one showing the least development in the various traits we have listed. Of the forms known today, a species of *Haploposthia* seems to be a useful form. We will here use *H. rubra* (Westblad 1945, 1948) (Fig. 13-12).

Now is the time to discuss fully the alternate view of the Acoela, that they are in reality not primitively simple but are the products of regressive evolution (Remane 1963a; Ax 1961, 1963). Ax has presented the most detailed discussion of this point and in speaking to his arguments we speak to others who share his point of view.

Ax's approach is through the use of homologies. "Methodologically, the only approach is through homology analysis. We must first of all carefully analyze the morphological developmental steps of single organs and apparatuses within the different systematic groups, and then we can determine from the various combinations of primitive and derived characters the phylogenetic position of orders and suborders in a natural system." (Ax 1961, p. 3. Original in German.) In criticism of this approach, we can only agree to the importance of homology—and Ax as a student of Remane is completely familiar with Remanian concepts. But to understand how Ax arrives at his interpretations of what is primitive and what derived, we must try to follow his thinking in specific instances. (a) The primary structure (primäre Struktur) of the female gonads is

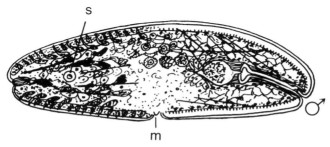

Fig. 13-12 Frontal section of *Haploposthia rubra*. The mouth (m) opens into the inner parenchyma, surrounding which is the peripheral parenchyma with variously differentiated parts, including statocyst (s). There is only a male gonopore (♂). (Redrawn, by permission, from Westblad (1945).)

a uniform ovary with entolecithal eggs (1961, p. 4). This is given as a simple declaration of fact, no ontogenetic or morphoclinal facts are adduced to support it. (b) Spiral cleavage is the primitive cleavage pattern because (in a footnote on p. 4–5) it is improbable to derive it from the modified cleavage patterns seen in the triclads and also argue for another separate origin of spiral cleavage in annelids and molluscs—he is implying all spiral cleavage is homologous and arose once. This type of thinking utilizes the major criterion of compositional similarity to establish homology and also uses the second accessory criterion of several different forms showing a similar trait. Though it does not use ontogenetic or morphocline data with the Turbellaria to argue for the plesiomorphic nature of spiral cleavage, it is a valid general argument. But being plesiomorphic in several phyla does not say where it first arose. However, it commits Ax to the position that acoel duet cleavage is derived from quartet cleavage ("the common ancestral type" Ax 1963, p. 206). This last conclusion is arrived at not from ontogenetic or morphocline data in the Turbellaria but from broader evolutionary suppositions. This is an unacceptable procedure: phylogenetic conclusions proceed from smaller groups to larger ones, and not the reverse (see earlier, Chapter 2). (c) Ax uses the same approach on the protonephridia and concludes from homologies with protonephridia in nemertines and aschelminths, that the acoel condition is derived by reductive loss from an ancestral one with protonephridia. (d) Arguing against Bresslau's (1933) derivation of a ciliated digestive lumen in the macrostomatids, catenulids, and polyclads as an improbable case of triple, parallel evolution, Ax chooses to propose a ciliated intestine as primitive, and the acoel solid, inner parenchyma as a regressed situation. Why Bresslau's situation has to be seen as three different lines of evolution derived from the acoeloid condition is not clear. In any case, Ax's method is to argue in probabilistic terms, i.e., what seems to him a simpler course of events, and not from any criteria of homology. He gets around his own argument for improbability by deciding there was a hypothetical ancestor with a ciliated digestive cavity. But how he decides this is ancestral to the acoels and not the other way around is simply not clear.

These examples show three different modes of defining primitive states: (a) simple declaration of so-called fact; (b) reference to phylogenetic presuppositions from other groups outside the one under consideration; and (c) somewhat subjective decisions as to the probable course of evolution with reference to a hypothetical form. None of these is justifiable phylogenetic procedure.

If we find Ax's procedure unacceptable in establishing the postulated specialized nature—largely through reductions—of the acoels, are there procedures that are acceptable that can be used? Earlier in this volume (Chapter 3) and elsewhere (Hanson 1963a) the problem of reduction has been discussed. In widely accepted examples of reduction, as in parasites (e.g., cestodes) and sessile forms (e.g., tunicates), it is seen that reduction of characters follows on a change of niche with three specific consequences. (1) The resulting change in selection pressures means certain traits are no longer selected for and mutation pressure breaks them down, hence, there is regression or reduction of them. (2) But not all traits are reduced—an unspecialized organism cannot survive. In fact, certain preexisting traits become even more specialized (the genitalia in cestodes, the pharyngeal basket in tunicates). (3) And lastly, the ontogeny of forms showing reduction often reveals evidence of the more complicated ancestral form.

The questions now are: Do the acoels show reduction in a few characters that are demonstrably not of selective value in the environment they are living in? Do they show special development of other characters? Are there recapitulative evidences of more complex ancestral traits? If the acoels are indeed examples of regressed evolution, we would predict affirmative answers to all of the foregoing questions. If, on the other hand, their organization represents primitive simplicity, the answers will be in the negative. As has been published earlier, though in somewhat different format, the answers are all negative (Hanson 1963a). The so-called reduction of the acoels—no intestine; no protonephridia; relatively simple genitalia, especially in the smallest forms with only one gonopore; no basement membrane; no gonadal tissue—eggs and sperm develop directly in the parenchyma; no oviducts or sperm ducts; duet spiral cleavage (these come from comparing *Haploposthia* to Ax's (1961, p. 48–51) archetype)—are not traits correlated with a given mode of life. Except for the genitalia, which do show varying degrees of complexity, the other traits are common to all acoels. They are common, therefore, to herbivores, carnivores, ectocommensals; to benthic and pelagic forms; to forms that are sand dwelling and those that live among algae. The acoeloid type of organization cannot be accounted for as a common adaptation to these varied modes of life. It is rather a primitive mode of organization that these worms have subsequently exploited in different directions.

Further, there are no specializations in the acoels, with the possible exception of duet-cleavage, that are not found in other turbellarian flatworms. In fact the specializations in the acoels are often more highly developed in the other worms. There is no evidence that the adaptive zone of the acoels is distinct from that of the other turbellarian worms nor that their mode of life would suggest the loss of traits found in the other Turbellaria.

Lastly, acoel embryogenesis does not show any signs of more complex characters being present in ancestral forms. Formerly, the belief that adult acoels were syncytial led to the view that cellularized blastomeres recapitulated an ancestral state that was largely lost in the adults (Ax 1963). The awareness that all Acoela thus far studied by electron microscopy are cellular as adults essentially demolishes the problems posed by that point of view.

We can only conclude, along with others who have long maintained such a point of view (Graff 1882; Bresslau 1933; Hyman 1951; Ivanov 1971; to name only the major workers), that the relative simplicity of the Acoela is not a regressed but a primitive condition. A form such as *Haploposthia rubra* (Fig. 13–12) is plesiomorphic to the Turbellaria as well as to the whole flatworm phylum.

PHYLOGENETIC ANALYSIS

A preliminary study of evolutionary relationships will be attempted here. It is preliminary in the sense that it does not employ the most desirable approach, which is the analysis of the homologies in small homogeneous groups and identification of their respective plesiomorphs and phylogenies, with subsequent attention to larger and larger sets of plesiomorphs. Rather, for practical reasons that guided our earlier formulation of dendrograms, we will start with what seem to be rather stable taxonomic groups and take selected species from there for critical comparison. This approach, it must be reemphasized is not ideal. There is no good assurance that the best plesiomorphs have been selected to represent each relatively homogeneous group, and in some cases comparisons are better made between the plesiomorph of one group and a derived form of a second group, rather than with the plesiomorph of that group. This we saw to be clearly the case with the ciliates. However, there are practical reasons for attempting this less than ideal excursion into turbellarian phylogeny. First, it illustrates in a concrete way what semes can be compared and how the trends in the group may be interpreted. It provides, then, a beginning for later, more exact work. Second, it is the most explicit way to formulate already existing phylogenetic speculations on the Turbellaria, speculations which are often based on even less trustworthy methodological foundations than the present attempt. Third, this attempt at phylogenetic analysis offers itself as a specific target for further criticism and improvement; it can be of heuristic value.

Phyletic relationships. Table 13–1 summarizes the data describing the available pertinent facts about the species being used for comparison. In that turbellarian systematics has been in a fluid state in recent years, it is also necessary to comment briefly on the basis for identifying so-called homogenous groups. This was done by comparing three different systematic schemes and by selecting a well-known (or in some cases the only known) form to represent a given group (Table 13–2). Clearly the decision to use well-known forms is no guarantee of finding sound plesiomorphs. However, well-known forms have the important advantage of providing sufficient data to allow numerous, effective comparisons.

From the data in Table 13–1 there can be brought together (Table 13–3) a

Table 13-1 Semic traits of ten species of flatworms selected to represent the various orders of the Turbellaria, according to Hyman's (1959) classification.

A. *Haploposthia rubra* (Order Acoela) (Fig. 13-1A)
   1. Body about 1000μm long and 200-300μm wide; mid-ventral mouth; single, posterior male gonopore
   2. Outer ciliated epidermis covering the parenchyma which is separable into outer parenchyma containing statocyst, brain, nerve cords, frontal gland, muscle fibers, gametocytes, seminal vesicle, penis, and male genital tract
   3. Secretory elements consisting of well-developed frontal gland and epidermal gland cells
   4. Musculature around the opening of the frontal gland with fibers lying in the parenchyma; especially well developed are the anterior retractor fibers
   5. Nervous fibers present as a subepidermal plexus with six cords—three pairs: dorsal, lateral and ventral; anterior brain lying just ahead of the statocyst
   6. Statocyst is only identified sensory structure
   7. No excretory organelles
   8. Inner parenchyma carries out digestion on food enclosed in vacuoles
   9. Both male and female gametes are formed. Only male organelles present as seminal vesicle, penis, and genital tract; latter opens at posterior end of worm
   10. Mating not observed; probably by copulation and exchange of sperm
   11. Cleavage not observed; probably spiral cleavage into duets
   12. Post-cleavage development not observed
   13. Feeding not observed (marine)

B. *Macrostomum appendiculatum* (Order Macrostomatida) (Fig. 13-1B)
   1. Flattened body 1-3 mm in length; ventral mouth anteriorly placed; separate male and female gonopores opening on posterior ventral surface; a spatulate, adhesive, tail with papillae
   2. Epidermis with cilia and (sensory?) bristles; parenchyma with well-developed organ systems; intestine with gastrodermis attached to simple pharynx
   3. Eosino- and cyanophilous epidermal glands, including rhabdites
   4. Subepidermal and parenchymal muscles
   5. Brain just anterior to pharynx; two lateral nerve cords and a few anterior commissures
   6. Paired eyespots dorsal to brain
   7. Two protonephridia
   8. Small, simple pharynx and ciliated gastrodermal lining of intestine
   9. Oocytes develop in two lateral strands, dorsal to spermatocytes. Mature eggs lie anterior to female gonopore which lies ¾ of the way posterior on the ventral surface. Spermatocytes start developing as two lateral strands more anteriorly than oocytes and pass below oocytes and ova to area of male gonopore. Here there is a seminal vesicle and bursa as well as a cuticularized penis. Male gonopore opens on ventral surface about in middle of spatulate tail.
   10. Mating not reported; probably by copulation and exchange of sperm
   11. Entolecithal egg and spiral cleavage closer to the acoel duet formation than to the polyclad quartet pattern
   12. Postcleavage organogenesis shows a provisional outer membrane of four cells and then its replacement by a true ectoderm which becomes epidermal. Inner cells differentiate into endoderm and then gastrodermis. The genitalia are the last to develop
   13. Feeding not reported (fresh and possibly brackish water)

C. *Stenostomum ventronephridium* (Order Catenulida) (Fig. 13-1C)
   1. Cylindrical body 0.5-1.0 mm in length with ventral mouth close to the ante-

Table 13-1 *(Cont.)*
C. *Stenostomum ventronephridium (Cont.)*

    rior end which also carries two lateral ciliated (sensory) pits; constriction of body at level of mouth separates anterior prostomium from rest of body. Nephridiopore terminates caudally

2. Ciliated epidermis with (sensory?) bristles; gastrodermal intestine connected to mouth by well-developed pharynx; parenchyma with various organ systems
3. Both types of epidermal glands present as well as glands around the pharynx; rhabdites in conspicuous clusters at anterior end
4. Subepidermal and parenchymal musculature
5. Brain just anterior to pharynx; eight pairs of cords extend from the brain
6. Sensory pits with associated cells are paired lateral depressions on prostomium
7. Protonephridium with two longitudinal canals extending the length of the body
8. Mouth leads to well-developed pharynx, which connects to ciliated intestine
9. One dorsal testis close to mouth, an efferent duct, cuticular penis, and genital antrum (channel leading to the exterior); the ovary is a single, ventral organ lying posterior to the mouth and lacks ducts or pore
10. Mating apparently by hypodermic impregnation
11. Entolecithal egg
12. Postcleavage development not reported; asexual reproduction is common
13. Feeds on ciliates, which are broken up by muscular action of intestine and then digested further and phagocytized (freshwater)

D. *Proporoplana jenseni* (Order Archoophora) (Fig. 13-1D)
1. Slender, cylindrical body tapering at each end, especially posteriorly, about 1.0 mm. in length; mouth at anterior tip and male gonopore at posterior tip
2. Ciliated epidermis; ciliated pharyngeal tube with muscular, plicate pharynx leading to gastrodermal intestine; parenchyma with various organ systems
3. At least a frontal gland is present
4. Musculature not described except for presence in pharynx
5. Large brain and six nerve cords, which seem to be subepidermal
6. No special sensory structures reported
7. No details on excretory organs
8. Mouth leads through pharyngeal tube to plicate pharynx and then to intestine reported to be syncytial
9. Testes lie ventrally toward posterior end and connect to a seminal vesicle and then a duct leads externally. Ova mature along dorsal side of parenchyma in midbody; no ducts
10. Mating not reported
11. No information on eggs
12. No information on organogenesis
13. Food habits not known (marine)

E. *Notoplana alcinoi* (Order Polycladida) (Fig. 13-1E)
1. Flattened, ovoid body slightly broader at anterior end; midventral mouth; male and female gonopores opening on ventral midline posterior to mouth with female opening posterior to male opening
2. Ciliated epidermis with cilia aligned in longitudinal rows; gastrodermis lines highly developed gastrovascular cavity; parenchyma solid with well-developed organ systems and considerable intercellular material
3. Eosino- and cyanophilous glands present, often sunken into parenchyma
4. Subepidermal muscle, well-developed on ventral side; parenchymal muscles too

Table 13–1 *(Cont.)*
E. Notoplana alcinoi *(Cont.)*
   5. Well-developed, bilobed brain and extensive nerve plexus of cords and commissures
   6. Eyespots numerous, inverse type; some associated with brain and others on the ends of two short, blunt stalks
   7. Excretory organs not studied in this form; a compound protonephidium in related forms
   8. Mouth leads to pharyngeal cavity within which there is a large, flaired pharynx. This connects to highly branched intestine extending throughout the parenchyma.
   9. Male parts: gonopore leads to antrum with penis, which connects internally to seminal prostate and thence to seminal vesicle. Female parts: gonopore leads to internal ducts, which terminate in a bursa. Sperm and eggs differentiate in testes and ovaries in the parenchyma.
   10. Copulation with exchange of sperm
   11. Spiral cleavage; quartet formation
   12. Progressive differentiation of stereogastrula; gastrodermis differentiates internally and connects to mouth. Genitalia are the last organ systems to differentiate.
   13. Prey variable, ingestion by pharynx which can extend out through mouth (marine)
F. *Plagiostomum album* (Order Holocoela) (Fig. 13–1F)
   1. Slender, round worms, with tapered ends, several mm in length; mouth at anterior tip; single common gonopore near posterior tip
   2. Ciliated epidermis; gastrodermal intestine, said to be syncytial, surrounded by parenchyma
   3. Eosino- and cyanophilous glands present
   4. Subepidermal and parenchymal musculature
   5. Anterior brain and six nerve cords
   6. Two eyespots
   7. Protonephridium present
   8. Large bulbuous pharynx connects anterior mouth to intestine, which is single large sac.
   9. Single common gonopore on ventral side of posterior tip; this leads to a common antrum, into which an ovovitelline duct connects dorsally and an introverted penis structure with glands is attached anteriorly. These structures lead to a prostatic vesicle and then a seminal vesicle.
   10. Mating not reported
   11. Ectolecithal egg with modified spiral cleavage
   12. Organogenesis not observed
   13. Feeding habits not reported (freshwater, typically)
G. *Prorhynchus* sp. (Order Lecithoepitheliata) (Fig. 13–1G)
   1. Cylindrical worm with tapering ends, 1–5 mm in length; mouth combined with male gonopore at anterior tip; female gonopore slightly anterior of midventral; two lateral nephridiopores
   2. Ciliated epidermis; gastrodermis forms intestine and conspicuous, muscular pharynx
   3. Glands throughout epidermis
   4. Subepidermal and parenchymal muscle
   5. Brain and eight cords with commissures
   6. Two eyespots and ciliated pits

Table 13–1 *(Cont.)*
G. *Prorhynchus sp. (Cont.)*
   7. Two protonephridial ducts opening laterally
   8. Mouth is followed by a slender tube-like vestibule connecting to a muscular pharynx, which connects directly to the intestine.
   9. Sperm develop in midbody and are conducted anteriorly along ventral side to a complex copulatory apparatus furnished with a penis. This then leads to a male antrum, which connects to the vestibule leading to the mouth. Eggs develop in the ovaries, which lie in the posterior ventral half of the worm. The oviduct leads to a female antrum connected to a gonopore.
   10. Mating not reported
   11. Eggs not known
   12. Organogenesis not known
   13. Feeding habits not known for this species
H. *Bothrioplana semperi* (Order Seriata) (Fig. 13–1H)
   1. Long body (about 5 mm), flattened, and squared off at anterior end; mouth just posterior to midventral with excretory pore anterior to it and a common (male and female) gonopore posterior to it; two lateral pairs of sensorial pits at anterior end
   2. Ciliated epidermis; gastrodermal lining of a branched and outpocketed intestine with attached plicate pharynx; solid parenchyma with well-differentiated organs
   3. Eosino- and cyanophilous glands present
   4. Subepidermal and parenchymal muscles
   5. Brain with eight nerve cords, the ventral ones better developed, and many, regular, circular commissures
   6. Bristles at anterior end as well as two pairs of lateral, sensory pits
   7. Protonephridia joining to common lateral collecting tubules and these join in midventral region to a nephridiopore.
   8. Plicate pharynx lying in a cavity just inside the mouth and connecting to an intestine, which has one anterior branch and two posterior branches lying on either side of the pharynx, which then fuse posterior to the pharynx; conspicuous diverticula of intestine
   9. Male gametes rarely mature, and reproduction is thought to be predominantly parthenogenetic. Female gametes develop laterally from gonopore. Yolk cells appear in well-developed vitelline organs lying anterior to gonopore.
   10. Mating not reported and may be nonexistent in these highly parthenogenetic worms
   11. Ectolecithal egg
   12. Organogenesis not described in detail
   13. Apparently bottom scavengers ingesting various kinds of food (fresh-water)
I. *Procerodes lobata* (Order Tricladida) (Fig. 13–1I)
   1. Planarian worms (about 10 mm) with flattened bodies; somewhat triangular anterior end, due to lateral lobes; rounded posterior; mouth in midline of ventral side and about ¾ of the way to posterior end; common gonopore posterior to mouth; dorsal and ventral nephridiopores
   2. Ciliated epidermis; gastrodermal lining of three-branched intestine with diverticula; solid parenchyma with well-formed organs
   3. Eosino- and cyanophilous glands; highly developed ventrally
   4. Subepidermal and parenchymal musculature; well developed around pharynx
   5. Large brain; two ventral nerve cords cross-linked by commissures
   6. Paired anterior eyespots; anterior lobes or auricles apparently with sensory function

Table 13–1 *(Cont.)*
I. Procerodes lobata *(Cont.)*
   7. Extensive development of flame cells and collecting tubules with dorsal and ventral pores
   8. Three branched digestive system (one anterior and two posterior) with outpocketings; plicate pharynx, directed posteriorly in body
   9. Many testes lying between intestinal diverticula; penis structure lying in a genital antrum, which connects to common gonopore; ovaries lying next to testes and connected to oviducts, which are paired and fuse just posterior to male copulatory structures; also attached at this point is a seminal bursa and glands that provide shell material for the eggs. These structures connect to the exterior via the common gonopore.
   10. Mating not reported
   11. Eggs are ectolecithal in this order.
   12. Development is very similar to fresh-water planarians, i.e., with transitory outer membrane and pharynx and subsequent appearance of ectodermal precursor of epidermis and internal differentiation of gastrodermis and new pharynx and structures derived from parenchyma.
   13. Bottom scavenger (marine)

J. *Mesostoma ehrenbergi* (Order Neorhabdocoela) (Fig. 13–1J)
   1. Flattened worm a few mm in length with tapering anterior end; mouth ventral and slightly anterior to midpoint; common gonopore just posterior to mouth
   2. Epidermis ciliated and with many rhabdites; central intestine connected to rosette type of pharynx; well-developed parenchyma
   3. Profusion of rhabdites in anterior parts of epidermis; also usual glandular material
   4. Subepidermal and parenchymal muscles
   5. Heavily developed brain; four heavy cords, extending anteriorly, two pairs (dorsal and ventral) of heavy cords extending posteriorly; few commissures
   6. Sensory bristles are sparsely distributed over the body surface. A pair of lateral, sensory pits lie at anterior end, and other apparently sensory organelles. Paired eyespots
   7. Two pairs of long, somewhat convoluted excretory canals (one pair anterior and one posterior) terminate internally with protonephridial flame cells and fuse centrally in worm before going to a nephridiopore in the pharyngeal vestibule.
   8. The mouth leads to a small pharyngeal vestibule in which a muscular, rosette-type pharynx is present. The pharynx leads to an unbranched intestine which runs anteriorly and posteriorly.
   9. Testes lie at both lateral margins of the worm and connect by ducts to a seminal vesicle and then to a penis structure. Latter is close to the common gonopore. A small centrally located ovary is distinct from the vitelline glands. Yolk cells from latter join with ova near a uterus and, after fertilization, the eggs and yolk cells, enclosed in a cocoon, are released from gonopore.
   10. Mating is a copulation with exchange of sperm.
   11. Ectolecithal eggs of two types: dormant, thick-shelled eggs appear in the spring; thin-shelled or subitaneous eggs develop immediately upon fertilization and may do so within the parent's uterus. Dormant eggs develop external to the parent.
   12. Organogenesis apparently similar to triclad's
   13. Probably a bottom feeder (fresh-water)

Table 13-2 Major taxonomic subdivisions according to Hyman (1951, 1959), Ax (1961, 1963), and Beauchamp (1961a).

Hyman (1951, 1959)	Ax (1961, 1963)	Beauchamp (1961)
		Archoophores[1]
Acoela	Acoela or Acoeloida	Acoeliens
	Nemertodermatida	Némertodermiens
Macrostomatida	Macrostomatida	Macrostomiens
Catenulida		Caténuliens
Gnathostomulida[2]		Gnathostomuliens[2]
		Hofsteniens[3]
Archoophora[4]	Proplicostomata	Proporoplaniens
Polycladida	Polycladida[5]	Polyclades
Acotylea		Acotylés
Cotylea		Cotylés
Holocoela	Prolecithophora[6]	
Lecithoepitheliata	Lecithoepitheliata	Périlécithophores
Seriata	Seriata	Protriclades
	Proseriata	Cyclocoeles
		Crossocoeles
Tricladida	Tricladida	Triclades
Maricola		Maricoles
Paludicola		Paludicoles
Terricola		Terricoles
Neorhabdocoela	Neorhadbocoela	Eulécithophores
Dalyellioida	Dalyellioida	Dalyelliens
Typhloplanoida	Typhloplanoida	Mesostomiens
Kalyptorhynchia	Kalyptorhynchia	Calyptorhynchiens
		Plagostomiens[7]
Temnocephalida	Temnocephalida	Témnocéphales[8]

[1] In the sense of Karling (1940) and Westblad (1948).

[2] Ax (1963), for sound reasons, does not include the Gnathostomulida as Turbellaria at present.

[3] Included in the Lecithoepitheliata by Hyman and in the Acoela by Ax.

[4] In the sense of Reisinger (1935).

[5] Acoela through Polycladida belong in "Stadiengruppe" Archoophora (Karling 1940; Westblad 1948).

[6] Prolecithophora through Temnocephalida belong in "Stadiengruppe" Neoophora (Westblad 1948).

[7] Included in the Holocoela by Hyman and in the Prolecithophora by Ax.

[8] Treated as a separate class by Beauchamp.

Table 13-3 Homologous relationships among selected Turbellaria.

	Body		Organ Systems						Reproduction				Feeding
	1	2	3	4	5	6	7	8	9	10	11	12	13
A. *Haploposthia rubra*	1A	2A	3A	4A	5A	6A	7A	8A	9A	10A	11A	?	?
B. *Macrostomum appendiculatum*	1B	2B	3B	4B	5B	6B	7B	8B	9B	10B	11B	12B	?
C. *Stenostomum ventronephridium*	1C	2B	3B	4B	5C	6C	7C	8C	9C	10C	11B	?	13C
D. *Proporoplana jenseni*	1A	2B	3A	4D	5D	6D	?	8D	9C	10A	?	?	?
E. *Notoplana alcinoi*	1B	2B	3A	4E	5E	6E	?	8E	9E	10A	11B	12E	13E
F. *Plagiostomum album*	1F	2B	3A	4B	5D	6B	7F	8F	9E	?	11F	?	?
G. *Prorhynchus* sp.	1G	2B	3A	4B	5G	6G	7B	8G	9G	?	?	?	?
H. *Bothrioplana semperi*	1H	2B	3A	4B	5G	6C	7H	8H	9H	?	11F	?	13H
I. *Procerodes lobata*	1H	2B	3I	4E	5I	6I	7F	8H	9I	?	11F	12I	13H
J. *Mesostoma ehrenbergi*	1H	2B	3B	4B	5J	6G	7J	8F	9J	10A	11J	12I	13H

Key to Semes:
1. Body form: 1A: rounded form, one gonopore; 1B: flat body, two gonopores; 1C: head-like anterior end, one gonopore; 1F: round body, one common gonopore; 1G: round body, mouth serves as male gonopore; 1H: flat body, common gonopore.
2. Body organization: 2A: no gastrodermis; 2B: all three body layers present.
3. Secretory structures: 3A: frontal gland and epidermal gland cell; 3B: epidermal glands including rhabdites; 3I: special ventral development.
4. Musculature: 4A: epidermal and parenchymal muscle fibers; 4B: subepidermal fibers; 4D: pharyngeal musculature; 4E: subepidermal and pharyngeal musculature.
5. Nervous system: 5A: six nerve cords; 5B: brain and two lateral cords; 5C: brain and eight pairs of cords; 5D: brain and six cords; 5E: bilobed brain and nerve plexus; 5G: brain and eight cords; 5I brain and two ventral cords; 5J: well-developed brain with four anterior cords and five pairs of cords extending posteriorly.
6. Sensory elements: 6A: statocyst; 6B: paired eyespots; 6C: sensory pits; 6D: no special structure; 6E: many eyespots; 6G: eyespots and ciliated pits; 6I: eyespots and lobes.
7. Excretory system: 7A: none; 7B: two protonephridia; 7C: paired canals; 7F: compound protonephridia; 7H: common collecting tubules and nephridiopore; 7J: nephridiopore in pharyngeal vestibule.
8. Digestive system: 8A: no gastrovascular cavity; 8B: simple pharynx and ciliated gastrodermis; 8C: pharynx well-developed and ciliated gastrodermis; 8D: plicate pharynx; 8E: flared pharynx and branched intestine; 8F: bulbous pharynx; 8G: pharyngeal vestibule; 8H: branches and diverticula in intestine.
9. Reproductive system: 9A: only gametes; 9B: male genitalia; 9C: male and female genitalia; 9E: complex genitalia; 9G: complex genitalia (different from (E); 9H: parthenogenesis; 9I: compound testes and ova; 9J: complex genitalia (different from 9E and 9G).
10. Mating, nature of gametes, etc.: 10A: copulation and exchange of sperm (not seen in species A, B, and D but seen in related forms); 10C: mating apparently by hypodermic impregnation.
11. Cleavage patterns: 11A: probably spiral cleavage into duets; 11B: entolecithal egg, spiral cleavage like acoel pattern; 11F: ectolecithal egg, modified spiral cleavage; 11J: ectolecithal eggs of two types.
12. Organogenesis: 12B: provisional outer membrane; 12E: stereogastrula and other special features; 12I: various special details (see TABLE 13-1).
13. Food habits: 13C: feeds on ciliates, ingested whole; 13E: extensible pharynx, scavenger; 13H: bottom scavenger.

summary of the known conditions of the thirteen semes under consideration as they are distributed among the ten selected species. And from these data the comparisons needed to determine plesio-, apo-, and neosemes is readily deduced and used to calculate the phyletic distances (R values) for all pairs of species (Table 13–4). These values are then shown in Fig. 13–13.

No dendrogram is plausibly extracted from the maze of phyletic distances developed here. First, note that plesiosemes increase in frequency as comparisons move away from the acoel plesiomorph (*Haploposthia rubra*). This implies that there was rapid evolution beyond the acoels to forms exploiting ecological niches different from the acoels (otherwise the acoels might well have been eliminated by the later forms). Second, it is impossible to show on the two dimensions of a page the correct quantitative relations inherent in the R values between these turbellarian species. For example, species C, *Stenostomum ventronephrium* is equidistant from species D, F, and G, and in fact about the same distance from species H and J, too. It differs from all of these significantly, but in different ways. This is a common situation here. This seems to argue for radiative evolution above the level of the acoels, and with the limited sample studied here it is impossible to argue convincingly for one evolutionary pathway over another. How does one decide how one goes from the plesiomorph to *Stenostomum*? Obviously, our sampling has been too simplistic. A sound turbellarian phylogeny will have to start from smaller taxonomic units and build more carefully to the level of the higher ones. The limitations of the analysis attempted here are explicit at this point.

Table 13–4 Phyletic comparisons between selected Turbellaria.

Homologies between semes 1 2 3 4 5 6 7 8 9 10 11 12 13	Semes compared t	Plesiosemes p	Aposemes a	Neosemes n	Phyletic distance R
A and B a a a a a a a n a a p ? ?	11	1	9	1	31.2
C a a a a a a a n a a a ? ?	11	0	10	1	38.2
D p a p a a a ? a a p ? ? ?	9	3	6	0	16.7
E a a p a a a ? a a p a ? ?	10	2	8	0	26.4
F a a p a a a n a a ? a ? ?	10	1	8	1	27.4
G a a p a a a n a a ? ? ? ?	9	1	7	1	23.7
H a a p a a a n a a ? a ? ?	10	1	8	1	27.4
I a a a a a a n a a ? a ? ?	10	0	9	1	34.3
J a a a a a a n a a p a ? ?	11	1	9	1	31.2
B and C a p p p a a a a a a p ? ?	11	4	7	0	18.5
D a p a a a a ? a a p ? ? ?	9	2	7	0	22.6

Table 13-4 Phyletic comparisons between selected Turbellaria.

	Homologies between semes 1 2 3 4 5 6 7 8 9 10 11 12 13	Semes compared t	Totals Plesiosemes p	Aposemes a	Neosemes n	Phyletic distance R
E	p p a a a a ? a a p p a ?	11	4	7	0	18.5
F	a p a p a a a a ? a ? ?	10	2	8	0	26.4
G	a p a p a a p a a ? ? ? ?	9	3	6	0	16.7
H	a p a p a a a a ? a ? ?	10	2	8	0	26.4
I	a p a a a a a a a ? a a ?	11	1	10	0	37.3
J	a p p a a a a a a a a ?	12	3	9	0	27.8
C and D	a p a a a ? a p a ? ? ?	9	2	7	0	22.6
E	a p a a a ? a a a p ? a	11	2	9	0	30.3
F	a p a p a a ? a a ? a ? ?	9	2	7	0	22.6
G	a p a p a a ? a a ? ? ? ?	8	2	6	0	18.8
H	a p a p a p ? a a ? a ? a	10	3	7	0	20.3
I	a p a a a a ? a a ? a ? a	10	1	9	0	33.3
J	a p p p a a ? a a a a ? a	11	3	8	0	24.0
D and E	a p p a a a ? a a p ? ? ?	9	3	6	0	16.7
F	a p p a p a ? a a ? ? ? ?	8	3	5	0	13.1
G	a p p a a a ? a a ? ? ? ?	8	2	6	0	18.8
H	a p p a a a ? a a ? ? ? ?	8	2	6	0	18.8
I	a p a a a a ? a a ? ? ? ?	8	1	7	0	25.4
J	a p a a a a ? a a p ? ? ?	9	2	7	0	22.6
E and F	a p p a a a ? a p ? a ? a	10	3	7	0	20.3
G	a p p a a a ? a a ? ? ? ?	8	2	6	0	18.8
H	a p p a a a ? a a ? a ? a	10	2	8	0	26.4
I	a p a p a a ? a a ? a a a	11	2	9	0	30.3
J	a p a a a a ? a a p a a a	12	2	10	0	34.2
F and G	a p p p a a a a ? ? ? ?	9	3	6	0	16.7
H	a p p p a a a a ? p ? ?	10	4	6	0	15.0
I	a p a a a p a a ? p ? ?	10	3	7	0	20.3
J	a p a p a a a p a ? a ? ?	10	3	7	0	20.3
G and H	a p p p p a a a a ? ? ? ?	9	4	5	0	11.7
I	a p a a a a a a ? ? ? ?	9	1	8	0	29.3
J	a p a p a p a a a ? ? ? ?	9	3	6	0	16.7
H and I	p p a a a a a p a ? p ? p	11	5	6	0	13.7
J	p p a p a a a a a ? a ? p	11	4	7	0	18.5
I and J	p p a p a a a a a a ? a p p	12	5	7	0	16.9

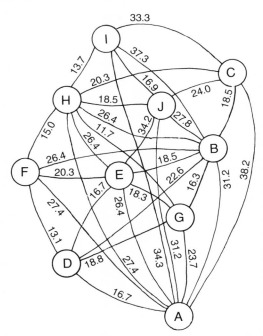

Fig. 13-13 Phyletic relationships between selected Turbellaria. Diagram of phyletic distances.

The foregoing results make very suspect the phylogeny proposed by Ax (1963) (Fig. 13-14). We have already commented on the unsupported idea that some form other than an acoel is the plesiomorph. The further comment needed is to challenge the linearity of progression in Ax's proposal. If we look at the lineages whose origins are unclear, i.e., indicated by Ax by a gap and emphasized by an asterisk in Fig. 13-14, and raise the possibility that in effect many or even all of these groups arose independently from an ancestral (acoeloid) stock, then we obtain a so-called dendrogram not unlike the one given by our own analysis (Fig. 13-13). This view of turbellarian evolution is also supported by Karling's proposals regarding the developmental potential of these worms as seen in the variety of solutions achieved for pharyngeal, genital, and adhesive organs, cited earlier. The main reason for Ax's vertical extension of turbellarian phylogeny comes from the artifact introduced by his two categories Archoophora and Neoophora (see left axis of Fig. 13-14). There is no convincing reason that the former be below (precede?) the latter except in ancestral form. From there on the two traits could show their own further development without the Neoophora being implicated as a later or higher development than the various highly evolved Archoophora. It would therefore seem more consonant with careful analysis to think of many different lines arising simultaneously from acoels more complex than *Haploposthia* and exploiting the various developmental potentials of the epi- and gastro-

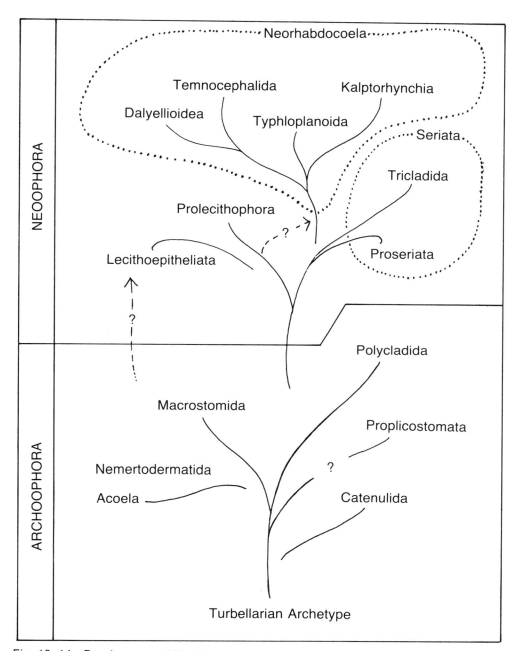

Fig. 13–14 Dendrogram of the Turbellaria according to Ax. (Redrawn, by permission, from Ax (1963).)

dermis with the intermediate parenchyma. The evolutionary pattern is "buisonnate" as in the zooflagellates, but here the plesiomorph is still arguably extant. Apparently the Turbellaria have hit upon some extraordinary organizational and developmental principle that is being thoroughly and independently exploited by many different evolutionary experiments.

Origins. There are three main proposals for the origin of flatworms in general and the turbellarians in particular. These are: (a) the Turbellaria came from coelomate forms showing spiral cleavage in their development; (b) they came from planuloid or ctenophore forms and hence have a coelenterate ancestry; and (c) they evolved from ciliatelike forms and therefore have a protozoan ancestry.

The proposed origin from bilateral, coelomate forms has had a long history (Remane 1936b; Ax 1961, 1963) and necessitates the point of view that significant reduction of certain specializations characterizes the whole flatworm phylum and not just the acoels. Such a view is predicated on a prior conclusion that the primitive Bilateria were coelomate and is labelled the enterocoel theory of derivation of coelomic cavities (Remane 1963b). This theory is refuted, first by the careful analysis of Hartman (1963), using in large part Remanian criteria of homology, and second by Clark's (1963, 1964) careful review and analysis of the origins of cavities and segmentation in the Metazoa. Clark's work convincingly demonstrates the probability that the coelom was not a primary condition in the Metazoa but a derived one, which came about as a result of increased size and locomotory habits in originally acoelomate forms. Such forms were initially solidly constructed with a triploblastic body, whose middle layer was a somewhat fluid mesenchyme, much like the turbellarian parenchyma. On Clark's view, the flatworms are primitively acoelomate, and not a product of evolutionary regression from the coelomate condition. Such a view has been held by others for years (see Hyman 1951; Hadzi 1953, 1963; Marcus 1958; Ivanov 1971) but never so cogently argued as by Clark.

Returning to the earlier point that the Turbellaria as a group show reductions, this can be answered in the same terms used to discuss the question of acoel reduction, and the conclusion is the same. There is no evidence for selection pressures that would result in loss of the coelom, the circulatory system, and closure of the anal end of the intestine leaving only a mouth opening (Ax 1961, 1963). It is further to be noted that Ax's phylogeny of the Turbellaria does not proceed by progressive reduction from the more complex forms to least complex ones, which is a logical extension of the idea of reduction. Rather, turbellarian evolution starts from the postulated reduced form (the turbellarian archetype or "Urform") and then proceeds to become complicated in all lines except the Acoela, which are the only ones presumably continuing the reductionist trend. There is no evidence or argumentation given to account for such hypothetical evolutionary trends in terms of possible selective advantages or even in terms of ecological function. The whole discussion seems to depend really on two points: (a) preserving the consequences of the enterocoel theory, and (b) that internal fertilization is a derived trait, derived from external fertilization. The first point we have already commented on and further details, as previously stated, are found in Hartman

(1963) and Clark (1963, 1964). The question of internal fertilization is a new point and needs explicit discussion now.

Actually, the question of internal fertilization contains two points: the one already alluded to—whether internal or external fertilization is the original mode of fertilization—and then the question as to whether or not there is a primitive type of sperm. Remane (1954, 1957, 1963b), Jägersten (1959), and Ax (1961, 1963) contend that external fertilization is the primitive mode and, following Franzen's (1956) comprehensive survey of sperm types, argue that external fertilization always shows the primitive type of sperm, given by Franzen as having "... a short, rounded or oval head, a middle piece containing four mitochondrial spheres, and a tail consisting of a long filament." (Franzen 1956, p. 461). (Fig. 13-15). Jägersten's (1959, p. 94) comment is, "It is probably seldom that a character can be designated as primitive with greater certainty than in this case."

Unfortunately, it is never safe to make categorical (or near categorical) conclusions as to what is primitive. "Primitive" is a relative term and applicable in the sense of ancestral to succeeding forms. It therefore depends on the phylogenetic analysis which determined the ancestral relationship. What is Franzen's analysis? In the first sentence of his comments on spermiogenesis from a phylogenetic view, Franzen candidly states "... the author has started from the assumption that the primitive type of sperm of the Metazoan is characterized by a short, rounded or oval head ... [etc.]" (p. 461). What is the basis for this assumption? A clue is given at the outset of the review of "Spermiogenesis and Biology of Fertilization": "In the Cnidaria, sperms that are obviously of the primitive type have been described by several authors." (p. 362). It seems that Franzen is satisfied that the Cnidaria are primitive (ancestral to the Metazoa?) and therefore their sperm is also primitive. No allowance is made for possible evolutionary mosaicism, for different degrees of evolutionary development of different parts of the organism. *In toto*, if Franzen's methodology for finding primitive traits is being properly understood, it comes down to the assumption that the Cnidaria are indeed ancestral to the Metazoa (which is debatable, if not even doubtful). Furthermore, whenever Franzen finds in other groups that the sperm has a round head, four mitochondria in the middle piece, and a long tail filament, those sperm are also "primitive," even though they occur in such complex forms as annelids, molluscs, sipunculids, brachiopods, enteropneusts, echinoderms, and the Acrania.

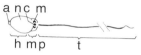

Fig. 13-15 Diagram of a "primitive" animal sperm. a—acrosome, c—centriole, h—head, m—mitochrondrion, mp—middle piece, n—nucleus, t—tail. (Redrawn, by permission, from Ax (1963). Original by Franzen.)

What is especially interesting, Franzen stresses and is echoed in this view by Jägersten, Remane, and Ax, is that external fertilization occurs whenever the "primitive" sperm are found. Hence, external fertilization is "primitive" too.

In response to this line of reasoning, the following points must be made. First, the correlation of sperm type and external fertilization could be the result of the same cell type performing the same function in different forms, i.e., functional convergence. Convergence will also account for its wide distribution; wherever sperm must move through the external medium, this so-called primitive type may well be present as the most efficient answer to that functional need. Note Franzen's claim that "The most important argument for the opinion that this is the primitive type of sperm is found in its occurrence in so numerous and widely distant groups." (p. 463) Note also Remane's first accessory criterion when formulated negatively, as Remane originally stated it: the probability of homology of a character decreases with the frequency of the occurrence of these characters in clearly unrelated forms. This criterion is applicable here, and when the probability of homology decreases, the probability of convergence increases in proportion. Second, there is the claim (Remane 1954, 1963a; Franzen 1956; Jägersten 1955, 1959; Ax 1961, 1963) that nowhere in the evolutionary history of animals has internal fertilization ever changed to the external mode, though the reverse is known to be true. The absence of instances of the first case depends on one's initial phylogenetic viewpoint. The enterocoel theory of invertebrate evolution allows the ordering of forms such that the stem line includes only forms with external fertilization. This is the fundamental evidence adduced by proponents for the primitive nature of external fertilization. The unacceptability of the enterocoel theory removes, therefore, that so-called evidence that no cases of transition from internal to external fertilization are known. A priori it is possible to suppose that the mode of fertilization evidenced by a species will reflect the reproductive behavior most advantageous to it; it can therefore be either internal or external, and there seem to be no irreversible barriers in going from one to the other.

And finally, third, the complex turbellarian sperm (Fig. 13-16), which Franzen (1956) and others think is so improbable as a "primitive" type, can be seen simply as a functional adaptation to the various modes of copulation and storage of gametes found in the free-living flatworms. This diversity may also be important as a reproductive barrier to interbreeding, an interpretation not unlike Karling's (1963) suggestion that the diversity of gonopores might be a device to hinder interspecific mating. Furthermore, the diversity of turbellarian sperm speaks to the possibility that it could well be ancestral to other sperm types; its evolutionary potential seems obviously great. Indeed, referring to Hendelberg's (1969, 1970) analysis, one would argue that the primitive turbellarian sperm is biflagellate, and monoflagellate sperm is a reduced situation. Hendelberg (1970) notes that "at the beginning of the pre-

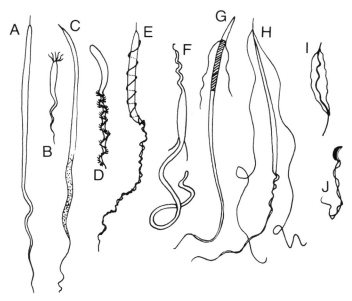

Fig. 13-16 Various types of turbellarian sperm. A. Found in acoels. B. *Paraphanostoma submaculatum*. C. *Convoluta convoluta* D. *Convoluta fulvomaculata*. E. *Haploposthia viridis*. F. *Macrostomum*. G. *Prostheceraeus vittatus*. H. *Dendrocoelum lacteum*. I. *Plagiostomum*. J. *Paravortex cardii* (Redrawn, by permission, from Ax (1963). Original from various sources.)

sent century the theory was put forward that the occurrence of two flagella in many turbellarian spermatozoa is a primitive feature, surviving since the period before the metazoan stage. It must be stressed, however, that the occurrence of two flagella does not imply that the turbellarian spermatozoa has not evolved from the primitive, monoflagellate metozoan type." (p. 369)

In conclusion, the argument that external fertilization is primitive and shows a primitive type of sperm is fictitious. In regard to the original point which originated this discussion—namely, why do certain workers insist on viewing the flatworms as derived from coelomate ancestors?—neither arguments for regressive evolution, nor phylogenetic biases (enterocoel theory), nor generalizations regarding primitive modes of fertilization and gametic structure, support the view of a coelomate ancestory of the Platyhelminthes.

Next, we examine the so-called theories of the coelenterate origin of the Turbellaria. There are two such proposals: that the polyclads arose from ctenophores, which carries the corollary that the rest of the flatworms arose from the polyclads; and that the Acoela arose from a cnidarian planuloid ancestor, and that the rest of the flatworms descended from acoels. These ideas have been commented on by Hyman (1951). The ctenophore-polyclad theory, developed originally by Kowalevsky and especially by Lang (1884), is succinctly set aside by Hyman as follows:

"Lang's theory of the ctenophores as intermediate forms between the radiates and the polyclad flatworms had a wide and extended acceptance but is to be rejected on the following grounds. First, it is argued by all students of ctenophores that the Platyctenea [which includes the ctenophores Lang thought were the key intermediates] are simply highly aberrant ctenophores without phylogenetic significance. Second, the embryology of the two groups [ctenophores and polyclads] is in fact very different, and the cleavage patterns follow very diverse plans, that of the ctenophores being biradial, and that of the polyclads spiral. Third, in the theory of Lang, the polyclads are necessarily regarded as the most primitive existing Bilateria; this is a mistaken idea, since it is now clear the order Acoela occupies that position, and the polyclads stem from acoel ancestors." (Hyman 1951, p. 7)

Hyman favors the planuloid-acoeloid approach to the problem and develops the argument in her own terms, though she follows Graff's (1882) lead in recognizing the primitive nature of the acoels. Hyman's method is superficially that of looking for homologies. After summarizing the chief characters of the acoel order in something less than a page of descriptive comment, she then concludes, "It is evident that there is no difficulty in passing from a planuloid type of ancestor to an acoeloid form (Fig. 2) [Fig. 13–17]. Epidermis, muscular system, interior mass, nervous system, and sex cells are, even in present Acoela, distinctly at a coelenterate stage or but slightly advanced from this stage. Consequently we accept the theory of the origin of the Bilateria from a planuloid ancestor by way of an acoeloid form." (Hyman 1951, p. 9).

Let us look more closely at this whole argument. First, the similarities mentioned, with regard to epidermis, muscular system, interior mass, nervous system, and sex cells are the result of vague comparisons between the Acoela as a group and the Cnidaria as a group, not the cnidarian planula in particular and certainly not the planula of a particular plesiomorphic species of cnidarian. The similarities are then a loosely defined amalgam of resemblances culled out of two large groups that are not units of evolution as species are. The methodology neither rigorously identifies homologies nor looks for them in the critical evolutionary groups—a plesiomorphic species of the derived group and some species of the ancestral group. However, such criticism really only says the method is incorrect and not that the conclusions are also incorrect, however crudely they were derived. Let us now examine this second aspect of the problem.

Assuming that the similarities presented by Hyman indeed represent homologous relationships, what of the traits she doesn't mention? Do they represent aposemic traits or neosemic traits, or traits in nowise related? Can these discontinuities be bridged, if not by intermediate forms, at least by selectionist arguments that make the transition from the cnidarians by way of a planula to the acoels an evolutionarily plausible transition? First we will

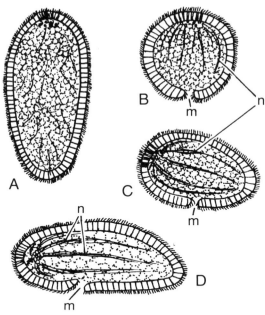

Fig. 13-17 Diagrams illustrating possible origin of acoels from a planuloid ancestor. A. Planula larva, mouthless with peripheral epidermis and inner entodermal mass. B. Hypothetical intermediate with mouth (m) and nerve cords (n). C. Further development of hypothetical intermediate; the body is elongating in a sagittal plane and the "nervous center" is shifting to new anterior end. D. An acoeloid condition. (Redrawn, by permission, from Hyman (1951) *The Invertebrates. Platyhelminthes and Rhynchocoela.* McGraw-Hill Book Company, Inc.)

mention the lack of similarities present and then turn to their discussion in selectionist terms.

The cnidarians and acoels differ in their organizational symmetry and location of the nerve center. Symmetry is radial as compared to bilateral, and the nervous center is aboral in planulae and anterior in acoels. These differences Hyman tries to account for by intermediate forms (Fig. 13-17) which show a shortening of the oral-aboral axis of the planula and its lateral extension, with shift of nervous material, to provide an acoeloid body. Not mentioned by Hyman are the following differences: the planula epidermis has one cilium per cell whereas the acoels have many (Ax 1963); male copulatory organs are present in all acoels but unknown in cnidarians; copulation and internal fertilization are universal in acoels, but copulation is unknown in the cnidarians, and their sperm are shed free in the surrounding medium, however fertilization is often internal; cleavage is of a very indeterminate type in the Cnidaria and is of the duet-spiral type in acoels; and the secretory organelles are different in acoels from those in cnidarians.

There is no way to bridge the differences in epithelial ciliation through intermediates. It might be argued that increased ciliation appears in the acoels because of their heavy dependence on it for locomotion. Similarly, though no intermediates bridge the differences in secretory structures, the acoel's dependence on secretions for protection and food capture are so adaptively important as to allow the proposal that they might even be innovations. Certainly some major food-capture device is necessary to substitute for cnidarian nematocysts and tentacles. More on this shortly. The transition to copulation and internal release of sperm can be considered, in the absence of intermediates, to be plausible as a result of selection favoring protection of gametes. Thus far, selectionist arguments have helped the planula-acoeloid theory.

However, no intermediates and no selectionist argument are available for going from indeterminate cleavage to duet-spiral cleavage. The reasons for those differences remain a mystery.

As for the postulated changes in body form and placement of internal parts, such as the nervous system, selectionist thinking argues against the planula-acoeloid proposals. In Hyman's arguments it is unclear how the hypothetical organisms (Fig. 13–17, B and C) intermediate to planula and acoel would function, specifically how they would feed. The planula does not feed; it never has a mouth. Cnidaria feed as adults when food is captured by nematocyst-bearing tentacles and is brought to the mouth. It makes no sense in selectionist terms to argue that an organism can simply forego one mode of feeding—the product of generations of adaptive evolution—and simply by forming a new hole in the larval ectoderm adopt a new mode of feeding. Organisms simply don't work that way. The acoel mouth works because it is part of a functional, complex whole, wherein musculature around the mouth, secretory and nervous materials are all brought into cooperative action. To argue that changes in the body axes of the planula and shifts in nervous tissue and sensory cells occurred *so that* a mouth could become functional is impossible in selectionist terms. Changes do not occur so as to anticipate selective advantages, except as preadaptations. The preadaptive role of change in body axis and tissue locations is simply not clear. These hypothetical changes proposed by Hyman destroy the continuity of feeding functions and other adaptations, too, that must be preserved if the evolutionary transition is to be plausible.

In its present form, the planula-acoeloid theory is unacceptable for the following reasons: it is methodologically unsound (homologies have not been properly identified), and differences between the forms being compared cannot in all cases be bridged by selectionist arguments making clear the differences as effective evolutionary innovations in the derived group. If there were a planula capable of feeding on algae and protozoa without the help of nematocysts and tentacles, and with other features homologous to the acoeloid condition seen in *Haploposthia*, then the theory would deserve further careful consideration.

There remains now the proposal deriving the acoels from ciliatelike ancestors. For it to be successful, it must fairly meet the kind of criticisms directed to the preceding theories. It must be methodologically sound in its analysis of homologies and it must account for the lack of homologies in convincing selectionist terms. The history of this idea is much older than its recent attribution to Hadzi (1949, 1953, 1963) and Steinböck (1954). This idea has been variously formulated by Jhering (1877), Kent (1880–1882), and Sedgwick (1895). These historical facts are summarized elsewhere (Hanson 1958) and need no further comment here. The analysis of ciliate-acoel homologies in terms of the Remanian criteria was carried out by Hanson (1963a). That analysis will be summarized here and revised in the light of two new points. First, there are the findings from electron microscopy that acoels are completely cellular and not syncytial as was previously thought to be the case. Second, the study of homologies is made more critical by the use of plesiomorphs, so that now we are explicitly making comparisons between *Haploposthia rubra* and the ciliate form most like it, which is thought to be *Stephanopogon mesnili*, the ciliate plesiomorph.

The most important information presently blocking acceptance of the ciliate-acoel theory is the cellular nature of the acoels. Though there has been no electron microscopy on the *Haploposthia* or any of the simpler acoels (grouped in the Opisthandropora-Abursalia of Westblad (1948)) to establish directly their cellular construction, generalization from those acoels thus far studied clearly leads one to expect a cellular organization for the smaller and relatively less complex acoeloid worms. The essential problem posed by this cellular organization is that it denies fulfillment of the compositional relationship in the search for homologies between acoels, represented by *Haploposthia rubra*, and ciliates such as *Stephanopagon mesnili*. When the older view of the syncytial or plasmodial nature of the acoels was utilized (Hanson 1963a), the arguments for homologies, in the light of Remanian criteria, appeared as follows: General body organization in both forms presents broad similarities in bilateral symmetry, anterior-posterior polarization, body surface ciliature, and presence of a simple mouth, and broadly similar internal subdivisions—ciliate ectoplasm, outer and inner endoplasm, on the one hand, and acoel epidermis, outer and inner parenchyma on the other. Both forms show diploid nuclei, but there are few in *Stephanopogon mesnili* and many in *Haploposthia*. In terms of organelle structure, relatively simple mouths open into the interior where food materials come to lie in food vacuoles. Contractile fibers lie at the base of the epidermis in *Haploposthia* and, though apparently absent in *S. mesnili*, are certainly prominent in other gymnostomes such as *Dileptus*, where they are also peripherally located. Secretory and extrusible elements are present in both forms and are concerned with food capture and possibly other roles. The glandular materials of the acoels are formed in the parenchyma, and the trichocysts are formed in the endoplasm (Yusa 1963; Ehret and de Haller 1963) and not from the infraciliature (Lwoff 1950 and earlier papers; cited by Remane 1963a, as a dissimilarity in ciliate and acoel elements). However, much more needs to be learned about the fine structure of the secretory organelles in acoels before critical comparisons can be made with gymnostome elements. The acoel nervous and sensory structures are not present in the ciliates. Both groups show much the same behavioral responses, and perhaps coordinatory fibers have evolved from ciliary rootlet systems in ciliates, and perhaps the epidermal nerve net of acoels also was derived from ciliary rootlets. But this is speculation inappropriate to comparisons between the forms designated for comparison and at present similarities in nervous and sensory structures do not really exist between ciliates and acoels; much less are homologies present. The reproductive system of the simpler acoels is much more complex than any comparable element in the gymnostomes and homologies cannot be found. In summary, regarding the structural parts of the organisms being compared, certain general similarities were found, some (general body organization, nuclear make-up, peripheral secretory and contractile elements) are more convincing than others (feeding elements, reproductive system). Clearly the major Remanian criterion was not met in any case. At best the second and third subsidiary criteria could be invoked.

Functional semes were also examined. The function of contractile, secretory-extrusible, and nervous-sensory systems are not studied adequately from a biochemical or physiological point of view in either the simpler acoels or gymnostomes so as to permit meaningful comparisons; hence, these traits have not been compared. Comparisons were made of feeding and reproductive processes. The holozoic ingestion of unicellular algae and protozoans is very similar in both groups, as well as their intracellular mode of digestion utilizing food vacuoles. Asexual reproduction is typical of ciliates but probably absent in the simpler acoels. Sexual reproduction is the rule with acoels, though details of the process are unknown for *Haploposthia*. However, certain generalizations from other acoels can be applied, i.e., the presence of cellularized male and female gametes which undergo a prezygotic reduction, copulation of hermaphroditic worms, exchange of sperm, internal fertilization, egg laying, duet cleavage and embryogenesis. Sexual processes are not reported for *Stephanopogon,* but judging from other gymnostomes, conjugation between cells is present, with the usual steps of prezygotic reduction, formation of migratory and stationary gametic nuclei, internal fertilization after exchange of so-called male pronuclei, and then separation with reformation of redifferentiated structures. In the two types of sexual processes, only the steps of coming together of two organisms, prezygotic reduction, exchange of migratory elements, and internal fertilization are alike. (It is to be noted these steps occur nowhere else in the Protista.) Cellularization of the acoel gametes and acoel egg laying, cleavage, and organogenesis have no counterpart in gymnostomous forms.

Again, to summarize, feeding and the sexual behavior of the compared forms are similar in several important details and suggest homologies but are not sufficiently alike as to allow one to say the major relationships of homologies are fulfilled. The events in acoel reproduction subsequent to fertilization have no real counterparts in the ciliates.

Because of these general similarities, and because of the ciliate-acoel proposal, authorities such as Remane (1956, 1963a) and Jägersten (1959) have argued the similarities, such as they are, can only be convergences. Let us look at this criticism in detail before going further. For if it is a valid one, it precludes the necessity of going to the next and final step in appraising the ciliate-acoel theory, which is to see if selectionist arguments can bridge the differences between the groups being compared.

Convergence becomes a convincing argument when one can see evidence for similar selective pressures on the forms thought to be convergent, especially in terms of exploitive traits, and when the second and third accessory criteria of Remane are seen *not* to apply. When one looks at the ecology of *Stephanopogon* and *Haploposthia*, it is clear we are looking, in both cases, at small forms which are benthic marine predators, feeding on algal forms smaller than themselves, and possibly on protozoans too. In these terms the argument for convergence seems plausible: both forms must move around, locate their prey, entangle it, and ingest it, all under similar circumstances and, hence, they might well come to resemble each other in ciliary locomotion, feeding organelle, and in secretory elements. Nevertheless, convergence is not indicated by the broader comparisons that are also part of the picture. The structures and functions just suggested as appropriate to the ways of life of *Stephanopogon* and *Haploposthia* are found throughout the gymnostomes and acoels, and, in fact, can be thought to be characteristic of ciliates and turbellarians. That is, the general structural and functional exploitive traits of *Stephanopogon* and *Haploposthia* are not peculiar adaptations of these forms, nor even of the larger taxa of which they are a part (Gymnostomatida and Acoela, respectively) but represent primitive or plesiomorphic modes of organization and exploitive activity of the larger monophyletic groups of which they are the plesiomorphs. Their type of organization is not secondarily convergent but rather a primary mode of organization adaptable to many modes of life. Rather than arguing that the similarities in *Haploposthia* and *Stephanopogon* are the result of convergence, we must argue that they do indeed somehow share a common ancestry.

This conclusion is borne out by looking at the application of the Remanian accessory criteria. The second criterion argues for convergence when one formulates it as saying that the probability of homology decreases when a similar structure or function is only found in a few forms. However, the feeding, locomotory, secretory, and muscular structures are common to all turbellarians and ciliates—some thousands of species—and not to just a few forms in each group. The use of this criterion suggests homology, not convergence. The same is true for the third criterion, which would favor convergence if there were only a few traits having the same distribution. However, we find similarities in the exploitive traits just mentioned and also in general body organization, in diploid nuclei and chromosome cycles, and in copulation and conjugation. The concurrence of these several important traits in the forms being compared again argues for homology rather than convergence. Therefore, the argument for convergent similarities in acoels and gymnostomes is not convincing, and we come to the last step in the analysis, the plausibility of selectionist arguments in bridging the differences between the type of organism represented by *S. mesnili* and that represented by *H. rubra*.

The place to start this discussion is to realize we are looking at the possibility of a common ancestor for forms like *S. mesnili* and *H. rubra*. This brings us to the ciliate eotype and the question to be asked next is this: What may be the first step, of selective advantage, that leads to a divergence from the path of evolution which lead to the gymnostome ciliates? One answer might be cellularization, for this is a major underlying difference between ciliates and turbellarians. If we start at this point, then there are two points which demand critical attention. (a) What is the origin of this cellularization, especially in the context of being selectively advantageous? (b) Does such cellularization help understand the other differences between acoelous and gymnostomatous forms?

Cellularization is an aid to intraorganismic specialization. As a device for compartmentalizing functional units of nucleus and cytoplasm, it allows exploitation of diverse patterns of differential gene activity within one organism, which appears today as the fundamental basis of differentiation (Markert and Ursprung 1971). In the ciliates, where there is typically one big polyploid nucleus to serve all the cytoplasm, the differential genic activity of this nucleus (Beale 1954; Siegel and Cohen 1963; Hanson and Kaneda 1968) is probably all or none. That is, certain genes act at one time and not again until the appropriate mechanism reactivates them. Gene products come out of the nucleus in a given sequence, and consequently the cytoplasm has a limited set of materials to work with. Variability is limited to what the cortex can do with this array of raw materials and to what external stimuli may do to change a pattern of expression already underway. There are advantages to such a system. Control of gene action is well regulated and probably capable of rapid response to external conditions—a highly adaptive situation. The sequence of the genes appropriate to fission results in many rapid fissions, and with all the essential genetic material gathered into the macronucleus, karyokinesis can readily segregate this precious genetic material in synchrony with cytokinesis.

However, it is not difficult to see advantages accruing to the alternate type of organization, i.e., instead of one large polyploid nucleus, to have many diploid, small nuclei. And when the cell so organized is already on the road to exploiting increase in size, due in large part to a permanent mouth, which aids its predatory proclivities, and is also exploiting complexity of cytoplasmic structure in its superficial layer (the cortex, probably composed of ectoplasm and at least part of the outer, gel-like endoplasm), then an organizational base that complements these tendencies is of selective advantage. Cellularization is such an organizational pattern, conducive to both increase in size and further specialization.

Such cellularization may have initially arisen simply as a result of developing permanent boundaries of endoplasmic reticulum around nuclei lying in the central part of the organism or located eccentrically at the inner edge of the outer endoplasm. This cellularization could well have included nuclei destined to differentiate into gametes as it would enhance their differentiation. When at least one such gamete included cytoplasm with its nucleus so as to permit regeneration of an adult, and when that development is initiated by fusion with another, probably smaller, migratory gamete whose nucleus also contributes to the genetic information of the new individual, that is the time sexual reproduction is achieved. The selective advantage of sexual reproduction over the ciliate separation of sexuality and reproduction, is the classical one of increased genetic variability in natural populations. Such variability becomes especially important when there is the opportunity of exploiting new organizational modes with significant potentials for further diversification.

Are such potentials present in the forms being discussed here — a form like the ciliate eotype or plesiomorph? The answer is: Yes, on three counts. First, we are dealing with a system that carries extraordinary morphogenetic capabilities in its cortical or peripheral structures (e.g. Tartar 1961, 1967; Sonneborn 1963, 1970a; Hanson 1967a). The ciliates demonstrate the capabilities of this system even when working within the relatively confined situation of having only one significant source of nuclear products. Second, and differing from the ciliate situation, is the true embryogeny that has been inserted into the normal life of these organisms postulated to have sexual reproduction. Such a situation provides greatly increased opportunities for new organizational modes through altering late, middle, or early stages of embryonic development, as students of the biogenetic law have long made clear. Third, and this depends on the zygote's products retaining their cellular organization, cleavage would achieve a wide distribution of nuclei throughout the body of the embryo, and hence the adult would come to have widely distributed nuclei. This placing of nuclei in as different parts of the body as possible increases the variety of local environments around the different nuclei, which is reinforced by presence of cell membranes, and provides, therefore, the opportunities of different gene action in different cells.

This would be the point of emergence of the eumetazoan eotype. Here there would be a bilaterally symmetrical, multicellular animal capable of that cellular heterogeneity that is the forerunner of tissue differentiation. Hence, cellularization of the gametes, which initially was a by-product of that general body cellularization appearing in response to pressures for further specialization, provided immediate and important additional momentum to the second or alternate line of ciliate eotype development which utilized a multinucleate body. Increased genotypic variability, combined with ontogenic plasticity, and provided with a mechanism enhancing intraorganismic cellular differences in organisms already tending to various specializations, all this sets the stage for emergence of metazoan forms of great evolutionary potential.

There remains, however, to discuss the problem of those traits that are not homologous between gymnostomelike and acoeloid forms. Cellularization, the most important difference, is accounted for as a selectively advantageous innovation. Nervous tissue and sensory cells may be thought of as specializations especially dependent on cellular organization because of the important role played by membranes in their function. Ciliates, it may be argued, can never develop a true nervous system because of the absence of internal membranes appropriately organized. The coordinated behavior and ability to sense various environmental stimuli may represent the peak of such behavior in single cells, which have only fibers, ectoplasm-endoplasm interfaces, and variously developed endoplasmic reticulum or a single cell membrane (Naitoh and Eckert 1969; Eckert and Naitoh 1972) as a physical basis for impulse transmission. These tendencies, in the context of cellular forms also using complex behavioral patterns as part of their predatory habit, could well have led to selection favoring the development of nervous and sensory cells. The last acoel trait lacking an homologous counterpart in the ciliates, is embryogenesis. Its emergence as a selectively advantageous consequence of cellularized gametes has already been discussed.

Hence, overall, the differences between the gymnostomes and acoels lie in just those characters where the differences can be seen as innovations of selective advantage. This is exactly what should be apparent if the ciliate-acoel theory is to be considered valid, for these are the traits wherein acoel-turbellarian biology is diverging from zooflagellate-ciliate biology. The absence of homologies specifically identifies where and how the acoels are innovating, and these innovations can be seen to be exploited further in realizing their potential. This is clear in the extensive development and specialization of nerve tissue—cords, commissures, and brains—and of extensive types of sensory cells and sensory behavior. We see it in the extensive cellular basis of specialization of system after system beyond the level found in the ciliates—the secretory system, the musculature, the emergence of a digestive system and an excretory system. And we see it in the rapid increase in size found in the Turbellaria. We see it, lastly, in the rapid development of the genitalia and the extensive use their development makes of internal cellular specializations. Also, the importance of sexual reproduction is evidenced by the fact that genital evolution proceeds more rapidly in the acoels than any other trait; it is indeed the most important basis for taxonomic separation of the different acoel species, being the most complex and variable of all acoel characters. These facts document that evolution of efficient sexual reproduction through copulation and embryogenesis was of major selective advantage to these forms, to be followed more slowly by various changes in exploitive traits and by changes in homeostatic mechanisms to handle increases in body complexity and size.

Phylogenetic conclusions. In sum, the origin of the turbellarian flatworms seems at present to be best understood by seeing them emerge from the ciliate eotype; they share a common ancestor with the ciliate plesiomorph. However, since these worms hit on such a remarkably rich complex of traits—especially cellularity, sexual reproduction, and embryogenesis—we must expect them to diverge from the ciliate ancestor and from present day ciliates in just those traits depending on these innovations. Since the traits implicate reproductive and homeostatic functions as well as exploitive ones and depend on a new structural basis for all of these, it is not surprising that clear identification of homologies is difficult. The best we can say in that regard is that many suggestive similarities are present, but no rigorously definable homologies. However, a consistent selectionist argument makes clear why such a situation is to be expected and why we cannot expect a more rigorously delimited phylogenetic argument.

Before closing this chapter it is necessary to recapitulate comments on points that are thought to mitigate strongly against the ciliate-acoel theory. The problem of seeing internal fertilization as the primitive condition in the Bilateria has been discussed above. External fertilization is not necessarily a primitive trait, nor internal fertilization a derived one. From the foregoing phylogenetic analysis of the Turbellaria it appears that internal fertilization is the primitive mode for bilaterally symmetrical multicellular animals. This is not weakened by Jägersten's (1959) comment that copulation is found in flagellates and in turbellarians but conjugation occurs in ciliates. "According to current opinion the gametes in the evolutionary line from *Flagellata* to *Ciliata* have lost their cellular nature; in other words, the sexual process has here changed from copulation to conjugation. This is evidently also Hadzi's view. The consequence of Hadzi's thesis, however, is that at the differentiation of the acoel type from the ciliate one, new free gametes would have been formed, and thereby the process of copulation would have reappeared." (p. 92–93.) This criticism rests on the mistaken idea that since cellular fusion of haploid cells during sexual processes in zooflagellates is called copulation, and transient pairing of hermaphroditic worms to exchange male gametes is also called copulation, that the two sets of processes are alike (even homologous?). They are highly dissimilar (nonhomologous). Ciliate conjugation is probably largely homologous to turbellarian copulation, despite its different name, because of the similarities in the two processes. The possible isomorphisms of processes are to be compared in the study of homology and not the isomorphism of labels.

And last of all is the question why acoel cleavage should follow the spiral-duet pattern if it recapitulates nothing, as the ciliate-acoel theory implies. Earlier (Hanson 1958, 1963a) this question was answered by simply suggesting that the spiral-duet pattern may well have been the only way the early acoel eggs could develop, namely, that pattern was simply a consequence of the mode of organization of the cells. They would have to show some mode initially, and spiral cleavage was it. More recent work on acoels (Costello 1960, 1961) shows that cleavage patterns follow on the position of the centrioles. Hence, spiral cleavage seems to be a direct mechanical consequence of centriole topography, perhaps of the mechanics of centriole duplication. The relatively large acoel eggs apparently use a spindle and aster fiber apparatus to cleave the egg and its blastomeres and thus set the pattern for much subsequent eumetazoan cleavage (Rappaport 1971). This, however, raises the problem of where the centriole came from, since ciliates show no such body nor do they show loss of nuclear membranes during mitosis. There is no complete answer to this at present. It is to be noted that cleavage division is quite different from that in more differentiated cells. Preliminary observations of nuclei in regenerating adult acoels show that no centrioles were observed and the nuclear membrane seemed to remain intact (Hanson 1961). The same has been reported for adult tissue mitosis in *Stenostomum* (Stern 1925). These results need to be checked with electron microscopy. If true, they lead to the following suggestion; the centriole has indeed reappeared or it never disappeared. It was earlier present in the zooflagellate ancestor of the ciliate eotype and may have persisted through that eotype to the flatworms playing an important role in aster and spindle formation and in defining fission planes. It makes itself visible during cleavage when an effective fission mechanism is necessary but is not apparent when cells are smaller, in the adult condition. In other words, it is present really only as a functional adaptation and does not pose the phylogenetic problem some have attributed to it (Grassé 1951a; Remane 1963a).

The Turbellaria present us with a group still preserving an extraordinary range of diversity extending from the simpler acoels to a wide variety of more complex forms. Their rapid development of complex internal organ systems, patterns of sexual reproduction connected to embryogenesis, and the basic bilateral symmetry of their triploblastic bodies all show the emergence of patterns and potential for extended metazoan evolutionary development.

Part Three

Perspectives

Perspectives

THE two preceding Parts have, respectively, developed and then applied a phylogenetic theory and its analytical methods. What has been accomplished? In answer, this final Part evaluates two things: the methodology and the conclusions deriving from it regarding the origin and early evolution of animals.

Chapter 14
Phylogenetic Methodology

AT the outset we recall that phylogenetic goals are twofold: to recreate the course of evolutionary history and to elucidate the evolutionary innovations that make that history possible.

Evolutionary History

Recreation of organismic history is fundamentally a probelm of identifying genetic relationships among species, and to achieve that, the following procedure was developed. It can be divided into three parts for our present purposes. First, there is the general description of the species under comparison and the finding of relatively homogeneous groupings of them in terms of similar semic traits. Second, homologies are searched for and, if present, located among the semes, a plesiomorph identified and paired comparisons made among all the species to be examined. Third, the comparisons are used to calculate phyletic distances, which allow formulation of dendrograms, and these are evaluated by reference back to the original descriptive data. What are the strengths and limitations of this procedure?

SEMES

Nowhere in the application of the phylogenetic methodology used here (Chapters 4–13), not even where the methodology was pursued in greatest depth, as with the Acantharia, the Ciliophora, and the Turbellaria, was the actual analysis in tight congruence with the ideal procedure outlined in Chapter 3. In each chapter it was soon clear from the general characteristics used to describe the structural and functional traits of each group of organisms that there was unevenness in our knowledge. Regarding some traits nothing was known, as, for example, zygote formation in acantharians, or selective factors in the guts of termites, or the functional basis of much nematocyst morphology, and so on. In other areas there was plentiful information, as with ciliate ciliary patterns, sponge spicules and skeletons, radiolarian tests, and so forth. And between these extremes were cases of intensive study of a limited number of forms, such as mastigont and kinetide patterns in certain zooflagellates and ciliates, regeneration in certain cnidarians, life cycles in certain foraminiferans, and the like.

556 *Perspectives*

This unevenness would have been expected on *a priori* grounds and was simply brought into sharp relief when the methodology demanded that the total biology of each species was to be scrutinized (Rule II, Chapter 3). However, that admonition regarding "the total biology" could not, in practical terms, be taken literally, since it would include, endlessly, everything knowable about a given species. It was, therefore, formulated as some twenty possible major semes (Table 3–2), but even this reformulation of the information content of a species was never fully realized. The actual and the ideal analyses never coincided.

However, a perfect coincidence is not necessary, because this approach only demands that a sufficient number of semes (empirically estimated to be not less than ten) be available for comparative purposes and that the semes not be derived from an intensive study of only a limited aspect of the biology of the forms under study. (One could imagine that intensive work on reproductive processes in the turbellarians could provide separate semes on mating behavior, fertilization, egg laying, cleavage, organogenesis, and maturation of young worms. These plus asexual reproduction and regenerative processes could provide the majority of a ten-seme study and would mean that the resultant phylogeny would be based largely on developmental features to the exclusion of other relevant aspects of turbellarian evolution. This would be an unacceptable basis for elucidating the history of the Turbellaria.) Therefore, if numerous semes reflecting various aspects of the biology of the forms being studied, are at hand, a phylogenetic analysis is possible, of course qualified by the nature of the data used. In the studies carried out in the preceding chapters there was always a balance between structural and functional semes and always a variety of aspects represented within these categories. The only major sources of semic information not utilized were the structures of natural populations and molecular data. There was reference to breeding patterns and to some molecular information, but never enough was on hand to allow for its systematic, comparative use.

All of this leads to two conclusions. One is that more data are needed, but they should be concentrated on semic traits, and they should complement data already on hand. This will maximize the usefulness of new data and efficiently focus research efforts. Second, it is possible, even with the data on hand, to find an appropriate number of semic traits and proceed with them into preliminary studies of homologies and actual phylogenetic analyses.

COMPARATIVE STUDIES

Homologies. The reformulated criteria of Remane are free of ambiguities in identifying homologous relationships. They also, as claimed for them, provide a sound means of distinguishing homologies from convergences. The most notable achievement of these criteria comes in establishing homologous relations in aposemic traits; the serial relationship can identify genetic relationships even between highly dissimilar structures and functions. This is the essential means of following the course of evolutionary change.

The central operation in determining homologies is the establishment of a certain minimum set of point-to-point similarities. The earlier discussion (Chapter 3) proposed that a minimum of four points in three-dimensional space be examined for similarities between structures and five to seven or more be examined in a linear (temporal) sequence for functions which are being compared. These guidelines were observed throughout, but as guidelines rather than as rigorous rules of practice. For nowhere was there recourse to diagrammatic abstractions, as of stomial parts in ciliates or cellular details in choanoflagellates and sponge choanocytes, to establish point-to-point identities. The basic reason for the looser application of these comparative procedures lies in the awareness that biological variation, even within a species, renders absolute identity an unobtainable ideal. Hence, the investigator is forced to use his or her judgment in determining whether or not the apparent similarities justify the claim of homology. We used, in the preceding chapters, the rule of thumb that if four out of five morphological points were quite similar, then a homologous relationship was indicated.

This is the point where the greatest subjectivity enters into the analytical procedure being espoused here. A first safeguard against distortions from subjective judgments is a clear statement of the features being compared. What trait is being studied? What are the component parts which provide the point-to-point comparison? If this is specified clearly, the opportunity for others to examine the judgment of the investigator is made real, and the objectivity of scientific inquiry through specified operations is made possible.

The second safeguard against improper point-to-point comparisons is to improve the whole technique of making such studies. It may be that a direction indicated by a mathematically gifted colleague, Professor Robert A. Rosenbaum, can provide a lead which will be rewarding. He points out the following: There are three types of identity among entities (= semes). (1) There can be an entity a such that $a = a$. (This is a *reflexive* relationship.) (2) There can be entities a and b such that if $a = b$, then $b = a$. (This is a *symmetrical* relationship between a and b.) (3) There can be a, b, and c, such that if $a = b$ and $b = c$, then $a = c$. (This syllogistic relationship is called *transitive*.) The transitive relationship is the one of special interest, especially in cases of limited or weak transitivity. Let us assume that rather than precise identity between the entities there is limited identity, signified by $\sim_1 < \sim_2 < \sim_3$. (The first symbol, $\sim_1$, expresses greater similarity than the second, and the second greater similarity than the third.) Hence the following relationship can hold $a \sim_1 b$ and $b \sim_1 c$, but $a \not\sim_1 c$. Namely, a can resemble b, and b resemble c to the same degree, but a does not resemble c to that same degree. This is another way of expressing the serial relationship. In phylogenetic terms, a

and c are not so similar as to be termed homologous entities, but with the insertion of b, homology is established. In such a case the relation between a and c may be expressed as $a \sim_2 c$. And greater degrees of divergence, e.g., $\sim_3$, can be identified and described when more intermediates are used in a given series. The point is that the concept of *limited transitivity* might be logically developed into an effective language for phylogenetic studies right at the point of its greatest analytic weakness—the determination of similarities in point-to-point comparisons.

Before leaving the discussion of homologies, we should mention two further points that emerged from studies on the amebas and ciliates, respectively. With the amebas we found that pseudopodia are semic, i.e., sufficiently complex to allow for point-to-point comparisons, but we found that distinguishing between certain points in the ameboid three-dimensional matrix was difficult. In these forms the anterior and posterior points of the pseudopod were easily identifiable, but to distinguish right from left or top from bottom meant using the substratum as a reference point outside the organism. In the absence of that reference point it was not possible to compare pseudopodia in a way to test the possibility of homology. Here was a seme that was untestable for homologous relationships and therefore phylogenetically meaningless.

The special point in the ciliates exemplifies an unexpected mode for establishing serial relationships that are homologous. This happened twice: the transition from gymnostome ciliates to suctorians by way of *Actinobolina* involved progressive loss of stomial parts and emergence of tentacles. Similarly, the transition from gymnostomes to hymenostomes by way of *Pseudomicrothorax* occurred through simultaneous loss of nassa and appearance of a ciliated prostomium. The lesson in both cases is that the serial relationship can be established not just by change in one structure or function, but by progressive changes in two providing there is an intermediate which demonstrates an overlapping of the phasing out of one feature and the phasing in of another.

Plesiomorph. The plesiomorph is a specially important feature of the phylogenetic procedure used here for two reasons. It establishes a reference point for determining directions of aposemic or neosemic change and, being a form that actually exists or existed, it precludes the abstractions and vagaries that inevitably accompany arguments which postulate hypothetical forms. Its weakness is that it is not and should not be expected to be an ideal representative of the ancestral form of a given group. (That form we call the eotype; we could not escape completely the need to deal in hypothetical ancestors.) But it seems correct to say that the specialties that occur in plesiomorphs are usually recognizable as such, e.g., the lack of ciliation on one surface of *Stephanopogon mesnili*. Rarely do we find such convincing plesiomorphs as the acoel worm *Haploposthia rubra*, even though here its male genitalia are a kind of unique specialization.

Comparisons between plesiomorphs and other forms and between these others are straightforward except, again, for a certain subjectivity arising from having to decide when a trait is plesiosemic or aposemic relative to another. The practice used in the preceding chapters was somewhat arbitrary. The guideline of 80% similarity (four out of five points alike) was used for plesiosemy and less similarity than that (homology established by intermediates, of course) meant aposemy, or even neosemy where no homology existed. Where the differences between compared semes was close to 80% we tended to a conservative judgment and treated the situation as plesiosemic. A fuller treatment of this problem, perhaps using the approach of limited transitivity, would be most helpful and important.

PHYLETIC DISTANCE AND DENDROGRAMS

Phyletic distance is measured by R values and these are calculated by

$$R = \left(\frac{-p + (2a)^2 + (3n)^2}{t} \right) + 1$$

where p, a, and n are the number of plesio-, apo-, and neosemic relationships existing between two species, and t is the total of semic categories used in the comparisons. The formula deserves only two further comments now. First, as stated previously (Chapter 3), the coefficients and exponents of a and p are arbitrary weightings arrived at empirically. They are therefore open to further manipulation and surely improved weightings can be found. As we saw especially in Chapter 10, the present formula results in two kinds of inconsistencies. In one we can find a pair of species whose homologous relationships can be expressed as 2-7-2 (two plesio-, seven apo-, and two neosemes) and in another, using the same semes, the homologous relationship is 2-6-3 (Table 10–4, species A and C and species A and H). The latter pair with greater divergence yet shows a smaller R value. In the other kind of inconsistency, equivalent homologous relationships, e.g., 6-6-0 and 5-5-0 (Table 10–4, species G and J and species B and K), are not expressed as equivalent R values. As stated earlier, the problem comes from the effect of squaring the larger values of aposemes. This is seen graphically in Fig. 14–1. As shown in Part A of that figure, the larger values of plesiosemes and neosemes are much larger than the lower ones. This result of squaring the weighted values makes larger numbers of these semes take on added importance as measures of differences between species. That is a desirable effect. The problem comes when we realize that with a circumscribed number of semes of this sort (we are ignoring the plesiosemes for the moment, for they have a negligible effect) increase in

560 Perspectives

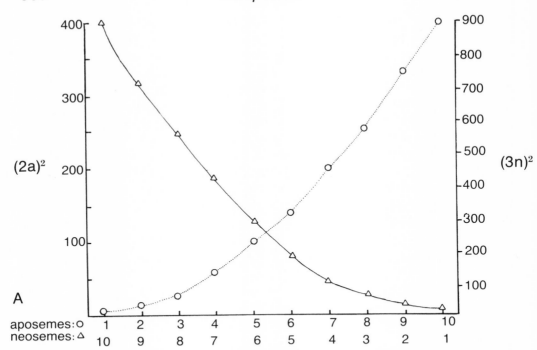

Fig. 14–1 The weighting of aposemes and neosemes. A. The progressive change in apo- and neosemic values as their numbers increase. B. The sum of a constant total number (ten) of apo- and neosemes. Note that when there are seven or more aposemes the value of apo- and neosemes increases even though the number of neo-

one category is loss from the other one. In Part B is plotted the sum of aposemic and neosemic values ranging from zero aposemes plus ten neosemes to the reverse situation. It is seen that where there are seven or more aposemes the combined values of apo- plus neosemes increases despite the fact that the number of neosemes—representing the most extreme differences, i.e., lack of homology—actually decreases. When there are six or less aposemes (four or more neosemes), then the combined values progressively increase to reflect the greater and greater divergence measured by increase in neosemes. We can only repeat that an improved system of weightings should be found, either by further empirical refinement or through some mathematical sequence of numbers that better fits our needs. That is a task left at present for those with a mathematical turn of mind.

The second further consideration of the problem of weightings might be to resolve the category of aposemes into its subcategories of hyperseme, hyposeme, polyseme, and episeme (Table 3–1). These categories were not used because there was little reference to them in the phylogenetic analyses and discussions developed in Chapters 5–13. Nonetheless, they might become useful for a more sensitive or accurate determination of R values.

Large phyletic distances need interpretation both in constructing and understanding dendrograms. When these distances are consistently greater than

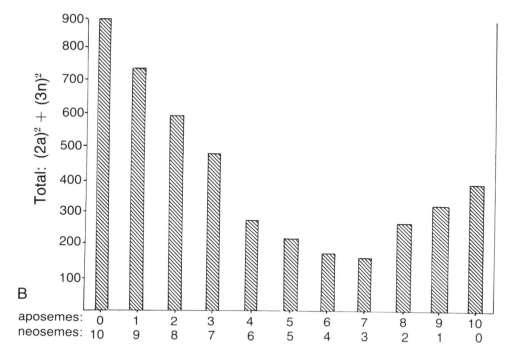

semes is decreasing. When there are six or less aposemes, and, conversely, four or more neosemes, the combined values of apo- and neosemes increases as the neosemes increase.

fifteen or so, it means significant evolution has taken place between the forms (species) under comparison. It is desirable to find intermediates so as to fill in the phylogenetic space between the distant forms. This is clearly exemplified by the large values of R between the acoel *Haploposthia* and other turbellarians. This situation tells us that much evolutionary change has occurred between the turbellarian plesiomorph and the other free-living flatworms. It further suggests that more complex acoels were probably ancestral to the other turbellarian groups, and therefore it would be informative to work out the phylogeny of the acoels and search among them for the origins of the other orders. This would seem to be a plausible way to resolve the various phyletic distances among the turbellarian species studied (Fig. 13–13) into an informative dendrogram.

Origins. Understandably, the most problematic aspect of phylogeny is the origin of any major group like a phylum. For reasons made clear in the various preceding chapters, one cannot expect to homologize readily a plesiomorph of one major group with a form in some other distinct group. There has been too much evolutionary change necessitated in moving from one adaptive zone to another to expect a full string of intermediates to persist as a connection between the two groups. Furthermore, we alluded in Chapter 1 to those morphologically similar forms occupying an adaptive peak defined by constraints of energy utilization rather than any obvious ecological adaptation. Might it be that adaptive zones have, as a broad underlying feature, certain limitations of organization imposed by whatever molecules (especially proteins) that can effectively serve the needs of survival there? This suggests that borders between adaptive zones are real barriers of energy expenditures that deny long-term survival, forcing forms who straddle them to return to their previous adaptive zone or move to another one of lower energy needs or die out. If something like this exists it would help explain the very clear and often wide separations existing between major groups of organisms.

Finally, in this same highly speculative vein, the pressures to evolve efficient use of available energy might have some connection to the waxing and waning of numbers seen in the geological record in the absence of significant morphological change or evidence of environmental change. Paleontologists are aware of these fluctuations (Muller 1955; Wiedmann 1969) and have often speculated on their causes. There is no consensus on the latter points.

Evolutionary Innovations

When homologies are identified directly by the positional and compositional criteria, there is little or no evidence of evolutionary change. When serial relationships must be invoked to establish homology, then there is evidence of aposemic development of the trait in question, and change is clearly in evidence. And when homologies are present in certain traits, establishing a common ancestor sometime in the past, then the nonhomologous semes are obvious evidence of divergence and, therefore, of evolutionary innovation in one or more of the species in question. These aposemic and neosemic evidences of evolutionary change only become apparent when we study as much of the biology of the forms being analyzed as is possible; in that larger context, evolutionary trends quite readily stand out with a clarity not otherwise obtainable. In ciliates, for example, we see conservatism in the uniformity of conjugation, less conservatism in the nuclear situation, and even less in the wonderfully malleable cortex and its multifarious formations and specializations. Exclusive emphasis on any one of these traits would result in three different views of ciliate evolution; together they fill out a mosaic pattern that is informative as regards the varying selection pressures to which these cells are responding.

THE LIMITS OF PHYLOGENETIC ANALYSIS

What are the limits of phylogenetic analyses based on Remanian criteria for identifying homologies? The operational limit is the absence of homology. For when this occurs, there is no evidence of genetic relationship and, therefore, no basis for deducing lines of descent. Such absences of homologies become painfully evident at the level of phyla when one looks for homologies between a plesiomorph of a given phylum and some other form outside that phylum. What we come to are evidences of possible homologies, as between sponge choanocytes and the collared flagellates, or between ciliate conjugation and flatworm copulation. And then speculation, tempered and limited by what is consistent with natural selection, attempts to make plausible the absence of homology in terms of adaptively selective changes between postulated ancestral and descendant forms. Beyond these speculative plausibilities phylogenetic analysis cannot go. The results are sometimes discouraging: no plausible ancestor for the cnidarian plesiomorph *Halammohydra* was found. And sometimes they are encouraging: there are sound reasons for understanding the differences between the acoel *Haploposthia rubra* and the ciliate *Stephanopogon mesnili* and for concluding that they could have evolved from a common ancestor much like the hypothetical ciliate eotype.

The only possible technical breakthrough that could help here seems to lie in molecular data. Thus far, unfortunately, molecular evolutionists have not seen their data as truly complementing more conventional phylogenetic studies. The latter have been largely used to confirm the validity of molecularly based phylogenies (Chapter 3). And conversely, conventional evolutionists concerned with phylogeny have not seen molecular data as being helpful. Might it not be that comparisons of selected proteins, e.g., cytochromes from plesiomorphic sponges compared to cytochromes of choanoflagellates, would help us to choose which sponge was phylogenetically closest to the collared flagellates? Might not similar work with *Halammohydra* and possible ancestors of the Cnidaria narrow the problem to a significant degree? The point is this: rather than seeing molecular and conventional approaches as competitors in drawing similar conclusions over similar sets of species, let us use the fuller range of information from conventional studies (augmented wherever appropriate by molecular data) to the limit of designating plesiomorphs, at the highest possible taxonomic levels, and their possible ancestors. Then bring molecular techniques to bear on the conservatism of macromolecules to find the phyletic distances between such plesiomorphic forms and their putative ancestors. In the absence of prior conventional studies, molecular phylogenists might have no good idea of where to look for plausible ancestors and, second, they might be hard put to interpret the difference they find among their comparisons. But in the context of conventional studies, plausible suggestions are present to guide the molecular studies, and if one of the possible demosponge plesiomorphs, *Oscarella* or *Halisarca*, shows more amino acid sequences similar to those in *Proterospongia* than the other plesiomorph, then a rational choice between the two candidates is possible.

In brief, phylogeny based on homologies above the molecular level can guide molecular studies so that the two levels of phylogenetic study will be truly complementary.

TERMINATION AND CONTINUITY IN EVOLUTIONARY DEVELOPMENT

A final area of discussion is to inquire into the basis for varying degrees of evolutionary success found among different groups of organisms. Why are some dead ends and others on the main line of evolutionary development? What distinguishes the two categories?

At present, the operations used for finding groups with limited or relatively unlimited evolutionary development are those needed to construct dendrograms, for position on the dendrogram identifies one group as being the terminus of evolutionary development and another as being on or close to a productive path of evolutionary change. Can we examine the groups lying in these two positions, choose the best examples and inductively find features characterizing each position? That would surely be one way to proceed. But a final overview of our phylogenetic conclusions comes in the next and final chapter, and so we are not yet ready to survey the overall dendrogram of protozoan and early metazoan evolution.

Is there another approach? Can we deductively infer, from our knowledge of how evolution proceeds by the mechanism of selective preservation of random mutations, what might be the bases for limited, on the one hand, and extensive, on the other hand, evolutionary development? The effort will be worth making if we can find *a priori* those characteristics that we expect will identify the two groups and then, in the next chapter, test our expectations against the dendrogramic summary of our phylogenetic conclusions.

Functional correlates of evolutionary development. A common reason given for limited evolutionary progress of a group of organisms is the somewhat vague notion of specialization or overspecialization. What does that mean? What is specialized? The whole organism or part of it? To explore the problem further we will recast it in selectionist terms, or better, in terms of those functional capabilities of a species that determine its evolution. In fact, we will argue that evolution that emphasizes exploitive functions will result in evolutionary dead ends, whereas those emphasizing homeostatic or reproductive functions are not limited in that fashion. After making those arguments, we will bring the discussion to a focus by proposing that dead-end groups can be characterized as largely cladogenetic forms, whereas the evolutionary more innovative groups are anagenetic.

All species must specialize to some degree to compete effectively for survival, but where the transition comes from successful specialization to doomed overspecialization we have no way of knowing today. The whole concept of overspecialization comes *a posteriori*. It is a hypothesis that explains from hindsight and has never been tested from *a priori* predictions. It presumably invokes a causal mechanism of the following sort: One species becomes adaptively specialized to others that it uses as food. Elimination of the latter dooms the former if the former cannot change its adaptation fast enough to other nutritional resources, or if it comes up against forms already well adapted to the only possible alternative food resources. The general situation inherent in this crude example is that if the environment remains constant, then highly-adapted forms will continue to survive, but when changes come, organisms must change too; and if they can't change enough, they die. They appear to have been overspecialized with regard to a certain mode of survival. We can view exploitive adaptations as representing an interface between organism and environment. They become so molded to the environment's resources that there is a unique fit between resource and response, and the organism becomes locked into that environment. And in that process, overspecialization becomes patent when the environment changes and reveals a vulnerable species or group of species.

Maintaining a constant internal environment is a homeostatic function common to all living things with the possible exception of viruses. Certainly all life whose organizational basis is cellular has to work to preserve a certain range of internal conditions which are essential to survival. The work is performed by cell organelles in unicellular organisms and by groups of cells (and their organelles) organized as tissues, organs, or organ systems in multicellular forms. The issue here is to inquire as to the role of homeostatic mechanisms in evolutionary changes, and, more specifically, to ask whether these mechanisms play any limited role in evolution.

As a first approximation, our answer would be that with very large size (redwood trees, large dinosaurs, and whales), homeostatic mechanisms in all probability do set limits to effective functioning. Circulatory, respiratory, endocrine, and nervous systems all may have limits above which they cannot properly meet the needs of very large organisms. (That upper limit may vary from system to system.) But within the size range encompassed by protozoans and lower metazoans, size is in all probability not a limiting factor. The limitations would be of two other sorts. Is there an organizational basis for achieving a given homeostatic function? Are there selection pressures favoring such functions? The latter is probably the key question for at the cellular level the occurrence of membranes, fibrils, microtubules, and other macromolecular aggregates, all supply a structural basis of quite extraordinary potential as evidenced by known organelle systems.

This structural plasticity endows a cell with the potential of meeting various selectionist pressures and, in contrast with exploitive traits where a close response to environmental resources occur, the limits for homeostatic functions are simply what best serves the needs of the unicellular or multicellular organism. As those needs change, the homeostatic responses can also change, using the mechanism of selection for advantageous mutations. This is not to say that external environmental conditions cannot change faster than the ability of an organism to cope with them in terms of maintaining a constantly viable internal environment. That can happen. But given gradual change, if the exploitive characteristics can continue to utilize the externally available resources, then the homeostatic functions can be expected to make needed adjustments too. The reverse is, however, not necessarily an expectation. For in this latter case we can refer to our previous examples of overspecialization and propose that there may be reduced availability of a food organism, and though the digestive system of the predator could handle new types of food, the exploitive mechanisms may not be able to provide them in a competitively effective way.

Homeostatic mechanisms don't bring species into competition with each other, though the more efficient functioning of such mechanisms may be selectively advantageous when competition arises over exploitation of a common pool of resources. Hence homeostatic parts of unicellular and smaller multicellular animals are not really limiting factors in their evolution.

The situation with reproductive functions is not unlike that just discussed. Reproductive potential is present in all organisms except for relatively rare cases of sterility due to disease, endocrine misfunction, abnormal development, or hybrid incompatibilities. And among patterns of reproduction are asexual and sexual ones, with the latter being more common. Sexual reproduction can vary between inbreeding and outbreeding. In none of these situations is the reproductive process a limiting aspect on the evolutionary potential of an organism. It can affect that potential through decreasing or increasing the genetic variability available to a population, but it cannot eliminate all variability, since random mutation is unavoidable; nor can it, with present mechanisms, force an organism to be more variable than is appropriate in selectionist terms. Selection apparently best operates in a relatively conservative context as judged from the relative rarity of mutations and the genetic stability of the mitoses that build multicellular systems and clones of unicellular ones.

The nature of the reproductive process utilized by a species reflects such factors as the variability of its niche and the length of time it has been exploiting its adaptive zone. Neither of these locks the species into a limited degree of specialization. (It seems plausible sex could reemerge in the naked amebas if it were selectively advantageous to explore new genetic combinations again.) And on the other hand, outbreeding or inbreeding or no breeding can be of advantage to organisms such that, given exploitive functions that achieve survival, the reproductive functions will then evolve so as to provide that stability that will be minimally disruptive. Neither extreme, nor possible intermediates, are limiting to the evolutionary development of a species. Reproductive functions are able to affect rates of evolution and will themselves probably evolve rapidly where a premium on a given type of reproduction is selectively advantageous, but they are not limiting factors in evolutionary change.

Cladogenetic vs. anagenetic tendencies. The foregoing discussion on limiting parameters in evolution comes to a clearer focus in the context of cladogenetic and anagenetic tendencies in evolution. First, the terms will need clarification, for they have been used with somewhat different meanings by different biologists, and then we can turn to their use in phylogenetic studies of evolutionary innovation. Cladogenesis and anagenesis were apparently first introduced by Rensch (1954, 1960). Cladogenesis referred to the origin of new organs and new "Bauplanen" (structural or body plans) and to the subsequent phyletic branching that follows from it. Anagenesis referred to long term evolutionary trends, earlier called orthogenetic trends and subsequently termed orthoselective (Simpson 1953a). Simpson gives passing reference to these terms, essentially equating cladogenesis to species diversification and anagenesis to species transformation. Huxley (1957) uses the term in essentially that sense too. We prefer to go back to a meaning closer to Rensch's original ones rather than follow more recent definitions.

The problem with the more recent views is that species diversification and transformation can occur from the same sort of evolutionary processes of mutation and selection, and to differentiate them on the basis of whether one or more than one species comes from the original stock seems excessively arbitrary (Bock 1965). The simplest course seems to be to accept the now widely accepted terms of diversification and transformation as applicable to events at the species level. For groups above the species level, where the larger evolutionary tendencies become apparent, there we will use cladogenetics and anagenesis as they will be now defined. *Cladogenesis* is the evolutionary diversification of a group of organisms around an exploitive theme, such as the flagellated chamber of sponges, the filopodial predation of radiolarians and acantharians, or the tentaculated polyp-medusa form of cnidarians. *Cladogenetic groups are those showing aposemic or neosemic change with regard to exploitive functions.* Variation is limited to one or a few exploitive themes. *Anagenesis* is evolutionary diversification of a group of organisms through development of new modes of homeostatic or reproductive functions, such as nuclear organization in the ciliates and cellular organization and its consequences in the turbellarians. *Anagenetic groups are those showing significant aposemic or neosemic change in connection with homeostatic or reproductive functions.* Variation is most importantly expressed as the emergence of new homeostatic and new reproductive modes. Anagenetic changes are only meaningful in the context of successful exploitation of the environment, and therefore a certain amount of cladogenetic proliferation will predictably accompany new anagenetic innovations. Hence, anagenetic groups will show some cladogenesis and cladogenetic groups, especially if they show extensive aposemy as in the case of the colonial siphonophores in the Cnidaria, will verge on neosemic development that might appear anagenetic. The difference between cladogenetic and anagenetic groups is therefore one of emphasis rather than clear difference.

The importance of these groups in phylogeny can be seen by examining the consequence of their origin and evolutionary development.

Cladogenetic groups, and here "groups" means any monophyletic set of species as determined through the phylogenetic procedures outlined earlier, will probably show initial tachytelic evolution as a result of rapid evolution to the more efficient forms of their basic exploitive trait. This tachytely, involving as it does essentially the same mode of living—the same niche practically—for the forms exploiting it, means they will be in competition with each other and probably replacing each other rapidly. Hence fossil evidence of this evolution will be scarce or nonexistent as occurs in tachytelic development (Simpson 1953a) and the final forms, depending on how far they have developed aposemically, may show little resemblance to their ancestor. It is, therefore, likely that the origins of cladogenetic groups will be hard to find, because the needed Remanian criterion of homology, which depends on serial relationships, may well be very difficult to apply. Indeed, since that criterion is the only relevant one for aposemes, the establishment of homologies may pass by default, and we will find on our hands a group not clearly related to any other. Additionally, within a group, a good deal of parallel evolution can be expected because of continued selection pressures to find all possible permutations of the basic successful mode of environmental exploitation.

Furthermore, cladogenetic groups, because of their evolutionary dependence on an exploitive theme, may well lock themselves into a close dependence on a given set of environmental resources. They thus become evolutionary dead ends and the termini of their phylogenetic developments.

Similarly now for anagenetic groups: their origin and probable position on the phylogenetic tree. To achieve anagenetic changes we need to consider traits that will not be bounded by adaptation to a given environment, and therefore we need to look past exploitive to homeostatic and reproductive traits. It is hard to predict exactly what the early fate of a new homeostatic or reproductive seme will be, for it depends on the efficacy of the exploitive characters with which it is associated. However, one guess might be that new internal capabilities might allow the organisms so endowed to explore new organizational modes, possibly even new Bauplanen, and wherever these are associated with efficient exploitive traits and an unoccupied or inefficiently occupied niche, there could be established a new species and its descendents. The consequence could be a wide-ranging variety of new homeostatic and reproductive traits and extensive development of old ones with some cladogenetic development associated with each one. And in that new niches would be involved here, the newer forms would not always replace old ones. The result would be a sequence of forms which preserved the evolutionary history of the group. Anagenetic groups can therefore be expected to be informative regarding their evolutionary developments.

There will be few anagenetic groups for the converse of the reasons that there are many cladogenetic ones. There are few major reproductive or homeostatic solutions to be used. Once found, however, the anagenetic groups should stand on the main lines of evolutionary development in phylogenetic dendrograms.

There now emerge four criteria that can be used to distinguish predominantly cladogenetic from predominantly anagenetic groups. The cladogenetic forms should be characterized by (1) variation on an exploitive theme, which will result in (2) a significant amount of parallel evolution, and which will also result in (3) difficulties in finding a plesiomorph and in (4) a large gap between suspected plesiomorphs and possible ancestral forms. By contrast, anagenetic forms can be expected to show (1) development of more than one theme, and these will be the less limiting ones related to homeostatic and reproductive functions, which in turn will mean (2) significant linear or orthoselective evolution rather than parallel evolution and also (3) convincing success in finding plesiomorphs and (4) plausible connections between plesiomorphs and possible ancestors.

This allows the following conclusion in terms of evolutionary innovation. The overarching task of the phylogenist is to find the anagenetic innovations and the anagenetic groups; they are the keys to evolutionary history.

Chapter 15
Phylogenetic Conclusions

THE plesiomorphs are the critical element in ascertaining phyletic relations between the higher taxa. They are critical because they are the most concrete representative available for defining the starting point of the evolutionary development of a given group of forms, and because they also represent the link to ancestral forms. Therefore, if the plesiomorph is not convincing, the needed reference point for describing evolutionary development within a group becomes elusive, and the perception of that development becomes blurred. And, looking to ancestral forms, the absence of a good plesiomorph renders the sought-for homologies with a plausible ancestor—a difficult pursuit under most, if not all, conditions—extremely problematical. The first issue, therefore, in pulling together final phylogenetic conclusions at the level of the highest taxa will be to review the limitations encountered in identifying the plesiomorphs of the major protozoan and metazoan groups studied here. Following that, tentative phyletic relations will be suggested, and from those we shall comment on the larger trends and innovative steps that seem apparent. That will set the stage for the final section, wherein conclusions from this study will be compared with the phylogenetic positions taken by other workers.

Origins of Major Groups

In Table 15-1 are given the major groups of organisms dealt with in this volume, broken down into subgroups where necessary to achieve what may be monophyletic groupings, and their proposed plesiomorphs and possible ancestral origins. Only in the cases of the Acantharia, Ciliophora, and Turbellaria is there confidence in the proposed plesiomorphic species. For all the rest, there are varying degrees of uncertainty. The situation can be reviewed as follows.

ZOOFLAGELLATES

In the zooflagellates, the choanoflagellates may be represented by a form from the genus *Codonosiga* as a plesiomorph. However, there was no information other than aposemic trends on which to choose a plesiomorph, and even this basis was tenuous, since it depended on a few semes whose directionality was only readable in terms of change in complexity. But whether

Table 15-1 Proposed plesiomorphs

Proposed Plesiomorph	Representative of Possible Ancestral Form
Zooflagelates	
Choanoflagellates: *Codonosiga* sp.	Chrysomonad: *Cyrtophora pedicellata*
Kinetoplastids:	
Bodonids: *Bodo saltans*	Euglenoid: *Peranema*
Trypanosomatines: *Leptomonas* sp.	
Trichomonads: *Monocercomonas* sp.	Apochlorotic, flagellated algas; however, loss of many intermediate forms leading to present-day array of complex, symbiotic zooflagellates renders identification of ancestral form(s) highly problematical.
Other symbiotic zooflagellates: five or more plesiomorphs	
Rhizoflagellates	
Foraminiferans: member of the Lagynidae	Probably polyphyletic. Ancestors among the various flagellated apochlorotic algas.
Various naked and shelled amebas	
Xenophyophorids	
Slime molds	
Heliozoans (Adansonian assemblage)	
Actinophryida: *Vampyrellidium* sp. ?	A rhizoflagellate
Desmothoracida: ?	A testate rhizoflagellate
Centrohelida: *Oxnerella* sp. ?	Helioflagellate: *Dimorpha mutans*
Helioflagellida: *Dimorpha mutans* ?	Testate ameba
Proteomyxida: ?	Ameboflagellate ?
Labyrinthulida: ?	Ameboflagellate ?
Radiolarians	
Actissa	
Thalassolampe	Marine dinoflagellate ?
Procyttarium	
Acantharians	
Acanthochiasma rubescens	Chrysomonad: *Chrysamoeba radians* ?
Ciliates	
Stephanopogon mesnili	Zooflagellate; multinucleate, hypermastigate form; but none is free-living
Sponges	
Demosponges: *Oscarella* sp. or *Halisarca* sp.	?
Calcisponges: *Leucosolenia* sp.	Choanoflagellate: *Proterospongia*
Hexactinellids: ?	
Cnidarians	
Otohydra vagans	?
Turbellarians	
Haploposthia rubra	Ciliate: *Stephanopogon mesnili*

this meant more complex was more evolved or the reverse could not be clearly determined on present evidence. *Codonosiga* represents one of the simpler collared flagellates.

The bodonids and the trypanosomatids pose a special problem. Though their common possession of kinetoplasts, striated pellicular structure, paraflagellar rods, and cytostomal tubes clearly imply homologous relationships, there are other important differences. The chief one is that the bodonids are free-living and the trypanosomatids are symbiotic. *Bodo saltans* could be the plesiomorph of its group and a leptomonad form the starting point of trypanosomatid evolution. Where is the common origin of these two forms? There is no good answer beyond the vague suggestion that symbiotic forms arose from free-living ones. The pellicular fiber system and paraflagellar rod might join the bodonids to the euglenoid forms, but this is not convincing. If later work does confirm this guess, it may well separate the bodonids and trypanosomatids from the rest of the zooflagellates whose ancestry could lie with another group, possibly the apochlorotic chrysomonads.

As discussed in Chapter 5, there are many separate groups of zooflagellates possessing two or more flagella apiece, and they are symbiotic. The discrete differences between these groups, largely in relation to specializations of the kinetide (mastigont), led us to suggest that there may have been radiative evolution from some free-living ancestral forms. These latter were believed to have shown considerable evolutionary diversity but then were wiped out, perhaps through unsuccessful competition with ciliates, with the result that only those flagellates escaping from competition by exploitation of special niches, namely symbiotic relationships, survived as a variety of relict groups. If these speculations are on the right track, there is no plesiomorph available for all these various forms. Each group will have its own plesiomorph, as with *Monocercomonas* for the trichomonad forms. Hence there is an enormous gulf between these various plesiomorphs and the now very distant free-living flagellated ancestral form, which could have been some colorless, unicellular, flagellated alga.

AMEBOID FORMS

We bypassed such problematical forms as the three groups of slime molds (which may or may not have a common ancestor) and the xenophyophorids. Looking at the various naked and shelled ameboid forms, except for the foraminiferans, we could come to no conclusions regarding the sure identification of monophyletic groups and therefore could say little that seemed sound regarding plesiomorphs and their origins beyond suggesting the obvious, i.e., that apochlorotic algae with ameboid tendencies were plausible ancestral forms. The chrysomonads have such forms.

As for the foraminiferans, their phylogeny is largely based on test morphology, and so the plesiomorph would lie at one end of the aposemic development shown in that structure. Glaessner (1947, 1963) favors a member of the superfamily Astrorhizidae but points out that of the four types of fossil tests found in this group, "It is hardly possible to decide which of these four is the most primitive." (1947, p. 91) Furthermore, he says nothing about the ancestor of this group. Le Calvez (1953), looking at live forms in addition to fossils, suggests the family Lagynidae as containing the simplest foraminiferans and, by extension, as possibly containing the plesiomorph of the group. These forms, showing minimal test formation, precede the Astrorhizidae in Le Calvez's classification and are derivable, he thinks, from testate amebas such as the Allogromidae. Le Calvez does not suggest a specific species as representative of what we would call a plesiomorph, and we will not attempt to do so here.

HELIOZOANS

The sun-animalcules are as difficult to deal with phylogenetically as the ameboid rhizoflagellates. For the labyrinthulids we have no suggestions for a plesiomorph nor is there a plausible ancestral group as yet. The Centrohelida do not provide a convincing plesiomorph, but many of their major features, notably the central granule, can be found in the helioflagellate form, *Dimorpha mutans*. And *D. mutans* itself might be derived from some ameboflagellate arising from a colorless alga. The Actinophryida could arise from an ameboid form—a rhizoflagellate—as could the Desmothoracida, providing in the latter case that a test was present. The testate amebas would meet that prerequisite. We have no suggestions for a plesiomorph or specific ancestral form for the proteomyxids.

RADIOLARIANS

Three different genera have been suggested as containing a radiolarian plesiomorph, i.e., *Actissa*, *Thalassolampe*, and *Procyttarium*. Earlier discussion reduced three possible candidates as ancestral forms to two, i.e., a dinoflagellate or a colorless chrysomonad. Both proposals face serious difficulties. The earlier hopes of Zimmerman (1930) and Biecheler (1935) that forms like *Gymnaster* and *Plectodinium* might show their perinuclear capsules to be homologous with the capsule of radiolarians seems not to stand up (Hollande, Cachon, and Cachon 1970). The later workers, propose, rather, that the cell wall of those dinoflagellates possessing thecae can be homologized to radiolarian capsules. This is doubtful if only on compositional grounds: thecae are usually made of cellulose, and capsules are thought to be glucoprotein.

The origin of radiolarians from filopodial chrysomonads such as *Leukochrysis* is difficult to accept, because only the filopodia make a plausible connection to the radiolarians. There is nothing in these bottom-crawling chrysomonads which anticipates capsule formation or a floating, planktonic mode of life.

Our best guess would favor the suppositions of the French workers in feeling that marine dinoflagellates might have evolved into radiolarians. The large gap between the radiolarians and possible ancestral forms emphasizes that the radiolarians evolved in the pre-Cambrian seas and have been ex-

ACANTHARIANS

Acanthochiasma rubescens is a quite convincing plesiomorph for this group. This is because of a morphocline analysis that conjoins several aposemic changes in this species and because of ontogenetic evidence.

But devolving from *A. rubescens* to a plausible ancestor poses sharp difficulties. Only general possibilities emerge, such as ameboid forms and colorless algae. If, as we have suggested, the early ontogeny of the acantharians contains a bottom-dwelling naked ameba, then the ameboid rhizoflagellates could be ancestral. The other possibility is to look for phytoflagellates with filopodial tendencies, such as *Chrysamoeba radians* (a chrysomonad). This form is not apochlorotic, though its single plastid suggests that apochlorotic forms might in fact arise spontaneously through lack of division of that organelle, for a variety of possible reasons, when the cell divides.

The acantharians present much the same evolutionary picture as the radiolorians except for their apparently much more recent appearance (there are no convincing fossil forms). If the recency of their origin is real, it might explain why the plesiomorph is so convincing, in the view that it has not been replaced yet by competition with subsequent forms but the latter have simply evolved further. This suggestion immediately raises the question as to why we might not also expect some of the ancestors of acanthochiasmid forms to also still be present. The answer could be that tachytelic evolution occurred up to the point of the plesiomorphic form, as invasion of a new adaptive zone proceeded rapidly through loss of traits no longer of selective value and through fuller development of those adaptations that were insuring survival. This could result in a rapid succession of forms until a definitive acantharian appeared, whereupon evolutionary rates became horotelic and proceeded further at a more moderate pace through dispersion (to avoid competition) and slight differences arising in response to selection for geographically determined differences.

In other words, the acantharians are well on their way to becoming another predominantly cladogenetic group like the radiolarians, but being younger there is still much of their history still intact in their plesiomorph and subsequent forms.

CILIATES

The ciliate plesiomorph *Stephanopogon mesnili* is a most convincing one. Several morphoclines come to a focal point at this species as well as certain ontogenetic evidence. Furthermore, it shows its own special adaptations, which help us understand its survival while other ciliates continue to evolve further in a variety of other directions.

Ciliates are most closely allied, through *S. mesnili*, to the zooflagellates. If our speculation regarding the loss of many free-living zooflagellate forms is correct, the difficulty in finding similarities between *S. mesnili* and known zooflagellates is understandable. The occurrence of kinetidal organization and multiple kinetides, multinuclearity, sexuality, and diploidy in the zooflagellates speak to the possibility of these forms as being antecedent to the ciliates. Certainly no other forms are more plausible. This postulated relationship is the first one encountered in this study where one major group of protozoans is derived from another protozoan group. Except for similar possibilities among minor groups within the heliozoans and ameboid forms, all other major protozoan groupings seem to have arisen from apochlorotic algae.

PORIFERA

The demosponges offer two equally likely choices for plesiomorph in the genera *Oscarella* and *Halisarca*. To this is added the calcisponge plesiomorph, *Leucosolenia*. (No plesiomorph was proposed for the hyalosponges.) These three candidates represent termini of morphoclines in sponge organization.

All of them are thought to be related, albeit distantly, to collared flagellates of the type represented by *Proterospongia*. That enigmatic form deserves much more study, including fine structure work on its choanocytes, fuller description of its development, and continuing search for a marine representative. The rediscovered *Proterospongia* are freshwater dwellers (Ertl 1968), and this is an unlikely place for the ancestor of the sponges to be located.

CNIDARIANS

After considerable difficulty and with real diffidence, *Otohydra vagans*, an actinulid hydrozoan, was proposed as the cnidarian plesiomorph. This assumes that other phylogenies that favor some group other than the actinulids, or even hydrozoans, as containing a representative of the cnidarian stem form, are here set aside. But as earlier discussions made clear (Chapter 12), the position advocated here is not convincingly better than those of other workers.

The origin of the cnidaria, regardless of what plesiomorph one chooses, seems at present to be totally obscure. Those who are willing to venture guesses (Jägersten 1955, 1959; Tuzet 1963; and others), invariably invoke a series of hypothetical intermediates to go from protozoan (Jägersten) or poriferan (Tuzet) ancestors to the Cnidaria. Such speculations, as has been emphasized before (Chapter 3), are untestable by normal phylogenetic procedures, and are also usually not developed with selectionist arguments kept in mind. They are morphological musings and meaningless (untestable) as scientific hypothesis.

TURBELLARIANS

Though *Haploposthia rubra* is a convincing plesiomorph for the free-living flatworms, it may well be that an even better one will be found. The reason is that the Acoela are still not a thoroughly studied group. (For example, in one summer's work at the Bermuda Biological Station this author turned up six species of acoels that were apparently undescribed, and this was done as an effort incidental to the main research program. A concerted collecting effort could probably have tripled the number of new forms.) Nonetheless, *H. rubra* does stand at the end of several morphoclines and therefore represents a convincing reference point for studying turbellarian evolution.

An acoel such as *H. rubra* is derivable from a ciliated form that could also have been ancestral to *Stephanopogon mesnili*. Though homologies are hard to come by, general similarities between the turbellarian and ciliate plesiomorphs do exist. Also, one can make plausible arguments to show why the differences existing between the two species could be bridged by selectively advantageous changes culminating in forms like the two different species and from them going on to the large groups of organisms represented by each of these species as plesiomorphs. For the ciliates, the basic organizational pattern was multinuclearity made up of a highly polyploid nucleus along with one or more diploid ones and enclosed in a highly differentiated cortex. For the turbellarians, multinuclearity is followed by cellular boundaries around diploid nuclei, thus achieving multicellularity, with conspicuous epidermally and subepidermally-located differentiations. Both patterns have advantages and have obviously been evolutionarily successful.

Phylogenetic Conclusions

The possible phylogenetic relationships that have emerged from the present study, in terms of the larger groupings of organisms, are summarized in Fig. 15–1. Proceeding from left to right across the figure, we see first that there are four groups of algae that may possibly have been ancestral to one or more groups of protozoans. However, it is also possible that all the protozoa arose from apochlorotic chrysomonads, but, if so, there were surely multiple independent origins from this one ancestral group. (We know that there are various colorless chrysomonads surviving today (Table 4–1).) The grouping labelled "Unidentified unicellular algal forms" refers to possible ancestors of certain protozoans that are so poorly defined as to make it quite presumptuous to put them in any given algal group. It might be that these obscure ancestors were apochlorotic Chlorophyta or even Cryptophyta. The latter is quite un-

PHYLOGENETIC CONCLUSIONS

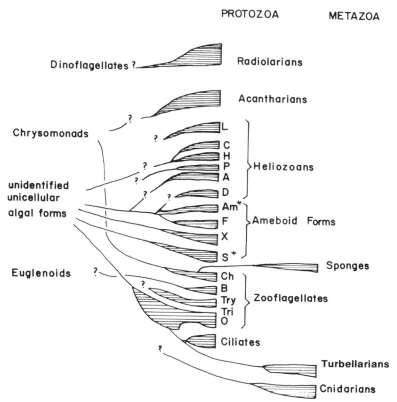

Fig. 15-1 Dendrogram of major protozoan and certain metazoan groups. A—Actinophryida, Am—naked and shelled amebas, B—Bodonida, C—Centrohelida, Ch—Choanoflagellata, D—Desmothoraca, F—Foraminiferida, H—Helioflagellidae, L—Labyrinthulida, O—other zooflagellates, P—Proteomyxida, S—slime molds, Tri—Trichomonadida, Try—Trypanosomatida, X—Xenophyophorida,*—possibly polyphyletic.

likely though, for no protozoans show the distinctive vestibular pocket lined with trichocysts that characterizes cryptomonads (Hollande 1952d).

To help resolve the many problems of algal ancestry of protozoans, the techniques of molecular phylogeny might be fruitfully brought to bear in the manner described in the preceding chapter. For example, it would be most interesting to compare the amino acid sequence of a cytochrome from *Actissa*, or other possible radiolarian plesiomorph, with comparable sequences for certain dinoflagellates or chrysomonads thought possibly to be ancestral to the radiolarians. And similarly for other problem areas.

In any case there are sound reasons for considering the protozoans to be polyphyletic in origin, arising either from various groups of chrysomonads or from them and certain other algal phyla (divisions). Among these various origins, the ameboid forms and heliozoans, in particular, seem especially to have had multiple origins. And except for one or two possible cases where one ameboid group gave rise to another—the desmothoracans and foraminiferans arising from shelled amebas—these groups have not evolved beyond their own specialized forms. To be sure, three groups with tests, namely, the radiolarians, acantharians, and foraminiferans, have evolved a significant diversity within their own organizational-functional (epigenotypic) pattern, but there is no real evidence that they have given rise to significant further evolution beyond their own group. As an inductive generalization it appears quite certain that the ameboid forms, speaking broadly now to include forms with and without tests and with all kinds of pseudopodia, are evolutionary dead ends. Their diversification seems to have been predominantly cladogenetic. One other possibility of long standing (Brucke 1861; Mereschkowsky 1910; Wallin 1923; reviewed by Minchin 1915; Wilson 1928) and redeveloped recently by Margulis (1970), for accounting for the polyphyletic origin of the ameboid protozoans is to propose a multiplicity of symbiotic associations of prokaryotic cells. This approach seems an unnecessary complication, since there appear to be many plausible ancestries for these protozoans from the colorless algae, as we and others before us, notably Pringsheim (1963) (who is nowhere mentioned by Margulis), have stressed many times. Furthermore, we also prefer to exhaust the possibilities of phylogenetic analysis through analysis of homologies and descent through continuity of cell lineages before invoking the hypothesis of symbiotic associations, a hypothesis which is still open to serious criticism (Raff and Mahler 1972).

The flagellated protozoans seem to have had a different long-range evolutionary outcome than the ameboid forms. From them there emerged the ciliates and probably also the symbiotic Sporozoa (an important group we have regretfully omitted from our analysis—see Chapter 4). The zooflagellates also show a polyphyletic origin like the ameboid forms, and, indeed, may well have shared a common ancestry with certain of them. But whereas the bodonids seem to have had a limited evolutionary development, the trypanosomatids have had considerable success as symbionts, and from the choanoflagellates and somewhere in the ancestry of the remaining zooflagellates, two further major groups have emerged—the sponges and the ciliates, respectively.

Though the sponges seem to many, but not all, workers to be an evolutionary dead end, the ciliate line seems to have continued on in two paths—one culminating in the ciliates proper and the other going on to give rise to the turbellarians and other metazoans (exclusive of the Cnidaria and probably, also, the Ctenophora and Mesozoa).

INNOVATIONS

It will be useful to continue here with the categories that emerged in the preceding discussion and deal, successively, with the pseudopodial forms (broadly defined to include all ameboid forms), the zooflagellates and ciliates, which will be grouped as kinetidal forms, and then to turn to the metazoans, where the sponges and cnidarians can be treated together and with final attention reserved for the turbellarians.

Pseudopodial forms. These are all predominantly cladogenetic organisms. They show the characteristics listed earlier (Chapter 14) that allow for identification of cladogenetic tendencies, i.e., diversification centering around an exploitive trait or theme, evidence of significant amounts of parellel evolution, plesiomorphs are difficult to find, and a significant gap exists between plausible plesiomorphs and their ancestors.

For the pseudopodial forms, the pseudopod in all its various manifestations is the exploitive trait that supplies the common thematic basis for evolutionary variety in this group. The occurrence of pseudopods determines the polarity, symmetry, and general organizational pattern of these cells. Within this pattern, nuclei and test formation are the other features that show significant development. Correlated with increase in cell size is the appearance of either multiple nuclei or a large polyploid or polyenergid nucleus. This seems a natural development, wherein increased amounts of genetic material would be needed to meet the needs of increased cytoplasmic volume. The precise basis of this karyoplasmic ratio is still unknown despite decades of intermittent research on the subject.

The presence of tests is also correlated, but not invariably, with larger size. Another correlation is between tests and diversification, as in the Acantharia, Radiolaria, and especially the Foraminiferida. It would seem that when such a structural feature appears, this is capitalized upon to evolve more complex internal organizations, such as the flotation apparatus, special layers of the cell body, and internal compartmentalizations (the foraminiferan locula). In other words, when something approaching a new organizational plan appears, these cells use it to the extent possible within the pseudopodial mode of life to which they are committed by prior adaptation. This is the limit of their anagenetic development except for one other possibility.

These three groups, especially the foraminiferans, all show variously complex life cycles. These may be responses to maintenance of sexuality based on relatively simple gametes on the one hand, and achievement of complex, vegetative cells on the other. That is, only by life cycles can the complex development from zygote to large, differentiated adult cell be achieved.

All of this gives one the feeling that a breakthrough to multicellularity might have readily occurred here, especially in those forms that kept their multiple nuclei separate and did not compound them into a single polyenergid structure. And perhaps there were evolutionary experiments in the direction of multicellularity, but they never succeeded for as yet obscure reasons. Coordinated pseudopodial activity is not amenable to cellular subdivision? Spherical symmetry is not conducive to active predation, which is the preeminent adaptive zone open to multicellular consumers? There is no development of nervous coordination to anticipate integrated multicellularity? Other speculations are also possible, but none is answerable at present.

Overall, then, the pseudopodial forms have been successful as bottom-dwelling or floating predators, relying on their pseudopodia for movement in many cases and for food capture in all cases. Within such an adaptive zone they have achieved significant increase in size, cellular differentiation, some nuclear complexity in the direction of multinuclear and polyenergid systems, and complex life cycles. Within this compass, though, they seem to have exhausted, or at least found limits to, their evolutionary potential; for though they have thousands and thousands of species which utilize this mode of life — their cladogenetic development is impressive — there seems to be no evidence of evolution into major new modes of structure and function. As concluded earlier, this is not unexpected in predominantly cladogenetic forms.

Kinetidal forms. The kinetosome has associated with it impressive morphogenetic powers, as demonstrated in the zooflagellates and ciliates. These organellar differentiations emerge as mastigonts or kinetides and provide the cells so equipped with an extraordinary potential for evolutionary development in terms of various internal structures (zooflagellates especially) and cortical differentiations (ciliates especially).

Within the zooflagellates, the major trend has been towards compounding and complexing the mastigont or kinetide. That is, kinetides are treated as structural units that are polymerized, but they are also made more complex by addition of various units, such as costas, crestas, axostyles, and so forth, and then these units can be polymerized or compounded. These two themes of evolution among these forms was earlier recognized as an important one (Dogiel 1927). It also occurs with the nuclei to a lesser degree, as seen in their compounding in the multinuclear forms. Increase in size also occurs here, as do complex life cycles in the forms still showing sexuality.

Within the ciliates, starting with an already compounded kinetidal surface, evolutionary pressures proceeded to vary the regularity of kinetidal rows and to join kinetides into special membranelles, as in adoral membranelles, or cirral tufts. This compounding of kinetides produced a new level of organellar structure. Associated with the cortex is the permanent feeding structure or cytostome. The ciliates are the only major group of protozoans to have a permanent ingestatory organelle. Only the symbiotic trypanosomatines are a possible exception to this claim for uniqueness in the ciliates. In any case, the complexity of the ciliate cytostome vastly surpasses its possible analog in the flagellates. The ciliate cytostome is a truly anagenetic advance over the condition found in other protozoans. For the first time a free-living protozoan predator is equipped with a structure that is to be associated with consumers from this point on, i.e., a mouth. Somewhere in the transition from presumed zooflagellate ancestor to ciliate eotype, the ameboid uptake of food, after becoming fixed in that part of the body not having flagella, then evolved a permanent structure and the cytostome became a reality. And apparently being of selective advantage, it persisted with locomotory needs being served by the cilia. Eventually, however, some of the cilia also became aids in feeding. Perhaps the propensity of the cortex to differentiate various kinetidal structures was a preadaption to also differentiating the cytostome, since fiber systems are an important part of the feeding organelle.

In addition to the remarkable fate of the cortical kinetides and the cytostome, the ciliates also show nuclear diversification on the theme of diploid micronuclei and polyenergid macronuclei. They also show size increases and complex life cycles.

In both the kinetidal and ameboid groups there has been increase in size, appearance of compound nuclear structures, and elaborate life cycles. The differences between the groups is that ameboid forms give up on kinetosomal based differentiations in favor of pseudopodia, whereas the zooflagellates and ciliates exploited the kinetosome.

Is the kinetosome the key to emergence of multicellularity? No, for multicellularity arose among the brown, red, and green algae, with the latter giving rise to the invasion of land by plants. And fungi are also multicellular. Among these forms the kinetosome and its derivatives have not played a conspicuous evolutionary role.

The previous question needs rephrasing. Is the kinetosome a preadaptation to the emergence of multicellular animals? Yes, but in two different ways. The kinetosome and its flagella are absolute prerequisites for the flagellated chambers of sponges. The emergence of the turbellarians is a different story, and we shall approach it by moving our speculations from the zooflagellates to the ciliates. In both groups the kinetide is clearly the basis of complex cytoplasmic differentiations.

In the ciliates the compounding and complexing of kinetides is associated with the cytostome. The functional coordination of these placed a premium on coordinatory devices, and further developments could only come from some basis for even more differentiation that would not preclude continued integrated activity of all parts. Internal cellularization would seem to have been the solution.

Internal cellularization is quite different from the coming together of discrete cells to form a coordinated whole (Hanson 1958). This latter mechanism was the almost certain mode or origin of all multicellular systems except for the Turbellaria and their successors, i.e., the rest of the Bilateria. In these other forms, arising from integration of a colony, the individual cells had to suppress their individuality and evolve cellular specializations and interconnections appropriate to transforming a colony into a superorganism with its own individuality. On the other hand, emergence of multicellularity from a multinucleated and highly differentiated unicell meant the coordination and connections of parts could persist but would be amplified by the further differentiation possible now from cellular units. This solution to the emergence of multicellularity keeps the integrity of the organism intact and keeps it moving evolutionarily in the direction it was already going.

The ciliates slipped off this evolutionary path by opting for a macronucleus. This precluded internal cellularization, but there was retained the advantages of unicellularity with rapid uniparental reproduction interspersed with occasional bouts of sexuality. It was clearly a successful option allowing for almost anagenetic development of the potentialities of the kinetidal complex and the associated cytostome. The real anagenetic step for ciliates, however, was emergence of the cytostome relative to the zooflagellate condition, but in that this was an exploitive advance, it would seem that eventually the ciliates were doomed to run out their potential unless something new was done. As we have said, that new thing was internal cellularization, a totally new way of organizing the internal—homeostatic and reproductive—economy of this line of evolution. Before examining the consequences of that within the Turbellaria, we need to examine the Porifera and the Cnidaria.

Sponges and cnidarians. These are two predominantly cladogenetic groups. The flagellated chamber and the tentaculated polyp and medusa are their bases, respectively, for exploiting their environments. There is considerable parallel evolution within each group. In both, plesiomorphs are hard to find, and there is a large gap between those plesiomorphs and possible ancestors. In fact, for the Cnidaria we could find no plausible ancestor.

The major trend in the sponges is, of course, the compounding of flagellated chambers as one proceeds from ascon to sycon to leucon. Associated with that are developments of various cell types between the chambers and of the skeletal parts. The embryology of these forms is also an important development, but it is hard to define trends here; rather, there are several variously complex ways of passing from the zygote to the adult.

In the cnidarians, the three different groups of the hydrozoans, scyphozoans, and anthozoans show emphasis on different tendencies, but these are not readily resolved into trends, as was obvious when one examined possible phylogenetic relations among them. The only development which aimed at superseding the polyp-medusa body plan was that of colony formation among the siphonophores, where individual polyps become contributing units within the colonial superorganism.

The cnidarians and sponges are in all probability evolutionary dead ends; they have an effective historical development in their own terms but will most probably never give rise to forms capable of taking the flagellated chamber and polyp-medusa themes into new adaptive zones and there evolving major new groups of organisms.

Turbellarians. These flatworms are an anagenetic group. Within their taxonomic confines they show the emergence of new organ systems such as an intestine and an excretory or protonephridial system, and organ systems already present in their plesiomorph show such considerable subsequent aposemic development (reflected in large phyletic distances—large R values) that those traits almost seem anagenetic too. We attribute this to the emergence of multicellularity and the exploitation of cytodifferentiation through embryogenesis. This was selectively advantageous because the ciliated ancestor of the turbellarians was very plausibly a multinucleated ciliate whose kinetidal equipment aided its functioning as a free-living predator and contributed to its evolving complexity. In that sense, the kinetosome-kinetide was a preadaptation to turbellarian functional and structural complexity. The transitional organisms gave up the rapid unicellular reproduction of their ciliate forebears for the genetic variability of sexual reproduction, which was useful in pursuing the epigenetic potentials of embryogenesis. All of these—genetic variability, cellular construction, developmental plasticity—together endowed the flatworms with an extraordinary evolutionary potential, which in hindsight we see has been realized in the diversity of eumetazoan forms.

In addition to the symbiotic trematodes and flukes it is probable, as others argue (e.g., Hyman 1940, 1951), that the nemertines and other worms and the rest of bilaterally symmetrical Metazoa arose from the Turbellaria. The turbellarians do show all the basic features of the Bilateria, namely, diploidy, sexual reproduction, embryogenesis, predaceous mode of life with a permanent mouth for ingesting food, and cephalization. And as we have commented earlier, the ciliates are the only protozoans that anticipate many of these traits, i.e., diploidy, sexuality by cross fertilization, permanent cytostome, and some tendency—through polarization of the cell-body—towards cephalization. The anagenetic trends of the kinetidal forms does continue into the turbellarians and on beyond them. This is the main line of early animal evolution (Fig. 15–1).

Other Phylogenetic Approaches

The foregoing conclusions differ in varying degrees from those of other workers. In the last two or three decades there has been renewed interest in phylogenetic problems, and the perennial questions of origins of major groups or of major evolutionary innovations have received their full share of attention. Corliss (1972) has carefully summarized the status of many of these. Among them are the origins of protozoans and multicellular animals. This section will critically review the suggestions of various workers in an attempt to sort out the alternatives that face the contemporary investigator in this area. Following our own procedural bias, we will first comment on the methodologies used by other workers and then turn to the products of their inquiries, i.e., their respective phylogenetic proposals.

METHODOLOGIES

Since in Chapter 3 there were given extensive critiques of those methodologies labelled morphological phylogenetics, the New Systematics, numerical taxonomy, and molecular phylogeny, we need not repeat those comments here. What is more to the point is the nature of the comparative process used by others, i.e., is there appeal to similarity as evidence of descent from a common ancestor? How is this similarity established? By reference to homology? What is the treatment of possible convergence and parallel evolution? And also important will be the nature of the material used for comparison, i.e., are hypothetical forms used? How much of the biology of the groups being compared is brought under scrutiny? What are the units of comparison? Single organisms? Species? Larger taxa? And so on. Another approach to these questions is to ask whether the three basic guidelines or rules used in our analysis (I. Species are the unit of comparison. II. Compare the whole biology of the forms in question. III. All conclusions must be consistent with selectionist principles.) have also been observed by other investigators.

Those who have written on the broader issues of the origin of multicellular animals (Hyman 1940; Baker 1948; Hardy 1953; Nursall 1959, 1962; Greenberg 1959; Kerkut 1960; Reutterer 1969) rarely spend time on critical reviews of methodologies and often turn directly to an enumeration of the mechanisms whereby multicellularity might occur and then proceed to ponder those possibilities. For example Reutterer cites four such possibilities:

1. Origin through association of free-swimming flagellates into a colony (the cell-state theory).
2. Origin through suppression after cell division, therefore by means of corm formation (the corm theory).
3. Origin through cellularization of a multinucleate cytoplasm of a free-swimming (or crawling) protozoan (the plasmodial theory).
4. Origin through fusion of bottom crawling ameboid protozoans to form a pseudoplasmodium (the ameboid theory).

With such an approach the problems of phylogeny become a search for forms that will illustrate one or more of the modes in question. This completely distorts the reality of evolution as a process that proceeds by natural selection through speciation to the formation of diverse groupings of organisms. In its place is a game of hide-and-seek, wherein the investigator seeks out those forms which are hiding from the preconceived evolutionary roles into which he chooses to insert them. And when such insertions are made, there is little or no possibility of evaluating the conclusions in the light of such fundamental biological principles as hereditary descent, adaptively significant changes, species transformation and diversification, and the like, because these basic foundational principles of evolution have been totally disregarded in most cases. On methodological grounds, much of the phylogenetic speculation of recent years can be dismissed as theoretically unjustifiable.

One commendable outcome of the foregoing procedure must be admitted. It frees the investigator from preconceptions allowing him or her the always-healthy option of probing for new points of view. As a consequence Nursall (1962) was willing to propose a very polyphyletic origin of metazoan groups—an intriguing antidote to the more conventional schemes of very linear evolution favored by Hadzi (1953, 1963) and Marcus (1958). Greenberg (1959) also has been willing to look for new origins of multicellularity in animals and has broken out of the mold of neo-Haeckelian thinking.

OTHER PHYLOGENETIC CONCLUSIONS

Other studies which have proceeded in more concrete terms often employ, though often only vaguely, comparisons that use homologous relations (Hyman 1940; de Beer 1954; Remane 1954; Jägersten 1955, 1959; Marcus 1958; Hadzi 1953, 1963; Hanson 1958; Dillon 1962, 1963; Steinböck 1963a; Corliss 1968). Among these workers some have offered their conclusions in the most tentative terms. For example, Hyman (1940, p. 252) says, "We may consider it plausible to suppose that the Metazoa arose from an axiate hollow spherical flagellated colony in which there occurred first a differentiation into somatic and reproductive cells and then a differentiation into locomotor-perceptive and nutritive types, through the wandering of the latter into the interior." There is not much to be done with such a vague supposition other than to marshal arguments against its plausibility by counter proposals, as has been done in the preceding chapters. And the same applies to others who also follow the Neo-Haeckelian view represented by Hyman, e.g., Remane (1954) and de Beer (1954).

Dillon (1962, 1963) represents a special case, for he has used one set of data to the exclusion of most others. He has taken cytological findings, largely relating to modes of cell division, and has developed phylogenetic conclusions on that basis. The narrowness of his data base has been criticized elsewhere (Hanson 1964); the essential fault lies in the exclusion of so much data of other sorts, such as gross cell morphology, biochemical data, life cycles, and so forth. Specifically, he ignores the guideline that demands that full consideration be given to our total biological knowledge of the organisms being compared.

Other workers such as Jägersten, Marcus, Hadzi, and Steinböck spend varying amounts of attention on particular groups. Marcus (1958) touches on the origin of the Metazoa in the context of a very broad reconsideration of phylogenetic problems, and Hadzi (1963) and Steinböck (1963a) do somewhat the same. Marcus and Steinböck are specialists on the flatworms, and Hadzi has specialized on the Cnidaria. For those who follow the history of debates in this area, there is the knowledge that Hadzi and Steinböck were promoting the idea of origin of acoels from ciliates prior to Hanson's (1958) first review of the topic from the side of protozoa. The difference between these three workers is that Hadzi and Steinböck have simply derived the Acoela from a ciliate ancestry, without explicit reference to a specific group or species of ciliate, and Hanson argues for the origin of the acoels and ciliates from a common ancestral form (the ciliate eotype, or protociliate as it has also been referred to (Hanson 1958, 1963a; Jägersten 1959).

Marcus takes a different view (Fig. 15-2), which he readily admits derives from his "zoological descent from Karl Heider." Marcus (1958, p. 24) adds with warm sincerity, "I cannot equal Heider in wisdom, but try to remember how serenely he pondered the matter of phylogeny." Also Marcus speaks respectfully of Remane's "rigor in establishing homologies", and presumably that bespeaks homologies as being the underlying theoretical principle for Marcus' proclivities in following Heider. The placement of the Platyhelminthes (Fig. 15-2) is justified by Marcus in describing them ". . . as an offshoot of the lower Spiralia. Most of their organs are simplified [i.e., reduced], not primitive, because an ancestral bilaterian must have had the fundamental features of the Archicoelomata, viz., three coeloms, mouth, anus, vessels, and perhaps tentacles." (p. 31–32). This argument for reduction in the Turbellaria we have already answered in response to Ax's claims along identical lines (Chapter 13). And the insistence on an archicoelomate ancestry is part of Marcus' deference to the enterocoel theory, which has been demolished by Hartman (1963). Therefore, the placement of the flatworms well up in the phylogenetic tree cannot be justified.

Marcus' treatment of phylogeny within the protozoans is very sketchy and reflects more the taxonomic treatment of these forms than any thorough phylogenetic analysis. His reference to French workers and the *Traité de Zoologie* implies an endorsement of evolutionary relations on which the present volume, too, has been heavily dependent. However, the complexities of the conclusions contained in the French masterwork are not at all evident in Marcus' summarizing diagram (Fig. 15-2). Little further can be said, then, on those problems.

Jägersten (1955) epitomizes the use of hypothetical forms in developing phylogenetic conclusions; he postulates fifteen such forms to go from an ancestral, pelagic *Blastaea* (a hollow colonial sphere of flagellated cells) to the Cnidaria, ancestral protostomes and deuterostomes, Porifera, and Planuloidea (Mesozoa) (Fig. 15-3). In this scheme Jägersten starts from suppositions as to

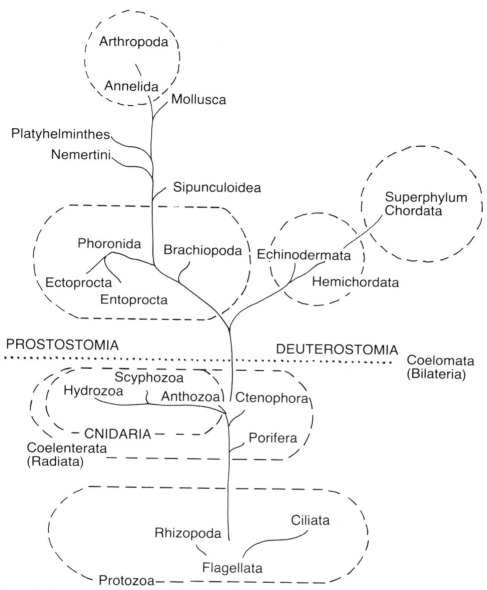

Fig. 15-2 Marcus' phylogenetic proposals (Redrawn, by permission, from Marcus (1958).)

the nature of a primitive invertebrate and includes embryological (i.e., Neo-Haeckelian) conclusions deriving from those groups. This approach totally disregards species as units of evolution, much if not most data other than embryological, and it ignores natural selection beyond considering how the hypothetical forms might move and take up food. In Chapter 13 we dealt in some detail with Jägersten's assumptions regarding primitive forms and the

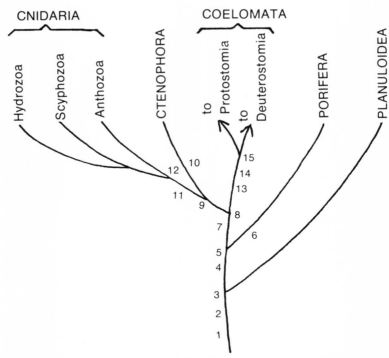

Fig. 15-3 Jägersten's proposal regarding metazoan origins following his Bilaterogasterea concept. (Redrawn, by permission, from Jägersten (1955).)

related idea of external fertilization being prior to internal fertilization. We saw that those were quite arbitrary inductive generalizations arising from his *a priori* assumptions as to how the evolution of body plans has arisen in the metazoans. It is not a theoretical approach based on homologies and the mechanism of evolution by natural selection. As Kohn (1973) has said recently in reviewing Jägersten's (1972) recent book, "Science fiction, according to a current dictionary, is 'fiction in which scientific discoveries and developments form an element of plot or background.' Books such as this one, avidly propounding a particular theory of early metazoan evolution, belong to this genre." (p. 790).

Another approach to the specific problem of the origin of multicellular animals is to use special forms as possible intermediates. In one sense this has already been done in our examination of the sponges, cnidarians, and turbellarians. Additionally, one could also examine the Mesozoa and Ctenophora and the curious Placozoa in the form of *Trichoplax adhaerens* being studied by Grell (1971a, 1971b, 1972, 1973) and Grell and Benwitz (1971). Neither the mesozoans nor the ctenophores have ever been seriously espoused as key intermediates in the emergence of multicellular animals. The invariably symbiotic nature of the mesozoans identifies them as a specialized group, possibly showing certain important reductions as a result of their symbiosis, and altogether they represent, with a high degree of probability, an evolutionary dead end. Things are not so clear with the comb jellies. They are free-swimming marine forms, at one time classified with the Cnidaria and other radially symmetrical forms as Coelenterata. Their separation into their own phylum indicates their unique status. Their origin seems a mystery, like that of the cnidarians, and they also seem a cladogenetic group but of limited success and without any evolutionary developments deriving from them.

Grell's analysis of *Trichoplax adhaerens* seems a stong candidate for one of Jägersten's hypothetical forms. Grell suggests that *T. adhaerens* might be similar to *Bilateroblastaea* (Fig. 15-3, form 3), for the organism is a flagellated, multicellular form, somewhat flattened and bottom crawling. It does not possess the anterior tuft of sensory cilia, the segregated reproductive cells, nor the polarity nor symmetry postulated by Jägersten. But it does have an outer epithelium that is convex on the upper (dorsal) surface and concave on the lower (ventral) surface. Grell reports that *T. adhaerens* reproduces not only by dividing in two but by swarmer formation. In both cases the reproductive product is multicellular. There is a kind of embryonic development that restores the reduced size to normal (Grell 1971a, 1971b). Grell has also initiated fine structure work, which shows the cell layers and demonstrates the presence of cells in the internal space between the two epithelial layers (Grell and Benwitz 1971). Though Grell recognizes the possible resemblance of *T. adhaerens* to an intermediate form in the transition between protists and metazoans, he makes no specific suggestions as to what these antecedent and descendant forms might be in order to complete the evolutionary sequence. His references to Haeckel's Gastraea and to Bütschli imply that his phylogenetic procedure is to accept a predetermined mode of transition from protist to metazoan, i.e., the neo-Haeckelian progression from colonial flagellates (what Reutterer calls the cell-state theory), and then suggest how *T. adhaerens* makes that progression a bit more real.

An alternative view of this placozoan is that it is a form of as yet obscure affinities. (See *Traité de Zoologie* for many such examples, included as *Incertae sedis*.). Only through the study of homologies will its evolutionary relationships to other forms be resolved.

A final set of brief comments now relates to criticisms of some ideas set forth in this volume. The viewpoint presented here has been anticipated in earlier publications (Hanson 1958, 1963a), and though the present work should be criticized on its own, some response is appropriate to the comments directed at its antecedents.

Jägersten (1959), Günther (1962), and Remane (1963a) have all stated that the proposed similarities between ciliates and acoels are, in Günther's words, "... of a physiological or ecological nature or are not actually present ....," and he goes on to say "... they above all do not remove the difficulties produced by this postulated derivation by the physiological nuclear duality and the 'conjugation' of the ciliates in this connection." (p. 552. Original in German.) Our only answer can be the comments made earlier in Chapter 13, where we urge the view that the acoels arose from a ciliate eotype lacking a nuclear dualism, and that conjugation is similar, to the verge of homology, with acoel mating behavior on the basis of point-to-point comparison of the two processes. The different labels applied to these sexual processes are misleading. Presumably, Günther's "physiological" and "ecological" similarities refer to convergent similarities. This point, too, we have covered earlier (Chapter 13) and reject it as being inconsistent with known causes of convergence and our understanding of acoel and ciliate biology.

A different kind of criticism is represented by Kerkut (1960), Pitelka (1963), and Barnes (1963). These people deal in broad generalities, and in that each approaches phylogenetic problems from a different viewpoint, they will be responded to individually. At the close of his chapter on the "Origin of the Metazoa," Kerkut asks rhetorically, "What conclusion then can be drawn concerning the possible relationship between the Protozoa and the Metazoa?" and his answer is, "The only thing that is certain is that at present we do not know this relationship. Almost every possible (as well as many impossible) relationship has been suggested, but the information available to us is insufficient to allow us to come to any scientific conclusion regarding the relationship. We can, if we like, *believe* that one or other of the various theories is the more correct, but we have no real evidence." (p. 49) Kerkut is substantially correct; our only demurrer is that phylogenetic analysis based on Remanian criteria in conjunction with modern evolutionary theory does supply a sound methodological basis for garnering evidence. Though this methodology still does not answer all our questions, especially regarding the origins of the higher taxa, it eliminates many proposals, narrows down remaining possibilities, and makes much more explicit the nature of the questions waiting to be answered. We have a better basis for evaluating evidence and conclusions, and that is a real gain.

Pitelka's (1963) position is stated as follows: "The author finds it impossible to visualize with Hadzi and Hanson (see Hanson, 1958) the subdivision of a ciliatelike organism into cellular compartments as a step in metazoan evolution. Comparison of the protozoan with the total metazoan organism does not help, because no known systems of structures in metazoan tissues seem related, morphologically, to ciliate intracellular fiber systems. But certainly the ultrastructure of lower metazoan ciliated epithelia and of the morphologically simple acoel turbellaria needs to be investigated." (p. 231) Her final sentence deserves to be reemphasized by pointing out again that all fine structure work to date has been on the larger, more complex acoels and not on forms similar to *Haploposthia*. Further, if visualization of subdivision of a ciliate is difficult to achieve, it might be because the process is being conceived as an all or none process. First, evolution wouldn't work that way (Bock 1965). Second, we are proposing that cellularization of gametes was the first step, something achievable within the ciliates as shown by *Cycloposthium bipalmatum* (Dogiel 1925), and then progressively included more and more of the organism as the advantages of further differentiation were realized in the emerging acoels. Such a gradual process may be easier to accept than a sudden morphological transformation. Another more subtle objection to internal cellularization is that its proposed occurrence in the transition from ciliate eotype to acoel is the only place where such a mechamism is thought to occur; all other emergences of multicelluarity are by integration of a protistan colony (Hanson 1958). And, therefore, why consider this exceptional proposal to be valid?

There are two reasons for accepting its validity as a possible mechanism for achieving multicellularity. First, it is the *a posteriori* consequence of searching for an ancestor to the turbellarian plesiomorph. That is, it is not an *a priori* scheme to which we are fitting possible candidates, but it arises from a rigorous search for phyletic relationships. Secondly, it makes sense in selectionist terms. Ciliate evolution was tending towards more and more complex predators with permanent ingestatory organs. The acoels and their descendents have enlarged upon that evolutionary trend.

Lastly, Barnes (1963) illustrates a general set of criticisms found in many textbooks on invertebrate biology. In this particular case, Barnes assumes that radial symmetry is a primitive situation and that colonial integration is not only the predominant but the only mode of achieving multicellularity. Our answers to both points are obvious by now. But to be explicit: there is no *a priori* justification for any pattern of symmetry to be primitive, nor is there any justification to assume *a priori* what are or are not the only possible modes of producing multicellular systems. Phylogenetic analysis proceeds by establishing homologies that then define phyletic relationships, and once these are reasonably clear, then and only then must the investigator decide what is primitive and what mechanisms may have prevailed to account for a certain evolutionary innovation.

Overview

Traditionally phylogeny and systematics have been closely joined, and typically a systematist, after considerable experience with a certain group of organisms, finally essays an evolutionary interpretation. In this volume our starting point has not been systematics but concern with the larger problems of evolutionary history. We have, obviously, been dependent on the prior work of systematists, especially in attempting to find reasonably homogeneous, i.e., monophyletic, groups. Having completed our phylogenetic analysis, are there any comments to be made regarding implications for the work of systematists? There are these. First, phylogeny is totally dependent on the identification of species, which is the bread and butter of systematic work. Second, given identified species, phylogeny and systematics can go their own way. This will not be acceptable to those who follow the philosophy of the New Systematics, for, almost by definition, such a person is committed to finding evolutionary relationships within the materials of his or her systematic studies. For a numerical taxonomist the converse holds; by definition phylogeny is a separate effort from the primary one of classification through quantitative measures of similarity. The phylogenetic efforts in this volume, though coming from one much more sympathetic to the work of the New Systematics than to numerical taxonomy, have little to suggest of use to systematists. Rather, the strong feeling is that it is a hindrance to rigorous phylogenetic work to worry about its taxonomic or systematic consequences. Hence we have made no proposals for new taxa or changes in taxonomic levels and the like. We leave that to the systematists; our sole aim has been the elucidation of evolutionary history and of the biological innovations that have emerged within the course of that historical development. And third, it does seem, however, that those groups of organisms whose history is so baffling at present are best treated taxonomically by explicitly disavowing historical connotations in their taxonomy. For example, the heliozoans and ameboid rhizoflagellates (excluding the Foraminiferida) are best treated in an Adansonian manner to achieve an efficient classification. When Pokorny (1967) refers to the Proteomyxida as "... a junk drawer of neglected creatures ..." she is being realistic and implying that pure techniques of classification, devoid of evolutionary speculation, will probably be of most service for this group. One also wonders whether certain zooflagellates, poriferans, and cnidarians are not also best handled by the techniques of numerical taxonomy, too.

The conclusion might be that systematists consider two different taxonomic procedures depending on the nature of the material at hand. Given many small groups with dubious homologous relationships among them, or given a larger group that is strongly cladogenetic, in both situations an Adansonian philosophy might well be effective. Where, by contrast, homologies are clear and anagenetic trends are also to be inferred, a systematics closely allied to phylogeny could be usefully applied. A reader being forewarned as to which approach is being used would also be forearmed in trying to understand the evolutionary biology of the group in question.

If we try to apply such an overview to the origin and early evolution of animals, we find the following: probably everywhere that apochlorotic algae have arisen we see the attempt of the former phototrophs to survive as osmotrophs or phagotrophs. The latter are, by definition, consumers and therefore animals. Therefore there has been a multiple origin of unicellular animals or protozoans. Many of them became specialized as ameboid cells, and this provided a considerable diversity of forms with blunt lobopodia or slender filopodia and often devoid of sexuality, presumably as a consequence of selection for a stable genotype. This diversity is scattered among the ameboid rhizoflagellates and heliozoans. In several groups test formation was achieved, and in three cases—radiolarians, acantharians, and foraminiferans—it resulted in extraordinary elaborations of the test along with increased size or number of nuclei, increase in cell size, and complex life cycles. Each of these three groups appears to have extended the potentialities of their development in a remarkable fashion but did not give rise to any new groups that were successful in their own right. They are all largely cladogenetic, evolutionary dead ends, and except for the testate forms are perhaps best treated systematically in Adansonian terms.

The main line of animal evolution seems to reside with those apochlorotic algae that retained their flagella and developed with it a variety of other microtubular and fibrillar structures known collectively as a kinetide. Compounding and complexing of the kinetide followed along with increase in number of nuclei and in the size of the cell itself. With emergence of a permanent mouth, the transition from zooflagellates to ciliates became apparent. At this point the ciliates went on to evolve dependence on a somatic macronucleus and a complexly differentiated cortex. Another line of development exploited multinuclearity by pushing on to multicellularity. These were the forerunners of the acoels, and there emerged at this point the bilaterally symmetrical forms with sexual reproduction, embryogenesis, tissue differentiations, and cephalization that mark the eumetazoans as a whole. There is clear evidence of significant anagenetic development in the zooflagellate-ciliate-turbellarian line of evolution. For these groups systematic treatment incorporating evolutionary conclusions is appropriate.

Regardless of the systematics used in organizing our knowledge of the protozoans and early metazoans, there remains a host of problems for further phylogenetic analysis. Such work is needed to reject, modify, or verify the views developed here. Phylogeny is still biology's major unfinished task.

References

Abram, D., and H. Koffler 1964 In vitro formation of flagella-like filaments and other structures from flagellin. J. Mol. Biol. 9:168–85.

Afzelius, B. A. 1963 The nucleus of Noctiluca scintillans. J. Cell Biol. 19:229–38.

Albaret, J.-L. 1970 Observations cytologiques sur les nyctothères des genres Prosicuophora de Puytorac et Oktem et Sicuophora de Puytorac et Grain, ciliés parasites de batraciens anoures d'Afrique noire. Descriptions de deux espèces nouvelles. Protistologica 6:183–98.

Alexieff, A. 1924 Comparison entre spermatozoides et flagellés. Arch. Protistenk. 49:104–11.

Allen, G. 1957 Reflexive catalysis, a possible mechanism of molecular duplication in prebiological evolution. Am. Nat. 91:65–78.

Allen, R. D. 1961 Ameboid movement. In The Cell eds. J. Brachet and A. E. Mirsky. Vol. II pp. 135–216. Academic Press, Inc., New York.

Allen, R. D. 1967 The structure, reconstruction, and possible functions of the components of the cortex of Tetrahymena pyriformis. J. Protozool. 14:553–65.

Allen, R. D. 1969 The morphogenesis of basal bodies and accessory structures of the cortex of the ciliated protozoan Tetrahymena pyriformis. J. Cell Biol. 40:716–33.

Allen, R. D. 1970 Comparative aspects of ameboid movement. Acta Protozool. 7:291–99.

Allen, R. D. 1971 Fine structure of membranous and microfibrillar systems in the cortex of Paramecium caudatum. J. Cell Biol. 49:1–20.

Allen, R. D., and N. Kamiya, eds. 1964 Primitive Motile Systems in Cell Biology. Academic Press, Inc., New York.

Allen, S. L. 1967 Chemical genetics of Protozoa. In Chemical Zoology. ed. M. Florkin and B. T. Scheer. Vol. I. The Protozoa ed. G. W. Kidder. pp. 617–94. Academic Press, Inc., New York.

Ammermann, D. 1964 Riesenchromosomen in der Makronukleusanlage des Ciliaten Stylonychia spec. Naturwissenschaften 51:249.

Ammermann, D. 1965 Cytologische und genetische Untersuchungen an dem Ciliaten Stylonychia mytilus Ehrenberg. Arch. Protistenk. 108: 109–52.

Amon, J. P., and F. O. Perkins 1968 Structure of Labyrinthula sp. zoospores. J. Protozool. 15:543–46.

Anderson, E. 1967 Cytoplasmic organelles and inclusions in Protozoa. In Research in Protozoology. ed. T.-T. Chen. Vol. I. pp. 1–40. Pergamon Press, New York.

Anderson, E., and H. W. Beams 1959 The cytology of Trichomonas as revealed by the electron microscope. J. Morphol. 104:205–36.

Anderson, E., and H. Beams 1960 The fine structure of the Heliozoan Actinosphaerium nucleofilum. J. Protozool. 7:190–99.

Anderson, W., and G. C. Hill 1969 Division and DNA synthesis in the kinetoplast of Crithidia fasciculata. J. Cell Sci. 4:611–20.

Andresen, N., C. Chapman-Andresen, and J. R. Nilsson 1968 The fine structure of Pelomyxa palustris. C. R. Trav. Lab. (Carlsberg) 36:285–317.

Anfinsen, C. B. 1955 The Molecular Basis of Evolution. John Wiley and Sons, New York.

Angell, R. W. 1967 The process of chamber formation in the foraminifer *Rosalina floridana* (Cushman). J. Protozool. 14:566–74.

Angelopoulos, E. 1970 Pellicular microtubules in the family Trypanosomatidae. J. Protozool. 17:39–51.

Anigstein, L. 1912 Zwei neue marine Ciliaten. Arch. Protistenk. 24:127–41.

Antipa, G. A., and E. B. Small 1971 A redescription of *Conchophthirius curtis* Engelmann, 1862, (Protozoa, Ciliaten). J. Protozool. 18:491–503.

Antonov, A. S. 1971 DNA: Origin, evolution and variability. In *Molecular Evolution. I. Chemical Evolution and the Origin of Life*. eds. R. Buvet and C. Ponnamperuma. pp. 420–24. American Elsevier Pub. Co., Inc., New York.

Apelt, G. 1969 Fortpflanzungsbiologie, Entwicklungszyklen und vergleichende Frühentwicklung acoeler Turbellarian. Mar. Biol. (Berlin) 4:267–325.

Arnott, S. 1971 The structure of transfer RNA. In *Progress in Biophysics and Molecular Biology*. eds. J. A. V. Butler and D. Noble. Vol. 22, pp. 181–213. Pergamon Press, New York.

Avers, C. J., M. W. Rancourt, and F. H. Lin 1965 Intracellular mitochondrial diversity in various strains of *Saccharomyces cerevisiae*. Proc. Nat. Acad. Sci. U.S.A. 54:527–35.

Ax, P. 1961 Phylogenie und Verwandtschaftsbeziehungen der Turbellarian. Ergeb. Biol. 24:1–68.

Ax, P. 1963 Relationships and phylogeny of the Turbellaria. In *The Lower Metazoa* eds. E. C. Dougherty, Z. N. Brown, E. D. Hanson, and W. D. Hartman. pp. 191–224. University of California Press, Berkeley.

Ax, P. 1971 Zur Systematik und Phylogenie der Trigonodyominae (Turbellaria, Neorhabdocoela). Akad. Wiss. Lit. Abh. Math. Naturwiss. Kl. (Mainz) 4:141–220.

Ax, P., and E. Schulz 1959 Ungeschlechtliche Fortpflanzung durch Paratomie bei acölen Turbellarien. Biol. Zentralb. 78:615–22.

Baccetti, B., ed. 1969 *Comparative Spermatology*. Accad. Naz. Lincei, Rome; Academic Press, Inc., New York.

Baddington, M. J., and D. F. Mettrick 1971 Seasonal changes in the chemical composition and food reserves of the freshwater triclad *Dugesia tigrina* (Platyhelminthes: Turbellaria). J. Fish Res. Board Canada 28:7–14.

Bagby, R. M. 1966 The fine structure of myocytes in the sponges *Microciona prolifera* (Ellis and Solander) and *Tedania ignis* (Duchassaing and Michelotti). J. Morphol. 118:167–82.

Bagby, R. M. 1970 The fine structure of pinacocytes in the marine sponge *Microciona prolifera* (Ellis and Solander). Z. Zellforsch. mikrosk. Anat. 105:579–94.

Baker, J. R. 1948 The status of the protozoa. Nature 161:548–51 and 587–89.

Baker, J. R. 1963 Speculations on the evolution of the family Trypanosomatidae Doflein, 1901. Exp. Parasitol. 13:219–33.

Balamuth, W. 1940 Regeneration in protozoa: a problem in morphogenesis. Quart. Rev. Biol. 15:290–337.

Balamuth, W. 1942 Studies on the organization of *Ciliate protozoa*. II. Reorganization processes in *Licnophora macfarlandi* during binary fission and regeneration. J. Exp. Zool. 91:15–43.

Baldwin, E. 1966 *An Introduction to Comparative Biochemistry*. 4th edition. Cambridge University Press, Cambridge.

Baldwin, E. 1963 Biochemistry and evolution. In *The Nature of Biological Diversity*, ed. J. M. Allen. pp. 45–68. McGraw-Hill Book Co., Inc., New York.

Baldwin, E. 1967 *Dynamic Aspects of Biochemistry*. 5th edition. Cambridge University Press, Cambridge.

Ball, G. H. 1969 Organisms living on and in Protozoa. In *Research in Protozoology* ed. T. T. Chen. Vol. 3, pp. 565–718. Pergamon Press, New York.

Band, R. N. 1962 Nutritional and related biological studies on the free-living soil amoeba *Hartmannella rhysodes*. J. Gen. Microbiol. 21:80–95.

Bandy, O. L. 1960 General correlation of foraminiferal structure with environment. Int. Cong., Session 21, Norden, part 22, pp. 7–19. Berlingske Bogtrykkeri, Copenhagen.

Bardele, C. F. 1968 *Acineta tuberosa* I. Der Feinbau des adulten Suktors. Arch. Protistenk. 110:403–21.

Bardele, C. F. 1972 A microtubule model for ingestion and transport in the suctorian tentacle. Z. Zellforsch. Mikrosk. Anat. 126:116–34.

Bardele, C. F., and K. G. Grell 1967 Elektronmikroskopische Beobachtungen zur Nahrungsaufnahme bei dem Suktor *Acineta tuberosa* Ehrenberg. Z. Zellforsch. 80:108–23.

Bardele, C. 1972 Cell cycle, morphogenesis and ultrastructure in the pseudoheliozoan *Clathrulina elegans*. Z. Zellforsch. 130:219–42.

Barghoorn, E. and J. W. Schopf 1966 Microorganism three billion years old from the Precambrian of South Africa. Science 152:758–63.

Barnes, R. D. 1963 *Invertebrate Zoology*. W. B. Saunders Co., Philadelphia.

Barrington, E. J. W. 1968 Phylogenetic perspectives in vertebrate endocrinology. In *Perspectives in Endocrinology*. eds. E. J. W. Barrington and C. B. Jorgensen. pp. 1–46. Academic Press, London.

Batham, E. J., and C. F. A. Pantin 1950 Muscular and hydrostatic action in the sea-anemone *Metridium senile* (L). J. Exp. Biol. 27:264–89.

Batisse, A. 1969 Les structures pédonculaires dans les genres *Tokophrya* Büschli et *Choanophrya* Hartog (Ciliata, Suctorida). Protistologica 5:387–412.

Baylor, E. R., and W. H. Sutcliffe, Jr. 1963 Dissolved organic matter in seawater as a source of particulate food. Limnol. Oceanog. 8:369–71.

Beale, G. H. 1954 *The Genetics of Paramecium aurelia*. The University Press, Cambridge.

Beale, G. H., J. K. L. Knowles, and A. Tait 1972 Mitochondrial genetics in *Paramecium*. Nature 235:396–97.

Beams, H., and R. L. King 1941 Some physical properties of the protoplasm of the protozoa. In *Protozoa in Biological Research* ed. G. N. Calkins and F. M. Summers pp. 43–110. Columbia University Press, New York.

Beauchamp, P. de 1961a Généralités sur les Plathelminthes. In *Traité de Zoologie*. ed. P.-P. Grassé. Vol. IV. pp. 23–33. Masson et Cie., Paris.

Beauchamp, P. de 1961b Classe de Turbellariés. In *Traité de Zoologie*. ed. P.-P. Grassé. Vol. IV. pp. 35–212. Masson et Cie., Paris.

Becker, A., and J. Hurwitz 1971 Current thoughts on the replication of DNA. *Prog. Nucleic Acid Res. Mol. Biol.* 11:423–59.

Beckwith, J. R., and D. Zipser, eds. 1970 *The Lactose Operon*. Cold Spring Harbor Laboratory, Cold Spring Harbor, N.Y.

Bĕlař, K. 1923 Untersuchungen an *Actinophrys sol*. I. Die Morphologie des Formwechsels. Arch. Protistenk. 46:1–96.

Bĕlař, K. 1926 Der Formwechsel der Protistenkerne. Ergeb. Fortschr. Zool. 6:235–654.

Benazzi, M. 1960 Evoluzione cromosomica e differenciamento razziole e specifico nei triclади. In *Evoluzione e Genetica* Colloquio Internaz. (1959) Prob. att. Sci. Cult. Acc. naz. Lincei 47:273–97.

Benazzi, M. 1963 Genetics of reproductive mechanisms and chromosome behavior in some fresh-water triclads. In *The Lower Metazoa*. ed. E. C. Dougherty, Z. N. Brown, E. D. Hanson, and W. D. Hartman. pp. 405–22. University of California Press, Berkeley.

Benazzi, M., E. Gianinni, and T. Puccinelli 1970 Karyological research on the American planarians *Dugesia dorotocephala* and *Dugesia tigrina*. J. Biol. Psychol. 12:81–82.

Bennett, H. S. 1956 The concepts of membrane flow and membrane vesiculation as mechanisms for active transport and ion pumping. J. Biophys. Biochem. Cytol. 2 (Suppl.): 99–103.

Berlyn, M. 1967 Gene-enzyme relationships in histidine biosynthesis in *Aspergillus nidulans*. Genetics 57: 561–70.

Berlyn, M., S. I. Ahmed, and N. H. Giles 1970 Organization of polyaromatic biosynthetic enzymes in a variety of photosynthetic organisms. J. Bacteriol. 104:768–74.

Berrill, N. J. 1949 Developmental analysis of Scyphomedusae. Biol. Devel. 24:395–410.

Berrill, N. J. 1961 *Growth, Development and Pattern*. W. H. Freeman and Co., San Francisco.

Berthold, W. U. 1971 Untersuchungen über die sexuelle Differenzierung der Foraminifere *Patellina corrugata* Williamson mit einem Beitrag zum Entwicklungsgang und Schalenbau. Arch. Protistenk. 113:147–84.

Biecheler, B. 1935 Sur un dinoflagellé à capsule perinucléaire, *Plectodinium*, n. gen. *nucleovolvatum* n. sp. et sur les relations des Peridiniens avec les Radiolaires. C. R. Acad. Sci. (Paris) 198:404–6.

Bilbaut, A., and M. Pavans de Cecatty 1971 Les différenciations cellularies de type choanocytaire chez l'Octocoralliaire *Veretillum cynomorium* Pall. C. R. Acad. Sci. [Ser. D.] (Paris) 272:3053–6.

Blackwelder, R. E. 1967 *Taxonomy. A Text and Reference Book*. John Wiley and Sons, Inc. New York.

Block, K. 1965 The biological synthesis of cholesterol. Science 150:19–28.

Blum, H. F. 1951 *Time's Arrow and Evolution*. Princeton University Press, Princeton.

Bock, W. J. 1963 Evolution and phylogeny in morphologically uniform groups. Am. Nat. 97:265–85.

Bock, W. J. 1965 The role of adaptive mechanisms in the origin of higher levels of organization. Syst. Zool. 14: 272–87.

Bohr, N. 1933 Light and life. Nature 131:421–23, 457–59.

Böhm, R. 1878 Helgoländer Leptomedusen. Z. Naturwiss. Jena. 12:68–203.

Bonner, D. M., J. A. DeMoss, and S. E. Mills 1965 The evolution of an enzyme. In *Evolving Genes and Proteins*, ed. V. Bryson and H. J. Vogel. pp. 305–18. Academic Press, Inc., New York.

Bonner, J. T. 1954 The development of cirri and bristles during binary fission in the ciliate *Euplotes eurystomus*. J. Morphol. 95:95–108.

Borgert, A. 1909 Untersuchungen über die Fortpflanzung der Tripyleen Radiolarien, speziell von *Aulacantha scolymantha* H. Theil II. Arch Protistenk. 14:134–261.

Borkott, H. 1970 Geschlechtliche Organization, Fortpflanzungsverhalten und Ursachen der sexuellen Vermehrung von *Stenostomum sthenum* nov. spec. (Turbellaria, Catenulidae) mit Beschreibung von 3 neuen *Stenostomum*-arten. Z. Morphol. Tiere 67:183–262.

Borojevic, R., and C. Lévi 1964 Étude au microscope électronique des cellules de l'éponge: *Ophlitaspongia seriata* (Grant), au cours de la réorganisation après dissociation. Z. Zellforsch. 64:708–25.

Borojevic, R. 1970 Différenciation cellulaire dans l'embryogenèse et la morphogenèse chez les Spongiaires. In *Biology of the Porifera* ed. W. G. Fry. No. 25 Symp. Zool. Soc. London pp. 467–90. Academic Press, Inc., New York.

Borst, P., and A. M. Kroon 1969 Mitochondrial DNA: Physicochemical properties, replication, and genetic function. Int. Rev. Cyt. 26:107–90.

Bouillon, J. 1968 Introduction to coelenterates. In *Chemical Zoology*, eds. M. Florkin and B. T. Scheer. Vol. II. *Porifera, Coelenterata*, and *Platyhelminthes*. pp. 81–147. Academic Press, Inc., New York.

Bovee, E. C. 1960 Studies on the Helioflagellates. V. Morphology and fission of *Dimorpha floridanis*, n. sp. Arch. Protistenk. 104:503–14.

Bovee, E. C. 1970 The lobose amebas. I. A key to the suborder Conopodina Bovee and Jahn 1966 and descriptions of thirteen new or little known

Mayorella species. Arch. Protistenk. 112:178–227.
Bovee, E. C., and T. L. Jahn 1965 Mechanisms of movement in taxonomy of Sarcodina. II. The organization of subclasses and orders in relationship to the classes Autotractea and Hydraulea. Am. Midland Nat. 73:293–98.
Bovee, E. C., and D. E. Wilson 1961 An amoeboid stage in the feeding behavior of the helioflagellate protozan *Dimorpha floridanis*. Am. Zool. 1:345 (abst.)
Boyden, A. 1947 Homology and analogy. Am. Midland Nat. 37:648–69.
Boyden, A. A. 1953 Comparative evolution with special reference to primitive mechanisms. Evol. 7:21–30.
Boyer, B. C. 1971 Regulative development in a spiralian embryo as shown by cell deletion experiments on the acoel *Childia*. J. Exp. Zool. 176: 97–105.
Bradbury, P. C. 1965 The infraciliature and argyrome of *Opisthonecta henneguyi* Fauré-Fremiet. J. Protozool. 12:345–63.
Bradley, Jr., T. B., R. C. Wohl, and R. F. Rieder 1967 Hemoglobin Gun Hill: deletion of five amino acid residues and impaired heme-globin binding. Science 157:1581–83.
Brady, H. B. 1884 *Report of the scientific results of the exploring voyage of H.M.S. Challenger*. Vol. IX (Zoology) pp. 1–814. London.
Brändle, E. 1971 Bedeutung der kolonialen Komponenten für die Bildung und Differenzierung der Medusen von *Podocoryne carnea* M. Sars. Roux' Arch. 166:254–86.
Bresslau, E. 1933 Turbellaria. In *Handbuch der Zoologie* eds. W. Kukenthal and T. Krumbach. Vol. II, Part i. Gruyter and Co., Berlin.
Brien, P. 1968 The sponges, or Porifera. In *Chemical Zoology* ed. M. Florkin and B. T. Scheer. Vol. II. *Porifera, Coelenterata, and Platyhelminthes*. pp. 1–30. Academic Press, Inc., New York.
Brien, P. 1973 Les Démosponges. Morphologie et reproduction. In *Traité de Zoologie*. ed. P.-P. Grassé. Tome III, fasc. I. pp. 133–461. Masson et Cie., Paris.
Britten, R. J., and D. E. Kohne 1967 Nucleotide sequence repetition in DNA. 1965–66 Carnegie Inst. Wash. Year Book 65:78–106.
Britten, R. J. 1968 Repeated sequences in DNA. Science 161:529–40.
Brooks, W. K. 1886 The life history of the hydromedusae: a discussion of the origin of the medusae and of the significance of metagenesis. Mem. Boston Soc. Nat. Hist. 3:359–430.
Brower, J. van Z. 1958 Experimental studies of mimicry in some North American butterflies. II. *Battus philenor* and *Papilio troilus, P. polyxenes* and *P. glaucus*. Evolution 12:123–36.
Brown, D. D., and I. B. Dawid 1968 Specific gene amplification in oocytes. Science 160:272–80.
Brown, D. D., and J. B. Gurdon 1964 Absence of ribosomal RNA synthesis in the anucleolate mutant of *Xenopus laevis*. Proc. Nat. Acad. Sci. U.S.A. 51:139–46.
Brucke, C. 1861 Die Elementarorganismen. Wiener Sitzber. 46
Bryan, J., and L. Wilson 1971 Are cytoplasmic microtubules heteropolymers? Proc. Natl. Acad. Sci. U.S.A. 8:1762–66.
Buchsbaum, R. 1948 *Animals without Backbones*. University of Chicago Press, Chicago.
Bullock, T. H., and G. A. Horridge 1965 *Structure and Function of the Nervous System of Invertebrates*. W. H. Freeman, San Francisco.
Burton, P. R. 1970 Fine structure of mitochondria of *Spirostomum ambiguum* as seen in sectioned and negatively-stained preparations. J. Protozool. 17:295–99.
Burton, P. R., and D. G. Dusanic 1968 Fine structure and replication of the kinetoplast of *Trypanosoma lewisi*. J. Cell Biol. 39:318–31.
Bush, L., and D. J. Zinn 1970 *Halammohydra schulzei*: first actinulid recorded from western Atlantic. Trans. Am. Microsc. Soc. 89:431–33.

Bütschli, O. 1880–1882 Protozoa. In *Bronn's Klassen und Ordnungen des Thier-Reichs.* Winter'sche Verlagshandlung, Leipzig und Heidelberg.

Bütschli, O. 1887–1889 *Protozoa; Bronn's Klassen und Ordnungen des Tierreichs.* 1:1–2035. Leipzig.

Bütschli, O. 1906 Uber die chemische Natur der Skelettsubstanz der Acantharia. Zool. Anz. 30:784–89.

Bütschli, O. 1908 Chemische Natur der Skelettsubstanz des *Podoactinelius* und der Acantharia überhaupt. In *Deutschen Südpolar Expedition 1901–1903.* ed. E. von Drygalski. Vol 9. pp. 238–57. Reimer Verlag, Berlin.

Butzel, H. M., Jr. and H. Horwitz 1965 Excystment in *Didinium.* J. Protozool. 12:413–16.

Butzel, H. M., Jr. and A. B. Bolten 1968 The relationship of the nutritive state of the prey organism *Paramecium aurelia* to the growth and encystment of *Didinium nasutum.* J. Protozool. 15:256–58.

Cachon-Enjumet, M. 1961 Contribution à l'étude des Radiolaires Phaeodariés. Arch. Zool. exp. gen. 100:151–238.

Cachon-Enjumet, M. 1964 L'évolution sporogénétique des Phaeodariés (Radiolaires). C. R. Acad. Sci. (Paris) 259:2677–79.

Cachon, J., and M. Cachon 1969 Révision systématique des Nassellaires Plectoidea à propos de la description d'un nouveau reprèsentant, *Plectogonidium deflandrei* nov. gen., nov. sp. Arch. Protistenk. 111:236–51.

Cachon, J., and M. Cachon 1971 Le système axopodial des Radiolaires Nassellaires. Arch. Protistenk. 113: 80–97.

Cachon, J., M. Cachon, and G. Ferru 1968 Rapports du squellete et du système axopodial chez les Radiolaires Nassellaires. C. R. Acad. Sci., Paris, Sèr. D 267:1602–4.

Cairns, J. 1963a The chromosome of *E. coli.* Cold Spring Harbor Symp. Quant. Biol. 28:43–46.

Cairns, J. 1963b The bacterial chromosome and its manner of replication as seen by autoradiography. J. Mol. Biol. 6:208–13.

Calder, D. R. 1971 Nematocysts of polyps of *Aurelia, Chrysaora,* and *Cyanea,* and their utility in identification. Trans. Am. Microsc. Soc. 90:269–74.

Calkins, G. N. 1901 *The Protozoa.* MacMillan Company, New York.

Calkins, G. N. 1926 *The Biology of the Protozoa.* Lea and Febiger, Philadelphia and New York.

Calvin, M. 1961 *Chemical Evolution.* Univ. of Oregon Press, Eugene.

Campbell, A. S. 1926 The cytology of *Tintinnopsis nucula* (Fol) Laackman with an account of its neuromotor apparatus, division, and new intranuclear parasite. Univ. Calif. Pub. Zool. 29:179–236.

Campbell, A. S. 1954a Radiolaria. In *Treatise on Invertebrate Paleontology.* ed. R. C. Moore. pp. D11–D163. Geol. Soc. Am. and Univ. Kansas Press, Lawrence.

Campbell, A. S. 1954b Tintinnia. In *Treatise on Invertebrate Paleontology* ed. R. C. Moore. Protista 3 (Protozoa, Chiefly Radiolaria and Tintinnia) pp. D166–D180. Geological Soc. America and Univ. Kansas Press, Lawrence.

Canella, M. F. 1957 Studie Ricerche sui Tentaculifer nel quadro della Biologia generale. Ann. Univ. Ferrara Sect. 3, Biol. Anim. 1:1–716.

Canella, M. F. 1958 Biologia degli infusori e ipotetici raffronti con i metazoi. Estratto dal Monitore Zoologico Italiano 65:164–89.

Canella, M. F. 1964 Strutture buccali, infraciliatura, filogenesi e sistematica dei ciliofori. Fatti, ipotesi, speculazioni. Ann. U. Ferrara (N.S.) Sezione III Biol. Anim. 2:119–88.

Canella, M. F. 1970 Sur les organelles ciliaries de l'appareil buccal des hyménostomes et autres ciliés. Ann. U. Ferrara (N.S.) Sezione III Biol. Anim. III (supp.) 1–235.

Cánovas, J. L., L. N. Ornston, and R. Y. Stanier 1967 Evolutionary significance of metabolic control systems. Science 156:1695–99.

Carasso, N., P. Favard, and

S. Goldfischer 1964 Localization à l'échelle des ultrastructures d'activités de phosphatases en rapport avec les processus digestifs chez un Cilié (*Campanella umbellaria*). J. Microscopie 3:297–322.

Carter, G. S. 1954 On Hadzi's interpretations of animal phylogeny. Syst. Zool. 3:163–67.

Carter, G. S. 1961 *A General Zoology of the Invertebrates* 4th ed. Sidgwick and Jackson, London.

Carter, G. S. 1965 Phylogenetic relations of the major groups of animals. In *Ideas in Modern Biology*. ed. J. A. Moore. Proc. XVI Int. Cong. Zool. Vol. 6 pp. 427–45. The Natural History Press, Garden City, N.J.

Casper, D. L. D., and A. Klug 1962 Physical principles in the construction of regular viruses. Cold Spring Habor Symp. Quant. Biol. 27:1–24.

Caullery, M. 1952 *Parasitism and Symbiosis*. Sidgwick and Jackson, London.

Chapman-Andresen, C. 1962 Studies on pinocytosis in amebae. C. R. Trav. Lab. Carlsberg. 33:73–264.

Chapman-Andresen, C. 1967 Studies on endocytosis in amoebae. The distribution of pinocytotically ingested dyes in relation to food vacuoles in *Chaos chaos*. I. Light microscopic observations. C. R. Trav. Lab. Carlsberg 36:161–87.

Chapman-Andresen, C. 1972 Membrane activity in fresh-water amebae. J. Protozool. 19:225–31.

Chapman, D. M. 1966 Evolution of the Scyphistoma. In *The Cnidaria and their Evolution*. ed. W. J. Rees. No. 16 Symp. Zool. Soc. London pp. 51–75. Academic Press, Inc., New York.

Chapman, G. 1966 The structure and functions of the mesoglea. In *The Cnidaria and their Evolution*. ed. W. J. Rees. No. 16 Symp. Zool. Soc. Lond. pp. 147–68. Academic Press, Inc., New York.

Chatton, E. 1920 Existence chez les Radiolaires de Peridiniens parasites considérés comme formes de reproduction de leurs hôtes. C. R. Acad. Sci. (Paris) 170:413–15.

Chatton, E. 1924 Sur les connexions flagellaires des éléments flagellés. Centrosome et mastigosomes. La cinétide, unité cinétoflagellaire. Cinétides simples et cinétides composées. C. R. Soc. Biol. 91:574–80.

Chatton, E. 1942 Le problème de la continuité du cinétome chez les Hypotriches. Nouvelles recherches sur l'*Euplotes crassus*. Bull. Biol. France-Belg. 76:314–35.

Chatton, E. 1953 Ordre des amoebiens nus au Amoebaea. In *Traité de Zoologie*, ed. P.-P. Grassé Tome I, fasc. II. pp. 5–91. Masson et Cie., Paris.

Chatton, E., and A. Lwoff 1935 Les ciliés apostomes. I. Aperçu historique et général; étude monographique des genres et des espèces. Arch. Zool. exp. gén. 77:1–453.

Chatton, E., and A. Lwoff 1936 Technique pour l'étude des protozoaires, specialement de leurs structures superficielles (cinétome et argyrome). Bull. Soc. Franç. Micr. 5:25–39.

Chatton, E., and T. Séguéla 1940 La continuité génétique des formations ciliaires chez les ciliés hypotriches. Le cinétome et l'argyrome au course de la division. Bull. Biol. de France-Belg. 74:349–442.

Chatton, E., A. Lwoff, and M. Lwoff 1931 La dualité figurée, substantielle et génétique, des corps basaux des cils chez les infusoires. Granules infraciliaire et corpuscule ciliaire. C. R. Soc. Biol., Paris 107:560–64.

Chatton, E., A. Lwoff, M. Lwoff, and J. Monod 1931 Sur la topographie, la structure et la continuité génétique du système ciliare de l'infusoire *Chilodon uncinatus*. Bull. Soc. Zool. Fr. 56:367–74.

Chatton, E., A. Lwoff, M. Lwoff, and L. Tellier 1929 L'infraciliature et la continuité génétique des blépharoplastes chez l'acinétien *Podophrya fixa* O. F. Müller. C. R. Soc. Biol. (Paris) 100:1191–96.

Cheissin, E. M., and G. I. Poljansky 1963 On the taxonomic system of the Protozoa. Acta Protozool. 1:327–52.

Chen, T.-T. 1951 Conjugation in

Paramecium bursaria. IV. Nuclear behavior in conjugation between old and young clones. J. Morphol. 88: 295–359.

Chen-Shan, L. 1970 Cortical morphogenesis in *Paramecium aurelia* following amputation of the anterior region. J. Exp. Zool. 174:463–78.

Child, F. 1967 The chemistry of protozoan cilia and flagella. In *Chemical Zoology* ed. M. Florkin and B. Scheer. Vol. I. *The Protozoa* ed. G. W. Kidder. pp. 381–93. Academic Press, Inc., New York.

Cho, P. L. 1971 Cortical patterns in two syngens of *Glaucoma.* J. Protozool. 18:180–83.

Clark, R. B. 1963 The evolution of the celom and metameric segmentation. In *The Lower Metazoa.* eds. E. C. Dougherty, Z. N. Brown, E. D. Hanson, and W. D. Hartman. pp. 91–107. University of California Press, Berkeley.

Clark, R. B. 1964 *Dynamics in Metazoan Evolution.* Clarendon Press, Oxford.

Cleland, R. 1964 The problem of species in *Oenothera.* Amer. Nat. 78: 5–28.

Clément-Iftode, F., and G. Versavel 1967 *Teutophrys triscula* (Chatton, de Beauchamp) cilié planctonique rare. Protistologica 3: 457–64.

Cleveland, L. R. 1935 The centriole and its role in mitosis as seen in living cells. Science 81:598–600.

Cleveland, L. R. 1938 Longitudinal and transverse fission in two closely related flagellates. Biol. Bull. 74:1–40.

Cleveland, L. R. 1947 The origin and evolution of meiosis. Science 105: 287–89.

Cleveland L. R. 1949 Hormone-induced sexual cycles of flagellates. I. Gametogenesis, fertilization, and meiosis in *Trichonympha.* J. Morphol. 86:215–28.

Cleveland, L. R. 1965a Hormone induced cycles of flagellates. XIV. Gametic meiosis and fertilization in *Macrospironympha.* Arch. Protist. 101:99–168.

Cleveland, L. R. 1965b Brief accounts of the sexual cycles of the flagellates of *Cryptocercus.* J. Protozool, 3:161–80.

Cleveland, L. R. 1965c Fertilization in *Trichonympha* from termites. Arch. Protist. 108:1–5.

Cleveland, L. R. 1966 Fertilization in *Mixotricha.* Arch. Protist. 109:37–38.

Cleveland, L. R., A. W. Burke, Jr., and P. Karlson 1960 Ecdysone induced modifications in the sexual cycles of the protozoa of *Cryptocercus.* J. Protozool. 7:229–39.

Cleveland, L. R., S. R. Hall, E. P. Sanders, and J. Collier 1934 The wood-feeding roach *Cryptocercus,* its protozoa, and the symbiosis between protozoa and roach. Mem. Am. Acad. Arts Sci. 17:187–342.

Collin, B. 1912 Étude monographique sur les Acinétiens. II. Morphologie, physiologie, systématique. Arch. Zool. exp. gén. 51:1–457.

Conner, R. 1967 Transport phenomena in Protozoa. In *Chemical Zoology.* eds. M. Florkin and B. T. Scheer. Vol I. *The Protozoa.* ed. G. W. Kidder. pp. 309–50. Academic Press, Inc., New York.

Corliss, J. O. 1953a Silver impregnation of ciliated protozoa by the Chatton-Lwoff technique. Stain Technol. 28:97–100.

Corliss, J. O. 1953b Comparative studies on holotrichous ciliates in the Colpidium-Glaucoma-Leucophrys-Tetrahymena group. II. Morphology, life cycles and systematic status of strains in pure culture. Parasitology 43:49–87.

Corliss, J. O. 1953c Protozoa and systematics. Yale Sci. Mag. 28:14–17, 36, 38, 40.

Corliss, J. O. 1956 On the evolution and systematics of the ciliated protozoa. Syst. Zool. 5:68–91, 121–40.

Corliss, J. O. 1957 Concerning the "cellularity" or acellularity of the Protozoa. Science 125:988–89.

Corliss, J. O. 1958a The phylogenetic significance of the genus *Pseudomicrothorax* in the evolution of holotrichous ciliates. Acta Biol. Acad. Sci. Hung. 8:367–88.

Corliss, J. O. 1958b The systematic

position of *Pseudomicrothorax dubius*, ciliate with a unique combination of anatomical features. J. Protozool. 5: 184–93.

Corliss, J. O. 1959 Some basic problems in the study of evolution of the ciliated protozoa. Proc. XV Int. Cong. Zool., London (1958) pp. 179–82.

Corliss, J. O. 1960 Comments on the systematics and phylogeny of the protozoa. Syst. Zool. 8:169–90.

Corliss, J. O. 1961 *The Ciliated Protozoa: Characterization, Classification, and Guide to the Literature.* Pergamon Press, New York.

Corliss, J. O. 1967 Systematics of the phylum Protozoa. In *Chemical Zoology* ed. M. Florkin and B. T. Scheer. Vol. I. *The Protozoa* ed. G. W. Kidder. pp. 1–20. Academic Press, Inc., New York.

Corliss, J. O. 1968 The value of ontogenetic data in reconstructing protozoan phylogenies. Trans. Am. Micro. Soc. 87:1–20.

Corliss, J. O. 1972 The ciliate protozoa and other organisms: Some unresolved questions of major phylogenetic significance. Am. Zool. 12:739–53.

Corliss, J. O. 1973 Guide to the literature on *Tetrahymena*: A companion piece to Elliott's "General Bibliography". Trans. Am. Microsc. Soc. 92:468–91.

Corliss, J. O. 1974a The changing world of ciliate systematics: Historical analysis of past efforts and a newly proposed phylogenetic scheme of classification for the protistan phylum Ciliophora. Syst. Zool. 23:91–138.

Corliss, J. O. 1974b Remarks on the composition of the large ciliate class Kinetofragmophora de Puytorac et al., 1974, and recognition of several new taxa therein, with emphasis on the primitive order Primociliatida n. ord. J. Protozool. 21:207–20.

Corliss, J. O. 1974c Time for evolutionary biologists to take more interest in protozoan phylogenetics? Taxon 23:497–522.

Corliss, J. O. 1974d Classification and phylogeny of the Protists. In *Actualités Protozoologiques*. ed. P. de Puytorac and J. Grain. Vol. I. pp. 251–64. University of Clermont, France.

Corliss, J. O. 1975 Taxonomic characterization of the suprafamilial groups in a revision of recently proposed schemes of classification for the phylum Ciliophora. Trans. Amer. Microsc. Soc. 94:224–67.

Costello, D. P. 1960 The giant cleavage spindle of the egg of *Polychoerus carmelensis*. Biol. Bull. 119:285.

Costello, D. P. 1961 The orientation of the centrioles in dividing cells and its significance. Biol. Bull. 121:368.

Cowden, R. 1970 Connective tissue in six marine sponges. Z. Mikro. Anat. Forsch. 82:557–69.

Cracroft, J. 1974 . Phylogenetic models and classification. *Syst. Zool.* 23:71–90.

Crick, F. H. C. 1963 The recent excitement in the coding problem. Prog. Nucleic Acid Res. 1:164–218.

Crow, J., and M. Kimura 1970 *An Introduction to Population Genetics Theory.* Harper and Row, New York.

Curtis, A. S. G. 1970 Problems and some solutions in the study of cellular aggregation. In *Biology of the Porifera*. ed. W. G. Fry. No. 25 Symp. Zool. Soc. London. pp. 335–52. Academic Press, Inc., New York.

Cushman, J. A. 1948 *Foraminifera: Their Classification and Economic Use.* 4th edition. Harvard University Press, Cambridge, Mass.

Dadd, R. H. 1970 Arthropod nutrition. In *Chemical Zoology.* Vol. V *Arthropoda*, Part A. eds. M. Florkin and B. T. Scheer. pp. 35–95. Academic Press, Inc., New York.

Dangeard, P.-A. 1900 Étude de la karyokinèse chez l'*Amoeba hyalina* sp. nov. Le Botaniste 7:49–83.

Daniels, E. W. and E. P. Breyer 1967 Ultrastructure of the giant amoeba *Pelomyxa palustris*. J. Protozool. 14:167–79.

Daniel, W. A., C. F. T. Mattern, and B. M. Honigberg 1970 Fine structure of the mastigont system in *Tritrichomonas muris* (Grassi). J. Protozool. 18:575–86.

Darwin, C. 1859 *On the Origin of*

Species by Natural Selection. In *On the Origin of Species. A Facsimile of the First Edition.* (1964) Introduction by E. Mayr. Harvard University Press, Cambridge, Mass.

Darwin, C. 1872 *The Origin of Species* 6th edition. John Murray, London.

Darwin, F. 1887 *The Life and Letters of Charles Darwin.* Vol. I. D. Appleton and Company, New York.

Dayhoff, M. O., ed. 1969 *Atlas of Protein Sequence and Structure.* Vol. 4. National Biomedical Research Foundation. Silver Spring, Md.

Dayhoff, M. O. 1971 Evolution of proteins. In *Molecular Evolution I. Chemical Evolution and the Origin of Life.* ed. R. Buvet and C. Ponnamperuma. pp. 392–419. American Elsevier Publishing Co., Inc., New York.

Dayhoff, M. O., and R. V. Eck 1969 Evolution of the globins. In *Atlas of Protein Sequence and Structure.* Vol. 4. ed. M. O. Dayhoff. pp. 17–24. National Biomedical Research Foundation, Silver Spring, Maryland.

Dayhoff, M. O., and P. J. McLaughlin 1969 Transfer RNA. In *Atlas of Protein Sequence and Structure,* Vol. 4. ed. M. O. Dayhoff. pp. 89–94. National Biomedical Research Foundation, Silver Spring, Md.

Dayhoff, M. O., and C. M. Park 1969 Cytochrome c: Building a phylogenetic tree. In *Atlas of Protein Sequence and Structure.* Vol. 4. ed. M. O. Dayhoff. pp. 7–16. National Biomedical Research Foundation, Silver Spring, Maryland.

de Bary, A. 1879 *Die Erscheinung der Symbiose.* Strassburg.

de Beer, G. R. 1940 *Embryos and Ancestors.* Clarendon Press, Oxford.

de Beer, G. R. 1954 The evolution of the Metazoa. In *Evolution as a Process.* ed. J. S. Huxley, A. C. Hardy, and E. B. Ford. pp. 24–33. Allen and Unwin, London.

de Duve, C. 1964 From cytases to lysosomes. Fed. Proc. 23:1045–49.

de Duve, C. 1969 The lysosome in retrospect. In *Lysosomes in Biology and Pathology.* eds. J. Y. Dingle and H. B. Fell pp. 3–42. Amer. Elsevier Pub. Co., New York.

de Duve, C., B. C. Pressman, R. Gianetto, R. Wattiaux, and F. Appelmans 1955 Tissue fractionation studies. 6. Intracellular distribution patterns of enzymes in rat-liver tissue. Biochem. J. 60:604–17.

Deflandre, G. 1928 Le genre *Arcella* Ehrenberg. Morphologie-Biologie. Essai phylogénétique et systématique. Arch. Protistenk. 64:152–287.

Deflandre, G. 1952 Classe des Radiolaires. In *Traité de Paleontologie.* ed. J. Piveteau. Tome I. pp. 303–13. Masson et Cie., Paris.

Deflandre, G. 1953a Généralités [Superclass Rhizopodes]. In *Traité de Zoologie* ed. P.-P. Grassé. Tome I, fasc. II. pp. 3–4. Masson et Cie., Paris.

Deflandre, G. 1953b Ordre des Aconchulina, de Saedeleer, Athalamia, Haeckel. In *Traité de Zoologie* ed. P.-P. Grassé. Tome I, fasc. II. pp. 92–6. Masson et Cie., Paris.

Deflandre, G. 1953c Ordres des Testacealobosa, des Testaceafilosa, des Thalamia ou Thecamoebiens (auct.). In *Traite de Zoologie* ed. P.-P. Grassé. Tome I, fasc. II. pp. 97–148. Masson et Cie., Paris.

Deflandre, G. 1953d Radiolaires fossiles. In *Traité de Zoologie* ed. P.-P. Grassé. Tome I, fasc. II. pp. 389–436. Masson et Cie., Paris.

Deflandre, G. 1963 *Pylentonema*, nouveau genre de Radiolaire du Viséen: Sphaerellaire ou Nassellaire? C. R. Acad. Sci. (Paris) 257:3981–84.

Deflandre, G., and P.-P. Grassé 1953 Généralités [sous-embranchement des Actinopodes]. In *Traité de Zoologie.* Tome I. fasc. 2. pp. 267–68. Masson et Cie., Paris.

Demerec, M. 1965 Homology and divergence in genetic material of *Salmonella typhimurium* and *Escherichia coli.* In *Evolving Genes and Proteins* eds. V. Bryson and H. F. Vogel. pp. 505–10. Academic Press, New York.

Demerec, M., and Z. Hartmann 1956 Trypotophan mutants in *Salmonella typhimurium.* Carnegie Inst. Wash. Publ. 612:5–33.

Deroux, G., and J. Dragesco 1968 Nouvelles données sur quelques ciliés holotriches cyrtophores à ciliature ventralle. Protistologica 4:365–403.

de Saedeleer, H. 1929a Recherches sur les choanocytes; l'origine des Spongiaires. Ann. Soc. Roy. Zool. Belg. 60:16–21.

de Saedeleer, H. 1929b Remarques relatives au précédent travail de W. N. Ellis "Recent Researches on the Choanoflagellates." Ann. Soc. Roy. Zool. Belg. 60:89–95.

de Terra, N. 1966 Leucine incorporation into the membranellar bands of regenerating and non-regenerating Stentors. Science 153:543–44.

de Terra, N. 1967 Macronuclear DNA synthesis in *Stentor:* Regulation by a cytoplasmic initiator. Proc. Natl. Acad. Sci. U.S.A. 57:607–14.

Dewey, V. C. 1967 Lipid composition, nutrition, and metabolism. In *Chemical Zoology*, eds. M. Florkin and B. T. Scheer. Vol. I *The Protozoa*, ed. G. W. Kidder. pp. 161–274. Academic Press, Inc., New York.

Didier, P. 1971 Contribution à l'étude comparée des ultrastructures corticales et buccales des ciliés hymenostomes peniculiens. Thèse présentée à L'U.E.R. des Sciences Exactes et Naturelles de l'Université de Clermont-Ferrand.

Diller, W. F. 1936 Nuclear reorganization processes in *Paramecium aurelia*, with descriptions of autogamy and "hemixis." J. Morphol. 59:11–67.

Dillon, L. S. 1962 Comparative cytology and the evolution of life. Evol. 16:102–17.

Dillon, L. S. 1963 A reclassification of the major groups of organisms based upon comparative cytology. Syst. Zool. 12:71–82.

Dippell, R. V. 1954 A preliminary report on the chromosomal constitution of certain variety 4 races of *Paramecium aurelia*. Proc. IX Intl. Cong. Gen., Bellagio, Italy, 1953, in Caryologia 6 (suppl.): 1109–11.

Dippell, R. V. 1962 The site of silver impregnation in *Paramecium aurelia*. J. Protozool. 9 (suppl.): 24.

Dippell, R. V. 1963 Nucleic acid distribution in the macronucleus of *Paramecium aurelia*. J. Cell Biol. 19:20A.

Dippell, R. V. 1968 The development of basal bodies in *Paramecium*. Proc. Natl. Acad. Sci. U.S.A. 61:461–68.

Dippell, R. V. and S. E. Sinton 1963 Localization of macronuclear DNA and RNA in *Paramecium aurelia*. J. Protozool. 10 (suppl.): 22–23.

Dobell, C. 1911 Principles of protistology. Arch. Protistenk. 23:269–310.

Dobell, C. 1917 On *Oxnerella maritima*, n. gen., n. sp., a new heliozoan, and its method of division; with some remarks on the centroplast of the Heliozoa. Quart. J. Microsc. Sci. 62:515–38.

Dobell, C. 1932 *Antony von Leeuwenhoek and his 'Little Animals'*. 1958 edition. Russell and Russell, Inc., New York.

Dobrzanska-Kaczanowska, J. 1963 Comparaison de la morphogenèse des Ciliés: *Chilodonella uncinata* (Ehrbg.), *Allosphaerium paraconvexa* sp. n. et *Heliochona scheuteni* (Stein). Acta Protozool. 1:353–94.

Dobzhansky, T. 1950 Mendelian populations and their evolution. Am. Nat. 84:401–18.

Dobzhansky, T. 1965 *Genetics and the Origin of Species*. 3rd Edition Revised. Columbia University Press, New York.

Dobzhansky, T. 1970 *Genetics and the Evolutionary Process*. Columbia University Press, New York.

Doflein, F., and E. Reichenow 1927–1929 *Lehrbuch der Protozoenkunde*. 5th ed. G. Fishcher, Jena.

Dogiel, V. 1925 Die Geschlechtsprozesse bei Infusorien. Arch. Protistenk. 50:283–442.

Dogiel, V. 1927 Polymerization als ein Prinzip der progressen Entwicklung der Protozoen. Biol. Zentralbl. 49:451–69.

Dogiel, V. A., J. I. Poljanskij, and E. M. Chejsin 1965 *General Protozoology*, 2nd edition. Clarendon

Press, Oxford.
d'Orbigny, A. D. 1826 Tableau méthodique de la classe des Céphalopodes. Ann. Sci. Nat. 7: 245–314.
Dorey, A. E. 1965 The organization and replacement of the epidermis in acoelous turbellarians. Quart. J. Microsc. Sci. 106:147–72.
Dougherty, E. 1953 Problems of nomenclature for the growth of organisms of one species with and without associated organisms of other species. J. Parasitol. 42:259–61.
Dougherty, E. C. 1955 Comparative evolution and the origin of sexuality. Syst. Zool. 4:145–69.
Dougherty, E. 1963 Nutrient media for axenic organisms, with special reference to Micrometazoa. In *The Lower Metazoa.* eds. E. Dougherty, Z. N. Brown, E. D. Hanson, W. D. Hartman. pp. 315–26. University of California Press, Berkeley.
Dragesco, J. 1960 Ciliés mésopsammiques littoraux. Systématique, morphologie, écologie. Thèses présentées à la Faculté des Sciences de l'Université de Paris. Serie A, No. 3577; No. d'ordre: 4449. pp. 5–356.
Dragesco, J. 1963 Compléments à la connaissance des Ciliés mésopsammiques de Roscoff. I. Holotriches. Cahiers Biol. Marine. 4: 91–119.
Dragesco, J. 1966 Ciliés libres de Thonon et ses environs. Protistologica 2:59–95.
Dragesco, J. and M. Tuffrau 1967 *Neobursaridium gigas* Balech, 1941, cilié, holotriche hymenostome pantropical. Protistologica 3:133–46.
Dreyer, F. 1892 Die Prinzipen der Gerüstbildung bei Rhizopoden, Spongien, und Echinodermen. Ein Versuch zur mechanisch Erklärung organischer Gebilde. Jen. Z. Bd. 26.
Duboscq, O., and O. Tuzet 1937 L'ovogenese, la fécondation et les premiers stades du développement des Éponges calcaires. Arch. Zool. exp. gén. 79:157–316.
Dujardin, F. 1835 Observations sur les rhizopodes et les infusoires. C. R. Acad. Sci. (Paris) No. 1835:338–40.
DuPraw, E. J. 1968 *Cell and Molecular Biology.* Academic Press, New York.
DuPraw, E. J. 1970 *DNA and Chromosomes.* Holt, Rinehart, and Winston, Inc., New York.
Dupy, Blanc, J. 1969 Étude par cytophotometrie des teneurs en ADN nucléaire chez trois espèces de paramécies, chez différentes variétés d'un même espèce, et chez différents types sexuels d'un même variété. Protistologica. 5:297–308.
Eckert, R. and Y. Naitoh 1972 Bioelectric control of locomotion in ciliates. J. Protozool. 19:237–43.
Ehret, C. F. 1960 Organelle systems and biological organization. Science 132:115–23.
Ehret, C. F. 1967 Paratene theory of the shapes of cells. J. Theor. Biol. 15: 263–72.
Ehret, C. F. and G. de Haller 1963 Origin, development and maturation of organelle systems of the cell surface in *Paramecium.* J. Ultrastruct. Res. 6 (suppl.): 1–42.
Ehret, C. F. and E. L. Powers 1957 The organization of gullet organelles in *Paramecium bursaria.* J. Protozool. 4:55–59.
Ehrlich. P. R. 1961 Has the biological species concept outlived its usefulness? Syst. Zool. 10:167–76.
Ehrlich, P. R. 1964 Some axioms of taxonomy. Syst. Zool. 13:109–23.
Ehrlich, P. R. 1965 Numerical taxonomy. Papua and New Guinea Sci. Soc. Ann. Rep. Proc. 17:10–14.
Ehrlich, P. R. and R. W. Holm 1963 *Process of Evolution.* McGraw-Hill Book Company, New York.
Ehrlich, P. R. and P. H. Raven 1969 Differentiation of populations. Science 165:1228–32.
Elliott, H. M., ed. 1973 *The Biology of Tetrahymena.* Dowden, Hutchinson, and Ross. Stroudsberg, Pennsylvania.
Ellis, W. N. 1929 Recent researches on Choanoflagellata (Craspedomonadines) (fresh-water and marine) with description of new genera and species. Ann. Soc. Roy. Zool. Belg. 60:49–88.

Ephrussi, B. 1949 Action de l'acriflavine sur les levures. In *Unités Biologiques Douées de Continuité Génétique.* Edition du Centre Nat. Rech. Sci. Paris.

Ephrussi, B. 1951 Remarks on cell heredity. In *Genetics in the 20th Century.* ed. L. C. Dunn. pp. 241–62. MacMillan, N.Y.

Ertl, M. 1968 Uber das Vorkommen von *Protospongia haeckeli* Kent in der Donau und einige Bemerkungen zur Taxonomie dieser Art. Arch. Protistenk. 111:18–23.

Estabrook, G. F. 1972 Theoretical models in systematic and evolutionary studies. Prog. in Theoret. Biol. 2:23–86.

Estève, J.-C. 1970 Distribution of acid phosphatase in *Paramecium caudatum:* Its relation with the process of digestion. J. Protozool. 17:29–35.

Evans, F. R., and J. O. Corliss 1964 Morphogenesis in the hymenostome ciliate *Pseudocohnilembus persalinus* and its taxonomic and phylogenetic implications. J. Protozool. 11:353–70.

Fauré-Fremiet, E. 1930 Growth and differentiation of the colonies of *Zoothamnium alternans* (Clap. and Lach.) Biol. Bull. 58:28–51.

Fauré-Fremiet, E. 1950a Morphologie comparée et systématique des Ciliés. Bull. Soc. Zool. Fr. 75:109–22.

Fauré-Fremiet, E. 1950b Mécanismes de la morphogénèse chez quelques ciliés gymnostomes hypostomiens. Arch. Anat. Microsc. Morphol. Exp. 39:1–14.

Fauré-Fremiet, E. 1950c Morphologie comparée des ciliés holotriches trichostomes. An. Acad. Bras. Cienc. 22:257–61.

Fauré-Fremiet, E. 1952 La diversification structurale des ciliés. Bull. Soc. Zool. Fr. 77:274–81.

Fauré-Fremiet, E. 1953 La bipartition énantiotrope chez les Ciliés. Oligotriches. Arch. Anat. Microsc. Morphol. Exp. 42:202–25.

Fauré-Fremiet, E. 1954a Réorganisation du type endomixique chez les Loxodidae et chez les *Centrophorella.* J. Protozool. 1:20–27.

Fauré-Fremiet, E. 1954b Les problèmes de la différentiation chez les protistes. Bull. Soc. Zool. Fr. 79: 311–29.

Fauré-Fremiet, E. 1954c Morphogénèse de bipartition chez *Urocentrum turbo* (cilié holotriche). J. Embryol Exp. Morphol. 2:227–38.

Fauré-Fremiet, E. 1958 The origin of the Metazoa and the stigma of the phytoflagellates. Quart. J. Microsc. Sci. 99:123–29.

Fauré-Fremiet, E. 1967a Chemical aspects of ecology. In *Chemical Zoology,* ed. M. Florkin and B. T. Scheer. Vol. I. *The Protozoa,* ed. G. W. Kidder. pp. 21–54. Academic Press, Inc., New York.

Fauré-Fremiet, E. 1967b La régénération chez les Protozoaires. Bull. Soc. Zool. 92:249–72.

Fauré-Fremiet, E. 1969a Remarques sur la systématique des ciliés Oligotrichida. Protistologica 5:345–52.

Fauré-Fremiet, E. 1969b A propos de *Deltopylum rhabdoides.* Acta Protozool. 7:25–26.

Fauré-Fremiet, E. and J. André 1967 Étude au microscope éléctronique du cilié *Pseudomicrothorax dubius* Maupas. J. Protozool. 14:464–73.

Fauré-Fremiet, E. and J. André 1968 Structure corticale d'une amibe édaphique. Protistologica 4:195–207.

Fauré-Fremiet, E., P. Favard, and N. Carasso 1962 Etude au microscope éléctronique des ultrastructures d'*Epistylis anastotica* (cilié péritriche). J. Microscopie 1:287–312.

Fauré-Fremiet, E., and M. Cl. Garnier 1970 Structure fine du *Strombidium sulcatum* Cl. et L. (Ciliata Oligotrichida). Protistologica 6:207–23.

Favard, P., and N. Carasso 1964 Etude de la pinocytose au niveau des vacuoles digestives de ciliés peritriches. J. Microscopie 3:671–96.

Fawcett, D. 1966 *The Cell, Its Organelles and Inclusions.* W. B. Saunders Co., Philadelphia.

Febvre, J. 1970 Les myonèmes d'Acanthaires. Essai d'interpretation de leur ultrastructure et leur cinétique. J. Protozool. 17 (suppl.): 34 (abst.).

Febvre, J. 1971 Le myonème d'Acanthaire: Essai d'interpretation ultrastructure et cinétique. Protistologica 7:379–91.

Finks, R. M. 1970 The evolution and ecologic history of sponges during palaeozoic times. In *Biology of the Porifera.* ed. W. G. Fry. No. 25 Symp. Zool. Soc. London pp. 3–22. Academic Press, Inc., New York.

Finley, H. 1952 Sexual differentiation in peritricous ciliates. J. Morph. 91:569–605.

Finley, H. 1969 What the peritrichs have taught us. J. Protozool. 16:1–5.

Finley, H. E., D. MacLaughlin, and D. M. Harrison 1959 Nonaxemic and axenic growth of *Vorticella microstoma.* J. Protozool. 6:201–5.

Fisher, R. A. 1930 *The Genetical Theory of Natural Selection.* Clarendon Press, Oxford.

Fitch, W. M., and E. Margoliash 1967 Construction of phylogenetic trees. Science 155:279–84.

Fjerdingstad, E. J. 1961a The ultrastructure of the choanocyte collars in *Spongilla lacustris* (L). Z. Zellforsch. 53:645–57.

Fjerdingstad, E. J. 1961b Ultrastructure of the collar of the choanoflagellate *Codonosiga botrytis* (Ehrenb.). Z. Zellforsch. 54:499–510.

Florkin, M., and B. T. Scheer (eds.) 1968 *Chemical Zoology.* Vol. II. *Porifera, Coelenterata, and Platyhelminthes.* Academic Press, Inc., New York.

Frankel, J. 1960a Morphogenesis in *Glaucoma chattoni.* J. Protozool. 7:362–76.

Frankel, J. 1960b Effects of localized damage on morphogenesis and cell division in a ciliate, *Glaucoma chattoni.* J. Exp. Zool. 143:175–94.

Frankel, J. 1962 The effects of heat, cold, and p-fluorophenylalanine on morphogenesis in synchronized *Tetrahymena pyriformis* Gl. Compt. Rend. Trav. Lab. Carlsberg 33:1–52.

Frankel, J. 1973 Dimensions of control of cortical patterns in *Euplotes:* The role of preexisting structure, the clonal life cycle, and the genotype. J. Exp. Biol. 183:71–94.

Franzen, A. 1956 On spermiogenesis, morphology of the spermatozoon, and biology of fertilization among invertebrates. Zool. Bid. Uppsala 31:355–482.

Freedman, L. J., and E. D. Hanson 1970 Further observations on the life cycle of *Leptomonas Karyophilus.* J. Protozool. 17 (suppl.): 18.

Friz, C. T. 1970 The protein amino acid compositions of *Amoeba proteus, A. discoides, A. dubia,* and *Pelomyxa carolinensis.* J. Protozool. 16:460–66.

Fry, W. G. ed. 1970 *Biology of the Porifera.* No. 25. Symp. Zool. Soc. London. Academic Press, Inc., New York.

Fuge, H. 1969 Electron microscopic studies on the intraflagellar structures of trypanosomes. J. Protozool. 16:460–66.

Fulton, C. 1971 Centrioles. In *Origin and continuity of cell organelles.* eds. J. Reinert and H. Ursprung. pp. 170–221. Springer Verlag, New York.

Fulton, J. D. 1969 Metabolism and pathogenic mechanisms of parasitic protozoa. In *Research in Protozoology,* ed. T.-T. Chen. pp. 389–504. Pergamon Press, Inc., New York.

Furgason, W. H. 1940 The significant cytostomal pattern of the "Glaucoma-Colpidium group," and a proposed new genus and species, *Tetrahymena geleii.* Arch. Protistenk. 94:224–26.

Gabriel, A. 1971 Évolution de l'incorporation de leucine tritiée dans la blastème et du taux de mitoses, au cours de la régénération de jeunes Planaires traitées par l'actinomycin D. C. R. Acad. Sci. (Paris) Sér. D. 272:2017–20.

Gall, J. G. 1959 Macronuclear duplication in the ciliated protozoan *Euplotes.* J. Biophys. Biochem. Cytol. 5:295–308.

Galloway, J. J. 1933 A Manual of the Foraminifera. Principia Press, Inc. Bloomington, Ind.

Garstang, W. 1946 The morphology and relations of the Siphonophora. Quart. J. Micro. Sci. 87:103–93.

Gatenby, J. B. 1920 The germ cells,

fertilisation and early development of *Grantia (Sycon) compressa*. J. Linn. Soc. (Zool.) 34:261–97.

Gatenby, J. B. 1927 Further notes on the gametogenesis and fertilisation of sponges. Quart. J. Microsc. Sci. 71: 173–88.

Gause, G. F. 1934 *The Struggle for Existence*. Williams and Wilkins, Baltimore.

Geiman, Q. M. 1964 Comparative physiology: Mutualism, symbiosis, and parasitism. Ann. Rev. Physiol. 26: 75–108.

Gelei, J. von 1930 "Echte" freie Nervenendigungen (Bemerkungen zu den Receptoren der Turbellarian). 2. Morph. Ökol. Tiere 18:786–98.

Gelei, J. von 1934 Der feinere Bau des Cytopharynx von *Paramecium* und seine systematische Bedeutung. Arch. Protistenk. 82:331–62.

Gelei, J. von 1935 Infusorien in Dienste der Forschung und des Unterrichtes. Biol. Zentr. 55:57–74.

Gelei, J. von 1950 Die Morphogenese der Einzeller mit Rücksicht auf die morphogenetischem Prinzipien von Sewertzoff. Acta biol. Acad. Sci. Hung. 1:69–134.

Ghiselin, M. T. 1969 *The Triumph of the Darwinian Method*. University of California Press, Berkeley.

Gibbons, I. R. and A. V. Grimstone 1960 On flagellar structure in certain flagellates. J. Biophys. Biochem. Cytol. 7:697–716.

Gibor, A. 1967 Inheritance of cytoplasmic organelles. In *Formation of Cell Organelles*. ed. K. B. Warren. pp. 305–16. Academic Press, New York.

Gibor, A., and S. Granick 1964 Plastids and mitochondria: Inheritable systems. Science 145:820–97.

Giese, A. C. 1973 *Blepharisma: The Biology of a Light-sensitive Protozoan*. Stanford University Press, Stanford.

Giles, N. H., M. E. Case, C. W. H. Partridge, and A. I. Ahmed 1967 A gene cluster in *Neurospora crassa* coding for an aggregate of five aromatic synthetic enzymes. Proc. Natl. Acad. Sci. U.S.A. 58:1453–60.

Gillies, C., and E. D. Hanson 1963 A new species of *Leptomonas* parasitizing the macronucleus of *Paramecium trichium*. J. Protozool. 10:467–73.

Gillies, C., and E. D. Hanson 1968 Morphogenesis of *Paramecium trichium*. Acta Protozool. 6:13–31.

Gilmour, J. S. L. 1940 Taxonomy and Philosophy. In *The New Systematics*. ed. J. S. Huxley pp. 461–74. Clarendon Press, Oxford.

Girgla, H. 1971 Cortical anatomy and morphogenesis in *Homalozoon vermiculare* (Stokes). Acta Protozool. 8:355–62.

Glaessner, M. F. 1947 *Principles of Micropaleontology*. John Wiley and Sons, Inc., New York.

Glaessner, M. F. 1963 Major trends in the evolution of the Foraminifera. In *Evolutionary Trends in the Foraminifera*. ed. G. H. R. von Koenigswald, J. D. Emeis, W. L. Bruning, and C. W. Wagner. pp. 9–24. Amsterdam Elsevier Publishing Company.

Goldschmidt, R. B. 1952a Evolution as viewed by one geneticist. Am. Sci. 40:84–98.

Goldschmidt, R. B. 1952b Homeotic mutations and evolution. Acta Biotheoretica (Leiden) 10:87–104.

Golikova, M. N. 1964 Polytene chromosomes in the developing macronucleus of a ciliate. Citologija 6:250–53.

Golinska, K. 1972 Studies on stomatogenesis in *Dileptus* (Ciliata, Holotricha) in the course of division processes. Acta Protozool. 9:283–97.

Golinska, K., and M. Doroszewski 1964 The cell shape of *Dileptus* in the course of division and regeneration. Acta Protozool. 2:59–67.

Goodwin, T. W. 1968 Pigments of Coelenterata. In *Chemical Zoology* eds. M. Florkin and B. T. Scheer. Vol. II. *Porifera, Coelenterata, and Platyhelminthes*. pp. 149–55. Academic Press, Inc., New York.

Goss, R. 1969 *Principles of Regeneration*. Academic Press, Inc., New York.

Graff, L. von 1882 *Monographie der*

Turbellarien. I. Rhabdocoelida. Leipzig.

Grain, J. 1966 Étude cytologique de quelques Ciliés Holotriches endocommensaux des Ruminants et des Equidés. Protistologica 2(1):59–141; 2(2):5–51.

Grain, J. 1968 Les systèmes fibrillaires chez *Stentor igneus* Ehrenberg et *Spirostomum ambiguum* Ehrenberg. Protistologica 4:27–35.

Grain, J. 1969 Le cinétosome et ses dérivés chez les Ciliés. Ann. Biol. 8: 53–97.

Grain, J. 1970 Structure et ultrastructure de *Lagynophora fusidens* Kahl, 1927. Protistologica 6:37–52.

Grain, J. 1972 Étude ultrastructurale d'*Halteria grandinella* O. F. M. (Cilié Oligotriche) et considérations phylogenetiques. Protistologica 8(2):179–97.

Grain, J. and K. Golinska 1969 Structure et ultrastructure de *Dileptus cygnus* Claparède et Lachman, 1859, cilié holotriche gymnostome. Protistologica 5:269–91.

Granick, S. 1959 The chloroplasts: Inheritance, structure and function. In *The Cell. Biochemistry, Physiology, Morphology.* eds. J. Brachet and A. E. Mirsky. Vol. II. p. 489–619. Academic Press, Inc., New York.

Grassé, P.-P., ed. 1952a *Phylogénie Protozaires: Généralités, Flagellés. Traité de Zoologie.* Tome I, fasc. I. Masson et Cie., Paris.

Grassé, P.-P. 1952b (Embranchement des Protozoaires) Généralités. *Traité de Zoologie.* ed. P.-P. Grassé. Tome I, fasc. I. pp. 37–132. Masson et Cie., Paris.

Grassé, P.-P. 1952c (Classe des Zooflagellés) Généralités. *Traité de Zoologie.* ed. P.-P. Grassé. Tome I, fasc. I. pp. 574–78. Masson et Cie., Paris.

Grassé, P.-P. 1952d Ordre des Trypanosomides. *Traité de Zoologie.* ed. P.-P. Grassé. Tome I, fasc. I. pp. 602–68. Masson et Cie., Paris.

Grassé, P.-P. 1952e Super-ordre des Opalines. In *Traité de Zoologie*, ed. P.-P. Grassé. Tome I, fasc. II. pp. 983–1004. Masson et Cie., Paris.

Grassé, P.-P., ed. 1953 *Protozaires: Rhizopodes, Actinopodes, Sporozoaires, Cnidosporidies. Traité de Zoologie* Tome I, fasc. II. Masson et Cie., Paris.

Grassé, P.-P. 1956a L'ultrastructure de *Pyrsonympha vertens* (Zooflagellata Pyrsonymphina): les flagelles et coaptation avec le corps, l'axostyle contractile, le paraxosyle, le cytoplasme. Arch. Biol. 67:595–611.

Grassé, P.-P. 1956b L'appareil parabasale et de Golgi sont un même organite. Leur ultrastructure, leurs modes de sécrétion. C. R. Acad. Sci. (Paris) 242:858–61.

Grassé, P.-P., and G. Deflandre 1952 Ordre des Bicoecidea. In *Traité de Zoologie* ed. P.-P. Grassé, Tome I fasc. I. pp. 599–601. Masson et Cie., Paris.

Greenberg, M. J. 1959 Ancestors, embryos, and symmetry. Syst. Zool. 8:212–21.

Grell, K. 1953a Die Chromosomen von *Aulacantha scolymantha.* Arch. Protistenk. 99:1–54.

Grell, K. G. 1953b Die Konjugation von *Ephelota gemmipara* R. Hertwig. Arch. Protistenk. 98:287–326.

Grell, K. G. 1953c Die Struktur des Macronucleus von *Tokophrya.* Arch. Protistenk. 98:466–68.

Grell, K. G. 1954 Der Generationswechsel der polythalamen Foraminifere *Rotaliella heterocaryotica.* Arch. Protistenk. 100:268–86.

Grell, C. 1956 Protozoa and algae. Ann. Rev. Microbiol. 10:307–28.

Grell, K. 1958 Untersuchungen über die Fortpflanzung und Sexualität der Foraminiferan, III. *Glabratella sulcata.* Arch. Protistenk. 102:449–72.

Grell, K. G. 1966 Amoeben der Familie Stereomyxidae. Arch. Protistenk. 109:147–54.

Grell, K. 1967 Sexual reproduction in Protozoa. In *Research in Protozoology* ed. T.-T. Chen. Vol. II. pp. 147–213. Pergamon Press, Inc., New York.

Grell, K. 1968 *Protozoologie* 2. Aufl. Springer, Berlin.

Grell, K. G. 1971a *Trichoplax adhaerens* F. E. Schulze und die

Entstehung der Metazoen. Naturwissenschaften Rundschau 24:160–61.

Grell, K. G. 1971b Embryonalentwicklung bei *Trichoplax adhaerens* F. E. Schulze. Naturwissenschaften 58:570.

Grell, K. 1972 Eibildung und Furchung von *Trichoplax adhaerens* F. E. Schulze (Placozoa). Z. Morphol. Tiere 73:297–314.

Grell, K. 1973 *Trichoplax adhaerens* and the origin of the Metazoa. *Actualité's Protozoologiques*. IVe. Cong. Int. Protozoologie. Paul Couty, Clermont-Ferrand.

Grell, K. G., and G. Benwitz 1971 Die Ultrastruktur von *Trichoplax adhaerens* F. E. Schulze. Cytobiologie 4:216–70.

Grell, K., and A. Ruthman 1964 Über die Karyologie des Radiolars und Feinstruktur seiner Chromosomen. Chromosoma 15:185–211.

Gremigni, V., and M. Benazzi 1969 Ricerche isochimiche e ultrastruturali sull'ovogenese dei Tricladi: I. Inclusi deutoplasmatici in *Dugesia lugubris* e *Dugesia benazii*. Atti Accad. Naz. Lincei Rend. CL Sec. Fes. Mat. Natur. Sez. III 47:101–8.

Grimstone, A. V. 1959 Cytology, homology, and phylogeny—a note on "organic design." Am. Nat. 93:273–82.

Grimstone, A. V. 1961 Fine structure and morphogenesis in protozoa. Biol. Rev. 36:97–150.

Grimstone, A. V., and L. R. Cleveland 1965 The fine structure and function of the contractile axostyles of certain flagellates. J. Cell Biol. 24:387–400.

Grimstone, A. V., and T. R. Gibbons 1966 The fine structure of the centriolar apparatus and associated structures in the complex flagellates *Trichonympha* and *Pseudotrichonympha*. Philos. Trans. Soc. Lond. 250:215–42.

Grobstein, C. 1964 *The Strategy of Life*. W. H. Freeman and Co., San Francisco.

Gruber, A. 1881 *Dimorpha mutans*. Eine Mischforme von Flagellaten und Heliozoen. Z. Wiss. Zool. 36:445.

Guilcher, Y. 1951 Contribution à l'étude des ciliés gemmipares, chonotriches et tentaculifères. Ann. Sci. Nat. Zool. (sér. 11) 13:33–132.

Günther, K. 1962 Systematik und Stammesgeschichte der Tiere. 1954–1959. Fortsch. Zool. 14:269–547.

Guttman, H. and F. G. Wallace 1964 Nutrition and physiology of the Trypanosomatidae. In *Biochemistry and Physiology of Protozoa*. ed. S. H. Hutner. Vol. III pp. 460–94. Academic Press, Inc., New York.

Hadzi, J. 1949 Die Ableitung der Knidarien von den Turbellarian und einige Folgerungen dieser Ableitung. C. R. XIIIe Cong. Zool. Paris, 1948.

Hadzi, J. 1953 An attempt to reconstruct the system of animal classification. Syst. Zool. 2:145–54.

Hadzi, J. 1963 *The Evolution of the Metazoa*. Pergamon Press, Inc. New York.

Haeckel, E. 1862 Die Radiolarien (Rhizopoda Radiara). Eine Monographie (mit Atlas). Reimer Verlag, Berlin.

Haeckel, E. 1866 *Generelle Morphologie der Organismen*. I. Allgemeine Anatomie der Organismen. II. Allgemeine Entwicklungsgeschichte der Organismen. G. Reimer Verlag, Berlin.

Haeckel, E. 1874 Die Gastraeatheorie, die phylogenetische Classifikation des Theirreichs und die Homologie der Keimblätter. Jenaische f. Naturwiss. 8:1–55.

Haeckel, E. 1887 Die Radiolarien. II. Grundriss einer allgemeinen Naturgeschichte der Radiolarien (mit Atlas). Reimer Verlag, Berlin.

Haeckel, E. 1888a Die Radiolarien. Theil III. Die Acantharien oder Actipyleen Radiolarien. Reimer Verlag, Berlin.

Haeckel, E. 1888b Die Radiolarien. IV. Die Phaeodarien oder Cannopyleen Radiolarien (mit Atlas). Reimer Verlag, Berlin.

Haeckel, E. 1894 Systematisches Phylogenie. Reimer Verlag, Berlin.

Haecker, V. 1908 Die Radiolarien in der Variations-und Artbildungslehre. Z.

Induct. Abstam. Vererbungslehre 2: 1–17.

Hafleigh, A. S., and C. A. Williams, Jr. 1966 Antigenic correspondence of serum albumins among the primates. Science 151:1530–35.

Haldane, J. B. S. 1932 *The Causes of Evolution*. Harper and Brothers, Publishers, New York.

Hall, R. P. 1953 *Protozoology*. Prentice-Hall, Inc. Englewood Cliffs, N.J.

Hall, R. P. 1965 *Protozoan Nutrition*. Blaisdell Publishing Company, New York.

Hall, B. D., and S. Spiegelman 1961 Sequence complimentarity of T2-DNA and T2-specific RNA. Proc. Natl. Acad. Sci. U.S.A. 47:137–46.

Hammen, C. S., and M. Florkin 1968 Composition and intermediary metabolism – Porifera. In *Chemical Zoology*. eds. M. Florkin and B. T. Scheer Vol. II. *Porifera, Coelenterata, and Platyhelminthes*. pp. 53–64. Academic Press, Inc., New York.

Hammond, D. M. 1937 The neuromotor system of *Euplotes patella* during binary fission and conjugation. Quart. J. Microsc. Sci. 79:507–57.

Hand, C. 1959 On the origin and phylogeny of the coelenterates. Syst. Zool. 8:191–202.

Hand, C. 1961 Present state of nematocyst research: Types, Structure and Function. In *The Biology of Hydra: 1961*. eds. H. M. Lenhoff and W. F. Loomis pp. 187–97. University of Miami Press, Coral Gables, Florida.

Hand, C. 1963 The early worm: A planula. In *The Lower Metazoa*. eds. E. C. Dougherty, Z. N. Brown, E. D. Hanson, and W. D. Hartman. pp. 33–39. University of California Press, Berkeley.

Hand, C. 1966 On the evolution of the Actiniaria. In *The Cnidaria and their Evolution*. ed. W. J. Rees. No. 16 Symp. Zool. Soc. London. pp. 135–46. Academic Press, Inc., New York.

Handler, P., ed. 1970 *Biology and the Future of Man*. Oxford University Press, New York.

Hanson, E. D. 1958 On the origin of the Eumetazoa. Syst. Zool. 7:16–47.

Hanson, E. D. 1961 *Convoluta sutcliffei*, a new species of acoelous Turbellaria. Trans. Am. Microsc. Soc. 80:423–33.

Hanson, E. D. 1962 Morphogenesis and Regeneration of Oral Structures in *Paramecium aurelia:* An Analysis of Intracellular Development. J. Exp. Zool. 150:45–68.

Hanson, E. D. 1963a Homologies and the ciliate origin of the Eumetazoa. In *The Lower Metazoa*. eds. E. C. Dougherty, Z. N. Brown, E. D. Hanson, and W. D. Hartman. pp. 7–22. University of California Press, Berkeley, Calif.

Hanson, E. D. 1963b Stages of scientific inquiry. Connecticut Med. 27:186–95.

Hanson, E. D. 1964 What are the basic principles of classification based on comparative cytology? Syst. Zool. 13:44–46.

Hanson, E. D. 1966 Evolution of the cell from primordial living systems. Quart. Rev. Biol. 41:1–12.

Hanson, E. D. 1967a Protozoan Development. In *Chemical Zoology* eds. M. Florkin and B. T. Scheer Vol. I. *Protozoa*, ed. G. W. Kidder. pp. 395–539. Academic Press, Inc. New York.

Hanson, E. D. 1967b Regeneration in acoelous flatworms: the role of the peripheral parenchyma. Roux' Arch. 159:298–313.

Hanson, E. D. 1969 Major patterns of ciliate stomatogenesis. In *Progress in Protozoology* Proc. IIIrd Int. Cong. Protozool. p. 97. Pub. House "Nauka," Leningrad.

Hanson, E. D. 1972 *Animal Diversity* 3rd edition. Prentice-Hall, Inc., Englewood Cliffs, New Jersey.

Hanson, E. D. 1975 Major evolutionary trends in animal protists. J. Protozool. 23:4–12.

Hanson, E. D., C. Gillies and M. Kaneda 1969 Oral structure, development and nuclear behavior during conjugation in *Paramecium aurelia*. J. Protozool. 16:197–204.

Hanson, E. D., and M. Kaneda 1968

Evidence for sequential gene action within the cell cycle of *Paramecium*. Genetics 60:793–805.

Hanson, E. D., T. Nakabayashi, M. Ishibashi, and S. Inoki 1963 Resistance to the drug propamidine in *Leishmania donovani*. Biken's J. 6:1–7.

Hanson, E. D., and R. M. Ungerleider 1973 The formation of the feeding organelle in *Paramecium aurelia*. J. Exp. Zool. 185:175–88.

Hanson, N. R. 1958 *Patterns of Discovery*. The University Press, Cambridge.

Hardy, A. C. 1953 On the origin of the Metazoa. Quart. J. Microsc. Sci. 94:441–43.

Hartmann, M. 1911 *Die Konstitution der Protistenkerne und ihre Bedeutung für die Zellenlehre*. Fischer, Jena.

Hartmann, M. 1913 Morphologie and Systematik der Amöben. *Handbuch der pathogenen Mikroorganismen*. Vol. 7 2nd ed. Jena.

Hartman, W. D. 1963 A critique of the enterocele theory. In *The Lower Metazoa* eds. E. C. Dougherty, Z. N. Brown, E. D. Hanson, and W. D. Hartman. pp. 55–77. University of California Press, Berkeley.

Hartman, W. D. 1971 Calcarea, Calcinea, Calcaranea (Vol. 2); Demospongia (Vol. 3); Hexactinellida (Vol. 6); Parazoa (Vol. 9); Porifera (Vol. 10). *McGraw-Hill Encyclopedia of Science and Technology*. McGraw-Hill Book Co., Inc. New York.

Hartman, W. D., and T. F. Goreau 1970 Jamaican coralline sponges: their morphology, ecology and fossil relatives. In *Biology of the Porifera* ed. W. G. Fry. No. 25 Symp. Zool. Soc. London pp. 205–43. Academic Press, Inc., New York.

Hatschek, B. 1888 *Lehrbuch der Zoologie*. Jena.

Hawes, R. S. J. 1963 The emergence of asexuality in the Protozoa. Quart. Rev. Biol. 38:234–42.

Hay, E. D. 1968 Structure and function of the nucleolus in developing cells. In *The Nucleus* eds. A. J. Dalton and F. Haguenau. pp. 1–79. Academic Press, Inc., New York.

Hayes, M. L. 1938 Cytological studies in *Dileptus anser*. Trans. Am. Microsc. Soc. 57:11–25.

Hayes, W. 1968 *The Genetics of Bacteria and their Viruses*. 2nd ed. Wiley, New York.

Heal, O. W. 1963 Morphological variation in certain Testacea (Protozoa, Rhizopoda). Arch. Protistenk. 106:351–68.

Heath, H. 1928 A sexually mature turbellarian resembling Müller's larva. J. Morph. 45:187–203.

Heckmann, K. 1965 Total Konjugation bei *Urostyla hologama* n. sp. Arch. Protistenk. 108:55–62.

Heckmann, K. 1967 Age-dependent intraclonal conjugation in *Euplotes crassus*. J. Exp. Zool. 165:269–78.

Hegner, R. 1920 The relations between nuclear number, chromatin mass, cytoplasmic mass, and shell characteristics in four species of the Genus *Arcella*. J. Exp. Zool. 30:1–95.

Hempel, C. G. 1966 *Philosophy of Natural Science*. Prentice-Hall, Inc., Englewood Cliffs, N.J.

Hendelberg, J. 1969 On the development of different types of spermatozoa from spermatids with two flagella in the Turbellaria with remarks on the ultrastructure of the flagella. Zool. Bid. Uppsala 38:1–50.

Hendelberg. J. 1970 On the number and ultrastructure of the flagella of flatworm spermatozoa. In *Comparative Spermatology*. Proc. 1st Int'l Symp. (1969) ed. B. Bocetti pp. 367–74. Accd. Naz. Lincei, Academic Press, Inc. New York.

Henderson, I. F., W. D. Henderson, and J. H. Kenneth 1963 *A Dictionary of Biological Terms*. Oliver and Boyd, London.

Hennig, W. 1950 *Grundzüge einer Theorie der phylogenetischen Systematik*. Deutscher Zentralverlag, Berlin.

Hennig, W. 1955 Phylogenetic systematics. Ann. Rev. Ent. 10:97–116.

Hennig. W. 1966 *Phylogenetic Systematics*. trans. by D. D. Davis and R. Zangerl. University of Illinois Press, Urbana.

REFERENCES

Hentschel, E. 1923–1925 Parazoa. Porifera, Schwämme. In *Handbuch der Zoologie*. eds. E. Kukenthal and T. Krumbach. Bd. I. pp. 307–418. de Gruyter and Co., Berlin.

Hertwig, O. 1875–1878 Beiträge zur Kenntnis der Bildung, Befruchtung, und Teilung des tierischen Eies, I–IV. Morphol. Jb. 1, 3, und 4.

Hertwig, R. 1876 *Zur Histologie der Radiolarien*. Leipzig.

Hertwig, R. 1879 *Der Organismus der Radiolarien*. Fischer Verlag, Jena.

Hertwig, R. 1889 Ueber der Conjugation der Infusorien. Abh. bayer. Akad. Wiss. 17:150–233.

Hertwig, R. 1899 Ueber Kerntheilung, Richtungskörperbildung, und Befruchtung von *Actinosphaerium Eichhornii*. Abh. bayer. Akad. Wissen. (Math.-Physik. Classe) 19:631–734.

Hertwig, R. and E. Lesser 1874 Ueber Rhizopodon und denselben nahestehenden Organismen. Arch. mikro. Anat. 10 (suppl.): 35–243.

Heywood, V. H., and J. McNeill 1964 *Phenetic and Phylogenetic Classification*. Systematics Association, London.

Hill, D., and J. W. Wells 1956 Cnidaria—General features. In *Treatise on Invertebrate Paleontology* (F.) *Coelenterata*. ed. R. C. Moore. pp. F5–F9. Geol. Soc. of America, Univ. Kansas Press, Lawrence, Kansas.

Hinegardner, R. T. 1968 Evolution of cellular DNA content in teleost fishes. Amer. Nat. 102:517–23.

Hinegardner, R. T., and J. Engelberg 1963 Rationale for a universal genetic code. Science 142:1083–85.

Hirsch, J. G. 1962 Cinemicrophotographic observations on granule lysis in polymorphonuclear leucocytes during phagocytosis. J. Exp. Med. 116:827–34.

Hoare, C. A. 1964 Morphological and taxonomic studies on mammalian trypanosomes. X. Revision of the systematics. J. Protozool. 11:200–207.

Hoare, C. A. 1967 Evolutionary trends in mammalian trypanosomes. Adv. Parasitol. 5:47–91.

Hollande, A. 1952a Classe des Cryptomonadines. In *Traité de Zoologie*. ed. P.-P. Grassé. Tome I, fasc. I, pp. 239–84. Masson et Cie., Paris.

Hollande, A. 1952b Classe des Chrysomonadine. In *Traité de Zoologie*. ed. P.-P. Grassé Tome I, fasc. I. pp. 471–570. Masson et Cie., Paris.

Hollande, A. 1952c Ordre des Choanoflagellés ou Craspédomonadines. In *Traité de Zoologie*. ed. P.-P. Grassé Tome I. fasc. I. pp. 239–84. Masson et Cie., Paris.

Hollande, A. 1952d Ordre des Bodonides. In *Traité de Zoologie*. ed. P.-P. Grassé. Tome I, fasc. I. pp. 669–93. Masson et Cie., Paris.

Hollande, A., and M. Cachon-Enjumet 1959 La polyploidie du noyau végétatif des Radiolaires. C. R. Acad. Sci., Paris. 248:2641–43.

Hollande, A., and M. Cachon-Enjumet 1960 Cytologie, évolution, et systématiques des Sphaerellaires (Radiolaires). Arch. Mus. d'Hist. Nat. 7e série. 7:1–134.

Hollande, A., M. Cachon, and J. Valentin 1967 Infrastructure des axopodes et organisation général de *Sticholonche zanclea* Hertwig (Radiolaire Sticholonchidea). Protistologica 3: 155–66.

Hollande, A., and M. Enjumet 1955 Sur l'évolution et la systématique des Labyrinthulidae. Étude de *Labyrinthuea algeriensis* nov. sp. Ann. Sci. Nat. Zool. 17:356–68.

Hollande, A., J. Cachon, and M. Cachon-Enjumet 1965 Les modalités de l'enkystement présporogénétique chez les Acanthaires. Protistologica 1:91–104.

Hollande, A., J. Cachon and M. Cachon 1969 La dinomitose atractophorienne à fuseau endonucleaire chez les Radiolaires Thalanophysidae. Son homologie avec la mitose des Foraminifères et avec celle des levures. C. R. Acad. Sci., Paris. Ser. D 269: 179–82.

Hollande, A., J. Cachon, and M. Cachon 1970 La signification de la membrane capsulaire des Radiolaires et ses rapports avec la plasmalemme et les membranes du réticulum endoplasmique. Affinités entre

Radiolaires, Héliozoaires et Péridiniens. Protistologica 6:311–18.

Hollande, A. and J. Carruette-Valentin 1971 Les atractophores, l'induction du fuseau de la division cellulaire chez les Hypermastigines. Étude infrastructurale et révision systématique des *Trichonymphines* et des *Spirotrichonymphines*. Protistologica 7:5–100.

Hollande, A., and J. Valentin 1967 Interprétation des structures dites "centriolaire" chez les Hypermastigines symbiontes des Termites et du *Cryptocercus*. C. R. Acad. Sci. Paris. 264:1868–71.

Hollande, A., and J. Valentin 1969 Appareil de Golgi, pinocytose, lysosomes, mitochondries, bactéries symbiontiques, atractophores, et pleuromitose chez les Hypermastigines du genre *Joenia*. Affinités entre Joeniides et Trichomonadines. Protistologica 5:39–86.

Holley, R. W., J. Apgar, G. Everett, J. Madison, M. Marquisse, S. Merrill, J. Penswick, and A. Zamir 1965 Structure of ribonucleic acid. Science 147:1462–65.

Holt, P. A. and G. B. Chapman 1971 The fine structure of the cyst wall of the ciliated protozoan *Didinium nasutum*. J. Protozool. 18:604–14.

Holt, P. A., and J. O. Corliss 1973 Pattern variability in microtubular arrays associated with the tentacles of *Actinobolina* (Ciliatea: Gymnostomatida). J. Cell Biol. 58:213–19.

Holz, G. G., Jr. 1960 Structural and functional changes in a generation in *Tetrahymena*. Biol. Bull. 118:84–95.

Honigberg, B. M. 1963 Evolutionary and systematic relationships in the flagellate order Trichomonadida Kirby. J. Protozool. 10:20–63.

Honigberg, B. M. 1967 Chemistry of parasitism among some Protozoa. In *Chemical Zoology*. eds. M. Florkin and B. T. Scheer. Vol. I *The Protozoa* ed. G. W. Kidder. pp. 695–814. Academic Press, Inc., New York.

Honigberg, B. M., W. Balamuth, E. C. Bovee, J. O. Corliss, M. Gojdics, R. P. Hall, R. R. Kudo, N. D. Levine, A. R. Loeblich, Jr., J. Weiser, and D. H. Wenrich 1964 A revised classification of the phylum Protozoa. J. Protozool. 11:7–20.

Honigberg, B. W., C. F. T. Mattern, and W. A. Daniel 1971 Fine structure of the mastigont system in *Tritrichomonas foetus* (Riedmüller). J. Protozool. 18:183–98.

Horn, W. 1929 Ueber die Zukunft der Insekten. Systematik Anz. Schädlingsk. 5:40–45.

Horowitz, N. H. 1945 On the evolution of biochemical syntheses. Proc. Nat. Acad. Sci. U.S.A. 31:153–57.

Horridge, G. A. 1969 Statocysts of medusae and evolution of stereocilia. Tissue Cell 1:341–53.

Houssay, D. and M. Prenant 1970 *Thecamoeba spheronucleus*. Données physiologique et étude morphologique en microscopie optique et et électronique. Arch. Protistenk. 112:228–51.

Hovasse, R. 1965 Ultrastructure comparée des axopodes chez les Héliozoaires des genres *Actinosphaerium, Actinophrys*, et *Raphidiophrys*. Protistologica 1:81–88.

Howard, A., and S. Pelc 1951 Nuclear incorporation of P^{32} as demonstrated by autoradiographs. Exp. Cell Res. 2:178–87.

Howell, J. V. 1960 *Glossary of Geology and Related Sciences*. Am. Geol. Inst., Washington, D.C.

Hoyer, B. H., B. J. McCarthy, and E. T. Bolton 1964 A molecular approach in the systematics of higher organisms. Science 144:959–67.

Hull, D. L. 1967 Certainty and circularity in evolutionary taxonomy. Evol. 21:174–89.

Hull, D. L. 1968 The operational imperative: Sense and nonsense in operationism. Syst. Zool. 17:438–57.

Humphreys, T. 1970 Biochemical analysis of sponge cell aggregation. In *Biology of the Porifera* ed. W. G. Fry. No. 25 Symp. Zool. Soc. London pp. 325–34. Academic Press, Inc., New York.

Huneeus, F. C. and P. F. Davison 1970

Fibrillar proteins from squid axons. I. Neurofilament proteins. J. Mol. Biol. 52:415–28.

Hutchinson, G. E. 1959 Homage to Santa Rosalia, or Why are there so many different kinds of animals? Amer. Nat. 93:145–59.

Huth, W. 1913 Zur Entwicklungsgeschichte der Thalassicollen. Arch. Protistenk. 30:1–124.

Hutner, S. H. and L. Provasoli 1951 The phytoflagellates. In *Biochemistry and Physiology of Protozoa.* Vol. I. ed. A. Lwoff. pp. 29–128. Academic Press, Inc., New York.

Hutner, S. H., H. Baker, O. Frank, and D. Cox 1972 In *Int'l. Encyclopedia of Food and Nutrition* ed. H. M. Sinclair. Vol. 18 *Nutrition of Lower Organisms* ed. R. N. Fiennes. pp. 85–177. Pergamon Press, Oxford.

Huxley, J. (ed.) 1940 *The New Systematics.* Clarendon Press, Oxford.

Huxley, J. 1943 *Evolution. The Modern Synthesis.* Harper and Brothers, Publishers, New York.

Huxley, J. 1957 Three types of evolutionary process. Nature 180:454–55.

Huxley, J. S. and G. R. de Beer 1934 *The Elements of Experimental Embryology.* Cambridge Univ. Press, Cambridge.

Hyman, L. H. 1940 *The Invertebrates.* Vol. I. *Protozoa through Ctenophora.* McGraw-Hill Book Company, New York.

Hyman, L. H. 1940–1959 *The Invertebrates.* Vol. I–V. McGraw-Hill Book Co., Inc., New York.

Hyman, L. H. 1951 *The Invertebrates.* Vol. II. *Platyhelminthes and Rhynchocoela.* McGraw-Hill Book Company, Inc., New York.

Hyman, L. H. 1959 *The Invertebrates.* Vol. V. *Smaller Coelomate Groups.* McGraw-Hill Book Company, New York.

Ingram, V. M. 1957 Gene mutations in human haemoglobin: the chemical difference between normal and sickle cell haemoglobin. Nature 180:326–28.

Inoki, S., K. Nakanishi, and T. Nakabayashi 1959 Observations on *Trichomonas vaginalis* by electron microscopy. Bikens J. 2:21–24.

Inoki, S., T. Nakabayashi, S. Fukukita, and H. Osaki 1960 Studies on immunological variation in *Trypanosoma gambiense.* V. Antigenic constitution of the relapse type strain in respect to its reversibility to the original type. Bikens J. 3:339–50.

Inoki, S., M. Ohno, K. Kondo, and H. Sakamato 1961 Electron microscopic observations on the "costa" as one of the organelles of *Trichomonas foetus.* Bikens J. 4:63–65.

Ivanic, M. 1934 Bau, Lebensweise und Entwicklungsgeschichte, nebst Bemerkungen über die systematische Stellung von *Vampyrellidium vagans* Zopf. La Cellule 43:147–74.

Ivanov, A. V. 1971 K voprusu o primitivnosti beskishechnykh resnichnykh chervei-Acoela. [Primitivism of Acoela.] Zool. Zh. 50:621–32.

Jägersten, G. 1955 On the early phylogeny of the Metazoa. The Bilaterogastraea theory. Zool. Bid. Uppsala 40:321–54.

Jägersten, G. 1959 Further remarks on the early phylogeny of the Metazoa. Zool. Bid. Uppsala 33:79–108.

Jägersten, G. 1972 *Evolution of the Metazoan Life Cycle; A Comprehensive Theory.* Academic Press, Inc., New York.

Jahn, T. L. and E. C. Bovee 1964 Protoplasmic movements and locomotion of Protozoa. In *Physiology and Biochemistry of the Protozoa.* Vol. III ed. S. H. Hutner. pp. 62–129. Academic Press, New York.

Jahn, T. L. and E. C. Bovee 1965a Movement and locomotion of microorganisms. Ann. Review Microbiol. 19:21–58.

Jahn, T. L. and E. C. Bovee 1965b Mechanisms of movement in taxonomy of Sarcodina. I. As a basis for a new major dichotomy into two classes, *Autotractea* and *Hydraulea.* Am. Midl. Nat. 73:30–40.

Jahn, T. L. and E. C. Bovee 1967

Motile behavior of Protozoa. In *Research in Protozoology*. ed. T.-T. Chen. Vol. 1. pp. 41–200.

Jahn, T. L. and R. A. Rinaldi 1959 Protoplasmic movement of the foraminiferan *Allogromia laticolloris*, and theory of its mechanism. Biol. Bull. 117:100–118.

Jahn, T. L., and F. F. Jahn 1949 *How to Know the Protozoa*. William C. Brown Co., Dubuque, Iowa.

Jankowski, A. W. 1964 Morphology and evolution of Ciliophora. III. Diagnoses and phylogenesis of 53 sapropelebionts, mainly of the order Heterotrichida. Arch. Protistenk. 107:185–294.

Jankowski, A. W. 1967 [A new system of ciliate Protozoa (Ciliophora).] Akad. Nauk SSSR, Trudy Zool. Inst. 43:3–54. (In Russian. Translated by E. S. Shafer. Copy kindly supplied by Drs. J. O. Corliss and E. B. Small, University of Maryland, U.S.A.)

Jankowski, A. W. 1973 [Taxonomic revision in the subphylum Ciliophora Doflein, 1901.] Zool. Zh. 52:165–75. (In Russian with English summary. Translation provided by Drs. J. O. Corliss and E. B. Small, University of Maryland, U.S.A.)

Jareño, M. A., P. Alonso, and J. Pérez-Silva 1970 Induced autogomy in 2 species of *Stylonychia*. J. Protozool. 17:384–88.

Jenkins, M. M. 1970 Sexuality in *Dugesia dorotocephala*. J. Biol. Psychol. 12:71–80.

Jennings, H. S. 1908 Heredity, variation and evolution in Protozoa. II. Heredity and variation in size and form in *Paramecium* with studies in growth, environmental action and selection. Proc. Am. Phil. Soc. 47:393–546.

Jennings, H. S. 1916 Heredity, variation and the results of selection in the uniparental reproduction of *Difflugia corona*. Genetics 1:407–534.

Jennings, H. S. 1923 *Behavior of the Lower Organisms*. Columbia University Press, New York.

Jennings, H. S. 1941 Inheritance in Protozoa. In *Protozoa in Biological Research*. eds. G. N. Calkins and F. M. Summers. pp. 710–71. Columbia Univ. Press, New York.

Jennings, J. B. 1957 Studies on feeding, digestion and food storage in free-living flatworms (Platyhelminthes: Turbellaria). Biol. Bull. 112:63–80.

Jennings, J. B. 1963 Some aspects of nutrition in the Turbellaria, Trematoda, and Rhynchocoela. In *The Lower Metazoa*. eds. E. C. Dougherty, Z. N. Brown, W. D. Hanson, and W. D. Hartman. pp. 345–353. University of California Press, Berkeley.

Jennings, J. B. 1968 Nutrition and Digestion. In *Chemical Zoology* eds. M. Florkin and B. T. Scheer. Vol. II. *Porifera, Coelenterata, and Platyhelminthes*. pp. 303–26. Academic Press, Inc., New York.

Jensen, D. D. 1959 A theory of the behavior of *Paramecium aurelia* and behavioral effects of feeding, fission, and ultraviolet microbeam irradiation. Behavior 15:82–122.

Jerka-Dziadosz, M. 1964 *Urostyla cristata*, sp. n. (Urostylidae, Hypotrichida); the morphology and morphogenesis. Acta Protozool. 2:123–28.

Jerka-Dziadosz, M. 1972 Cortical development in *Urostyla*. I. Comparative study on morphogenesis in *U. cristata* and *U. grandis*. Acta Protozool. 10:73–99.

Jerka-Dziadosz, M. and J. Frankel 1969 An Analysis of the Formation of Ciliary Primordia in the Hypotrich Ciliate *Urostyla* weissei. J. Protozool. 16:612–37.

Jhering, H. von 1877 *Vergleichende Anatomie des Nervensystems und Philogenie der Mollusken*. Leipzig.

Jinks, J. L. 1964 Extrachromosomal inheritance. Prentice-Hall, Inc., Englewood Cliffs, N.J.

Johannsen, W. 1911 The genotype conception of heredity. Am. Nat. 45:129–59.

Johnson, W. H. 1956 Nutrition of Protozoa. Ann. Rev. Microbiol. 10:193–212.

Jollos, V. 1924 Untersuchungen über Variabilität und Vererbung bei *Arcella*.

Arch. Protistenk. 49:307–74.
Jones, W. C. 1962 Is there a nervous system in sponges? Biol. Rev. 37:1–50.
Jörgensen, E. 1924 Mediterranean Tintinnidae. Rep. Danish oceanogr. Exped. 1908–1910 Medit. 2 (Biol.) pp. 1–110.
Jörgensen, E. and A. Kahl 1933 Tintinnidae (Nachträge). In *Die Tierwelt der Nord- und Ostsee*. eds. G. Grimpe and E. Wagler. Lief. 23 (Teil II, c2) pp. 27–28. Leipzig.
Joyon, L. and J.-P. Mignot 1969 Données recentes sur la structure de la cinétide chez les protozoaires flagellés. Année Biol. 4e. serie 8:1–52.
Joyon, L., J.-P. Mignot, M. R. Kattar, and G. Brugerolle 1969 Compléments à l'étude des Trichomonadida et plus particulièrement de leur cinétide. Protistologica 5:309–26.
Jukes, T. H. 1965 Coding triplets and their possible evolutionary implications. Biochem. Biophys. Res. Commun. 19:391–96.
Jukes, T. H. 1966 *Molecules and Evolution*. Columbia Univ. Press, New York.
Kaczanowski, A. 1965 About a presumable phylogenetic link between *Mesnilella* (Astomata) and *Conchostoma longissimum* Fauré-Fremiet (Trichostomata). Acta Protozool. 3: 225–32.
Kaczanowska, J. and D. Kowalska 1969 Studies on topography of the cortical organelles of *Chilodonella cucullus* (O.F.M.). I. The cortical organelles and intraclonal dimorphism. Acta Protozool. 6:1–15.
Kahl, A. 1930–1935 Urtiere oder Protozoa. I. Wimpertiere oder Ciliata (Infusoria). In *Die Tierwelt Deutschlands* Dahl. F. 1. Allgemeiner Teil und Prostomata (1930) pp. 1–180. 2. Holotricha (1931) pp. 181–398. 3. Spirotricha (1932) pp. 399–650. 4. Peritricha Chonotricha (1935) pp. 651–886. Fischer, Jena.
Kahl, A. 1931 Über die verwandtschaftlichen Beziehungen der Suctorien zu den prostomen Infusorien. Arch. Protistenk. 73:423–81.
Kahl, A. 1934 Suctoria. In Tierwelt der Nord- und Ostee eds. G. Grimpe and A. Remane. pp. 184–226 (published 1940) Akademische Verlagsgesellschaft, Leipzig.
Kalley, J. P. and T. Bisalputra 1970 *Peridinium trochoideum:* The fine structure of the theca as shown by freeze-etching. J. Ultrastructure Res. 31:95–108.
Kaneda, M. 1961 Fine structure of the macronucleus of the gymnostome ciliate, *Chlamydodon pedarius*. Jap. J. Genet. 36:223–34.
Kaneda, M. and E. D. Hanson 1974 Growth Patterns and Morpho-Genetic Events in the Cell Cycle of *Paramecium aurelia*. In *Paramecium: A Current Survey*. ed. W. J. van Wagtendonk. pp. 219–62. Elsevier Publishing Company, Amsterdam.
Kaplan, N. O. 1965 Evolution of dehydrogenases. In *Evolving Genes and Proteins*. eds. V. Bryson and H. T. Vogel. pp. 243–78. Academic Press, Inc., New York.
Karling, T. 1963 Some evolutionary trends in turbellarian morphology. In *The Lower Metazoa*. eds. E. C. Dougherty, Z. N. Brown, E. D. Hanson, and W. D. Hartman. pp. 225–33. University of California Press, Berkeley.
Kates, J. R. and L. Goldstein 1964 A comparison of the protein composition of three species of amoebae. J. Protozool. 11:30–35.
Kato, K. 1940 On the development of some Japanese polyclads. Jap. J. Zool. 8:537–73.
Kavanau, L. J. 1963 A new theory of amoeboid locomotion. J. Theoret. Biol. 4:124–41.
Kenk, R. 1970 Freshwater triclads (Turbellaria) of North America. IV. The polypharyngeal species of Phagocota. Smithson. Contr. Zool. 80:1–17.
Kent, S. 1880–1882 *Manual of the Infusoria, including a Description of all known Flagellate, Ciliate, and Tentaculiferous Protozoa, British and Foreign, and an Account of the Organization and Affinities of the Sponges*. (3 vols.) Bogue, London.

Kerkut, G. A. 1960 *Implications of Evolution*. Pergamon Press, New York.

Kidder, G. W. 1967 Nitrogen: Distribution, Nutrition, and Metabolism. In *Chemical Zoology* eds. M. Florkin and B. T. Scheer. Vol. I. *Protozoa*. ed. G. W. Kidder. pp. 93–159. Academic Press, New York.

Kidder, G. W., and V. C. Dewey 1951 The biochemistry of ciliates. In *Biochemistry and Physiology of Protozoa*. ed. A. Lwoff. Vol. I. pp. 323–400. Academic Press, Inc., New York.

Kiefer, B. I., H. Sakai, A. Solari, and D. Mazia 1966 The molecular unit of the microtubules of the mitotic apparatus. J. Mol. Biol. 20:75–79.

Kimball, R. F. 1942 The nature and inheritance of mating types in *Euplotes patella*. Genetics 27:269–85.

Kimball, R. F., and S. W. Perdue 1962 Quantitative cytochemical studies on *Paramecium*. V. Autoradiographic studies of nucleic acid synthesis. Exp. Cell Res. 27:405–15.

Kimball, R. F., and D. M. Prescott 1962 Deoxyribonucleic acid synthesis and distribution during growth and amitosis of the macronucleus of *Euplotes*. J. Protozool. 9:88–92.

King, J. L. and T. H. Jukes 1969 Non-Darwinian evolution. Science 164:788–98.

Kinosita, H., and A. Murakami 1967 Control of ciliary motion. Physiol. Rev. 47:53–82.

Kirby, H. 1937 Host-parasite relations in the distribution of Protozoa in termites. Univ. Calif. Publ. Zool. 41:189–212.

Kirby, H. 1944 Some observations on cytology and morphogenesis in flagellate Protozoa. J. Morph. 75:361–421.

Kirby, H. 1947 Flagellate and host relationships of trichomonad flagellates. J. Parasitol. 33:214–28.

Kirby, H. 1949 Systematic differentiation and evolution of termites in flagellates. Revista Soc. Mex. Hist. Nat. 10:57–79.

Kirby, H. 1950 *Materials and Methods in the Study of Protozoa*. pp. 1–72. Univ. Calif. Press, Berkeley.

Kitching, J. A. 1964 The axopods of the sun animalcule, *Actinophrys sol* (Heliozoa). In *Primitive Motile Systems in Cell Biology*. eds. R. D. Allen and N. Kamiya pp. 445–56. Academic Press, Inc., New York.

Kitching, J. A. 1967 Contractile vacuoles, ionic regulation, and excretion. In *Research in Protozoology*. ed. T.-T. Chen. Vol. 1. pp. 307–36. Pergamon Press, New York.

Klein, B. M. 1926 Ergebnisse mit einer silber Methode bei Ciliaten. Arch. Protistenk. 56:243–79.

Kluyver, A. J., and C. B. van Niel 1956 *The Microbe's Contribution to Biology*. Harvard University Press, Cambridge.

Koenigswald, G. N. R. von, J. D. Emeis, W. L. Buning, and C. W. Wagner (eds.) 1963 *Evolutionary Trends in Foraminifera*. Elsevier Publishing Company, Amsterdam.

Kofoid, C. A. 1930 Factors in the evolution of the pelagic Ciliata, the Tintinnoinea. Contrib. Mar. Biol. Stanford Univ., pp. 1–39.

Kofoid, C. A. and A. S. Campbell 1929 A conspectus of the marine and fresh-water Ciliata belonging to the sub-order Tintinnoinea, with descriptions of new species principally from the Agassiz Expedition to the eastern tropical Pacific 1904–1905. Univ. Calif. Publ. Zool. 34:1–403.

Kohn, A. 1973 Review of *Evolution of the Metazoan Life Cycle* by G. Jägersten. Science 179:789–90.

Koidl, B. 1970 Zur Regeneration des Nervensystems von *Bothrioplana semperi* Braun. Zool. Anz. 185:75–84.

Koopowitz, H. 1970 Feeding behavior and the role of the brain in the polyclad flatworm *Planocera gilchristi*. Anim. Behav. 18:31–35.

Kormos, J. 1958 Phylogenetische Untersuchungen und Suctorien. Acta Biol. Acad. Sci. Hungaricae 9:9–23.

Kowalevsky, A. 1866 Entwicklungsgeschichte der einfachen Ascidien. Mem. Acad. Sci. St.-Pétersbourg ser. 7. 10:1–19.

Kozloff, E. 1965 New species of acoel turbellarians from the Pacific Coast. Biol. Bull. 129:151-66.

Kudo, R. R. 1954 *Protozoology* 4th edition. Charles C. Thomas, Publisher, Springfield, Illinois.

Kudo, R. R. 1965 *Protozoology* 5th edition. Charles C. Thomas, Publisher, Springfield, Illinois.

Kühn, A. 1914 Entwicklungsgeschichte und Verwandtschaftsverhältnisse der Hydrozoen. I. Die Hydroiden. Erg. Fortschr. Zool. 4:1-284.

Kühn, A. 1926 *Morphologie der Tiere im Bildern*. II Heft. Berlin.

Kühn, A. 1955 *Vorlesungen über Entwicklungsphysiologie*. Springer-Verlag, Berlin.

Kümmel, G. and J. Brandenberg 1961 Die Reussengeisselzellen (Cyrtocyten). Z. Naturforsch. 16:692-97.

Kung, C. and R. Eckert 1972 Genetic modification of electric properties of an excitable membrane. Proc. Nat. Acad. Sci. U.S.A. 69:93-97.

Kuźnicki, L. 1970 Mechanisms of the motor responses of *Paramecium*. Acta Protozool. 8:83-118.

Lacassagne, M. 1968 Les anisorhizes atriches, nouveau type de nématocystes stomocnides présents chez deux formes d'*Halammohydra* (Hydrozoaires, Actinulides). C. R. Acad. Sci. Sér. D. 266:2090-92.

Lanfranchi, A. 1969 Nuovi otoplanisi (Turbellaria Proseriata) delle coste della liquria e della toscana. Boll. Zool. 36:167-88.

Lang, A. 1884 Die Polycladen (Seeplanarien). *Fauna und Flora Golfes von Neapel.* 11 Monograf. Leipzig.

Lavier, G. 1942-1943 L'évolution de la morphologie dans le genre *Trypanosoma*. Ann. Parasitol. 19: 168-96.

Laubenfels, M. W. de 1955 Porifera. In *Treatise on Invertebrate Paleontology*. ed. R. C. Moore. Part E. *Archaeocyatha and Porifera*. pp. E21-E112. Geol. Soc. Amer. and University of Kansas Press, Lawrence, Kansas.

Le Calvez, J. 1938 Recherches sur les Foraminifères. Arch. Zool. exp. gen. 80:163-333.

Le Calvez, J. 1953 Ordre des Foraminifères. In *Traité de Zoologie*. ed. P.-P. Grassé. Tome I. fasc. II. pp. 149-265. Masson et Cie., Paris.

Lee, J. T. and M. E. McEnery 1970 Autogamy in *Allogromia laticollaris* (Foraminifera). J. Protozool. 17:184-95.

Leedale, G. F. 1967 Euglenida/Euglenophyta. Ann. Rev. Microbiol. 21:31-48.

Leff, J. and N. I. Krinsky 1967 A mutagenic effect of visible light mediated by endogenous pigments in *Euglena gracilis*. Science 158:1332-35.

Lehninger, A. 1965 *The Mitochondrion*. W. A. Benjamin, Inc., New York.

Lehninger, A. 1970 *Biochemistry: The Molecular Basis of Cell Structure and Function*. Worth, New York.

Lenhoff, H. M. and W. F. Loomis 1961 *The Biology of Hydra and Some Other Coelenterates* 1961. Univ. of Miami Press, Coral Gables, Florida.

Lenhoff, H. M. 1968 Chemical perspectives on the feeding response, digestion, and nutrition of selected coelenterates. In *Chemical Zoology* eds. M. Florkin and B. T. Scheer Vol. II *Porifera, Coelenterata, and Platyhelminthes*. pp. 157-221. Academic Press, Inc., New York.

Leone, C. A. (ed.) 1964 *Taxonomic Biochemistry and Serology*. Ronald Press, New York.

Lévi, C. 1956 Étude des *Halisarca* de Roscoff. Embryologie et systématique des démosponges. Arch. Zool. exp. gén. 93:1-184.

Lévi, C. 1963 Gastrulation and larval phylogeny in sponges. In *The Lower Metazoa* eds. E. C. Dougherty, Z. N. Brown, E. D. Hanson, and W. D. Hartman. pp. 375-82. Univ. California Press, Berkeley, Calif.

Lévi, C. 1970 Les Cellules des éponges. In *Biology of the Porifera* ed. W. G. Fry. No. 25 Symp. Zool. Soc. Lond. pp. 353-64. Academic Press, Inc., New York.

Levine, N. D. 1962 Protozoology today. J. Protozool. 9:1-6.

Lewis, E. B. 1945 The relation of

repeats to position effect in *Drosophila melanogaster*. Genetics 30:137–66.

Lewis, K. R. and B. John 1970 *The Organization of Heredity*. Amer. Elsevier Publ. Co., Inc., New York.

Lewontin, R. C. (editor) 1968 *Population Biology and Evolution*. Syracuse University Press, Syracuse, N.Y.

Lewontin, R. C. 1970 The units of selection. Ann. Rev. Ecol. Syst. 1:1–18.

Li, C. C. 1955 *Population Genetics*. University of Chicago Press, Chicago.

Lilly, D. M. 1967 Growth factors in Protozoa. In *Chemical Zoology*. eds. M. Florkin and B. M. Scheer. Vol. I *Protozoa*. ed. G. W. Kidder. pp. 275–307. Academic Press, Inc., New York.

Linnaeus, C. 1758 Systema naturae per regna tria naturae, secundum classes, ordines, genera, species cum characteribus, differentiis, synonymes, locis. Editio decima, reformata, Tom. I. Laurentiae Salvii, Holmiae.

Linstedt, K. J. 1971 Biphasic feeding response in a sea anemone: control by asparagine and glutathione. Science 173:333–34.

Loeblich, A. R. and H. Tappan 1961 Suprageneric classification of the Rhizopoda. J. Paleontol. 35:245–330.

Loeblich, A. R. and H. Tappan 1964 Sarcodina. Chiefly "Thecamoebians" and Foraminiferida. In *Treatise on Invertebrate Paleontology*. ed. R. C. Moore. Part C. Protista 2, vol. I and II. Geol. Soc. Am. and Univ. of Kansas Press, Lawrence, Kansas.

Loening, U. E. 1968 Molecular weights of ribosomal RNA in relation to evolution. J. Mol. Biol. 38:355–65.

Lom, J. 1964 The morphology and morphogenesis of the buccal ciliary organelles in some peritrichous ciliates. Arch. Protistenk. 107:131–62.

Lom, J. and N. Kozloff 1969 Ultrastructure of the cortical regions of ancistrocomid ciliates. Protistologica 5:173–92.

Lowell, R. D. and A. L. Burnett 1969 Regeneration of complete Hydra from epidermal explants. Biol. Bull. 137:312–20.

Lubinsky, G. 1957 Studies on the evolution of the Ophryoscolecidae (Ciliata: Oligotricha). III. Phylogeny of the Ophryoscolecidae based on their comparative morphology. Canad. J. Zool. 35:141–59.

Luck, D. J. L. 1965 Formation of mitochondria in *Neurospora crassa*. Amer. Nat. 99:241–53.

Lund, E. J. 1917 Reversibility of morphogenetic processes in *Bursaria*. J. Exp. Zool. 24:1–33.

Luporini, P. 1970 Life cycle of autogamous strains of *Euplotes minuta*. J. Protozool. 17:324–28.

Lwoff, A. 1923 Sur un infusoire cilié homocaryote à vie libre. C. R. Acad. Sci. Paris. 177:910.

Lwoff, A. 1926 Le cycle nucléaire de *Stephanopogon mesnili* Lw. (cilié homocaryote). Arch. Zool. exp. gén. 78:117–32.

Lwoff, A. 1943 *L'Evolution Physiologique. Étude des pertes de fonctions chez les microorganismes*. Hermann, Paris.

Lwoff, A. 1950 *Problems of Morphogenesis in Ciliates*. John Wiley and Sons, Inc., New York.

Lwoff, M. 1951 The nutrition of parasitic flagellates (Trypanosomidae, Trichomonadinae). In *Biochemistry and Physiology of Protozoa* Vol. I. ed. A. Lwoff. p. 129–76. Academic Press, Inc., New York.

Lwoff, A. and H. Dusi 1935 La suppression expérimentale des chloroplastes chez *Euglena mesnili*. C. R. Soc. Biol. 119:1092.

MacGinnitie, G. E. and M. MacGinnitie 1949 *Natural History of Marine Animals*. McGraw-Hill Book Company, New York.

Machemer, H. 1969a Eine 2-Gradienten-Hypothese für die Metachronieregulation bei Ciliaten. Arch. Protistenk. 111:100–28.

Machemer, H. 1969b Regulation der Cilienmetachronie bei der "Fluchreaktion" von *Paramecium*. J. Protozool. 16:764–71.

Mackie, G. O. 1965 Conduction in the nerve-free epithelia of siphonophores. Amer. Zool. 5:439–53.

Mackinnon, D. L. and R. S. J. Hawes 1961 *An Introduction to the Study of Protozoa*. Oxford University Press, Oxford.

Madison, J. T., G. A. Everett, and H. Kung 1966 Nucleotide sequence of a yeast tyrosine transfer RNA. Science 153:531–34.

Marcus, E. 1945 Sobre Catenulido Brasileiros. Sao Paulo.

Marcus, E. 1954 Turbellaria brasileiros (11). Pop. Avulsos Dept. Zoologia, Sao Paulo, Brasil. 11, 24:419–89.

Marcus, E. 1958 On the evolution of the animal phyla. Quart. Rev. Biol. 33:24–58.

Marcus, E. and W. MacNae 1954 Architomy in a species of *Convoluta*. Nature 173:130.

Margoliash, E. and E. L. Smith 1965 Structural and functional aspects of cytochrome *c* in relation to evolution. In *Evolving Genes and Proteins*. eds. V. Bryson and H. J. Vogel. pp. 221–42. Academic Press, Inc., New York.

Margulis, L. 1970 *Origin of Eukaryotic Cells*. Yale University Press, New Haven, Conn.

Markert, C. and H. Ursprung 1971 *Developmental Genetics*. Prentice-Hall, Inc. Englewood Cliffs, New Jersey.

Marmur, J., S. Falkow, and M. Mandel 1963 New approaches to bacterial taxonomy. Ann. Rev. Microbiol. 17:329–72.

Maslin, T. P. 1952 Morphological criteria of phyletic relationships. Syst. Zool. 1:49–70.

Mast, S. O. 1926 Structure, movement, locomotion, and stimulation in Amoeba. J. Morph. 41:347–425.

Mattern, C. F. T., B. W. Honigberg, and W. A. Daniel 1967 The mastigont system of *Trichomonas gallinae* (Rivalta) as revealed by electron microscopy. J. Protozool. 14:320–29.

Mattern, C. F. T., B. M. Honigberg, and W. A. Daniel 1969 Structure of *Hypotrichomonas acosta* (Moskowitz) (Monocercomonadidae, Trichomonadida) as revealed by electron microscopy. J. Protozool. 16:668–85.

Maupas, E. 1889 La rajeunissement karyogamigue chez les ciliés. Arch. Zool. exp. gén. 7:149–517.

Mayr, E. 1940 Speciation phenomena in birds. Am. Nat. 74:249–78.

Mayr, E. 1942 *Systematics and the Origin of Species*. Columbia University Press, New York.

Mayr, E. 1961 Cause and effect in biology. Science 134:1501–6.

Mayr, E. 1963 *Animal Species and Evolution*. Belknap Press, Harvard University Press, Cambridge.

Mayr, E. 1969 *Principles of Systematic Zoology*. McGraw-Hill Book Company, New York.

Mayr, E., G. Linsley, and R. L. Usinger 1953 *Methods and Principles of Systematic Zoology*. McGraw-Hill Book Company, New York.

Mereschkowsky, C. 1910 Theorie der zwei Plasmaarten als Grundlage der Symbiogenesis, einer neuen Lehre von der Entstehung der Organismen. Biol. Centralbl. 30:278–303, 321–47, 353–67.

Messier, P.-E. 1971 Sub-pellicular microtubules in *Crithidia fasciculata*. J. Protozool. 18:223–31.

Milder, R. and M. P. Deane 1969 The cytostome of *Trypanosoma cruzi* and *T. conorhini*. J. Protozool. 16:730–37.

Millecchia, L. L., and M. A. Rudzinska 1972 The permanence of the infraciliature in Suctoria: An electronmicroscopic study of pattern formation in *Tokophrya infusionum*. J. Protozool. 19:473–83.

Miller, R. P. 1970 Sperm migration prior to fertilization in the hydroid *Gonothyrea loveni*. J. Exp. Zool. 75:493–504.

Minchin, E. A. 1912 *An Introduction to the Study of the Protozoa*. Edw. Arnold, London.

Minchin, E. A. 1915 The evolution of the cell. Rept. Brit. Ass. Adv. Sci. 85:437–64.

Monod, J. 1971 *Chance and Necessity*. (trans. A. Wainhouse) Alfred A. Knopf, New York.

Monod, J., J.-P. Changeux, and F. Jacob 1963 Allosteric proteins and cellular control systems. J. Mol. Biol. 6:306–29.

Moore, R. C. 1956 Introduction. In *Treatise on Invertebrate Paleontology.* Part F. *Coelenterata.* pp. F2–F4. Geol. Soc. Amer., Univ. Kansas Press, Lawrence, Kansas.

Morat, G. 1970 Études cytophotometriques et autoradiographiques de l'appareil nucleaire de *Colpidium campylum* au cours du cycle cellulaire. Protistologica 6:83–95.

Moroczewski, J. 1965 Taxocénoses des *Testacea* de quelques petits bassin de terrains inondables de la Narew. Acta. Protozool. 3:189–213.

Moroff, T. and G. Stiasny 1909 Über Bau und Entwicklung von *Acanthometron pellucidum* J. M. Arch. Protistenk. 16:207–36.

Morowitz, H. 1968 *Energy Flow in Biology.* Academic Press, New York.

Morowitz, H. 1971 Energy flow and biological organization. (A paper presented at the International Biophysics Meeting, Sept. 1968, Cambridge, Mass.)

Moses, M. J. 1964 The nucleus and chromosomes: A cytological perspective. In *Cytology and Cell Function* ed. G. H. Bourne. pp. 423–558. Academic Press, Inc., New York.

Mosevich, T. M. 1968 The fine structure of the macronucleus in *Ichthyophthirius multifiliis* at different stages of the life cycle. Protistologica 4:469–76.

Mugard, H. 1948 Régulation du nombre des cinéties au course du cycle de croissance et de division chez un Cilié *Ichthyophthirius multifiliis* Fouquet. Arch. Anat. Microsc. 37: 204–13.

Mugard, H. 1949 Contribution à l'étude des infusoires hymenostomes histiophages. Ann. Sci. Nat. Zool. (ser. 11) 10:171–268.

Muller, H. J. 1932 Some genetic aspects of sex. Amer. Nat. 66:118–38.

Muller, H. J. 1955 Life. Science 121: 1–9.

Muller, H. J. 1958 Human values in relation to evolution. Science 127: 625–29.

Muller, J. 1858 Ueber die Thalassicollen, Polycystinen, und Acanthometren des Mittlemeeres. Abh. Akad. Wiss., Berlin. pp. 1–62.

Müller, W. A. 1969a Die Steuerung des morphogenetischen Fliessgleichgewichts in den Polypen von *Hydractinia echinata.* II. Biologischexperimentalle Untersuchungen. Roux' Arch. 163: 334–56.

Müller, W. A. 1969b Die Steuerung des morphogenetischen Fleissgleichgewichts in den Polypen von *Hydractinia echinata.* II. Chemischanalytische Untersuchungen. Roux' Arch. 163:357–74.

Myers, E. H. 1936 The life-history of *Spirillina vivipara* Ehrenberg, with notes on morphogenesis, systematics, and distribution of the Foraminifera. J. R. Microsc. Soc. 56:120–46.

Myers, E. H. 1943 Life activities of the Foraminifera in relation to marine ecology. Am. Phil. Soc., Proc. 86: 439–58.

Myers, E. H. and W. S. Cole 1957 Foraminifera. In *Treatise on Marine Ecology and Paleoecology.* ed. J. W. Hedgpeth. Vol. 1 pp. 1075–81. Geol. Soc. Am., Mem. 67.

Naef, A. 1919 *Idealistische Morphologie und Phylogenetik.* Fischer Verlag, Jena.

Naef, A. 1931 Phylogenie der Tiere. In *Handbuch der Vererbungswissenschaft* eds. E. Baur and M. Hartmann. Vol. III, 1 (Liefg. 13)

Naitoh, Y. and R. Eckert 1969 Ciliary orientation: Controlled by cell membrane or by intracellular fibrils? Science 166:1633–35.

Nanney, D. L. 1966 Cortical integration in *Tetrahymena:* An exercise in cytogeometry. J. Exp. Zool. 161:307–18.

Nanney, D. L. 1967 Cortical slippage in *Tetrahymena.* J. Exp. Zool. 166: 163–70.

Nanney, D. L. 1968 Cortical patterns in cellular morphogenesis. Science 160:496–502.

Needham, A. E. 1959 The origination of life. Quart. Rev. Biol. 34:189–209.

Netzel, H. 1971 Die Schalenbildung

bei der Thekamöben Gattung *Arcella* (Rhizopoda, Testacea). Cytobiologie 3:89–92.

Netzel, H. 1972a Die Bildung der Gehäusewand bei der Thekamöbe *Centropyxis discoides* (Rhizopoda, Testacea). Z. Zellforsch. 135:45–54.

Netzel, H. 1972b Die Schalenbildung bei *Difflugia oviformis* (Rhizopoda, Testacea). Z. Zellforsch. 135:55–61.

Netzel, H. 1972c Morphogeneses des Gehäuses von *Euglypha rotunda* (Rhizopoda, Testacea). Z. Zellforsch. 135:63–69.

Netzel, H. and H. H. Heunert 1971 Die Zellteilung bei *Arcella vulgaris* var. *multinucleata* (Rhizopoda, Testacea). Arch. Protistenk. 113:285–92.

Newton, W. D. 1970 Gastrulation in the turbellarian *Hydrolimax grisea* (Platyhelminthes: Plagiostomidae): Formation of the epidermal cavity, inversion and epiboly. Biol. Bull. 139:539–48.

Nielsen, M. H., J. Ludvik, and R. Nielsen 1966 On the ultrastructure of *Trichomonas vaginalis*. J. Micros. 5:229–50.

Nilsson, J. R. 1962 Observations on *Neobursaridium gigas* Balech, 1941 (Ciliata Heterotrichida). J. Protozool. 9:273–76.

Nilsson, J. R. 1969 The fine structure of *Neobursaridium gigas*. C. R. Trav. Lab. Carlsberg 37:49–76.

Nilsson, J. R. 1970 Suggestive structural evidence for macronuclear "subnuclei" in *Tetrahymena pyriformis* Gl. J. Protozool. 17:539–48.

Noirot-Timothée, C. 1958 L'ultrastructure de la limite ectoplasme-endoplasme et des fibres formant le caryophore chez les ciliés du genre *Isotricha* Stein (holotriches trichostomes). C. R. Acad. Sci. Paris 247:692–95.

Noirot-Timothée, C. 1968 Les sacs parasomaux sont des sites de pinocytose. Étude expérimentale à l'aide de thorotrast chez *Trichidinopsis paradoxa* (Ciliata Peritricha). C. R. Acad. Sci. Paris Sér. D. 268:2334–36.

Noirot-Timothée, C. 1969 Les affinités des Entodiniomorphida (Ciliata). Prog. Protozool. Third Int. Cong. Protozool. p. 371. Pub. House "Nauka," Leningrad.

Noirot-Timothée, C. and J. Lom 1965 L'ultrastructure de l'"Hoplocinetie" des cilies peritriches. Comparaison avec la membrane undulante des hymenostomes. Protistologica 1:33–40.

Noland, L. E. and M. Gojdics 1967 Ecology of Free-living Protozoa. In *Research in Protozoology*. ed. T.-T. Chen Vol. 2, pp. 215–66. Pergamon Press, New York.

Norrevang, A. 1964 Choanocytes in the skin of *Harrimania kupferi* (Enteropneusta). Nature 204:398–99.

Noble, E. R., W. L. McRary, and E. T. Beaver 1953 Cell division in trypanosomes. Trans. Am. Microsc. Soc. 72:236–48.

Novikoff, A. B., E. Essner, and N. Quintana 1964 Golgi apparatus and lysosomes. Fed. Proc. 23:1010–22.

Nursall, J. R. 1959 The origin of the Metazoa. Trans. Roy. Soc. Canada 53: 1–5.

Nursall, J. R. 1962 On the origins of the major groups of animals. Evol. 16:118–23.

Odum, H. T. 1941 Notes on the strontium content of sea water celestine radiolarian and strontionite snail shells. Science 114:211–13.

Olive, L. 1970 The Mycetozoa: A revised classification. Bot. Rev. 36: 59–89.

Oosawa, F. and S. Higashi 1967 Statistical thermodynamics of polymerization and polymorphism of protein. Prog. Theor. Biol. 1:79–164.

Oosawa, F., M. Kosai, S. Hatano, and S. Asakura 1966 Polymerization of actin and flagellin. In *Principles of Biomolecular Organization*. eds. G. E. W. Wolstenholme and C. M. O'Connor. pp. 273–303. Little, Brown and Co., Boston, Mass.

Oparin, A. I. 1962 *Life: Its Nature, Origin, and Development*. Academic Press, Inc., New York.

Organ, A. E., E. C. Bovee, and T. L. Jahn 1967 The mechanism of the nephridial apparatus of *Paramecium multimicronucleatum*. II. The filling of the water expulsion vesicle. J. Protozool. 14 (suppl.): 13 (abst.).

Orgel, L. E. 1968 Evolution of the genetic apparatus. J. Mol. Biol. 38: 381–93.

Orlov, A. (ed.) 1962 *Fundamentals of Paleontology*. General Part. Protozoa. (Trans. from Russian (1959) by the Israel Program for Scientific Translations. Jerusalem)

Ovchinnikova, L. P. 1970 Variability of DNA content in micronuclei of *Paramecium bursaria*. Acta Protozool. 7:211–20.

Owen, R. 1848 Report on the archetype and homologies of the vertebrate skeleton. Reports 16th Meeting British Assoc. Adv. Sci. pp. 169–340.

Ozeki, Y., T. Ono, S. Okubo, and S. Inoki 1970 Electron microscopy of DNA released from ruptured kinetoplasts of *Trypanosoma gambiense*. Biken J. 13:387–93.

Ozeki, Y., V. Sooksri, T. Ono, and S. Inoki 1971 Studies on the ultrastructure of kinetoplasts of *Trypanosoma cruzi* and *Trypanosoma gambiense* by autoradiography enzymatic digestion. Biken J. 14: 97–118.

Page, F. C. 1967 Taxonomic criteria for limax amoebae, with descriptions of 3 new species of *Hartmannella* and 3 of *Vahlkampfia*. J. Protozool. 14:499–521.

Page, F. C. 1968 Generic criteria for *Flabellula*, *Rugipes*, and *Hyalodiscus*, with descriptions of species. J. Protozool. 15:9–26.

Pantin, C. F. A. 1951 Organic Design. Adv. Sci. 8:138–50.

Pantin, C. F. A. 1956 The origin of the nervous system. Pubbl. Staz. Zool. Napoli 28:171–81.

Pantin, C. F. A. 1966 Homology, analogy, and chemical identity in the Cnidaria. In *The Cnidaria and their Evolution*. ed. W. J. Rees. No. 16 Symp. Zool. Soc. London. pp. 1–16. Academic Press, Inc., New York.

Papi, F. 1957 Sopra un nuovo Turbellario arcooforo di particolare significato filetico e sull posizione della fam. Hofsteniidae nel sistema dei Turbellari. Pubbl. Staz. Zool. Napoli 30:132–48.

Pappas, G. D. 1956 The fine structure of the nuclear envelope of *Amoeba proteus*. J. Biophys. Biochem. Cyt. 2 (suppl.): 431–34.

Párducz, B. 1952 Ú; gyorsfestö eljárás a véglénykutatás és okatás szolgálatában. Ann. hist-nat. Mus. Hung. 2:5–12.

Párducz, B. 1957 Reizphysiologische Untersuchungen an Ziliaten. VII. Das Problem der vorbestimmten Leitungsbahnen. Acta Biol. Acad. Sci. Hung. 8:219–51.

Párducz, B. 1959 Reizphysiologische Untersuchungen an Ziliaten. VIII. Ablauf der Fluchtreaktion bei allseitiger und anhaltender Reizung. Ann. hist-nat. Mus. Hung. 51:227–46.

Párducz, B. 1967 Ciliary movement and coordination in ciliates. Int. Rev. Cytol. 21:91–128.

Parke, M., I. Manton, and B. Clarke 1955 Studies on marine flagellates. II. Three new species of *Chrysochromulina*. J. mar. biol. Ass. U. K. 34:579–609.

Parker, G. H. 1910 The reaction of sponges, with a consideration of the origin of the nervous system. J. Exp. Zool. 8:1–41.

Parker, G. H. 1919 *The Elementary Nervous System*. Lippincott, Philadelphia.

Pascher, A. 1918 Flagellaten und Rhizopoden und ihre gegenseitigen. Arch. Protistenk. 38:1–88.

Pascher, A. 1927 *Volvocales-Phytomonadinae*, Süsswasserflora Deutschlands, Osterreichs und der Schweiz. Jena.

Pascher, A. 1931 Systematische Übersicht über die mit Flagellaten in Zusammenhang stehender Algenreihen und Versuch einer Einreihung dieser Algenstämme in de Stämme des Pflanzenreiches. Beih. Bot. Zbl. II 48:317–32.

Passano, L. M. 1963 Primitive nervous systems. Proc. Nat. Acad. Sci. U.S.A. 50:306–13.

Passano, K. N. and L. M. Passano 1971 The endodermal nerve net of Scyphozoa. J. Morph. 133:105–24.

Pätau, K. 1937 Ueber die Natur der sog. Anisosporen der Radiolarien. Verh. Deutsch. Zool. Ges. Leipzig. 39:93–98.

Patterson, J. T. and W. S. Stone 1952 *Evolution in the Genus Drosophila.* MacMillan, New York.

Paulin, J. J. and J. O. Corliss 1969 Ultrastructural and other observations which suggest suctorian affinities for the taxonomically enigmatic ciliate *Cyathodinium.* J. Protozool. 16:216–23.

Pavans de Cecatty, M. 1955 Le système nerveux des Éponges calcaires et silicieux. Ann. Sci. Nat., Zool. (ser. 11) 17:203–88.

Pavan, M. and M. V. Dazzini 1971 Toxicology and pharmocology — Arthropoda. In *Chemical Zoology* Vol. VI *Arthropoda* Part B. eds. M. Florkin and B. T. Scheer. pp. 365–409. Academic Press, Inc., New York.

Pavans de Cecatty, M., Y. Thiney, and R. Garrone 1970 Les bases ultracturales des communications intercellularies dans les oscules de quelques éponges. In *Biology of the Porifera* ed. W. G. Fry. No. 25 Symp. Zool. Soc. London pp. 449–66. Academic Press, Inc., New York.

Pedersen, K. J. 1961a Studies on the nature of the planarian connective tissue. Zellforsch. 53:569–608.

Pedersen, K. J. 1961b Some observations on the fine structure of planarian protonephridia and gastrodermal phagocytes. Zellforsch. 53:609–28.

Pedersen, K. J. 1964 The cellular organization of *Convoluta convoluta,* an acoel turbellarian. A cytological, histochemical and fine structural study. Zellforsch. 64:655–87.

Pedersen, K. J. 1966 The organization of the connective tissue of *Discocelides langi* (Turbellaria, Polycladida). Zellforsch. 71:94–117.

Penard, E. 1902 *Faune Rhizopodique du Bassin du Léman.* Henry Kündig, Libraire de l'Institut, Genève.

Penard, E. 1904 *Les Héliozoaires d'Eau Douce.* Kündig et Fils, Genève.

Péres-Silva, J., and P. Alonso 1966 Demonstration of polytene chromosomes in the macronuclear anlage of oxytrichous ciliates. Arch. Protistenk. 109:65–70.

Perkins, F. O., R. W. Ramsey, and S. F. Street 1971 The ultrastructure of fishing tentacle muscle in the jellyfish *Chrysaora quinquecirrha:* A comparison of contracted and relaxed states. J. Ultrastruct. Res. 35:431–50.

Petersen, J. B., and J. B. Hansen 1954 Electron microscope observations on *Codonosiga botrytis* (Ehr.) James-Clark. Batan. Tidssker. 51:281–91.

Petersen, J. B., and J. B. Hansen 1960 Elektronenmikroskopische Untersuchungen von zwei Arten der Heliozoen-Gattung *Acanthocystis.* Arch. Protistenk. 104:547–53.

Petrunkevitch, A. 1952 Macroevolution and the fossil record of the Arachnida. Amer. Sci. 40:99–122.

Phleger, F. B. 1960 *Ecology and Distribution of Recent Foraminifera.* Johns Hopkins Press, Baltimore.

Pickens, L. E. R. and R. J. Skaer 1966 A review of researches on nematocysts. In *The Cnidaria and their Evolution.* ed. W. J. Rees. No. 16 Symp. Zool. Soc. London. pp. 19–50. Academic Press, Inc., New York.

Pirie, N. W. 1937 The meaninglessness of the terms life and living. In *Perspectives in Biochemistry.* eds. J. Needham and D. Green. pp. 11–22. Cambridge University Press, Cambridge.

Pitelka, D. R. 1963 *Electron Microscopic Structure of Protozoa.* Pergamon Press, New York.

Pitelka, D. R. 1965 New observations on the cortical ultrastructure in *Paramecium.* J. Microsc. 4:373–94.

Pitelka, D. R. 1969 Fibrillar systems in Protozoa. In *Research in Protozoology.* ed. T.-T. Chen. Vol 3. pp. 282–388. Pergamon Press, New York.

Pitelka, D. R. 1970 Ciliate ultrastructure: Some problems in cell biology. J. Protozool. 17:1–10.

Pokorny, K. S. 1967 *Labyrinthula.* J. Protozool. 14:697–708.

Pokorny, V. 1958 *Grundzüge der zoologishchen Mikropaläontologie.*

Deutscher Verlag des Wissenschaften, Berlin.

Poljansky, G. and I. B. Raikov 1961 Nature et origine du dualisme nucléaire chez les infusoires ciliés. Bull. Soc. Zool. Fr. 84:402–11.

Popofsky, A. 1909 Die Radiolarien der Antarktic. In *Deutsche Südpolar-Expedition 1901–1903*. ed. E. von Drygalski. Vol X (Zool. II) pp. 183–305. Reimer Verlag, Berlin.

Popofsky, A. 1913a Die Sphaerellarien des Warmwassersgebietes. In *Deutsche Südpolar-Expedition 1901–1903*. ed. E. von Drygalski. Vol. XIII (Zool. V) pp. 73–160. Reimer Verlag, Berlin.

Popofsky, A. 1913b Die Nassellarian des Warmwassersgebietes. In *Deutsche Südpolar-Expedition 1901–1903*. ed. E. von Drygalski. Vol. XIV (Zool. VI). pp. 217–416. Reimer Verlag, Berlin.

Porter, E. D. 1960 The buccal organelles in *Paramecium aurelia* during fission and conjugation with special reference to kinetosomes. J. Protozool. 7:211–17.

Porter, E. D. 1962 A theory of morphogenetic migration in *Paramecium aurelia*. J. Protozool. 9 (suppl.): 27.

Porter, K. R. 1964 Cell fine structure and biosynthesis of intercellular macromolecules. Biophys. J. 4:167–96.

Porter, K. R., and M. A. Bonneville 1963 *An Introduction to the Fine Structure of Cells and Tissues*. Lea and Febiger, Philadelphia.

Powers, D. A. 1970 A numerical taxonomic study of Hawaiian reef corals. Pacific Sci. 24:180–86.

Preer, J. R., Jr. 1969 Genetics of the Protozoa. In *Research in Protozoology* ed. T.-T. Chen. Vol. 3. pp. 129–278. Pergamon Press, Oxford.

Prescott, D. M. and R. F. Kimball 1961 Relation between RNA, DNA, and protein synthesis in the replicating nucleus of *Euplotes*. Proc. Natl. Acad. Sci. U.S.A. 47:686–93.

Preston, T. M. 1969 The form and function of the cytostome-cytopharynx of the culture forms of the elasmobranch haemoflagellate *Trypanosoma raiae* Laveran and Mesnil. J. Protozool. 16:320–33.

Pringsheim, E. G. 1952 On the nutrition of Ochromonas. Quart. J. Microsc. Sci. 93:71–96.

Pringsheim, E. G. 1963 *Farblose Algen*. Fischer Verlag, Stuttgart.

Prout, T. 1964 Structural reduction in evolution. Am. Nat. 98:239–49.

Provasoli, L., S. H. Hutner, and A. Schatz 1948 Streptomycin-induced chlorophyll-less races of *Euglena*. Proc. Soc. Exptl. Biol. Med. 69:279–82.

Puytorac, P. de 1959a Quelques observations sur l'évolution et les origines des cilies Astomatida. Proc. XV Int. Cong. Zool. London (1958) pp. 649–51.

Puytorac, P. de 1959b Structures et ultrastructures nucléolaires et périnucléolaires du macronoyau intermitotique des Ciliés Haptophryidae. C. R. Acad. Sci. Paris. 249:1709–11.

Puytorac, P. de 1967 Aspects de l'ultrastructure du cilié *Lembadion leucas* (Maskell). Protistologica 3: 269–74.

Puytorac, P. de 1968a Sur l'ultrastructure de la cavité buccale des Ciliés Hysterocinetida Diesing. C. R. Acad. Sci. Paris Sér. D. 266:1508–10.

Puytorac, P. de 1968b Rémarques à propos de l'ultrastructure du suçoir d'*Ancistrocoma myae* (K. et B.), Cilié Rhynocodea. C. R. Acad. Sci. Paris. Sér. D. 268:820–22.

Puytorac, P. de 1970 Definitions of ciliate descriptive terms. J. Protozool. 17:358.

Puytorac, P. de and J. Grain, eds. 1974 *Actualités Protozoologiques*. 4e. Int. Cong. Protozoologie. University of Clermont, France.

Puytorac, P. de and M. R. Kattar 1969 Observations sur l'ultrastructure de cilié *Helicoprorodon multinucleatum* Dragesco 1900. Protistologica 5:549–60.

Puytorac, P. de and T. Njiné 1970 Sur l'ultrastructure des *Loxodes* (Ciliés holotriches). Protistologica 6:427–44.

Puytorac, P. de and A. Savoie 1968 Observations cytologiques et biologiques sur *Prorodon palustris* nov.

sp. Protistologica 4:53–60.
Puytorac, P. de, M. Roque, and M. Tuffrau 1966 Étude cytologique du cilié *Philaster digitiformis* Fabre-Domergue. Protistologica 2:5–15.
Raabe, Z. 1964a Remarks on the principles and outline of the system of *Protozoa*. Acta Protozool. 2:1–18.
Raabe, Z. 1964b The taxonomic position and rank of *Peritricha*. Acta Protozool. 2:19–32.
Raabe, Z. 1967 Ordo Thigmotricha (Ciliata, Holotricha). Acta Protozool. 5:1–36.
Raabe, Z. 1968 Two new species of *Thigmotricha* (ciliata, Holotricha) from *Theodoxus fluviotitis*. Acta Protozool. 6:169–73.
Raabe, Z. 1969 Les processus morphogénétiques chez les Ciliés thigmotriches. In *Progress in Protozoology* Proc. IIIrd Int. Cong. Protozool. p. 83. Publ. House "Nauka," Leningrad.
Raabe, Z. 1970a Ordo *Thigmotricha* (*Ciliata-Holotricha*). II. Familia *Hemispheirdae*. Acta Protozool. 7:117–80.
Raabe, Z. 1970b Ordo *Thigmotrich* (*Ciliata-Holotricha*) III. Familiae *Ancistrocomidae* et *Sphenophryidae*. Acta Protozool. 7:387–463.
Radzikowski, S. 1965 Changes in the heteromeric macronucleus in division of *Chilodonella cucullulus* (Müller), Acta Protozool. 3:233–38.
Raff, R. A. and H. R. Mahler 1972 The non-symbiotic origin of mitochondria. Science 177:575–82.
Raikov, I. B. 1958 Der Formwechsel des Kernapparates einiger niederer Ciliaten. Arch. J. Protistenk. 103:129–92.
Raikov, I. B. 1959 Der Formwechsel des Kernapparates einiger niederer Ciliaten. II. Die Gattung *Loxodes*. Arch. Protistenk. 104:1–42.
Raikov, I. B. 1963a On the origin of nuclear dualism in ciliates. In *Progress in Protozoology*. Proc. 1st Int. Congress Protozool. (1961) Prague. pp. 253–58.
Raikov, I. B. 1963b The nuclear apparatus of *Remanella multinucleata* Kahl (Ciliata, Holotricha). Acta Biol. Hung. 14:221–29.
Raikov, I. B. 1969 The macronucleus of ciliates. In *Research in Protozoology*. ed. T.-T. Chen. Vol. 3, pp. 1–128. Pergamon Press, Oxford.
Raikov, I. B., E. M. Cheissin and E. G. Buse 1963 A photometric study of DNA content of macro- and micronuclei in *Paramecium caudatum*, *Nassula ornata* and *Loxodes magnus*. Acta Protozool. 1:285–300.
Randall, J. T. and J. M. Hopkins 1962 On the stalks of certain peritrichs. Philos. Trans. R. Soc. Lond. Series B 245:59–79.
Rappaport, R. 1971 Cytokinesis in animal cells. Int. Rev. Cytol. 31:169–213.
Raskin, G. 1925 Uber die Axopodien der Heliozoa und die Greiftentakeln der Ephelotidae. Arch. Protistenk. 52:207–16.
Rasmont, R. 1959 L'ultrastructure des choanocytes d'éponges. Ann. Sci. Nat. Zool. 12:253–62.
Rasmont, R. 1968 Nutrition and Digestion. In *Chemical Zoology* ed. M. Florkin and B. T. Scheer. Vol. II *Porifera, Coelenterata, and Platyhelminthes*. pp. 43–51. Academic Press, Inc., New York.
Rasmont, R., J. Bouillon, P. Castiaux, and G. Vandermeersche 1957 Structure submicroscopique de la collerette des choanocytes d'éponges. C. R. Acad. Sci. Paris 255:1571–74.
Read, C. P. 1968 Intermediary metabolism of flatworms. In *Chemical Zoology* ed. M. Florkin and B. T. Scheer. Vol. II *Porifera, Coelenterata, and Platyhelminthes*. pp. 327–57. Academic Press, Inc., New York.
Rees, W. J. ed. 1966a *The Cnidaria and Their Evolution*. No. 16 Symp. Zool. Soc. Lond. Academic Press, Inc., New York.
Rees, W. J. 1966b The evolution of the Hydrozoa. In *The Cnidaria and Their Evolution*. ed. W. J. Rees. No. 16 Symp. Zool. Soc. London. pp. 199–221. Academic Press, Inc., New York.
Reid, R. E. H. 1970 Tetraxons and demosponge phylogeny. In *Biology of the Porifera*. ed. W. G. Fry. No. 25

Symp. Zool. Soc. Lond. pp. 63–89. Academic Press, Inc., New York.

Reiff, I. 1968 Die genetische Determination multipler Paarungstypen bei dem Ciliaten *Uronychia transfuga* (Hypotricha, Euplotidae). Arch. Protistenk. 110:372–97.

Reisinger, E. 1923 Turbellaria Strudelwürmer. In *Biologie der Tiere Deutschlands*. ed. P. Schulze. Lief. 6, Teil 4.

Reisinger, E. 1935 Ergebnisse einer von E. Reisinger and O. Steinböck mit Hilfe des Rask-orsted Fonds durchgeföhrten Reise in Grönland 1926. 6. *Proporoplana jenseni* n. gen. n sp., ein morphologisch bedentsamer Turbellarientyp. Vidensk. Medd. Dansk. Naturh. Feren. Kbn. 98:243–59.

Reisinger, E. 1959 Anoromogenetische und parasitogene Syncytienbildung bei Turbellarian. Protoplasma 50:627–43.

Reisinger, E. 1961 Morphologie des Coelenteraten, acoelomaten und pseudocoelomaten Würmer. Fort. Zool. 13:1–82.

Reisner, A. H., J. Rowe, and H. M. Macindoe 1968 Structural studies on the ribosomes of *Paramecium:* Evidence for a "primitive" animal ribosome. J. Mol. Biol. 32:587–610.

Remane, A. 1927 *Halammohydra*, ein eigenartiges Hydrozoan der Nord- und Ostsee. Z. Morph. Ökol. Tiere. 7: 643–67.

Remane, A. 1954 Die Geschichte der Tiere. In *Die Evolution der Organismen*. ed. G. Heberer. 2nd ed. pp. 340–422. Fischer Verlag, Stuttgart.

Remane, A. 1955 Morphologie als Homologienforschung. Zool. Anz. Suppl. 18:159–83.

Remane, A. 1956 *Die Grundlagen des naturlichen Systems der vergleichenden Anatomie und Phylogenetik*. 2. Aufl. Geest und Portig K. G., Leipzig.

Remane, A. 1957 Zur Verwandtschaft und Ableitung der niederen Metazoen. Zool. Anz. Suppl. 21:179–96.

Remane, A. 1963a The evolution of the Metazoa from colonial flagellates vs. plasmodial ciliates. In *The Lower Metazoa*. eds. E. C. Dougherty, Z. N. Brown, E. D. Hanson, and W. D. Hartman. pp. 23–32. University of California Press, Berkeley.

Remane, A. 1963b The enterocelic origin of the celom. In *The Lower Metazoa*. eds. E. C. Dougherty, Z. N. Brown, E. D. Hanson, and W. D. Hartman. pp. 78–90. University of California Press, Berkeley, Calif.

Rensch, B. 1954 *Neuere Probleme der Abstammungslehre*. 2. Aufl. Enke Verlag, Stuttgart.

Rensch, B. 1960 *Evolution Above the Species Level*. Columbia University Press, New York.

Reschetnjak, V. V. 1955 Vertical distribution of Radiolaria of the Kurilo-Kamschatka depression. Trudy Zool. Inst. Acad. Sci. U.S.S.R. 21:94–101.

Reutterer, A. 1969 Zum Problem der Metazoenabstammung. Z. Zool. Syst. Evol. 7:30–53.

Reynoldson, T. P., and L. S. Bellamy 1970 The status of *Dugesia lugubris* and *D. polychroa* (Turbellaria, Tricladida) in Britain. J. Zool. (London) 162:157–77.

Riser, N. W. and M. Y. Morse eds. 1974 *Biology of the Turbellaria*. McGraw-Hill Book Company, New York.

Robertson, J. D. 1959 The ultrastructure of cell membranes and their derivatives. Biochem. Soc. Sum. 16:3–43.

Rogers, W. P. 1962 *The Nature of Parasitism*. Academic Press, Inc., New York.

Romanes, G. J. 1897 Darwin, and after Darwin. Vol. 3. Open Court, Chicago.

Romer, A. S. 1949 *The Vertebrate Body*. W. B. Saunders Co., Philadelphia.

Roodyn, D. B., and D. Wilkie 1968 *The Biogenesis of Mitochondria*. Methuen and Co., London.

Roque, M. 1956a L'évolution de la ciliature buccale pendant l'autogamie et la conjugaison chez *Paramecium aurelia*. C. R. Acad. Sci. Paris 242: 2592–95.

Roque, M. 1956b La stomatogénèse pendant l'autogamie, la conjugaison et la division chez *Paramecium aurelia*. C. R. Acad. Sci. Paris 243:1564–65.

Roque, M. 1961 Recherches sur les infusoires ciliés: les hymenostomes peniculiens. Bull. Biol. 45:439–519.

Roque, M., and P. de Puytorac 1967 Infraciliature d'un nouvel Ophryoglenidae: *Ichthyophthiriodes browni* n. g., n. sp. Protistologica 3:465–73.

Roque, M., P. de Puytorac, and J. Lom 1967 L'architecture buccale et la stomatogénèse d'*Ichthyophthirius multifilliis* Fouquet, 1976. Protistologica 3:79–90.

Roque, M., P. de Puytorac, and A. Savoie 1965 *Ophryoglena bacterocaryon*, sp. nov., Cilié heterotriche péniculien (cytologie, ultrastructure, cycle). Arch. Zool. exp. gén. 105:309–94.

Rose, S. M. 1970 *Regeneration*. Appleton Century Crofts.

Rose, P. G. and A. L. Burnett 1970 The origin of secretory cells in *Cordylophora caspia* during regeneration. Roux' Arch. 165:192–216.

Rosenbaum, J. and F. M. Child 1967 Flagellar regeneration in protozoan flagellates. J. Cell Biol. 34:345–64.

Roskin, G. I. 1957 Nervous system of sponges. Adv. Mod. Biol. Moscow (Usp. Sovr. Biol.) 43:199–207 (In Russian).

Ross, A. 1968 The substructure of centriole subfibers. J. Ultrastruct. Res. 23:537–39.

Ross, H. H. 1964 Review of *Principles of Numerical Taxonomy*. Syst. Zool. 13:106–8.

Rouiller, C. and E. Fauré-Fremiet 1957 Ultrastructure reticulée d'une fibre squelettique chez un cilié. J. Ultrastruct. Res. 1:1–13.

Rudzinska, M. A. 1970 The mechanism of food intake in *Tokophrya infusonium* and ultrastructural changes in food vacuoles during digestion. J. Protozool. 17:626–41.

Rudzinska, M. A., P. A. D'Alesandro, and W. Trager 1964 The fine structure of *Leishmania donovani* and the role of the kinetoplast in the leishmania-leptomonad transformation. J. Protozool. 11:166–91.

Rudzinska, M. A. and W. Trager 1957 Intracellular phagotrophy by malaria parasites: an electron microscope study of *Plasmodium lophurae*. J. Protozool. 4:190–99.

Rudzinska, M. A. and W. Trager 1962 Intracellular phagotrophy in *Babesia rodhaini* as revealed by electron microscopy. J. Protozool. 9:279–88.

Rumjantzew, A. and E. Wermel 1925 Untersuchungen über den Protoplasmabau von *Actinosphaerium eichhorni*. Arch. Protistenk. 52:217–64.

Russell, F. S. 1953 *The Medusae of the British Isles: Anthomedusae, Leptomedusae, Trachymedusae, and Narcomedusae*. Cambridge Univ. Press, Cambridge.

Ruthman, A. and K. G. Grell 1964 Die Feinstruktur des intracapsularen Cytoplasmas bei dem Radiolar *Aulacantha scolymantha*. Z. Zellforsch. 63:97–119.

Ryley, J. F. 1967 Carbohydrates and respiration. In *Chemical Zoology* ed. M. Florkin and B. T. Scheer Vol. I. *The Protozoa* ed. G. W. Kidder. pp. 55–92. Academic Press, Inc. New York.

Sager, R. and F. J. Ryan 1961 *Cell Heredity*. John Wiley and Sons, Inc., New York.

Said, R. 1950 The distribution of Foraminifera in the northern Red sea. Contributions Cushman Found. Foram. Res. 1:9–29.

Saidova, Kh. M. 1960 Raspredelenie foraminifer v donnykh otlozheniyakh Okhotskogo Morya. Akad. Nauk. SSSR, Instituta Okeanologii, Trudy 32:96–157.

Sakai, H. 1966 Studies on sulfhydryl groups during cell division of sea-urchin eggs. VIII. Some properties of mitotic apparatus proteins. Biochim. Biophys. Acta 112:132–45.

Samuels, R. 1957a Studies of *Tritrichomonas batrachorum* (Perty) I. The trophic organism. J. Protozool. 4:110–18.

Samuels, R. 1957b Studies of *Tritrichomonas batrachorum* (Perty) II. Normal mitosis and morphogenesis. Trans. Am. Microsc. Soc. 76:295–307.

Schaeffer, A. A. 1920 *Ameboid Movement*. Princeton University Press,

Princeton.

Schaeffer, A. A. 1926 *Taxonomy of the Amebas*. Carnegie Institution of Washington (Publ. No. 345). 24:1–116.

Schaudinn, F. 1896 Ueber das Centralkorn der Heliozoen, ein Beitrag zum Centrosomenfrage. Verhandl. deutsch. Zool. Ges. 6:113–31.

Scherfel, A. 1901 Kleiner Beitrag zur Phylogenie einiger Gruppen niederer Organismen. Bot. Z. 59:143–58.

Schewiakoff, W. 1926 Die Acantharia des Golfes von Neapel. *Fauna e Flora del Golfo di Napoli* 37. Monografia. Pubbl. Sta. Zool. a. Napoli.

Schilke, K. 1970 Zur Morphologie und Phylogenie der Schizorhynchia (Turbellaria, Kalyptorhynchia). Z. Morph. Tiere 67:118–71.

Schindewolf, O. H. 1942 Entwicklung im Lichte der Paläontologie. Der Biologie 11:113–25.

Schindewolf, O. 1950 *Grundfragen der Paläontologie*. Schweizerbart, Stuttgart.

Schmidt-Nielsen, B. 1964 Organ systems in adaptation: the execretory system. In *Handbook of Physiology. Section 4 Adaptation to the Environment*. eds. D. B. Dill, E. F. Adolph, and C. G. Wilber. Am. Physiol. Soc., Washington, D.C.

Schneider, L. 1960 Elektronenmikroskopische Untersuchungen über das Nephridialsystem von *Paramecium*. J. Protozool. 7:75–90.

Schneider, L. 1964 Elektronenmikroskopische Untersuchungen der Ernahrungsorganellen von *Paramecium*. Der Cytopharynx. Z. Zellforsch. 62:198–224.

Schröder, O. 1908 Eine gestrielte Acanthometride (*Podactinelius sessilis* Ol. Schr. n.o.n.sp.) In *Deutschen Südpolar Expedition 1901–1903*. ed. E. von Drygalski Vol. IX pp. 225–36. Reimer Verlag, Berlin.

Schulze, F. E. 1875 Rhizopodienstudien. IV. Arch. Mikros. Anat. 11:329–53.

Schuster, F. 1963 An electron microscope study of the amoeboflagellate, *Naegleria gruberi* (Schordinger). I. The amoeboid and flagellate stages. J. Protozool. 10:297–320.

Schwartz, V. 1935 Versuche über Regeneration und Kerndimorphismus bei *Stentor coeruleus* Ehrbg. Arch. Protistenk. 85:100–139.

Schwartz, V. 1958 Chromosomen in Makronucleus von *Paramecium bursaria*. Biol. Zentr. 77:347–64.

Scriven, M. 1959 Explanation and prediction in evolutionary theory. Science 130:377–82.

Sedgwick, A. 1895 Further remarks on the cell-theory, with a reply to Mr. Bourne. Quart. J. Microsc. Sci. 38:331–37.

Sehnal, F. 1971 Endocrines of Arthropods. In *Chemical Zoology Vol. VI Arthropoda Part B*. eds. M. Florkin and B. T. Scheer. p. 308–46. Academic Press, Inc., New York.

Seilern-Aspang, F. 1957 Die Entwicklung von Macrostomum. Zool. Jahrb., Abt. Anat. 76:311–30.

Seshachar, B. R., K. N. Saxena, and H. Girgla 1971 Some factors governing feeding behavior of *Homalozoon vermiculare* (Ciliophora, Holotricha). J. Protozool. 18:90–95.

Setlow, R. B. and E. C. Pollard 1962 *Molecular Biophysics*. Addison-Wesley Pub. Co., Reading, Mass.

Sewertzoff, A. N. 1931 Morphologischen Gesetzmässigkeiten der Evolution. Jena.

Sharp, R. G. 1914 *Diplodinium ecaudatum* with an account of its neuromotor apparatus. Univ. Calif. Publ. Zool. 13:42–122.

Shelanski, M. L. and E. W. Taylor 1968 Properties of the protein subunit of central-pair and outer-doublet microtubules of sea urchin flagella. J. Cell Biol. 38:304–15.

Siegel, R. W. 1961 Direct and indirect evidence that free-living ciliates conjugate in nature. J. Protozool. 8:27–29.

Siegel, R. W. and L. W. Cohen 1963 A temporal sequence for genic expression: cell differentiation in *Paramecium*. Am Zool. 3:127–34.

Sigal, J. 1952 Foraminifères. In *Traité de Paleontologie*. ed. J. Piveteau. Tome 1. pp. 133–78, 192–301. Masson et Cie.,

Paris.

Simpson, G. G. 1953a *The Major Features of Evolution*. Columbia University Press, New York.

Simpson, G. G. 1953b *Evolution and Geography*. Condon Lectures. Oregon State Systems of Higher Education, Eugene, Oregon.

Simpson, G. G. 1961 *Principles of Animal Taxonomy*. Columbia University Press, New York.

Simpson, G. G. 1964 Organisms and molecules in evolution. Science 146:1535-38.

Singh, B. N. 1951 Nuclear division in Amoebeae and its bearing on classification. Nature 167:582-84.

Singh, B. N. 1952 Nuclear division in nine species of small free-living amoebae and its bearing on the classification of the order Amoebida. Phil. Trans. R. Soc. Lond. ser. B. 236:405-61.

Skaer, R. J. 1961 Some aspects of the cytology of *Polycelis niger*. Quart. J. Microsc. Sci. 102:295-317.

Sleigh, M. A. 1962 *The Biology of Cilia and Flagella*. Pergamon Press, London.

Small, E. 1968 The Scuticociliatida, a new order of the class Ciliatea (Phylum Protozoa, Subphylum Ciliophora). Trans. Am. Microsc. Soc. 86:345-70.

Sneath, P. H. A. 1957 Some thoughts on bacterial classification. J. Gen. Microbiol. 17:184-200.

Sneath, P. H. A. 1958 Some aspects of Adansonian classification and of the taxonomic theory of correlated features. Ann. Microbiol. Enzymol. 8:261-68.

Sneath, P. H. A. 1964 Comparative biochemical genetics in bacterial taxonomy. In *Taxonomic Biochemistry and Serology*. ed. C. A. Leone. pp. 565-85. Ronald Press, New York.

Sokal, R. R. and P. H. A. Sneath 1963 *Principles of Numerical Taxonomy*. W. H. Freeman and Company, San Francisco.

Soldo, A. J. and W. J. van Wagtendonk 1969 The nutrition of *Paramecium aurelia*, stock 299. J. Protozool. 16:500-506.

Sollas, I. B. J. 1906 Porifera (Sponges). In *The Cambridge Natural History*. eds. S. F. Harmer and A. E. Shipley. pp. 163-242. MacMillan and Co., London.

Sonneborn, T. M. 1947 Recent advances in the genetics of *Paramecium* and *Euplotes*. Ad. Gen. 1:263-358.

Sonneborn, T. M. 1954 The relation of autogamy to senescence and rejuvenescence in *Paramecium aurelia*. J. Protozool. 1:38-53.

Sonneborn, T. M. 1957 Breeding systems, reproductive methods, and species problems in Protozoa. In *The Species Problem*. ed. E. Mayr. pp. 155-324. AAAS Publication, Washington, D.C.

Sonneborn, T. M. 1960 The gene and cell differentiation. Proc. Natl. Acad. Sci. U.S.A. 46:149-65.

Sonneborn, T. M. 1963 Does preformed cell structure play an essential role in cell heredity? In *The Nature of Biological Diversity* ed. T. M. Allen. pp. 165-221, McGraw-Hill Book Co., New York.

Sonneborn, T. M. 1964 The differentiation of cells. Proc. Natl. Acad. Sci. U.S.A. 51:915-29.

Sonneborn, T. M. 1965a Nucleotide sequence of a gene: first complete specification. Science 148:1410.

Sonneborn, T. M. 1965b Degeneracy of the genetic code: Extent, nature, and genetic implication. In *Evolving Genes and Proteins* (eds.) V. Bryson and H. J. Vogel p. 377-97. Academic Press, Inc., New York.

Sonneborn, T. M. 1970a Gene action in development. Proc. R. Soc. Lond. B. 176:347-66.

Sonneborn, T. M. 1970b Methods in *Paramecium* research. In *Methods in Cell Physiology*. ed. D. Prescott. Vol. 4 pp. 241-339. Academic Press, Inc., New York.

Spirin, A. S. and L. P. Gavrilova 1969 *The Ribosome* Springer Verlag, Berlin.

Stebbins, G. L. 1950 *Variation and Evolution in Plants*. Columbia University Press, New York.

Stebbins, L. 1960 The comparative evolution of genetic systems. In *Evolution After Darwin*. Vol. I. *The*

Evolution of Life. ed. S. Tax. pp. 197–226. University of Chicago Press, Chicago.

Steinböck, O. 1954 Regeneration azöler Turbellarien. Zool. Anz. (Suppl.) 18:86–94.

Steinböck, O. 1958 Schlusswort zur Diskussion Remane-Steinböck. Zool. Anz. (Suppl.) 21:196–218.

Steinböck, O. 1963a Origin and affinities of the lower Metazoa: The "aceloid" ancestry of the Metazoa. In *The Lower Metazoa*. eds. E. C. Dougherty, Z. N. Brown, E. D. Hanson, and W. D. Hartman. pp. 40–54. University of California Press, Berkeley.

Steinböck, O. 1963b Regenerations- und Konplantations-versuch an *Amphiscolops* spec. (Turbellaria Acoela). Roux' Arch. 154:308–53.

Steinert, G. 1958 Etudes sur le determinisme de la morphogénèse d'un trypanosome. Exp. Cell Res. 15:560–69.

Steinert, G. 1960 Mitochondria associated with the kinetonucleus of *Trypanosoma mega*. J. Biophys. Biochem. Cytol. 8:542–46.

Steinert, G. and A. B. Novikoff 1960 The existence of a cytostome and the occurrence of pinocytosis in the trypanosome (*Trypanosoma mega*). J. Biophys. Biochem. Cytol. 8:563–70.

Stent, G. 1971 *Molecular Genetics. An Introductory Narrative*. W. H. Freeman and Co., San Francisco.

Stephens, R. E. 1968 Reassociation of microtubule protein. J. Mol. Biol. 33:517–19.

Stern, C. 1924 Untersuchungen über Acanthocystideen. Arch. Protistenk. 48:436–91.

Stern, C. 1925 Die Mitose der Epidermiskerne von *Stenostonum*. Z. Zellforsch. 2:121–28.

Stout, J. D. 1960 Morphogenesis in the ciliate *Bresslaua vorax* Kahl and the phylogeny of the Colpodidae. J. Protozool. 7:26–35.

Stuart K. and E. D. Hanson 1967 Acriflavin induction of dyskinetoplasy in *Leptomonas karyophilus*. J. Protozool. 14:39–43.

Sturtevant, A. H. 1937 Essays on evolution. I. On the effects of selection on mutation rate. Quart. Rev. Biol. 12:464–67.

Sueoka, N. 1965 On the evolution of informational macromolecules. In *Evolving Genes and Proteins*. eds. V. Bryson and H. J. Vogel. pp. 479–96. Academic Press, Inc., New York.

Summers, F. M. 1941 The protozoa in connection with morphogenetic problems. In *Protozoa in Biological Research*. eds. G. N. Calkins and F. M. Summers. pp. 772–817. Columbia University Press, New York.

Sutcliffe, W. H., Jr., E. R. Baylor, and D. W. Menzel 1963 Sea surface chemistry and Langmuir circulation. Deep-Sea Res. 10:223–43.

Swanson, C. 1957 *Cytology and Cytogenetics*. Prentice-Hall, Inc. Englewood Cliffs, New Jersey.

Swedmark, B. and G. Teissier 1966 The Actinulida and their evolutionary significance. In *The Cnidaria and their Evolution*. ed. W. J. Rees No. 16 Symp. Zool. Soc. Lond. pp. 119–32. Academic Press, Inc., New York.

Tartar, V. 1954 Anomalies of regeneration in *Paramecium*. J. Protozool. 1:11–17.

Tartar, V. 1961 *The Biology of Stentor*. Pergamon Press, New York.

Tartar, V. 1967 Morphogenesis in Protozoa. In *Research in Protozoology* ed. T.-T. Chen. Vol. 2, pp. 1–115. Pergamon Press, New York.

Tatton, A. K. 1954 Siphonophora of the Indian Ocean. "Discovery" Reports 27:1–162.

Taylor, C. V. 1941 Fibrillar systems in Ciliates. In *Protozoa in Biological Research*. eds. G. N. Calkins and F. M. Summers. pp. 191–270. Columbia University Press, New York.

Taylor, J. H. 1957 The time and mode of duplication of chromosomes. Amer. Nat. 91:209–21.

Taylor, A. E. R., and D. G. Godfrey 1969 A new organelle of bloodstream salivarian trypanosomes. J. Protozool. 16:466–70.

Thiel, H. 1966 The evolution of the Scyphozoa. A Review. In *The Cnidaria and their Evolution*. ed. W. J. Rees. No.

16 Symp. Zool. Soc. Lond. pp. 77–116. Academic Press, Inc., New York.
Thompson, D'A. W. 1942 *On Growth and Form*. MacMillan Company, New York.
Thompson, J. D., Jr. and J. O. Corliss 1958 A redescription of the holotrichous ciliate *Pseudomicrothorax dubius* with particular reference to its morphogenesis. J. Protozool. 15:175–84.
Threadgold, L. T. 1967 *The Ultrastructure of the Animal Cell*. Pergamon Press, Oxford.
Tilney, L. G. 1968 Studies on the microtubules of Heliozoa. IV. The effect of colchicine on the formation and maintenance of the axopodia and the redevelopment of pattern in *Actinosphaerium nucleofilum* (Barrett). J. Cell Sci. 3:549–62.
Tilney, L. G. 1968 The assembly of microtubules and their role in the development of cell form. In *The Emergence of Order in Developing Systems*. ed. M. Locke pp. 63–102. Academic Press, Inc., New York.
Totton, A. K. 1954 Siphonophora of the Indian Ocean. "Discovery" Rep. 27: 1–162.
Trager, W. 1963 Differentiation in Protozoa. J. Protozool. 10:1–6.
Trager, W. and S. M. Krassner 1967 Growth of parasitic Protozoa in tissue cultures. In *Research in Protozoology*. ed. T.-T. Chen. Vol. 2, pp. 357–82. Pergamon Press, New York.
Trégouboff, G. 1953a Caractères généraux. In *Traité de Zoologie*. ed. P.-P. Grassé. Tome I, fasc. II. pp. 269–70. Masson et Cie., Paris.
Trégouboff, G. 1953b Classe des Acanthaires. In *Traité de Zoologie*. ed. P.-P. Grassé. Tome I, fasc. II. pp. 271–320. Masson et Cie., Paris.
Trégouboff, G. 1953c Classe des Radiolaires. In *Traité de Zoologie*. ed. P.-P. Grassé. Tome I, fasc. II. pp. 321–88. Masson et Cie., Paris.
Trégouboff, G. 1953d Classe des Heliozoaires. In *Traité de Zoologie*. ed. P.-P. Grassé. Tome I, fasc. II. pp. 437–89. Masson et Cie., Paris.
Tucker, J. B. 1966 Fine structure and morphogenesis of the pharyngeal basket of the ciliate *Nassula*. J. Protozool. 13 (suppl.): 30.
Tucker, J. B. 1968 Fine structure and function of the cytopharyngeal basket in the Ciliate *Nassula*. J. Cell Sci. 3:493–574.
Tucker, J. B. 1971 Development and deployment of cilia, basal bodies, and other microtubular organelles in the cortex of the ciliate *Nassula*. J. Cell Sci. 9:539–67.
Tuffrau, M. 1952 La morphogénèse de division chez les Colpodidae. Bull. Biol. de France-Belg. 86:309–20.
Tuffrau, M. 1964a Différenciations fibrillaires et coordinations motrices chez quelques Hypotriches. J. Protozool. 11 (suppl.): 158:50.
Tuffrau, M. 1964b La morphogénèse de bipartition et les structures neuromotrices dans le genre *Aspidisca* (Ciliés Hypotriches). Revue de quelques éspèces. Cah. Biol. Mar. 5:173–99.
Tuffrau, M. 1967a Les structures fibrillaires somatiques et buccales chez les ciliés heterotriches. Protistologica 3:369–94.
Tuffrau, M. 1967b Perfectionnements et pratique de la technique d'impregnation au protargol des infusoires ciliés. Protistologica 3:91–98.
Tuffrau, M. 1969 L'origine du primordium buccale chez les Ciliés Hypotriches. Protistologica 5:227–37.
Tuomikoski, R. 1967 Notes on some principles of phylogenetic systematics. Ann. Ent. Fenn. 33:137–47.
Tuzet, O. 1963 The phylogeny of sponges according to embryological, histological, and serological data, and their affinities with the Protozoa and Cnidaria. In *The Lower Metazoa*. eds. E. C. Dougherty, Z. N. Brown, E. D. Hanson, and W. D. Hartman. pp. 129–48. Univ. California Press, Berkeley.
Tuzet, O. 1973 Éponges calcaires. In *Traité de Zoologie*. ed. P.-P. Grassé. Tome III, fasc. I. pp. 27–132. Masson et Cie., Paris.
Tuzet, O., and M. Pavans de Cecatty 1952 Les cellules nerveuses de

Grantia compressa pennigera Haeckel (Éponge calcaire hétérocoele). C. R. Acad. Sci., Paris. 235:3088–90.

Tuzet, O., and M. Pavans de Cecatty 1953 Les cellules nerveuses et neuromusculaires de l'Éponge: *Cliona celata* Grant. C. R. Acad. Sci., Paris. 236: 2342–4.

Tuzet, O., and M. Pavans de Cecatty 1958 Le spermatogenèse, l'ovogenèse, la fécondation et les premiers stades du dévelopement chez *Hippsongia communis* Link. (= *H. equina* O. S.) Bull. Biol. 92:331–48.

Tyler, S., and R. M. Rieger 1975 Uniflagellate spermatozoa in *Nemertoderma* (Turbellaria) and their phylogenetic significance. Science 188:730–32.

Ueda, K. 1961 Structure of plant cells with special reference to lower plants. VI. Structure of chloroplasts in algae. Cytologia 26:344–58.

Umylina, T. M. 1971 Kariotipy Baikal'skiph planarii roda Bdellocephala de Mann (Turbellaria, Tricladida, Paludicola) [Karyotypes of Baikal planarians of the genus Bdellocephala de Mann (Turbellaria, Tricladida, Paludicola)]. Zool. Zh. 50:130–33.

Vacelet, J. 1970 Les éponges pharetionides actuelles. In *Biology of the Porifera* ed. W. G. Fry. No. 25 Symp. Zool. Soc. Lond. pp. 189–264. Academic Press, Inc. New York.

Vacelet, J. 1971 Ultrastructure et formation des fibres de spongine d'éponges cornées Verongia. J. Microscopie 10:13–32.

Valkonov, A. 1928 Protistenstudien. III. Die Stielbildung bei Desmothoraken nebst einigen Worten über die Stellung dieser Gruppe in System. Arch. Protistenk. 64:446–56.

Valkonov, A. 1940 Die Heliozoen und Proteomyxien. Artbestand und sonstige kritische Bemerkungen. Arch. Protistenk. 93:225–54.

van Potter, R. 1971 *Bioethics: Bridge to the Future.* Prentice-Hall, Inc., Englewood Cliffs, N.J.

van Wagtendonk, W. J. 1955 Encystment and excystment of Protozoa. In *Biochemistry and Physiology of Protozoa.* eds. S. H. Hutner and A. Lwoff. Vol. II, pp. 85–90. Academic Press, Inc., New York.

van Wagtendonk, W. J., and A. T. Soldo 1970 Methods in the axenic culture of *Paramecium aurelia.* In *Methods of Cell Physiology.* ed. D. M. Prescott Vol. IV, pp. 117–30. Academic Press, Inc., New York.

van Wagtendonk, W. ed. 1974 *Paramecium: A Current Survey.* Elsevier Scientific Publishing Company, Amsterdam.

Vickerman, K. 1969 The fine structure of *Trypanosoma congolense* in its bloodstream phase. J. Protozool. 16: 54–69.

Vickerman, K. 1962 The mechanisms of cyclical development in trypanosomes of the *Trypanosoma brucei* sub group: an hypothesis based on ultrastructural observations. Trans. R. Soc. Trop. Med. Hyg. 56:487–95.

Villeneuve, F. 1937 Sur la structure de *Cienkowska Mereschkowskyi* Cienk. et *Actinolophus pedunculatus* Schulze, Héliozoaires des eaux saumâtres de sète. Arch. Zool. exp. gén. 78:243–50.

Vivier, E. and J. Andre 1961 Données structurales et ultrastructurales nouvelles sur la conjugaison de *Paramecium caudatum.* J. Protozool. 8:416–26.

Vivier, E., B. Legrand and A. Petitprez 1969 Recherches cytochimiques et ultrastructurales sur des inclusions polysaccharidiques et calciques du Spirostome; leurs relations avec la contractilité. Protistologica 5:145–59.

Volkonsky, M. 1931 *Hartmannella castellani* Douglas et classification des Hartmannelles. Arch. Zool. exp. gén. 72:317–39.

Waddington, C. H. 1960 Evolutionary Adaptation. In *The Evolution of Life.* ed. S. Tax, p. 381–402. University of Chicago Press, Chicago.

Wagner, R. T. and H. K. Mitchell 1964 *Genetics and Metabolism.* 2nd edition. John Wiley and Sons, New York.

Wallace, B. 1966 *Chromosomes, Giant Molecules, and Evolution.* W. W.

Norton and Co., Inc., New York.
Wallin, I. E. 1923 The mitochondria problem. Am. Nat. 57:255–61.
Waterman, T. H. 1961 Comparative physiology. In *The Physiology of the Crustacea*. Vol. II. *Sense Organs, Interaction, Behavior*. ed. T. H. Waterman. pp. 521–93. Academic Press, Inc., New York.
Watson, J. D. 1970 *The Molecular Biology of the Gene*. 2nd edition. W. A. Benjamin, Inc., New York.
Watson, J. D. and F. H. C. Crick 1953a A structure for deoxyribose nucleic acid. Nature 171:737–38.
Watson, J. D. and F. H. C. Crick 1953b Genetical implications of the structure of deoxyribonucleic acid. Nature 171:964–67.
Weill, R. 1934 Contribution à l'étude des Cnidaires et de leurs nématocystes. I. Recherches sur les nématocystes (Morphologie-Physiologie-Développement). Trav. Stat. Zool. Wimereux 10: 1–347.
Wenrich, D. H. 1929 The structure and behavior of *Actinobolus vorax* n. sp. (Protozoa, Ciliata). Biol. Bull. 56: 390–401.
Wenrich, D. H. 1954 Sex in Protozoa: A comparative review. In *Sex in Microorganisms*. eds. D. H. Wenrich, T. F. Lewis, and J. R. Raper, pp. 134–265. AAAS Publication, Washington, D.C.
Wessels, N. K., B. S. Spooner, J. T. Wrenn, and K. M. Yamada 1971 Microfilaments in cellular and developmental processes. Science 171:135–43.
Wessenberg, H. 1961 Studies on the life cycle and morphogenesis of Opalina. Univ. California Publ. Zool. 61:315–70.
Wessenberg, H. and G. Antipa 1968 Studies on *Didinium nasutum*. I. Structure and ultrastructure. Protistologica 4:427–47.
Wessenberg, H. and G. Antipa 1970 Capture and ingestion of Paramecium by *Didinium nasutum*. J. Protozool. 17:250–70.
Westblad, E. 1937 Die Turbellarien-Gattung *Nemertoderma* Steinböck. Acta Soc. Fauna Flora fenn. 60:45–89.
Westblad, E. 1945 Studien über Skandinavische Turbellaria Acoela. III. Ark. Zool. 36A:1–56.
Westblad, E. 1948 Stüdien über Skandinavische Turbellaria Acoela. V. Ark. Zool. 41A:1–82.
Westfall, J. 1971 Ultrastructure of synapses in a primitive coelenterate. J. Ultrastruct. Res. 32:237–46.
White, M. J. D. 1954 *Animals Cytology and Evolution*. 2nd edition. Cambridge University Press, Cambridge.
Wichterman, R. 1953 *The Biology of Paramecium*. Blakiston, New York.
Widersten, B. 1968 On the morphology and development in some cnidarian larvae. Zool. Bid. Uppsala 37:139–79.
Wiedmann, J. 1969 The heteromorphs and ammonoid extinction. Biol. Rev. 44:563–602.
Wigg, D., E. C. Bovee, and T. L. Jahn 1967 The evacuation mechanism of the water expulsion vesicle ("contractile vacuole") of *Amoeba proteus*. J. Protozool. 14:104–8.
Williams, M. B. 1970 Deducing the consequence of evolution: A mathematical mode. J. Theo. Biol. 29:343–85.
Willier, B. H., L. H. Hyman, and S. A. Rifenburgh 1925 A histochemical study of intracellular digestion in triclad flatworms. J. Morphol. 40:299–340.
Wilson, E. B. 1928 *The Cell in Development and Heredity* 3rd edition. MacMillan Company, New York.
Wilson, H. V. 1907 On some phenomena of coalescence and regeneration in sponges. J. Exp. Zool. 5:245–58.
Wise, B. N. 1965 The morphogenetic cycle in *Euplotes eurystomus* and its bearing on problems of ciliate morphogenesis. J. Protozool. 12:626–48.
Woese, C. R. 1967 *The Genetic Code*. Harper and Row, New York.
Wohlman, A. and R. D. Allen 1968 Structural organization associated with pseudopod extension and contraction during cell locomotion in *Difflugia*. J. Cell Sci. 3:105–14.
Wolfe, J. 1969 Structural aspects of amitosis: a light and electron

microscope study of the isolated macronuclei of *Paramecium aurelia* and *Tetrahymena pyriformis*. Chromosoma 23:59–72.

Wolfe, J. 1970 Structural analysis of basal bodies of the isolated oral apparatus of *Tetrahymena pyriformis*. J. Cell Sci. 6:679–700.

Wolff, E. 1962 Recent researches on the regeneration of planaria. In *Regeneration*. ed. D. Rudnick. pp. 53–84. Ronald Press Co., New York.

Wolska, M. 1966a Division morphogenesis in the genus *Didesmis* Fior. of the family *Buetschliidae* (Ciliata, Gymnostomata). Arch. Protozool. 4:15–18.

Wolska, M. 1966b Study on the family *Blepharocorythidae* Hsiung. I. Preliminary remarks. Acta Protozool. 4:97–103.

Wolska, M. 1967a Study on the family *Blepharocorythidae* Hsiung. II. *Charonina ventriculi* (Jameson). Acta Protozool. 4:279–83.

Wolska, M. 1967b Study of the family *Bleopharocorythidae* Hsiung. III. *Raabena bella* gen. n., sp. n. from the intestine of the Indian elephant. Acta Protozool. 4:285–90.

Wolska, M. 1971a Studies on the family *Blepharocorythidae* Hsiung. V. A review of the genera and species. Acta Protozool. 9:23–40.

Wolska, M. 1971b Studies on the family *Blepharocorythidae* Hsiung. VI. Phylogenesis of the family and the description of the new genus *Circodinium* gen. n. with the species *C. minimum* (Gassousky, 1918). Acta Protozool. 9:171–94.

Wolstenholme, D. R. and I. B. Dawid 1968 A size difference between mitochondrial DNA molecules of urodele and anuran amphibia. J. Cell Biol. 39:222–28.

Wood, A. 1948 The structure of the wall of the test in the Foraminifera; its value in classification. Quart. J. Geol. Soc. 104:229–55.

Wood, A. E. 1950 Porcupines, paleogeography, and parallelism. Evol. 4:87–98.

Woodard, J., B Gelber, and H. Swift 1961 Nucleoprotein changes during the mitotic cycle in *Paramecium aurelia*. Exp. Cell Res. 23:258–64.

Woodger, J. H. 1945 On biological transformation. In *Essays on Growth and Form Presented to D'Arcy Wentworth Thompson*. ed. W. E. Le Gros Clark and P. B. Medawar. pp. 94–120. Clarendon Press, Oxford.

Woodruff, L. L. and G. Baitsell 1911 Rhythms in the reproductive activity of Infusoria. J. Exp. Zool. 11:339–59.

Woodruff, L. L. and R. Erdmann 1914 A normal periodic reorganization process without cell fusion in Paramecium. J. Exp. Zool. 17:425–518.

Wright, S. 1932 The roles of mutation, inbreeding, crossbreeding, and selection in evolution. Proc. Sixth Intern. Cong. Genetics 1:356–66.

Wright, S. 1949a Adaptation and selection. In *Genetics, Paleontology, and Evolution*. ed. G. L. Jepsen, E. Mayr, and G. S. Simpson. pp. 365–89. Princeton University Press, Princeton, New Jersey.

Wright, S. 1949b Population structure in evolution. Proc. Am. Phil. Soc. 93:471–78.

Yusa, A. 1957 The morphology and morphogenesis of the buccal organelles in *Paramecium* with reference to their systematic significance. J. Protozool. 4:128–42.

Yusa, A. 1963 An electron microscope study on regeneration of trichocysts in *Paramecium caudatum*. J. Protozool. 10:253–62.

Ziegler, B. and S. Rietschel 1970 Phylogenetic relationships of fossil calcisponges. In *Biology of the Porifera*. ed. W. G. Fry. No. 25 Symp. Zool. Soc. Lond. pp. 23–40. Academic Press, Inc., New York.

Zimmerman, A. 1930 Neue und wenig bekannte Kleinalgen von Neapel. I-B. Zeit. Botan. 23:419–42.

Zimmerman, W. 1959 Die Methoden der Phylogenetik. In *Die Evolution der Organismen*. ed. G. Heberer. pp. 25–102. Fischer Verlag, Stuttgart.

Zuckerkandl, E., and L. Pauling
  1965 Evolutionary divergence and convergence in proteins. In *Evolving Genes and Proteins*. eds. V. Bryson and H. J. Vogel. p. 97–166. Academic Press, Inc., New York.

Zuelzer, M. 1909 Bau und Entwicklung von *Wagnerella borealis* Meresch. Arch. Protistenk. 17:135–202.

Zumstein, H. 1900 Zur Morphologie und Physiologie der *Euglena gracilis* Klebs. Jahrb. wiss. Botan. 34:149–98.

Author Index

Abram, 41, 54
Adanson, 114
Afzelius, 42
Ahmed, 40
Albaret, 340
Alexieff, 306, 307
Allen, 54, 56, 67, 141, 145, 167, 204, 210, 226, 328, 329, 333, 346, 347
Alonso, 338, 361
Ammerman, 338, 373
Amon, 263
Anderson, 159, 161, 162, 168, 203, 246, 251, 252, 337, 338
André, 207, 352, 354, 376, 391
Andresen, 206
Anfinsen, 36
Angell, 215, 230
Angelopoulos, 159
Anigstein, 384
Antipa, 130, 329, 331, 341, 343, 347, 366
Antonov, 29
Arnott, 30
Avers, 51
Ax, 105, 496, 506, 511, 513, 519, 521, 522, 523, 530, 534, 535, 536, 537, 538, 539, 541, 586
Baccetti, 63
Baddington, 510
Bagby, 427
Baitsell, 355
Baker, 192, 193, 194, 584
Balamuth, 349
Baldwin, 39, 63
Band, 211
Bandy, 210
Bardele, 261, 331, 340, 341, 367
Barghoorn, 281
Barnes, 590, 591

Barrington, 39
Batham, 471
Batisse, 337
Beale, 58, 546
Beams, 159, 203, 246, 247, 251, 252
Beauchamp, 493, 495, 496, 497, 500, 508, 509, 511, 519, 530
Beaver, 167
Becker, 42
Beckwith, 44
Bělař, 135, 206, 243, 248, 249, 253, 256, 259, 260, 261
Bellamy, 519
Benazzi, 508, 512
Bennett, 50
Benwitz, 588, 589
Berlyn, 40
Berrill, 64, 473
Berthold, 213, 215
Biecheler, 288, 573
Bilbaut, 464
Bisalputra, 288
Blackwelder, 114
Block, 52
Blum, 6
Bock, 87, 105, 567, 591
Böhm, 483, 485
Bohr, 418
Bolten, 353
Bolton, 27, 29
Bonner, 40, 348
Bonneville, 49
Borgert, 275
Borkott, 512
Borojevic, 428, 448
Borst, 51
Bovee, 57, 165, 202, 204, 205, 209, 221, 222, 223, 226, 245, 259, 260, 278, 333, 343
Boyden, 74, 135, 146

Boyer, 511
Bradbury, 406
Bradley, 33
Brady, 233
Brahe, 72
Brändle, 475
Bresslau, 522, 524
Breyer, 206, 207
Brien, 420, 436, 448
Britten, 28
Brooks, 483, 485
Brower, 278
Brown, 43
Brucke, 51, 578
Bryan, 54
Buchsbaum, 61
Bullock, 61, 132, 428, 464, 470, 473, 501, 502, 503, 510
Burnett, 475
Burton, 168, 338
Buse, 339, 368
Bush, 455
Bütschli, 242, 243, 254, 265, 292, 320, 326, 369, 437, 589
Butzel, 353
Cachon, 269, 270, 271, 272, 274, 275, 276, 285, 288, 289, 298, 299, 573
Cachon-Enjumet, 265, 272, 274, 275, 298, 299
Cairns, 24
Calder, 466, 467, 478
Calkins, 263, 283, 290
Calvin, 38
Campbell, 279, 280, 282, 284, 285, 310, 324, 388, 389, 391, 398
Canella, 347, 357, 379, 381, 382, 391, 392, 394, 410, 413
Cánovas, 40
Carasso, 333, 341

638

AUTHOR INDEX

Carruette-Valentin, 157, 159, 161, 167, 168
Carter, 486
Casper, 41
Caullery, 163
Changeux, 32
Chapman, 337, 467, 468, 472, 481
Chapman-Andresen, 127, 206, 207, 210, 213
Chatton, 157, 204, 205, 206, 238, 239, 265, 345, 346, 347, 348, 352, 356, 374, 378, 379, 387, 391, 393
Cheissin (*see also* Chejsin) 339, 368, 377, 410
Chejsin (*see also* Cheissin) 148, 154, 163, 215, 266, 276, 279, 281, 282
Chen, 139, 144
Chen-Shan, 348
Child, 56, 342
Cho, 331
Christenson, 330
Clark, 61, 63, 151, 471, 510, 536, 537
Cleland, 147
Clément-Iftode, 365
Cleveland, 42, 136, 138, 145, 156, 158, 166, 167, 168, 169, 173, 174, 183
Cohen, 356, 546
Cole, 210, 212
Collins, 393, 394, 407
Conner, 210, 211, 213, 344
Connor, 166
Corliss, 103, 132, 153, 326, 330, 334, 335, 336, 339, 345, 349, 350, 352, 357, 363, 364, 367, 374, 375, 377, 379, 381, 382, 384, 385, 389, 391, 394, 396, 405, 406, 407, 408, 409, 410, 411, 415, 583, 585
Costello, 509, 550
Cowden, 428
Crick, 24, 26
Crow, 10, 73
Curtis, 439
Cushman, 228, 235
Cuvier, 7
Dadd, 38
d'Alesandro, 172
Dangeard, 202

Daniel, 157, 158, 159, 160, 162
Daniels, 206, 207
Darwin, 13, 70, 71, 72, 73, 74, 108, 110
Davison, 54
Dawid, 43, 51
Dayhoff, 30, 31, 32, 34
Dazzini, 39
Deane, 160
de Bary, 163
de Beer, 63, 65, 585
de Duve, 50
DeFlandre, 182, 200, 204, 207, 214, 215, 238, 239, 265, 271, 281, 284, 285, 310, 324, 380
de Haller, 347, 544
Demerec, 39
DeMoss, 40
Deroux, 371
de Saedeleer, 181, 452
de Terra, 345
Dewey, 166, 211, 214, 344
Didier, 358
Diller, 355
Dillon, 585
Dippell, 46, 56, 144, 167, 338, 346, 347
Dobell, 125, 132, 253, 254, 255, 256, 326, 327, 379, 381
Dobrzhanska-Kaczanowska, 339, 351, 353, 384, 389, 390, 405
Dobzhansky, 5, 9, 11, 16, 45, 72, 73, 109, 145
Doflein, 244
Dogiel, 148, 154, 164, 215, 266, 276, 279, 281, 282, 591
d'Orbigny, 277
Dorey, 494, 496, 497, 499, 500, 501
Doroszewski, 350, 351, 371
Dougherty, 135, 146, 211
Dragesco, 338, 339, 352, 365, 371, 372, 381, 412
Dreyer, 276
Duboscq, 435, 436, 438, 452
Dujardin, 227
DuPraw, 24, 41, 42, 43, 44, 46
Dupy-Blanc, 338

Dusanic, 168
Dusi, 151
Eck, 34
Eckert, 132, 343, 548
Ehret, 328, 347, 348, 544
Ehrlich, 9, 12, 116, 146
Ellis, 181
Engelberg, 30, 31, 58
Enjumet, 263
Ephrussi, 51
Erdmann, 355
Ertl, 452
Essner, 50
Estève, 338, 341
Evans, 345, 349
Everett, 30
Falkow, 28
Fauré-Fremiet, 207, 273, 326, 330, 333, 334, 340, 342, 343, 350, 352, 355, 356, 357, 364, 369, 370, 371, 373, 374, 375, 377, 379, 382, 388, 389, 391, 398, 406, 409, 410, 413, 414, 418
Favard, 333, 341
Fawcett, 42
Febvre, 295, 296, 297, 301
Ferru, 276
Finks, 446
Finley, 372, 379, 392, 410
Fisher, 72, 109, 146, 147, 148
Fitch, 119, 120
Fjerdingstad, 161, 440
Florkin, 434, 455
Frankel, 348, 349, 351, 352, 353
Franzen, 537, 538
Freedman, 191
Friz, 240
Fry, 421
Fuge, 159, 176
Fulton, 56, 166, 346
Furgason, 337, 364, 366, 375, 407
Gabriel, 513
Gall, 345
Galloway, 228, 231, 233, 235
Garnier, 343
Garonne, 434
Garstang, 485
Gatenby, 435

Gavrilova, 57
Gegenbaur, 270
Geiman, 163, 164
Gelber, 339, 345, 368
Gelei, 63, 334, 345, 501, 505
Gerard, 419
Ghiselin, 72
Giannini, 508
Gibbons, 54, 157, 161, 175
Gibor, 52
Giles, 40
Gillies, 171, 191, 347, 348, 349, 354, 376
Gilmour, 115, 116
Girgla, 342, 350
Glaessner, 234, 235, 236, 237, 572
Godfrey, 162
Goethe, 109
Gojdics, 342, 373
Gold, 389
Goldfischer, 341
Goldschmidt, 17, 70
Goldstein, 240
Golikova, 338
Golinska, 329, 333, 335, 341, 343, 350, 351, 371
Goodwin, 459
Goreau, 419, 439
Goss, 513
Götte, 512
Graff, 524, 540
Grain, 54, 328, 329, 333, 341, 343, 358, 396, 408
Granick, 52, 57
Grassé, 154, 155, 156, 158, 159, 160, 162, 168, 169, 170, 171, 176, 182, 183, 190, 193, 194, 198, 199, 200, 203, 217, 220, 222, 232, 240, 257, 266, 267, 268, 271, 285, 294, 295, 296, 300, 321, 323, 380, 415, 550
Greenberg, 584, 585
Grell, 42, 135, 136, 139, 154, 217, 218, 227, 272, 275, 276, 333, 338, 341, 367, 372, 588, 589
Gremigni, 512
Grimstone, 54, 56, 157, 158, 161, 175, 346
Grobstein, 6

Gruber, 259
Guilcher, 354, 383, 384, 389, 394, 405
Günther, 112, 589, 590
Gurdon, 43
Guttman, 192
Hadzi, 105, 486, 536, 543, 549, 585, 586, 590
Haeckel, 63, 105, 108, 191, 242, 265, 266, 268, 270, 271, 274, 279, 280, 282, 283, 287, 290, 292, 297, 319, 322, 490, 589
Haecker, 273, 276, 278, 279, 281, 282
Hafleigh, 36, 37
Haldane, 109
Hall, 31, 128, 154, 166, 263, 344
Hammen, 434
Hammond, 348
Hand, 465, 466, 478, 488, 489
Handler, 8
Hansen, 160, 254
Hansgirg, 128
Hanson, 10, 15, 21, 56, 58, 63, 67, 71, 72, 81, 82, 135, 146, 152, 166, 171, 172, 191, 192, 214, 330, 336, 345, 346, 347, 348, 349, 350, 352, 354, 357, 376, 379, 413, 415, 416, 418, 490, 496, 509, 513, 522, 523, 543, 546, 547, 550, 581, 585, 586, 589, 590, 591
Hardy, 584
Harrison, 392, 410
Hartman, 419, 420, 427, 437, 439, 444, 448, 488, 536, 586
Hartmann, 39, 244, 257
Hatschek, 454
Hawes, 135, 146, 192, 203
Hay, 42
Hayes, 146, 369
Heal, 212
Heath, 512
Heckmann, 356, 372
Hegner, 215
Heider, 586
Hempel, 71, 72
Hendelberg, 507, 508, 538

Henderson, 98
Hennig, 74, 89, 92, 98, 99, 104, 105, 110, 111, 112, 116, 195, 278, 446
Henstchel, 421, 431, 437
Herder, 586
Hertwig, 79, 80, 135, 137, 238, 242, 245, 248, 249, 265, 270, 271, 272, 277, 280, 282, 283, 297, 301, 319, 354, 355
Heunert, 215
Heywood, 114
Higashi, 54
Hill, 168, 466, 474, 481, 484
Hinegardner, 29, 30, 31, 58
Hirsch, 50
Hoare, 170, 171, 191, 192, 193, 194
Hollande, 157, 158, 159, 161, 164, 167, 168, 169, 181, 182, 263, 265, 269, 270, 271, 272, 274, 275, 288, 289, 298, 299, 452, 573, 575
Holley, 25, 30
Holm, 9
Holt, 337, 367
Holz, 345
Honigberg, 154, 155, 157, 158, 159, 160, 162, 163, 166, 168, 176, 180, 182, 194, 195, 198, 199, 204, 218, 220, 221, 222, 239, 242, 245, 246, 259, 260, 262, 290, 359, 375, 383, 393, 415
Hopkins, 333
Horn, 111
Horowitz, 37, 38
Horridge, 61, 132, 428, 464, 470, 473, 486, 488, 501, 502, 503, 510
Horwitz, 353
Houssay, 206, 226
Hovasse, 254, 256
Howard, 45
Howell, 98
Hoyer, 27, 29
Hull, 79, 116, 117, 146
Humphreys, 439
Huneeus, 54
Hurwitz, 42
Hutchinson, 8

Huth, 272, 274
Hutner, 127, 128, 151, 166, 344
Huxley, 18, 65, 71, 108, 112, 567
Hyman, 63, 64, 65, 127, 131, 306, 410, 415, 420, 421, 422, 423, 424, 425, 426, 427, 429, 430, 432, 433, 435, 437, 438, 439, 451, 453, 454, 455, 457, 458, 460, 461, 463, 464, 467, 470, 471, 483, 485, 489, 493, 495, 497, 500, 501, 502, 504, 505, 507, 509, 510, 511, 524, 530, 536, 539, 540, 541, 542, 583, 584, 585
Ingram, 35
Inoki, 51, 159, 172
Ivanic, 247, 253
Ivanov, 524, 536
Jacob, 32
Jägersten, 105, 449, 487, 488, 490, 537, 538, 545, 549, 575, 585, 586, 587, 588, 589
Jahn, 57, 165, 209, 210, 221, 222, 223, 226, 278, 330, 333, 343
Jankowski, 326, 349, 350, 352, 382, 387, 388, 391, 394, 395, 396, 405, 408, 409, 410, 411
Jareño, 361
Jenkins, 513
Jennings, 132, 208, 214, 343, 345, 510, 515
Jensen, 61, 132
Jerka-Dziadosz, 348, 353
Jhering, 543
Jinks, 134
Johannesen, 88
John, 40
Johnson, 211
Jollos, 215
Jones, 428
Jörgensen, 398
Joyon, 157, 158, 160, 176
Jukes, 23, 26, 36, 119
Kaczanowska, 348, 371, 378
Kaczanowski, 392
Kahl, 326, 334, 362, 367, 369, 377, 379, 380, 384, 385, 386, 387, 388, 389, 391, 392, 393, 394, 397, 398, 407, 411
Kalley, 288
Kamiya, 204, 210
Kaneda, 330, 336, 339, 345, 347, 349, 352, 354, 376, 546
Kaplan, 35, 36
Karling, 506, 517, 518, 530, 534, 538
Kates, 240
Kato, 512
Kattar, 331
Kavanau, 210
Kenk, 519
Kenneth, 98
Kepler, 72, 73
Kent, 452, 543
Kerkut, 584, 590
Kidder, 211, 214, 344
Kiefer, 54
Kimball, 132, 345
Kimura, 73
King, 36, 119, 247
Kinosita, 343
Kirby, 145, 160, 173, 174, 194, 195, 199, 345
Kitching, 165, 203, 333, 343
Klein, 345
Klug, 41
Kluyver, 5
Knowles, 58
Koenigswald, 235
Koffler, 41, 54
Kofoid, 330, 398
Kohn, 588
Kohne, 28
Koidl, 513
Koopowitz, 510
Kormos, 394, 407
Kowalevsky, 129, 539
Kowalska, 348, 341, 378
Kozloff, 358, 392, 519
Krassner, 342
Krinsky, 151
Kroon, 51
Kudo, 55, 135, 154, 160, 338, 386, 387, 388, 390, 391, 415
Kühn, 64, 244, 245, 485
Kung, 30, 343
Kuznicki, 343
Lacassagne, 466, 484
Lanfranchi, 519
Lang, 539, 540
Laubenfells, 421, 425, 427, 428, 429, 443, 446
Lavier, 193
Le Calvez, 211, 212, 215, 223, 230, 235, 299, 309, 572, 573
Lee, 217
Leedale, 151
Leeuwenhoek, 125, 127, 326, 343
Leff, 151
Legrand, 338
Lehninger, 32, 49, 51
Lemmerman, 128
Lenhoff, 61
Leone, 37
Lesser, 238, 242
Leuckart, 454
Lévi, 419, 420, 426, 428, 435, 437, 438, 444, 445, 446, 448, 449, 450, 451
Levine, 221, 228
Levins, 73
Lewis, 40, 67
Lewontin, 9, 23, 66, 73, 95
Li, 10
Lilly, 210, 214, 344
Lin, 51
Linnaeus, 202
Linsley, 111, 112, 114, 116, 128, 131, 151, 153
Linstedt, 472
Loeblich, 211, 212, 215, 219, 227, 228, 229, 230, 231, 233, 234, 235, 237
Loening, 30
Lom, 336, 337, 351, 352, 353, 358, 364, 372, 386, 392, 412
Loomis, 61
Lowell, 475
Lubinsky, 395, 397
Luck, 51
Ludvik, 158
Lund, 332, 352, 353, 372, 412
Luporini, 361
Lwoff, 38, 56, 128, 131, 151, 338, 345, 346, 347, 353, 356, 369, 374, 375, 377, 380, 382, 384, 387, 390, 391, 393, 544

MacArthur, 73
MacGinnitie, 455
Machemer, 343
Macindoe, 30, 56
Mackie, 481
Mackinnon, 192, 203
MacNae, 513
Madison, 30
Mahler, 578
Mandel, 28
Manton, 151
Marcus, 64, 105, 495, 513, 519, 536, 585, 586, 587
Margoliash, 34, 86, 119, 120
Margulis, 51, 578
Markert, 546
Marmur, 28
Maslin, 99, 100, 195
Mast, 209
Mattern, 157, 158, 159, 160, 162
Maupas, 352
Mayr, 8, 9, 10, 11, 12, 13, 14, 17, 22, 65, 72, 92, 95, 98, 109, 111, 112, 114, 116, 132, 135, 145, 146, 147, 148, 153
McCarthy, 27, 29
McEnery, 217
McLaughlin, 31, 392, 410
McNeil, 114
McRory, 167
Menzel, 273
Mereschkowsky, 51, 578
Messier, 159
Mettrick, 510
Meyers, 212
Mignot, 157, 176
Milder, 160
Millecchia, 347
Miller, 473
Mills, 40
Minchin, 51, 415, 578
Mitchell, 10
Monod, 31, 32
Moore, 481, 482
Moraczewski, 212, 215
Morat, 339
Moroff, 299, 301
Morowitz, 6, 8
Moses, 42
Mosevich, 340
Mugard, 348, 386
Muller, 5, 132, 135, 562

Müller, 290, 293, 475, 512
Murakami, 343
Myers, 210, 212, 217
Naef, 80, 105, 109
Naitoh, 132, 343, 548
Nakabayashi, 51, 159
Nakanishi, 51, 159
Nanney, 331
Needham, 37
Netzel, 208, 214, 215
Newton, 512
Nielsen, 158
Nilsson, 206, 339
Njiné, 370, 377, 378, 380, 392
Noble, 167
Noirot-Timothée, 337, 340, 342, 352, 353, 396, 405
Noland, 330, 342, 373
Norrevang, 465
Novikoff, 50, 160, 164
Nursall, 584, 585
Odum, 292, 293
Olive, 217
Oosawa, 41, 54
Oparin, 37
Organ, 343
Orgel, 67
Orlov, 324
Ornston, 40
Ovchinnikova, 338
Owen, 74, 110
Ozeki, 161
Page, 203, 204, 205, 206, 226
Pantin, 9, 61, 469, 471, 486, 488
Papi, 515, 517
Pappas, 42
Parducz, 343
Park, 34
Parke, 151
Parker, 433, 469
Pascher, 129, 149, 150, 239, 240
Passano, 61, 469
Pätau, 265
Patterson, 144
Paulin, 394, 407
Pauling, 33, 34, 35, 36, 89
Pavan, 39
Pavans de Cecatty, 428, 434, 435, 464
Pedersen, 494, 496, 497, 501, 502, 505, 515

Pelc, 45
Penard, 205, 243, 244, 245, 248, 251, 254, 258, 261
Perdue, 345
Péres-Silva, 338, 361
Perkins, 47, 263, 465
Petersen, 160, 254
Petitprez, 338
Petrunkevitch, 17
Phleger, 212
Pickens, 489
Pirie, 5, 21
Pitelka, 50, 54, 55, 157, 158, 159, 160, 162, 175, 176, 177, 191, 194, 200, 207, 328, 333, 337, 339, 340, 343, 415, 590
Pokorny, 233, 263, 592
Poljanskij, 148, 154, 163, 215, 266, 276, 279, 281, 282, 338, 377, 410
Pollard, 40
Popofsky, 265, 266, 280, 282, 284, 285
Porter, 49, 50, 354, 377
Potter, 7
Powers, 328, 491
Preer, 141
Prenant, 206, 226
Prescott, 345
Preston, 160
Pringsheim, 128, 149, 150, 151, 152, 240, 288, 578
Prout, 87
Provasoli, 127, 128, 151
Puccinelli, 508
Puytorac, 329, 331, 333, 340, 347, 348, 351, 352, 364, 370, 371, 372, 377, 378, 380, 386, 392, 417
Quintana, 50
Raabe, 330, 352, 354, 362, 363, 377, 392, 406, 410
Radzikowski, 339
Raff, 578
Raikov, 42, 332, 338, 339, 355, 368, 369, 373, 376, 379, 380, 381, 417
Ramsey, 465
Rancourt, 51
Randall, 333
Rappaport, 550
Raskin, 246

AUTHOR INDEX

Rasmont, 426, 434, 440
Raven, 12
Rees, 455, 480, 483, 485
Reichenow, 244
Reid, 446, 449, 450
Reiff, 373
Reisinger, 496, 505, 530
Reisner, 30, 56
Remane, 74, 75, 76, 77, 78, 81, 85, 88, 89, 94, 105, 109, 110, 111, 112, 116, 118, 178, 253, 258, 305, 357, 366, 455, 486, 487, 488, 517, 518, 521, 536, 537, 538, 544, 545, 550, 585, 586, 589
Rensch, 17, 18, 567
Reschetnjak, 273, 279
Reutterer, 584
Reynoldson, 519
Rieder, 33
Rietschel, 446
Rifenburgh, 510
Rinaldi, 210
Robertson, 49
Rogers, 163
Romanes, 13
Romer, 79, 82
Roodyn, 51
Roque, 331, 348, 351, 352, 372, 376, 386, 412
Rose, 475
Rosenbaum, 56, 557
Roskin, 428
Ross, 46, 116
Roth, 247
Rouillier, 333
Rowe, 30, 56
Rudzinska, 132, 172, 333, 341, 347
Rumjantzew, 246, 247, 251
Russell, 457, 459, 460
Ruthman, 272, 275
Ryan, 25, 150
Ryley, 166, 211, 214
Sager, 25, 150
Said, 211
Saidova, 212
Sakai, 46
Samuels, 156, 167
Savoie, 352, 371, 372, 412
Saxena, 342
Schaeffer, 205, 209, 220, 224

Schatz, 151
Schaudinn, 243, 256, 262
Scheer, 455
Scherfel, 150
Schewiakoff, 290, 291, 292, 294, 295, 296, 297, 298, 299, 300, 301, 302, 304, 305, 308, 310, 311, 317, 318, 319, 321, 325, 383
Schilke, 519
Schindewolf, 17, 485
Schmidt-Nielsen, 39, 63
Schneider, 333, 343
Schnitz, 128
Schopf, 281
Schröder, 320
Schulz, 513
Schulze, 223, 431
Schuster, 46
Schwartz, 338, 352
Scriven, 72
Sedgwick, 543
Séguéla, 348, 352
Sehnal, 39
Seilern-Aspang, 511
Seshachar, 342
Setlow, 40
Sewertzoff, 63, 65, 232
Shelansky, 54
Shigenaka, 247
Sibley, 332
Siegel, 145, 356, 546
Sigal, 231, 235
Simpson, 13, 14, 15, 16, 17, 18, 19, 20, 22, 74, 75, 84, 87, 108, 109, 110, 111, 112, 116, 567
Singh, 205
Sinton, 338
Skaer, 489, 494
Sleigh, 342
Slobodkin, 73
Small, 346, 347, 352, 366, 378, 383, 386, 391, 406, 410
Smith, 34, 86
Sneath, 80, 109, 113, 114, 115, 116, 117, 118, 119, 146, 264, 320
Sokal, 80, 109, 113, 114, 115, 116, 117, 118, 119, 146, 264, 320
Soldo, 344
Sollas, 419, 421, 427, 437

Sonneborn, 12, 25, 30, 56, 58, 66, 67, 135, 136, 137, 140, 141, 142, 143, 144, 145, 146, 147, 148, 172, 174, 215, 217, 241, 346, 347, 354, 355, 356, 373, 413, 546
Spiegelman, 31
Spirin, 57
Stanier, 40
Stebbins, 11, 135, 146, 147
Steinböck, 496, 513, 543, 585, 586
Steinert, 51, 160, 161, 164, 172
Stent, 44, 57
Stephens, 41, 54
Stern, 243, 255, 260, 263, 509, 550
Stiasny, 299, 301
Stone, 144
Stout, 389
Street, 465
Stuart, 192
Sturtevant, 134
Sueoka, 28
Summers, 356
Sutcliffe, 273
Swammerdam, 326
Swanson, 45
Swedmark, 455, 474, 483, 485, 486
Swift, 339, 345, 368
Tait, 58
Tappan, 211, 212, 215, 219, 227, 228, 229, 230, 231, 233, 234, 235, 237
Tartar, 56, 332, 345, 350, 352, 353, 376, 413, 546
Tatton, 485
Taylor, 44, 54, 162
Teissier, 455, 477, 483, 485, 486
Thiel, 473
Thiney, 434
Thompson, 9, 75, 208, 214, 276, 293, 367, 385, 391
Threadgold, 41, 50
Tilney, 54, 203, 246
Trager, 132, 172, 342
Trégouboff, 39, 223, 244, 245, 246, 251, 253, 254, 258, 259, 260, 261, 262, 265, 266, 268, 269, 271,

Trégouboff (cont.)
  273, 274, 279, 280, 290,
  291, 292, 294, 297, 301,
  308, 320
Tucker, 341
Tuffrau, 335, 337, 345, 352,
  366, 372, 386, 387, 389,
  395, 412
Tuomikowski, 69
Tuzet, 421, 426, 427, 428,
  435, 436, 438, 452, 575
Ueda, 51
Umylina, 508
Ungerleider, 349
Ursprung, 546
Usinger, 111, 112, 114, 116,
  153
Vacelet, 440, 443
Valentin, 158, 161, 164, 272
Valkonov, 244, 245, 250,
  261, 262, 263
Van Niel, 5
van Wagtendonk, 344, 353
Versavel, 365
Vickerman, 158, 162, 173
Villeneuve, 257
Vivier, 338, 354, 376
Volkonsky, 205
von Siebold, 127
Waddington, 12, 16, 65
Wagner, 10
Wallace, 45, 72, 73, 192
Wallin, 51, 578
Waterman, 82
Watson, 24, 25, 26, 57
Weills, 466
Wells, 466, 474, 481, 484
Wenrich, 135, 367
Wermel, 246, 247, 251
Wessenberg, 329, 331, 341,
  343, 415
Westblad, 503, 521, 530,
  543
Westfall, 464, 465
White, 45
Wichterman, 334, 355, 373
Widersten, 474
Wiedmann, 562
Wigg, 57, 165, 333, 343
Wilkie, 51
Williams, 36, 37, 72
Willier, 510
Wilson, 259, 439
Wise, 330, 348, 349, 353
Wilson, 41, 42, 47, 49, 50,
  51, 54
Woese, 58
Wohl, 33
Wohlman, 54
Wolfe, 333
Wolff, 513
Wolska, 53, 88, 91, 96, 97,
  353, 388, 391, 396, 397,
  405
Wolstenholme, 51
Wood, 87, 229, 235
Woodard, 339, 345, 368
Woodger, 89
Woodruff, 355
Wright, 11, 109
Yusa, 352, 544
Ziegler, 446
Zimmerman, 111, 288, 573
Zinn, 455
Zipser, 44
Zuckerkandl, 33, 34, 35, 36,
  89
Zuelzer, 253, 256, 258
Zumstein, 149

Organism Index

Acantharia (acantharians), 154, 155, 202, 221, 223, 246, 247, 250, 271, 273, 287, 289, 290 *et seq.*, 327, 359, 369, 383, 432, 555, 567, 570, 571, 574, 577 *et seq.*, 593
Acanthochiasma rubescens, 294, 296, 310, 311, 314, 317, 322, 323, 571, 574;
   A. fusiformis, 295
Acanthochiasmidae, 298, 306, 319, 321
Acanthocyrta haeckeli, 295
Acanthocystidea, 244
Acanthocystis, 243, 256, 259;
   A. aculeata, 243, 254, 256, 262
Acantholithium stellatum, 312, 314, 316, *et seq.*
Acanthometra pellucida, 312, 314, 316, *et seq.*, 320
Acanthometridae (acanthometrids), 318 *et seq.*
Acanthometron, 299
Acanthophlegma krohni, 296, 311, 314, 317
Acanthophlegmidae, 319
Acanthospira spiralis, 294
Acanthostaurus purpurascens, 296, 299, 300
Acetabularia, 256
Acoela (acoels), 489, 494, 495, 496, 497, 500 *et seq.*, 504 *et seq.*, 515, 517 *et seq.*, 530, 532, 534 *et seq.*, 539 *et seq.*, 548, 550, 561, 563, 576, 586, 589 *et seq.*, 593
acoelomate animals, 516
Aconchulina, 221
Acrania, 537

Acrasiales, 238
Acrasida, 222
Actinelia (actinelids), 320, 322, 323
Actinobolina, 367, 368, 558;
   A. vorax, 367
actinomycetes, 30
Actinolophus, 258
Actinophrydea, 244
Actinophryida (actinophryidans), 245 *et seq.*, 254, 255, 263, 571, 573, 577
Actinophryidia, 244
Actinophrys, 243, 246 *et seq.*;
   A. sol, 135, 243, 248, 249, 251, 253;
   A. pedunculata, 257
Actinopoda (actinopods), 270
Actinopodea, 221, 222
Actinopodia, 222
Actinosphaerium, 246 *et seq.*, 255;
   A. eichorni, 248, 249, 251, 253;
   A. nuleofilum, 251, 252
Actinulida, 485
Actissa (= *Procyttarium*), 285, 287, 571, 573;
   A. princeps, 282, 283, 289
Aequorea, 460
Afronta, 507
Aglantha, 460
Aglaura hemistoma, 480
algae, 30, 48, 51, 52, 55, 56, 128, 129, 131, 149, 152, 176, 223, 230, 240, 262, 273, 297, 340, 341, 359, 361, 369, 379, 381, 384 *et seq.*, 399, 400, 414, 494, 509, 523, 544, 545, 571, 572, 576, 577, 581;
   apochlorotic (colorless),

198, 288, 571, 574, 578, 592;
   blue-green, 30, 385;
   brown, 149, 581;
   flagellated, 265;
   green, 51, 56, 256, 581;
   red, 581;
   symbiotic, 278, 459
Allogromia, 211;
   A. laticollaris, 217
Allogromidae, 573
Alternatia, 222
amebas (ameboid forms), 12, 54, 137, 154, 155, 166, 202 *et seq.*, 308, 327, 377, 380, 384, 558, 566, 571 *et seq.*, 577, 578, 581, 584
ameboid rhizoflagellates, 592, 593
ameboflagellate, 571
Amoeba, 42;
   A. discoides, 240;
   A. proteus, 202, 203, 240;
   A. stigmatica, 150
Amoebaea, 221, 238
Amphibia (amphibians), 18, 79, 84, 171, 194
Amphidiscophora, 420
Amphilithidae, 319
Amphilithium concretum, 312, 314, 316, 317
Amphilonche elongata, 295
Amphinema, 460
Amphiscolops, 507;
   A. langerhansi, 509, 513
Ancistrina, 363
Ancistrumina, 362
Annelida (annelids), 3, 39, 70, 86, 192, 212, 522, 537, 587
Anthophysa vegetans, 150
Anthozoa (anthozoans), 454 *et seq.*, 459, 465, 468, 471, 473 *et seq.*, 477,

645

Anthozoa (cont.)
479, 480 et seq., 486,
488, 582, 587, 588
Aphrothoraca
(aphrothoracans), 243, 254,
255, 257, 259, 260
apochlorotic (colorless)
forms (see algae,
chrysomonads)
Apostomatida
(apostomatids), 356,
387, 393, 397, 400, 406
et seq., 410, 411
Arachnosphaera myriacantha, 267, 282
Arcella, 207, 215
Archaeopteryx, 17, 22
Archicoelomata, 586
Archilopsis unipunctata,
508
Archoophora
(archoophorans), 506,
526, 530, 534, 535
armadillos, 72
Arthracantha, 292, 317 et
seq.
Arthropoda (arthropods),
20, 36, 64, 192, 587
artiodactyls, 17
aschelminths, 522
Ascobolus, 40
Astomatida (astomatids),
352, 392, 406
astomes, 352, 377
Astracantha parodoxa, 282
Astrolithidae, 318, 319
Astrolithium bulbiferum,
294
Astrorhizidae, 572, 573
Athalamia, 221
Atlanticetta craspedota,
282
Aulacantha, 275, 276;
A. scolymantha, 267, 275
Aurelia, 454
Automalos hamatus, 508
Autotractea, 221
Azoosporidae, 245, 262
Bacillus subtilis, 28
bacteria (bacterial), 9, 24,
28 et seq., 34, 39, 40,
41, 133, 152, 164, 172,
183, 210, 213, 230, 251,
273, 281, 297, 340, 341,
359, 361, 369, 381, 385,
387, 389, 414, 490
bacteriophages, 7
Balantidium, 350
Barbulanympha sp., 186,
188
Bdellocephala, 508
bedbug, 169
beetle, 185
Bicoeca lacustris, 182, 184
Bicosoecida (Bicoecidea,
bicosoecids), 178, 182,
184, 198, 200
Bilateria (bilaterians), 493,
510, 536, 540, 549, 581,
583, 586, 587
Bilateroblastaea, 589
bird, 17, 104, 105, 171, 194
Blastaea, 586
Blastocrithidia, 171
blood-flagellates, 38, 51,
165, 166
Bodo, 157, 169;
B. saltans, 156, 160, 182,
184, 194, 571, 572
Bodomorphidae, 182, 184
Bodomorpha minima, 182,
184
Bodonida (bodonids), 571,
572, 577, 578
Bodonidae, 182, 184
Bodonidea, 182, 184
bodonines, 157, 176, 178,
191, 194, 198, 200
bony fishes, 20, 84
Bothrioplana, 506, 513;
B. semperi, 494, 528, 531
Botryopyle dictyocephalus,
282
bottle-nosed dolphin, 85
Bougainvillia, 460
Boveria, 362
Brachiopoda (brachiopods),
20, 63, 537, 587
Bresslaua, 352, 389
Bryophyta, 52
Bursaria, 352, 371, 378,
406, 412;
B. truncatella, 332, 341,
352, 372
butterflies, 278
Calcarea (calcareous
sponges, calcisponges),
419, 426, 427, 431, 433,
436 et seq., 441 et seq.,
571, 575
Calcaronea, 420
Calcinea, 420
Calcispongiae, 420
Calonympha, 415
Calonymphidae
(calonymphids), 157,
160, 182, 195, 197
Campanella umbellaria,
364
Camptonema, 246 et seq.;
C. mutans, 253
Candida, 120
*Carcharodopharynx
arcanus*, 517
Carposphaera nodosa, 282
Castanella balfouri, 282
Catenulida (catenulids),
504, 505, 513, 522, 525,
530, 535
Catharina florea, 394
Cementella loricata, 282
Centrohelida
(centrohelidans), 245,
253 et. seq., 260, 263,
321, 571, 573, 577
Centrohelidia, 244
cephalopods, 227
Centropyxis, 215
Ceractinomorpha, 420
Cercobodo neimi, 182, 184
Cercobodonidae, 182, 184
cestodes, 154, 493, 522, 523
cetaceans, 287
chaetognaths, 185
Chalarothoraca
(chalarothoracans), 243,
254, 255, 257
Challengeron armata, 82
Chaos chaos, 202
Chaunacantha, 299, 317 et
seq.
chicken, 28, 120
Childia, 511
Chilodon, 348, 352
Chilodonella, 348;
C. uncinatus, 351

ORGANISM INDEX

Chilomitus, 197
chiton, 20
Chlamydodon mnemosyne, 371
Chlamydomonas, 56, 86, 150
Chlamydophora, 243, 254
Chlamydophrys sp., 203
Chlorophyta, 51, 576
Choanoflagellata (choanoflagellates), 160, 161, 164, 168, 176, 178, 181, 182, 184, 190, 198, 200, 452, 490, 557, 563, 570, 571, 577, 578
chondrichthyan fish, 84
Chonotrichida (chonotrichs), 352, 370, 378, 383, 384, 389, 405, 408, 411
Chordata (chordates), 3, 20, 63, 70, 129, 488, 586
Chromulina, 150
Chrysamoeba pyrenoidifera, 319;
  C. radians, 571, 574
Chrysarachnion insidians, 150
Chrysaora, 465, 468
Chrysodendron ramosum, 150
Chrysomonadida (chrysomonads), 151, 159, 181, 190, 200, 240, 263, 288, 289;
  colorless, 322, 325, 389, 571 *et seq.*, 576, 577;
  cyrtophorine, 181
Chrysophyceae, 288
Chrysophyta, 288
Cienkowskya, 258;
  C. mereschkowski, 257
Ciliophora (ciliates, ciliated protozoa), 54 *et seq.*, 66, 85, 98, 103, 132, 133, 135, 137, 139, 144, 145, 154, 155, 166, 172, 174, 180, 191, 192, 200, 210, 211, 240, 241, 248, 260, 309, 326 *et seq.* 489, 491, 492, 531, 536, 543 *et seq.*, 555, 557, 558, 562, 563, 567, 570 *et seq.*, 574 *et seq.*, 580 *et seq.*, 586, 587, 589 *et seq.*, 593
Clathrina, 422, 435, 445
Clathrulina, 262;
  C. elegans, 261
Clathrulinidea, 244
Climacammina, 233
Climacostomum, 336
Cnidaria (cnidarians), 154, 439, 454 *et. seq.*, 503, 510, 537, 540, 541, 542, 555, 563, 567, 568, 571, 575 *et seq.*, 577, 582, 586, 587 *et seq.*, 592
cnidarian polyps, 399
Cnidosporidia (cnidosporidians), 154, 387, 489
coccolithophores, 154
Codonodendron ocellatum, 150
Codonosiga, 165, 181;
  C. botryotis, 82, 184, 190;
  Codonosiga sp., 571
Coelosomides, 350, 405, 407;
  C. marina, 384, 385, 390, 398, 399, 403
Coelacantha ornata, 282
Coelenterata (coelenterates), 154, 387, 454, 539, 540, 587, 589
Coelodendrum gracillimum, 282
coelomate metazoans, 437
Coelomata, 587, 588
Coleaspis coronata, 313, 314, 317, 323
collared flagellates, 181, 199, 575
colorless (apochlorotic) forms:
  algae, 152, 240, 578;
  chrysomonads, 322, 325, 573;
  flagellates, 155, 322
Colpoda, 35, 336, 352, 389;
comb jellies (see Ctenophora), 454, 589
Conacon foliaceus, 311, 312, 314, 317
Conaconidae, 319
Conchidium, 275;
  C. argiope, 267
Conchophthirius, 366
condylarths, 17
Conradocystis dinobryonis, 150
Convoluta, 507, 512;
  C. convoluta, 502, 539;
  C. fulvomaculata, 539;
  C. roscoffensis, 496, 498, 511;
  C. sutcliffei, 504, 509
copepods, 297, 387, 393, 399, 472
Coracatharina, 394
Corallistes, 430
corals, 63, 454, 459, 488
Coronympha octonaria, 182, 184
Craspedomonadaceae, 150
Cribrogenerina, 233
Cribrostomum, 233
cricket, 189
Crithidia, 171, 192;
  C. euryophthalmi, 169, 171
Crocodilus niloticus, 170
Crustacea (crustaceans), 212, 230, 297, 381, 384, 399, 509, 519
Cryptocercus, 145, 169, 179;
  C. punctulata, 156
cryptomonads, 377, 384, 489, 577
Cryptophyta, 576
Ctenophora (ctenophores), 454, 455, 482, 489, 536, 539, 540, 578, 587, 588, 589
Cyanea, 468
Cyathodinium, 394
Cyclia, 222
Cycloposthium bipalmatum, 591
Cyclops, 467

Cyrtophora pedicellata, 181, 190, 571
Cystidium princeps, 282
dalyelloids, 502
decapod crustaceans, 212
Demospongiae (demosponges), 419, 421, 425, 428, 431, 435, 436, 437, 441, 442, 444 *et seq.*, 449, 450, 452, 563, 571, 575
Demothoraca (demothoracans), 577, 588
Dendrocoelum lacteum, 539
Dendrosoma, 85
Dendromonas virgaria, 150
Dermocystidium marinum, 47
Desmothoraca, 243, 244, 259
Desmothoracida, 245, 261 *et seq.*, 571, 573
Deuterostomia (deuterostomes), 586 *et seq.*
Devescovinidae (devescovinids), 182, 184, 194, 195, 197
Devescovina duboscqui, 182, 184
 D. *striata*, 306
diatoms, 230, 297, 377, 384, 385, 389
Dictophimus sphaerocephalus, 267, 282
Dictyacantha tetragonopa, 313, 314, 317
Dictyanthidae, 319
Didesmis, 396, 405, 411;
 D. *ovalis*, 397
Didinium, 129, 130, 331, 341;
 D. *nasutum*, 330
Difflugia, 202, 215, 227, 241;
 D. *corona*, 208, 214;
 D. *oviformis*, 214
Dileptus, 331, 335, 336, 341, 342, 365, 371;
 D. *anser*, 339, 350, 351, 369;
 D. *estuarinum*, 339, 543
Dimorpha, 259, 263, 321;

D. *floridanis*, 245, 259, 260;
D. *mutans*, 259, 260, 571, 573
Dimorphella elegans, 260
Dinophyceae (dinoflagellates), 288, 289, 322, 380, 389, 489, 571, 573, 577
Diodon, 75
Diploconidae, 319
Diploconus fasces, 313, 314, 317
Discocelides langi, 502
Discorbis orbicularis, 231
Distomatina, 186, 188
dog, 120
dolphins, 84
Donatomonas cylindrica, 150
donkey, 120
Dorataspidae, 319
Dorataspis loricata, 295
Drosophila, 7, 63, 83;
 D. *melanogaster*, 44;
 D. *persimilis*, 83;
 D. *pseudoobscura*, 83
duck, 120
Dugesia (see, also, planarians), 508, 512;
 D. *dorotocephala*, 513
Echinodermata (echinoderms), 20, 39, 86, 454, 455, 537, 587
Echinosphaerium nucleofilum, 247, 252
Ectoprocta, 587
elephant, 37, 388, 399
Elephas maximus, 388
Eleuthropyxis, 150
Elphidium crispum, 231, 232
enteropneusts, 465, 537
Entodiniomorphida (entodiniomorphids), 353, 388, 395 *et seq.*, 405, 406, 408, 411 *et seq.*
Entoprocta, 587
Epidinium parvicaudatum, 330
Epipyxis, 150
Escherichia coli, 7, 31

Eucomonympha imla, 186, 188
Eucomonymphidae, 186, 188
Euglena (euglenoids), 176, 194, 200, 571, 572, 577;
 E. *gracilis*, 149, 151
Euglypha, 208, 227
eulecithophores, 511
Eumetazoa (eumetazoans), 418, 454, 455, 491, 593
Euplotes, 132, 339, 345, 348, 349, 350, 352, 353, 395;
 E. *eurystomus*, 330;
 E. *minuta*, 348
Euryophthalmus convivus, 169
Exogena, 411
false heliozoans, 244, 245
ferns, 30
Filida, 221, 222
Filoreticulosia, 222
Filosa, 221, 224, 238
Filosia, 222
finches, 71
fish, 28, 29, 79, 84 *et seq.*, 171, 194, 278, 286, 472
Flagellata (flagellates), 42, 55, 85, 128, 129, 133, 136, 155, 210, 240, 260, 262, 263, 275, 307, 321, 322, 379, 411, 414, 549, 563, 584, 587
flatworms, 39, 64, 86, 88, 154, 439, 455, 493 *et. seq.*, 561, 563, 576, 586
flukes, 493, 583
Foaina, 200
Foraminiferida (foraminiferans), 133, 136, 202, 207, 211 *et seq.*, 219, 221, 224, 227 *et seq.*, 250, 305, 309, 555, 571 *et seq.*, 577 *et seq.*, 592, 593
frogs, 185, 200, 415
Frontonia, 406, 409;
 F. *invellatum*, 358;
 F. *marina*, 385, 391, 399
fruit fly (see also,

ORGANISM INDEX

Drosophila), 83
fungi, 34, 37, 40, 56, 131, 151, 152, 221, 581
gammarids, 384
Geleia, 381
Geodia, 430
Giardia intestinalis, 186, 188
Gigartaconidae, 319
Gigarton fragilis, 312
Glabratella sulcata, 136, 139
Glaucoma, 331, 352
Glossatella, 398, 400, 406, 409;
    G. pisicola, 391, 392;
    Glossatella sp., 386, 392, 399, 400
Glossina, 193;
    G. papilis, 170
Gnathostomulida (gnathostomulids), 506, 530
Gonionemus, 60, 464, 465
Gonthyrea, 473
Gorgonia (gorgonians), 454, 458
Graffizoon lobata, 513
Grantia, 430
Granuloreticulida, 222
Granuloreticula, 221
Granuloreticulosa, 221, 224
Granuloreticulosia, 222
Grebnickiella, 133
guinea pig, 29
Gymnaster, 573;
    G. pentasterias, 288
Gymnostomatida (gymnostomes), 103, 335, 341, 342, 348, 352, 365, 366, 369, 370, 377 *et seq.*, 383, 384, 389, 391, 392, 394, 397, 407, 408, 410 *et seq.*, 543 *et seq.*, 548, 558
Gymnoeraspedidae, 182, 184
Gyratrix, 78
Halammohydra, 55, 484 *et seq.*, 563
Halisarca, 435, 448, 449, 450, 563, 575;
    H. dujardini, 436;

Halisarca sp., 571
hamster, 29
Haploposthia, 521, 523, 534, 543 *et seq.*, 561, 590;
    H. rubra, 495, 521, 524, 525, 531, 532, 543, 546, 559, 563, 571, 576;
    H. viridis, 539
Heliochona, 389, 398;
    H. scheuteni, 383, 384, 389, 399, 405
Heliocoprorodon, 331
Helioflagellida, 571
Helioflagellata, 244
Helioflagellidae, 245, 259 *et seq.*, 262, 576
Heliozoa (heliozoans), 154, 155, 202, 221, 223, 242 *et seq.*, 290, 305, 321, 322, 327, 571, 573, 575, 592, 593
Heliozoia, 287, 289, 290
Hemichordata, 587
Hemispeira, 362
Herpetomonas, 171
Heterolagynion, 150
Heterotrichida (heterotrichs), 331, 335 *et seq.*, 352, 359, 360, 362, 366, 381, 387, 395, 398, 407, 408, 410 *et seq.*
Hexactinellida (hexactinellids, hyalosponges), 419, 422, 427, 429, 431, 441, 442, 444, 448, 449, 571, 575
Hexalaspidae, 319
Hexamastix, 167, 168, 197
Hexasterophora, 420
Hircinia, 432
Hofstenia, 501, 504;
    H. pardii, 506, 515, 517
Holacantha, 299, 317, 318, 319, 320
Holocoela, 527, 530
Homotrema rubrum, 13, 230
Holomastigotes, 198;
    H. elongatum, 186, 188
Holomastigotidae, 186, 188

Holophrya, 379, 380
Holotrichia (Holotricha, holotrichs), 331, 348, 349, 360 *et seq.*, 387, 392, 393, 408, 411
holothurian, 362
Homalozoon, 342, 350
Homoeonema, 85
Hoplonymphidae, 186, 188
horse, 120
human, 29, 34, 44, 120, 180, 185
Hyalonoema, 430
hyalosponges (*see* Hexactinellida)
Hybocodon, 460
Hydra, 61, 454, 457, 467, 472, 473, 488;
    H. littoralis, 458
Hydraulea, 221, 222
Hydrolimax, 512
Hydrozoa (hydrozoans), 454, *et seq.*, 461, 462, 467, 471, 473, *et seq.*, 477, 479, 480 *et seq.*, 488, 575, 577, 582, 587, 588
Hymenostomatida (hymenostomes), 335, 349, 352, 358, 363, 372, 378, 385, 391, 405 *et seq.*, 558
hypermastigids, 180, 198, 199
hypermastigote, 275
Hyporadiolarida, 222
Hypotrichida (hypotrichs), 335, 336, 338, 347, 348, 353, 360, 361, 372, 381, 387, 395, 407, 408, 411
Hypotrichomonadinae, 197
Hypotrichomonas (hypotrichomonads), 158, 197
Hysterocinetidae, 364
icthyosaurs, 84 *et seq.*
Icthyophthirioides, 348;
    I. browni, 331
Icthyophthririus multifiliis, 348, 351
Idionympha perissa, 186, 188
Ignatocoma salbellarum, 358, 359

insects, 34, 38, 39, 86, 112, 170, 171, 173, 180, 185, 191 et seq., 200
invertebrates, 61, 85, 191, 194, 503, 587, 591
Iridia, 217;
  I. lucida, 215, 216, 228, 231
Isotricha, 340
Joenia, 164;
  J. annectens, 186, 188
Joeniidae, 186, 188
Kahliella, 412;
  K. acrobates, 387, 391, 395, 398, 399
kalyptorhynchids, 502
kangaroo, 120
Kephyrion, 150
Karotomorpha bufonis, 182, 184
Karotomorphidae, 182, 184
Keratosa, 420, 450
Kinetoplastida (kinetoplastids), 157, 178, 191, 198, 199, 571
Kybotion eremita, 150
Labyrinthula algeriensis, 263
Labyrinthulia, 222
Labyrinthulida (labyrinthulids), 202, 218, 263, 571, 573, 577
Labyrinthulidae, 245, 263
Labyrinthuloidea, 244
Lacrymaria, 342
Lagynidae, 571, 573
Lagynion, 150
Latimeria, 20
Lecithoepitheliata, 527, 530, 535
leeches, 171, 192, 193, 194
Leishmania, 171;
  L. donovani, 172
Leptodiscus, 85
Leptomonas (leptomonads), 171, 194, 572;
  L. karyophilus, 191, 192;
  Leptomonas sp., 571
Leucandra, 433
Leucosolenia, 422, 426, 430, 445, 448, 449, 575;
  L. botryoides, 436;
  Leucosolenia sp., 571

Leukochrysis, 150, 288, 289, 573
Leukopyxis asymmetrica, 150
Lillium, 44
Lingula, 14, 20
Lithopteridae (lithopterids), 318, 319
Lithopterida fenestrata, 312, 313, 314, 317
Lituolacea, 235
lizards, 185, 200
Lobida, 221, 222
Lobosa, 221, 224, 238
Lobosia, 222
Lophomonadidae, 186, 188
Lophomonadina, 186, 188
Lophomonas, 198;
  L. blattarum, 186, 188
Loxodes, 369, 381, 382;
  L. rostrum, 355
Loxophyllum pseudosetigerum, 365;
  L. kahli, 365
Lyrocephalus, 77
Macrorhynchus, 78
Macrostomatida (macrostomatids), 508, 522, 525, 530, 535
Macrostomum, 511, 519, 520, 539;
  M. appendiculatum, 495, 525, 531
Mallomonas, 152;
  M. apochromatica, 150
Mammalia (mammals), 17, 20, 79, 84, 85, 171, 193, 194, 405
Mastigamoebidae (mastigamebas), 239
Meseres, 412;
  M. cordiformis, 388, 391, 397, 399, 400
Mesostoma (mesostomes), 505, 506;
  M. ehrenbergi, 495, 529, 531
Mesozoa (mesozoans), 578, 586, 588
Metacoronympha senta, 200
Metaphyta (see also, plants), 174
Metazoa (metazoans), 9, 48,
53, 56, 60, 61, 62, 64, 96, 125, 129, 133, 136, 153 et seq., 174, 213, 248, 286, 306, 321, 344, 419, 435, 437, 439, 491, 503, 536, 537, 539, 565, 570, 577, 578, 583, 585, 586, 588 et seq., 593
Metoposaurus, 77
Metopus, 408
Metridium, 458
Microjoenia fallax, 86, 188
Microjoenidae, 186, 188
Micrometazoa, 399
Miliolacea, 235
Minona evelinae, 508
Miracella ovulum, 282
Mollusca (molluscs), 20, 39, 63, 86, 227, 392, 488, 522, 537, 587
Monadida, 244, 245
Monadodendron distans, 150
Monas, 150
Monaxida, 420
monkey, 120
Monocelida, 518
Monocelidae (monocelids), 504, 507, 508
Monocelis, 78, 496;
  M. fusca, 508
Monocercomonadidae, 182, 184, 194, 196
Monocercomonas (monocercomonads), 168, 195, 196, 197, 572;
  M. colubrorum, 182, 184;
  M. moskowitzi, 197;
  Monocercomonas, sp. 571
Monoraphis, 428
Monosiga, 150
moth, 120
mouse, 29
Muggiaea, 461
Mycetozoia (mycetozoans), 222, 238
Mycetozoida, 222
Naegleria, 46
Nasellaria (nasellarians), 269, 272, 276, 281, 284, 285
Nassula, 367, 405, 410, 411, 412;

N. ornata, 338, 383, 384, 390, 398, 399, 403
Nausithae, 454
Nemertodermatida, 535
Nemertini (nemertines), 522, 583, 587
Neobursaridium, 352, 378, 412;
  N. gigas, 372
Neoophora, 534, 535
Neorhabdocoela (neorhabdocoels), 506, 510, 511, 529, 530, 535
Neurospora, 7, 40, 120
Noctiluca, 42
Nodosariacea, 235
Notila, 415
Notoplana alcinoi, 495, 525, 526, 531
Nyctotherus, 340
Obelia, 460
ocean sunfish, 75
Ochromonas, 150, 151, 152;
  O. malhemensis, 151
octopus, 37
Odontostomatida (odontostomes), 338, 395, 408, 411
Oikomonas, 150
Oligotrichida (oligotrichs), 370, 388, 397, 400, 408, 410, 411
opalinids, 415
Ophlitospongia seriata, 428
Ophryoglena, 371
ophryoglenids, 371, 372
Orbulina universa, 223
Oroscena regalis, 282
Orthogoriscus, 75
Oscarella, 448, 449, 563, 575;
  Oscarella sp., 571
Ostrea, 20
Otocelis, 507
Otohydra, 485, 490;
  O. vagans, 486, 488, 489, 571, 575
Oxnerella, 256, 259;
  O. maritima, 254, 255;
  Oxnerella sp., 571
Oxymonadidae, 186, 188
Oxymonadina, 186, 188
Oxymonas grandis, 156,
  186, 188
Palmenia, 507
Paracicerina maristol, 517
Paramecium (paramecium, paramecia), 7, 56, 66, 130, 133, 136, 141, 143, 147, 217, 328, 329, 334, 336, 337, 340, 341, 343 *et seq.*, 349, 352, 361, 364, 372, 373, 391;
  P. aurelia, 103, 133, 135 *et seq.*, 330, 332, 339, 347, 348, 354 *et seq.*, 376;
  P. bursaria, 143, 144, 348, 356;
  P. caudatum, 328, 329, 333, 344;
  P. multimicronuleatum, 141;
  P. trichium, 134, 191, 347, 348
Paraphanostoma submaculatum, 539
Parapodophrya, 393, 394, 398;
  P. soliformis, 386, 387, 391, 399
parasitic flagellates, 51
parasitic flatworms, 87
Paravortex cardii, 539
Pelagea, 463
Pelodinium reniforme, 388, 395, 398, 399, 408
Pelomyxa palustris, 206
Penardia mutabilis, 203
penguin, 120
Pentatrichomonas, 197
Pentatrichomonoides, 197
Pentatrichomonoidinae, 197
Peraclistus oophagus, 508
Peranema, 176, 194, 571
Perca fluviatilis (perch), 386, 392
Peritrichia, 392, 393
Peritrichida (peritrichs), 103, 326, 334 *et seq.*, 352, 353, 356, 359, 362 *et seq.*, 366, 370, 372, 378, 386, 394, 400, 406, 408 *et seq.*
Phaeodaria (phaeodarians), 269, 273, 281, 283, 284, 285
Phaeodina sp., 282
Phaeometra, 283
Phaeophyta (*see also*, algae), 149
Phalansteriidae, 182, 184
Phalansterium, 181;
  P. consociatum, 182, 184, 190
Pheronema, 422
Phialella, 460
Philaster, 366, 392, 406, 409, 412;
  P. digitiformis, 386, 391, 399
Phoronida, 587
Phractopelta dorataspis, 313, 314, 317;
  P. tessarapsis, 295
Phractopeltidae, 319
phycomycetes, 263
Phyllacantha, 318, 319
Phyllostauridae, 319
Phyllostaurus siculus, 294, 313, 314, 317, 318
Physalia pelagica, 455
phytoflagellate, 198
Phytomonas, 171, 180
phytoplankton, 297
pig, 120
pigeon, 120
pinnipeds, 287
Piroplasmea, 218, 222
Placozoa, 588
Plagiostomum, 539;
  P. album, 495, 527, 531
planarians, 493, 496, 497, 501, 502, 504, 508, 513, 519
Planocera inquiena, 511
plants, 28, 30, 52, 129, 133, 147, 171, 180, 414, 581
Planuloidea, 586, 588
Plasmodiophorales, 217, 238
Plasmodium, 133
Platyctenea, 540
Platyhelminthes, 64, 493, 586, 587
Platytheca micropora, 150
Plectodinium, 573;
  P. nucleovolvatum, 288

Plegmosphaera myriacantha, 282
Pleuronema, 366;
   P. coronatum, 330
pleuronematines, 337
Pleuraspis costata, 313
Plumularia, 458
Podactinelius sessilis, 320, 321
Podophrya collini, 354
Polycelis, 508
Polychoerus carmelensis, 509
Polycladida (polyclads), 494, 496, 497, 500 *et seq.,* 506, 510 *et seq.,* 517, 522, 526, 530, 535, 539, 540
Polymastigidae, 182, 184
Polymastix melonthae, 182, 184
porcupine fish, 75
porcupines, 86
Porifera (sponges), 20, 63, 70, 154, 276, 286, 419 *et seq.,* 454, 490 *et seq.,* 503, 510, 555, 557, 563, 567, 571, 575, 577, 578, 581, 582, 586 *et seq.,* 592
porpoises, 86
Poteriochromonas, 150
Proboveria, 350
Procerodes lobata, 495, 528, 529, 531
Procyttarium (see *Actissa*), 283, 285, 571, 573
Podactinelius sessilis, 292
Prolecithophora (prolecithophorans), 506, 508, 535
Proplicostomata, 535
Proporoplana jenseni, 495, 526, 531
Proporus, 507
Prorhynchus, 78;
   Prohynchus sp., 495, 517, 525, 528, 531
Prorodon, 335, 352, 371, 379, 389
Proseriata, 535
Prostheceraeus vittatus, 539

proteomyxans, 223
Proteomyxia, 244
Proteomyxida (proteomyxids), 262, 263, 571, 573, 577, 592
Proteomyxidea, 244
Proteomyxidia, 245
Proteromonadidae, 182, 184
Protermonadina, 182, 184
Proteromonas lacertaeviridis, 182, 184
Proterospongia, 452, 453, 563, 571, 575;
   P. haeckeli, 452
Protista (protistans, protists), 379, 545, 589, 591
Protocruzia, 397, 406, 407, 410, 412;
   P. adhaerans, 387, 391, 394, 398, 399, 407
Protospira, 363
Protostomia (protostomes), 586, 587, 588
protozoa (protozoans), 9, 12, 30, 53, 56, 85, 86, 96, 108, 125, 127 *et seq.,* 155, 162, 166, 172, 174, 191, 210, 221, 223, 240, 242, 262, 263, 273, 297, 305, 324, 340, 344, 361, 381, 400, 452, 509, 510, 536, 544, 545, 565, 570, 575 *et seq.,* 580, 583, 584, 586, 587, 590, 592, 593
Protrichomonas, 197
pseudocoelomate animals, 516
Pseudocohnilembus, 349
Pseudolithidae, 318, 319
Pseudolithium, 318;
   P. bifidum, 312, 314, 316 *et seq.*
Pseudomicrothorax, 349, 352, 367, 371, 378, 405, 407, 412, 558;
   P. dubius, 367, 371, 385, 391, 398, 399
Pseudoprorodon, 350, 389
Pseudospora, 262
Pseudotrichomonas, 197
Pseudotrichonympha, 161

Pseudotrypanosoma, 197
Ptygura, 85, 86
Pyrrophyta (see Dinophyceae), 288
Pyrsonympha, 176;
   P. flagellata, 186, 188
Pyrsonymphidae, 186, 188
Pyrsonymphina, 186, 188
Raabena bella, 388, 391, 396, 397, 398, 399, 405
rabbit, 120
Radiata, 454, 587, 589
Radiolaria (radiolarians), 39, 154, 155, 202, 221, 223, 242, 246, 247, 257, 265 *et seq.,* 290, 293, 305, 309, 321, 322, 324, 327, 389, 432, 555, 567, 571, 573 *et seq.,* 577 *et seq.,* 593
Radiolarida, 222
Raphidiophrys, 255;
   R. elegans, 256
rat, 189
Remanella, 369, 382;
   R. multinucleata, 103, 332, 355, 365, 379
reptile, 17, 79, 84, 85, 171, 185, 194, 200
Retortomonadidae, 186, 188
Retortomonadina, 186, 188
Retortomonas gryllotalpae, 186, 188
rhabdocoels, 486, 508
rhesus monkey, 28
Rhizochrysididae, 288
Rhizochrysis crassipes, 150
rhizoflagellates, 154, 155 *et seq.,* 202 *et seq.,* 571, 573, 574, 592, 593
Rhizomastigida, 239, 245
Rhizonympha jahieri, 186, 188
Rhizonymphidae, 186, 188
Rhizopodea (rhizopods), 221, 222, 242, 262
roaches, 145, 155, 164, 173, 174, 179, 180, 189, 199, 200, 415
rodent, 86
rose corals, 459
Rotaliella heterocaryotica, 217

ORGANISM INDEX 653

rotifers, 85, 86, 248, 385
ruminants, 193, 333
Saccharomyces, 40, 120
Salivaria, 193
salmon, 28, 29
Salpingoeca gracilis, 182, 184, 190
Salpingoecidae, 182, 184
sand shark, 85
Sarcodina, 204, 221, 222
Sarsia, 460
Sclerospongiae (coralline or sclerosponges), 419, 439, 442
screw worm, 120
Scuticociliatida (scuticociliates), 352, 366, 378, 386, 391, 406, 407, 410 et seq.
Scyphozoa (scyphozoans), 454, 455, 465, 471, 473, 475, 477, 479 et seq., 488, 582, 587
sea anemones, 454, 457, 458, 463, 475, 487
sea cucumber, 362
sea fan, 458, 459, 471
sea pens, 471
sea squirts, 129
sea urchin, 54
sea whips, 459
Seriata, 506, 528, 530, 535
sharks, 84, 86
silicoflagellates, 154
Siphonophora (siphonophores), 185, 461, 462, 481, 490, 568, 582
Siphonophorida, 482
Sipunculoidea (sipunculids) 537, 587
slime molds, 154, 202, 209, 214, 217, 221, 244, 571, 572, 577
snake, 120
Sphaenacantha, 318, 319
Sphaerellaria (sphaerellarians), 271
Sphenophrya dosiniae, 354
Spiralia, 63, 586
Spirillinacea, 236
Spirillina vivipara, 217
Spirochona, 389

Spiromonadidae, 182, 184
Spiromonas augusta, 182, 184
Spirophrya, 407;
 S. *subparasitica*, 387, 391, 393, 399, 407
Spirostomum, 342
Spirotrichia (Spirotricha), 393, 408, 411
Spriotrichida (spirotrichs), 103, 336, 352, 378, 394, 395, 411, 412
Spirotrichonympha, 168, 198;
 S. *flagellata*, 186, 188;
 S. *polygyra*, 156, 167
Spirotrichonymphidae, 186, 188
Spirotrichonymphina, 186, 188
sponges (see Porifera)
Spongiaria, 454
Spongilla, 439
Sporozoa (sporozoans), 47, 154, 215, 327, 578, 588
Spumellaria (spumellarians), 268, 271, 277, 281, 284, 285
squids, 20
starfish, 63, 105
Stauraconidae, 319
Stauracon pallidus, 294, 312, 314, 317
Stauraconthidae, 319
Staurapsis stauracantha, 295
Staurojoenidae, 186, 188
stegacephalons, 77
Stenostomum, 509, 550;
 S. *ventronephridium*, 495, 525, 526, 531, 532;
 S. *sthenum*, 512
stenostomid, 513
Stentor, 85, 86, 332, 335, 336, 341, 342, 345, 352, 369;
 S. *ambiguum*, 369;
 S. *auricula*, 369;
 S. *coeruleus*, 103, 339;
 S. *roeseli*, 369
Stephalia, 461
Stephanocyphus, 454

Stephanopogon, 338, 377, 378, 379, 398, 405, 410, 411, 415, 416, 544, 545;
 S. *colpoda*, 338
 S. *mesnili*, 338, 365, 369, 377, 378, 379, 383, 384, 390, 398, 399, 400, 403, 414, 543, 546, 558, 563, 571, 574 et seq.
Stercoraria, 193
Sticholonche zanclea, 271
Stokesiella lepteca, 150
Streblomastigidae, 186, 188
Streblomastix strix, 186, 188
Stylochromonas minuta, 50
Stylopyxis, 150
Stylosphaera lithatractus, 267
Suctoria (suctorians), 85, 331, 333, 341, 347, 354, 359, 362 et seq., 367, 368, 370, 372, 378, 394, 406, 407, 410, 414, 558
Suctorida, 352, 386, 393, 408, 411
Sycon, 422, 423
Symphacantha, 299, 317, 318, 319, 320
Synchromonas pallida, 150
Syncoryne, 85
Synura, 380
tapeworms, 87, 88, 493
Tedania, 422
Temnocephalida, 493
Teratonympha mirabilis, 186, 188
Teratonymphidae, 186, 188
termites, 145, 155, 164, 173, 174, 175, 180, 185, 189, 199, 200, 415, 555
Testaceafilosa, 221
Testacealobosa, 221
Tetractinellida, 420
Tetractinomorpha, 420
Tetrahymena, 56, 133, 328, 329, 331, 339, 344 et seq., 349, 350, 352, 364, 367, 371, 391;
 T. *pyriformis*, 103, 329, 351, 356, 358

tetrahymenines, 337, 366
tetrapods, 18
Tetratrichomastix, 197
Tetratrichomonas, 197
Teutophrys, 365
Textularia, 233
Thalamia, 221
Thalanophysa, 275
Thalassicola nucleata, 268
Thalassolampe, 285, 289, 571, 573;
  T. primordialis, 283
Thalassophysa sanguinolenta 270
Thallasothamnus pinetum, 282
Thaumatomonadidae, 182, 184
Thaumatomonas lauterhorni, 182, 184
Thecamoebae, 212
Thigmotrichida (thigmotrichs), 352, 354, 359, 362 et seq., 370, 378, 391, 392, 406, 408, 411
ticks, 194
Tillina, 389
Tintinnida (tintinnids), 388, 398, 397, 408, 410
Tintinnopsis nucula, 388, 389, 391, 398, 399
Tokophrya, 347
Trachelocerca, 369, 382;
  T. phoenicopterus, 373
Tracheophyta, 52
trematodes, 154, 493
Tricercomitus, 197
Tricladida (triclads), 502, 505, 506, 508, 510, 511, 517, 519, 522, 528, 530, 535
Trichodinopsis paradoxa, 342
Trichomitus, 197
Trichomonadida, 194, 195, 196, 198, 199, 415, 577
Trichomonadidae (trichomonadids), 157, 159, 176, 180, 182, 184, 197, 198
Trichomonadina, 182, 184

Trichomonadinae, 197
Trichomonas (trichomonads), 160, 161, 165, 167, 168, 177, 197, 306, 571, 572;
  T. gallinae, 158, 159;
  T. muris, 162;
  T. vaginalis, 51, 158, 182, 184
Trichonympha, 136, 138, 161, 164, 165, 166;
  T. campanula, 186, 188
Trichonymphidae (trichonymphids), 161, 164, 186, 188
Trichonymphina, 186, 188
Trichoplax adhaerens, 588, 589
Trichostomatida (trichostomes), 352, 365, 377, 378, 381, 384, 389, 396, 405, 408, 411, 412
Tridyctiopus elegans, 268
Triomitopsis, 197
Triplagicantha abietina, 282
Tritrichomonadinae, 197
Tritrichomonas (tritrichomonads), 158, 160, 197;
  T. augusta, 158;
  T. batrachorum, 156, 167
  T. cricetis, 158
true heliozoans, 244, 245
Trypanophidae, 182, 184
Trypanophis grobbeni, 182, 184
Trypanosoma (trypanosomes), 133, 157, 159, 164 et seq., 171, et seq., 193, 194;
  T. brucei, 162, 173;
  T. congolense, 162;
  T. cruzi, 167;
  T. gambiense, 156, 172, 182, 184;
  T. grayi, 170, 171, 172;
  T. mega, 160, 165, 172;
  T. alae, 160;
  T. vivax, 162

Trypanosomatida (trypanosomatids), 572, 577, 578
Trypanosomatidae, 160, 177
Trypanosomatina (trypanosomatines), 157, 169, 176, 178, 180, 191, 192, 198, 200, 571, 581
Trypanosomidae, 182, 184
Trypanosomidea, 182, 184
tsetse flies, 192, 193
Tubularia (tubularians), 457
tuna, 120
Tunicata (tunicates), 20, 88, 129, 523
Turbellaria (turbellarians), 154, 212, 381, 489, 493 et seq., 555, 556, 561, 567, 570, 571, 576, 577, 581, 583, 588, 591
turtle, 120
Tuscarora nationalis, 268
Urinympha, 415
Urocentrum, 406
Uronychia, 395;
  U. transfuga, 373
Urostyla, 348;
  U. hologama, 372
Vahlkampfia inornata, 203
Vampyrella, 245, 246, 262
Vampyrellidium, 246 et seq.;
  Vampyrellidium sp., 571
Veretillum cynomorium, 464
vertebrates, 16 et seq., 22, 34, 39, 42, 53, 54, 85, 86, 104, 119, 171, 180, 191, 193, 194, 273, 415
viruses, 21, 24, 37, 41, 133, 146
Vorticella, 326, 341;
  V. campanula, 103
Wagnerella, 254, 258;
  W. borealis, 253, 256, 257
water-flea, 326
waterlouse, 326
whale, 34
wheat, 34
woodpeckers, 86

worms, 63, 64, 491, 493, 494, 496, 509, 511, 512, 514, 515, 518, 523, 534, 544, 549, 583
Xenophyophorida (xenophyophorids), 202, 218, 222, 238, 571, 572, 577
Xiphacantha quadridentata, 295, 313, 314, 317, 323
yeast, 51
Zea, 7
zoochlorellae, 243, 502
zooflagellates, 137, 154, 155 et seq., 162, 306, 309, 327, 328, 344, 357, 362, 370, 380, 408, 415 et seq., 452, 489, 490, 491, 492, 535, 548, 550, 555, 570, 571, 572, 575, 577, 578, 580, 581, 582, 592, 593
zooplankton, 286
Zoosporidae, 245, 262
Zoothamnium alternans, 356

Subject Index

aboral field or structures, 336, 337, 362, 374
acantharians (*see also* Organism Index), 290 *et seq.*;
 functions of, 297 *et seq.*;
 homologies of, 302 *et seq.*;
 origins of, 321 *et seq.*;
 phylogenetic analysis of, 310 *et seq.*;
 phylogenetic conclusions, 324 *et seq.*;
 phylogeny of, 302 *et seq.*;
 structure of, 290 *et seq.*
accessory criterion (*see* subsidiary criterion), 178, 199, 522, 538, 544, 545
acellular condition, 132
acoelomate forms, 536
ACTH, 36
actinula, 474, 479, 480, 483 *et seq.*, 575
actomyosin, 465
Adansonian taxonomy, 114 *et seq.*, 245, 264, 290, 398, 420, 442, 454, 491, 571, 592, 593
adaptation, 5, 17, 18, 71, 108, 172, 179, 262, 273, 308, 327, 342, 378, 380, 400, 416, 488, 523, 538, 542, 545, 550, 562, 563, 574, 575, 579
adaptive peaks, 148, 562
adaptive radiation, 284, 455, 476
adaptive zone, 18, 19, 106, 281, 289, 324, 451, 452, 523, 561, 562, 566, 574, 580, 583
adhesive disk, 333, 457
adhesive structures, 518, 534

adoral depression, 334, 335, 336, 337, 349
adoral field, 336, 337, 359
adoral groove, 334, 337
adoral membranelle, 330, 580
adoral zone, 334, 336;
 membranelles of, 335, 336
adrenocorticotrophic hormone (*see* ACTH)
ageing, 140, 355, 356
alveoli, 329, 333, 358
amebocytes, 422, 427, 428, 433 *et seq.*, 438, 440, 452
ameboid forms (*see also* Organism Index), 202 *et seq.*;
 functions of, 208 *et seq.*;
 homologies of, 223 *et seq.*;
 phylogenetic conclusions, 238 *et seq.*;
 semic traits of, 220 *et seq.*;
 structures of, 203 *et seq.*;
 taxonomy of, 220 *et seq.*
ameboid motion, 204, 208, 209 *et seq.*, 218, 226;
 contraction-hydraulic system, 204, 205, 209, 223;
 gel conversion, 209;
 shearing flow system, 204, 209, 223
amino acids, 30, 31, 33, 131, 166, 211, 240, 344;
 gross analysis, 32;
 invariant sites, 35;
 residue, 33 *et seq.*;
 sequence of, 7, 25, 31, 32, 34, 39, 67, 84, 121, 240, 563, 577;
 substitution pattern, 33
amitosis, 339

amphiblastula, 435 *et seq.*, 439, 444, 445, 448
anagenesis, 18 *et seq.*, 490, 564, 567 *et seq.*, 579, 580, 582, 583, 592, 593;
 criteria for, 569;
 potential of, 19
analogy, 84, 177, 336
anal pore (*see* cytoproct)
ancestral character (*see also* plesioseme), 92
ancestors or ancestral forms, 69, 98, 153, 179, 239, 240, 277, 306, 321 *et seq.*, 357, 376, 380, 383, 397, 408, 414 *et seq.*, 451, 452, 455, 484, 485, 487, 490, 522, 523, 534, 536, 537, 539 *et seq.*, 543, 545, 546, 549, 550, 558, 561, 563, 568, 569, 570, 573, 575, 578, 579, 580, 582 *et seq.*, 586
animalcule, 127, 326, 327, 573
apochlorotic condition (*see also* colorless algae), 151
aposeme (*see also* episeme, hyperseme, hyposeme, polyseme), 92, 94, 97, 98, 100, 101, 103, 104, 107, 108, 113, 118, 122, 153, 177, 180, 191, 195, 198, 258, 277, 280, 282, 285, 289, 302, 304 *et seq.*, 309 *et seq.*, 315, 323, 374, 375, 398, 400, 401 *et seq.*, 405, 440, 443, 445 *et seq.*, 449, 477 *et seq.*, 484, 491, 513, 515 *et seq.*, 532, 533, 540, 557 *et seq.*, 562, 567, 568, 570, 572

SUBJECT INDEX 657

appendages, 494
archaeocytes, 427, 428, 435, 438, 439, 440, 441
archetype, 534, 536
arrested change, 19
ascon, 422, 423, 424, 425, 426, 434, 436, 437, 441, 442, 445, 446, 447, 448, 453, 582
asemic traits, 96
asexual organisms (see also ecospecies), 240, 241
asexual reproduction (see also fission), 133 et seq., 147, 173, 215, 248, 255, 274, 297, 309, 310, 324, 345, 360, 393, 407, 438, 440, 441, 458, 464, 475 et seq., 478, 511, 513, 544, 556;
 architomy, 513;
 paratomy, 513
asparagine, 472
assimilation, (see also osmotrophy, saprotrophy), 127
aster, 45, 47, 168
aster fibers, 338, 509, 550
astropyle, 268;
 macropyle, 275;
 micropyle, 275
atractophore, 161, 167, 168, 175, 275
autogamy, 133, 136, 137, 141, 143, 144, 217, 353, 354, 361, 373
autotrophy, 239
axoflagellum, 272
axoneme, 246, 247, 249, 251, 253, 256, 259, 261, 296, 299, 310
axoplast, 271, 276, 277
axopodia, 161, 181, 203, 210, 217, 218, 242, 246, 247, 249 et seq., 255, 256, 258 et seq., 271 et seq., 276 et seq., 286, 292, 299, 300, 303, 305, 306, 308, 310 et seq., 314, 321, 322
axostyle, 157, 158, 160, 167, 176, 177, 178, 183, 187, 195, 197, 200, 580
basal body (see also kinetosome), 46, 55, 56,
 157, 261, 328, 333, 426, 481
bases, 166, 344
base sequence, 24, 25, 48, 66, 121, 240
Bauplan, 449, 477, 478, 567, 568
Beagle, voyage of, 70
bell, 460, 478
benthic forms (see also bottom dwelling forms), 211, 217, 523, 531, 545
bilaterogastraea, 487, 488, 588
biochemical pathways, 96
biochemical unity, 37
biocoenoses, 212
biogenetic law (recapitulation), 63, 99, 105, 167, 191, 195, 231, 235, 262, 276, 281, 307 et seq., 322, 324, 344, 353, 376, 417, 482, 513, 520, 523, 547, 550
biosynthetic capacities, loss of, 131
biosynthetic pathways, 37 et seq.
blastocoel, 64
blastomeres, 63, 64, 435, 436, 511, 512, 519, 523, 550
blastula, 435, 444;
 diplo-, 435 et seq., 444, 445, 447
Blätterkreuz, 293, 294, 313
blepharoplast (see also kinetosome), 55
body form, 421 et seq., 440, 442 et seq., 447, 455 et seq., 478, 494, 520, 531;
 asconoid (see ascon);
 leuconoid (see leucon);
 rhagonoid (see rhagon);
 syconoid (see sycon)
body layers (organization), 476 et seq., 494 et seq., 514, 515, 531, 546
body plan (see also Bauplan), 95, 96, 113, 567, 588;
 pentaradiate, 20
bottom dwelling forms (see also benthic forms), 378, 414, 490, 573, 580,
 584, 589
bradytely, 14, 15
brain, 501 et seq., 510, 516, 525 et seq., 531, 548
breeding patterns, 65, 96, 139;
 inbreeding, 135, 140 et seq., 173, 174, 217, 249, 565;
 outbreeding (panmixia), 133, 140 et seq., 173, 217, 565
bristles, 494, 504, 514
buccal cavity (see prostomial cavity), 334, 336, 396
buccal overture (see prostomial opening)
budding, 133, 134, 256 et seq., 345, 352 et seq., 357, 359, 360, 364, 370, 371, 374, 378, 384, 385, 387 et seq., 393, 394, 399, 400, 438, 458, 459, 464, 474 et seq., 479
calcium-aluminum silicate, 292, 304, 310, 322
calcium carbonate (calcareous), 419, 428, 432, 443, 446
calymma, 242, 265 et seq., 270, 271, 273, 277, 278, 282, 288
canalization, 12, 65
capsule, 274, 279, 280 et seq., 284 et seq., 288 292, 300, 302, 303, 305, 307, 311 et seq.;
 perforated, 265, 268, 295, 297, 314, 320, 323, 573
carbohydrate, 39, 192, 211
carbon rich aggregated particles, 273
caryonide, 140, 142, 143
caryosome, 206, 219
casual change (see episodic change), 19
cavities, 536
cell, 6, 22, 40 et seq., 59, 67, 131, 215, 228, 229, 263, 265, 274, 278, 290, 291, 298, 310, 316, 327, 329, 333, 335, 340, 343, 348, 386, 389, 398, 413 et seq., 428, 437, 463,

cell (cont.)
    464, 471, 477, 491, 496
    et seq., 503, 505, 506,
    514, 517, 518, 547, 550,
    565, 579, 585, 586;
  diversity of, 58 et seq.;
  organizational units of
    life, 8
cell body, 327, 330, 341,
    342, 353, 357, 374, 398,
    579
cell cycle, 45, 58, 59, 331,
    359
cell division (see also
    fission), 128, 173, 205,
    250, 254, 439, 584, 585
cell life cycle, 59
cell membrane (see also
    plasmalemma, plasma
    membrane), 52 et seq.,
    162, 206, 210, 213, 269,
    270, 287, 547, 565
cell organelles, 6, 41, 57 et
    seq., 58, 66, 67, 121,
    132, 160 et seq., 179,
    206 et seq., 213, 226,
    230, 239, 271, 326, 331,
    337, 338, 346, 348, 353,
    357, 360, 361, 368, 373,
    376, 392, 414, 427, 565;
  organelle assembly, 301
cell shape, 327, 333
cell size, 579, 593
cell types, 426, 440 et seq.
cellular differentiation, 580,
    583
cellularization, 496, 545,
    547 et seq., 581;
  internal, 581, 582, 591;
  of gametes, 591;
  through integration of a
    colony, 581
cell walls, 490
central capsule, 242, 268 et
    seq., 274, 277, 289
central granule, 253, 255,
    259, 260, 320
centriole, 45 et seq., 55,
    161, 167, 185, 195, 218,
    238, 256, 275, 338, 369,
    417, 427, 440, 509, 537,
    550
centromere, 275
centroplast, 253 et seq.

centrosome, 218, 239, 248,
    256, 259, 261, 338, 369
cephalization, 129, 583, 593
chance resemblance, 84
channel system, 419, 421,
    439, 443, 451
character phylogeny, 111,
    112
chemoheterotrophy, 37, 38,
    131, 149, 151, 273
chloroplast (see plastids),
    499
choanocytes (collar cells),
    422, 423, 426, 428, 434
    et seq., 440, 441, 447,
    451, 452, 464, 490, 557,
    563, 575;
  monotypic, 428;
  polytypic, 428
choanosome, 429, 436
chorological data (see also
    distributional data), 487
chorological progression, 99
chromatin (see chromo-
    somes)
chromidia, 206
chromocytes, 427, 440
chromosomes, 10, 24, 40, 43
    et seq., 46, 48, 58, 66,
    132, 133, 144, 162, 167,
    173, 219, 248, 249, 251,
    255, 256, 258, 260, 272,
    275, 276, 309, 338, 359,
    369, 507, 508, 547
ciliary capsule (see ciliary
    unit)
ciliary patterns (see kinety
    patterns)
ciliary row (see kinety), 500
ciliary units, 328 et seq.,
    331, 346 et seq., 354,
    358, 359, 362, 374
ciliates (see also Organism
    Index), 326 et seq.;
  functions of, 340 et seq.;
  homologous relationships
    of, 357 et seq.;
  origins of, 403 et seq.;
  phylogenetic analysis of,
    363, 374, 382 et seq.;
  phylogenetic conclusions,
    409 et seq.;
  phylogeny of, 357 et seq.;
  structures of, 327 et seq.

cilium, -a, 46, 49, 53, 54, 55,
    61, 128, 132, 246, 329,
    330, 333, 337, 342, 343,
    346 et seq., 348, 354,
    362, 374, 386, 387, 414,
    463, 474, 475, 482, 486,
    489, 494, 496 et seq.,
    507, 509, 510, 521, 541,
    542, 545, 558, 581, 589
cinétide (see kinetide)
circulation, 471, 473
circulatory system (see also
    organ systems), 62, 434,
    511, 565
cirrus, -i, 329, 330, 387, 394,
    407, 580
cistron, 26, 40, 44
cladogenesis, 18 et seq.,
    323, 325, 490, 564, 567
    et seq., 574, 578 et seq.,
    582, 589, 592, 593;
  criteria for, 569;
  potential of, 19
cleavage, 63, 214, 345, 435,
    473, 477, 486, 509, 511,
    512, 515, 519, 520, 525
    et seq., 531, 541, 542,
    545, 547, 550, 556;
  duet formation, 511, 522,
    523, 525, 531, 541, 542,
    544, 550;
  quartet formation, 511,
    522;
  spiral, 511, 512, 521 et
    seq., 531, 540 et seq.,
    550
cline, 99
clone, 140, 141
cnidarians (see also
    Organism Index), 454
    et seq.;
  functions of, 472 et seq.;
  homologous semes of, 476
    et seq.;
  origins of, 489 et seq.;
  phylogenetic analysis of,
    488 et seq.;
  plesiomorph of, 481 et
    seq.;
  structures of, 455 et seq.
cnidoblast, 464, 467
cnidocil, 465
cnidocysts, 489
code (see genetic code)

codon, 24, 26, 28, 30, 31, 36
coeloblastula, 473
coelom, 536, 586
coelomate forms, 536, 537
coenosarc, 457
collagen, 477
collar, 161, 162, 167, 179, 181, 185, 423, 426, 436, 440, 464, 563
collar cells (*see* choanocytes)
collencytes, 428
colloblasts, 489
collozoum stage, 274
colonies, 356, 452, 453, 456 *et seq.*, 461, 480, 481, 489 *et seq.*, 568, 581, 582, 584 *et seq.*, 591; integration of, 591
colony formation, 131, 248, 274
colorless algae (*see* algae), 287
commensalism, 163, 164, 362, 405, 523
commissures (*see* nervous cross connectives), 502, 513, 516, 548
comparative biochemistry, 344
comparisons, 556 *et seq.*
compositional criterion or relationship, 80, 83, 84, 94, 97, 98, 176, 177, 190, 227, 264, 288, 362, 440, 442, 477 *et seq.*, 514 *et seq.*, 522, 543, 562, 573
compounding of structures, 163, 168, 200, 580, 593
conjugation, 133, 135 *et seq.*, 140, 141, 143, 353 *et seq.*, 363, 372 *et seq.*, 380, 384 *et seq.*, 393, 399, 400, 413 *et seq.*, 544, 546, 549, 562, 590
conservative characters, 69, 108, 111, 280, 377, 382, 476, 477, 519, 520, 562, 563, 566
constraints on living systems, 8, 21
constraints of energy utilization, 562

consumers, 128, 131, 416, 580, 592
contractile fibrils, 333, 342
contractile fibers, 543
contractile vacuoles (*see also* expulsive vesicles), 57, 132, 165, 202, 213, 225, 243, 250, 332, 333, 343, 344, 346, 359, 360, 367, 368, 400
convergence, 40, 69, 84 *et seq.*, 88, 95, 118, 223, 240, 264, 284, 288, 322, 336, 357, 362, 443, 538, 545, 546, 556, 584, 590; compositional, 86; positional, 86
copulation (*see also* mating), 135, 518 *et seq.*, 525 *et seq.*, 531, 538, 541, 542, 546, 549, 550, 563
copulatory apparatus, 501, 507, 517, 518, 541
cormidia, 461
cortex, 331 *et seq.*, 337, 340, 341, 359, 365, 370, 413, 424, 425, 546, 547, 562, 576, 580, 581
cortical morphogenesis, 344, 345, 372
cortical unit (*see* ciliary unit)
costa, 157, 158, 177, 185, 195, 197, 580
craspedon (*see* velum)
cresta, 160, 185, 199, 580
crystals, 57, 202, 206, 274, 338, 359, 369
cysts, 169, 185, 207 *et seq.*, 218 *et seq.*, 227, 249, 250, 262, 337, 344, 353, 360
cytochrome, 32, 34, 35, 86, 119, 563, 577
cytogamy, 353, 355, 373
cytokinesis (*see also* cell division, fission), 351, 546
cytopharynx, 160, 334, 335
cytoplasm, 48 *et seq.*, 128, 209, 225, 230, 239, 244, 249, 271, 276, 340, 341, 427, 546, 547, 579, 584

cytoproct, 333, 341, 342, 346, 359, 360, 368, 373, 388, 400
cytoskeleton, 54, 333
cytostome, 160, 161, 179, 191, 334 *et seq.*, 340 *et seq.*, 350, 353, 359, 364 *et seq.*, 376 *et seq.*, 384 *et seq.*, 399, 403, 407, 413, 414, 416, 572, 580 *et seq.*
dactylozooids, 458, 459, 461, 462
defined culture medium, 344
dehydrogenase, 35, 36, 37
deme, 10, 11, 12, 14, 16, 65, 71, 95, 143, 147, 173
dendrogram, 102, 103, 107, 108, 118, 119, 122, 153, 317 *et seq.*, 383, 403, 404, 408 *et seq.*, 524, 532, 535, 555, 559 *et seq.*, 564, 569, 577
deoxyribonucleic acid (*see* DNA)
derived character (*see also* aposeme), 92
dermal pores, 423, 427
development, 63 *et seq.*, 66, 82, 113, 133, 211, 214, 215, 236, 249, 258, 277, 280, 281, 282, 289, 298, 302, 306, 308, 309, 324, 346, 349, 376, 413, 421, 437, 473, 475, 479, 480, 483, 486, 487, 512, 514, 518, 520, 547, 575; conservative aspects of, 64, 66; innovative aspects of, 66; postplanula, 479, 480; postswarmer, 304, 306, 312 *et seq.*; preplanula, 477 *et seq.*
differentiation, 475, 512, 515, 546
digestion, 509, 515, 519, 525 *et seq.*
digestive cavity (*see* gastrovascular cavity)
digestive system (*see also* organ system), 60, 127, 129, 172, 470 *et seq.*,

digestive system (*cont.*) 505 *et seq.*, 512, 513, 517, 519, 520, 531, 548, 566
diploblastic, 490
diploidy, 48, 132, 135, 162, 173, 217, 231, 277, 309, 338, 368, 369, 376, 379, 381, 414, 415, 543, 546, 547, 575, 576, 583
distributional data (*see also* chorological data), 280, 375
DNA, 10, 24 *et seq.*, 31, 42 *et seq.*, 51, 52, 65, 66, 96, 121, 151, 161, 168, 179, 227, 338, 368, 373;
  hybridization of, 26 *et seq.*
dorsal nerve tube, 129
dorsal nerve cord, 70
ear, 79
early evolution of animals, 493
ecology, 211, 218, 227, 236, 238, 307, 311, 326, 340, 342, 358, 373, 407, 512, 545, 562, 589
ecospecies, 146, 147, 149, 241
ectoderm, 64, 437, 512, 542
ectomesenchyme, 428, 435 *et seq.*
ectoplasm, 243, 244, 246, 250, 253, 254, 258, 259, 261, 262, 265, 266 *et seq.*, 269 *et seq.*, 291, 292, 298, 301 *et seq.*, 305, 307, 310 *et seq.*, 318, 340, 360, 368, 543, 547, 548
eggs, 63, 427, 435, 473, 496, 511, 512, 516 *et seq.*, 523, 544, 545, 550, 556;
  ectolecithal, 511, 512, 519, 520, 527 *et seq.*, 531;
  entolecithal, 511, 519, 522, 525, 531
ejectosomes, 489
elastic fibers, 294, 302, 303, 307, 311, 312, 313, 314
embryo, 54, 98, 437, 509, 512, 515, 547

embryogenesis (*see* embryology)
embryology (*see also* development), 12, 420, 437, 520, 540, 544, 547 *et seq.*, 550, 582, 583, 593
encystment, 344, 345, 353
endocrine system (*see also* organ systems), 61, 132, 565
endoderm, 64, 437, 475, 483, 520, 541
endomixis, 353, 355, 374
endoplasm, 243, 244, 246, 250, 253, 254, 258, 261, 265, 268 *et seq.*, 278, 286, 288, 291, 292, 298, 301 *et seq.*, 305, 307, 310 *et seq.*, 319, 340, 359, 360, 369, 414, 543, 544, 547, 548
endoplasmic organelles, 337 *et seq.*
endoplasmic reticulum, 49, 50, 206, 269, 337, 375, 427, 547, 548
endoral kinety, 337, 353, 359, 364
endosomes, 339
energy levels, 8
enterocoel theory, 488, 536 *et seq.*, 586
enzymes, 32, 36, 50, 58, 341, 434, 465, 510, 515;
  functions of, 32
environmental variability, 10
eotype, 98, 414, 416, 449, 452, 453, 485, 491, 546 *et seq.*, 558, 563, 581, 586, 590, 591;
  eumetazoan, 547
ephyra, 474, 479
epidermis, 64, 454, 459, 462, 464, 465, 468, 469, 471, 475, 476, 496 *et seq.*, 501, 502, 509, 511, 512, 514, 518, 525 *et seq.*, 531, 540, 541, 543, 576
epigenesis (*see* development), 64, 66, 259, 300, 302, 309, 583

epigenetic system 12, 16, 20, 21, 22, 578
epiplasm, 328
episeme, 93, 100, 560
episodic change, 19
epistomial disc (field), 335, 336, 405
epithelio-muscular tissue, 463, 465, 477
epithelium, 494, 496, 505, 506, 517, 518, 542, 590
eukaryotes, 24, 39, 40, 41, 43, 44, 51, 54, 58, 132, 162, 206, 309, 344
eumetazoans (*see* Eumetazoa), 583;
  evolutionary development of, 455;
  origin of, 455
evolution, 5 *et seq.*;
  convergence (*see* convergence);
  conservative nature of, 3, 14, 16;
  dead ends, 107, 377, 453, 455, 493, 564, 569, 578, 582, 589, 593;
  development, 233, 393, 492, 517, 550, 564 *et seq.*, 580;
  history, 108, 455, 555 *et seq.*;
  innovative nature of, 3, 14, 16, 17, 20 *et seq.*, 562 *et seq.*;
  major tactic of, 67;
  mosaicism, 22, 382, 537, 562;
  parallel (*see* parallel evolution);
  plasticity, 108;
  potential, 107, 110, 121, 228, 237, 238, 392, 538, 560, 580;
  process, 5 *et seq.*, 418;
  radiation, 318;
  rates of (*see* bradytely, horotely, tachytely), 14 *et seq.*;
  regression (*see* reduction);
  relationships (*see also* dendrogram, phyletic relationships), 348, 363, 421, 524, 589;

SUBJECT INDEX

transspecific, 17 et seq.
evolvability, 5
excretion, 471, 473, 510
excretory pore (see also
  nephridiopore), 505
excretory system, (see also
  organ systems), 62, 132,
  505, 514, 520, 531, 548,
  583
excystment, 344, 353
exoskeleton, 471, 491
exploitive functions, 6, 10,
  18, 19, 20, 69, 105, 106,
  113, 120, 121, 128, 131,
  164 et seq., 209 et seq.,
  273 et seq., 297, 310,
  340 et seq., 377, 413,
  433 et seq., 451, 472,
  491, 509 et seq., 531,
  545, 546, 549, 564, 565,
  567, 569, 579
expulsive vesicles, 57
external morphology, 493 et
  seq.
extracellular secretions and
  accumulations, 53, 337
eyespots, 493, 494, 503,
  504, 514, 516, 521, 531
feeding ciliature, 331, 362
feeding habits, 374
feeding organelle (see also
  cytostome, prosto-
  mium), 327, 333 et seq.,
  347, 357, 374, 545, 557,
  558
feeding polyp (see
  gastrozooid)
fertilization, 63, 134, 135,
  138, 173, 214, 232, 249,
  250, 298, 355, 357, 414,
  435, 437, 441, 473, 477,
  486, 511, 515, 519, 536
  et seq., 541, 544, 545,
  549, 556, 583, 588;
  anisogamy, 136, 138, 394;
  isogamy, 135, 136, 299,
  394, 414;
  oogamy, 135
fiber systems, 55
fins, 84
filopodia, 203, 210, 213,
  218, 221, 223, 224, 228,
  230, 239, 242, 244, 246,
  261, 264, 265, 281, 286

et seq., 303, 305, 322,
  323, 325, 567, 573, 574,
  593
fission (see also asexual
  reproduction), 133, 134,
  136, 150, 166, 173, 195,
  248, 250, 251, 276, 304,
  314, 344 et seq., 353,
  354, 360, 369, 370, 374,
  376, 380, 386 et seq.,
  399, 400, 407, 414, 415,
  417, 475, 476, 546, 550;
  binary fission, 134, 214,
    248, 255, 258, 260, 261,
    274, 275, 297, 298, 302,
    306, 309 et seq.;
  enantiomorphic, 370, 388,
    389, 399, 400;
  homothetogenic, 344,
    345, 348, 357, 360, 370,
    374, 380, 384 et seq.,
    393, 399, 400, 413, 416;
  longitudinal, 134, 166,
    344, 415;
  multiple, 134, 136, 214,
    215, 217, 255, 274, 291;
  symmetrigenic, 134, 160,
    344, 370, 386, 415, 416;
  transverse, 344, 513
fitness, 22, 107
flagellated chamber, 20, 70,
  421 et seq., 425, 426,
  428, 434, 438 et seq.,
  446, 447, 451, 567, 581,
  582
flagellum, 41, 46, 49, 54, 55,
  61, 86, 128, 132, 152,
  155 et seq., 206, 238 et
  seq., 245, 259, 261, 274,
  275, 292, 306 et seq.,
  311, 321, 323, 362, 422,
  426, 427, 434, 437, 438,
  445, 463, 464, 474, 491,
  508, 521, 538, 539, 572,
  581, 586, 589, 593;
  bacterial, 54
flotation apparatus, 273,
  290, 293 et seq., 301 et
  seq., 305, 307, 314, 324,
  579
food capture (see predation)
food reserves, 465, 472, 510
food vacuoles, 50, 127, 228,
  251, 252, 254, 266, 269,

273, 278, 291, 310, 333,
  334, 337, 340, 341, 353,
  371, 373, 414, 490, 510,
  515, 517, 525, 543
form and function, 7, 22
fossil record, 14, 16, 20, 310
fossils, 63, 66, 69, 70, 72,
  73, 99, 113, 227, 228,
  233, 236, 258, 281, 285,
  310, 320, 324, 375, 439,
  443, 444, 446, 481 et
  seq., 568, 572 et seq.
founder principle, 14
functional correlates of evolutionary development,
  564 et seq.
Galapagos Archipelago, 70,
  72
gametes, 63, 135, 136, 139,
  206, 214, 217, 249, 291,
  297, 302, 306, 322, 435,
  437, 441, 446, 477, 480,
  482, 507, 515, 518, 519,
  531, 538, 542, 544, 545,
  547 et seq., 579
gamete nuclei, 355, 372,
  386, 414, 416, 544
gametocytes, 427
gametogenesis, 435, 441
gamogony, 228
ganglion cells (see nerve
  cells)
gastrodermis, 64, 454, 460,
  462, 464, 465, 468, 470
  et seq., 475, 476, 490,
  494, 497, 506, 510, 514,
  515, 517, 520, 525 et
  seq., 531
gastrovascular (digestive)
  cavity, 60, 454, 460,
  462, 465, 470, 472, 477,
  488, 490, 494, 497, 505,
  506, 510, 512, 515, 522,
  531
gastrozooid, 459, 461, 462
gastrula, 437, 447, 486, 512,
  515;
  stereogastrula, 449, 473,
    477, 482, 512, 531
G-C content, 28, 29
gelatinous layer (material),
  265, 291, 294, 296, 299,
  302, 303, 307 et seq.,
  309, 311 et seq., 320, 322

gemmule, 438, 439, 441
gene pool, 9 et seq., 14, 15, 17, 21, 67, 69, 145, 148, 356, 359
genes, 6, 7, 10, 15, 17, 214;
  action of, 301, 475, 546, 547;
  products of, 546
genic flow, 10 et seq., 546
genetic code, 8, 26, 30, 58, 119;
  degeneracy of, 26, 30
genetic control, 39 et seq.
genetic drift, 10
genetics, 211, 237, 413
genitalia, 83, 507, 508, 512, 513, 515, 517 et seq., 521, 523, 535, 548, 549, 559
genital openings (orifices), 494, 496, 514
genomes, 272, 276
geological character precedence, 99
geological record (see also fossil record), 562
germ cells, 63
germ layer, 64
gland cells, 427, 464, 465, 497, 499, 505, 514, 516, 525 et seq.
globins, 32
glutathione, 472
glycocalyx, 206, 207, 210
glycogen, 57
Golgi apparatus (see also parabasal), 50, 176, 206, 252, 269, 307, 338, 427, 498
gonads (see also reproductive structures), 460, 471, 479, 507, 512, 521, 523
gonophore, 461, 462
gonopore (see also genital opening), 495, 507, 508, 515, 517, 521, 523, 525 et seq., 531, 538
gonozooid, 459, 462
granulo-fibrillar network, 328, 333
growth factors, 211
gullet (see cytostome and prostomium)

haploidy, 48, 63, 132, 134, 135, 162, 173, 217, 231, 249, 276, 277, 309, 354, 414
haplokinety, 364, 366, 386
haptocysts, 331, 341, 359
Hardy-Weinberg law, 65, 95
heliozoans, (see also Organism Index), 242 et seq.;
  functions of, 246 et seq.;
  homogeneous groupings of, 242 et seq.;
  homologies in, 250 et seq.;
  organization of, 246 et seq.
hemixis, 353, 355
hemoglobin, 32, 34, 35, 39, 86, 510;
  Gun Hill, 33;
  sickle cell, 35
hermaphroditism, 471, 486, 507, 544, 549
heterogeneous groups, 105, 106
hexamerous body organization, 456, 457, 470, 477, 483
histology, 426 et seq., 463 et seq.
homeostatic functions, 6, 10, 18, 19, 69, 105, 106, 113, 121, 131, 133, 165 et seq., 209, 213 et seq., 231, 273, 297, 340, 343, 434, 472 et seq., 509 et seq., 549, 565 et seq., 582
homogeneous groupings, 104 et seq., 106, 316, 555
homogenetic, 75, 110, 112, 113
homokaryotic forms, 377, 380, 382
homology, 74 et seq., 88, 91, 106, 110 et seq., 118, 120 et seq., 153, 175, 195, 199, 218, 223, 226, 228, 233, 236, 238, 241, 250, 251, 253 et seq., 260, 261, 263, 264, 277 et seq., 292, 302, 305, 307, 314, 315, 322, 325, 336, 349, 354, 357 et seq., 382, 399, 401 et seq., 405, 413, 414, 419, 439, 442, 443, 445, 446, 455, 478, 484, 490, 513 et seq., 519, 524, 531 et seq., 536, 538, 540, 542, 543, 545, 546, 549, 555 et seq., 559, 562, 570, 572, 578, 584 et seq., 580 et seq.;
  circular definition of, 79, 113;
  definition of, 82;
  serial, 72;
  special, 74
homonomy, 518
homoplasy (see also analogy, chance resemblance, convergence, mimicry, parallelism), 84
horotely, 14 et seq., 286, 323, 574
hydranth, 456, 457, 458, 479
hydrocaulus, 457, 458
hyperseme, 93, 100, 560
hyposeme, 93, 97, 100, 288, 306, 560
hypostome, 456, 469, 486
hypothetical forms, 69, 542, 558, 575, 584, 586
hypothetical intermediates, 490
immunological studies, 32, 36
informational content of organisms, 23 et seq.;
  of prokaryotes, 40
infraciliary lattice, 328, 333, 340
infraciliature, 374, 375, 544
ingestion, 341
ingestors, 518 et seq., 531, 544, 546
inner cell body (see Weichkörper), 291, 318, 324
innovative characters, 69, 74, 93, 104, 106, 107 et seq., 112, 113, 121, 122, 162, 163, 193, 280, 382, 413, 416, 451, 491, 519,

SUBJECT INDEX

542, 549, 564, 567, 569, 570, 578 *et seq.*, 583, 591, 592
intestine (*see also* gastrovascular cavity), 494, 505, 506, 512, 515, 517, 520, 521, 523, 525 *et seq.*, 536, 583
interstitial cells, 464, 465
inversion (sponge embryo), 438
isolating mechanism, 16
jellyfish (*see also* medusa), 454, 455, 465, 476
karyoplasmic ratio, 579
keratin, 428, 432, 433, 443, 448
ketoadipyl CoA, 40
kinetide (*see also* mastigont), 157 *et seq.*, 162, 165 *et seq.*, 177, 178, 194, 195, 197 *et seq.*, 328, 347, 349, 362, 417, 496, 500, 555, 572, 575, 579 *et seq.*, 583, 593
kinetodesma, 328, 329, 358
kinetochore, 144
kinetoplast, 51, 161, 162, 167, 170, 171, 179, 185, 571
kinetosomal territory (*see* ciliary unit)
kinetosome, 46, 54 *et seq.*, 155, 157, 159 *et seq.*, 167, 168, 170, 178, 181, 194, 195, 200, 328, 329, 333, 346 *et seq.*, 358, 360 *et seq.*, 364, 376, 395, 413 *et seq.*, 427, 497, 580, 581, 583
kinety, 328 *et seq.*, 330, 331, 333, 337, 346 *et seq.*, 354, 360, 374, 376, 384 *et seq.*, 395, 414, 496
kinety patterns, 328 *et seq.*, 336, 357, 359, 362, 364, 368, 398, 399, 555
larvae, 88, 273, 278, 354, 364, 368, 425, 435 *et seq.*, 440, 442, 444 *et seq.*, 447, 448, 468, 474, 477, 478, 484, 487, 512, 541, 542

leucoflagellates (*see also* apochlorotic algae, colorless algae), 128
leucon, 424, 425, 433, 434, 437, 444, 445, 447, 448, 451, 453, 582;
  aphodal condition, 424;
  diplodal condition, 424;
  eurypylous condition, 424, 425
leucoplast, 263
life cycles, 133, 162, 169, 173, 179 *et seq.*, 191, 202, 214 *et seq.*, 235, 238, 239, 245, 261, 345, 348, 355 *et seq.*, 387, 415, 474, 475, 483, 555, 579, 581, 585, 593
lineage allomorphosis, 22
lipid, 35, 39, 52, 57, 192, 211, 228, 273, 278, 286, 291, 308, 310, 338, 369
lobes, 493, 494
lobopodia, 203, 218, 221, 223, 224, 238, 239, 246, 256, 308, 593
locomotion, 128 *et seq.*, 495, 509, 516, 542
loculum, 208, 229, 230, 231, 579
lophophores, 20
loricas, 333, 337, 359, 374, 388, 389, 398
lysosome, 50, 338, 341
macrocytes, 428
macronucleus, 134, 137, 217, 228, 287, 299, 327, 338 *et seq.*, 340, 354, 355, 357, 360, 361, 367 *et seq.*, 373, 374, 376, 377, 379 *et seq.*, 384 *et seq.*, 394, 396, 399, 400, 407, 417, 546, 581;
  heteromeric, 339;
  homomeric, 339
manubrium, 460, 462, 469, 481
mastigonemes, 157, 159, 175
mastigont (*see also* kinetide), 157, 195, 200, 362, 417, 555, 572, 580;
  karyomastigont, 183, 200
mating (*see also* copulation), 514, 515, 519, 520, 525 *et seq.*, 531, 545
meaningful comparisons, 92
medusa, 454, 457, 459, 468, 469, 472, 474 *et seq.*, 485 *et seq.*, 490, 567, 582, 583;
  hydromedusae, 456, 459, 460, 462, 468, 469;
  scyphomedusae, 456, 460, 462, 463, 469, 471, 481
meiosis, 41, 45 *et seq.*, 63, 134, 173, 249, 354, 355, 359, 361, 435;
  one-step, 145
mesenchyme, 423, 424, 429, 494, 514, 536
mesoderm, 64
mesoglea, 276, 422, 427, 428, 433, 452 *et seq.*, 459, 462, 464, 465, 467 *et seq.*, 469, 472, 476, 480, 483, 491
messenger RNA (*see* mRNA)
metabolic pathways, 214, 219, 344
metabolic products, 338
microfibrils, 329, 333, 340, 565, 593
microfilaments, 53 *et seq.*, 176, 498, 502
micronucleus, 134, 137, 217, 228, 299, 327, 338, 339, 354 *et seq.*, 359 *et seq.*, 368, 369, 373, 374, 376, 379, 380, 384 *et seq.*, 399, 400, 581
microtubules, 41, 47, 53 *et seq.*, 157, 158, 160, 161, 175, 176, 181, 203, 246, 247, 254, 259, 261, 271, 272, 275, 276, 279, 328, 329, 340, 341, 358, 367, 426, 496, 497, 565, 593
mimicry, 84, 324;
  Batesian, 324;
  Müllerian, 324
mitochondria, 24, 50 *et seq.*, 58, 132, 134, 161, 162, 179, 206, 252, 307, 328, 337, 338, 375, 427, 498, 502, 537

mitosis, 41, 45 *et seq.*, 168, 185, 189, 205 *et seq.*, 218, 219, 238, 248, 249, 251, 253, 274, 338, 355, 359, 369, 373, 374, 509;
  mesomitosis, 206, 219, 226, 238, 380;
  metamitosis, 206, 219, 226;
  pleuromitosis, 168, 275;
  promitosis, 205, 219, 226
mitotic apparatus, 46, 53
molecular evolution, 109, 119 *et seq.*;
  molecular semes, 120;
  molecular phylogemy, 121
molecular information, 23 *et seq.*, 556, 562, 563
molecular phylogeny, 577, 584
monophyletic group, 101, 106, 154, 238, 263, 281, 446, 489, 545, 568, 572, 591
morphocline, 99, 100, 106, 195, 212, 289, 324, 375, 377 *et seq.*, 396, 447 *et seq.*, 481, 483 *et seq.*, 487, 520 *et seq.*, 574 *et seq.*
morphological phylogenetics, 109 *et seq.*, 584
morula, 436, 444, 486
mouth, 456, 460, 465, 469, 470, 472, 475, 477, 478, 481, 485, 490, 494, 495, 497, 505, 509, 512, 514, 515, 519 *et seq.*, 531, 536, 542, 543, 547, 581, 583, 586, 593
mucocysts, 331, 359
Müller's law, 293, 300, 304, 310, 320
Müller's vesicles, 332, 333
multicellular systems, 59 *et seq.*, 454, 455
multicellularity, 418, 453, 493, 548, 565, 566, 576, 579 *et seq.*, 585, 586, 589, 591, 593;
  origin of, 453, 584
multinuclearity, 205, 309, 575, 576, 580, 583, 584, 593
muscle cells (musculature), 498, 501, 502, 505, 506, 509, 510, 514, 516, 518, 520, 521, 525 *et seq.*, 531, 540, 542, 546, 548
muscle fibers, 501, 525 *et seq.*, 531
muscular system (*see also* organ systems), 61, 129, 468, 477, 478, 480
mutation, 10, 12, 22, 36, 38, 70, 87, 107, 134, 172, 564 *et seq.*;
  macromutation, 17
mutational distance, 119
mutation pressure, 87
mutualism, 163, 164, 199
myocytes, 427, 440
myoglobin, 32, 33, 34
myonemes, 291, 294 *et seq.*, 299 *et seq.*, 305, 307, 311 *et seq.*, 316, 318, 332, 333, 374, 468

nassa, 341, 353, 367, 370, 376 *et seq.*, 384, 391, 399, 412, 558
natural classification, 115
natural selection (*see* selection)
nemadesma, 335, 365
nematocysts, 57, 460, 462, 464, 465 *et seq.*, 469, 472, 477, 478, 480, 484, 485, 488, 490, 491, 542, 555
nematocytes, 465, 466
Neo-Haeckelian view, 585, 587, 589
neontology, 13, 98, 113, 481
neoseme, 93, 94, 97 100, 101, 103, 106, 107, 113, 122, 179, 311, 315, 323, 374, 398, 400, 401 *et seq.*, 409, 440, 480, 520, 532, 533, 540, 558 *et seq.*, 562, 567, 568
neoteny, 512
nephridiopore (*see also* excretory pore), 505, 517, 526 *et seq.*, 531
nerve cells, 428, 464
nerve net, 61, 472, 478, 480, 485, 513, 544
nervous-coordinatory system (*see also* nervous system), 434
nervous cords, 501, 502, 510, 513, 516, 525 *et seq.*, 531, 541, 548
nervous cross connectives (*see* commissures), 502
nervous system (*see also* organ systems), 61, 129, 501 *et seq.*, 510, 513, 514, 516, 518, 521, 531, 540, 542, 544, 548, 565
New Systematics, 108, 109, 111 *et seq.*, 114, 153, 357, 442, 584, 592
niche, 9, 16, 18, 21, 39, 58, 65, 67, 88, 89, 108, 122, 129, 131, 148, 219, 220, 241, 287, 373, 374, 381, 416, 478, 485, 523, 532, 568, 569, 574
nuclear membrane, 41, 168, 206, 219, 246, 248, 251, 256, 258, 259, 274, 275, 338, 339, 509, 550
nucleation site, 54
nucleic acids, 23, 24 *et seq.*, 39, 146, 240
nucleoli, 42 *et seq.*, 162, 206, 219, 338, 368, 380, 435
nucleoplasm, 42, 338, 369
nucleus (*see also* macronucleus, micronucleus), 24, 41 *et seq.*, 162, 168, 170, 171, 179, 181, 192, 200, 202, 205 *et seq.*, 215, 218, 220, 225, 226, 228, 233, 238, 239, 243, 244, 246, 248, 250, 252 *et seq.*, 261, 265, 270 *et seq.*, 277, 281, 285 *et seq.*, 291, 298 *et seq.*, 307, 309, 320, 327, 338 *et seq.*, 353, 368, 374, 379, 380, 382, 413 *et seq.*, 427, 435, 440, 464, 498, 501, 508, 537, 543, 544, 546, 547, 562, 567, 576, 579, 580, 590, 593
numerical taxonomy, 109, 113 *et seq.*, 121, 491,

584, 592;
evaluation of affinity, 114
nutrition (*see also* assimilation, autotrophy, chemoheterotrophy, osmotrophy, phagocytosis, phagotrophy, phototrophy, pinocytosis, predation, saprotrophy), 214, 344, 433;
bacterial, 344;
metazoan, 344;
protozoan, 344
nutritive cells, 464, 465, 477
octomerous body organization, 457, 470, 477
ontogenetic character precedence (*see* biogenetic law), 99
ontogeny, 233, 307, 309, 310, 318, 322, 324, 375, 376, 446, 447, 451, 482 *et seq.*, 485, 489, 518 *et seq.*, 523, 548, 574, 575
operational rigor, 117
operon, 40, 44
oral disk, 456
oral lobes, 456, 462, 463
oral structures (*see* feeding organelle)
organic design, 9; limitations of, 86
organogenesis, 519, 520, 525 *et seq.*, 531, 545, 556
organ systems (*see also* circulatory, digestive, endocrine, excretory, integumentary, muscular, nervous (coordinatory-sensory), reproductive, respiratory, skeletal systems), 59 *et seq.*, 95, 96, 121, 512, 550, 565, 583
origin of life, 122
origins, 357, 403 *et seq.*, 407, 418, 493, 534, 536 *et seq.*, 549, 561 *et seq.*, 568, 570 *et seq.*, 578, 583, 586, 588, 592;
of multicellular animals, 583 *et seq.*;

of protozoa, 583, 592
orthogenesis, 567
orthoselection, 567, 569
osculum, 70, 422 *et seq.*, 427, 431
osmoregulation, 343
osmotrophy (*see* assimilation saprotrophy), 131, 151, 163, 592
OTU, 114, 116, 117, 118, 146
ovary, 517, 522
ovum (*see* eggs), 473
ovocyte, 435, 441
paleontology, 13, 98, 113 (*see also* fossil record), 227, 235, 259, 280, 282, 285, 289, 481, 482, 487, 562
papaine, 36
parabasal (*see also* Golgi apparatus), 157 *et seq.*, 176 *et seq.*, 181, 183, 187, 195, 197, 200, 427, 440
paradesmose, 427, 440
parallel evolution, 69, 84, 87, 88, 148, 233, 281 *et seq.*, 289, 316, 322, 323, 357, 371, 372, 406, 443, 445, 446, 451, 484, 491, 518, 522, 568, 569, 579, 582, 584
parasites (parasitism), 133, 163, 164, 192, 202, 211, 217, 230, 238, 262, 327, 377, 522
parasomal sac, 328, 329, 342, 346, 348, 358
parenchyma, 494, 496, 497, 501, 502, 504 *et seq.*, 509, 511, 514, 516, 518, 520 *et seq.*, 523 *et seq.*, 531, 543, 544
parenchymella, 435 *et seq.*, 444, 445, 448
paroral kinety, 330, 337, 349, 350, 353, 359, 364, 366, 372, 377, 412
peak-valley model, 11
pedogamy, 133, 135, 249, 250
pedogenesis, 65

pelta, 160, 179, 185, 199
peniculus (*see* prostomial membranelles)
penis, 496, 507, 511, 518, 525 *et seq.*
perisarc, 457, 458, 480
peristome, 334, 359, 399;
peristomial ciliature, 371;
peristomial depression, 335, 336, 365, 389
phagocytosis, 127, 210, 340, 510, 515, 519
phagotrophy, 128, 131, 152, 163 *et seq.*, 217, 240, 263, 273, 373, 592
pharyngeal basket (*see* nassa)
pharynx, 501, 505, 506, 509, 510, 512, 515, 517 *et seq.*, 525 *et seq.*, 531, 534
phase space, 226
phenetic classification, 204, 226
phenotypic variation, 214
phototrophy, 38, 149, 152, 240, 592
phyletic distance, 94, 95, 101 *et seq.*, 107, 119, 122, 199, 311 *et seq.*, 315, 317, 320, 357, 383, 400, 401 *et seq.*, 404, 405, 409, 449, 488, 532 *et seq.*, 555, 559 *et seq.*, 563, 583;
guidelines for interpreting, 102 *et seq.*
phyletic relationships (*see also* evolutionary relationships), 55, 198 *et seq.*, 317, 374, 383 *et seq.*, 404, 420, 439, 488, 524 *et seq.*, 534, 570, 576, 591
phylogenetic analysis, 68 *et seq.*, 211, 217, 476, 488 *et seq.*, 513, 522, 524 *et seq.*, 537, 549, 556, 562 *et seq.*;
procedural guidelines, 68 *et seq.*;
Rule I, 68, 106, 113, 584;
Rule II, 69, 95, 105,

phylogenetic (*cont.*)
113, 556, 584;
Rule III, 70, 113, 584;
procedural summary, 106 *et seq.*;
procedures, 94 *et seq.*
phylogenetic conclusions, general, 570 *et seq.*
phylogenetic methodology, general, 555 *et seq.*
phylogenetic overview, 591 *et seq.*
phylogenetic tree (*see also* dendrograms), 103, 234, 378, 383, 568
phylogeny, 3, 5, 23, 37, 52, 67, 162, 202, 213, 218, 219, 231, 233, 245, 282, 289, 302 *et seq.*, 319, 345, 347, 350, 357, 375, 382, 389, 392, 394, 405, 408 *et seq.*, 418, 421, 449, 450, 474, 476 *et seq.*, 489, 524, 532, 534, 536, 556, 561, 563, 572, 584, 586, 592, 593;
animal, 52;
plant, 52
pinacocytes, 427, 428, 440
pinocytosis, 165, 209, 211, 341, 342, 515, 519
pinocytotic vesicles, 50, 127
planula, 473 *et seq.*, 475, 477, 479 *et seq.*, 486, 541, 542
planuloid form, 536, 537, 540, 541
plasmalemma (*see* cell membrane), 206, 210, 213, 219
plasma membrane (*see also* cell membrane), 127, 132, 261, 328, 329, 343, 368, 375
plastids, 24, 30, 51 *et seq.*, 58, 149, 150, 151, 323, 574;
leucoplast, 150
plastogamy, 217, 248, 250
plesiomorph, 98 *et seq.*, 104, 106, 107, 111, 122, 152, 153, 181 *et seq.*, 194, 195, 198, 199, 259,
260, 263, 277, 281 *et seq.*, 285 *et seq.*, 307, 309 *et seq.*, 316, 318, 322 *et seq.*, 362, 363, 375 *et seq.*, 383, 384, 389 *et seq.*, 403, 407, 408, 410, 412, 413, 415, 439, 446 *et seq.*, 451, 452, 476, 481 *et seq.*, 513, 516, 520 *et seq.*, 524, 532, 534, 535, 540, 543, 545, 547, 549, 555, 558 *et seq.*, 561, 563, 569 *et seq.*, 579, 582, 583, 591
plesiosome, 92, 94, 97, 98, 100, 101, 104, 106 *et seq.*, 113, 118, 121, 122, 153, 276, 280, 288, 311, 315, 398, 400, 401 *et seq.*, 441, 445, 477, 478, 480, 513, 515 *et seq.*, 532, 533, 559
plesiotype, 98
pneumatophore, 462
podocone, 268, 269, 272, 285
point-to-point comparisons, 89 *et seq.*, 121, 177, 178, 225, 226, 358, 362, 373, 514, 557, 558, 590
polar capsules, 489
polarization, 421, 436, 579, 583, 589
polyenergid, 257, 272, 274, 282, 285 *et seq.*, 338, 579 *et seq.*
polykinety, 336, 359, 364, 366, 386
polymerization (*see* compounding of structures), 580, 593
polymorphism, 421, 459
polyp, 454 *et seq.*, 467, 469, *et seq.*, 474 *et seq.*, 487, 491, 567, 582, 583;
anthozoan, 456
polypeptide, 24, 25, 31 *et seq.*, 119, 121
polypeptide sequences, 491
polyphyletic, 200, 238, 239, 392, 571, 577, 578, 585
polyploidy, 132, 257, 265, 272, 287, 309, 338, 339,
368, 369, 376, 377, 380, 508, 546, 547, 576, 579
polyseme, 93, 94, 100, 560
populations, 65 *et seq.*, 556, 566;
genetic variability of, 10, 547, 548
pore cell, 427
pores, 427, 441
porocytes, 423, 426, 440
positional criterion or relationship, 75, 76, 80, 86, 94, 97, 98, 177, 224 *et seq.*, 288, 362, 440 *et seq.*, 477, 478, 514 *et seq.*, 562
postzygotic processes, 358, 361, 373, 374
preadaptation, 16, 22, 288, 542, 581, 583
predation, 129, 131, 212, 247, 273, 278, 281, 286, 288, 289, 322, 324, 341, 342, 361, 472, 514, 516, 519, 531, 542, 545, 566, 567, 574, 580, 583, 591
prestomial parts, 335, 365, 371, 388, 396
prezygotic processes, 358, 364, 373, 374
primitive forms, 69, 87, 98, 99, 153, 253, 377, 379 *et seq.*, 394, 414, 449, 455, 486 *et seq.*, 508, 520, 521, 524, 536 *et seq.*, 545, 572, 586, 587, 591
primordial animal, 155
probability of correspondence, 91
proboscis, 335, 336
producers, 128, 131
prokaryotes, 24, 40, 41, 44, 51, 578
prostomial cavity, 334, 335, 364, 376, 387, 391, 407
prostomial membranelles, 335
prostomial opening, 334, 335, 337, 347, 368
prostomium, 336, 337, 341, 348, 349, 352, 353, 359, 362, 364, 365 *et seq.*, 371, 376, 377, 385 *et*

SUBJECT INDEX 667

seq., 395, 558;
membranelles of, 336, 366
protein, 6, 28, 30, 31 et seq., 35, 39, 54, 57, 58, 66, 67, 96, 128, 148, 226, 240, 368, 510, 563;
physical properties, 32
protocnidarian, 483
protonephridium, 62, 505, 510, 511, 516, 517, 521 et seq., 525 et seq., 531, 583
protozoans (*see also* protozoa), 127 et seq.;
as ingestors, 127 et seq., 130;
definition of, 127;
origins of, 149 et seq.;
reproduction of, 133 et seq.;
sexuality among, 133 et seq.;
species structure of, 137 et seq.;
species problem in, 145 et seq.
pseudoheliozoans, 244
pseudogamy, 509
pseudometazoan, 326, 418
pseudopodia (*see also* axopodia, filopodia, lobopodia, reticulopodia), 61, 128, 132, 152, 161, 162, 183, 203 et seq., 205, 209, 210, 213, 218, 220, 221, 223, 224, 226 et seq., 230, 239, 240, 243 et seq., 263, 264, 273, 290 et seq., 300, 307 et seq., 314, 320, 322, 324, 558, 578 et seq., 581
quantum change, 19
R (*see* phyletic distance)
radial canals, 456, 460, 462, 469
radiate animals, 454
radii, 456, 460;
location of, 457
radiolarians (*see also* Organism Index), 265 et seq.;
functions of, 273 et seq.;

origin of, 285 et seq.;
plesiomorph of, 280 et seq.;
phylogeny of, 277 et seq.;
structure of, 265 et seq.
recapitulation (*see* biogenetic law)
receptor cells, 503, 516
reduction (regressive evolution), 69, 84, 86 et seq., 100, 197, 200, 235, 287, 288, 324, 377, 381, 509, 516, 521, 522 et seq., 536, 539, 586, 589
regeneration, 438, 439, 464, 475, 513, 514, 555, 556
regressive evolution (*see* reduction)
relict group, 201, 416
Remanian criteria (*see also* accessory criterion, compositional criterion, positional criterion, serial criterion), 75 et seq., 107, 177, 190, 195, 218, 223, 226, 258, 357, 364, 365, 375, 536, 543, 544, 556, 562, 568, 590;
continuity criterion, 75;
reformulation of, 80 et seq.;
special quality criterion, 75, 76
reproduction (*see also* asexual and sexual reproduction), 531, 556
reproductive functions, 6, 10, 18, 19, 69, 105, 106, 121, 166 et seq., 209, 214 et seq., 231, 270, 274 et seq., 297 et seq., 340, 344 et seq., 382, 435 et seq., 439, 441, 473 et seq., 489, 511 et seq., 531, 549, 566 et seq., 582
reproductively isolated populations, 71, 80
reproductive structures (*see also* gonads), 456
reproductive system (*see also* organ systems), 62, 471, 507 et seq., 514, 520, 531, 544

respiration, 471, 473
respiratory system (*see also* organ systems), 62, 511, 565
reticulopodia, 203, 292, 297, 299, 300, 301, 303, 305, 306, 308, 310 et seq.
retractor muscles, 477, 478
rhabdites, 497, 514, 526, 531
rhabdoids, 497, 500, 516
rhagon, 425, 437, 445, 446, 447
ribonucleic acid (*see* RNA)
ribosomal RNA (*see* rRNA)
ribosomes, 30, 49, 56 et seq., 58, 206, 337
ring canal, 460
RNA, 24, 27, 28, 30 et seq., 42, 43, 57, 96, 368, 512;
mRNA, 25, 30, 40, 45;
rRNA, 30, 43;
tRNA, 25, 30, 31
sagittocysts, 489, 500, 514
saprotrophy (*see also* assimilation, osmotrophy), 131, 151, 152
scleroblast, 276, 427, 438
scopula, 333, 387, 392, 393
scyphistoma, 454, 476, 479
secretion, 509
secretory elements, 497 et seq., 514, 519, 531, 541 et seq., 548
segmentation, 536
selection, 6, 7, 10, 12, 18, 28, 39, 48, 56, 65, 68, 70 et seq., 87, 95, 107, 108, 110 et seq., 116, 119 et seq., 129, 131, 136, 144, 147, 163, 164, 172, 174, 193, 197, 200, 212, 215, 223, 237 et seq., 241, 281, 286, 323, 356, 357, 377, 381, 382, 413, 417, 418, 451, 491, 518, 523, 536, 540 et seq., 545, 546, 548, 549, 555, 562 et seq., 574, 576, 581, 583, 584, 587, 588, 591
self-fertilization (*see also* autogamy), 355
semantide, 89

semaphoront, 89
semes (*see also* aposemes, neosemes, plesiosemes), 88 *et seq.*, 92, 95, 100, 104, 106, 118, 119, 121, 156, 175, 179, 190, 199, 207, 218 *et seq.*, 250, 253, 277, 279, 302 *et seq.*, 311, 315, 357 *et seq.*, 373, 374, 383, 384, 398, 400 *et seq.*, 405, 439 *et seq.*, 442, 513 *et seq.*, 519, 525 *et seq.*, 531, 532, 544, 555 *et seq.*, 562, 570, 583;
  comparisons of, 97;
  major animal, 96;
  molecular, 96;
  types of, 92 *et seq.*
seminal bursa, 507, 514, 517, 518
seminal receptacle, 517, 518
seminal vesicle, 507
sensory and nervous system (*see also* nervous system), 468 *et seq.*, 473, 478 *et seq.*, 485, 488
sensory organs, 493, 494, 519, 531, 544, 548
sensory pits, 494, 505, 518, 526, 531
sensory structures (sense organs or sensory cells), 459, 460, 462, 464, 465, 468, 477, 480, 481
septum, 457, 470, 477, 478, 484, 485
serial criterion or relationship, 80 *et seq.*, 87, 88, 94, 97, 98, 112, 180, 190, 305, 365, 398, 417, 442, 478, 557, 558, 562, 568
sessile forms, 87, 523
sexual behavior (*see* conjugation, copulation, mating)
sexuality (*see also* sexual processes, sexual reproduction), 575, 579, 582, 593
sexual processes, 12, 169, 173, 174, 202, 214, 215, 231, 236, 240, 344, 345, 353 *et seq.*, 357, 359, 361, 373 *et seq.*, 413, 590
sexual reproduction, 215, 248 *et seq.*, 258, 260, 261, 265, 274, 297, 298, 306, 308, 309, 310, 421, 435, 453, 458, 473, 477, 480, 511 *et seq.*, 519, 547, 549, 550, 566, 583, 593
shell, 202, 214, 261, 265, 277, 279, 281, 286, 287, 289, 303, 313
sibling species, 83
silver impregnation technique, 346, 347, 501
silica, 265, 266, 273, 276, 286, 288, 292, 419, 428, 429, 431 *et seq.*, 446, 448
similarities, 114;
  determination of, 81, 82, 90;
  of subfunctions, 91
siphonoglyph, 470, 477
skeletal plates (platelets), 333, 374
skeletal system (*see also* organ systems), 63, 129
skeleton, 242, 243, 245, 246, 254, 258, 565 *et seq.*, 267, 270, 271, 276, 278 *et seq.*, 284 *et seq.*, 291, 292, 298, 299, 301 *et seq.*, 307, 308, 311, 314, 316, 322, 324, 419, 427, 432, 433, 439, 440, 442 *et seq.*, 447, 448, 451, 453, 459, 471 *et seq.*, 480, 491, 555, 582;
  calcareous, 459, 471, 480;
  hydraulic, 471;
  proteinaceous, 459, 471, 480
skull, 77, 79
somatic ciliature (*see* surface ciliature)
speciation, 10, 112, 113, 121
species, 5, 8, 9, 12 *et seq.*, 17, 23, 66, 68, 69, 71, 73, 80, 82 *et seq.*, 93 *et seq.*, 117, 122, 139, 145 *et seq.*, 173, 212, 220, 221, 228, 233, 244, 245, 261, 266, 274, 286, 298 *et seq.*, 306, 310, 311, 314, 316 *et seq.*, 320, 326, 327, 331, 333, 348, 354, 356, 359, 362, 363, 368, 379, 382 *et seq.*, 391, 394, 399, 400, 409, 421, 444, 460, 494, 508, 514, 518, 521, 525, 532, 540, 546, 549, 555 *et seq.*, 559, 561, 564 *et seq.*, 568, 570, 574, 576, 580, 584, 586, 587, 592;
  biological concept of, 139, 141, 146 *et seq.*;
  diversification of, 13, 18, 73, 112, 233, 567, 585;
  diversity of, 8;
  in phylogenetic studies, 83;
  total number of, 8;
  transformation, 13, 18, 73, 74, 233, 567, 585
sperm, 63, 86, 306, 307, 427, 435, 437, 473, 507, 508, 511, 517 *et seq.*, 523, 525 *et seq.*, 531, 537 *et seq.*, 541, 542
spermatozoa (*see* sperm)
spermiogenesis, 435, 508, 537
spicules, 265, 266, 276, 277, 279, *281*, 285 *et seq.*, 292, 294, 300, 301, 303 *et seq.*, 311 *et seq.*, 316, 318, 320, 419 *et seq.*, 427, 428, 431 *et seq.*, 438 *et seq.*, 442 *et seq.*, 447 *et seq.*, 451, 459, 555;
  desmas, 419, 431;
  megascleres, 428 *et seq.*, 443, 444;
  microscleres, 428, 429, 431, 444, 450;
  monaxonic, 429, 443, 450;
  polyaxonic, 431;
  spheres, 431;
  triaxonic, 429
spindle, 45, 47, 161, 168,

SUBJECT INDEX 669

185, 189, 195, 238, 248, 249, 256, 259, 260, 275, 338, 369, 417, 550
spirocysts, 465, 478, 488
sponges (*see also* Organism Index), 419 *et seq.*;
 classification of, 420;
 functions of, 433 *et seq.*;
 homologous relationships of, 440 *et seq.*;
 origins of, 451 *et seq.*;
 phylogenetic analysis of, 449 *et seq.*;
 phylogeny of, 439 *et seq.*;
 semes of, 440 *et seq.*;
 structures of, 421 *et seq.*
spongin, 419, 432, 433, 440, 448
spongioblast, 433
spongocoel, 422, 425, 429, 434
spore, 263, 275
stalks, 333, 360, 370, 374, 387
starch, 57
stasigenesis, 18, 20
statocyst, 469, 471, 486, 489, 494, 501, 503, 504, 513 *et seq.*, 521, 525 *et seq.*, 531
stem forms, 69, 98, 105, 106, 538, 575
stomatogenesis, 344 *et seq.*, 348 *et seq.*, 358, 360, 371, 372, 374, 376, 378, 384, 385, 388, 391, 399, 400, 404, 405, 407, 410, 412, 413;
 parafissional opisthe, 350, 351, 361, 371, 378, 385, 399, 404;
 prefissional opisthe, 350, 351, 361, 371, 373, 378, 386 *et seq.*, 399, 404 *et seq.*, 412;
 prefissional proter and opisthe, 351, 352, 361, 371, 374, 378, 384 *et seq.*, 391, 399, 404, 412;
 postfissional proter and opisthe, 351, 352, 361, 371, 384, 399, 404, 412
stomial parts (*see* feeding organelle)

stomodeum, 470
striated fibrils, 157, 159, 177, 328, 333, 358
strontium sulfate, 292, 320
subsidiary criterion (accessory criterion), 177, 178, 190, 253, 258, 304, 305, 362, 442
superorganism, 581, 582
surface ciliature, 330, 331, 348, 349, 350, 359, 379, 385, 386, 398, 417
swarmer, 265, 272, 274, 297, 298, 300, 302, 304, 306, 308, 309 *et seq.*, 322, 386, 387, 393, 394, 589
sycon, 423 *et seq.*, 434, 437, 442, 445, 447, 448, 582
symbiosis (symbionts), 59, 87, 154, 163 *et seq.*, 171, 172, 173, 178, 179, 180, 189, 191, 192, 195, 198, 199, 200, 201, 202, 206, 213, 218, 219, 227, 265, 278, 291, 308, 311, 333, 338, 340, 342, 356, 359, 361, 362, 369, 373, 392, 399, 400, 415, 472, 493, 494, 499, 571, 572, 578, 580, 583, 588, 589
symbiotic algae, 273
symbiotic association of prokaryotic cells, 51, 578
symmetry, 95, 96, 129, 276, 360, 379, 421, 454 *et seq.*, 480, 483, 485, 541, 579, 589, 591;
 asymmetry, 421;
 axial, 445;
 bilateral, 129, 266, 285, 456, 457, 470, 477, 486, 491, 493, 514, 541, 543, 548, 550, 583, 593;
 biradial, 456;
 pentaradiate, 105;
 radial, 129, 243, 261, 266, 281, 285, 378, 380, 386, 421, 422, 454 *et seq.*, 477, 541, 591;
 spherical, 262, 264, 272, 285, 322, 580
symplesioseme, 277

synaposemy, 446, 481
synclone, 140, 141, 144
syncytium, 496, 497, 523, 527, 543
systematics, 591, 592
tachytely, 15, 16, 19, 20, 119, 323, 568
tarsal bones, 17
taxonomy (*see also* systematics), 114;
 definition of, 111;
 phenetic, 113, 114;
 phyletic, 113
tentacles, 70, 331, 333, 341, 359, 360, 364, 367, 377, 386, 387, 393, 394, 407, 414, 454 *et seq.*, 462, 464, 465, 469, 470, 472, 473, 475, 477, 478, 481, 483, 485, 486, 489, 493, 494, 542, 558, 567, 582, 586
testes, 512, 517
tests, 202, 207 *et seq.*, 211, 214, 215, 219 *et seq.*, 225 *et seq.*, 233, 235 *et seq.*, 241, 262, 266, 284, 297, 432, 555, 572 *et seq.*, 578, 579, 593;
 as ballast, 213, 237;
 calcareous, 213;
 endogenous (autogenous), 207;
 exogenous (heterogeneous), 207, 221;
 formation of, 218, 219
tetramerous body organization, 456, 457, 462, 477, 483, 489
thecas, 337, 359, 573
thesocytes, 427, 440
thermoregulation, 22
three dimensional comparisons, 91
thigmotactic surface, 333
tissue differentiation, 593
tissue level of organization, 427
tissues (*see also* histology), 6, 59, 121, 427, 454, 471, 491, 493, 494, 496, 497, 502, 515, 548, 565
toxicysts, 331, 340, 342, 359, 365, 367, 414, 489

transfer RNA (*see* RNA)
transformation series, 99, 280
transitive relationship, 557 *et seq.*;
  limited or weak, 557, 558
trends, 17, 19, 60, 110, 194, 280, 281, 375 *et seq.*, 392, 394, 396 *et seq.*, 412, 413, 420, 433, 440, 476, 481, 483, 485, 505, 516, 517, 520, 570, 582
trichocyst, 57, 328, 331, 332, 346, 354, 359, 377, 489, 544, 577
triploblastic forms, 536, 550
trophocyte, 435, 438
turbellarians (*see also* Organism Index), 493 *et seq.*;
  functions of, 509 *et seq.*;
  homologies of, 513 *et seq.*;
  phylogenetic analysis of, 525 *et seq.*;
  phylogenetic conclusions, 549 *et seq.*;
  phylogeny of, 513 *et seq.*;
  structures of, 493 *et seq.*
umbrella (*see* bell)
undulating membrane, 157 *et seq.*, 176, 183, 195, 197
undulating membranelle (*see* paroral kinety)
unicellularity, 131 *et seq.*, 265, 413, 515, 565, 566
unity of living systems, 8
unitary origin of life, 67
vacuoles, 265
velum, 459, 460
vestibulum (*see* adoral depression)
volvocine colony (coenobium), 490
Weichkörper (*see* inner cell body), 291, 302, 303, 304, 305, 311, 312, 313, 359, 369, 442
weighted characters, 101 *et seq.*, 114, 122, 409, 559, 560
yolk glands, 507, 517
zooflagellates (*see also* Organism Index), 155 *et seq.*;
  ancestry of, 178;
  functions of, 163 *et seq.*;
  homologies of, 175 *et seq.*;
  phylogenetic analysis of, 181 *et seq.*;
  phylogenetic conclusions, 199 *et seq.*;
  phylogeny of, 174 *et seq.*;
  structures of, 157 *et seq.*
zoogeography, 72, 73
zoospore, 256, 262, 263, 274, 276, 277, 281, 299, 300
zootrophy (*see* phagotrophy)
zygote, 139, 216, 231, 274, 298, 299, 300, 308, 354, 435, 473, 481, 547, 555, 582